STEM CELL BIOLOGY AND GENE THERAPY

STEM CELL BIOLOGY AND GENE THERAPY

Edited by

PETER J. QUESENBERRY
GARY S. STEIN
University of Massachusetts
Worcester, Massachusetts

BERNARD G. FORGET
SHERMAN M. WEISSMAN
Yale University School of Medicine
New Haven, Connecticut

WILEY-LISS
A JOHN WILEY & SONS, INC., PUBLICATION
New York • Chichester • Weinheim • Brisbane • Singapore • Toronto

This book is printed on acid-free paper. ♾

Published simultaneously in Canada.

While the authors, editors and publisher believe that drug selection and dosage and the specification and usage of equipment and devices, as set forth in this book, are in accord with current recommendations and practice at the time of publication, they accept no legal responsibility for any errors or omissions, and make no warranty, expressed or implied, with respect to material contained herein. In view of ongoing research, equipment modifications, changes in governmental regulations and the constant flow of information relating to drug therapy, drug reactions, and the use of equipment and devices, the reader is urged to review and evaluate the information provided in the package insert or instructions for each drug, piece of equipment, or device for, among other things, any changes in instructions or indication of dosage or usage and for added warnings and precautions.

Library of Congress Cataloging-in-Publication Data:

Stem cell biology and gene therapy / edited by Peter J. Quesenberry, . . . [et al.].
p. cm.
Includes index.
ISBN 0-471-14656-0 (alk. paper)
1. Gene therapy. 2. Hematopoietic stem cells. 3. Hematopoietic stem cells—Therapeutic use. I. Quesenberry, Peter J.
RB155.8.S84 1998
616'.042–dc21 97-48252

Printed in the United States of America.

10 9 8 7 6 5 4 3 2 1

CONTENTS

PREFACE

During the past several years there have been significant conceptual and experimental advances in stem cell biology and gene therapy. Consequently, this is a topic of importance to a broad spectrum of investigators and physicians pursuing biological regulatory mechanisms and treatment of human disease. At one time, stem cell investigations were principally confined to identification and characterization of primitive cells which commit to the hematopoietic lineage. In a restricted sense, such studies provided a viable basis for defining regulatory parameters of hematopoietic cell differentiation. But significantly, from a broader biological perspective, a paradigm was established for studying progenitor cells which are competent to develop and sustain the phenotypic properties in a series of cells and tissues. Insight into fundamental properties of stem cells is being translated to therapeutic applications. Here again, the hematopoietic stem cell has paved the way for emerging applications of stem cell biology to treatment of disease.

In response to the requirement for a volume which covers stem cell biology and gene therapy from both biological and clinical standpoints, we have organized this book into four sections. The first section presents fundamental regulatory mechanisms which are operative in stem cells. Consideration is given to the strengths and limitations of protocols for stem cell isolation. The validity of markers for the primitive status of stem cell populations is critically evaluated. Regulatory mechanisms which support competency for proliferation and cell cycle progression as well as down regulation of growth that accompanies differentiation are reviewed. Cytokines and growth factors which are key modulators of growth control and microenvironmental (stromal) influences are discussed within the contexts of physiological control and as an arsenal of factors to selectively influence the extent to which stem cells engage in renewal, proliferation and differentiation.

The second section is devoted to a consideration of stem cells from a therapeutic perspective. Preparations of stem cells are addressed in relation to retention of biological activities, ability to sustain lineage potentials following ex-vivo expansion and capabilities for both engraftment and differentiation following return to the in vivo environment. These are important considerations for the therapeutic potential of stem cells as vehicles for gene therapy. Additionally, insight into regulatory components of tissue remodelling can be appreciated.

Section three addresses delivery systems for gene therapy. It is here that the rate limiting steps for many applications of DNA-mediated treatment of human disease reside. An overview of the most promising vectors is presented. Consideration is given to levels of activity, sustained expression and options for conditional activity. Immunization by gene transfer and antisense strategies are considered.

The final section focuses on clinical applications for gene therapy. There is a systematic assessment of approaches to eliminate defective genes or replace or supplement activities of defective genes by manipulating cells at the molecular level. Our intention was not to be inclusive. Rather, we have provided examples to illustrate how gene therapy offers viable alternatives to conventional treatments and options for therapy where currently success has been minimal. Examples include but are not restricted to metabolic disorders, coagulation disorders, anemias, pulmonary diseases, cancer, cardiovascular disease, and neurological disorders.

Because gene therapy is a rapidly advancing field, it would not be realistic to comprehensively cover the basic biological and clinical parameters. Rather, we have selected components of regulatory mechanisms that are central to clinical applications. The therapeutic applications were chosen as representative of the potential which gene therapy provides for treatment of human disease. The challenges now faced are numerous. These are being met by refinements in vector systems as well as gene modifications and improvements in stem cell engraftment strategies.

THE EDITORS

CONTRIBUTORS

Giridhar R. Akkaraju, Department of Molecular Genetics and Biochemistry, University of Pittsburgh School of Medicine, Pittsburgh, PA 15261

George F. Atweh, Professor of Medicine, Division of Hematology, Mount Sinai School of Medicine, One Gustav Levy Place, New York, NY 10029

A. Bahnson, University of Pittsburgh Medical Center, Pittsburgh, PA 15261

E. Ball, University of Pittsburgh Medical Center, Pittsburgh, PA 15261

V. Bansal, University of Pittsburgh Medical Center, Pittsburgh, PA 15261

J. A. Barranger, University of Pittsburgh Medical Center, Pittsburgh, PA 15261

M. Beeler, University of Pittsburgh Medical Center, Pittsburgh, PA 15261

Martha C. Bohn, Children's Memorial Institute for Education and Research, Department of Pediatrics, Northwestern University Medical School, Chicago, IL 60614

Xandra O. Breakfield, Massachusetts General Hospital East, Molecular Neurogenetics Unit, 13th Street, Building 149, Charlestown, MA 02129

Antony W. Burgess, Ludwig Institute for Cancer Research, Melbourne 3050, Australia

Michael J. Caufield, Department of Virus and Cell Biology, Merck Research Labs, West Point, PA 19454

Derek Choi-Lundberg, Department of Neurobiology and Anatomy, Box 603, 601 Elmwood Ave, University of Rochester School of Medicine and Dentistry, Rochester, NY 14642

Kenneth W. Culver, Director, Gene Therapy Research and Clinical Affairs, Codon Pharmaceuticals, Gaithesburg, MD 20877

Charles S. Day, Department of Orthopedic Surgery, Musculoskeletal Research Center, Children's Hospital of Pittsburgh, University of Pittsburgh School of Medicine, Pittsburgh, PA 15261

J. Dunigan, University of Pittsburgh Medical Center, Pittsburgh, PA 15261

M. Eljanne, University of Pittsburgh Medical Center, Pittsburgh, PA 15261

Bernard G. Forget, Professor of Medicine, Hematology Section, Yale University School of Medicine, 333 Cedar Street, New Haven, CT 06510

Jonathan C. Fox, University of Pennsylvania, Department of Medicine, Room 809c, Stellar-Chance Laboratories, 422 Curie Boulevard, Philadelphia, PA 19104–6100

Baruch Frenkel, Department of Orthopedics & Institute for Genetic Medicine, University of Southern California School of Medicine, 2250 Alcazar St., Los Angeles, CA 90033

Joseph C. Glorioso, Department of Molecular Genetics and Biochemistry, University of Pittsburgh School of Medicine, Pittsburgh, PA 15261

William F. Goins, Department of Molecular Genetics and Biochemistry, University of Pittsburgh School of Medicine, Pittsburgh, PA 15261

Michael M. Gottesman, Laboratory of Cell Biology, National Cancer Institute, NIH, Building 37, Room 1B22, 37 Convent Drive, Bethesda, MD 20892-4255

Katherine A. High, The Children's Hospital of Philadelphia, 34th Street and Civic Center Boulevard, Room 310 Abramson Research Center, Philadelphia, PA 19104

Anthony D. Ho, Department of Medicine, University of California, San Diego, La Jolla, CA 92093-0671

Johnny Huard, Department of Orthopedic Surgery, Musculoskeletal Research Center, Children's Hospital of Pittsburgh, University of Pittsburgh School of Medicine, Pittsburgh, PA 15261

Dennet R. Hushka, 1708 William & Mary Common, Hillsboro, NJ 08876

David Jacoby, Massachusetts General Hospital East, Molecular Neurogenetics Unit, 13th Street, Building 149, Charlestown, MA 02129

Karen Johnston, Massachusetts General Hospital East, Molecular Neurogenetics Unit, 13th Street, Building 149, Charlestown, MA 02129

Stefan Karlsson, Gene Therapy Center, Molecular Medicine Section and The Department of Medicine, University Hospital, Lund SE-22362, Sweden

A. Kemp, University of Pittsburgh Medical Center, Pittsburgh, PA 15261

David Krisky, Department of Molecular Genetics and Biochemistry, University of Pittsburgh School of Medicine, Pittsburgh, PA 15261

Gabriele Kroner-Lux, University of North Carolina at Chapel Hill, Gene Therapy Center, CB#7352, 7109 Thurston-Bowles, Chapel Hill, NC 27599

J. Lancia, University of Pittsburgh Medical Center, Pittsburgh, PA 15261

Ping Law, Department of Medicine, University of California, San Diego, La Jolla, CA 92093

Xinqiang Li, Department of Biology, University of California, San Diego, La Jolla, CA 92093

Jane B. Lian, Department of Cell Biology and Cancer Center, University of Massachusetts Medical Center, 55 Lake Avenue North, Worcester, MA 01655

Thomas Licht, Laboratory of Molecular Biology, National Cancer Institute, NIH, Building 37, Room 416, 37 Convent Drive, Bethesda, MD 20892–4255; and University of Ulm, Department of Internal Medicine III, Robert-Kock-Strasse 8, D-89081 Ulm, Germany

Margaret A. Liu, Department of Virus and Cell Biology, Merck Research Labs, West Point, PA 19454

S. Lucot, University of Pittsburgh Medical Center, Pittsburgh, PA 15261

J. Mannion-Henderson, University of Pittsburgh Medical Center, Pittsburgh, PA 15261

Peggy Marconi, Department of Molecular Genetics and Biochemistry, University of Pittsburgh School of Medicine, Pittsburgh, PA 15261

Jeffrey A. Medin, Section of Hematology/Oncology, University of Illinois at Chicago, Chicago, IL 60607-7173

J. Mierski, University of Pittsburgh Medical Center, Pittsburgh, PA 15261

T. Mohney, University of Pittsburgh Medical Center, Pittsburgh, PA 15261

Malcolm A.S. Moore, James Ewing Laboratory of Developmental Hematopoiesis, Memorial Sloan Kettering Cancer Center, 1275 York Avenue, New York, NY 10021

Susan K. Nilsson, Peter MacCallum Cancer Institute, A' Beckett St., Melbourne 3000, Australia

M. Nimgaonkar, University of Pittsburgh Medical Center, Pittsburgh, PA 15261

Tom Oligino, Dept. of Molecular Genetics & Biochemisty, University of Pittsburgh School of Medicine, Pittsburgh, PA 15261

Ira Pastan, Laboratory of Molecular Biology, National Cancer Institute, NIH, Building 37, Room 416, 37 Convent Drive, Bethesda, MD 20892-4255

Peter Pechan, Massachusetts General Hospital East, Molecular Neurogenetics Unit, 13th Street, Building 149, Charlestown, MA 02129

Archibald S. Perkins, Departments of Medicine, Human Genetics, and Pathology, Yale University School of Medicine, 336 BCMM, 295 Congress Avenue, New Haven, CT 06510

Ruth Pettengell, Department of Hematology, St. George's Hospital Medical School, Cranmer Terrace, London SW17 0RE, United Kingdom

Peter J. Quesenberry, University of Massachusetts Medical Center, Cancer Center, 373 Plantation Street, Biotech II, Suite 202, Worcester, MA 01605

E.O. Rice, University of Pittsburgh Medical Center, Pittsburgh, PA 15261

John Richter, Gene Therapy Center, Molecular Medicine Section and The Department of Medicine, University Hospital, Lund, Sweden

Richard Jude Samulski, University of North Carolina at Chapel Hill, Gene Therapy Center, 7109 Thurston-Bowles, Chapel Hill, NC 27599

S. Schierer-Fochler, University of Pittsburgh Medical Center, Pittsburgh, PA 15261

Thomas Shenk, Howard Hughes Medical Institute, Department of Molecular Biology, Princeton University, Princeton, NJ 08544-1014

Arun Srivastava, Division of Hematology/Oncology, Departments of Medicine, Microbiology & Immunology, Medical Science Building 255, Walther Oncology Center, Indiana University School of Medicine, 635 Barnhill Drive, Indianapolis, IN 46202-5120

Gary S. Stein, Department of Cell Biology and Cancer Center, University of Massachusetts Medical Center, 55 Lake Avenue North, Worcester, MA 01655

Janet L. Stein, Department of Cell Biology and Cancer Center, University of Massachusetts Medical Center, 55 Lake Avenue North, Worcester, MA 01655

W. Swaney, University of Pittsburgh Medical Center, Pittsburgh, PA 15261

N. Takiyama, University of Pittsburgh Medical Center, Pittsburgh, PA 15261

Christopher E. Walsh, University of North Carolina at Chapel Hill, Gene Therapy Center, 7109 Thurston-Bowles, Chapel Hill, NC 27599

Sherman M. Weissman, Department of Medicine, Human Genetics, and Pathology, Yale University School of Medicine, 336 BCMM, 295 Congress Avenue, New Haven, CT 06510

John C. Van Gilder, Chairman of the Division of Neurosurgery, University of Iowa School of Medicine, Iowa City, IA 52242

André J. Van Wijnen, Department of Cell Biology and Cancer Center, University of Massachusetts Medical Center, 55 Lake Avenue North, Worcester, MA 01655

Flossie Wong-Staal, Department of Biology, University of California, San Diego, La Jolla, CA 92093

1

STEM CELL SYSTEMS: BASIC PRINCIPLES AND METHODOLOGIES

SUSAN K. NILSSON AND PETER J. QUESENBERRY

Peter MacCullum Cancer Institute, Melbourne, Australia (S.K.N.), Cancer Center, University of Massachusetts Medical Center, Worcester, MA 01605 (P.J.Q.)

CONCEPTS AND METHODS IN HEMATOPOIESIS

Hematopoietic cell research has a long and distinguished history. At the present time it is probably the best understood proliferation differentiation system in mammalian biology. The first beginnings of insight into this area involved the utilization of various dyes and stains to identify different subsets of cells in the peripheral blood and other tissues. These studies showed a variety of cell types that appeared to be largely produced in the bone marrow. Studies on bone marrow morphology indicated a relatively ordered sequence of differentiation in different cell lineages and eventual transit into the peripheral blood system.

Studies at Brookhaven after World War II utilizing tritiated thymidine to mark cells and map cell cycle established the kinetics of cell production in the marrow and distribution in the peripheral blood. These studies, again, indicated an orderly progression of proliferation and then differentiation in the marrow with release to the blood and a distribution of cells within the peripheral bloodstream between a marginating and circulating pool for granulocytes.

Stem Cell Biology and Gene Therapy, Edited by Peter J. Quesenberry, Gary S. Stein, Bernard G. Forget, and Sherman M. Weissman
ISBN 0-471-14656-0

Much of the early work focused on erythropoiesis because of the availability of the reticulocyte as a marker for newly produced red cells and the utilization of radioactive ^{59}Fe as a marker for newly produced red cells. In addition, it was found that the induction of polycythemia by hypertransfusion or hypoxia resulted in a complete shut off of murine erythropoiesis in the marrow, thus providing a model for the testing of substances and their effects on the hematopoietic system (Cotes and Bangham, 1981; Jacobson et al., 1957). This led to the demonstration of a humoral agent inducing differentiation into the erythroid pathway (Borsook et al., 1954; Cotes and Bangham, 1981; Erslev, 1953; Jacobson et al., 1957; Plzak et al., 1955), erythropoietin.

The more recent history of hematopoiesis has involved the study of hematopoietic progenitor stem cells and their regulatory influences. These studies have led to the definition of a variety of precursor stem cells and hematopoietic regulators, several of which are now in clinical use.

HISTORY

The study of hematopoietic cell systems initially involved a number of approaches utilizing the whole animal and marrow cell production after different cytotoxic insults. Granulocyte repopulation and red cell repopulation were studied, as were the responses to various injected agents. These types of studies represented a beginning, but in fact gave little insight into the hematopoietic regulatory systems. The existence of hematopoietic repopulating cells was established in the early 1950s in studies on lethally irradiated mice (Ford et al., 1956; Lindsley et al., 1955; Mitchison, 1956; Nowell et al., 1956), but perhaps the first real break in this area was the description of the colony-forming unit-spleen CFU-S by Till and McCulloch (1961). This was the first clonal hematopoietic stem cell assay. Extensive studies with CFU-S established that the cell that formed nodules on the spleen of irradiated animals had extensive, although heterogenous, renewal capacity, had a tremendous proliferative and differentiation capacity, and was capable of differentiating into erythroid, granulocyte, and megakaryocyte lineages (Curry and Trentin, 1967; Fowler et al., 1967; Lewis and Trobaugh, 1964; Siminovitch et al., 1963). This cell was relatively dormant, although it could be induced into cell cycle (Becker et al., 1965; Iscove et al., 1970; Richard, et al., 1970), and provided the first model system for hematopoietic stem cells. Subsequent work indicated that the CFU-S, as monitored at 8–9 days after irradiation and infusion of marrow cells, was in fact a heterogeneous population and probably was made up predominately of progenitors. Different populations of cells were monitored with longer intervals postirradiation, although even at days 12–14 CFU-S did not precisely monitor the long-term engrafting stem cell, as assessed in *in vivo* models (Bertoncello et al., 1985; Jones et al., 1990; Magli et al., 1982; Mulder and Visser, 1987; Ploemacher and Brons, 1988a–c, 1989).

IN VITRO CLONAL PROGENITOR ASSAYS

The next big step in the field was the elucidation of *in vitro* clonal assays for granulocyte-macrophage progenitors by Bradley and Metcalf (1966) and Pluznik and Sachs (1965). These progenitors responded to exogenous stimulators with the formation of clones of granulocytes and macrophages in soft agar culture. Subsequent studies characterized a relatively large number of clonal progenitors in agar, methylcellulose, or plasma clot. These included burst-forming unit erythroid, CFU erythroid (Gregory, 1976; Iscove and Sieber, 1975; McLeod et al., 1974), colony-forming cell granulocyte-macrophage, colony-forming cell granulocyte, colony-forming cell macrophage, CFU megakaryocyte, burst-forming unit megakaryocyte (Bradley and Metcalf, 1966; Briddell et al., 1989; Gregory, 1976; Long et al., 1985; Monette et al., 1980; Pluznik and Sachs, 1965), GEMM-CFU, and others. Mast cell and eosinophil lineages were also determined, and work by Ogawa and colleagues showed that with single cell transfer there existed a relatively large number of progenitors that showed different lineage predilections *in vitro* (Nakahata and Ogawa, 1982; Ogawa et al., 1985; T. Suda et al., 1983, 1984; J. Suda et al., 1984). In general, these progenitor classes were in active cell cycle and had limited renewal and proliferative potentials. These classes of cells are perhaps best defined now by cytokine responsiveness and time in culture.

PRIMITIVE CLONAL PROGENITORS AND STROMAL BASED ASSAYS

The next important phase of development of the clonal assay involved the characterization of multifactor responsive progenitors. This field was led by the initial work of Dr. Ray Bradley, who first characterized synergisms and then described, along with Dr. Hodgson, the high proliferative potential colony-forming cell (HPP-CFC) (Bradley and Hodgson, 1979; Hodgson and Bradley, 1979; McNiece et al., 1986, 1988a,b). This is perhaps, at present, the best assay for a primitive progenitor and is characterized by multifactor sensitivity, dormancy, and tremendous proliferative and renewal potential.

Work in this area has evolved, with studies showing that increasing the number of growth factors may increase the lineage potential of the assayed cells so that seven to nine growth factors may be optimal, especially when working with relatively purified stem cells. Additional work utilizing serum-free systems and purified cells suggests, that, in fact, for expression of this cell, a relatively large number of growth factors (three or more) are needed and that increasing growth factors increases the proliferative potential. Other primitive progenitors that have been assayed in vitro as potential surrogates for engrafting cells include the blast CFU of Ogawa (Nakahata and Ogawa, 1982; Ogawa et al., 1985, T. Suda et al., 1983, 1984; J. Suda et al., 1984), which is a highly primitive cell with great proliferative and differentiation potential.

In addition, a fairly large amount of work has characterized stromal-based systems in which hematopoietic progenitors adhere to and grow on adherent marrow stromal cells. The long-term culture-initiating cell (LTC-IC) and cobblestone-forming area assay are two of these assays (Ploemacher et al., 1989; Slovik et al., 1984; Sutherland et al., 1989, 1990). These assays do appear to monitor relatively primitive cells, although recent work with human cells comparing the LTC-IC with cells engrafting and proliferating in immunodeficient NOD-SCID animals suggests that LTC-IC may monitor a more differentiated and less primitive hematopoietic stem cell.

IN VITRO STROMAL-BASED CULTURE SYSTEMS

A major step forward was the elucidation by Michael Dexter and colleagues of the murine "Dexter" long-term marrow culture system (Dexter et al., 1977). This culture system produced ongoing hematopoiesis over time, in the obligate presence of adherent stromal cells. The classic Dexter system needed hydrocortisone and stroma, and, in the presence of this plus horse serum, ongoing granulocyte, macrophage, and megakaryocyte production and the support of a wide variety of progenitor/stem cells, including CFU-S, diffusion chamber CFU, megakaryocyte CFU, HPP-CFC, and granulocyte-macrophage CFU, occurred (Doukas et al., 1985, 1986; Levitt and Quesenberry, 1980; McGrath et al., 1987; Quesenberry et al., 1984). A variation of this system, the Whitlock-Witte system, allowed for B-cell growth (Whitlock and Witte, 1982). These stromal-based systems provided the first *in vitro* models for the microenvironment and were equally of interest as a potential hunting ground for new stromal-based growth factors, a number of which have been isolated from different stromal systems, including steel factor and interleukin-7.

CYTOKINES

The number of cytokines discovered to be active on the hematopoietic system continues to expand. Initially, erythropoietin for erythropoiesis and the colony-stimulating factors for macrophages and granulocytes were defined. Subsequently, multi-CFS or interleukin-3 for multilineage progenitors was characterized and since then a relatively large number of factors have been characterized and cloned and some are in clinical trial or clinical practice. It is estimated that over 60 factors are active on the hematopoietic system, many of which may stimulate both early stem cells for proliferation and differentiation and the function of differentiated end cells. These factors frequently synergize, and they may inhibit or stimulate proliferation, differentiation, apoptosis, function, or motility. Tables 1 and 2 present a noninclusive list of these factors with an attempt to order them via their initially described action or their level of action on hematopoiesis.

TABLE 1 Cytokines Active on Lymphohematopoietic Cells (Interleukins, Colony-Stimulating Factors, and Erythropoietin)*

Cytokine	Initially Described or Highlighted Bioactivity
Erythropoietin	Red cell formation
Granulocyte colony-stimulating factor	Granulocyte colony formation, stem cell release and stimulation
Granulocyte-macrophage colony-stimulating factor	Granulocyte and macrophage colony formation
Colony-stimulating factor-1	Macrophage colony formation
Interleukin-1	B and T cell regulator, endogenous pyrogen, and inducer of other factors
Interleukin-2	T-cell growth factor
Interleukin-3 (Multicolony-stimulating factor)	Stimulator granulocyte, macrophage, eosinophil, mast cell, and megakaryocyte colony formation and interacts with erythropoietin to stimulate erythroid bursts
Interleukin-4	B-cell proliferation and immunoglobulin secretion
Interleukin-5	B-cell differentiation and immunoglobulin secretion; eosinophil proliferation and maturation
Interleukin-6	B-cell differentiation and immunoglobulin secretion; multiple actions on hematopoietic stem progenitors
Interleukin-7	Stimulation of pre-B-cell production
Interleukin-9	Erythroid colony formation and stimulation of proliferation of a megakaryocyte cell line
Interleukin-10	Inhibits cytokine synthesis by T cells; increases cytotoxic T-cell numbers and function
Interleukin-11	B-cell, megakaryocyte, and stem cell stimulator
Interleukin-12	Natural killer cell stimulator; helper T-cell proliferation and interferon production by T and natural killer cells
Interleukin-13	Similar to interleukin-4; B-cell immunoglobulin production and isotype switching; stimulates T-cell proliferation and inhibits monocyte inflammatory cytokine production.
Interleukin-15	Similar to interleukin-2 actions

* Noninclusive

IN VIVO ENGRAFTMENT ASSAY

The gold standard for hematopoietic stem cells has been studies of *in vivo* renewal and engraftment. These were initially carried out in irradiated murine hosts. Early studies looked at radiation survival or renewal using limiting dilution approaches. These provided some of the initial insights into hematopoietic stem cells, but were inherently difficult to quantitate and have been

TABLE 2 Additional Cytokines Active on Lymphohematopoiesis*

Cytokine	Initially Described or Principal Bioactivity
Kit ligand (steel factor)	Synergistic with many other factors to stimulate progenitor and early multi-potential and erythroid cells
Basic fibroblast growth factor	Synergistic with other growth factors
Platelet-derived growth factor	Erythropoietic and granulopoietic progenitors
Hepatocyte growth factor	Synergistic activity on progenitors
Insulin-like growth factor II	Erythroid and granulopoietic progenitors
Leukemia inhibitory factor	Early stem cell and megakaryocyte progenitors

Others with various activities, predominantly inhibitory, include transforming growth factor-β, the interferons and prostaglandins, inhibin, macrophage-inflammatory protein-1α and other members of the chemokine family, the pentapeptide, and seraspenide. * Noninclusive

supplanted by other approaches. Harrison and colleagues (1988) have championed competitive repopulation in irradiated hosts in which one cell population competes with another and quantitative estimates of early or long-term repopulating are obtained. In these studies, stem cells that are responsible for long-term hematopoiesis appear to be dormant and rare and have tremendous renewal potential, while those for shorter term may include these cells but also a population of more rapidly proliferating cells. More recently, emphasis has been placed on the ability to engraft into nonmyeloablated nontreated hosts (Peters et al., 1995, 1996; Ramshaw et al., 1995; Rao et al., 1997; Stewart et al., 1993). Such engraftment appears to be quantitative, the final result being determined by stem cell competition between engrafting and donor stem cells. The latter model may actually give better engraftment at the stem cell level, as there is evidence that cytotoxic therapy or radiation may damage the microenvironment and actually impair engraftment (Shirota and Tavassoli, 1992). These types of assays have allowed one to estimate the meanings of some of the surrogate assays. Probably the best correlates with long-term renewal are the HPP-CFC and blast CFU with a major question mark remaining about the LTC-IC or cobblestone-forming area cells.

STEM CELL PURIFICATION

To appropriately study stem cell populations, many investigators have worked out techniques to separate populations of cells based on differentiation markers, physical characteristics, or special epitopes defining primitive cells (Baines et al., 1988; Bauman et al., 1986; Berman and Basch, 1985; Bertoncello et al., 1986; Boswell et al., 1984; Civin et al., 1984, 1987; Engh et al., 1978, 1981; Fitchen et al., 1987; Harris et al., 1984; Jordan et al., 1990; Katz et al., 1985; Lu et al., 1987; Mulder et al., 1984, 1985; Muller-Sieburg et al., 1986; Spangrude et al., 1988; Trask and Engh, 1980). Different approaches have been taken for both human and murine hematopoiesis. In general, cells with defined differentiated markers are removed by antibody staining using magnetic bead and/or fluorescent cell sorting; then specific antigens expressed on primitive cells are tagged with antibodies with various fluorochrome tags, and these are selectively separated by fluorescent activated cell sorting. In the mouse, C-kit, Sca and other epitopes have been used, while in the human antibodies distinguishing CD34 and CD38 cells have been frequently utilized. In addition, the use of rhodamine and Hoescht dyes to stain cells has allowed for the separation of relatively primitive populations. Dull staining rhodamine and Hoescht cells seem to mark a population of cells with long-term marrow repopulating capacity. The lineage-negative, rhodamine^{lo}/Hoescht^{lo} cells when assayed *in vitro* are approximately 90%–100% multifactor HPP-CFC. Studies of these highly purified cell populations have allowed the definition of early and long-term repopulating cells. It appears that the longest term repopulating cell also has significantly early repopulating potential but that there may be other cells with only early repopulating potential.

Work by Kaufman et al. (1994) has suggested the presence of facilitator cells, which may enhance engraftment or long-term expression of stem cells in allogeneic systems. More recently, work by Nilsson et al. (1997) has suggested the existence of an autologous facilitator cell. These highly purified populations of stem cells are of significant interest with regard to their potential use in gene therapy approaches or marrow stem cell expansion.

SPECIFIC CONSIDERATIONS REGARDING HUMAN VERSUS MURINE HEMATOPOIETIC STEM CELL SYSTEMS

Studies in the human system have been limited because no repopulation model exists. Human bone marrow transplantation has allowed for a definition of whole populations with the ability to repopulate, but specific subpopulations have not been studied in this mode. Immunodeficient murine models have been utilized as an attempt to have a surrogate *in vivo* assay for human hematopoietic stem cells. The SCID or NOD-SCID animals allow for engraftment of human cells (Lowry et al., 1996), and, although the differentiation lineages are skewed and there are problems with this model, this is probably the closest one comes to being able to assess human engrafting stem cells. The HPP-CFC and possibly the CFU blast are probably the best surrogate assays, but caution must be utilized in assessing such assays since there are clearly situations in which the *in vitro* studies may not mirror *in vivo* engraftment. A particular example of this are studies in murine species in which cytokine exposure of marrow *in vitro* has lead to an expansion of HPP-CFC with cycle induction while markedly decreasing long-term engrafting potential (Peters et al., 1995, 1996). The immunodeficient mouse models continue to evolve and hold a promise of being able to carry out more adequate assessments of human engrafting cells.

SUMMARY

Extensive information has been developed on both human and murine stem cell systems and their cytokine and stromal regulators. We continue to expand our insights into this system with a growing potential for application to gene therapy or to marrow stem cell expansion.

REFERENCES

Baines P, Mayani H, Bains M, Fisher J, Hoy T, Jacobs A (1988): Enrichment of CD 34 (My10)-positive myeloid and erythroid progenitors from human marrow and their growth in cultures supplemented with recombinant human granulocyte-macrophage colony-stimulating factor. *Exp Hematol* 16:785–789.

Bauman JGJ, Wagemaker G, Visser JWM (1986): A fractionation procedure of mouse bone marrow cells yielding exclusively pluripotent stem cells and committed progenitors. *J Cell Physiol* 128:133–142.

Becker AJ, McCulloch EA, Siminovitch L, Till JE (1965): The effect of differing demands for blood cell production on DNA synthesis by hemopoietic colony-forming cells of mice. *Blood* 26:296–308.

Berman JW, Basch RS (1985): Thy-1 antigen expression by murine hematopoietic precursor cells. *Exp Hematol* 12:1152–1156.

Bertoncello I, Bartelmez SH, Bradley TR, Stanley ER, Harris RA, Sandrin MS, Kriegler AB, McNiece IK, Hunter SD, Hodgson GS (1986): Isolation and analysis of primitive hemopoietic progenitor cells on the basis of differential expression of Qa-m7 antigen. *J Immunol* 136:3219–3224.

Bertoncello I, Hodgson GS, Bradley TR (1985): Multiparameter analysis of transplantable hemopoietic stem cells. I. The separation and enrichment of stem cells homing to marrow and spleen on the basis of rhodamine-123 fluorescence. *Exp Hematol* 13:999–1006.

Borsook H, Graybiel A, Keighley G, Windsor E (1954): Polycythemic response in normal adult rats to a nonprotein plasma extract from anemic rabbits. *Blood* 9:734–742.

Boswell HS, Wade PM Jr, Quesenberry PJ (1984): Thy-1 antigen expression by murine high proliferative capacity hematopoietic progenitor cells: Relation between sensitivity depletion by Thy-1 antibody and stem cell generation potential. *J Immunol* 133:2940–2949.

Bradley TR, Hodgson GS (1979): Detection of primitive macrophage progenitor cells in mouse bone marrow. *Blood* 54:1446–1450.

Bradley TR, Metcalf D (1966): The growth of mouse bone marrow cells *in vitro*. *Aust J Exp Biol Med Sci* 44:287–300.

Briddell RA, Brandt JE, Straneva JE, Srour EF, Hoffman R (1989): Characterization of the human burst-forming unit–megakaryocyte. *Blood* 74:145–151.

Civin CI, Banguenigo ML, Strauss LS, Loken MR (1987): Antigenic analysis of hematopoiesis. VI. Flow cytometric characterization of My 10-positive progenitor cells in normal bone marrow. *Exp Hematol* 15:10–17.

Civin CI, Strauss LC, Brovall C, Fackler MJ, Schwartz JF, Shaper JH (1984): Antigenic analysis of hematopoiesis. III. A hematopoietic progenitor cell surface antigen defined by a monoclonal antibody raised against KG-1a cells. *J Immunol* 133: 157–165.

Cotes PM, Bangham DR (1981): Bioassay of erythropoietin in mice made polycythaemic by exposure to air at a reduced pressure. *Nature* 191:1065–1067.

Curry JL, Trentin JJ (1967): Hemopoietic spleen colony studies. IV. Phytohemagglutinin and hemopoietic regeneration. *J Exp Med* 126:819–932.

Dexter TM, Allen TD, Lajtha LG (1977): Conditions controlling the proliferation of haemopoietic stem cells *in vitro*. *J Cell Physiol* 91:335–344.

Doukas MA, Niskanen E, Quesenberry PJ (1985): Lithium stimulation of granulopoiesis in diffusion chambers—A model of humoral, indirect stimulation of stem cell proliferation. *Blood* 65:163–168.

Doukas MA, Niskanen E, Quesenberry PJ (1986): The effect of lithium on stem cell and stromal cell proliferation *in vitro*. *Exp Hematol* 14:215–221.

Engh GJVD, Russell J, de Cicco D (1978): Surface antigens of hemopoietic stem cells: The expression of BAS, Thy-1, and H-2 antigens on CFU-S. In Baum SJ, Ledney GD (eds): Experimental Hematology Today. New York: Springer, p 9.

Engh GJVD, Trask B, Visser JWM (1981): Surface antigens of pluripotent and committed haemopoietic stem cells. In Neth R, Ballo RC, Graf T, Mannweiler K, Winkler K (eds): Modern Trends in Human Leukemia. IV. Haematology and Blood Transfusion, Vol 26. Berlin: Springer, p 305.

Erslev AJL (1953): Humoral regulation of red cell production. *Blood* 8:349–357.

Fitchen JH, Foon KA, Cline NJ (1987): The antigenic characteristics of hemopoietic stem cells. *N Engl J Med* 305:17–25.

Ford CE, Hamerton JL, Barnes DWH, Loutit JF (1956): Cytological identification of radiation-chimaeras. *Nature* 177:452–453.

Fowler JH, Wu AM, Till JE, McCulloch EA, Siminovitch L (1967): The cellular composition of hemopoietic spleen colonies. *J Cell Physiol* 69:65–71.

Gregory CJ (1976): Erythropoietin sensitivity as a differentiation marker in the hemopoietic system: Studies of three erythropoietic colony reponse in culture. *J Cell Physiol* 89:289–301.

Harris RA, Hogarth PM, Wadeson LJ, Collins P, McKenzie IFC, Penington DG (1984): An antigenic difference between cells forming early and late haematopoietic spleen colonies (CFU-S). *Nature* 307:638–641.

Harrison DE, Astle CM, Lerner C (1988): Number and continuous proliferative pattern of transplanted primitive immunohematopoietic stem cells. *Proc Natl Acad Sci USA* 85:822–826.

Hodgson GS, Bradley TR (1979): Effects of endotoxin and extracts of pregnant uterus on the receover of hemopoiesis after 5-fluorouracil. *Cancer Chemother Rep* 63:1761–1769.

Iscove NN, Sieber F (1975): Erythroid progenitors in mouse bone marrow detected by macroscopic colony formation in culture. *Exp Hematol* 3:32–43.

Iscove NN, Till JE, McCulloch EA (1970): The proliferative states of mouse granulopoietic progenitor cells. *Proc Soc Exp Biol Med* 134:33–36.

Jacobson LO, Goldwasser E, Plzak L, et al. (1957): Studies on erythropoiesis. IV. Reticulocyte response of hypophysectomized and polycythemic rodents to erythropoietin. *Proc Soc Exp Biol Med* 94:243–249.

Jones RJ, Wagner JE, Celano P, Zicha MS, Sharkas SJ (1990): Separation of pluripotent hematopoietic stem cells from spleen-colony-forming cells. *Nature* 347:188–189.

Jordan CT, McKearn JP, Lemischka IR (1990): Cellular and developmental properties of fetal hematopoietic stem cells. *Cell* 61:953–963.

Katz FE, Tindle R, Sutherland DR, Greaves MJ (1985): Identification of a membrane glycoprotein associated with haemopoietic progenitor cells. *Leuk Res* 9:191–198.

Kaufman CL, Colson YL, Wren SM, Watkins S, Simmons RL, Ildstad ST (1994): Phenotypic characterization of a novel bone marrow–derived cell that facilitates engraftment of allogeneic bone marrow stem cells. *Blood* 94:2436–2446.

Levitt L, Quesenberry PJ (1980): The effect of lithium on murine hematopoiesis in a liquid culture system. *N Engl J Med* 302:713–719.

Lewis JP, Trobaugh FE Jr (1964): Hematopoietic stem cells. *Nature* 204:589–590.

Lindsley DL, Odell TT Jr, Tausche FG (1955): Implantation of functional erythropoietin elements following total-body irradiation. *Proc Soc Exp Biol Med* 90:512–515.

Long MW, Gragowski LL, Heffner CH, Boxer LA (1985): Phorbol diesters stimulate the development of an early murine progenitor cell. The burst forming unit-megakaryocyte. *J Clin Invest* 76:431–438.

Lowry PA, Shultz LD, Greiner DL, Hesselton RM, Kittler ELW, Tiarks CY, Rao SS, Leif JH, Ramshaw HS, Quesenberry PJ (1996): Improved engraftment of human cord blood stem cells in NOD/LtSz-*scid/scid* mice following irradiation or multiple day injections into unirradiated recipients. *Biol Blood Marrow Trans* 2:15–23.

Lu L, Walker D, Broxmeyer HE, Hoffman R, Hu W, Walker E (1987): Characterization of adult human marrow hematopoietic progenitors highly enriched by two color cell sorting with My-10 and major histocompatibility class II monoclonal antibodies. *J Immunol* 139:1823–1829.

Magli MC, Iscove NN, Odartchenko N (1982): Transient nature of early haematopoietic spleen colonies. *Nature* 295:527–529.

McGrath HE, Liang C, Alberico T, Quesenberry PJ (1987): The effect of lithium on growth factor production in long-term bone marrow cultures. *Blood* 70:1136–1142.

McLeod DL, Shreeve MM, Axelrad AA (1974): Improved plasma culture system for production of erythrocytic colonies *in vitro*: Quantitative assay method for CFU-3. *Blood* 44:517–534.

McNiece IK, Bradley TR, Kreigler AB, Hodgson GS (1986): Subpopulations of mouse bone marrow high proliferative potential colony forming cells (HPP-CFC). *Exp Hematol* 14:856–860.

McNiece IK, Robinson BE, Quesenberry PJ (1988a): Stimulation of murine high proliferative potential colony forming cells by the combination of GM-CSF and CSF-1. *Blood* 72:191–195.

McNiece IK, Stewart FM, Decon DM, Quesenberry PJ (1988b): Synergistic interactions between hematopoietic growth factors as detected by *in vitro* mouse bone marrow colony formation. *Exp Hematol* 16:383–388.

Mitchison NA (1956): The colonisation of irradiated tissue by transplanted spleen cells. *Br J Exp Pathol* 37:239–247.

Monette FC, Kent RB, Weiner EJ, Jarris RF Jr, Oullette PL, Thorson JA, Zelick RD (1980): Cell-cycle properties and proliferation kinetics of late erythroid progenitors in murine bone marrow. *Exp Hematol* 8:484–493.

Mulder AH, Bauman JGJ, Visser JWM, Boersma WJA, Engh GJVD (1984): Separation of spleen colony forming units and prothymocytes by use of a monoclonal antibody detecting an H-2K determinant. *Cell Immunol* 88:401–410.

Mulder AH, Visser JWM (1987): Separation and functional analysis of bone marrow cells separated by rhodamine-123 fluorescence. *Exp Hematol* 15:99–104.

Mulder AH, Visser JWM, Engh GJVD (1985): Thymus regeneration by bone marrow cell suspensions differing in the potential to form early and late spleen colonies. *Exp Hematol* 13:768–775.

Muller-Sieburg CE, Whitlock CA, Weissman IL (1986): Isolation of two early B lymphocyte progenitors from mouse marrow: a committed pre-pre-B cell and a clonogenic Thy-1 lo hematopoietic stem cell. *Cell* 44:653–662.

Nakahata T, Ogawa M (1982): Hemopoietic colony-forming cells in umbilical cord blood with extensive capability to generate mono- and multipotential hemopoietic progenitors. *J Clin Invest* 70:1324–1328.

Nilsson S, Dooner M, Tiarks C, Heinz-Ulrich W, Quesenberry PJ (1997): Potential and distribution of transplanted hematopoietic stem cells in a nonablated mouse model. *Blood* 89:4013–4020.

Nowell PC, Cole LJ, Habermeyer JG, Roan PL (1956): Growth and continued function of rat marrow cells in x-radiated mice. *Cancer Res* 16:258–261.

Ogawa M, Pharr PN, Suda T (1985): Stochastic nature of stem cell functions in culture. In Cronkite E, Dainiak N, McCaffrey R, Palek J, Quesenberry P (eds): Hematopoietic Stem Cell Physiology. New York: Alan R. Liss, Inc., p 11.

Pearson-White S, Deacon D, Crittenden R, Brady G, Iscove N, Quesenberry PJ (1995): The *ski/sno* proto-oncogene family in hematopoietic development. *Blood* 86:2146–2155.

Peters SO, Kittler EL, Ramshaw HS, Quesenberry PJ (1995): Murine marrow cells expanded in culture with IL-3, IL-6, IL-11, and SCF acquire an engraftment defect in normal hosts. *Exp Hematol* 23:461–469.

Peters SO, Kittler ELW, Ramshaw HS, Quesenberry PJ (1996): *Ex vivo* expansion of murine marrow cells with interleukin-3, interleukin-6, interleukin-11, and stem cell factor leads to impaired engraftment in irradiated hosts. *Blood* 87:30–37.

Ploemacher RE, Brons NHC (1988a): Cells with marrow and spleen repopulating ability and forming spleen colonies on day 16, 12 and 8 are sequentially ordered on the basis of increasing rhodamine 123 retention. *J Cell Physiol* 136:531–536.

Ploemacher RE, Brons NHC (1988b): Isolation of hemopoietic stem cell subsets from murine bone marrow. I. Radioprotective ability of purified cell suspensions differing in the proportion of day 7 and day 12 CFU-S. *Exp Hematol* 16:21–26.

Ploemacher RE, Brons NHC (1988c): Isolation of hemopoietic stem cell subsets from murine bone marrow. II. Evidence for an early precursor of day 12 CFU-S and cells associated with radioprotective ability. *Exp Hematol* 16:27–32.

Ploemacher RE, Brons NHC (1989): Separation of CFU-S from primitive cells responsible for reconstitution of the bone marrow hemopoietic stem cell compartment following irradiation: Evidence for a pre-CFU-S cell. *Exp Hematol* 17:263–266.

Ploemacher RE, van der Slijs JP, Uoerman JSA, Brons NHC (1989): An *in vitro* limiting dilution assay of long-term repopulating hematopoietic stem cells in the mouse. *Blood* 74:2755–63.

Pluznik DH, Sachs L (1965): The cloning of normal "mast" cells in tissue culture. *J Cell Comp Physiol* 66:319–324.

Plzak LF, Fried W, Jacobson LO, Bethard WF (1955): Demonstration of stimulation of erythropoiesis by plasma from anemic rats using Fe^{59}. *J Lab Clin Med* 45:671–678.

Quesenberry PJ, Coppola MA, Gualtieri RJ, Wade PM Jr, Song ZX, Doukas MA, Shideler CE, Baker DG, McGrath HE (1984): Lithium stimulation of murine hematopoiesis in liquid culture: An effect mediated by marrow stromal cells. *Blood* 63:121–127.

Ramshaw HS, Rao SS, Crittenden RB, Peters SO, Weier HU, Quesenberry PJ (1995): Engraftment of bone marrow cells into normal unprepared hosts: Effects of 5-fluorouracil and cell cycle status. *Blood* 86:924–929.

Rao SS, Peters SO, Crittenden RB, Stewart FM, Ramshaw HS, Quesenberry PJ (1997): Stem cell transplantation in the normal nonmyeloablated host: Relationship between cell dose, schedule and engraftment. *Exp Hematol* 25:114–121.

Rickard KA, Shadduck RK, Howard DE, Stohlman F Jr (1970): A differential effect of hydroxyurea on hemopoietic stem cell colonies *in vitro* and *in vivo*. *Proc Soc Exp Biol Med* 134:152–156.

Shirota T, Tavassoli M (1992): Alterations of bone marrow sinus endothelium induced by ionizing irradiation: Implications in the homing of intravenously transplanted marrow cells. *Blood Cells* 18:197–214.

Siminovitch L, McCulloch EA, Till JE (1963): The distribution of colony-forming cells among spleen colonies. *J Cell Comp Physiol* 62:327–336.

Slovic FT, Abboud CN, Brennan JK, Lichtman MA (1984): Survival of granulocytic progenitors in the nonadherent and adherent compartments of human long-term marrow cultures. *Exp Hematol* 12:327–338.

Spangrude GJ, Heimfeld S, Weissman IL (1988): Purification and characterization of mouse hematopoietic stem cells. *Science* 241:58–62.

Stewart FM, Crittenden R, Lowry PA, Pearson-White S, Quesenberry PJ (1993): Long-term engraftment of normal and post-5-fluorouracil murine marrow into normal nonmyeloablated mice. *Blood* 81:2566–2571.

Suda J, Suda T, Ogawa M (1984): Analysis of differentiation of mouse hemopoietic stem cells in culture by sequential replating of paired progenitors. *Blood* 64:393–399.

Suda T, Suda J, Ogawa M (1984): Disparate differentiation in mouse hemopoietic colonies derived from paired progenitors. *Proc Natl Acad Sci USA* 81:2520–2524.

Suda T, Suda J, Ogawa M (1983): Single cell origin of mouse hemopoietic colonies expressing multiple lineages in variable combinations. *Proc Natl Acad Sci USA* 80:6689–6693.

Sutherland HJ, Eaves CJ, Eaves AC, Dragowska W, Lansdorp PM (1989): Characterization and partial purification of human marrow cells capable of initiating long-term hematopoiesis *in vitro*. *Blood* 74:1563–1570.

Sutherland HJ, Lansdorp PM, Henkelman DH, Eaves AC, Eaves CJ (1990): Functional characterization of individual human haematopoietic stem cells cultured at limiting dilution on supportive marrow stromal layers. *Proc Natl Acad Sci USA* 87: 3584–3588.

Till JE, McCulloch EA (1961): A direct measurement of the radiation sensitivity of normal mouse bone marrow cells. *Radiat Res* 14:213–222.

Trask B, Engh GJVD (1980): Antigen expression of CFU-S determined by light activated cell sorting. In Baum SJ, Ledney D, Van Bekkum DW (eds): Experimental Hematology Today. New York: Karger, p 299.

Whitlock CA, Witte ON (1982): Long-term culture of B lymphocytes and their precursors from murine bone marrow. *Proc Natl Acad Sci USA* 79:3608–3612.

2

CYTOKINE/GROWTH FACTOR RESPONSIVENESS OF EARLY HEMOPOIETIC PROGENITOR CELLS

ANTONY W. BURGESS
Ludwig Institute for Cancer Research, Melbourne 3050, Australia

INTRODUCTION

The availability of recombinant cytokines and purified subpopulations of early hemopoietic progenitor cells has led to an explosion of reports on the regulation of the self-renewal, proliferation, and differentiation processes in hemopoiesis (Ball et al., 1995; Nicholls et al., 1995; Levesque et al., 1996). However, the results from even the simplest *in vitro* systems involving purified progenitor cells, a single growth factor/cytokine, and serum-free conditions can be difficult to interpret. Many different criteria are used to purify the progenitor cells, and the biological assays used to define the characteristics of these early cells vary from long-term repopulation studies to adherence induction or colony growth. Consequently, it is often a challenge to know which early hemopoietic cells are being studied: stem cells (Ogawa, 1993; Lord and Dexter, 1995), multipotential but committed progenitor cells, bispecific lymphohemopoietic progenitors, or lineage-specific progenitors.

Clearly, stem cells exist and can be maintained in culture indefinitely, but the only clear example to date is the propagation of murine embryonal stem

Stem Cell Biology and Gene Therapy, Edited by Peter J. Quesenberry, Gary S. Stein, Bernard G. Forget, and Sherman M. Weissman
ISBN 0-471-14656-0

(ES) cells (Smith et al., 1988; Williams et al., 1988) in the presence of leukemia inhibitory factor (LIF). Even after years in culture, these totipotent cells retain their ability to contribute to the formation of all tissues. Although a single cytokine, LIF, appears to be required to prevent differentiation and maintain self-renewal of ES cells indefinitely, as yet there is no equivalent culture system for hemopoietic stem cells.

Mixed cultures of early murine progenitor cells and bone marrow stromal cells can produce committed progenitor cells for many months, but until now no combination of cytokines has appeared to permit indefinite self-renewal of hemopoietic stem cells. Interestingly, complex mixtures of cytokines appear to improve the production of early hemopoietic progenitor cells *in vitro,* but whether self-renewing divisions actually occur *in vitro* is still the subject of considerable debate (Lord and Dexter, 1995). Furthermore, it is still far from clear whether these *in vitro* actions of cytokine mixtures are relevant to the physiological self-renewal or differentiation of stem cells in animals.

The theme of this book centers around the latest attempts to introduce functional genes into the self-renewing hemopoietic systems; thus this chapter focuses on a description and analysis of the cytokines which regulate the early hemopoietic progenitor cells. Hopefully, an understanding of the biological effects of cytokines on these cells will improve our ability to introduce/repair genes in the self-renewing hemopoietic compartment. There is a brief description of "different" classes of early hemopoietic progenitor/stem cells and a discussion of the assays which might be used to predict the effectiveness of genetic manipulation strategies. The role of cytokine/growth factor networks for the regulation of self-renewal, proliferation, lineage commitment, differentiation, and function is also presented. Finally, the interactions between stromal/endothelial cells, the extracellular matrix systems, and the growth factor/cytokine networks have now been explored in detail, and some of the recent results of these studies are discussed.

Clearly, several cytokines/growth factors stimulate the production of other hemopoietic regulators. These "endogenous" regulators can act in synergy with the exogenous cytokines to influence the hemopoietic process. Whilst these cross-inductions can be detected and even characterized *in vitro,* these interactions are extraordinarily difficult to either detect or analyze in animals. One example relates to the analysis of the factors controlling the mobilization of peripheral blood "stem" cells (Duhrsen et al., 1988; Sockinski et al., 1988). Interestingly, regulators which were initially thought to be lineage specific, e.g., granulocyte colony-stimulating factor (G-CSF) (Burgess and Metcalf, 1980) or thrombopoietin (Ebbe, 1997), are known to interact with multipotential cells and even with cells in other lineages. Interestingly, even the early studies in hemopoiesis suggested that hemopoietic progenitor cells may be stimulated by a particular cytokine to produce endogenous cytokines capable of controlling the differentiation process (Sachs, 1985).

Individual cytokines/growth factors can have powerful effects on the production of hemopoietic cells. The effects of cytokines on early progenitor cells have been grouped into three classes (Ogawa, 1993): I, triggering from dormancy; II, maintenance of the proliferative stage; and III, synergy with cytokines from groups I or II (see Table 1) (Yonemura et al., 1996). The roles of several cytokines in each of these processes is explored in some detail in several sections of this chapter.

For many years there has been debate on the role of cytokines in determining lineage-specific cell production. One hypothesis holds that stem cells display lineage-specific cytokine receptors and that, when the corresponding cytokine is present, the stem cell differentiates along that lineage. The other hypothesis maintains that stem cells differentiate in a stochastic manner, producing lineage-specific progenitor cells which display particular cytokine/growth factor receptors. In the absence of the corresponding cytokine the progenitor cells die; in the presence of appropriate growth factor/cytokine, a program of proliferation and terminal differentiation ensues. Clearly, the earliest hemopoietic progenitor cells display cytokine receptors, but these do not appear to be lineage determinant or even necessary for lineage-specific differentiation (e.g., mice lacking c-mpl can still produce platelets) (Alexander et al., 1996).

The significance of inhibitory modulators of hemopoiesis has been difficult to establish. However, it has been reported that cytokines such as transforming growth factor-β (TGF-β) (Sitnicka et al., 1996b) and tumor necrosis factor-α (TNF-α) (Jacobsen et al., 1996) can inhibit the division(s) of stem cells (both *in vivo* and *in vitro*); however, whether there is a regulatory role for TGF-β or TNF-α in normal hemostasis is yet to be resolved.

The chapter concludes with a short summary of evidence relating to the possible role of autocrine cytokine/growth factors in the maintenance of the leukemic state. There is some evidence that modulation of stimulation such as granulocyte-macrophage (GM)-CSF or inhibition of its receptor can influence leukemic cell proliferation (Russell, 1992). Combinations of anticytokine

TABLE 1 Classes of Cytokines Acting on Early Hemopoietic Progenitor Cells[1]

CLASS		
I Triggering (G0→Cycling)	II Maintenance (Cell Cycle)	III Synergistic
Interleukin-6	Interleukin-3	Interleukin-3
Interleukin-11	Interleukin-4	Stem cell factor
Interleukin-12		
G-CSF	GM-CSF	Flt-3 ligand
Flt-3 ligand		
Leukemia inhibitory factor		
Thrombopoietin		

[1] See Ku et al. (1996).

agents and cytotoxic drugs offer new possibilities for more effective leukemia therapies.

EARLY PROGENITOR CELLS AND CYTOKINE RESPONSIVENESS

The early hemopoietic progenitor cell compartment consists of all of the cells between the self-renewing, multipotential lymphohemopoietic progenitor cells, multipotential myeloid progenitor cells, and bispecific progenitor cells. All of these progenitors have a capacity for proliferation, but their ability to repopulate irradiated recipient mice in the long term diminishes as the progenitor cells mature (Lord and Dexter, 1995).

It is not the purpose of this chapter to analyze the definition of the hemopoietic stem cell (Ogawa, 1993; Lord and Dexter, 1995); however, several biological characteristics are important for an analysis of cytokine action on the outcome of both transplantation and gene therapy. It is critical that cells used to repopulate a "defective" hemopoietic system (whether the defect is due to disease or ablation) be capable of long-term repopulation. It appears that the ability of primitive hemopoietic stem cells to repopulate (self-maintain) diminishes with increasing numbers of divisions (Hole et al., 1996; Young et al., 1996b), yet many techniques for effecting stable genetic alterations often require some cell cycling for the integration of the genetic vector. Clearly, ES cells can proliferate and self-renew in the presence of LIF, but even the most effective cytokine systems have little success in maintaining the long-term repopulating ability (LTRA) of primitive hemopoietic stem cells (Traycoff et al., 1995). More success appears to be possible by co-culturing the primitive hemopoietic stem cells with cytokines and stromal cells (Young et al., 1996b); however, whilst the *in vitro* proliferative and "self-renewal" capacity of hemopoietic cells cultured under these conditions was apparent, the LTRA of these cells is yet to be proved (Traycoff et al., 1995; Young et al., 1996b).

The effects of cytokines on these early progenitor cells (self-renewing) are quite profound (Table 1) (Ku et al., 1996). Some cytokines (e.g., interleukin [IL]-6, IL-11, or LIF) trigger noncycling stem cells into a proliferative phase (Ogawa, 1993; Shah et al., 1996); however, in many experiments this is associated with a loss of long-term repopulating cells. In particular, two cytokines, IL-3 and IL-12, stimulate the proliferative expansion of cells associated with the stem cell compartment, but there appears to be a significant reduction in the LTRA of the cells produced in these cultures. Although the definition of LTRA has not been standardized (Lord and Dexter, 1995), it is an important concept if our understanding of the role of cytokines and/or integrins in the regulation of self-renewal and differentiation is to progress.

Although human stem cells appear to be restricted to the $CD34^+$ population, there is some indication that mouse stem cells exist in both the $CD34^+$ and $CD34^{lo/-}$ subpopulation (Osawa et al., 1996). Whilst $CD34^-$ do not produce colonies with IL-3, when IL-3 and stem cell factor (SCF) are used together, there is a remarkable increase in the formation of large multilineage colonies. Subsequent *in vivo* analyses showed that $CD34^+$ murine progenitors produced early but not sustained engraftment, whereas the $CD34^{lo/-}$ subpopulations yielded delayed but long-term reconstitution. The existence of at least two classes of long-term repopulating cells is supported by genetic marking studies (Spangrude et al., 1995).

Many individual cytokines can influence the survival, proliferation, and differentiation of early progenitor cells; most success in generating early hemopoietic progenitor cells has come through the use of multiple cytokines or combinations of cytokines with stromal cells/stromal cell lines (Traycoff et al., 1995; Young et al., 1996b). While the effectiveness of stromal cell monolayers for maintaining hemopoiesis *in vitro* has been known for many years, the molecular basis of the important interactions between early hemopoietic cells and stromal cells is still far from clear. There is now interesting data to implicate integrin–extracellular matrix interactions in the maintenance of the self-renewal potential of stem cells (Levesque et al., 1996); indeed, the activation level of fibronectin receptors expressed by $CD34^+$ hemopoietic progenitor cells correlates well with their proliferative potential. Furthermore, the activation of the fibronectin receptors on early progenitors appears to be controlled by the cytokines which control the proliferation (Levesque et al., 1996): IL-3, GM-CSF, and SCF; the synergistic action of these cytokines on proliferation also extends to the induction of fibronectin-mediated adherence of early hemopoietic progenitor cells.

Improvements in bone marrow transplantation and/or the genetic manipulation of the hemopoietic system will require a better understanding of self-renewing, hemopoietic stem cells. For bone marrow transplantation, identification of the normal cells with LTRA should eventually allow the removal of neoplastic cells. Expansion of LTRA cells *in vitro* would allow the establishment of donor banks on a much wider scale. Whilst the proliferative potential of early bone marrow and peripheral blood progenitors is similar, in mice the fetal liver progenitors appear to be greater than adult progenitors. If cord blood progenitors can be expanded, these cells may have even greater potency for transplantation.

The ultimate source of primitive progenitor cells is arguably ES cells. It is not only possible to expand murine ES cells *in vitro;* several laboratories have also been able to generate hemopoietic stem cells from murine ES cells (Nakano et al., 1994; Lieschke and Dunn, 1995; Palacios et al., 1995; Hole et al., 1996). Once LIF is withdrawn from the ES cell cultures, differentiation is initiated, and after 4 days primitive hemopoietic progenitor cells can be detected. Interestingly, the hemopoietic cells only develop if the embryoid bodies remain intact. The repopulation potency of these cells is still under investigation, but the ability to manipulate ES cells in culture would make this source of hemopoietic cells attractive for genetic therapies. Although no exogenous cytokines are necessary for the production of murine hemopoietic progenitor cells from ES cells, polymerase chain reaction (PCR) analysis of the embryoid bodies has detected SCF message. Several hemopoietin receptors (e.g., epo, SCF, macrophage M-CSF, IL-3, G-CSF, and gp130) were also detected by reverse transcription PCR, but expression was weak (or nonexistent) until after the hemopoietic cells were detected (Hole et al., 1996).

If retroviruses are to be effective genetic vectors, it is essential that cell division of the hemopoietic stem cell can be stimulated without inducing differentiation. Whilst it is possible to express genes from retroviral vectors in some clones of progenitors, these clones invariably cease proliferating after 30 weeks (Tumas et al., 1996). Unless vectors which can infect quiescent hemopoietic stem cells can be developed, it will be essential to develop cytokine/stromal culture systems which can stimulate some cell division without inducing differentiation (Goodell et al., 1996). If we can improve our understanding of cytokine–receptor and integrin–extracellular matrix signaling, it is likely that we will be able to introduce genes into hemopoietic cells with self-renewing capacity.

STROMAL CELL MAINTENANCE OF LONG-TERM REPOPULATING STEM CELLS

The first success in producing significant numbers of hemopoietic progenitors in culture was achieved by using bone marrow stromal cells (Dexter et al., 1977). Dexter and his colleagues established murine bone marrow stromal layers and reseeded these with bone marrow cells. They were able to harvest white blood cells from these cultures for many months. There has always been some uncertainty about the role of the stromal cells in the maintenance of hemopoietic stem cells—are they a source of cytokines, integrin–extracellular matrix interactions, or metabolite control? Direct contact between the stroma and stem cells appears to be required for the maintenance of B-lymphocyte production (Manabe et al., 1994); however, the maintenance of $CD34^+$, lin^-, $HLA\text{-}Dr^-$ cells does not require direct interactions with stroma (Verfaillie, 1993). Indeed, by using purified $CD34^+$ cells from cord blood, more self-renewing, long-term culture-initiating cells are produced if direct contact with the stroma can be avoided (Abe et al., 1996).

From the earliest observations on stem cell proliferation and differentiation in the spleen, it was suggested that different microenvironments controlled the generation of lineage-specific progeny and the self-renewal of stem cells (McCulloch, 1970). It appeared that the stromal cells were heterogeneous. Indeed, it is probable that only a small proportion of the stromal cells established in Dexter cultures are capable of supporting hemopoiesis. Several groups have provided support for this concept by isolating stromal cell lines with different abilities for sustaining lymphoid and myeloid cells (Deryugina and Muller-Sieberg, 1993). By infecting murine fetal liver cells with a temperature-sensitive SV40 large-T antigen, many stromal cell lines have been developed (Wineman et al., 1996). Many of these clones did not support hemopoiesis at all; others allowed only limited stem cell persistence, and one cell line (S17) sustained high levels of primitive, long-term repopulating cells (Abe et al., 1996). In transplantation studies the stem cells maintained on S17 stromal cells compete well with primary stem cells from bone marrow. Whilst many cytokines (flt-3 ligand, SCF, IL-6, IGF-1, and TGF-β) appear to be produced by S17 cells, similar combinations of cytokines were produced by the stromal cell lines which failed to sustain stem cells. It is still unclear whether there is an uncharacterized soluble- or membrane-associated protein produced by S17 cells, which is critical for stem cell maintenance.

Although stromal cell cultures produce many cytokines (Quesenberry et al., 1994), the effectiveness of stromal cells for supporting the maintenance of early progenitor cells can be modulated by the addition of cytokines to the cultures (Szilvassy et al., 1996). Using several independently derived murine stromal cell lines (Sys-1, S17, and PA6), Szilvassy et al. observed that IL-3, SCF, or LIF increased the ability of the stromal cells to support stem cells in 4 week cultures. Thus, when purified stem cells (thy-1^{lo} Sca-1^{+} H2K^{hi}) isolated from the bone marrow of 5-fluorouracil–treated mice were co-cultured for 2 weeks with the stromal layer, in the absence of LIF, the long-term renewing cells decreased from almost 2% of the initial cell inoculum to 0.25% of the final cell suspension. When LIF (10 ng/ml) was added to this stromal cell/stem cell co-culture system, the long-term renewing cells were maintained at their original frequency even after 2 weeks.

Since LIF does not sustain cultures of purified hemopoietic stem cells in the absence of stroma, either LIF acts indirectly to induce the production of essential "stem cell factor(s)" by the stromal cells or both LIF and a stromal cell factor are required to sustain the long-term renewing cells. LIF does upregulate the production of several cytokines by stromal cells, including IL-1β, IL-2, IL-6, G-CSF, TGF-β, LIF, and SCF (Szilvassy et al., 1996). In particular, the SCF appears to be necessary for the effects of LIF in this stromal cell system.

The induction of cytokines/growth factors by other cytokines is a feature of the cytokine/growth factor regulatory network. The induction of growth factor/cytokine cascades not only makes the analysis of cytokine action difficult, but it has confounded our ability to define the physiology and/or mechanism of action of a particular growth factor. Even in the simplest cell biological system, it is difficult to interpret results in a cause-and-effect framework. G-CSF was first detected by its differentiation-inducing activity on murine myeloid leukemia cell line WEHI-3B(D^{+}) cells (Burgess and Metcalf, 1980). Further analysis has now shown that the effects of G-CSF on WEHI-3B(D^{+}) cells are density dependent (Bohmer and Burgess, 1988) and that the conditioned medium from high-density WEHI-3B(D^{+}) cells induces the differentiation. The identity of this factor is still to be determined, but clearly there is more to be learned about G-CSF–induced control of differentiation signals in myeloid cells.

Interestingly, an analysis of peripheral blood stem cells (PBSCs) has revealed a subpopulation of cells capable of inducing cytokine production from stromal cells (Mielcarek et al., 1996). In the presence of G-CSF, the kinetics of neutrophil recovery from bone marrow stem cells and PBSC transplants are essentially the same, and it is still not clear why PBSCs lead to a more rapid platelet recovery than bone marrow–derived stem cells. Since the repopulating cells from both sources appear to be similar, it has been suggested that there may be accessory cells associated with the PBSCs which improve the production of megakaryocytes and platelets. There are many more T lymphocytes associated with G-CSF–mobilized PBSCs, but these are not responsible for stimulating cytokine production by stromal cells. It is the $CD14^+$ monocytes which induce IL-6 and G-CSF secretion when added to stromal cells. $CD14^+$ monocytes secrete IL-1α and IL-1β, both of which are required for induction of cytokine synthesis by the stromal cells (Bagby, 1989; Dinarello, 1996). Whilst IL-6 is known to accelerate the reconstitution of platelets (D'Hondt et al., 1995), $CD34^+$/PBSCs support early recovery of platelets despite the depletion of $CD14^+$ monocytes. Of course, the $CD34^+$/PBSCs could contain $CD14^+$ precursors which expand and differentiate on transplantation, or the peripheral blood $CD34^+$ cells may be qualitatively different from $CD34^+$ stem cells from bone marrow. This conundrum might be resolved by identifying and removing the $CD14^+$ precursors (if they exist) or perhaps by testing IL-1α/IL-1β–deficient PBSCs from IL-1–deficient mice in a murine stem cell engraftment model (Zheng et al., 1995).

This discussion would not be complete without mentioning the putative inhibitory regulators of long-term repopulating cells produced by the stromal cells. TGF-β has an inhibitory effect on hemopoiesis both *in vitro* and *in vivo*. The presence of TGF-β1 mRNA has been detected in stroma, and the TGF-β1 protein is present in the media of long-term bone marrow cultures (Cashman et al., 1990). TGF-β1 has been shown to inhibit the early divisions of long-term repopulating cells (LTR-HSC) (Sitnicka et al., 1996b). Using a highly enriched population of LTR-HSC (obtained by sorting of lin^--depleted bone marrow cells for $rhodamine^{lo}$/$Hoecht^{lo}$ cells), Sitnicka et al. observed proliferation and differentiation in the presence of IL-3, IL-6, and SCF. After 8 days in culture, the LTR-HSC proliferated from 70 cells to 300,000, but the number of cells capable of forming high proliferative potential colonies (HPP-CFC) declined from 64 to 0. When TGF-β1 (1 ng/ml) was added to the conditioned medium, proliferation was reduced sevenfold, but there were actually more HPP-CFCs (90 cells) than at the start of the culture. Although there were no GM-CFCs at the start of these cultures, after 8 days they represented 4% of the total cell number. TGF-β1 (1 ng/ml) inhibited the production of GM-CFC by 50%; thus TGF-β appears to inhibit the later expansion divisions of progenitor cells more than the divisions between the LTR-HSCs and GM-CFCs.

In culture systems designed to expand human long-term initiating cells (Soma et al., 1996) or murine long-term repopulating cells, TGF-β inhibits stem cell survival. Indeed, if neutralizing antibodies against TGF-β are added to the 4 day cultures, almost 30% more long-term repopulating cells are recovered. TGF-β inhibits the stimulation of stem cells by the flt-3 ligand (Jacobsen et al., 1996). Differentiation toward myeloid and lymphoid precursors is inhibited with equal potency. Similarly, neutralizing antibodies against TGF-β stimulate (sevenfold) the clonal proliferation of lin$^-$ Sca-1$^+$ stem cells. This report also describes inhibitory effects of TNF-α on hemopoietic stem cells; however, TNF-α is considerably less potent than TGF-β.

There has been a long-standing interest in the molecular mechanisms which link the nervous and hemopoietic systems (Savino and Dardenne, 1995). In recent years, several neuropeptides have been shown to induce cytokine production by stromal cells and monocytes (Rameshwar et al., 1994). The substance P peptide induces the production of SCF by bone marrow stroma (Rameshwar and Gascon, 1995), and as a consequence, substance P is an indirect stimulator of early hemopoietic events. In contrast, the related neuropeptide NK-A inhibits the production of CFCs in short-term bone marrow cultures (Rameshwar and Gascon, 1996). Most of the inhibitory action of NK-A is attributable to the induction of TGF-β1 secretion from stromal cells. Since substance P and NK-A are the products of the same gene, it is conceivable that regulators of differential splicing are also hemopoietic regulators. Clearly, indirect regulation of hemopoiesis through the regulation of cytokine production by stromal cells provides a sensitive and powerful system for amplifying small changes in a single cytokine of the network. Furthermore, it is now clear that the stroma can act to initiate an appropriate hemopoietic response to signals emanating from other tissue systems.

Although competent bone marrow stromal layers can establish spontaneously *in vitro,* human stromal layers are often unreliable and/or short lived. Basic fibroblast growth factor (bFGF) not only increases the rate of establishment of the stromal layers, but improves the myelopoietic support provided by the stromal cells (Oliver et al., 1990; Wilson et al., 1991). Furthermore, human stromal cells established in the presence of bFGF are able to support bone marrow progenitor cell production for at least 2 months. Using k-FGF, Quito et al. (1996) were able to show that stromal cells could support human progenitor cell production for more than 8 months. Whilst the FGFs may influence the differentiation and/or proliferation of stem or hemopoietic progenitor cells directly, their major effect appears to be on the survival and quality of the stromal layer. It is likely that the FGFs increase the production of proteoglycan and extracellular matrix secretion by the stromal cells, and it is known that some of these proteoglycans (e.g., heparan sulfate) are required for cytokine expansion of long-term culture-initiating cells (Gupta et al., 1996). In part, the matrix components are required for the association of the hemopoietic stem cells with the stroma (Siczkowski et al., 1992; Bruno et al., 1993), but these polymers are highly charged and also interact with many of the cytokines known to influence the survival of long-term repopulating cells.

It is interesting to note that the CD34 cell surface marker is a highly glycosylated glycoprotein (a sialomucin), and when this is knocked out of the mouse genome there are decreased numbers of progenitor cells (Cheng et al., 1996). There is a delay in the establishment of myelopoiesis in both $CD34^{-/-}$ embryoid bodies and $CD34^{-/-}$ mouse embryos. In the bone marrow of adult mice the frequency of colony-forming cells is 3-4 fold reduced, but there is not significant change in marrow cellularity or the number and composition of either the white or red blood cells. It is important that the range of extracellular matrix proteins involved in the modulation of stem cell survival be investigated in depth. The right combination of these proteins and cytokines may eventually be able to replace the need for stromal cells when expanding stem cells *in vitro* or removing contaminating tumor cells by culture *in vitro.*

CYTOKINES AND STEM CELL EXPANSION AND MOBILIZATION

Until recently, most human hemopoietic culture systems only supported the production of mature cells from lineage-specific progenitors, and even complex combinations of cytokines had had limited success in sustaining self-renewing stem cells in these cultures. However, stimulation of stromal cells and rapid metabolite removal in continuous perfusion systems have improved the longevity of hemopoietic stem cells in culture (Schwartz et al., 1991a,b). When combinations of SCF, IL-1, IL-3, GM-CSF, and erythropoietin are added to these perfusion or bioreactor cultures of tumor bone marrow, it is possible to achieve a four- to eightfold stimulation of long-term culture-initiating cells (Koller et al., 1993). As mentioned earlier, ES cells can produce hemopoietic stem cells; however, at present, human ES cells are not readily available for *ex vivo* expansion. Umbilical cord blood stem cells have a greater proliferative capacity than their adult counterparts (Broxmeyer et al., 1991; Emerson, 1996).

Stem cells can be mobilized into the peripheral blood of humans by cytotoxic drugs (Juttner et al., 1988) or cytokines (Duhrsen et al., 1988; Bodine et al., 1996; Yamamoto et al., 1996). The normal frequency of $CD34^+$ cells is usually 40/100,000 white blood cells. This frequency increases to >1,000/100,000 after patients are treated with G-CSF or >300/100,000 after GM-CSF treatment (Ho et al., 1996). However, the proportion of early progenitors ($CD34^+$/$CD38^-$) in the blood of patients treated with G-CSF is only a quarter of the proportion of early progenitors in the blood of patients treated with GM-CSF. The combination of G-CSF and GM-CSF produces mobilized PBSCs with the characteristics of the $CD34^+$ cells found in umbilical cord blood. Furthermore, almost three times as many peripheral blood progenitor cells can be harvested using the GM-CSF/G-CSF combination in place of G-CSF or GM-CSF alone (Winter et al., 1996).

Whilst many cytokines induce a rapid mobilization of hemopoietic stem cells, there is usually a consequential short-term depletion of bone marrow progenitor cells (Bodine et al., 1996). However, 2 weeks after treating mice with G-CSF and SCF there was a 10-fold increase in the repopulating ability of bone marrow cells (Bodine et al., 1996). It will be interesting to explore whether routine "priming" of potential bone marrow or peripheral blood progenitor donors leads to higher yields of repopulating cells. Already it is clear that the pretreatment of patients with SCF, followed by G-CSF, increases the circulating peripheral blood progenitor cells 30-fold over G-CSF alone (Elwood et al., 1996). The quality of the repopulating cells and the timing of the priming doses, as well as the most effective cytokine combination, would need to be established.

One combination of cytokines—IL-1β, IL-3, IL-6, G-CSF, GM-CSF, and SCF—appears to be very effective for stimulating the production of $CD34^+$ cells from peripheral blood (Brugger et al., 1993). It is quite confusing to try and predict which combination will be effective for sustaining or expanding self-renewing stem cells. For example, there are a number of reports indicating that IL-3 and IL-1 drive differentiation (toward the production of lineage-specific progenitors) rather than self-renewal, so the repopulation potential of these cells is reduced (Yonemura et al., 1996). Similar observations were made by one group using either G-CSF or GM-CSF (Brugger et al., 1993), but most studies have found G-CSF and GM-CSF to augment/sustain self-renewal (Peters et al., 1995). Clearly, the actions of different cytokines are dependent on the presence of other cytokines. Thus SCF and IL-3, acting together or alone, do not sustain the self-renewing population; however, when either IL-6 or IL-1 is added to the cytokine mixture, stem cell survival is enhanced. As mentioned earlier, these studies are often complicated because an endogenous cytokine network is activated by specific stimulators; e.g., G-CSF stimulates the release of IL-1α and IL-6 from stromal cell subsets (Mielcarek et al., 1996).

What is the basis of this cytokine synergy? Perhaps there are several subsets of stem cells, each with multiple cytokine receptors—thus, the greater the range of cytokines, the higher the likelihood of stimulating more stem cell subsets. Alternatively, none of the individual cytokines/growth factors may be capable of sustaining the stem cell *per se,* but together they induce the secretion of unknown factor(s) from accessory cells (or even the $CD34^+$ cells themselves), and it is this factor which is responsible for stem cell maintenance or expansion. Yet another possibility lies in a potential requirement of stem cells for multiple signaling pathways (e.g., tyrosine kinase, JAKs, integrin induction, extracellular matrix secretion, gp130-associated signals, and phosphoinositol signals) to be activated before self-renewal divisions can be stimulated. Time and time again, researchers have reported increased stem cell production with increasingly complex mixtures of cytokines. A detailed analysis of the receptor (cytokine) classes and signaling systems triggered by these mixtures of cytokines may give some clue to the molecular basis of this synergy. If the combinations are involved in stimulating an as yet unknown stem cell regulator, it may be necessary to collect conditioned media from the stimulated cells, neutralize the known cytokines, and test for factors which modulate stem cell survival. Similarly, the adherence status of the stem cells needs to be determined; perhaps a particular combination of known cytokines is necessary for induction of the multiple processes required for self-renewal. Even at the single cell level, this will be a difficult problem to analyze, as induced autocrine factors may be quite potent effectors of self-renewal or differentiation.

FLT-3/FLK-2 LIGAND

Several years ago, a new family of cytokine receptors, flt-3 (also called flk-2), was described (Matthews et al., 1991; Rosnet et al., 1991). Flt-3 is clearly related to the M-CSF and SCF receptors called c-fms and c-kit respectively, so it was reasonable to expect that the ligand for flt-3 would also be a hemopoietic regulator (Lyman et al., 1993; Hannum et al., 1994). After identifying, cloning, and expressing the flt-3 ligand (FL), Lyman et al. (1993) demonstrated that it stimulated, albeit weakly, thymidine incorporation by murine fetal liver lymphomyeloid stem cells (i.e., AA4.1^{+}, Sca-1^{+}, lin^{lo}) (Rusten et al., 1996). Murine FL also stimulates thymidine incorporation into a CD34^{+} subpopulation from human bone marrow cells. Although the proliferative response by murine stem cells to FL was significant (eightfold background), the FL response was less than 10% of the response of the stem cells to SCF; the effects of FL are much stronger when it is used as a synergistic factor with SCF (Lyman et al., 1993).

The synergistic action of FL extends to co-stimulation of stem cells with other cytokines: IL-3, IL-6, G-CSF, GM-CSF; however, SCF appears to be a more effective synergistic agent than FL (Hannum et al., 1994; Rasko et al., 1995; Rusten et al., 1996). The committed murine progenitor cells (AA4.1^{+}, Sca-1^{+}, lin^{high}) do not respond to FL, and there is only a weak synergistic interaction between FL and GM-CSF or IL-3. SCF is a much stronger stimulator of committed myeloid progenitor cells than FL (Hannum et al., 1994). There are neither direct nor synergistic actions of FL on erythroid progenitor cells. In combination with IL-3, FL was as potent as SCF in stimulating myeloid colony formation. Again, the most effective stimulus for colony formation from committed myeloid progenitors appears to be a mixture of three cytokines: SCF, IL-3 and FL (Rusten et al., 1996). When added together with SCF, IL-6, IL-11, or IL-3, there is a strong proliferative response of myeloid progenitors to FL (Hirayama et al., 1995). FL appears to be the most potent stimulus for enhancing the survival of primitive murine stem cells (lin^{-}, kit^{+}, Sca-1^{+}) (Rasko et al., 1995). Survival of these cells was enhanced even further when both FL and SCF were added to the cultures. The lymphohemopoietic progenitor cells (post-5-fluorouracil, lin^{-}, Ly6a/E^{+}) (Hirayama et al., 1995) are also stimulated weakly by FL.

B-lymphoid progenitor cell production is also stimulated by FL. The strongest response was observed with both FL and IL-11; interestingly, if either IL-3 or IL1α is added to these B-lymphoid–generating cultures, no pre B-cells are generated (Hirayama et al., 1995). G-CSF or IL-6 can also combine with

FL to generate pre-B cells, and again this stimulation is inhibited completely by IL-3 or IL-1α. It has been suggested that FL synergizes with other cytokines because it shortens the G1 phase of the cell cycle (Hirayama et al., 1995); TGF-β appears to antagonize this effect of FL on G1. Consequently, the doubling time of IL-3–stimulated blast cell colonies can be reduced from 21 to 17 h with FL. In the presence of TGF-β, IL-3, and FL, the doubling time returns to 23 h, but if antisense TGF-β oligonucleotides are added to the IL-3/FL–stimulated blast cell colonies, the doubling time decreases to 13 h.

Clearly, the differential effects of cytokines on the maturation and self-renewal of stem cells need to be investigated carefully, but the combination of IL-3 (or SCF), FL, and TGF-β antagonists appears to be a strong, cell-free proliferative stimulus for early hemopoietic stem cells. (Ohishi et al., 1996).

FL can stimulate quiescent, human hemopoietic stem cells ($CD34^+$, $CD38^-$ cells) (Shah et al., 1996). The stimulatory effects of FL have been observed in co-culture assays with bone marrow stromal cells and $CD34^+$, $CD38^-$ progenitor cells. In the presence of SCF, IL-6, and IL-3, these cultures increase in cell number 30–100-fold after 3 weeks. When FL is added to the culture medium, cell numbers increase 300–2,000-fold (Shah et al., 1996). In the absence of stroma, the SCF/IL-6/IL-3 combination fails to stimulate proliferation, but if FL is also present, even in the absence of stroma, proliferation occurs and cell numbers increase 10–400-fold. Although the IL-6 in the SCF/IL-6/IL-3 combination only increases cell production marginally, the absence of either IL-3 or SCF reduces proliferation significantly (Shah et al., 1996). Whilst SCF/IL-3 stimulates the expansion of $CD34^+$, $CD38^-$ cells (1500-fold) in these stromal layer cultures, the expansion only continues for 50–60 days. If FL is present as well, there is a greater expansion (3,000-fold) and the $CD34^+$, $CD38^-$ cells persist for more than 100 days. Again, unless FL is present, the production of colony-forming cells stops by 60 days. Indeed, when FL/SCF/IL-3/IL-6 are present, the proliferative capacity of $CD34^+$, $CD38^-$ cells in this stromal co-culture system increases steadily up to 12 weeks in culture (Shah et al., 1996). Gene transfer into hemopoietic stem cells may be facilitated by co-culture systems such as this; in particular, since FL appears to stimulate the proliferation of quiescent stem cells, it should facilitate the introduction of genetic material into repopulating, hemopoietic cells.

In mice, daily injections of FL (10 μg/day) stimulate progenitor cell production and the mobilization of day 13 CFU-S (Brasel et al., 1996). After 10 days, the total number of blood cells increases ~3-fold, with a large increase in the monocyte population. Despite the daily injections, after 15 days the white cell numbers in the blood decrease; however, there is still a fourfold increase of white blood cell numbers in the spleen. Thus, despite the increased frequency of myeloid progenitor cells after 10 days of FL treatment, there is no profound increase in white blood cell numbers in the spleen or blood after 15 or 20 days; presumably, other cytokines limit the production of mature (functional) white blood cells. Note that the converse can also be true: in c-mpl$^{-/-}$ (Alexander et al., 1996) and tpo$^{-/-}$ mice (Carver-Moore et al., 1996) there is a profound reduction in the frequency of myeloid progenitor cells, but no change in the number of bone marrow or circulating neutrophils.

STEM CELL FACTOR (SCF)

Two of the earliest descriptions of genetic defects affecting hemopoiesis were associated with the steel (Sl) and the white spotting (W) mutations of mice (Russell, 1979). The absence of c-kit (W) or SCF (Sl) is associated with macrocytic anemia, mast cell deficiency, sterility, and pigment defects. Many years later, after the stem cell defect (W) and the environmental defect (Sl) had been explored in detail, molecular biologists identified the W mutation as an SCF receptor (c-kit) defect and the Sl mutation as an SCF (also called *kit ligand Steel Factor*) defect.

Although it is a weak stimulator of *in vitro* colony-forming cells, SCF synergizes with epo, G-CSF, GM-CSF, IL-1, IL-3, IL-6, IL-7, IL-11, and thrombopoietin to increase both the number and size of hemopoietic colonies. SCF appears to be a more potent stimulator of cells in the pre-CFC compartment (Olweus et al., 1996). In cultures containing SCF, epo, IL-3, IL-6, GM-CSF, and G-CSF, an enriched population of primitive murine hemopoietic cells ($CD34^{hi}$, $CD38^{lo}$, $CD50^{+}$), usually 1% of $CD34^{+}$ cells, was capable of generating $CD34^{+}$ cells for 14 days. When SCF is omitted from these cultures, the frequency of $CD34^{+}$ cells decreases from 28 to 17% (Olweus et al., 1996). In contrast, the omission of epo does not alter the production of $CD34^{+}$ cells; however, the optimal self-renewal of the $CD34^{hi}$, $CD38^{lo}$, $CD50^{+}$ cells in these cultures required both epo and SCF.

Blast cell colony-forming cells (BC-CFC) have long been used *in vitro* as a surrogate marker for quantitating hemopoietic stem cells (Migliaccio et al., 1996); however, analysis of BC-CFC subpopulations using flow cytometry and rhodamine staining indicates that the stem cells are only a small proportion of the BC-CFC (Migliaccio et al., 1996). A primitive stem cell population can be prepared by selecting for WGA^{+}, 15.1^{-}, rho^{-} cells. Almost 10% of these cells were capable of permanent reconstitution of W/W^{v} mice. When the WGA^{+}, $lin\text{-}15^{-}$, rho^{lo} cells are cultured in the presence of IL-3, SCF, or a mixture of the two cytokines, GM-CFC are generated (Migliaccio et al., 1996). SCF also combines with IL-3 to stimulate the proliferation and differentiation of myeloid cells. SCF also synergizes with a number of cytokines (e.g., IL-1β and LIF) for increasing the generation of HPP-CFU, CFC mix, GM-CFC, and BFU-E in culture (Imamura et al., 1996). When all three cytokines are present, there is a significant increase in the production of HPP-CFC and CFC mixed, so it is reasonable to assume that this combination of cytokines has a powerful stimulatory effect on the survival and expansion of the long-term repopulation cells.

Interestingly, SCF also appears to have a positive effect on the cell systems which receive and expand hemopoietic stem cells after transplantation

(Broudy et al., 1995). Pretreatment of mice with SCF and IL-11 resulted in improved recovery (1,000-fold) of GM-CFC from the spleen; furthermore, the pretreatment improved the survival of bone marrow transplanted, 5-fluorouracil–ablated mice (Broudy et al., 1996). The increased numbers of progenitor cells might be explained by improved seeding efficiencies of stem cells into the spleen or bone marrow (Broudy et al., 1996) and/or by an increased capacity of the stroma to stimulate the proliferation of seeded stem cells, but the molecular changes on stromal cells induced by SCF/IL-11 have yet to be identified.

GRANULOCYTE-COLONY STIMULATING FACTOR

G-CSF was discovered as a cytokine capable of inducing the differentiation of WEHI-3B(D^+) cells (Burgess and Metcalf, 1980). Its apparent lineage specificity, when acting on normal bone marrow, suggested that it might be a regulator of the later stages of neutrophil production. However, during the first clinical trial of human G-CSF, Duhrsen et al. (1988) observed that G-CSF stimulated the release of hemopoietic progenitor cells into the peripheral blood. While this action of G-CSF may well be indirect, it has been reported that purified hemopoietic stem cells (lin^-, $CD34^+$, $CD38^-$) proliferate for 14 days in the presence of SCF and G-CSF (Nishi et al., 1996). Both cytokines are necessary, as there is little or no proliferation in the presence of either cytokine alone. Again, since G-CSF can stimulate the production of other cytokines, some caution must be used when interpreting these results. These analyses need to be repeated at the single cell level under conditions where the effects of induced cytokine production are minimized.

The mobilization of peripheral blood progenitor cells is one of the most useful clinical applications of G-CSF (Prosper et al., 1996; Varas et al., 1996). These cells can be purged of tumor cells and used directly for hemopoietic reconstitution or manipulated *in vitro* for subsequent gene therapy. Whilst G-CSF stem cell mobilization is already used widely, improved yields of stem cells would facilitate both the routine use of these cells and the feasibility of gene therapy. By pretreating patients with IL-3, the yield of G-CSF–mobilized progenitor cells can be increased threefold (Huhn et al., 1996). Similar synergy has been observed by combining SCF and G-CSF (Nishi et al., 1996). In a murine model, the use of both cytokines more than doubled the yield of spleen colony-forming cells in the peripheral blood (Drize et al., 1996). By optimizing the timing of the SCF and G-CSF delivery, even larger increases in human PBSCs have been achieved (G.C. Begley, personal communication). In analyzing the utility of these multiple cytokine regimes, it will be important to monitor the long-term repopulation ability of the harvested PBSCs and/or stem cells after the *in vitro* manipulation. Whilst short-term reconstitution can be enhanced and the efficiency of gene transfer can be increased by multiple cytokine stimulation, up to now it has proved difficult to transfer new genes to the long-term repopulating cells.

INTERLEUKIN-3

IL-3 promotes the survival of primitive human hemopoietic cells (Brandt et al., 1994). A synthetic derivative of IL-3 (with 10-fold increased biological activity) accelerates the recovery of platelets and neutrophils after radiation-induced bone marrow aplasia (Farese et al., 1996). However, questions have been raised about the effects of both IL-1 and IL-3 on the self-renewing capacity of hemopoietic stem cells (Yonemura et al., 1996). Yonemura et al observed short-term (7 day) *in vitro* expansion (approximately twofold) of primitive hemopoietic progenitor cells; however, by 14 days the presence of IL-3 had reduced the proportion of Ly6a/E$^+$, c-kit$^+$ cells to 10% of control cultures. Similarly, the repopulating ability of primitive murine hemopoietic progenitor cells (Ly6a/E$^+$, c-kit) was abolished by culturing in the presence of either IL-3 or IL-1 (Yonemura et al., 1996). These cultures all contained SCF, IL-6, IL-11, and epo, so the effects of IL-3 and IL-1 on self-renewal capacity must dominate the proliferative and/or survival activities of the other cytokines. These observations are difficult to reconcile with many previous studies that report that IL-3 and SCF are required for the survival of dormant primitive progenitor cells (Bodine et al., 1989; Katayama et al., 1993; Kobayashi et al., 1996).

THROMBOPOIETIN

Thrombopoietin (tpo, also called mpl ligand) regulates the production and maturation of megakaryocytes and platelets (Ebbe, 1997). Discovered only recently, tpo was initially believed to be specific for the megakaryocyte/platelet lineage (Kaushansky, 1995). However, tpo also appears to act on primitive progenitor cell populations (Zeigler et al., 1994; Broudy et al., 1995; Farese et al., 1995; Debili et al., 1997). Indeed, c-mpl$^{-/-}$ (Alexander et al., 1996) and tpo$^{-/-}$ (Carver-Moore et al., 1996) deficient mice have reduced numbers of multipotential hemopoietic progenitor cells in both bone marrow and spleen. Despite the reduced numbers of myeloid progenitor cells, the mpl$^{-/-}$ and tpo$^{-/-}$ mice have normal neutrophil numbers. At present there are no data to indicate how the normal neutrophil numbers are achieved. It has also been shown that tpo stimulates the expansion of early hemopoietic cells *in vitro* (Kobayashi et al., 1996; Sitnicka et al., 1996a; Young et al., 1996a). Tpo acts in synergy with IL-3 to increase the production of primitive hemopoietic cells in culture.

Cytokines (growth factors, interleukins and colony-stimulating factors) regulate all aspects of the development of hemopoietic progenitor cells; however, the most potent modulator of the primtive hemopoietic cells appears to be bone marrow–derived stromal cells. Stromal cells provide multiple cytokines and extracellular matrix components; together with the appropriate hemopoietic stem cell receptor display, these two signaling systems appear to be required for the stimulation of self-renewal cell divisions. At present, it appears difficult to introduce new genes into long-term repopulating cells; however, the appropriate mixture of cytokines, extracellular matrix components, and differentiation inhibitors can be expected eventually to provide the appropriate conditions for gene transfer into stem cells.

AUTOCRINE STIMULATION IN MYELOID LEUKEMIAS

The uncontrolled production of myeloid cells which occurs in leukemia can be caused by many distinct mechanisms, including the activation of ras (Farr et al., 1988), abl (Ben-Neriah et al., 1986) or c-fms (Jacobs et al., 1990). There have been a number of reports linking the autocrine production of cytokines/growth factors to leukemogenesis (Cozzolino et al., 1989), but invariably these studies have been directed toward the characterization of these cells *in vitro.* Unfortunately, when cells are cultured *in vitro,* growth factor/cytokine production can be stimulated (Kaufman et al., 1988), so it is difficult to determine whether these autocrine regulators influence the initial or even later stages of leukemogenesis *per se* (Lang and Burgess, 1990).

Several types of mutations appear to influence the leukemogenesis process: The activity of nuclear proteins such as myc or myb (Sanchez-Garcia and Grutz, 1995) may be perturbed, increasing the ability of early progenitor cells to self-renew ("immortalize"). Once lesions such as this are complemented by excess growth factor/cytokine (e.g., autocrine) synthesis, cell production would be dysregulated. The combined effects of these "immortalizing" and "growth factor–related" lesions may lead to excess cells within different hemopoietic compartments, i.e., blast cell leukemias or chronic granulocyte leukemia.

Myelomonocytic leukemias are unusual in that they will proliferate *in vitro* without an exogenous source of GM-CSF. Thus, chronic myelomonocytic leukemia (CMML) cells are known to secrete sufficient GM-CSF and IL-6 to allow colony formation *in vitro* (Everson et al., 1989). IL-10 inhibits cytokine production. IL-10 has been used to demonstrate that autocrine GM-CSF and not G-CSF, IL-3, or IL-6 was responsible for the autonomous proliferation of CMML cells *in vitro* (Geissler et al., 1996).

The proliferative state of leukemic blast cells may be controlled indirectly by the cytokine-inductive effects of IL-1 (Dinarello, 1996). IL-1 increases the production of GM-CSF, which appears to be responsible for the proliferative effects of IL-1 (Russell, 1992).

IL-1 secretion has been associated with several leukemic cell types, including acute and chronic myeloid and acute and chronic lymphocytic leukemias.

The induction of IL-1β is associated with the leukemic state; however, its proliferative effects appear to be mediated by GM-CSF. Antisense oligonucleotides to GM-CSF mRNA reduce the proliferation of these cells (Russell, 1992). Thus, AML cell proliferation can be reduced by neutralizing antibodies to IL-1β (but not IL-1α), soluble IL-1 receptor, antisense oligonucleotides against IL-1–converting enzyme (Stosic-Grujicic et al., 1995), or, as mentioned above, antisense GM-CSF oligonucleotides.

It is presumed that autocrine GM-CSF facilitates the survival and/or proliferation of all types of myelomonocytic leukemia cells, but there is no evidence that it influences the self-renewal capacity of the leukemic progenitor cells. The increased self-renewal capacity of these cells is presumably due to mutations controlling the expression of immortalizing genes such as myb or myc.

REFERENCES

Abe T, Takaue Y, Kawano Y, Kuroda Y (1996): Effect of recombinant erythropoietin in interaction with stromal factors on cord blood hematopoiesis. *Blood* 87:3212–3217.

Alexander WN, Roberts AW, Nicola NA, Li R, Metcalf D (1996): Deficiencies in progenitor cells of multiple hematopoietic lineages and defective megakaryocytopoiesis in mice lacking the thrombopoietin receptor c-Mpl. *Blood* 87:2162–2170.

Bagby GC (1989): Interleukin-1 and hematopoiesis. *Blood Rev* 3:152–161.

Ball TC, Hirayama F, Ogawa M (1995): Lymphohematopoietic progenitors of normal mice. *Blood* 85:3086–3092.

Ben-Neriah Y, Daley GQ, Mes-Masson M-M, Witte ON, Baltimore D (1986): The chronic myelogenous leukemia-specific p210 protein is the product of the *bcr/abl* hybrid gene. *Science* 233:212–214.

Bodine D, Karlsson S, Nienhuis AW (1989): Combination of interleukins 3 and 6 preserves stem cell function in culture and enhances retrovirus-mediated gene transfer into hematopoietic stem cells. *Proc Natl Acad Sci USA* 86:8897–8901.

Bodine DM, Seidel NE, Orlic D (1996): Bone marrow collected 14 days after *in vivo* administration of granulocyte colony-stimulating factor and stem cell factor to mice has 10-fold more repopulating ability than untreated bone marrow. *Blood* 88:89–97.

Bohmer RM, Burgess AW (1988): Granulocyte colonly stimulating factor (G-CSF) does not induce differentiation of WEHI-3B(D^+) cells but is required for the survival of the mature progeny. *Int J Cancer* 41:53–58.

Brandt JE, Bhalla K, Hoffman R (1994): Effects of interleukin-3 and c-kit ligand on the survival of various classes of human hematopoietic progenitor cells. *Blood* 83:1507–1514.

Brasel K, McKenna HJ, Morrissey PJ, Charrier K, Morris AE, Lee CC, Williams DE, Lyman SD (1996): Hematologic effects of flt3 ligand *in vivo* in mice. *Blood* 88:2004–2012.

Broudy VC, Lin NL, Kaushansky K (1995): Thrombopoietin (c-mpl ligand) interacts synergistically with erythropoietin stem cell factor and interleukin-11 to enhance murine megakaryocyte colony growth and increases megakaryocyte ploidy *in vitro*. *Blood* 85:1719–1726.

Broudy VC, Lin NL, Priestley GV, Nocka K, Wolf NS (1996): Interaction of stem cell factor and its receptor c-*kit* mediates lodgment and acute expansion of hematopoietic cells in the murine spleen. *Blood* 88:75–81.

Broxmeyer HE, Kurtzberg J, Gluckman E, Auerbach AD, Douglas G, Cooper S, Falkenberg JH, Bard J, Boyse EA (1991): Umbilical cord blood hematopoietic stem and repopulating cells in human clinical transplantation. *Blood Cells* 17:313–329.

Brugger W, Mocklin W, Heimfeld S, Berenson RJ, Mertelsmann R, Kanz L (1993): *Ex vivo* expansion of enriched peripheral blood $CD34^+$ progenitor cells by stem cell factor, interleukin-1β (IL-1β), IL-6, IL-3, interferon-γ, and erythropoietin. *Blood* 81:2579–2584.

Bruno E, Luikart SD, Dixit V, Long MW, Hoffman R (1993): Heparan sulfate proteoglycan (HS-PG) modulates the adhesion of pluripotent hematopoietic stem cells to cytokines. *Exp Hematol* 21:1025.

Burgess AW, Metcalf D (1980): Characterization of a serum factor stimulating the differentiation of myelomonocytic leukemic cells. *Int J Cancer* 26:647–654.

Carver-Moore K, Broxmeyer HE, Luoh S-M, Cooper S, Peng J, Burstein SA, Moore MW, de Sauvage FJ (1996): Low levels of erythroid and myeloid progenitors in thrombopoietin– and c-*mpl*–deficient mice. *Blood* 88:803–808.

Cashman JD, Eaves AC, Raines EW, Ross R, Eaves CJ (1990): Mechanisms that regulate the cell cycle status of very primitive hematopoietic cells in long-term human marrow cultures. I. Stimulatory role of a variety of mesenchymal cell activators and inhibitory role of TGFβ_1. *Blood* 75:75–96.

Cheng J, Baumheuter S, Cacalano G, Carver-Moore K, Thibodeaux H, Thomas R, Broxmeyer HE, Cooper S, Mague N, Moore M, Lasky LA (1996): Hematopoietic defects in mice lacking the sialomucin CD34. *Blood* 87:479–490.

Cozzolino F, Rubartelli A, Aldinucci D, Sitia R, Torcia M, Shaw A, Di Guglielmo R (1989): Interleukin 1 as an autocrine growth factor for acute myeloid leukemia cells. *Proc Natl Acad Sci USA* 86:2369–2373.

Debili N, Cramer E, Wendling F, Vainchenker W (1997): *In vitro* effects of Mpl ligand on human hemopoietic progenitor cells. In Kuter DJ, Hunt P, Sheridan W, Zucker-Franklin D (eds): Thrombopoiesis and Thrombopoietins: Molecular, Cellular, Preclinical, and Clinical Biology. Totowa, NJ: Humana Press, Inc, pp 217–235.

Deryugina EI, Muller-Sieberg CE (1993): Stromal cells in long-term cultures: Key to the elucidation of hematopoietic development. *Crit Rev Immunol* 13:115–150.

Dexter TM, Allen TD, Lajtha LG (1977): Conditions controlling the proliferation of haemopoietic stem cells *in vitro*. *J Cell Physiol* 91:335–344.

D'Hondt V, Humblet Y, Guillaume T, Baatout S, Chatelain CMB, Longueville J, Feyens AM, De Greve J, Van Oosterom A, Von Graffenreid B, Donnez J, Symann M (1995): Thrombopoietic effects and toxicity of interleukin-6 in patients with ovarian cancer before and after chemotherapy: A multicentric placebo-controlled, randomized phase Ib study. *Blood* 85:2347–2353.

Dinarello CA (1996): Biologic basis for interleukin-1 in disease. *Blood* 87:2095–2147.

Drize N, Chertkov J, Samoilina N, Zander A (1996): Effect of cytokine treatment (granulocyte colony-stimulating factor and stem cell factor) on hematopoiesis and the circulating pool of hematopoietic stem cells in mice. *Exp Hematol* 24:816–822.

Duhrsen U, Villeval JL, Boyd J, Kannourakis G, Morstyn G, Metcalf D (1988): Effects of recombinant human granulocyte colony-stimulating factor on hematopoetic progenitor cells in cancer patients. *Blood* 72:2074–2081.

Ebbe S (1997): Foreword. In Kuter DJ, Hunt P, Sheridan W, Zucker-Franklin D (eds): Thrombopoiesis and Thrombopoietins: Molecular, Cellular, Preclinical, and Clinical Biology. Totowa, NJ: Humana Press, pp v–xi.

Elwood NJ, Zogos H, Willson T, Begley GC (1996): Retroviral transduction of human progenitor cells: Use of granulocyte colony-stimulating factor plus stem cell factor to mobilize progenitor cells *in vivo* and stimulation by Flt-3–Flk-2 ligand *in vitro*. *Blood* 88:4452–4462.

Emerson SG (1996): *Ex vivo* expansion of hematopoietic precursors, progenitors, and stem cells: The next generation of cellular therapeutics. *Blood* 87:3082–3088.

Everson MP, Brown CB, Lilly MB (1989): Interleukin-6 and granulocyte-macrophage colony-stimulating factor are candidate growth factors for chronic myelomonocytic leukemia cells. *Blood* 74:1472–1476.

Farese AM, Herodin F, McKearn JP, Baum C, Burton E, MacVittie TJ (1996): Acceleration of hematopoietic reconstitution with a synthetic cytokine (SC-55494) after radiation-induced bone marrow aplasia. *Blood* 87:581–591.

Farese AM, Hunt P, Boone T, MacVittie TJ (1995): Recombinant human megakaryocyte growth and development factor stimulates thrombocytopoiesis in normal nonhuman primates. *Blood* 86:54–59.

Farr CJ, Saiki RK, Erlich HA, McCormick F, Marshall CJ (1988): Analysis of *ras* gene mutations in acute myeloid leukemia by polymerase chain reaction and oligonucleotide probes. *Proc Natl Acad Sci USA* 85:1629–1633.

Geissler K, Ohler L, Fodinger M, Virgolini I, Leimer M, Kabrna E, Kollars M, Skoupy S, Bohle B, Rogy M, Lechner K (1996): Interleukin 10 inhibits growth and granulocyte/macrophage colony-stimulating factor production in chronic myelomonocytic leukemia cells. *J Exp Med* 184:1377–1384.

Goodell MA, Brose K, Paradis G, Connor AS, Mulligan RC (1996): Isolation and functional properties of murine hematopoietic stem cells that are replicating *in vivo*. *J Exp Med* 183:1797–1806.

Gupta P, McCarthy JB, Verfaillie CM (1996): Stromal fibroblast heparan sulfate is required for cytokine-mediated *ex vivo* maintenance of human long-term culture-initiating cells. *Blood* 87:3229–3236.

Hannum C, Culpepper J, Campbell D, McClanahan T, Zurawski S, Bazan JF, Kastelein R, Hudak S, Wagner J, Mattson J, Luh J, Duda G, Martina N, Peterson D, Meneon S, Shanafelt A, Meunch M, Keiner G, Namikawa R, Rennick D, Roncarolo M-G, Ziotnik A, Rosnet O, Dubreull P, Birnbaum D, Lee F (1994): Ligand for FLT3/FLK2 receptor tyrosine kinase regulates growth of haematopoietic stem cells and is encoded by variant RNAs. *Nature* 368:643–648.

Hirayama F, Lyman SD, Clark SC, Ogawa M (1995): The *flt*3 supports proliferation of lymphohematopoietic progenitors and early B-lymphoid progenitors. *Blood* 85:1762–1768.

Ho AD, Young D, Maruyama M, Corringham RET, Mason JR, Thompson P, Grenier K, Law P, Terstappen LWMM, Lane T (1996): Pluripotent and lineage-committed $CD34^+$ subsets in leukapheresis products mobilized by G-CSF, GM-CSF vs. a combination of both. *Exp Hematol* 24:1460–1468.

Hole N, Graham GJ, Menzel U, Ansell JD (1996): A limited temporal window for the derivation of multilineage repopulating hematopoietic progenitors during embryonal stem cell differentiation *in vitro*. *Blood* 88:1266–1276.

Huhn RD, Yurkow EJ, Tushinski R, Clarke L, Sturgill MG, Hoffman R, Sheay W, Cody R, Philipp C, Resta D, George M (1996): Recombinant human interleukin-3 (rhIL-3) enhances the mobilization of peripheral blood progenitor cells by recombinant human granulocyte colony-stimulating factor (rhG-CSF) in normal volunteers. *Exp Hematol* 24:839–847.

Imamura M, Zhu X, Han M, Kobayashi M, Hashino S, Tanaka J, Kobayashi S, Kasai M, Asaka M (1996): *In vitro* expansion of murine hematopoietic progenitor cells by leukemia inhibitory factor, stem cell factor, and interleukin-1β. *Exp Hematol* 24:1280–1288.

Jacobs A, Carter G, Ridge S, Hughes D, Geddes D, Bowen D, Holmes J, Clark RE, Padua RA (1990): Haematopoietic and molecular abnormalities in the myelodysplastic syndrome. In Sachs L, Abraham NG, Widermann CJ, Levine AS, Konwalinka G (eds): Molecular Biology of Haematopoiesis. Andover, Hampshire; Intercept, pp 571–577.

Jacobsen SEW, Veiby OP, Myklebust J, Okkenhaug C, Lyman SD (1996): Ability of flt3 ligand to stimulate the *in vitro* growth of primitive murine hematopoietic progenitors is potently and directly inhibited by transforming growth factor-β and directly inhibited by transforming growth factor-β and tumor necrosis factor-α. *Blood* 87:5016–5026.

Juttner CA, To HO, Ho JK, Bardy PG, Dyson PG, Haylock DN, Kimber RJ (1988): Early lympho-hematopoietic recovery after autografting using peripheral blood stem cells in acute non-lymphoblastic leukemia. *Transplant Proc* 20:40–42.

Katayama N, Clark SC, Ogawa M (1993): Growth factor requirement for survival in cell-cycle dormancy of primitive murine lymphohematopoietic progenitors. *Blood* 81:610–616.

Kaufman DC, Baer MR, Gao XZ, Wang ZQ, Preisler HD (1988): Enhanced expression of the granulocyte-macrophage colony stimulating factor gene in acute myelocytic leukemia cells following *in vitro* blast cell enrichment. *Blood* 72:1329–1332.

Kaushansky K (1995): Thrombopoietin: The primary regulator of platelet production. *Blood* 86:419–431.

Kobayashi M, Laver JH, Kato T, Miyazaki H, Ogawa M (1996): Thrombopoietin supports proliferation of human primitive hematopoietic cells in synergy with steel factor and/or interleukin-3. *Blood* 88:429–436.

Koller MR, Emerson SG, Palsson BO (1993): Large-scale expansion of human hematopoietic stem and progenitor cells from bone marrow mononuclear cells in continuous perfusion culture. *Blood* 82:378–384.

Ku H, Yonemura Y, Kaushansky K, Ogawa M (1996): Thrombopoietin, the ligand for the mpl receptor, synergizes with steel factor and other early acting cytokines in supporting proliferation of primitive hematopoietic progenitors of mice. *Blood* 87:4544–4551.

Lang RA, Burgess AW (1990): Autocrine growth factors and tumourigenic transformation. *Immunol Today* 11:244–245.

Levesque J-P, Haylock DN, Simmons PJ (1996): Cytokine regulation of proliferation and cell adhesion are correlated events in human CD34$^+$ hemopoietic progenitors. *Blood* 88:1168–1176.

Lieschke GJ, Dunn AR (1995): Development of functional macrophages from embryonal stem cells *in vitro*. *Exp Hematol* 23:328–334.

Lord BI, Dexter TM (1995): Which are the hematopoietic stem cells? *Exp Hematol* 23:1237–1241.

Lyman SD, James L, Vanden Bos T, de Vries P, Brasel K, Gliniak B, Hollingsworth LT, Picha KS, McKenna HJ, Splett RR, Fletcher FA, Maraskovsky E, Farrah T, Foxworthe D, Williams DE, Beckmann MP (1993): Molecular cloning of a ligand for the flt-3/flk-2 tyrosine kinase receptor: A proliferative factor for primitive hematopoietic cells. *Cell* 75:1157–1167.

Manabe A, Murti KG, Coustan-Smith E, Kumagai M, Behm FG, Raimond SD, Campana D (1994): Adhesion-dependent survival of normal and leukemic human B lymphoblasts on bone marrow stromal cells. *Blood* 83:758–766.

Matthews W, Jordan CT, Wiegand GW, Pardoll D, Lemischka IR (1991): A receptor tyrosine kinase specific to hematopoietic stem and progenitor cell-enriched populations. *Cell* 65:1143–1152.

McCulloch EA (1970): Control of hematopoiesis at the cellular level. In Gordon AS (ed): Regulation of Hematopoiesis. New York; Appleton, Century and Crofts, pp 133–159.

Mielcarek M, Roecklein BA, Torok-Storb B (1996): $CD14^+$ cells in granulocyte colony-stimulating factor (G-CSF)–mobilized peripheral blood mononuclear cells induce secretion of interleukin-6 and G-CSF by marrow stroma. *Blood* 87:574–580.

Migliaccio G, Baiocchi M, Adamson JW, Migliaccio AR (1996): Isolation and biological characterization of two classes of blast-cell colony-forming cells from normal murine marrow. *Blood* 87:4091–4099.

Nakano T, Kodama H, Honjo T (1994): Generation of lymphohematopoietic cells from embryonic stem cells in culture. *Science* 265:1098–1101.

Nicholls SE, Heyworth CM, Dexter TM, Lord JM, Johnson GD, Whetton AD (1995): IL-4 promotes macrophage development by rapidly stimulating lineage restriction of bipotent granulocyte-macrophage colony-forming cells. *J Immunol* 155:845–853.

Nishi N, Ishikawa R, Inoue H, Nishikawa M, Kakeda M, Yoneya T, Tsumura H, Ohashi H, Yamaguchi Y, Motoki K, Sudo T, Mori KJ (1996): Granulocyte-colony stimulating factor and stem cell factor are the crucial factors in long-term culture of human primitive hematopoietic cells supported by a murine stromal cell line. *Exp Hematol* 24:1312–1321.

Ogawa M (1993): Differentiation and proliferation of hematopoietic stem cells. *Blood* 81:2844–2853.

Ohishi K, Katayama N, Itoh R, Mahmud N, Miwa H, Kita K, Minami N, Shirakawa S, Lyman SD, Shiku H (1996): Accelerated cell-cycling of hematopoietic progenitors by the *flt3* ligand that is modulated by transforming growth factor-β. *Blood* 87:1718–1727.

Oliver LJ, Rifkin DB, Gabrilove JL, Hannocks MJ, Wilson EL (1990): Long-term culture of human bone marrow stromal cells in the presence of basic fibroblast growth factor. *Growth Factors* 3:231–236.

Olweus J, Terstappen WMM, Thompson PA, Lund-Johansen F (1996): Expression and function of receptors for stem cell factor and erythropoietin during lineage commitment of human hematopoietic progenitor cells. *Blood* 88:1594–1607.

Osawa M, Hanada K, Hamada H, Nakauchi H (1996): Long-term lymphohematopoietic reconstitution by a single CD34-low/negative hematopoietic stem cell. *Science* 273:242–244.

Palacios R, Golunski E, Samardis J (1995): *In vitro* generation of hematopoietic stem cells from an embryonic stem cell line. *Proc Natl Acad Sci USA* 92:7530–7534.

Peters SO, Kittler EL, Ramshaw HS, Quesenberry PJ (1995): Murine marrow cells expanded in culture with IL-3, IL-6, IL-11, and SCF acquire an engraftment defect in normal hosts. *Exp Hematol* 23:461–469.

Prosper F, Stroncek D, Verfaillie C (1996): Phenotypic and functional characterization of long-term culture-initiating cells present in peripheral blood progenitor collections of normal donors treated with granulocyte colony-stimulating factor. *Blood* 88:2033–2042.

Quesenberry PJ, Crittenden RB, Lowry P, Kittler EW, Rao S, Peters S, Ramshaw H, Stewart FM (1994): *In vitro* and *in vivo* studies of stromal niches. *Blood Cells* 20:97–104.

Quito FL, Beh J, Bashayan O, Basilico C, Basch RS (1996): Effects of fibroblast growth factor-4 (k-FGF) on long-term cultures of human bone marrow cells. *Blood* 87:1282–1291.

Rameshwar P, Ganea D, Gascon P (1994): Induction of IL-3 and granulocyte-macrophage colony-stimulating factor by substance P in bone marrow cells is partially mediated by human peripheral blood mononuclear cells. *J Immunol* 152:4044–4054.

Rameshwar P, Gascon P (1995): Substance P (SP) mediates production of stem cell factor and inteleukin-1 in bone marrow stroma: Potential autoregulatory role for these cytokines in SP receptor expression and induction. *Blood* 86:482–490.

Rameshwar P, Gascon P (1996): Induction of negative hematopoietic regulators by neurokinin-A in bone marrow stroma. *Blood* 88:98–106.

Rasko JE, Metcalf D, Rossner MT, Begley CG, Nicola NA (1995): The flt-3/flk-2 ligand: Receptor distribution and action on murine haemopoietic cell survival and proliferation. *Leukemia* 9:2058–2066.

Rosnet O, Marchetto S, deLapeyriere O, Birnbaum D (1991): Murine Flt3, a gene encoding a novel tyrosine kinase receptor of the PDGFR/CSF1R family. *Oncogene* 6:1641–1650.

Russell ES (1979): Hereditary anaemias of the mouse. *Adv Genet* 20:357.

Russell NH (1992): Autocrine growth factors and leukaemic haematopoiesis. *Blood Rev* 6:149–156.

Rusten LS, Luman SD, Veiby OP, Jacobsen SEW (1996): The FLT ligand is a direct and potent stimulator of the growth of primitive and committed human $CD34^+$ bone marrow progenitor cells in vitro. *Blood* 87:1317–1325.

Sachs L (1985): Regulatory proteins for growth and differentiation in normal and leukemic hematopoitic cells: Normal differentiation and the uncoupling of controls in myeloid leukemia. In Ford RJ, Maizel AL (eds): Mediators in Cell Growth and Differentiation. New York; Raven Press, pp 341–360.

Sanchez-Garcia I, Grutz G (1995): Tumorigenic activity of the BCR-ABL oncogenes is mediated by bcl-2. *Proc Natl Acad Sci USA* 92:5287–5291.

Savino W, Dardenne M (1995): Immune-neuroendocrine interactions. *Immunol Today* 16:318–322.

Schwartz R, Emerson SG, Clarke MF, Palsson BO (1991a): *In vitro* myelopoiesis stimulated by rapid medium exchange and supplementation with hematopoietic growth factors. *Blood* 78:3155–3161.

Schwartz R, Palsson BO, Emerson SG (1991): Rapid medium and serum exchange increases the longevity and productivity of human bone marrow cultures. *Proc Natl Acad Sci USA* 88:6760–6764.

Shah AJ, Smorgorzewska EM, Hannum C, Crooks GM (1996): Flt3 ligand induces proliferation of quiescent human bone marrow $CD34^{+}CD38^{-}$ cells and maintains progenitor cells *in vitro*. *Blood* 87:3563–3570.

Siczkowski M, Clarke D, Gordon MY (1992): Binding of primitive hematopoietic progenitor cells to marrow stromal cells involves heparan sulfate. *Blood* 80:912–919.

Sitnicka E, Lin N, Priestley GV, Fox N, Broudy VC, Wolf NS, Kaushansky K (1996a): The effect of thrombopoietin on the proliferation and differentiation of murine hematopoietic stem cells. *Blood* 87:4998–5005.

Sitnicka E, Ruscenti FW, Priestley GV, Wolf NS, Bartelmez SH (1996b): Transforming growth factor β_1 directly and reversibly inhibits the initial cell divisions of long-term repopulating hematopietic stem cells. *Blood* 88:82–88.

Smith AG, Heath JK, Donaldson DD, Wong GG, Moreau J, Stahl M, Rogers D (1988): Inhibition of pluripotential embryonic stem cell differentiation by purified polypeptides. *Nature* 336:688–690.

Sockinski MA, Cannistra SA, Elias A, Antman KH, Schnipper L, Griffin JD (1988): Granulocyte colony-stimulating factor expands the circulating haemopoietic progenitor cell compartment in man. *Lancet* 1:1194–1198.

Soma T, Yu JM, Dunbar CE (1996): Maintenance of murine long-term repopulating stem cells in *ex vivo* culture is affected by modulation of transforming growth factor-β but not macrophage inflammatory protein 1-α activities. *Blood* 87:4561–4567.

Spangrude GJ, Brooke DM, Tumas DB (1995): Long-term repopulation of irradiated mice with limiting numbers of purified hematopoietic stem cells: *In vivo* expansion of stem cell phenotype but not function. *Blood* 85:1006–1016.

Stosic-Grujicic S, Basara N, Milenkovic P, Dinarello CA (1995): Modulation of acute myeloblastic leukemia (AML) cell proliferation and blast colony formation by antisense oligomer of IL-1 beta converting enzyme (ICE) and IL-1 receptor antagonist (IL-1ra). *J Chemother* 7:67–71.

Szilvassy SJ, Weller KP, Lin W, Sharma AK, Ho AK, Tsukamoto A, Hoffman R, Leiby KR, Gearing DP (1996): Leukemia inhibitory factor upregulates cytokine expression by a murine stromal cell line enabling the maintenance of highly enriched competitive repopulating stem cells. *Blood* 11:4618–4628.

Traycoff CM, Kosak ST, Grigsby S, Srour EF (1995): Evaluation of *ex vivo* expansion potential of cord blood and bone marrow hematopoietic progenitor cells using cell tracking and limiting dilution analysis. *Blood* 85:2059–2068.

Tumas DB, Spangrude GJ, Brooks DM, Williams CD, Chesebro B (1996): High-frequency cell surface expression of a foreign protein in murine hematopoietic stem cells using a new retroviral vector. *Blood* 87:509–517.

Varas F, Bernad A, Bueren JA (1996): Granulocyte colony-stimulating factor mobilizes into peripheral blood the complete clonal repertoire of hematopoietic precursors residing in the bone marrow of mice. *Blood* 88:2495–2501.

Verfaillie CM (1993): Soluble factor(s) produced by human bone marrow stroma increase cytokine-induced proliferation and maturation of primitive hematopoietic progenitors while preventing their terminal differentiation. *Blood* 82:2045–2053.

Williams RL, Hilton DJ, Pease S, Willson TA, Stewart CL, Gearing DP, Wagner EF, Metcalf D, Nicola NA, Gough NM (1988): Myeloid leukaemia inhibitory factor maintains the developmental potential of embryonic stem cells. *Nature* 336:684–687.

Wilson EL, Rifkin DB, Gabrilove JL, Hannocks M, Wilson EL (1991): Basic fibroblast growth factor stimulates myelopoiesis in long-term human bone marrow cultures. *Blood* 77:954–960.

Wineman J, Moore K, Lemischka I, Muller-Sieburg C (1996): Functional heterogeneity of the hematopoietic microenvironment: Rare stromal elements maintain long-term repolulating stem cells. *Blood* 87:4082–4090.

Winter JN, Lazarus HM, Rademaker A, Villa M, Mangan C, Tallman M, Jahnke L, Gordon L, Newman S, Byrd K, Cooper BW, Horvath N, Crum E, Stadtmauer EA, Conklin E, Bauman A, Martin J, Goolsby C, Gerson SL, Bender J, O'Gorman M (1996): Phase I/II study of combined granulocyte colony-stimulating factor and granulocyte-macrophage colony-stimulating factor administration for the mobilization of hemopoietic progenitor cells. *J Clin Oncol* 14:277–286.

Yamamoto Y, Yasumizu R, Amou Y, Watanabe N, Nishio N, Toki J, Fukuhara S, Ikehara S (1996): Characterization of peripheral blood stem cells in mice. *Blood* 88(2): 445–454.

Yonemura Y, Ku H, Hirayama F, Souza LM, Ogawa M (1996): Interleukin 3 or interleukin 1 abrogates the reconstituting ability of hematopoietic stem cells. *Proc Natl Acad Sci USA* 93:4040–4044.

Young JC, Bruno E, Luens KM, Wu S, Backer M, Murray LJ (1996a): Thrombopoietin stimulates megakaryocytopoiesis, myelopoiesis, and expansion of $CD34^+$ progenitor cells from single $CD34^+$ Thy-$1^+$$Lin^-$ primitive progenitor cells. *Blood* 88:1619–1631.

Young JC, Varma A, DiGiusto D, Backer MP (1996b): Retention of quiescent hematopoietic cells with high proliferative potential during *ex vivo* stem cell culture. *Blood* 87:545–556.

Zeigler FC, de Sauvage F, Widmer HR, Keller GA, Donahue C, Schreiber RD, Malloy B, Hass P, Eaton D, Matthews W (1994): *In vitro* megakaryocytopoietic and thrombopoietic activity of c-mpl ligand (TPO) on purified murine hematopoietic stem cells. *Blood* 84:4045–4052.

Zheng H, Fletcher D, Kozak W, Jiang M, Hoffman K, Conn CA, Soszynski S, Grabiec C, Trumbauer ME, Shaw A, Kostura MJ, Stevens K, Rosen H, North RJ, Chen HY, Tocci MJ, Kluger MJ, Van der Ploog LHT (1995): Resistance to fever induction and impaired acute-phase response in interleukin-1 deficient mice. *Immunity* 3:9–19.

3

MOLECULAR MECHANISMS CONTROLLING THE CELL CYCLE AND PROLIFERATION–DIFFERENTIATION INTERRELATIONSHIPS

GARY S. STEIN, ANDRÉ J. VAN WIJNEN,
DENNET R. HUSHKA, BARUCH FRENKEL,
JANE B. LIAN, AND JANET L. STEIN

Department of Cell Biology and Cancer Center, University of Massachusetts Medical Center, Worcester, MA 01655 (G.S.S., A.J.V., J.B.L., J.L.S.), 1708 William and Mary Common, Hillsboro, NJ (D.R.H.), Department of Orthopedics and Institute of Genetic Medicine, University of Southern California School of Medicine, Los Angeles, CA (B.F.)

INTRODUCTION

Proliferation is a fundamental requirement for developmental establishment and renewal of tissues. Consequently, an understanding of stem cell biology and utilization of stem cells for gene therapy necessitates delineation of growth regulatory mechanisms that are operative both *in vivo* and *ex vivo*. Such insight into control of proliferation will facilitate the design of protocols for expression of stem cells, including under conditions where transfected genes are expressed with fidelity. Equally important, a basis will be provided for stringent regulation of the proliferative process under conditions where expres-

Stem Cell Biology and Gene Therapy, Edited by Peter J. Quesenberry, Gary S. Stein, Bernard G. Forget, and Sherman M. Weissman
ISBN 0-471-14656-0

sion of postproliferative cell and tissue-specific parameters are not compromised.

In this chapter, we explore the broad spectrum of signaling mechanisms that integrate and amplify growth-related regulatory cues that modulate the expression of genes requisite for proliferation and cell cycle control. The modularly organized promoter elements of genes that support the onset and progression of proliferation are presented as blueprints for growth control within the context of responsiveness to mediators of proliferative status. These encompass cytokines, growth factors, steroid hormones, and cell cycle regulatory factors that include but are not restricted to cyclins, cyclin-dependent kinases, cyclin inhibitors, and tumor suppressor proteins. Additionally, we address the activities of factors that mediate the postproliferative downregulation of cell cycle and cell growth regulatory pathways to support quiescence and expression of phenotypic genes at the onset of differentiation (Stein and Lian, 1993; Stein et al., 1990). Our rationale is that, from a biological perspective, to understand growth control we must define the stringent regulation of molecular parameters associated with competency for initiation of proliferation and progression through the cell cycle (Baserga and Rubin, 1993). Each component of the regulatory cascade indicates a step in physiological control (Nurse, 1994). There is growing appreciation for uncompromised operation of multiple cell cycle checkpoints that serve as surveillance mechanisms and invoke repair pathways or apoptosis under defined circumstances.

We refrain from providing a catalog of proliferation-associated regulatory factors. Our objective is to confine our consideration to those components of growth control that contribute insight into rate-limiting events operative at principal cell cycle transition points as well as at the onset and termination of proliferation. Because the volume of information on cell cycle and growth control has expanded dramatically, a single chapter cannot be comprehensive. Moreover, to serve the theme of this book, we emphasize functional interrelationships between physiological regulatory signals that contribute to cell cycle control and phenotypic competency. However, we provide references to a series of excellent review articles that broadly treat cell cycle regulatory concepts and serve as databases for an expanded segment of the relevant literature.

GROWTH REGULATORY MECHANISMS: A HISTORICAL PERSPECTIVE

A historical perspective of cell cycle control offers a conceptual and experimental basis for our current understanding of growth regulation at the cellular, biochemical, and molecular levels. The complexity of proliferation-related regulatory mechanisms is becoming increasingly evident. As a result, we are expanding our appreciation of interrelationships between cell cycle control and accommodation of requirements for proliferation under a broad spectrum of biological circumstances that range from early stage cleavage divisions during the initial periods of embryogenesis to compensatory proliferation during tissue renewal and remodeling.

Subdivision of the Cell Cycle Into Functional Stages

The cornerstone for investigations into mammalian cell cycle control is the documentation by Howard and Pelc (1951) nearly four decades ago that proliferation of eukaryotic cells, analogous to that of bacteria, requires discrete periods of DNA replication (S phase) and mitotic division (M) with a postsynthetic, premitotic period designated G2 and a postmitotic, presynthetic period designated G1 (Fig. 1). The foundation for pursuit of regulatory mechanisms associated with growth control and cell cycle progression was provided by an elegant series of cell fusion and nuclear transplant experiments (reviewed by Prescott, 1976; Heichman and Roberts, 1994). Consequential influences of cytoplasm from various stages of the cell cycle on nuclei from other periods demonstrated the following basic principles of cell cycle control: (1) the onset of DNA synthesis is determined by cytoplasmic factors present through-

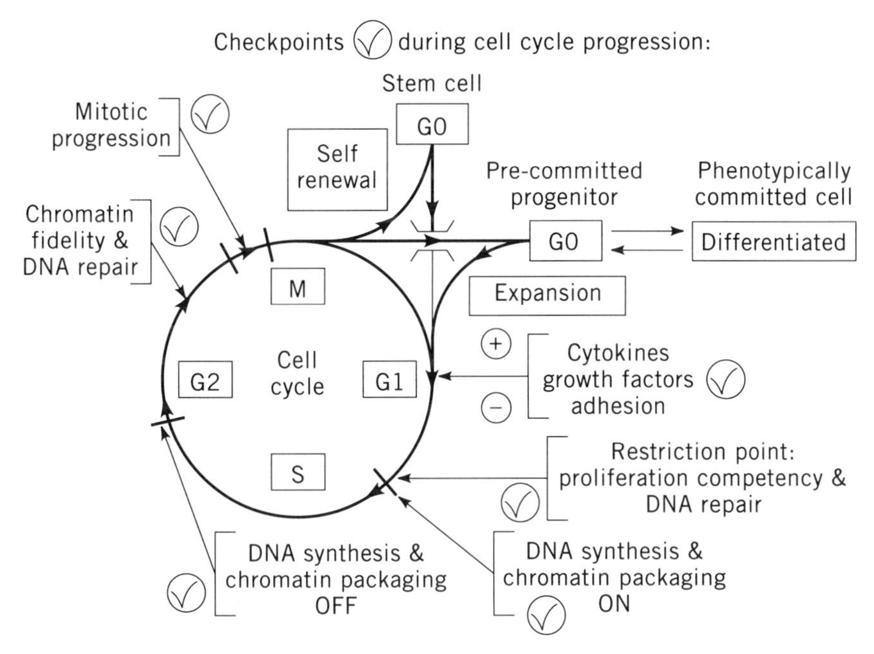

FIGURE 1. Checkpoints during cell cycle progression. The four stages of the cell cycle (G1, S, G2, and M) each contain several critical cell cycle checkpoints at which competency for cell cycle progression is monitored. Entry and exit of the cell cycle is controlled by growth regulatory factors (e.g., cytokines, growth factors, cell adhesion, and/or cell–cell contact) which determine self-renewal of stem cells and expansion of pre-committed progenitor cells.

out S phase but absent in pre-S phase; (2) a nuclear mechanism prevents re-replication of DNA without passage through mitosis; and (3) a dominant cytoplasmic factor in mitotic cells promotes mitosis in interphase cells irrespective of whether DNA replication has occurred. The broad biological relevance of cell cycle regulatory parameters is reflected by the phylogenetic conservation of control mechanisms in yeast, protozoans, echinoderms, amphibian oocytes, and mammalian cells.

The initial indication that modifications in gene expression are required to support entry into S phase and mitosis was obtained from inhibitor studies. First observed was the necessity of transcription and protein synthesis for DNA replication and mitotic division (Terasima and Yasukawa, 1966; Baserga et al., 1965). Restriction points late in G1 and G2 for competency to initiate S phase and mitosis were mapped (Sherr, 1993; Dowdy et al., 1993; Pardee, 1989). Subsequently, by the combined application of gene expression inhibitors and modulation of growth factor levels in cultured cells, a mitogen-dependent (growth factor/cytokine responsive) period was defined early in G1 in which competency for proliferation is established, and a late G1 restriction point was identified in which competency for cell cycle progression is attained (Pardee, 1989).

Cell Cycle Checkpoints

Checkpoints have been identified that govern passage through G1 and G2 (Fig. 1), where competency for cell cycle traverse is monitored (Nurse, 1994; Sherr, 1993, 1994; Dowdy et al., 1993). The first evidence for these checkpoints was provided by the observation of delayed entry into S phase or mitosis following exposure to radiation or carcinogens. Editing functions are operative, and decisions for continued proliferation, growth arrest, or apoptotic cell death are executed at these regulatory junctures (White, 1994). Here, long-standing fundamental questions deal with the requirement of proliferation for the onset of differentiation as well as the extent to which proliferation and postproliferative expression of cell and tissue phenotypic genes are functionally interrelated or mutually exclusive. Knowledge of control that is operative at cell cycle checkpoints is rapidly accruing. The complexity of the surveillance mechanisms that govern decisions for cell cycle progression is becoming increasingly apparent. We are now aware of multiple checkpoints during S phase that monitor regulatory events associated with DNA replication, histone biosynthesis, and fidelity of chromatin assembly. Mitosis is similarly controlled by an intricate series of checkpoints that are responsive to biochemical and structural parameters of chromosome condensation, mitotic apparatus assembly, chromosome alignment, chromosome movement, and cytokinesis.

Biochemical And Molecular Parameters Of Cell Cycle Control

Characterization of the biochemical and molecular components of cell cycle and growth control emerged from systematic analysis of conditional cell cycle mutants in yeast (Hartwell et al., 1974; Hartwell and Weinert, 1989). These studies were the foundation for the concept that cell cycle competency and progression are controlled by an integrated cascade of phosphorylation-dependent regulatory signals. Cyclins are synthesized and activated in a cell cycle–dependent manner and function as regulatory subunits of cyclin-dependent kinases (CDKs). The CDKs phosphorylate a broad spectrum of structural proteins and transcription factors to mediate sequential parameters of cell cycle control (Fig. 2). By complementation analysis, the genes for mammalian homologs of the yeast cell cycle regulatory proteins have been identified. *In vivo* overexpression, antisense, and antibody analyses have verified conservation of cell cycle–dependent regulatory activities and have validated functional contributions to control of cell cycle stage-specific events. The emerging concept is that the cyclins and CDKs are responsive to regulation by the phosphorylation-dependent signaling pathways associated with activities of the early response genes, which are upregulated following mitogen stimulation of proliferation (reviewed by Hunter and Pines, 1994; Hartwell and Weinert, 1989; MacLachlan et al., 1995) (summarized in Fig. 2). Cyclin-dependent phosphorylation is functionally linked to activation and suppression of both p53 and Rb-related tumor suppressor genes, which mediate transcriptional events involved with passage into S phase. The activities of the CDKs are downregulated by a series of inhibitors (designated CDIs) and mediators of ubiquitination, which signal destabilization and/or destruction of these regulatory complexes in a cell cycle–dependent manner. Particularly significant is the accumulating evidence for functional interrelationships between activities of cyclin–CDK complexes and growth arrest at G1 and G2 checkpoints, when editing and repair are monitored following DNA damage. It is at these times, and in relation to these processes, that apoptotic cell death is invoked as a compensatory mechanism.

During G1, expression of genes associated with deoxynucleotide biosynthesis are upregulated (e.g., thymidine kinase, thymidylate synthase, dihydrofolate reductase) in preparation for DNA synthesis (King et al., 1994; Pardee and Keyomarsi, 1992; Johnson, 1992). As cells progress through G1, regulatory factors required for initiation of DNA replication are sequentially expressed and/or activated. Following stimulation of quiescent cells to proliferate, expression of the fos/jun-related early response genes is induced early in G1, playing a pivotal role in activation of subsequent cell cycle regulatory events. In S phase, DNA replication is paralleled by and functionally coupled with histone gene expression, providing the necessary basic chromosomal proteins (H1, H4, H3, H2A, and H2B) for packaging newly replicated DNA into chromatin (Azizkhan et al., 1993; Plumb et al., 1983). During G2, regulatory factors for mitosis are synthesized, and modifications of chromatin structure to support mitotic chromosome condensation occur (reviewed by MacLachlan et al., 1995). Mitosis involves a sequential remodeling of genome architecture from uncondensed chromatin to highly condensed chromosomes and back to chromatin; assembly and subsequent disassembly of the mitotic apparatus; breakdown and reformation of the nuclear membrane; and modifications in activities

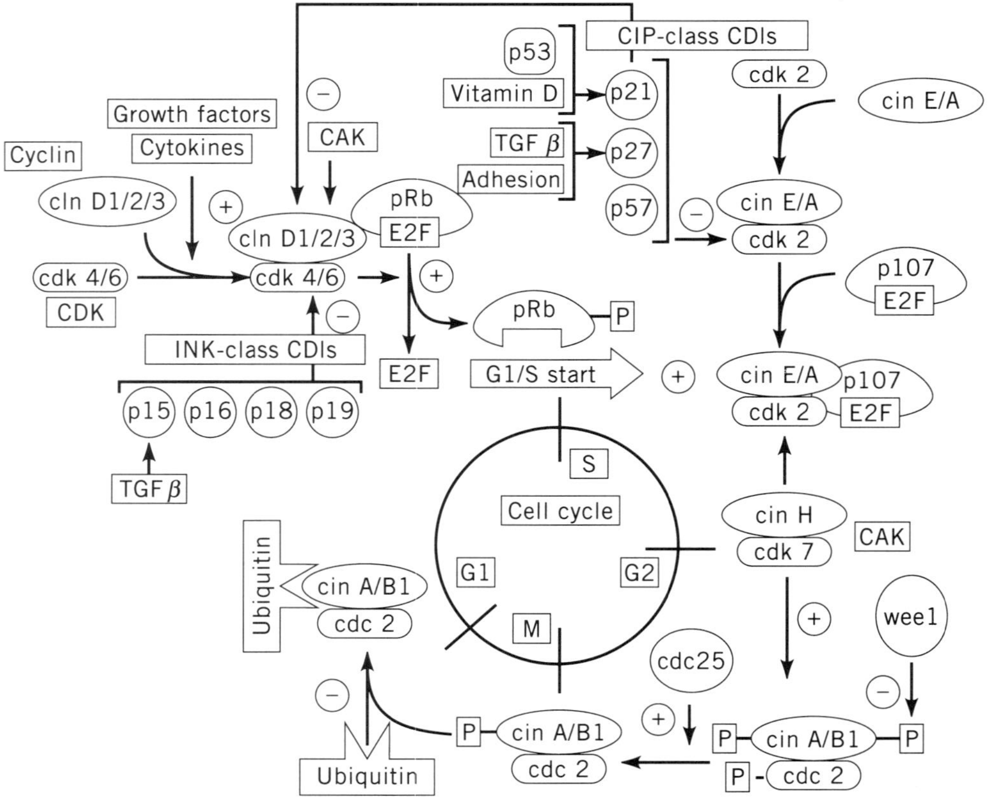

The cell cycle: a functionally integrated cascade of positive and negative regulatory factors

of factors required for reinitiation of cell cycle progression, quiescence, or differentiation.

As the sophistication of experimental approaches for dissection of promoter elements and characterization of cognate regulatory factors increases, there is an emerging recognition of cyclic modifications in occupancy of promoter domains and protein–protein interactions that control cell cycle progression. The changes observed in the site II cell cycle regulatory promoter element of the histone gene illustrate such changes in factor occupancy and activities that are functionally linked to activation, suppression, and subtle modifications in levels of expression that are proliferation dependent (van Wijnen et al., 1994, 1996; Vaughan et al., 1995; Stein et al., 1989; Holthuis et al., 1990; Pauli et al., 1987).

Recently, considerable attention has been directed to experimentally addressing transcriptional regulatory mechanisms that are associated with cell cycle and growth control. However, the importance of control at post-transcriptional levels should not be underestimated. For example, cell–cycle dependent modifications in histone mRNA processing and stability contribute to linkage of histone gene expression with DNA replication and the S phase of the cell cycle (Harris et al., 1991; Marzluff and Pandey, 1988; Morris et al., 1991; Pelz et al., 1991; Pardee, 1989). Compartmentalization of cell cycle regulatory factors and/or cognate gene transcripts, as well as phosphorylation of regulatory proteins, are components of post-transcriptional growth control (Zambetti et al., 1987; Birnbaum et al., 1997).

Multiple, Interdependent Cycles Operative During Proliferation

Several interdependent cycles are functionally linked to control of proliferation. The first is a stringently regulated sequential series of biochemical and molecular parameters that support genome replication and mitotic division. The second is a cascade of cyclin-related regulatory factors that transduce growth factor–mediated signals into discrete phosphorylation events, controlling expression of genes responsible for both initiation of proliferation and competency for cell cycle progression. Other cell cycle–related regulatory loops involve chromosome condensation, spindle assembly, metabolism, and

FIGURE 2. Regulation of the cell cycle by cyclins, CDKs, and CDIs. Cell cycle progression is determined by an integrated cascade of positive (e.g., cyclins) and negative (e.g., CDIs) cell cycle regulatory factors that influence the activities of cyclin-dependent kinases. CDKs mediate phosphorylation of pRb-related proteins, which results in activation of E2F and other transcription factors, as well as induction of G1/S phase–related gene expression by E2F-dependent and E2F-independent mechanisms. The activities of CDKs are also influenced by phosphorylation (e.g., CDK-activating kinase [CAK, Wee 1], dephosphorylation [cdc25], and ubiquitin-dependent proteolysis).

assembly of cdc2 and assembly/disassembly of DNA replication factor complexes (replicators and potential initiator proteins [Stillman, 1996; Hamlin et al., 1994; Muzi-Falconi and Kelly, 1995; Clyne and Kelly, 1995; Dubey et al., 1996; Gavin et al., 1995; Gossen et al., 1995; Leatherwood et al., 1996; Carpenter et al., 1996; Wohlgemuth et al., 1994]). It is becoming increasingly evident that each step in the regulatory cycles governing proliferation is responsive to multiple signaling pathways and has multiple regulatory options. The diversity in cyclin–CDK complexes accommodates control of proliferation under multiple biological circumstances and provides functional redundancy as a compensatory mechanism. The regulatory events associated with the proliferation-related cycles support control within the contexts of (1) responsiveness to a broad spectrum of positive and negative mitogenic factors; (2) cell–cell and cell–extracellular matrix interactions; (3) monitoring sequence integrity of the genome and invoking editing and/or apoptotic mechanisms if required; and (4) competency for differentiation.

Accommodation of Unique Cell Cycle Regulatory Requirements in Specialized Cells

Consistent with the stringent requirement for fidelity of DNA replication and DNA repair to execute proliferation, stage-specific modifications in control of cell cycle regulatory factors have been observed to parallel physiological changes and perturbations in growth control. Some striking examples of physiological changes are regulatory mechanisms that support developmental transitions during early embryogenesis, when DNA replication and mitotic division occur in rapid succession in the absence of significant G1 or G2 periods. In contrast, proliferation in somatic cells of the adult requires passage through a cell cycle with G1, S, G2, and mitotic periods that are operative and necessary. Often, a prolonged G1 period provides support for long-term quiescence of cells and tissues while retaining the competency to reinitiate proliferation for tissue remodeling and renewal.

The abrogated components of growth control in transformed and tumor cells are associated with and functionally linked to both the regulation and regulatory activities of cyclin–CDK complexes. Characteristic alterations have been associated with progressive stages of neoplasia and specific tumors (reviewed by Hunter and Pines, 1994; Hartwell and Weinert, 1989). Consequently, tumor cells are providing us with valuable insight into rate-limiting regulatory steps in cell cycle and cell growth control. In addition, we are increasing our opportunity to therapeutically rectify proliferative disorders in a targeted manner. Particularly challenging is the possibility of restoring fidelity of regulatory mechanisms operative at cell cycle checkpoints, when responses to apoptotic signals prevent accumulation and phenotypic expression of mutations associated with growth control perturbations.

FACTORS CONTROLLING PROLIFERATION AND CELL CYCLE PROGRESSION: THE REGULATORY AND REGULATED MECHANISMS

We present an overview of the principal components of positive and negative control that govern entrance into and progression through the cell cycle (reviewed by Norbury and Nurse, 1992). Emphasis is on the regulated and regulatory parameters of cell cycle control, especially initiation of proliferation and G1 and G2 functions that ensure that the factors required for the next phase have been synthesized and that the previous phase has been completed. Transitions through and between different phases of the cell cycle that are regulated by checkpoints are described when certain prerequisites must be met for progression to occur (Hartwell and Weinert, 1989). The factors that support activities at these cell cycle checkpoints are discussed in relation to ultimately ensuring that newly divided cells receive a full and intact complement of hereditary material.

Entry of Quiescent Cells Into the Cell Cycle

When quiescent cells (G0) are stimulated to proliferate and divide, they enter G1, the first phase of the cell cycle where the enzymes required for DNA replication are synthesized. Before a cell can progress through G1 and begin DNA synthesis (S phase), it must pass through a checkpoint in late G1, which is known as the *restriction point* (Pardee, 1989). At this restriction point, both positive and negative external growth signals are integrated into the cell cycle. If conditions are appropriate, the cell proceeds through the remainder of G1 and enters the S phase. Once the cell passes the restriction point, it is refractory to withdrawal of mitogens or to growth inhibitory signals and is committed to progressing through the remainder of the cell cycle unless it is subjected to DNA damage or metabolic disturbance (Pardee, 1989). In mammalian cells, progression through the cell cycle is regulated by the activity of a family of threonine/serine kinases designated *cyclin-dependent kinases* (Morgan, 1995). CDKs are regulated by both positive and negative phosphorylation and their reversible association with specific cyclins during defined phases of the cell cycle. In general, the levels of CDK proteins remain relatively constant during the cell cycle, whereas the expression of specific cyclins is confined to distinct phases of the cell cycle where they are quickly degraded after having completed their function.

The retinoblastoma protein (Rb), a tumor suppressor, is a member of a family of related proteins that include p107 and p130. Rb has been shown to have a critical role in the regulation of cell proliferation, particularly in progression through G1 (reviewed by Weinberg, 1995). Rb functions as a signal transducer, receiving both growth-promoting and -inhibitory signals and linking them to the transcriptional machinery required for cell cycle progression or cell cycle arrest. In quiescent cells or cells re-entering G1 from mitosis, Rb exists in an underphosphorylated or dephosphorylated state. As cells are stimulated to proliferate and enter the cell cycle, Rb becomes progressively phosphorylated. In early G1, Rb is phosphorylated by CDK4 and CDK6, in conjunction with their obligate catalytic partners the D-type cyclins (Meyerson and Harlow, 1994; Matsushime et al., 1994). When quiescent cells are exposed to growth factor signals, and depending on the cell type, one or more members of a family of delayed early response genes, the D-type cyclins are induced. Combinations of the D-type cyclins are expressed in a cell type–specific manner. This evidence suggests that D-type cyclins have a key role in sensing and integrating mitogenic signals to the cell cycle (Sherr, 1993). However, other growth factor–dependent events are required for the formation of active CDK4–cyclin D complexes and subsequent cell cycle entry as well as progression. Cyclin D and CDK4 ectopically overexpressed in fibroblasts do not assemble in active complexes in the absence of serum. However, the exact biochemical link between mitogenic signaling cascades and the expression of D-type cyclins have not been identified. The D-type cyclins have been shown to bind to Rb by a consensus amino acid motif, LXCXE, that functions to target CDK4 and CDK6 to Rb (Dowdy et al., 1993; Ewen et al., 1993). Rb phosphorylation by cyclin D–CDK4/6 complexes is first detected in mid-G1 and is maximal near the G1/S transition.

In its unphosphorylated state, Rb exists as a heterodimer with specific members of the E2F family of transcription factors (Weinberg, 1995). Other E2F family members bind specifically to either p107 or p130. Rb phosphorylation results in the dissociation and activation of E2F. E2F is a heterodimer composed of E2F and DP-related factors. Activation of the E2F family of transcription factors positively regulates genes required for entry into S phase and DNA synthesis that include cdc2, c-myc, cyclin A, dihydrofolate reductase (DHRF), DNA polymerase-α, thymidine kinase (CTK), thymidylate kinase (TMK), dihydroorotate synthase (CAD), and ribonucleotide reductase M2 (RNR) (reviewed by Nevins, 1992; LaThangue, 1994). Growth inhibitory signals that specifically arrest cells in G1 maintain Rb in a hypophosphorylated state as a heterodimic complex with E2F, repressing E2F activity and preventing the expression of genes required for DNA synthesis.

Progression through the late G1 restriction point is controlled by a CDK2–cyclin E complex. As cells reach the G1/S transition, Rb phosphorylation shifts from being mediated by cyclin D–CDK complexes to cyclin E–CDK2 complexes. The cyclin E gene is responsive to E2F regulation and is transiently expressed with mRNA levels being maximal at the G1/S transition. The E2F-responsive nature of cyclin E results in regulation of cyclin E activity by a positive feedback loop. The exact functions of Rb phosphorylation by these different CDK complexes are not known. However, CDKs may function to change Rb phosphorylation from a mitogen-dependent state (cyclin D–CDK4) to a mitogen-independent state (cyclin E–CDK2) (Sherr, 1996). Rb phosphorylation continues throughout the remainder of the cell cycle until anaphase;

these phosphorylations are mediated by cyclin A– and cyclin B-dependent kinases, depending on the cell cycle phase. Cyclin E is rapidly degraded by a ubiquitin-dependent proteolysis as cells enter S phase. CDK2 phosphorylation of cyclin E is required for this process.

Cyclin A synthesis is induced in late G1 by activated E2F. CDK2–cyclin A is required for the initiation of DNA replication and to support DNA synthesis throughout S phase. As discussed below, cyclin A–dependent kinases are involved in phosphorylating substrates that are components of replication origins. In addition, CDK2–cyclin A phosphorylates DP-1, which results in decreased E2F transactivation activity as cells enter S phase. Once a cell completes mitosis, Rb is dephosphorylated in anaphase by a type 1 phosphatase (Durfee et al., 1993).

At the onset of mitosis, chromosome condensation is mediated by cdc2–cyclin B-dependent phosphorylation of a number of substrates (Guadagno et al., 1993; Swenson et al., 1986; Pines and Hunt, 1987). Phosphorylation of histone H1 by cdc2–cyclin B modifies chromatin structure through alterations in nucleosome interactions (Minshull et al., 1989; Jerzmanowski and Cole, 1992). cdc2–cyclin B also contributes to chromosome condensation by phosphorylating and consequently activating caseine kinase, which is a topoisomerase II activator (Reeves, 1992). cdc2 has been functionally linked to control of mitosis by phosphorylation-dependent abrogation of lamin phosphorylation, which results in nuclear envelope breakdown. In addition, both cdc2–cyclin A and cdc2–cyclin B promote microtubule formation from centromeres (Nigg, 1992, 1993; Ohta et al., 1993).

The complexity of cdc2 regulation in relation to cell cycle control is illustrated by phosphorylation-dependent changes at the onset of mitosis. During G2, cdc2 is inactivated by phosphorylation of Tyr-15 by Wee-1. Initiation of mitosis is functionally coupled to inactivation of Wee-1 by phosphorylation, which is mediated by a series of kinases that include the nim1 kinase. cdc25 phosphatase dephosphorylates cdc2 at the onset of mitosis, and the cdc2–cyclin B complex phosphorylates and activates cdc25. Phosphorylation of Thr-161 by CDK-activating kinase (CAK) (MO15 or CDK7) in association with cyclin H is required for maximal cdc2 activity.

Cell Cycle Regulation in Continuously Dividing Cells

Compared with quiescent cells that enter the cell cycle, some cells undergo a high rate of expansion and do not exit the cell cycle after one division but undergo continuous division. For example, embryonic cells, cells in tissues that undergo a high rate of renewal such as in the lining of the intestine, the skin, hematopoietic linages, and cancer cells, continuously divide without exiting the cell cycle. Because the critical events that determine entry and progression through the cell cycle reside in G1, particularly at the restriction point, these cells may bypass some of the rate limiting events determining cell cycle competency (reviewed by Sherr, 1996). A classic example is the overexpression of cyclin D1 in a variety of tumors, including esophageal, hepatic, head and neck, breast, colorectal, and some sarcomas. Overexpression of cyclin D1 has been correlated with hyperplastic and neoplastic cell growth phenotypes in a number of these tissues that are characterized by reduced growth factor requirements and shortened cell cycle transit. Normal cells that undergo continuous division are able to respond to signaling pathways that prevent uncontrolled proliferation, while cancer cells abandon these controls and remain in the cell cycle (Sherr, 1996).

Negative Regulation of CDK Activity by CDK Inhibitors

In the past 5 years a number of proteins that bind to and inhibit the activity of CDKs have been identified and designated *CDK-inhibitor proteins* (CDIs; reviewed by Hunter and Pines, 1994) (see Table 1). These proteins are expressed in a tissue- and cell type–specific manner and have been associated with diverse responses such as growth arrest, differentiation, and apoptosis. The role of these proteins in these responses have been demonstrated in gene

TABLE 1 Properties of Cyclin-Dependent Kinase Inhibitors

Inhibitor	Specificity	Comments	Phenotype of Knockout Mouse
$p15^{INK4b}$	CDK4/CDK6	Inducible by TGF-β1 (Hannon and Beach, 1994)	
$p16^{INK4a}$	CDK4/CDK6	Constitutively expressed (Serrano et al., 1993)	Develop spontaneous tumors at early age, increased sensitivity to carcinogens (Serrano et al., 1996)
$p18^{INK4c}$	CDK4/CDK6	Guan et al (1994)	
$p19^{INK4d}$	CDK4/CDK6	Interacts with Nur77 (Chan et al., 1995)	
$p21^{Cip1}$	Universal	Inducible by p53/MyoD (Guo et al., 1995; El-Deiry et al., 1993)	Defective G1 checkpoint control (Deng et al., 1995)
$p27^{Kip1}$	Universal	Activated by TGF-β1, cell–cell contact (Polyak et al., 1994a,b)	Increased body size, organ hyperplasia, female sterility (Kiyokawa et al., 1996; Nakayama et al., 1996; Fero et al., 1996)
$p57^{Kip2}$	Universal	Associated with terminal differentiation (Matsuoka et al., 1995)	

knockout experiments. These proteins can be divided into two families based on structural homology and amino acid similarity. The Cip/Kip family, which includes p21, p27, and p57, all contain a region of homology that mediates their binding to CDKs and is required for their inhibition of CDK activity. The INK family, which includes p15, p16, p18, and p19, contain multiple ankyrin repeats. These two families differ in substrate specificity, with the Cip/Kip family targeting a wide range of CDK–cyclin complexes, and are therefore considered universal CDIs. The INK family appears to be specific for CDK4 and CDK6.

In addition to their distinct substrate specificity, CDIs are also grouped into two categories based on their expression, namely, the constitutive and the inducible. Some of the constitutively expressed CDIs, such as p16 and p27, have been shown to bind to cyclin–CDK complexes. The activity of these complexes depends on the stoichiometry of CDI binding. It is thought that the constitutively expressed CDIs act to prevent inappropriate cell cycle progression and that the levels of cyclins and CDKs must exceed a threshold level of constitutive CDI in order for cell cycle progression to occur. Recent studies suggest that $p27^{Kip1}$ may be required for restriction point control (Coats et al., 1996). The levels of $p27^{Kip1}$ are increased in fibroblasts deprived of serum mitogens and consequently arrested in G1. Serum stimulation of these cells results in cell cycle progression and decreased levels of $p27^{Kip1}$. Mice nulliyzgous for p27 demonstrate a number of phenotypes that include an overall body size that is one-third larger than the wild-type controls, female sterility that is associated with a defect in luteal cell differentiation, and pituitary adenomas. The increase in animal size is due to an increased number of cells, not to increased cell size. A variety of stimuli have been shown to induce different CDIs (see Table 1). $p21^{Cip1}$ is transcriptionally activated by the tumor suppressor p53 in response to DNA damage (El-Deiry et al., 1993). In addition, a number of lines of evidence suggest that $p21^{Cip1}$ may have a functional role in differentiation of a variety of cell types. MyoD, a skeletal muscle-specific bHLH transcription factor that induces terminal cell cycle arrest associated with skeletal muscle differentiation, has been shown to induce the levels of $p21^{Cip1}$ (Halevy et al., 1995). *In situ* hybridization of developing mouse embryos demonstrates that p21 mRNA is localized to tissues that primarily contain postmitotic differentiated cells (Parker et al., 1995). CDIs have been shown to be regulated at transcriptional and post-translational levels.

Roles for Cell Cycle Regulatory Factors in Differentiation

Because acquisition of tissue-specific phenotypes is normally associated with growth arrest, most of the attention to cell cycle regulatory factors within the context of differentiation has been focused on mechanisms that ensure exit from the cell cycle as a prerequisite for differentiation. These mechanisms include hypophosphorylation of Rb and decreased representation or activity of cyclin–CDK complexes, with recent emphasis on the upregulation of CDIs during differentiation. While this concept and its significance to carcinogenesis have been addressed in several recent reviews (Weinberg, 1995; MacLachlan et al., 1995; Kranenburg et al., 1995a; Marks et al., 1996), here we present representative evidence demonstrating that many factors controlling cell cycle progression acquire additional roles postproliferatively and are in fact active in the differentiation process of various cell types.

pRb and Related Proteins. The best characterized function of the Rb gene is the association of hypophosphorylated pRb with E2F transcription factors. Consequently, phosphorylation of pRb by CDKs during mid and late G1 results in the timely transcriptional activation of E2F-regulated genes (reviewed by Weinberg, 1995; Nevins, 1992; La Thangue, 1994). However, it is now clear that pRb function is not confined to its roles in the control of cell cycle progression and maintenance of quiescence. Numerous investigators have reported high abundance of pRb in a variety of postmitotic cell types, where it is mostly found hypophosphorylated (Yen et al., 1993; Kiyokawa et al., 1993; Szekely et al., 1993; Cordon-Cardo and Richon, 1994; reviewed in Weinberg, 1995). Perhaps the best evidence for postproliferative roles for pRb is provided by pRb gene ablation studies. Cell proliferation in early mouse embryos lacking functional pRb initially appears normal, possibly attributable to functional redundancy with the pRb-related proteins p107 and/or p130. However, at midgestation, abrogation of neural development and erythropoiesis occur, resulting in prenatal lethality (Jacks et al., 1992; Clarke et al., 1992; Lee et al., 1992, 1995; reviewed by Weinberg, 1995; Slack and Miller, 1996). Biochemical and cellular analyses of the pRb-deficient cells confirmed maturation defects and p53-dependent premature apoptosis of specific neuron populations (Lee et al., 1994; Morgenbesser et al., 1994). In addition, pRb plays a role in myoblast differentiation, as demonstrated by the abrogation of this process in pRb-deficient cells (Schneider et al., 1994; Novitch et al., 1996; reviewed by Wiman, 1993). Similarly, *in vitro* adipocyte differentiation in Rb-deficient fibroblasts is blocked and is restorable upon re-expression of pRb (Chen et al., 1996). Finally pRb has been implicated in sustaining the differentiation state of HL60 human leukemia cells (Yen and Varvayanis, 1994). pRb-induced differentiation has been attributed to its ability to interact with and modulate activity of transcription factors, such as MyoD (Gu et al., 1993), adipogenic inducers of the C/EBP family (Chen et al., 1996), NF-IL6, the glucocorticoid receptor (Singh et al., 1995), ATF-2, (reviewed by Sellers and Kaelin, 1996; Kouzarides, 1995; Chen et al., 1995), and possibly some newly identified (Buyse et al., 1995) and yet to be identified transcription factors. From the standpoint of new functions that cell cycle regulatory molecules acquire during differentiation, it is of particular interest that Rb family members may function postproliferatively as stable complexes with E2F transcription factors, which in some cases are observed predominantly in terminally differentiated cells (see, for example, Corbeil et al., 1995; Jiang et al., 1995; Shin et al., 1995).

Like pRb, the related proteins p107 and p130 interact with E2F transcription factors via their "pocket domain." Unlike pRb, they also interact with cyclin–CDK complexes via a sequence resembling a domain found in CDIs. However, as with pRb, mice lacking both p107 and p130 do not exhibit a generalized cell cycle defect, but rather specific defects in chondrocytic growth and limb development, with neonatal lethality (Cobrinik et al., 1996). Thus, while pRb plays a role in neural and hematopoietic development, p107 and p130 become critical later, during skeletal development. *In vitro,* upregulation of p130 in differentiating L6 cells has been implicated in myotube formation (Kiess et al., 1995a), and p107 has been recently found in a transcription factor complex bound to the bone-specific osteocalcin promoter (van Gurp et al., 1997). Thus, all three members of the Rb gene family seem to play tissue-specific differentiation-related roles in addition to their more traditional regulatory roles in cell cycle progression. Inhibition of apoptosis by pRb family members as a mechanism that maintains a differentiated phenotype is discussed elsewhere in this chapter.

Cyclin-Dependent Kinases. Whereas most CDKs normally function during active cell proliferation, two members of this family of enzymes, CDK5 and CDK7, are clearly different (Table 2). CDK5 is expressed in the central nervous system in a specific spatial and temporal fashion. *In situ* analyses indicate that it is excluded from mitotic neurons, and functional assays demonstrate postmitotic kinase activity associated with CDK5 in differentiating neurons (Tsai et al., 1993). Finally, manipulations of CDK5 and its p35 partner protein *in vitro* resulted in direct effects on neurite outgrowth in cortical cultures (Nikolic et al., 1996), and targeted deletion of CDK5 in mice resulted in defective brain development and perinatal lethality (Ohshima et al., 1996). Thus, CDK5 plays a critical role in central nervous system development. Unlike CDK5, the CAK CDK7 is abundant in diverse cell types, both proliferating and differentiated (Bartkova et al., 1996). Other CDKs (e.g., CDC2, CDK2, CDK4) have been occasionally observed in some postproliferative cells (Table 2), usually without, but in some cases with (Bartkova et al., 1996; Dobashi et al., 1996; Smith et al., 1997; Gao et al., 1995; Kranenburg et al., 1995a; Jahn et al., 1994), associated kinase activity. The requirement of these kinases for cell cycle progression makes it difficult to directly address their role in cell differentiation using gene ablation approaches.

Cyclins. There is increasing evidence for cell type–specific postproliferative retention and even upregulation of cyclins in differentiating cells (Table 2). However, in most cases no kinase activity is associated with these postproliferative cyclins, and therefore their function may be related to association with other, nonkinase, proteins, such as pRb (e.g., Chen et al., 1995) or, as recently suggested, nuclear hormone receptors (Zwijsen et al., 1997). In one case, the postproliferative upregulation of cyclin E has been shown to support an osteoblast differentiation-related kinase activity, suppressible by inhibitory

TABLE 2 Persistence or Upregulation of Cyclins and Cyclin-Dependent Kinases in Differentiating Cells

Cyclin A	NGF-induced PC12 pheochromocytoma cells (Buchkovich and Ziff, 1994)
	DMSO-induced HL60 cells (Burger et al., 1994)
Cyclin B	Lens fiber cells (Gao et al., 1995)
	Primary rat calvarial osteoblasts (Smith et al., 1995)
	DMSO-induced HL60 cells (Burger et al., 1994)
Cyclin D1	Hepatocytes (Loyer et al., 1994)
	Kidney cells (Godbout and Andison, 1996)
	Senescent WI-38 human diploid fibroblasts (Lucibello et al., 1993)
	NGF-induced PC12 pheochromocytoma cells (Dobashi et al., 1995; Yan and Ziff, 1995; Tamaru et al., 1994; van Grunsven et al., 1996)
	TPA-induced HL60 cells (Horiguchi-Yamada et al., 1994; Burger et al., 1994)
	Megakaryocytes (Dami, HEL, and K562 cell lines) (Willhide et al., 1995)
	P19 embryonal carcinoma (Kranenburg et al., 1995a)
Cyclin D2	Specific neural populations (Ross et al., 1996; Ross and Risken, 1994)
	P19 embryonal carcinoma (Kranenburg et al., 1995a)
Cyclin D3	L6, L8, G8, and C2C12 myoblasts (Kiess et al., 1995b; Rao and Kohtz, 1995; Jahn et al., 1994)
	HMBA-induced MEL cells (Kiyokawa et al., 1994)
Cyclin E	Intestinal epithelial cells (Chandrasekaran et al., 1996)
	Hepatocytes (Loyer et al., 1994)
	Primary rat calvarial osteoblasts (Smith et al., 1995, 1997)
	Senescent WI-38 human diploid fibroblasts (Lucibello et al., 1993)
	L6 myoblasts (Kiess et al., 1995a,b)
	NGF-induced PC12 pheochromocytoma cells (Dobashi et al., 1995)
cdc2	Lens fiber cells (Gao et al., 1995)
	Sertoli cells (Rhee and Wolgemuth, 1995)
	C2C12 (Jahn et al., 1994)
CDK2	Sertoli cells (Rhee and Wolgemuth, 1995)
	L6 myoblasts (Kiess et al., 1995a,b)
	NGF-induced PC12 pheochromocytoma cells (Yan and Ziff, 1995)
CDK4	Sertoli cells (Rhee and Wolgemuth, 1995)
	Intestinal epithelial cells (Chandrasekaran et al., 1996)
	L6 myoblasts (Kiess et al., 1995a,b)
	NGF-induced PC12 pheochromocytoma cells (Dobashi et al., 1996; Yan and Ziff, 1995)
	P19 embryonal carcinoma (Kranenburg et al., 1995a,b)
CDK5	Embryonic mouse neurons (Tsai et al., 1993)
	Embryonic *Xenopus* neurons (Gervasi and Szaro, 1995)
CDK7	Quiescent cells (various types) (Bartkova et al., 1996)
Pctaire 1	Sertoli cells (Rhee and Wolgemuth, 1995)
Pctaire 3	Sertoli cells (Rhee and Wolgemuth, 1995)

activity residing in proliferating osteoblasts (Smith et al., 1997). Additional evidence for involvement of cyclins in cell differentiation come from experiments manipulating their levels. Transgenic mice overexpressing cyclin D1 under the control of an immunoglobulin enhancer contained fewer mature B and T cells (Bodrug et al., 1994); The myeloid cell line 32D fails to differentiate in the presence of overexpressed cyclins D2 or D3, probably due to their interaction with pRb and/or p107 (Kato and Sherr, 1993). Cyclin D1 gene ablation in mice resulted in breast- and nervous tissue–specific defects (Fantl et al., 1995; Sicinski et al., 1995). Consistent with this result, overexpression of cyclin D1 in mouse mammary epithelial cells resulted in a more differentiated phenotype (Han et al., 1996). These findings may be attributed to the CDK-independent activation of the estrogen receptor by cyclin D1 (Zwijsen et al., 1997). In some other cases, however, cyclins may support the induction or maintenance of a differentiated phenotype by activation of CDKs (Bartkova et al., 1996; Dobashi et al., 1996; Smith et al., 1997; Gao et al., 1995; Kranenburg et al., 1995a; Jahn et al., 1994) or association with pRB/E2F differentiation-related complexes (Kiyokawa et al., 1994).

In summary, the similarities of cell cycle regulatory molecules among various tissues, as well as across species, suggest common mechanisms underlying cell cycle control in diverse cell types. However, it is becoming clear that specific cells have unique cell cycle requirements. This is evidenced by cell type–specific representation of cell cycle regulatory factors and by cell type–specific effects observed following over- and underexpression of these molecules. Moreover, as cells acquire specific phenotypic properties, some cell cycle regulatory factors persist or even become more abundant postproliferatively. Not only Rb family members but also cyclins and CDKs seem to play roles in cell type–specific differentiation processes. We are currently witnessing the initial steps in a growing field, investigating how these cell cycle regulatory molecules acquire new functions during differentiation, including interactions among themselves, with E2F and other transcription factor families, and with transcriptional elements of growth- and differentiation-related genes.

TRANSCRIPTIONAL CONTROL DURING THE CELL CYCLE

Transcriptional Activation and Suppression of Genes Involved in Nucleotide Metabolism at the Restriction Point of the G1/S Transition

During the G1/S phase transition, three critical events occur that prepare the cell for the duplication of chromatin. First, genes encoding enzymes involved in nucleotide metabolism are activated to ensure that cellular deoxynucleotide triphosphate pools are adequate for the onset of DNA synthesis. Second, multiprotein complexes at DNA replication origins are assembled that both regulate the initiation of DNA synthesis and prevent re-initiation at the same origin. Third, histone proteins are synthesized *de novo* to accommodate the packaging of newly replicated DNA into nucleosomes. Transcriptional activation of gene expression at the G1/S phase transition represents the initial rate-limiting step for cell cycle progression into S phase.

The restriction point prior to the G1/S phase transition integrates a multiplicity of cell signaling pathways that monitor growth factor levels, nutrient status, and cell–cell contact. This integration of positive and negative cell cycle regulatory cues culminates in the transcriptional upregulation of genes encoding enzymes and accessory factors that directly and indirectly control nucleotide metabolism and DNA synthesis (Fig. 3).

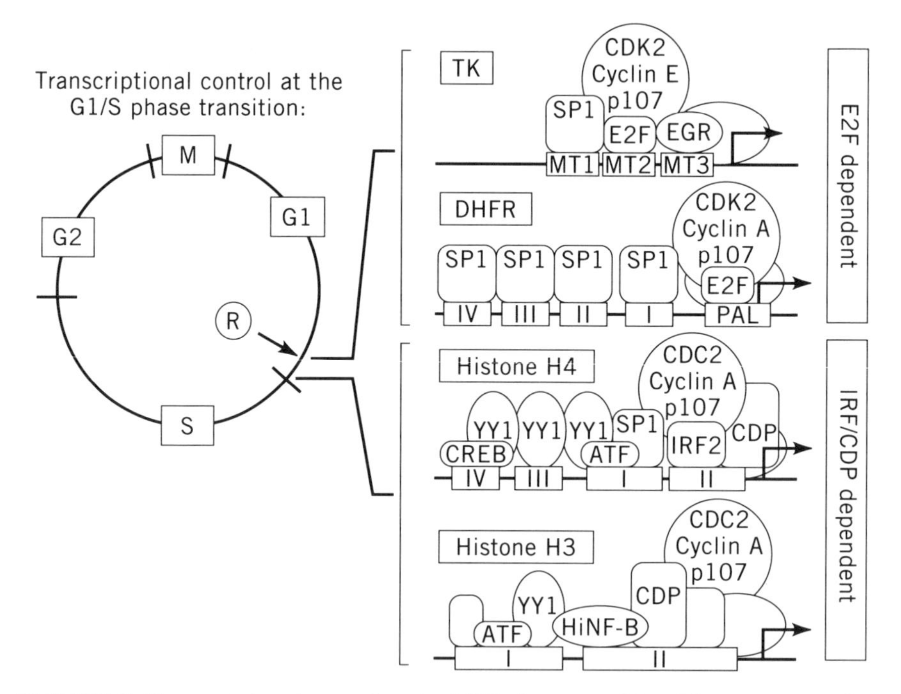

FIGURE 3. Transcriptional control at the G1/S phase transition. The genes encoding enzymes involved in nucleotide metabolism (e.g., TK and DHFR) and histone biosynthesis (e.g., H4 and H3) each are controlled by diverse arrays of promoter regulatory elements (open boxes) which influence transcriptional initiation by RNA polymerase II (grey ovals). E2F elements in the promoters of the TK and DHFR genes interact with heterodimeric E2F factors that associate with CDKs, cyclins, and pRb-related proteins. In contrast, histone genes are controlled by the site II cell cycle regulatory element, which interacts with CDP and IRF2 proteins. Analogous to E2F-dependent mechanisms, CDP interacts with cdc2, cyclin A, and pRb, whereas IRF2 performs an activating function similar to "free" E2F. The presence of binding sites for SP1 in the promoters of G1/S phase–related genes provides a shared mechanism for further enhancement of transcription at the onset of S phase.

Analysis of the thymidine kinase (TK) promoter and cognate promoter factors has revealed that maximal TK gene transcription involves at least three distinct *cis*-acting elements (MT1, MT2, and MT3) (Fridovich-Keil et al., 1993; Dou et al., 1994a,b; Dou and Pardee, 1996; Li et al., 1993; Good et al., 1995). These elements interact with cell cycle-dependent (e.g., Yi1 and Yi2) and constitutive (e.g., SP1) DNA-binding proteins. The Yi complexes interacting with the MT2 motif are associated with p107, as well as cyclin- and CDK-related proteins (Dou et al., 1994a,b; Dou and Pardee, 1996; Li et al., 1993; Good et al., 1995). The Yi complexes are analogous to or identical with E2F-related higher order complexes containing cyclins, CDKs, and pRb-related proteins. Interestingly, cyclins A and E may represent the labile and rate-limiting restriction point proteins, which were originally postulated based on results from early studies on cell growth control (Dou et al., 1993).

Each of the G1/S phase genes is controlled by different arrays of *cis*-acting promoter elements and cognate factors. One unifying theme among many promoters of the R-point genes is the presence of E2F and SP1 consensus elements. Thus, one mechanism by which the cell achieves coordinate and temporal regulation of these genes at the G1/S phase boundary is directly linked to the release of transcriptionally active E2F from inactive E2F–pRb complexes. The disruption of E2F–pRb is mediated by CDK4–CDK6-dependent phosphorylation of pRb in response to growth factor stimulation and cell cycle entry. Hence, the E2F-dependent activation of the R-point genes provides linkage between the onset of S phase and control of cell growth.

The E2F transcription factor represents a heterogenous class of heterodimers formed between one of five different E2F proteins (i.e., E2F-1 to E2F-5) and one of three distinct DP factors (DP-1 to DP-3). The various E2F factors may display preferences in promoter specificity, differ in the regulation of their DNA-binding activities during the cell cycle, and bind selectively to distinct pRb proteins (Meyers and Hiebert, 1995; Chen et al., 1995). The mechanism by which this multiplicity of E2F factors orchestrate transcriptional regulation of diverse sets of genes at the G1/S phase transition is only beginning to be understood. Apart from the role of "free" E2F in activating genes at the G1/S phase transition, promoter-bound complexes of E2F factors associated with pRb-related proteins, cyclin A, and CDK2 have active roles in repression of gene expression during early S phase (Krek et al., 1994).

E2F-responsive transcriptional modulation of R-point genes requires participation of the SP1 family of transcription factors (e.g., SP1 and SP3). For example, the TK promoter contains one E2F site and one SP1 site, and both are required for maximal transcriptional responsiveness at the G1/S phase boundary (Karlseder et al., 1996). This synergistic enhancement involves direct protein–protein interactions between E2F and SP1. Consistent with the critical role of SP1 in cell cycle control of gene expression, protein–protein interactions between SP1 and pRb can also occur, suggesting that pRb can modulate the activities of E2F and SP1 in concert. Analogous to the TK promoter, the DHFR promoter is regulated by four SP1 elements that, together with E2F, mediate transcriptional upregulation at the G1/S phase transition (Wade et al., 1995; Good et al., 1996; Schulze et al., 1994; Wells et al., 1996; Azizkhan et al., 1993; Schilling and Farnham, 1994, 1995; Wells et al., 1997). Interestingly, SP3 selectively represses SP1 activation of the DHFR promoter, but not the TK or histone H4 promoter (Birnbaum et al., 1995b). It appears that the cellular ratio of SP1 and SP3 levels may influence specific classes of cell cycle–regulated genes, but the physiological function of this regulatory mechanism remains to be elucidated.

Initiation of DNA Synthesis in Relation to Transcriptional Control at the G1/S Phase Transition

Conditions that establish competency for the initiation of DNA synthesis in vertebrates are monitored in part by the origin recognition complex (ORC) (Hickey and Malkas, 1997; Stillman, 1996). This complex appears to contain sequence-specific proteins that mark the location of DNA replication origins. Prior to S phase, the labile Cdc6p protein associates with ORC, which stages the subsequent binding of Mcm proteins ("licensing factors") to form large origin-bound pre-replication complexes. The mechanism by which these complexes facilitate the onset of template-directed synthesis of DNA remains to be established. However, activation of S phase–dependent CDKs is required for the initiation of DNA replication, but this event is also thought to prevent assembly of new pre-replication complexes (Stillman, 1996). This hypothesis provides a potential mechanism for stringent control of chromosomal duplication, which should occur only once during each somatic cell cycle. Thus, checkpoint controls at the onset of DNA synthesis serve to signal cellular competency for S phase entry and maintenance of the normal diploid genotype upon mitosis.

Once DNA synthesis has been initiated, replicative activity is confined to specific locations within the nucleus, referred to as DNA replication foci. DNA replication foci represent subnuclear domains that are thought to be highly enriched in multisubunit complexes ("DNA replication factories") containing enzymes involved in DNA synthesis, including DNA polymerases α and δ, PCNA, and DNA ligase (Hickey and Malkas, 1997; Leonhardt and Cardoso, 1995). The concentration of these factors at DNA replication foci that are associated with the nuclear matrix provides a solid-phase framework for understanding catalytic and regulatory components of DNA replication.

Coordinate Activation of Multiple DNA Replication-Dependent Histone Genes During the G1/S Phase Transition

The initiation of histone protein synthesis at the G1/S phase transition is tightly coupled to the start and progression of DNA synthesis. To prevent disorganization of nuclear architecture and chromosomal catastrophe during chromosome segregation at mitosis, it is critical that newly replicated DNA is packaged immediately into nucleosomes. Histones permit the precise packaging of 2 m of DNA into chromatin within each cell nucleus (diameter approximately 10 μm). This functional and temporal coupling poses stringent constraints on multiple parameters of histone gene expression, because somatic cells do not have storage pools for histone protein or histone mRNAs. The vast number of histone polypeptides that must be synthesized and the limited time of S phase allotted for this process necessitate a high histone protein synthesis rate. Mass production of each histone subtype occurs at an average rate of several thousand proteins per second throughout S phase. Moreover, because each 0.2 kb of DNA is packaged by nucleosomal octamers composed of histones H2A, H2B, H3, and H4, the stoichiometric synthesis of each of the histone subtypes is essential for efficient DNA packaging. Consequently, histone gene regulatory factors integrate a series of cell signaling pathways that monitor the onset of S phase and coordinate the expression of 50–100 distinct histone gene subtypes.

The first rate-limiting factor of histone gene expression is the enhancement of a low transcription rate that persists throughout the cell cycle (Plumb et al., 1983). Histone H4 gene transcription has been extensively studied, and a series of *cis*-acting elements and cognate factors have been identified by our laboratory (Vaughan et al., 1995; van Wijnen et al., 1994, 1996; Birnbaum et al., 1995a; Guo et al., 1995; Aziz et al., 1998; Stein et al., 1994). We first showed that genomic occupancy of histone gene promoter elements occurs throughout the cell cycle (Pauli et al., 1987), which was subsequently also shown for the R-point gene DHFR (Wells et al., 1996) by others. The constitutive occupancy of promoter regulatory elements is consistent with the concept that protein–protein interactions, post-translational modifications, and alterations in chromatin structure are important factors in modulating transcription of histone genes and other genes expressed during S phase (Chrysogelos et al., 1985, 1989; Moreno et al., 1986; Pemov et al., 1995; Ljungman, 1996). Similar to the R-point genes, the presence of SP1-binding sites is critical for maximum activation of histone genes. However, unlike the R-point genes, the majority of histone genes do not contain E2F elements. Rather, a sophisticated and E2F-independent transcriptional mechanism has evolved for coordinate activation of histone genes.

As with E2F-responsive genes, E2F-independent transcriptional control mechanisms must account for G1/S phase–dependent enhancement of transcription, as well as attenuation of gene transcription at later stages of S phase. The key cell cycle element for histone H4 genes is a highly conserved promoter domain designated site II, which encompasses binding sites for IRF2, the homeodomain-related "CCAAT displacement protein" CDP/cut, and the TATA-binding complex TFIID (Vaughan et al., 1995; van Wijnen et al., 1994, 1996; Aziz et al., 1998; Stein et al., 1994). IRF2 is required for maximal activation of histone gene transcription and appears to function at the G1/S phase boundary in a manner analogous to "free" E2F by enhancing cell cycle–dependent transcription rates by about threefold (Vaughan et al., 1995).

Phosphorylation of IRF2 *in vivo* occurs primarily on serine residues, which may be mediated by several ubiquitous kinases including casein kinase II, protein kinase A, and protein kinase C (Birnbaum et al., 1997). Interestingly, IRF2 activity does not appear to be directly linked to phosphorylation by mitogen-activated protein kinases or CDKs.

Involvement of the CDP/cut homeodomain protein in cell cycle control was initially established by the finding that this factor is a component of the HiNF-D complex (van Wijnen et al., 1996). In the multisubunit HiNF-D complex, the CDP/cut protein is associated with pRb, cyclin A, and CDK1/cdc2 (van Wijnen et al., 1994, 1997; Shakoori et al., 1995). HiNF-D may have a bifunctional role in H4 gene transcription. For example, binding of the HiNF-D complex to the H4 promoter is essential for maximal H4 gene promoter activity in cells nullizygous for IRF2. However, overexpression of the CDP/cut DNA-binding subunit of HiNF-D results in repression of H4 promoter activity (van Wijnen et al., 1996). CDP/cut in association with pRb, CDK1/cdc2, and cyclin A may perform a function very similar to that of the multiplicity of higher order E2F complexes. These CDP complexes bound to cyclins, CDKs, and pRb-related proteins attenuate the enhanced levels of histone gene transcription during mid S phase, when physiological demand for histone mRNAs begins to diminish.

Similar to the R-point genes, histone gene promoters have auxiliary elements (e.g., site I) that support transcriptional activation during the cell cycle. For example, histone H4 genes contain binding sites for YY1 and SP1. The interaction of SP1 with site I modulates the efficiency of H4 gene transcription by an order of magnitude (Birnbaum et al., 1995a). The binding of YY1 to multiple sites in the histone H4 promoter may facilitate gene–nuclear matrix interactions (Guo et al., 1995; Last et al., unpublished data). In addition, it has recently been shown that YY1 associates with the histone deacetylase rpd3. The possibility arises that post-translational modifications of histone proteins when bound as nucleosomes to the H4 promoter may parallel the modifications in chromatin structure that accompany modulations of histone H4 gene expression (Chrysogelos et al., 1985, 1989).

Stoichiometric synthesis of histone mRNAs and proteins requires coordinate control of histone gene expression at several gene regulatory levels. At the transcriptional level, coordinate activation of the five histone gene classes at the G1/S phase transition may be mediated by CDP/cut, which has been shown to interact with the promoters of all major histone gene subtypes. The association of CDP/cut with pRb, cyclin A, and CDK1/cdc2 as components of the HiNF-D complexes interacting with DNA replication-dependent histone genes provides direct functional linkage between transcriptional coordination of histone gene expression and cyclin/CDK signaling mechanisms that mediate cell cycle progression.

The secondary levels at which histone gene expression is coordinated occur by post-transcriptional mechanisms, including transcript elongation and 3′ end processing, that produce mature histone mRNAs. Histone mRNAs do not have polyA tails, but instead contain a unique histone-specific hairpin-loop structure. Histone 3′ end processing is mediated by U7 snRNP complexes (Marzluff and Pandey, 1988; Schumperli, 1988), and accessory proteins that recognize histone mRNA 3′ ends are being characterized. It has been postulated that histone mRNA 3′ end processing is a key step in histone gene expression (Marzluff and Pandey, 1988; Schumperli, 1988; Harris et al., 1991), but the regulatory events that connect activation of this process to cell cycle progression remain to be established. However, because all histone mRNAs have highly similar structural elements at the 3′ end, the recognition of these structures by regulatory factors (Wang et al., 1996; Martin et al., 1997) may represent an important mechanism by which mature histone mRNAs are produced, transported to specific cytoplasmic locations, translated, and degraded.

Selective Downregulation of Histone Gene Expression Upon Cessation of DNA Replication at the S/G2 Transition

When cells approach the S/G2 transition, most of the genome has been replicated, and the demand for histone proteins to package newly replicated DNA diminishes. Cells must ensure that histones do not accumulate in excess, as these highly basic proteins would likely interfere with cellular and particularly nucleic acid metabolism. Therefore, histone mRNAs are selectively degraded during late S phase in concert with the completion of DNA synthesis. Molecular mechanisms have been elucidated that account for post-transcriptional control of histone gene expression during S phase by modulating mRNA stability. The histone mRNA-specific stem-loop structure plays a key role in regulating histone mRNA turnover. This stem-loop motif is present in all mRNAs encoding the live cell cycle–regulated histone subtypes. Therefore, this structure is considered pivotal in maintaining the stoichiometric balance of the five histone classes and coupling with DNA replication. Selective degradation of histone mRNA when DNA synthesis is halted is mediated by a 3′ exonuclease and requires active translation of histone mRNA bound to polyribosomes (Marzluff and Pandey, 1988; Schumperli, 1988; Stein and Stein, 1984; Zambetti et al., 1987). Interestingly, mRNA destabilization does not occur when histone mRNA is targeted to membrane-bound ribosomes rather than to ribosomes associated with the nonmembranous cytoskeleton. Thus, histone mRNA degradation is a dynamic process that requires macromolecular complexes at specific subcellular locations.

It has been shown that histone proteins mediate histone mRNA degradation (Pelz and Ross, 1987) and that histone mRNA half lives are modulated during the cell cycle (Morris et al., 1991). Because histone mRNAs are most stable when free histone protein concentrations are minimal and highly unstable when histone proteins accumulate in excess, it appears that selective downregulation of histone gene expression is achieved by an autoregulatory mechanism.

Molecular Mechanisms Monitoring Chromosomal Integrity During the Cell Cycle

The integrity of the genome and fidelity of the encoded genetic information is monitored at key cell cycle stages to ensure that daughter cells receive a duplicate copy of the chromosomal complement. DNA damage checkpoints that decide between cell cycle progression, cell cycle inhibition, and apoptosis have been identified in G1, S, and G2. Several factors have been shown to be involved in DNA repair pathways, including the tumor suppressors p53 (Elledge, 1996) and IRF1 (Tanaka et al., 1994; Tamura et al., 1995), the ataxia telangiectasia (AT)–related proteins (Elledge, 1996), and the p53-inducible gene p21 (Sherr and Roberts, 1995). DNA signaling pathways must involve a sensor that is linked to damaged DNA. However, definitive proof of factors performing this sensoring task have not been identified in mammalian cells. One DNA damage repair pathway may involve activation of p53 by the "AT-mutated" (ATM) protein. This results in p53-dependent activation of p21 and subsequent inhibition of cell cycle stage-specific CDKs. However, alternative pathways involving the "AT and rad-related" (ATR) protein, as well as IRF1 exist. The importance of DNA repair pathways during the mammalian cell cycle is reflected by conservation of analogous mechanisms in yeast (Elledge, 1996).

Cell death by apoptosis is an important checkpoint that functions to eliminate cells that have undergone catastrophic genetic damage. An important component of this checkpoint is the p53 tumor suppressor gene, which is a transcriptional regulator of genes controlling both cell cycle progression and cell death (reviewed by Levine, 1997). p53 normally has a very short half life, but accumulates in response to DNA damage or other stressors. p53 function, by its ability to transactivate $p21^{Cip}$, is to ensure that in response to genetic damage cells arrest in G1, thereby allowing cells to repair their damaged DNA (reviewed by Sherr, 1996). If the DNA damage is too great, the cell will enter the apoptotic pathway.

Apoptosis is characterized by condensation of chromatin followed by fragmentation of the nucleus as a result of endonuclease digestion of DNA between nucleosomes. Biochemical events and molecular markers functionally related to the promotion or prevention of apoptosis have been identified. A number of cell cycle regulatory proteins (c-fos, c-myc, c-jun, cdc2) and transcriptional activators (p53, IRF1, IRF2) appear to be upregulated in response to apoptosis-inducing stimuli and have been implicated, therefore, as essential mediators of the process (reviewed by Pandey and Wang, 1995). The fact that the p53 gene is the most frequently mutated gene in human cancers demonstrates its key role in these pathways. In addition, p53-independent pathways leading to apoptosis have been identified that require IRF1 and IRF2 and that appear to be cell type specific (Tamura et al., 1995; Tanaka et al., 1994).

Several genes have been identified that function in the decision-making process between cell survival and cell death. bcl-2 blocks cell death in thymocytes induced by growth factor deprivation, glucocorticoids, and γ irradiation, but not death induced by all stimuli (Korsmeyer, 1992). bax, a homolog of bcl-2, heterodimerizes with and inactivates bcl-2, promoting cell death (Oltvai et al., 1993). A number of other bcl-2 family members, including alternatively spliced forms, that are involved in apoptosis have been identified (Boise et al., 1993). p53-dependent modulation of these genes may be a critical regulatory process that initiates apoptosis. For example, p53 is a direct transcriptional activator of the pro-apoptotic bax gene and can downregulate survival genes such as bcl-2 (Miyashita and Reed, 1995). A family of interleukin-1β–converting enzyme (ICE) proteinases that may be involved in the activation of endonucleases(s) responsible for degradation of DNA have been identified (Flaws et al., 1995; Miura et al., 1993). Other substrates that are cleaved by ICE-like proteases during apoptosis include the DNA repair enzyme polyADP-ribose polymerase and pRb (Jänicke et al., 1996).

Apoptosis is an important component of growth control during embryogenesis, organogenesis, and tissue morphogenesis, as well as in the maintenance of proliferation homeostasis in many adult tissues (Clarke, 1990). However, the signaling cascades contributing to the activation of the apoptotic pathway are not well defined. Molecular mechanisms mediating these responses may derive from specific pre-apoptotic signaling molecules, e.g., Msx expression in developing limbs, disruption of cell–extracellular matrix, and cell–cell interactions, that can influence cell death pathways.

CONTROL MECHANISMS THAT MODULATE THE ACTIVITIES OF CELL CYCLE REGULATORY FACTORS

Cell Cycle Stage-Specific and Ubiquitin-Dependent Turnover of Gene Regulatory Factors

The activation and inactivation of cell cycle regulatory factors at specific stages of the cell cycle occur at multiple levels and are often achieved by a combination of control mechanisms. As discussed above, CDK-mediated phosphorylation pathways represent an important level of control. For example, the DNA-binding activity of the histone gene regulatory HiNF-D complex is dramatically increased at the G1/S phase boundary in stem cells (Shakoori et al., 1995; van Wijnen et al., 1997). This increase in HiNF-D activity occurs, although the levels of the DNA-binding subunit CDP/cut remain constant. Concomitantly, levels of CDK1–cdc2 and cyclin A are increased, and pRb becomes hyperphosphorylated (Shakoori et al., 1995; van Wijnen et al., 1997), which is known to occur in a CDK-dependent manner.

The inactivation of regulatory factors represents an equally important cell cycle control mechanism. Recently, many studies have focused on the role of ubiquitin-dependent proteolysis of factors by the 22 S proteasome (Tanaka and Tsurumi, 1997). The ubiquitin-proteasome system involves a large number of enzymes mediating ubiquitin activation (E1), ubiquitin conjugation (E2), or ubiquitin ligation (E3) that modulate turnover of cell cycle regulatory proteins (reviewed by King et al., 1997). For example, degradation of G1 cyclins involves the CDC34 protein. CDC34 is a ubiquitin-conjugating enzyme and is conserved between yeast and vertebrates. The CDC34 gene is essential for the G1/S phase transition. Ubiquitin-dependent degradation is also involved in constitutive turnover of cyclins throughout G1, reflecting the labile nature of G1 competency factors such as cyclin E.

Ubiquitin-dependent degradation also performs a key regulatory function during the G2/M transition (King et al., 1997). The E2 enzymes encoded by UBC4 and UBC9 are involved in degradation of specific cyclins prior to the onset of mitosis. Similarly, completion of mitosis is regulated by the anaphase-promoting complex. This high molecular weight ubiquitination complex is essential for chromosome segregation, but specific molecular targets have not been identified. Apart from degradation of cyclins at specific stages during the cell cycle, ubiquitin- dependent proteolysis may also be important for regulating the activities of oncoproteins, including c-fos, c-jun, and IRF2.

Parameters of Nuclear Architecture Influencing Cell Cycle Control

Intracellular targeting of cell cycle regulatory proteins may contribute to cell cycle control by concentrating the activities of these factors to specific intracellular domains. For example, CDC2/cdk1 displays a dispersed subnuclear distribution, but is also highly concentrated at the centrosomes (Pockwinse et al., 1997). This subnuclear location may regulate the assembly of the mitotic spindle. Association of cyclin B with membranes may be functionally linked to the breakdown of the nuclear membrane and/or subsequent formation of the nuclear envelope after mitosis (Jackman et al., 1995). The presence of the tumor suppressor protein p53 in the cytoplasm or nucleus has been correlated with cell growth control and the neoplastic phenotype of the cell (Moll et al., 1996). The retinoblastoma protein pRb (Mancini et al., 1994), as well as transcription factors related to the c-fos and c-jun oncoproteins (van Wijnen et al., 1993), are associated with the nuclear matrix in a cell cycle– and/or cell growth–dependent manner. The association of DNA ligase I and DNA methyltransferase with nuclear matrix–associated DNA replication domains also provides a mechanism for integrating enzymatic and regulatory functions at specific locations (Leonhardt and Cardoso, 1995). Furthermore, the PML protein (Weis et al., 1994; Grande et al., 1996) and AML proteins (unpublished data) are redirected to distinct subnuclear domains as a consequence of structural modifications in these proteins due to chromosomal translocations in promyelocytic or myelogenous leukemias.

Because modifications in nuclear architecture are hallmarks of cancer cells, it is of considerable importance to define molecular principles governing the normal targeting of gene regulatory molecules to specific subnuclear domains. Recently, we have shown that the transcriptionally active AML-1B protein, a gene regulatory factor with a causative role in the progression of acute myelogenous leukemia, contains a 31 amino acid nuclear matrix–targeting signal (NMTS) (Zeng et al., 1997). The NMTS contributes to trans-activation *in vivo* and directs the hematopoietic and bone-specific family of AML/CBF regulatory factors to sites within the nucleus that are transcriptionally competent (Zeng et al., 1998). This NMTS is an autonomous protein domain that is absent in the transcriptionally inactive protein isoform AML-1. The NMTS is the frequent target of rearrangements due to chromosomal translocations that result in the separation of the NMTS, as well as a closely associated *trans*-activation domain, from the DNA-binding domain of AML-1B. Hence, stringent control of intranuclear targeting may represent an important mechanism for maintaining the normal cell growth phenotype of somatic cells.

CONCLUSION

The necessity for stringent control of cell cycle and growth regulatory mechanisms is becoming increasingly appreciated. There is acute awareness of the required fidelity of growth control to support proliferation during development and tissue remodeling throughout the life of an organism. Significant inroads are being made into elucidation of both stimulation of proliferation and exit from the cell cycle for establishment and maintenance of cells and tissues with specialized functions. Equally important, we are gaining insight into controls that are operative to sustain pools of stem cells with competency to expand in response to physiological demands and commit to specific phenotypes.

Studies carried out during the past several years have provided indications of key regulatory points during the cell cycle that support competency for proliferation and cell cycle progression. A series of checkpoints have been functionally defined that monitor fidelity of the complex and interdependent regulatory events that control each cell cycle transition and initiation as well as cessation of proliferation. The extensive repertoire of growth regulatory factors that have been identified and characterized offer valuable clues to control of growth and differentiation. We are beginning to understand interrelationships of growth regulatory mechanisms with components of cellular architecture. Additionally, we are learning effective ways to control proliferation under a broad spectrum of conditions. Consequently, new approaches are emerging for treatment of proliferation disorders that include but are not restricted to cancer. Our capabilities for controlling the growth properties of stem cells without compromising options for phenotype development are expanding. One can be confident that insight into parameters of gene expression that control stem cell proliferation and differentiation is forthcoming. It is realistic to expect enhanced options for development of targeted and physiologically regulated gene therapy that provides sustained effectiveness.

ACKNOWLEDGMENTS

The authors thank Elizabeth Bronstein for editorial assistance with the preparation of the manuscript. Studies described in this chapter were in part supported by grants from the National Institutes of Health (GM32010, AR45689, AR45688, AR39588).

REFERENCES

Aziz F, van Wijnen AJ, Vaughan PS, Wu S, Shakoori AR, Lian JB, Soprano KJ, Stein JL, Stein GS (1998): The integrated activities of IRF-2 (HiNF-M), CDP/cut (HiNF-D) and H4TF-2 (HiNF-P) regulate transcription of a cell cycle controlled human histone H4 gene: Mechanistic differences between distinct H4 genes. *Mol Biol Rep* 25:1–12.

Azizkhan JC, Jensen DE, Pierce AJ, Wade M (1993): Transcription from TATA-less promoters: Dihydrofolate reductase as a model. *Crit Rev Eukaryot Gene Exp* 3:229–254.

Bartkova J, Zemanova M, Bartek J (1996): Expression of CDK7/CAK in normal and tumor cells of diverse histogenesis, cell-cycle position and differentiation. *Int J Cancer* 66:732–737.

Baserga R, Estensen RD, Petersen RO (1965): Inhibition of DNA synthesis in Ehrlich ascites cells by actinomycin D. II. The presynthetic block in the cell cycle. *Proc Natl Acad Sci USA* 54:1141.

Baserga R, Rubin R (1993): Cell cycle and growth control. *Crit Rev Eukaryot Gene Exp* 3:47–61.

Birnbaum MJ, Wright KL, van Wijnen AJ, Ramsey-Ewing AL, Bourke MT, Last TJ, Aziz F, Frenkel B, Rao BR, Aronin N, Stein GS, Stein JL (1995a): Functional role for Sp1 in the transcriptional amplification of a cell cycle regulated histone H4 gene. *Biochemistry* 34:7648–7658.

Birnbaum MJ, van Wijnen AJ, Odgren PR, Last TJ, Suske G, Stein GS, Stein JL (1995b): Sp1 *trans*-activation of cell cycle regulated promoters is selectively repressed by Sp3. *Biochemistry* 34:16503–16508.

Birnbaum MJ, van Zundert B, Vaughan PS, Whitmarsh AJ, van Wijnen AJ, Davis RJ, Stein GS, Stein JL (1997): Phosphorylation of the oncogenic transcription factor interferon regulatory factor 2 (IRF2) *in vitro* and *in vivo*. *J Cell Biochem* 66:175–183.

Bodrug SE, Warner BJ, Bath ML, Lindeman GJ, Harris AW, Adams JM (1994): Cyclin D1 transgene impedes lymphocyte maturation and collaborates in lymphomagenesis with the myc gene. *EMBO J* 13:2124–2130.

Boise LH, González-Garcia M, Postema CE, Ding L, Lindsten T, Turka LA, Mao X, Nuñez G, Thompson CB (1993): bcl-x a bcl-2-related gene that functions as a dominant regulator of apoptotic cell death. *Cell* 74:597–608.

Buchkovich KJ, Ziff EB (1994): Nerve growth factor regulates the expression and activity of p33cdk2 and p34cdc2 kinases in PC12 pheochromocytoma cells. *Mol Biol Cell* 5:1225–1241.

Burger C, Wick M, Muller R (1994): Lineage-specific regulation of cell cycle gene expression in differentiating myeloid cells. *J Cell Sci* 107:2047–2054.

Buyse IM, Shao, G, Huang S (1995): The retinoblastoma protein binds to RIZ, a zinc-finger protein that shares an epitope with the adenovirus E1A protein. *Proc Natl Acad Sci USA* 92:4467–4471.

Carpenter PB, Mueller PR, Dunphy WG (1996): Role for a *Xenopus* orc2-related protein in controlling DNA replication. *Nature* 379:357–360.

Chan FKM, Zhang J, Cheng L, Shapiro DN, Winoto A (1995): Identification of human and mouse p19, a novel CDK4 and CDK6 inhibitor with homology to $p16^{ink4}$. *Mol Cell Biol* 15:2682–2688.

Chandrasekaran C, Coopersmith CM, Gordon JI (1996): Use of normal and transgenic mice to examine the relationship between terminal differentiation of intestinal epithelial cells and accumulation of their cell cycle regulators. *J Biol Chem* 271:28414–28421.

Chen P-L, Riley DJ, Chen Y, Lee WH (1996): Retinoblastoma protein positively regulates terminal adipocyte differentiation through direct interaction with C/EBPs. *Genes Dev* 10:2794–2804.

Chen P-L, Riley DJ, Lee W-H (1995): The retinoblastoma protein as a fundamental mediator of growth and differentiation signals. *Crit Rev Eukaryot Gene Exp* 5:79–95.

Chrysogelos S, Riley DE, Stein GS, Stein JL (1985): A human histone H4 gene exhibits cell cycle dependent changes in chromatin structure that correlate with its expression. *Proc Natl Acad Sci USA* 82:7535–7539.

Chrysogelos S, Riley DE, Stein GS, Stein JL (1989): Fine mapping of the chromatin structure of a cell cycle–regulated human H4 histone gene. *J Biol Chem* 264:1232–1237.

Clarke AR, Maandag ER, van Roon M, van der Lugt NMT, van der Valk M, Hooper ML, Berns A, te Riele H (1992): Requirement for a functional Rb-1 gene in murine development. *Nature* 359:328–330.

Clarke PHG (1990): Developmental cell death: Morphological diversity and multiple mechanisms. *Anat Embryol* 181:195–213.

Clyne RK, Kelly TJ (1995): Genetic analysis of an ARS element from the fission yeast *Schizosaccharomyces pombe. EMBO J* 14:6348–6357.

Coats S, Flanagan WM, Nourse J, Roberts JM (1996): Requirement of $p27^{Kip1}$ for restriction point control of the fibroblast cell cycle. *Science* 272:877–880.

Cobrinik D, Lee M-H, Hannon G, Mulligan G, Bronson RT, Dyson N, Harlow E, Beach D, Weinberg RA, Jacks T (1996): Shared role of the pRB-related p130 and p107 proteins in limb development. *Genes Dev* 10:1633–1644.

Corbeil HB, Whyte P, Branton PE (1995): Characterization of transcription factor E2F complexes during muscle and neuronal differentiation. *Oncogene* 11:909–920.

Cordon-Cardo C, Richon VM (1994): Expression of the retinoblastoma protein is regulated in normal human tissues. *Am J Pathol* 144:500–510.

Deng C, Zhang P, Harper JW, Elledge SJ, Leder P (1995): Mice lacking $p21^{CIP1/WAF1}$ undergo normal development, but are defective in G1 checkpoint control. *Cell* 82:675–684.

Dobashi Y, Kudoh T, Matsumine A, Toyoshima K, Akiyama T (1995): Constitutive overexpression of CDK2 inhibits neuronal differentiation of rat pheochromocytoma PC12 cells. *J Biol Chem* 270:23031–23037.

Dobashi Y, Kudoh T, Toyoshima K, Akiyama T (1996): Persistent activation of CDK4 during neuronal differentiation of rat pheochromocytoma PC12 cells. *Biochem Biophys Res Commun* 221:351–355.

Dou QP, Levin AH, Zhao S, Pardee AB (1993): Cyclin E and cyclin A as candidates for the restriction point protein. *Cancer Res* 53:1493–1497.

Dou QP, Molnar G, Pardee AB (1994a): Cyclin D1/cdk2 kinase is present in a G1 phase-specific protein complex Yi1 that binds to the mouse thymidine kinase gene promoter. *Biochem Biophys Res Commun* 205:1859–1868.

Dou QP, Pardee AB (1996): Transcriptional activation of thymidine kinase, a marker for cell cycle control. *Prog Nucleic Acid Res Mol Biol* 53:197–217.

Dou QP, Zhao S, Levin AH, Wang J, Helin K, Pardee AB (1994b): G1/S-regulated E2F-containing protein complexes bind to the mouse thymidine kinase gene promoter. *J Biol Chem* 269:1306–1313.

Dowdy SF, Hinds PW, Louie K, Reed SI, Arnold A, Weinberg RA (1993): Physical interaction of the retinoblastoma protein with human D cyclins. *Cell* 73:499–511.

Dubey DD, Kim SM, Todorov IT, Huberman JA (1996): Large, complex modular structure of a fission yeast DNA replication origin. *Curr Biol* 6:467–473.

Durfee T, Becherer K, Chen P-L, Yeh S-H, Yang Y, Kilburn AE, Lee W-H, Elledge SJ (1993): The retinoblastoma protein associates with the protein phosphatase type 1 catalytic subunit. *Genes Dev* 7:555–569.

El-Deiry WS, Tokino T, Velculescu VE, Levy DB, Parsons R, Trent JM, Lin D, Mercer E, Kinzler KW, Vogelstein B (1993): WAF1, a potential mediator of p53 tumor suppression. *Cell* 75:817–825.

Elledge SJ (1996): Cell cycle checkpoints: Preventing an identity crisis. *Science* 274:1664–1672.

Ewen ME, Sluss HK, Sherr CJ, Matsushime H, Kato J, Livingston DM (1993): Functional interactions of the retinoblastoma protein with mammalian D-type cyclins. *Cell* 73:487–497.

Fantl V, Stamp G, Andrews A, Rosewell I, Dickson C (1995): Mice lacking cyclin D1 are small and show defects in eye and mammary gland development. *Genes Dev* 9:2364–2372.

Fero ML, Rivkin M, Tasch M, Porter P, Carow CE, Firpo E, Polyak K, Tsai L-H, Broudy V, Perlmutter RM, Kaushansky K, Roberts JM (1996): A syndrome of multiorgan hyperplasia with features of gigantism, tumorigenesis, and female sterility in $p27^{Kip1}$-deficient mice. *Cell* 85:733–744.

Flaws JA, Kugu K, Trbovich AM, DeSauti A, Tilly KI, Hirshfield AN, Tilly JL (1995): Interleukin-1 beta-converting enzyme-related proteases (IRPs) and mammalian cell death: dissociation of IRP-induced oligonucleosomal endonuclease activity from morphological apoptosis in granulosa cells of the ovarian follicle. *Endocrinology* 136:5042–5053.

Fridovich-Keil JL, Markell PJ, Gudas JM, Pardee AB (1993): DNA sequences required for serum-responsive regulation of expression from the mouse thymidine kinase promoter. *Cell Growth Differ* 4:679–687.

Gao CY, Bassnett S, Zelenka PS (1995): Cyclin B, p34cdc2, and H1-kinase activity in terminally differentiating lens fiber cells. *Dev Biol* 169:185–194.

Gavin KA, Hidaka M, Stillman B (1995): Conserved initiator proteins in eukaryotes. *Science* 270:1667–1671.

Gervasi C, Szaro BG (1995): The *Xenopus laevis* homologue to the neuronal cyclin-dependent kinase (cdk5) is expressed in embryos by gastrulation. *Mol Brain Res* 33:192–200.

Godbout R, Andison R (1996): Elevated levels of cyclin D1 mRNA in the undifferentiated chick retina. *Gene* 182:111–115.

Good L, Chen J, Chen KY (1995): Analysis of sequence-specific binding activity of cis-elements in human thymidine kinase gene promoter during G1/S phase transition. *J Cell Physiol* 163:636–644.

Good L, Dimri GP, Campisi J, Chen KY (1996): Regulation of dihydrofolate reductase gene expression and E2F components in human diploid fibroblasts during growth and senescence. *J Cell Physiol* 168:580–588.

Gossen M, Pak DT, Hansen SK, Acharya JK, Botchan MR (1995): A *Drosophila* homolog of the yeast origin recognition complex. *Science* 270:1674–1677.

Grande MA, van der Kraan I, van Steensel B, Schul W, de The H, van der Voort HT, de Jong L, van Driel R (1996): PML-containing nuclear bodies: Their spatial distribution in relation to other nuclear components. *J Cell Biochem* 63:280–291.

Gu W, Schneider JW, Condorelli G, Kaushal S, Mahdavi V, Nadal-Ginard B (1993): Interaction of myogenic factors and the retinoblastoma protein mediates muscle cell commitment and differentiation. *Cell* 72:309–324.

Guadagno TM, Ohtsubo M, Roberts JM, Assioan RK (1993): A link between cyclin A expression and adhesion dependent cell cycle progression. *Science* 262:1572–1575.

Guan K-L, Jenkins CW, Li Y, Nichols MA, Wu X, O'Keefe CL, Matera AG, Xiong Y (1994): Growth suppression by p18, a $p16^{INK4/MTS1}$- and $p14^{INK4B/MTS2}$-related CDK6 inhibitor, correlates with wild-type pRb function. *Genes Dev* 8:2939–2952.

Guo B, Odgren PR, van Wijnen AJ, Last TJ, Nickerson J, Penman S, Lian JB, Stein JL, Stein GS (1995): The nuclear matrix protein NMP-1 is the transcription factor YY1. *Proc Natl Acad Sci USA* 92:10526–10530.

Guo K, Wang J, Andres V, Smith RC, Walsh K (1995): MyoD-induced expression of p21 inhibits cyclin-dependent kinase activity upon myocyte terminal differentiation. *Mol Cell Biol* 15:3823–3829.

Halevy O, Novitch BG, Spicer DB, Skapek SX, Rhee J, Hannon GJ, Beach D, Lassar AB (1995): Correlation of terminal cell cycle arrest of skeletal muscle with induction of p21 by MyoD. *Science* 267:1018–1021.

Hamlin JL, Mosca PJ, Levenson VV (1994): Defining origins of replication in mammalian cells. *Biochim Biophys Acta* 1198:85–111.

Han EK, Begemann M, Sgambato A, Soh JW, Doki Y, Xing WQ, Liu W, Weinstein IB (1996): Increased expression of cyclin D1 in a murine mammary epithelial cell line induces p27kip1, inhibits growth, and enhances apoptosis. *Cell Growth Differ* 7:699–710.

Hannon GJ, Beach D (1994): $p15^{INK4B}$ is a potential effector of TGF-β–induced cell cycle arrest. *Nature* 371:257–261.

Harris ME, Bohni R, Schneiderman MH, Ramamurthy L, Schumperli D, Marzluff WF (1991): Regulation of histone mRNA in the unperturbed cell cycle: Evidence suggesting control at two post-transcriptional steps. *Mol Cell Biol* 11:2416–2424.

Hartwell LH, Culotti J, Pringle JR, Reid BJ (1974): Genetic control of the cell division cycle in yeast. *Science* 183:46–51.

Hartwell LH, Weinert TA (1989): Checkpoints: Controls that ensure the order of cell cycle events. *Science* 246:629–634.

Heichman KA, Roberts JM (1994): Rules to replicate by. *Cell* 79:557–562.

Hickey RJ, Malkas LH (1997): Mammalian cell DNA replication. *Crit Rev Eukaryot Gene Exp* 7:125–157.

Holthuis J, Owen TA, van Wijnen AJ, Wright KL, Ransey-Ewing A, Kennedy MB, Carter R, Cosenza SC, Soprano KJ, Lian JB, Stein JL, Stein GS (1990): Tumor cells exhibit deregulation of the cell cycle histone gene promoter factor HiNF-D. *Science* 247:1454–1457.

Horiguchi-Yamada J, Yamada H, Nakada S, Ochi K, Nemoto T (1994): Changes of G1 cyclins, cdk2, and cyclin A during the differentiation of HL60 cells induced by TPA. *Mol Cell Biochem* 132:31–37.

Howard A, Pelc SR (1951): Nuclear incorporation of 32p as demonstrated by autoradiography. Exp Cell Res 2:178–187.

Hunter T, Pines J (1994): Cyclins and cancer II: Cyclin D and CDK inhibitors come of age. *Cell* 79:573–582.

Jackman M, Firth M, Pines J (1995): Human cyclins B1 and B2 are localized to strikingly different structures: B1 to microtubules, B2 primarily to the Golgi apparatus. *EMBO J* 14:1646–1654.

Jacks T, Fazeli A, Schmitt EM, Bronson RT, Goodell MA, Weinberg RA (1992): Effects of an Rb mutation in the mouse. *Nature* 359:295–300.

Jahn L, Sadoshima J, Izumo S (1994): Cyclins and cyclin-dependent kinases are differentially regulated during terminal differentiation of C2C12 muscle cells. *Exp Cell Res* 212:297–307.

Jänicke RU, Walker PA, Lin XY, Porter AG (1996): Specific cleavage of the retinoblastoma protein by an ICE-like protease in apoptosis. *EMBO J* 15:6969–6978.

Jerzmanowski A, Cole RD (1992): Partial displacement of histone H1 from chromatin is required before it can be phosphorylated by mitotic H1 kinase *in vitro*. *J Biol Chem* 267:8514–8520.

Jiang H, Lin J, Young SM, Goldstein NI, Waxman S, Davila V, Chellappan SP, Fisher PB (1995): Cell cycle gene expression and E2F transcription factor complexes in human melanoma cells induced to terminally differentiate. *Oncogene* 11:1179–1189.

Johnson LF (1992): G1 events and the regulation of genes for S-phase enzymes. *Curr Opin Cell Biol* 4:149–154.

Karlseder J, Rotheneder H, Wintersberger E (1996): Interaction of Sp1 with the growth- and cell cycle-regulated transcription factor E2F. *Mol Cell Biol* 16:1659–1667.

Kato JY, Sherr CJ (1993): Inhibition of granulocyte differentiation by G1 cyclins D2 and D3 but not D1. *Proc Natl Acad Sci USA* 90:11513–11517.

Kiess M, Gill RM, Hamel PA (1995a): Expression and activity of the retinoblastoma protein (pRB) family proteins, p107 and p130, during L6 myoblast differentiation. *Cell Growth Differ* 6:1287–1298.

Kiess M, Gill RM, Hamel PA (1995b): Expression of the positive regulator of cell cycle progression, cyclin D3, is induced during differentiation of myoblasts into quiescent myotubes. *Oncogene* 10:159–166.

King RW, Deshaies RJ, Peters JM, Kirschner MW (1997): How proteolysis drives the cell cycle. *Science* 274:1652–1659.

King RW, Jackson PK, Kirschner MW (1994): Mitosis in transition. *Cell* 79:563–571.

Kiyokawa H, Kineman RD, Manova-Todorova KO, Soares VC, Hoffman ES, Ono M, Khanam D, Hayday AC, Frohman LA, Koff A (1996): Enhanced growth of mice lacking the cyclin-dependent kinase inhibitor function of $p27^{Kip1}$. *Cell* 85:721–732.

Kiyokawa H, Richon VM, Rifkind RA, Marks PA (1994): Suppression of cyclin-dependent kinase 4 during induced differentiation of erythroleukemia cells. *Mol Cell Biol* 14:7195–7203.

Kiyokawa H, Richon VM, Venta-Perez G, Rifkind RA, Marks PA (1993): Hexamethylenebisacetamide-induced erythroleukemia cell differentiation involves modulation of events required for cell cycle progression through G1. *Proc Natl Acad Sci USA* 90:6746–6750.

Korsmeyer SJ (1992): Bcl-2 initiates a new category of oncogenes: Regulators of cell death. *Blood* 80:879–886.

Kouzarides T (1995): Transcriptional control by the retinoblastoma protein. *Semin Cancer Biol* 6:91–98.

Kranenburg O, de Groot RP, Van der Eb AJ, Zantema A (1995a): Differentiation of P19 EC cells leads to differential modulation of cyclin-dependent kinase activities and to changes in the cell cycle profile. *Oncogene* 10:87–95.

Kranenburg O, van der Eb AJ, Zantema A (1995b): Cyclin-dependent kinases and pRb: Regulators of the proliferation-differentiation switch. *FEBS Lett* 367:103–106.

Krek W, Ewen ME, Shirodkar S, Arany Z, Kaelin WG, Livingston DM (1994): Negative regulation of the growth-promoting transcription factor E2F-1 by a stably bound cyclin A–dependent protein kinase. *Cell* 78:161–172.

La Thangue NB (1994): DRTF1/E2F: An expanding family of heterodimeric transcription factors implicated in cell-cycle control. *TIBS* 19:108–114.

Leatherwood J, Lopez-Girona A, Russell P (1996): Interaction of Cdc2 and Cdc18 with a fission yeast ORC2-like protein. *Nature* 379:360–363.

Lee EY-HP, Chang C-Y, Hu N, Wang Y-CJ, Lai C-C, Herrup K, Lee W-H, Bradley A (1992): Mice deficient for Rb are nonviable and show defects in neurogenesis and haematopoiesis. *Nature* 359:288–294.

Lee EY, Hu N, Yuan SS, Cox LA, Bradley A, Lee WH, Herrup K (1994): Dual roles of the retinoblastoma protein in cell cycle regulation and neuron differentiation. *Genes Dev* 8:2008–2021.

Lee WH, Chen PL, Riley DJ (1995): Regulatory networks of the retinoblastoma protein. *Ann NY Acad Sci* 752:432–445.

Leonhardt H, Cardoso MC (1995): Targeting and association of proteins with functional domains in the nucleus: The insoluble solution. *Int Rev Cytol* 162B:303–335.

Levine AJ (1997): p53, the cellular gatekeeper for growth and division. *Cell* 88:323–331.

Li LJ, Naeve GS, Lee AS (1993): Temporal regulation of cyclin A-p107 and p33cdk2 complexes binding to a human thymidine kinase promoter element important for G1-S phase transcriptional regulation. *Proc Natl Acad Sci USA* 90:3554–3558.

Ljungman M (1996): Effect of differential gene expression on the chromatin structure of the DHFR gene domain *in vivo*. *Biochim Biophys Acta* 1307:171–177.

Loyer P, Glaise D, Cariou S, Baffet G, Meijer L, Guguen-Guillouzo C (1994): Expression and activation of cdks (1 and 2) and cyclins in the cell cycle progression during liver regeneration. *J Biol Chem* 269:2491–2500.

Lucibello FC, Sewing A, Brusselbach S, Burger C, Muller R (1993): Deregulation of cyclins D1 and E and suppression of cdk2 and cdk4 in senescent human fibroblasts. *J Cell Sci* 105:123–133.

MacLachlan TK, Sang N, Giordano A (1995): Cyclins, cyclin-dependent kinases and Cdk inhibitors: Implications in cell cycle control and cancer. *Crit Rev Eukaryot Gene Exp* 5:127–156.

Mancini MA, Shan B, Nickerson JA, Penman S, Lee WH (1994): The retinoblastoma gene product is a cell cycle-dependent, nuclear matrix-associated protein. *Proc Natl Acad Sci USA* 91:418–422.

Marks PA, Richon VM, Rifkind RA (1996): Cell cycle regulatory proteins are targets for induced differentiation of transformed cells: Molecular and clinical studies employing hybrid polar compounds. *Int J Hematol* 63:1–17.

Martin F, Schaller A, Eglite S, Schumperli D, Muller B (1997): The gene for histone RNA hairpin binding protein is located on human chromosome 4 and encodes a novel type of RNA binding protein. *EMBO J* 16:769–778.

Marzluff WF, Pandey NB (1988): Multiple regulatory steps control histone mRNA concentrations. *Trends Biochem Sci* 13:49–52.

Matsuoka S, Edwards MC, Bai C, Parker S, Zhang P, Baldini A, Harper JW, Elledge SJ (1995): $p57^{KIP2}$, a structurally distinct member of the $p21^{CIP1}$ Cdk inhibitor family, is a candidate tumor suppressor gene. *Genes Dev* 9:650–662.

Matsushime H, Quelle DE, Shurtleff SA, Shibuya M, Sherr CJ, Kato J-Y (1994): D-type cyclin-dependent kinase activity in mammalian cells. *Mol Cell Biol* 14:2066–2076.

Meyers S, Hiebert SW (1995): Indirect and direct disruption of transcriptional regulation in cancer: E2F and AML-1. *Crit Rev Eukaryot Gene Exp* 5:365–383.

Meyerson M, Harlow E (1994): Identification of G1 kinase activity for cdk6, a novel cyclin D partner. *Mol Cell Biol* 14:2077–2086.

Minshull J, Blow JJ, Hunt T (1989): Translation of cylcin mRNA is necessary for extracts of activated *Xenopus* eggs to enter mitosis. *Cell* 56:947–956.

Miura M, Zhu H, Rotello R, Hartwieg EA, Yuan J (1993): Induction of apoptosis in fibroblasts by IL-1 beta-converting enzyme, a mammalian homolog of the C. elegans cell death gene ced-3. *Cell* 75:653–660.

Miyashita T, Reed JC (1995): Tumor suppressor p53 is a direct transcriptional activator of the human bax gene. *Cell* 80:293–299.

Moll UM, Ostermeyer AG, Haladay R, Winkfield B, Frazier M, Zambetti G (1996): Cytoplasmic sequestration of wild-type p53 protein impairs the G1 checkpoint after DNA damage. *Mol Cell Biol* 16:1126–1137.

Moreno ML, Chrysogelos SA, Stein GS, Stein JL (1986): Reversible changes in the nucleosomal organization of a human H4 histone gene during the cell cycle. *Biochemistry* 25:5364–5370.

Morgan DO (1995): Principles of CDK regulation. *Nature* 374:131–134.

Morgenbesser SD, Williams BO, Jacks T, DePinho RA (1994): p53-dependent apoptosis produced by Rb-deficiency in the developing mouse lens. *Nature* 371:72–74.

Morris TD, Weber LA, Hickey E, Stein GS, Stein JL (1991): Changes in the stability of a human H3 histone mRNA during the HeLa cell cycle. *Mol Cell Biol* 11:544–553.

Muzi-Falconi M, Kelly TJ (1995): Orp1, a member of the Cdc18/Cdc6 family of S-phase regulators, is homologous to a component of the origin recognition complex. *Proc Natl Acad Sci USA* 92:12475–12479.

Nakayama K, Ishida N, Shirane M, Inomata A, Inoue T, Shishido N, Horii I, Loh DY, Nakayama K (1996): Mice lacking $p27^{Kip1}$ display increased body size, multiple organ hyperplasia, retinal dysplasia, and pituitary tumors. *Cell* 85:707–720.

Nevins JR (1992): E2F: A link between the Rb tumor suppressor protein and viral oncoproteins. *Science* 258:424–429.

Nigg EA (1992): Assembly–disassembly of the nuclear lamina. *Curr Opin Cell Biol* 4:105–109.

Nigg EA (1993): Targets of cylcin-dependent protein kinases. *Curr Opin Cell Biol* 5:187–193.

Nikolic M, Dudek H, Kwon YT, Ramos YF, Tsai LH (1996): The cdk5/p35 kinase is essential for neurite outgrowth during neuronal differentiation. *Genes Dev* 10: 816–825.

Norbury C, Nurse P (1992): Animal cell cycles and their control. *Annu Rev Biochem* 61:441–470.

Novitch BG, Mulligan GJ, Jacks T, Lassar AB (1996): Skeletal muscle cells lacking the retinoblastoma protein display defects in muscle gene expression and accumulate in S and G2 phases of the cell cycle. *J Cell Biol* 135:441–456.

Nurse P (1994): Ordering S phase and M phase in the cell cycle. *Cell* 79:547–550.

Ohshima T, Ward JM, Huh CG, Longenecker G, Veeranna, Pant HC, Brady RO, Martin LJ, Kulkarni AB (1996): Targeted disruption of the cyclin-dependent kinase 5 gene results in abnormal corticogenesis, neuronal pathology and perinatal death. *Proc Natl Acad Sci USA* 93:11173–11178.

Ohta K, Shiina N, Okamura E, Hisanaga S, Kishimoto T, Endo S, Gotoh Y, Nishida E, Sakai H (1993): Microtubule enucleating activity of centrosomes in cell-free extracts from *Xenopus* eggs: Involvement of phosphorylation and accumulation of pericentriolar material. *J Cell Sci* 104:125–137.

Oltvai ZN, Milliman CL, Korsmeyer SJ (1993): Bcl-2 heterodimerizes *in vivo* with a conserved homolog, Bax, that accelerates programmed cell death. *Cell* 74:609–619.

Pandey S, Wang E (1995): Cells en route to apoptosis are characterized by the upregulation of c-fos, c-myc, c-jun, cdc2, and Rb phosphorylation, resembling events of early cell-cycle traverse. *J Cell Biochem* 58:135–150.

Pardee AB (1989): G1 events and regulation of cell proliferation. *Science* 246:603–608.

Pardee AB, Keyomarsi K (1992): Cell multiplication. *Curr Opin Cell Biol* 4:141–143.

Parker SB, Eichele G, Zhang P, Rawls A, Sands AT, Bradley A, Olson EN, Harper JW, Elledge SJ (1995): p53-independent expression of $p21^{Cip1}$ in muscle and other terminally differentiating cells. *Science* 267:1024–1027.

Pauli U, Chrysogelos S, Stein, J, Stein G, Nick H (1987): Protein–DNA interactions *in vivo* upstream of a cell cycle regulated human H4 histone gene. *Science* 236:1308–1311.

Pelz SW, Brewer, G, Bernstein P, Hart PA, Ross J (1991): Regulation of mRNA turnover in eukaryotic cells. *Crit Rev Eukaryot Gene Exp* 1:99–126.

Pelz SW, Ross J (1987): Autogenous regulation of histone mRNA decay by histone proteins in a cell-free system. *Mol Cell Biol* 7:4345–4356.

Pemov A, Bavykin S, Hamlin JL (1995): Proximal and long-range alterations in chromatin structure surrounding the Chinese hamster dihydrofolate reductase promoter. *Biochemistry* 34:2381–2392.

Pines J, Hunt T (1987): Molecular cloning and characterization of the mRNA for cyclin from sea urchin eggs. *EMBO J* 6:2987–2995.

Plumb MA, Stein JL, Stein GS (1983): Coordinate regulation of multiple histone mRNAs during the cell cycle in HeLa cells. *Nucleic Acids Res* 11:2391–2410.

Pockwinse SM, Krockmalnic G, Doxsey SJ, Nickerson J, Lian JB, van Wijnen AJ, Stein JL, Stein, GS, Penman S (1997): Cell cycle independent interaction of CDC2 with the centrosome, which is associated with the nuclear matrix-intermediate filament scaffold. *Proc Natl Acad Sci USA* 94:3022–3027.

Polyak K, Kato J, Solomon MJ, Sherr CJ, Massague J, Roberts JM, Koff A (1994a): $p27^{Kip1}$, a cyclin–Cdk inhibitor, links transforming growth factor-β and contact inhibition to cell cycle arrest. *Genes Dev,* 8:9–22.

Polyak K, Lee M-H, Erdjument-Bromage H, Koff A, Roberts JM, Tempst P, Massague J (1994b): Cloning of $p27^{Kip1}$, a cyclin-dependent kinase inhibitor and a potential mediator of extracellular antimitogenic signals. *Cell* 78:59–66.

Prescott DM (1976): Reproduction of Eukaryotic Cells. New York: Academic Press.

Rao SS, Kohtz DS (1995): Positive and negative regulation of D-type cyclin expression in skeletal myoblasts by basic fibroblast growth factor and transforming growth factor beta. A role for cyclin D1 in control of myoblast differentiation. *J Biol Chem* 270:4093–4100.

Reeves R (1992): Chromatin changes during the cell cycle. *Curr Opin Cell Biol* 4:413–423.

Rhee K, Wolgemuth DJ (1995): Cdk family genes are expressed not only in dividing but also in terminally differentiated mouse germ cells, suggesting their possible function during both cell division and differentiation. *Dev Dynamics* 204:406–420.

Ross ME, Carter ML, Lee JH (1996): MN20, a D2 cyclin, is transiently expressed in selected neural populations during embryogenesis. *J Neurosci* 16:210–219.

Ross ME, Risken M (1994): MN20, a D2 cyclin found in brain, is implicated in neural differentiation. *J Neurosci* 14:6384–6391.

Schilling LJ, Farnham PJ (1994): Transcriptional regulation of the dihydrofolate reductase/rep-3 locus. *Crit Rev Eukaryot Gene Exp* 4:19–53.

Schilling LJ, Farnham PJ (1995): The bidirectionally transcribed dihydrofolate reductase and rep-3a promoters are growth regulated by distinct mechanisms. *Cell Growth Differ* 6:541–548.

Schneider JW, Gu W, Zhu L, Mahdavi V, Nadal-Ginard B (1994): Reversal of terminal differentiation mediated by p107 in Rb−/− muscle cells. *Science* 264:1467–1471.

Schulze A, Zerfass K, Spitkovsky D, Henglein B, Jansen-Durr P (1994): Activation of the E2F transcription factor by cyclin D1 is blocked by p161INK4, the product of the putative tumor suppressor gene MTS1. Oncogene 9:3475–3482.

Schumperli D (1988): Multilevel regulation of replication-dependent histone genes. *Trends Genet* 4:187–191.

Sellers WR, Kaelin WG (1996): xRB as a modulator of transcription. *Biochim Biophys Acta* 1288:M1–M5.

Serrano M, Hannon GJ, Beach D (1993): A new regulatory motif in cell cycle control causing specific inhibition of cyclin D/CDK4. *Nature* 366:704–707.

Serrano M, Lee H-W, Chin L, Cordon-Cardo C, Beach D, DePinho RA (1996): Role of the INK4a locus in tumor suppression and cell mortality. *Cell* 85:27–37.

Shakoori AR, van Wijnen AJ, Cooper C, Aziz F, Birnbaum M, Reddy GPV, Grana X, De Luca A, Giordano A, Lian JB, Stein JL, Quesenberry P, Stein GS (1995): Cytokine induction of proliferation and expression of CDC2 and cyclin A in FDC-P1 myeloid hematopoietic progenitor cells: Regulation of ubiquitous and cell cycle–dependent histone gene transcription factors. *J Cell Biochem* 59:291–302.

Sherr CJ (1993): Mammalian G1 cyclins. *Cell* 73:1059–1065.

Sherr CJ (1994): G1 phase progression: Cycling on cue. *Cell* 79:551–555.

Sherr CJ (1996): Cancer cell cycles. *Science* 274:1672–1677.

Sherr CJ and Roberts JM (1995): Inhibitors of mammalian G1 cyclin-dependent kinases. *Genes Dev* 9:1149–1163.

Shin EK, Shin A, Paulding C, Schaffhausen B, Yee AS (1995): Multiple change in E2F function and regulation occur upon muscle differentiation. *Mol Cell Biol* 15:2252–2262.

Sicinski P, Donaher JL, Parker SB, Li T, Fazeli A, Gardner H, Haslam SZ, Bronson RT, Elledge SJ, Weinberg RA (1995): Cyclin D1 provides a link between development and oncogenesis in the retina and breast. *Cell* 82:621–630.

Singh P, Coe J, Hong W (1995): A role for retinoblastoma protein in potentiating transcriptional activation by the glucocorticoid receptor. *Nature* 374:562–565.

Slack RS, Miller FD (1996): Retinoblastoma gene in mouse neural development. *Dev Genet* 18:81–91.

Smith E, Frenkel B, MacLachlan T, Giordano A, Stein JL, Lian JB, Stein GS (1997): Post-proliferative, cyclin E–associated kinase activity in differentiated osteoblasts: Inhibition by proliferating osteoblasts and osteosarcoma cells. *J Cell Biochem* 66:141–152.

Smith E, Frenkel B, Schlegel R, Giordano A, Lian JB, Stein JL, Stein GS (1995): Expression of cell cycle regulatory factors in differentiating osteoblasts: Post-proliferative upregulation of cyclins B and E. *Cancer Res* 55:5019–5024.

Stein GS, Lian JB (1993): Molecular mechanisms mediating proliferation/differentiation interrelationships during progressive development of the osteoblast phenotype. *Endocrine Rev* 14:424–442.

Stein GS, Lian JB, Owen TA (1990): Relationship of cell growth to the regulation of tissue-specific gene expression during osteoblast differentiation. *FASEB J* 4:3111–3123.

Stein GS, Lian JB, Stein JL, Briggs R, Shalhoub V, Wright K, Pauli U, van Wijnen A (1989): Altered binding of human histone gene transcription factors during the shutdown of proliferation and onset of differentiation in HL-60 cells. *Proc Natl Acad Sci USA* 86:1865–1869.

Stein GS, Stein JL (1984): Is human histone gene expression autogenously regulated? *Mol Cell Biochem* 64:105–110.

Stein GS, Stein JL, van Wijnen AJ, Lian JB (1994): Histone gene transcription: A model for responsiveness to an integrated series of regulatory signals mediating cell cycle control and proliferation/differentiation interrelationships. *J Cell Biochem* 54:393–404.

Stillman B (1996): Cell cycle control of DNA replication. *Science* 274:1659–1664.

Swenson KI, Farrell KM, Ruderman JV (1986): The clam embryo protein cyclin A induces entry into M phase and the resumption of meiosis in *Xenopus* oocytes. *Cell* 47:861–870.

Szekely L, Jin P, Jiang WQ, Rosen A, Wiman KG, Klein G, Ringertz N (1993): Position-dependent nuclear accumulation of the retinoblastoma (RB) protein during *in vitro* myogenesis. *J Cell Physiol* 155:313–322.

Tamura T, Ishihara M, Lamphier MS, Tanaka N, Oishi I, Aizawa S, Matsuyama T, Mak TW, Taki S, Taniguchi T (1995): An IRF-1–dependent pathway of DNA damage-induced apoptosis in mitogen-activated T lymphocytes. *Nature* 376: 596–599.

Tamaru T, Okada M, Nakagawa H (1994): Differential expression of D type cyclins during neuronal maturation. *Neurosci Lett* 168:229–232.

Tanaka N, Ishihara M, Kitagawa M, Harada H, Kimura T, Matsuyama T, Lamphier MS, Aizawa S, Mak TW, Taniguchi T (1994): Cellular commitment to oncogene-induced transformation or apoptosis is dependent on the transcription factor IRF-1. *Cell* 77:829–839.

Tanaka K, Tsurumi C (1997): The 26S proteosome: Subunits and functions. *Mol Biol Rep* 24:3–11.

Terasima T, Yasukawa M (1966): Synthesis of G1 protein preceding DNA synthesis in cultured mammalian cells. *Exp Cell Res* 44:669.

Tsai LH, Takahashi T, Caviness VS Jr, Harlow E (1993): Activity and expression pattern of cyclin-dependent kinase 5 in the embryonic mouse nervous system. *Development* 119:1029–1040.

van Grunsven LA, Billon N, Savatier P, Thomas A, Urdiales JL, Rudkin BB (1996): Effect of nerve growth factor on the expression of cell cycle regulatory proteins in PC12 cells: Dissection of the neurotrophic response from the anti-mitogenic response. *Oncogene* 12:1347–1356.

van Gurp MF, Hoffman H, Tufarelli C, Stein JL, Lian BJ, Neufeld EJ, Stein GS, van Wijnen AJ (1998): The CDP/*cut* homeodomain protein represses osteocalcin gene transcription via a tissue-specific promoter element: Formation of proliferation-specific protein/DNA complexes with the pRB-related protein p107 and cyclin A. *J Cell Biochem,* in press.

van Wijnen AJ, Aziz F, Grana X, De Luca A, Desai RK, Jaarsveld K, Last TJ, Soprano K, Giordano A, Lian JB, Stein JL, Stein GS (1994): Transcription of histone H4, H3 and H1 cell cycle genes: Promoter factor HiNF-D contains CDC2, cyclin A and an RB-related protein. *Proc Natl Acad Sci USA* 91:12882–12886.

van Wijnen AJ, Bidwell JP, Fey EG, Penman S, Lian JB, Stein JL, Stein GS (1993): Nuclear matrix association of multiple sequence specific DNA binding activities related to SP-1, ATF, CCAAT, C/EBP, OCT-1 and AP-1. *Biochemistry* 32:8397–8402.

van Wijnen AJ, Cooper C, Odgren P, Aziz F, De Luca A, Shakoori RA, Giordano A, Quesenberry PJ, Lian JB, Stein GS, Stein JL (1997): Cell cycle dependent modifications in activities of pRb-related tumor suppressors and proliferation-specific CDP/*cut* homeodomain factors in murine hematopoietic progenitor cells. *J Cell Biochem* 66:512–523.

van Wijnen AJ, van Gurp MF, de Ridder M, Tufarelli C, Last TJ, Birnbaum M, Vaughan PS, Giordano A, Krek W, Neufeld EJ, Stein JL, Stein GS (1996): CDP/*cut* is the DNA binding subunit of histone gene transcription factor HiNF-D: A mechanism for gene regulation at the G1/S phase cell cycle transition point independent of transcription factor E2F. *Proc Natl Acad Sci USA* 93:11516–11521.

Vaughan PS, Aziz F, van Wijnen AJ, Wu S, Harada H, Taniguchi T, Soprano K, Stein GS, Stein JL (1995): Activation of a cell cycle regulated histone gene by the oncogenic transcription factor IRF2. *Nature* 377:362–365.

Wade M, Blake MC, Jambou RC, Helin K, Harlow E, Azizkhan JC (1995): An inverted repeat motif stabilizes binding of E2F and enhances transcription of the dihydrofolate reductase gene. *J Biol Chem* 270:9783–9791.

Wang ZF, Whitfield ML, Ingledue TC 3rd, Dominski Z, Marzluff WF (1996): The protein that binds the 3′ end of histone mRNA: A novel RNA-binding protein required for histone pre-mRNA processing. *Genes Dev* 10:3028–3040.

Weinberg RA (1995): The retinoblastoma protein and cell cycle control. *Cell* 81:323–330.

Weis K, Rambaud S, Lavau C, Jansen J, Carvalho T, Carmo-Fonseca M, Lamond A, Dejean A (1994): Retinoic acid regulates aberrant nuclear localization of PML-RAR alpha in acute promyelocytic leukemia cells. *Cell* 76:345–356.

Wells J, Held P, Illenye S, Heintz NH (1996): Protein–DNA interactions at the major and minor promoters of the divergently transcribed DHFR and rep3 genes during the Chinese hamster ovary cell cycle. *Mol Cell Biol* 16:634–647.

Wells JM, Illenye S, Magae J, Wu CL, Heintz NH (1997): Accumulation of E2F-4/DP-1 DNA binding complexes correlates with induction of DHFR gene expression during the G1 to S phase transition. *J Biol Chem* 272:4483–4492.

White E (1994): p53, guardian of Rb. Nature 371:21–22.

Wilhide CC, Van Dang C, Dipersio J, Kenedy AA, Bray PF (1995): Overexpression of cyclin D1 in the Dami megakaryocytic cell line causes growth arrest. *Blood* 86:294–304.

Wiman KG (1993): The retinoblastoma gene: Role in cell cycle control and cell differentiation. *FASEB J* 7:841–845.

Wohlgemuth JG, Bulboaca GH, Moghadam M, Caddle MS, Calos MP (1994): Physical mapping of origins of replication in the fission yeast *Schizosaccharomyces pombe. Mol Biol Cell* 5:839–849.

Yan GZ, Ziff EB (1995): NGF regulates the PC12 cell cycle machinery through specific inhibition of the Cdk kinases and induction of cyclin D1. J Neurosci 15:6200–6212.

Yen A, Varvayanis S (1994): Late dephosphorylation of the RB protein in G2 during the process of induced cell differentiation. *Exp Cell Res* 214:250–257.

Yen A, Varvayanis S, Platko JD (1993): 12-*O*-tetradecanoylphorbol-13-acetate and staurosporine induce increased retinoblastoma tumor suppressor gene expression with megakaryocytic differentiation of leukemic cells. *Cancer Res* 53:3085–3091.

Zambetti G, Stein JL, Stein GS (1987): Targeting of a chimeric human histone fusion mRNA to membrane bound polysomes in HeLa cells. *Proc Natl Acad Sci USA* 84:2683–2687.

Zeng C, van Wijnen AJ, Stein JL, Meyers S, Sun W, Shopland L, Lawrence JB, Penman S, Lian JB, Stein GS, Hiebert SW (1997): Identification of a nuclear matrix targeting signal in the leukemia and bone-related transcription factor AML-CBF-α transcription factors. *Proc Natl Acad Sci USA* 94:6746–6751.

Zeng C, McNeil S, Nickerson J, Shopland L, Lawrence JB, Penman S, Hiebert S, Lian JB, van Wijnen AJ, Stein JL, Stein GS (1998): Intranuclear targeting of AML/CBFα regulatory factors to nuclear matrix-associated transcriptional domains. *Proc Natl Acad Sci USA* 95:1585–1589.

Zwijsen RML, Wientjens E, Klompmaker R, van der Sman J, Bernards R, Michalides RJAM (1997): CDK-independent activation of estrogen receptor by cyclin D1. *Cell* 88:405–415.

4

STEM CELL TRANSCRIPTION

SHERMAN M. WEISSMAN AND ARCHIBALD S. PERKINS
Department of Medicine, Human Genetics and Pathology, Yale University School of Medicine, New Haven, CT 06510

INTRODUCTION

Two decades ago, the emerging field of eukaryotic molecular biology began to cut its teeth on the complicated issue of tissue-specific gene transcription. Each wave of advancement since then, propagated by concomitant advances in technology, brought the description of new transcription factors and novel mechanisms of regulation. Sequence databases now teem with nearly countless zinc fingers proteins, helix-loop-helix proteins, homeodomain proteins, and other transcription factors. In parallel, developmental biologists have identified some of the mechanisms of cell fate determination and cell induction. The focus of this chapter is to encapsulate these advances as relevant to the question of what defines a stem cell: How is a stem cell population generated? How is it maintained? How do cells depart from the stem cell pool into committed derivatives? We focus on the hematopoietic system, since it is one about which a good deal is known, and draw upon others as needed. Despite the accumulation of a large body of facts and knowledge, we are still in the early phases of grappling with the questions posed above. In summarizing what is known in the field of transcriptional regulation of stem cells, we hope to provide the reader with an appreciation for what has been accomplished and a sense of what challenges lie ahead.

Stem Cell Biology and Gene Therapy, Edited by Peter J. Quesenberry, Gary S. Stein, Bernard G. Forget, and Sherman M. Weissman
ISBN 0-471-14656-0

BIOLOGY OF STEM CELLS

Anatomy

A discussion of transcriptional regulation of stem cells in hematopoiesis may begin with a consideration of the ontogeny of blood cell progenitors and the regulation of blood cell production. In birds, amphibians, and mammals there are two anatomical sites at which hematopoietic cells originate: an extraembryonic site and an intraembryonic site (Zon, 1995). These different anatomical sites subserve two different functions: The former supports early stage embryos at a time in development when the size of the embryo creates a limited demand on oxygen transport to the tissues, while the latter supports later-stage embryos and the fetus, where there is substantial tissue mass that demands considerable oxygen delivery. Thus, the initial phase of hemopoiesis is of limited capacity, while the latter, or definitive, hemopoietic system is more robust. Given the presence of the maternal immune system, the primary requirement made of intrauterine hematopoiesis is red blood cell production, although by *in vitro* assays it is clear that hematopoietic progenitors of the myeloid and lymphoid lineages are present (Auerbach et al., 1996).

In mammalian embryos, the yolk sac serves as the extraembryonic site, while in frogs this function is served by the ventral blood islands. The intraembryonic origin of hematopoiesis in mice is the para-aortic area near the mesonephros and genital ridge (AGM, for aorta-gonad-mesonephros) (Godin et al., 1993; Medvinsky et al., 1993; Muller et al., 1994). This shifts to the fetal liver at days 12–14 of gestation.

The contribution of cells from either site to embryonic hepatic hematopoiesis and adult hematopoiesis may vary at least quantitatively between species. For example, in the frog the ventral blood islands (VBI, the extraembryonic source) can make a small contribution even to adult erythropoiesis. The contribution of the VBI to hematopoietic cell populations does not simply decline with time: VBI-derived B cells and leukocytes actually increase markedly, although transiently, near the time of metamorphosis (Chen and Turpen, 1995).

In the mouse some controversy remains, but intraembryonic stem cells seem to be the major progenitors for adult hemopoiesis. Yolk sac hematopoiesis in the mouse is primarily erythropoietic and produces nucleated erythrocytes that are larger than adult erythrocytes. Interestingly colony-forming unit cells (CFUs) are detectable by 7–8 days of gestation in the yolk sac, although the earliest reported stem cells, as judged by transplantation, were seen at day 11 (Auerbach et al., 1996). Multipotential stem cells appear intraembryonically at embryonic day 9.5, and only appear somewhat later in the yolk sac, raising the possibility that these yolk sac cells are derived from the embryonic circulation (Dieterlen-Lievre et al., 1997). In addition, mature hematopoietic cells in the yolk sac appear with rapid kinetics, significantly faster than is seen during marrow replacement with definitive stem cells. These differences in kinetics suggest that in the early yolk sac mature cells develop from a pool of cells that are significantly more mature than true stem cells. Therefore, early hematopoiesis may bypass in part or whole the stem cell stage of development.

Concept of a Hematopoietic Stem Cell

By definition, a stem cell is able to reconstitute all hematopoietic lineages, including the formation of additional stem cells. One can assay for stem cells by a variety of methods, the most meaningful of which is the long-term reconstitution of an irradiated recipient, typically a mouse. Experimentally, stem cells can be isolated on the basis of the presence of certain cell surface markers, such as CD34 (Krause et al., 1994), Sca-1, and c-kit, and the absence of lineage markers (Spangrude et al., 1988). A recent report shows that cells that are c-kit and Sca-1 positive but CD34 negative can also engraft long term (Osawa et al., 1996). One ordinarily reconstitutes a mouse with no fewer than several hundred stem cells. However, the actual number of these cells needed to repopulate an irradiated mouse long term is very likely lower, and may be as low as one. It is hypothesized that two populations of cells are needed: (1) committed progenitors that provide initial engraftment and (2) stem cells that give long-term engraftment (Jones et al., 1990). With attention to this experimental difficulty, it becomes increasingly likely that there is a population of stem cells in normal (mouse) marrow such that a single cell can completely repopulate the hematopoietic system. Whether there are even more flexible stem cells that also can contribute to the endothelium is a matter of current discussion.

Stem Cell Renewal vs. Commitment

By necessity, some cell divisions must result in two progeny that assume different cell fates. A number of models exist for asymmetrical cell division in experimental organisms, such that the two progeny of a single division cycle are different from each other (Horvitz and Herskowitz, 1992). In some strains of *Saccharomyces cerevisiae,* after cell division the progeny of the mother cell switch mating type while the progeny of the daughter do not. This ensures that a population of cells contains similar numbers of cells of each mating type. The lineages of daughter cells are formally analogous to stem cells in that they continue to reproduce both themselves and cells of the alternate mating type. Mating type switching is mediated by an endonuclease that is only expressed at certain stages of the cell cycle and only in cells that will switch mating type. Current models suggest that the asymmetrical regulation of expression of this endonuclease is mediated by an asymmetrical concentration of a factor that is a negative regulator of transcription in daughter cells as compared to mother cells (Amon, 1996; Bobola et al., 1996; Jansen et al., 1996; Sil and Herskowitz, 1996).

In metazoan organisms, the origins of asymmetry have been extensively studied in *Drosophila* oogenesis and embryogenesis. Each oocyte precursor divides four times, generating 16 cells, of which 15 become nurse cells and one the oocyte. One basis of assymmetry may be the nonequivalent localization of the 16 cells with respect to one another that arises with these division. Subsequent asymmetrical localization of molecules in the oocyte and asymmetrical delivery of signaling molecules between nurse cells and oocyte produce the principal axes for embryonic development.

Positional cues play a role in *Caenorhabditis elegans* stem cell ontogeny: In vulvar development, the daughters of a given progenitor form a single layer of cells that all have similar developmental potential, but the fate that the progeny cells assume depends on their proximity to an external anchor cell that apparently sends a diffusible intercellular signal to these cells (Newman et al., 1996). In the frog, asymmetry is present in the oocyte and furthered by a postfertilization rotation of the cytoplasm, with an axis determined by the site of sperm entry. In the earliest cell divisions in mouse embryos, there is a regulated form of cell division such that the orientation of mitotic spindles is not random but involves rotations that ensure a specific relation of the axes of division of the earliest progeny cells.

Thus, two basic mechanisms—asymmetrical progeny or progeny localized in an asymmetrical environment—are theoretically possible in the case of hematopoietic stem cells (Morrison et al., 1997). The bone marrow stroma, which can influence blood cell development, may provide asymmetrical cues.

Intrinsic Vs. Instructive Models of Blood Cell Development

A panoply of cytokines, interleukins, growth factors, and interferons have been implicated in the proliferation, survival, differentiation, and activation of hematopoietic stem cells and their committed progenitors (Dexter and Spooncer, 1987; Metcalf, 1989; Whetton and Dexter, 1989). Their discovery and characterization led to the idea that these ligands, through interaction with specific receptors on the cell surface of early progenitors, may drive hematopoietic cells into different lineages and provide instructive cues for differentiation. However, experiments by Dexter and coworkers support an intrinsic model of differentiation (Fairbairn et al., 1993). It had been shown that cultured interleukin (IL)-3–dependent myeloid progenitors (FDCP-mix) can differentiate in the presence of certain cytokines, such as granulocyte colony-stimulating factor (G-CSF), but in the absence of any factor will undergo apoptosis. It was unclear, however, whether the G-CSF was providing an instructive signal that directed differentiation or a survival signal that allowed the innate differentiation program to manifest itself. By transferring in the antiapoptotic gene bcl-2, they were able to maintain the cells in the absence of factors and observed that many differentiation features emerged. These included changes in nuclear morphology and activation of specific transcripts associated with differentiation (Fairbairn et al., 1993). These results suggest that committed myeloid progenitor cells have an intrinsic ability to differentiate into mature cells and that the role of the growth factors is to enable the cell to express this program through supporting cell survival and preventing cell death. Consistent with this model are data obtained from gene targeting experiments in which mice bearing loss of either cytokine genes (Dranoff and Mulligan, 1994; Kopf et al., 1996; Leischke et al., 1994a,b) or genes for their receptors (Nishinakamura et al., 1996) continue to produce mature hematopoietic cells, albeit at a much lower number, and sometimes exhibiting abnormal function.

The importance of these models to stem cell biology and to transcriptional regulation of differentiation is fundamental. An instructive model suggests that signals initiated by the binding of ligands to cell surface receptors lead to transcriptional changes that are key determinants of stem cell maintenance and maturation. In an intrinsic model of hematopoietic cell differentiation, cytokine signaling serves mainly to counter cell death and does not *per se* lead to changes in transcription that are integral to the differentiation process. These transcriptional changes are mediated by cell autonomous mechanisms about which little information is available at present. Unlike muscle differentiation, in which MyoD as a single "master" gene can initiate a whole program of differentiation, there does not appear to be as of yet such master genes for hematopoiesis. Nonetheless, through a variety of approaches discussed below, numerous transcription factors that appear critical for blood cell development have been identified.

Cellular Commitment and Restriction of Differentiation Alternatives

One of the central questions concerning stem cell biology is the mechanism by which the stem cell becomes committed to a particular lineage and thus restricted in its pluripotency. These studies have been hampered by technical hurdles: Direct access to normal stem cells that are synchronized at a specific early step in differentiation is experimentally very difficult and uncertain. Hence conclusions have been based on analyses of mixed cell populations of primary bone marrow cells or on results obtained from cultured cell lines that may differ from their normal counterparts *vis à vis* patterns of gene expression and mechanisms of transcriptional regulation. Paradoxically, primitive cells may express genes that are hallmarks of mature cells (Hu et al., 1997), suggesting that the complex of regulatory events leading to initial expression of lineage-specific genes may be quantitatively and qualitatively different from the regulatory processes that determine expression of the gene at later stages of differentiation. After initial expression of genes that are characteristic of a given lineage, a series of further events must occur so that the relatively small number of incompletely differentiated precursor cells ultimately give rise to a large population of cells expressing a restricted subset of the gene products found in the early precursor. Thus there appears to be a loosening of gene expression, followed by a severe restriction in gene expression concomitant with an upregulation of selected transcripts. Among the steps that must be accomplished are lineage-specific silencing of selected genes expressed in precursors; cell proliferation to generate appropriate numbers of end-stage cells; regulation and suppression of apoptosis; and progressive amplification in the level of expression of specific gene products, such as the hemoglobin chains, and down regulation of other genes, such as those for various markers of the precursor cells. The expression of cell type–specific and housekeeping genes must be coordinated with the stage of maturation of a cell so that the cell does not prematurely enter into terminal differentiation. Although little is known about the mechanisms in play, regulation may hypothetically involve use of different transcription factors or complexes to regulate the same gene at different stages of development in a single lineage as well as in different lineages.

Coupling of Cell Cycle and Apoptosis to Differentiation: Can Differentiation Be Uncoupled From G1 Arrest?

The maintenance of the stem cell or progenitor cell phenotype requires suppression of apoptosis (Jacobson et al., 1997) and a block to differentiation. While active cycling is not required, these cells must retain the capacity to enter the cell cycle upon appropriate stimulation. As cells depart from the stem cell pool and become committed progenitors, a series of complex decision points are encountered within each cell cycle that are believed to influence the differentiation process. The advancement of a cell toward a differentiated phenotype is accompanied by a series of transcriptional changes that are associated with a slowing of the cell cycle. Certain transcripts, such as c-*myc,* that are implicated in cell cycle advancement are downregulated, while certain gene products, such as Rb, the retinoblastoma gene product, accumulate. Both of these changes are thought to be concordant with the slowing of the cell cycle: c-myc is known to activate the cdc25 gene (Galaktionov et al., 1996), a positive regulator of the cell cycle, and Rb, through association with the transcription factor E2F, is known to repress the activation of genes necessary for the G1 to S transition (P.L. Chen et al., 1995; Sanchez and Dynlacht, 1996). But is slowing of the cell cycle necessary for differentiation? Several recent experiments suggest that it is not. Certain mutants of Rb, identified by scanning mutagenesis of the protein, are unable to halt the progression of G1 to S, but are able to induce the differentiation of osteosarcoma cells (Sellers, WR et al., 1998). In addition, the overexpression of certain antiapoptotic genes, such as bcl-2, can help to maintain cell growth but do not appear to interfere with differentiation. These findings argue that the transcriptional regulatory pathways that control differentiation, and thus are likely also implicated in maintenance of stem cell phenotype, are distinct from those that control the cell cycle.

It is clear that stem cell pools must be resistant to apoptotic cell death. However, controlled cell death is a central component of developmental processes in hematopoiesis, best exemplified in the immune system. In both the B-cell and T-cell lineages, cells bearing receptors that interact too strongly with host antigens must be downregulated or eliminated. Selection of developing T cells is particularly complex. It involves both positive selection for cells that are able to recognize antigen in the context of self but not nonself-MHC antigens and also elimination of cells that recognize self-antigens too avidly. Developing thymocytes are sensitive to both radiation-induced and steroid-induced apoptosis, and the thymic epithelium itself may be steroidogenic (Zilberman et al., 1996). However, steroids may protect thymocytes against apoptosis induced by other agents. The interesting suggestion has been made that the balance between steroid levels and self-recognition–induced apoptosis in the thymus may serve to eliminate both thymocytes that entirely fail to recognize self, because of steroid apoptosis, and thymocytes that recognize self too strongly, because the stimulus to apoptosis is too strong to be countered by the steroid levels (Ashwell et al., 1996).

Feedback at the Organismal Level

As discussed above in the context of intrinsic vs. instructive models of hemopoiesis, extrinsic cues may not be required for cellular maturation. However, this does not mean that these extracellular signaling molecules do not have an important role in the organismal regulation of hematopoiesis. Some mechanism must exist to tell the marrow to increase the production of certain lineages in times of heightened demand or deficit. For the granulocytic series, it is not established what this signal is, although CSFs may play an important role. Both G-CSF and granulocyte-macrophage (GM)-CSF stimulate the production of granulocytes in adult animals and to this extent play a role similar to that of erythropoietin for red cells. However, gene knockout of these factors does not prevent the formation of mature granulocytes so that the factors are not necessary for either specification or significant expansion of this lineage (Metcalf et al., 1996). The mechanism by which the organism senses granulopenia, however, remains obscure.

It is well established that erythropoietin is essential for a positive feedback stimulus in the presence of anemia and decreased oxygen-carrying capacity. It is interesting, however, that a portion of yolk sac erythropoiesis in the mouse is independent of erythropoietin or the erythropoietin receptor. Conversely, productive intrahepatic erythropoiesis in the developing animal is strictly dependent on this signaling system (C.S. Lin, et al., 1996; Wu et al., 1995). An interesting parallel is that certain transcription factor gene knockouts (*vide infra*) selectively spare the initial yolk sac erythropoiesis while blocking subsequent hematopoietic development. However, these effects may not be truly analogous since the fetal livers from mice lacking erythropoietin or the erythropoietin receptor actually contain an increased number of erythroid (E) blast-forming units (E-BFU) and E-CFU, while the transcription factor knockouts exhibit decreased numbers of these progenitors. Nonetheless, these results highlight the distinction between extraembryonic and intraembryonic hematopoiesis.

Thrombopoietin and its receptor mpl regulate platelet levels by stimulating megakaryocyte production. However, mice with a knockout of the thrombopoietin gene still have some platelets and megakaryocytes (Carver-Moore et al., 1996; de Sauvage et al., 1996), indicating that, as with erythropoietin, the hormonal stimulation system is not obligatory for establishment of the relevant differentiation pathway. The feedback loop regulating thrombopoietin levels is not fully elucidated, but may involve removal of thrombopoietin by circulating platelets. Mice lacking the mpl receptor for thrombopoietin are also markedly deficient in platelets, but in addition show some deficiency of hematopoietic precursors. Further thrombopoietin can stimulate the proliferation of hematopoietic precursor cells *in vitro* (Ramsfjell et al., 1996) and rescue BFU-E colony formation from GATA-$1^{-/-}$ embryonal stem cells (Kieran et al., 1996), suggesting there may be more than one role for this system in hematopoietic regulation.

ROLE OF TRANSCRIPTION FACTORS IN STEM CELL REGULATION

General Transcription Apparatus

The general transcription apparatus responsible for forming mRNA includes a core RNA polymerase II that itself is a multipeptide complex and a variety of associated general transcription factors that may also have a role in other processes such as DNA repair (Orphanides et al., 1996). It is beyond the scope of this chapter to discuss these factors other than to note that there are several proteins of the general transcription apparatus that may serve as adaptors that are necessary for the function of specific types of transcription activation domains present on cell type specific factors, including those reviewed below (Hampsey and Reinberg, 1997; Laemmli and Tjian, 1996; Zawel and Reinberg, 1995).

Multiple Modalities of Transcription Factor Function

Transcription factors generally have a modular construction that greatly expedites analysis of their function (Calkhoven and Ab, 1996). The gene for a typical transcription factor may consist of one or more segments that encode protein domains that mediate transcription activation (Triezenberg, 1995), together with separable segments that encode domains with sequence-specific DNA-binding properties. Other segments may encode domains such as leucine zippers or poz domains that specifically interact with other proteins such as other transcription factors, and yet other segments that encode domains involved in the regulation of the DNA-binding or transcriptional activity of the overall factor. These gene segments may be excised and recombined to create factors that would, for example, contain a transcription activation domain from one factor attached to a sequence recognition and DNA-binding domain from another factor, even when one component is derived from yeast and the other from animal cells. This underlies a number of useful assays, including the yeast two hybrid system widely used for finding proteins that interact with a given target protein (Fields and Song, 1989).

Several sequence-specific DNA-binding protein motifs recur in multiple transcription factors, and representative crystal structures have been determined (Nelson, 1995). One common motif is the zinc finger (Knegtel et al., 1995; Pieler and Bellefroid, 1994), a structure that consists of a peptide with a tetra-complexed zinc. One finger may make contact with three successive residues in DNA, and it is even possible, to a limited extent, to construct proteins with desired sequence specificity by combining zinc fingers along established rules (e.g., Choo et al., 1994). Other motifs, such as the helix-turn-helix motif, or the homoeobox structure, depend for their sequence specificity on an alpha helical domain that binds DNA by inserting into the major groove of DNA. The homeodomain proteins, by themselves, typically have rather nonspecific DNA-binding ability. However, through heterodimerization with certain partners, this binding ability becomes much more specific (Chang et al., 1996). Proteins of the HMG group bind to DNA in the minor groove and cause varying degrees of helical bending.

Activation domains are also quite heterogeneous. One large class consists of acidic domains, but these may contain nonacidic amino acids such as phenylalanine that are necessary for the activation function (Cress and Triezenberg, 1991). Runs of proline or glutamine may also be hallmarks of transcription activation domains. These transcription activation domains may function by interaction either directly with general transcription factors such as the TATA-binding factor (TBF) or TFIIB, or through specific members of a group of coactivator proteins (TAFs) that themselves form a complex with TFIID. Regardless of whether specific protein contacts are known or not, the details of how transcription activation is achieved in these cases are obscure.

Formation and Maintenance of Gene-Specific Transcription Complexes

The process of establishing stable states of active or potentially active transcription units in a cell type is, not surprisingly, very complex and probably different for different genes. In some cases, such as the heat shock proteins, transcription factors and even RNA polymerase II may be located on promoters before activation (Quivy and Becker, 1996), while the determining factor, the HSF transcription factor, either only binds DNA after heat shock (Landsberger and Wolffe, 1995) or is prebound but becomes activated by phosphorylation. In other cases, such as the interferon gene, there appears to be highly cooperative interactions among several different factors necessary to form an activation complex. The presence of DNase-hypersensitive sites at the transcription initiation sites of genes presumably reflects the assembly of such protein complexes and is in general tissue specific, although the hypersensitive sites may be present even in the absence of the accumulation of gene transcripts. Activated states of genes also may be stabilized by covalent modifications such as DNA demethylation in mammalian cells. Interestingly, there are reports that demethylation may be an active process (Razin and Shemer, 1995) that does not require DNA replication but may involve RNA (Weiss et al., 1996).

Histones and Nucleosomes. Nucleosome placement may be critical for determining the activity of a promoter, as in the yeast PH05 promoter (Svaren and Horz, 1997), and can serve to bend the DNA in such a way as to bring in proximity transcription factors whose binding sites are widely separated on linear DNA. Recently there has been considerable progress in defining cellular systems involved in assembly of histones on DNA or in modifying the positioning of histone on the DNA. One system, developed by Stillman and his coworkers (see Verreault et al., 1996), will assemble histone octamers *de novo,* but only works on newly replicated DNA in a coupled system. Another remarkable system (Varga-Weisz et al., 1995) will not only assemble histones on preformed DNA but promotes sufficient histone mobility so that, for example, chainging the ionic strength of the medium results in changes in the internucleosome spacing of preformed chromatin. Two related systems have been identified in several organisms including yeast, where the abundant multiprotein RSC complex has also been found to be essential for mitotic growth (Cairns et al., 1996).

Acetylation of the N-terminal portion of core histones has been known for some time to be present particularly in active chromatin. Acetylation of different N-terminal lysines in histones H3 and H4 appear to serve different biological functions. Thus certain lysines in the N termini of histones H3 and H4 are acetylated in newly formed histones and this is part of the process of nuclear import and chromatin assembly. Other residues are acetylated specifically in histones in euchromatin, in some cases more specifically in genes that are being transcribed. Recently there has been considerable progress in characterizing and cloning histone acetyltransferases with different specificities and functions (Brownell and Allis, 1996; Parthun et al., 1996). The yeast transcription activation factor GCN5 and its *Tetrahymena* and human analogs have been shown to be histone acetylases (Brownell et al., 1996; Kuo et al., 1996; Wang et al., 1997), indicating that histone acetylation may be part of a mechanism used by transcription factors for gene activation. Mammalian analogs of yeast histone acetylases have been found, and the large coactivator protein CPB not only associates with other histone acetylases but apparently has an endogenous histone acetylase activity (Bannister and Kouzarides, 1996; Ogryzko et al., 1996). Even one of the TBF-associated TAF components of the TATA box–binding general transcription factor TFIID has been found to have histone acetylase activity (Mizzen et al., 1996). Specific histone deacetylases have also been identified (Taunton et al., 1996; Wolffe, 1996).

HMG Box Proteins. In addition to nucleosomes, architectural proteins, such as those containing HMG boxes, may establish the degree of DNA bending in the promoter region and may determine, or be essential for, the activity of a gene. The HMG box proteins fall into two classes: those, such as HMG-IY, that are relatively sequence nonspecific and generally expressed; and those, such as LEF-1, that show restricted patterns of cell type expression and more specificity in DNA binding.

Polycomb and Trithorax Group Proteins. Genetic analyses in *Drosophila* have shown that while the *initiation* of gene activation or inactivation may be mediated by one class of proteins, the *maintenance* of the state of the gene during development is dependent on a different class of proteins. It is envisioned that there are multiprotein systems for maintaining the activity (encoded by the trithorax group of genes) or inactivity (polycomb group) of target genes. In *Drosophila,* specific sequences termed PRE elements (Pirrotta et al., 1995; Zink and Paro, 1995) may be recognized by polycomb proteins and serve as initiation sites for assembly of these maintenance complexes that then can affect the activity of larger regions of DNA. Nucleic acid and protein homologies and limited biochemical studies indicate that homologous systems for stabilizing activated or silenced states operate in animal cells. Unfortunately, in no case is there an atomic level, or even a complete molecular level, understanding of the structure of the chromatin in a silenced DNA protein complex.

The mammalian homologs of *Drosophila* polycomb gene family proteins may be important in the regulation of hematopoiesis. The vav gene encodes a remarkable protein that contains a pleckstrin homology domain, SH3 and SH2 homology domain, regions of homology with guanine nucleotide exchange factors, sites for tyrosine phosphorylation, and nuclear localization signals. Vav is expressed exclusively in cells of the hematopoietic system and has been shown by gene knockout studies to play an important role in lymphocyte proliferation and signaling from antigen receptors. Recently vav has been shown to interact with the mammalian homolog of *Drosophila* enhancer of zeste protein, a putative member of the polycomb group (Hobert et al., 1996).

The Bmi-1 gene of mice encodes a homolog of the *Drosophila* posterior sex combs protein, and knockout of Bmi-1 causes severe defects in hematopoiesis (Vanderlugt et al., 1996). Overexpression of Bmi-1, via retroviral insertional activation, is associated with leukemia in mice (van Lohuizen et al., 1991), and transgenic mice may show homeotic-like anterior transformations of body structures (Alkema et al., 1997). The Bmi-1 protein has a RING finger domain (Borden and Freemont, 1996; Saurin et al., 1996; Lovering et al., 1993) that is important for its localization to specific nuclear domains (Cohen et al., 1996), which can be visualized as speckled dots with anti-Bmi-1 antibodies. Bmi-1 exists in the nucleus as a complex with Mel18 (Tagawa et al., 1990), M33, and Mph1, which share homology to the *Drosophila* polycomb (Mel18

[Tagawa et al., 1990] and M33 [Pearce et al., 1992]) and polyhomeotic (Mph1 [Alkema et al., 1997]) genes. The findings that this group of mammalian genes all bear similarity to *Drosophila* polycomb group genes, and can cause homeotic-like transformations when disrupted in mice, and that their proteins exist as a complex in defined nuclear structures together suggest that this represents a functional equivalent to the *Drosophila* complex and may exert its effects through modification of gene expression at multiple loci via alteration in chromatin structure.

MLL (All-1)

The Mll-1 or All-1 gene (Mbangkollo et al., 1995) was first recognized because of its dysregulation and involvement in chromosome translocations in various leukemias (Downing and Look, 1996; Joh et al., 1996; Rubnitz et al., 1996). The gene encodes an amino acid motif known as the SET domain that is also found in the trithorax group of *Drosophila* genes involved in maintaining or extending the expression range of homeobox genes (see above). Mice carrying a heterozygous disruption of the Mll gene show neurological abnormalities with posterior shifting of the anterior boundaries of expression of certain HOX genes (Yu et al., 1995). These mice showed decreased numbers of red cells, megakaryocytes and B cells, consistent with a role for this gene in hematopoiesis. Knockout of the Mll genes in embryonic stem cells in cultures produces cells that readily form immature hematopoietic colonies but fail to produce mature lineage restricted colonies (Fidanza et al., 1996). These results are all consistent with a role for the Mll gene in upregulating or maintaining the expression of HOX genes that is analogous to that of the *Drosophila* trithorax gene and with the suggestion that leukemic rearrangements of Mll produce a dominant-negative product that interferes with normal Mll function.

Non-DNA–Binding Proteins in Transcription Complexes. Proteins that do not themselves have DNA-binding motifs may regulate transcription by associating with specific combinations of transcription factors on DNA. These proteins presumably can mediate both positive and negative synergy between transcription factors, and their role in cellular physiology is only beginning to be appreciated. These proteins may be considered coordinator factors. Examples include the CBP/p300 proteins that interact with a number of transcription factors (Janknecht and Hunter 1996) the well-known cell cycle regulatory retinoblastoma protein and the related protein p107 (Bartek et al., 1996; P.L. Chen et al., 1995; Muller, 1995; Sanchez and Dynlacht, 1996). E2F and DP are transcription factors that have intrinsic DNA-binding activity. The E2F–DP heterodimeric complex can activate late G1 and S phase cell cycle–dependent transcription. However, the retinoblastoma protein, in its underphosporylated form, will bind to the E2F–DP1 complex, converting it from a transcriptional activator to a repressor. Conversely, the retinoblastoma protein can also bind to transcription factors and activate them, promoting differentiation (P.L. Chen et al., 1996).

A remarkable example of transcriptional regulation is that exerted by the action of a non-DNA–binding regulatory protein, through its association with promoter-binding proteins and promoter DNA of the MHC class II genes. Patients with the bare lymphocyte syndrome appear to be normal except for the consequences of a failure to express structurally normal MHC class II antigens. Mach and colleagues (1996) have elegantly elucidated the molecular defects in these patients. One group of subjects was particularly fascinating in that all factors generating *in vitro* gel shifts with regulatory elements for the class II genes were present and intact, indicating that all of the DNA-binding regulatory proteins for class II genes were present in lymphocytes from patients with the syndrome. However, complementation assays with cDNA libraries led to the cloning of a gene for a protein (CIITA) that bound to the promoter only in the presence of the complete set of other factors. This protein did not bind to DNA directly and did not interact strongly enough with any single factor to be detectable in *in vitro* protein–protein interaction assays or in the yeast two-hybrid system. Regulation of the new protein apparently is sufficient to account for many apsects of class II regulation, such as response to interferon-γ, or the shut down of class II genes in maturing plasmacytes.

It is striking that the CIITA protein is apparently needed only for the transcription of a very small set of genes related to class II function. The point of this illustration in the present context is that this protein would not necessarily have been detected by the conventional molecular biological procedures, which aim at finding factors that bind to and act at a given promoter. It was only the uncovering of a mutant that drew attention to the protein. This could be a unique example, but the possibility exists that other factors active in stem cell formation or differentiation at the transcriptional level could be similarly elusive.

Negative Regulatory Proteins. In addition to positive regulators of transcription, negative regulation can occur at multiple levels. The "coordinator" proteins referred to in the above paragraphs can act as negative regulators as well as positive regulators, as seen, for example, in the silencing of E2F-responsive promoters by the Rb–E2F complex. The Id gene encodes a widely expressed helix-loop-helix (bHLH) protein that lacks a DNA-binding domain (Benezra et al., 1990). It can dimerize with bHLH transcription factors via the HLH domain and neutralize them by preventing DNA binding (Sun et al., 1991). This has been studied particularly in muscle development (Weintraub, 1993), but similar effects are expected in hematopoiesis where several bHLH proteins, such as SCL/Tal1, are involved in lineage specification. The zinc finger protein MZF1 is preferentially expressed in hematopoietic cells and can prevent the induced expression of CD34 and c-myb during differentiation of embryonic stem cells (Perrotti et al., 1996). Several other DNA-binding proteins have been observed to negatively regulate some aspect of hematopoietic lineage expression, but their physiological role in general needs to be elucidated in more detail. Other negative regulatory transcription factors may be generated from alternatively spliced forms of mRNA of transcription-activating proteins.

Modes of Regulation of Transcription Factor Activity. The activities of transcriptional regulatory proteins are controlled by a variety of mechanisms, including cellular localization, post-translational modification, and proteolytic processing. Phosphorylation plays an important role in the regulation of transcriptional activity in many systems, including the hematopoietic system. Among other notable examples are the central role of phosphorylation of STAT factors in the transcriptional response to interferon and cytokines (Ihle, 1996), the role of phosphorylation of IκB and NF-κB in the inflammatory response (Z.J. Chen et al., 1996; Lee et al., 1997), and the role of phosphorylation in modulating the activity of NFAT transcription factors (Park et al., 1995; Ruff and Leach, 1995; Shaw et al., 1995) and therefore regulating the expression of cytokines such as IL-2 by T cells.

Attenuation of mRNA Transcription. In prokaryotes a common mechanism of regulation of mRNA level is at the level of transcription attenuation. Here transcription initiation may occur normally, but elongation of the transcript is reduced or completely blocked at certain regulated sites. In prokaryotes transcription termination may be controlled by secondary structural rearrangements that occur during translation of the nascent RNA or by proteins that bind to specific sites in the DNA. The former mechanism is not thought to be available to animal cells because translation occurs in the cytoplasm. However, a number of cases of potential transcriptional attenuation have been described for various genes, including myc (Roberts and Bentley, 1992). A recent report suggests that transcription of at least the first exon of embryonic β-globin chains may occur in adult erythroid cells, although transcription of downstream exons is depressed and essentially no embryonic globin is produced (Kollia et al., 1996).

Post-Transcriptional Regulation of mRNA Levels. At stages of differentiation at which lineage choice is determined by the relative level of various factors, post-transcriptional regulation of specific mRNA abundance could well prove as important as changes in the level of transcription factors. The area of post-transcriptional regulation of gene expression at the level of mRNA processing and stability, translational regulation, and protein stability during hematopoietic cell differentiation and development has not been as extensively studied as have the transcriptional regulators. Some striking examples of these effects are known, however. For example, as erythropoietic cells mature *in vitro* there may be a marked up regulation of the mRNA for the Tal1 transcription factor (see below) with a parallel decline in the levels of Tal1 protein

(Murrell et al., 1995). Globin mRNA becomes less stable when K562 cells are induced to differentiate along the megakaryocytic lineage by treatment with PMA (Lumelsky et al., 1991). A ribonucleoprotein assemblage on the 3′ untranslated region of α-globin mRNA confers stability to the mRNA (Wang et al., 1995; Weiss and Liebhaber, 1994), and a different complex appears to function on the 3′ untranslated region of β-globin mRNA (Russell and Liebhaber, 1996). In view of the known effects of 3′ mRNA sequences on translation in other systems, it would not be surprising to find that such ribonucleoprotein complexes modify translation efficiency as well as stability of mRNA.

TRANSCRIPTIONAL REGULATION IN HEMATOPOIESIS

Successive Expression: Transcriptional Cascades in Hematopoiesis

Methods for preparing pure populations of hematopoietic stem cells are still evolving, and there is probably more than one type of cell that is able to cause long-term generation of the lineages *in vivo,* as discussed elsewhere in this volume. The actual number of stem cells that can be obtained from an organism is very small. Normal stem cells cannot be expanded *in vitro* in pure populations synchronized with respect to their level of differentiation. As a result, information about the molecular content and gene regulation in stem cells has to be obtained or inferred from specialized approaches such as (1) single-cell polymerase chain reaction; (2) cultured cell line models; (3) gene knockout and transgene experiments; and (4) study of the transcriptional regulation of stem cell genes in other cell types. The data available from all these sources are for the most part limited and uneven, and the present chapter includes only selective examples of the type of information obtained by the above means and some discussion of gene control during the replication and early differentiation of hematopoietic stem cells.

Somatic cells of any lineage express at a significant level only a small fraction of the genes of an organism. For a gene such as globin there must be a first point in differentiation at which "significant" levels of this gene are expressed. It follows, therefore, that at that point the correct combination of transcription factors and chromatin structure that permit expression must also exist. Simplistically, one might have imagined that this expression is triggered by a unique combination of regulatory factors that mediate and create, for example, an "erythropoietic state" in the precursor cell and that expression of this group of regulatory factors mediates the entire process of erythropoietic differentiation. Not surprisingly, the actual process of erythroid differentiation is much more complex and multitiered.

One example of a potential transcription factor cascade is that provided by the factors GATA-1 and GATA-2 in erythroid differentiation (Weiss and Orkin, 1995a). Both transcription factor genes contain GATA-binding sites in their promoters and, in principle, might be subject to auto- and cross-regulation. GATA-2 is normally expressed earlier in hematopoietic differentiation than is GATA-1 and in a broader range of cells. In the absence of GATA-1, there is a 50-fold increase in the level of GATA-2 expression in erythroid cells (Weiss et al., 1994). On the other hand, GATA-1 expression occurs even in the absence of GATA-2 (F.Y. Tsai et al., 1994).

A different type of cascade is illustrated by the ets-related transcription factor PU.1 (SPI1). This factor is expressed in cells of the myelomonocytic and B-cell lineages and, along with C-EBPα and perhaps AML1 (CBFα2), provides cell type–specific activation of a number of genes in these lineages, including cytokine receptors (Smith et al., 1996; Zhang et al., 1996). Stimulation of these receptors in turn can promote differentiation and expansion of the hematopoietic cells and also activate other transcription factors. PU.1 also activates its own transcription (H. Chen et al., 1995), providing an additional measure of positive feedback.

Finally, in the absence of knowledge of a gene's function, it is dangerous to presume that, because knockout leads to the disappearance of a given lineage, the gene in question has a function unique to that lineage. A curious extreme example is the recently reported knockout of DNA ligase 1. This protein is thought to play an important role in lagging strand synthesis during DNA replication. However, homozygous knockout embryos lived to about day 16 postconception and showed no gross morphological abnormalities other than pallor and a small liver. There was a marked deficiency or complete lack of intraembryonic erythropoiesis, although initial results suggest that the total number of nonerythroid hematopoietic cells in the fetal liver of these animals was normal (Bentley et al., 1996). No explanation is currently available for the highly selective effect of knockout of this very generally expressed gene.

Transcription Factor Complexes and Interactions

The theme of combinatorial action of transcription factors to form active, or inactive, heterodimers arises so frequently as to be more the rule than the exception in transcriptional control, and the hematopoietic system presents a number of examples. For example, the selective expression of fetal (γ-globin) versus adult globin β-like chain mRNA is a relatively intensively studied example of the combinatorial action of transcription factors in regulating the expression of specific genes. GATA-1 is required for high level expression of both genes, but EKLF, the erythroid-specific Kruppel-like zinc finger transcription factor (see p. 113), is strictly required for expression of adult β-globin but not for fetal γ-globin production (Donze et al., 1995; Nuez et al., 1995; Perkins et al., 1995). Conversely there is a stage-specific selector element (Amrolia et al., 1995) in the fetal globin promoter that requires a heteromeric complex that contains a widely expressed transcription factor (CP2) and an additional stage-specific factor whose cloning has not yet been reported (Jane et al., 1995).

A number of dynamic processes may serve to reinforce differentiation along one lineage and repress alternative states of differentiation in hematopoietic cells. For example, the GATA-1 transcription factor that is expressed at high levels in differentiated erythroid cells activates its own transcription and may depress the transcription of GATA-2, which is found in less differentiated hematopoietic cells. In addition, among other genes, GATA-1 promotes transcription of the erythropoietin receptor, which presumably can transmit signals favoring erythropoietic development. We are only at the beginning of understanding such positive and negative crossover signals that occur at the transcriptional level during differentiation and development of the various hematopoietic lineages.

In addition to sequences proximal to the start of transcription, more distal sequences may be important for establishing the levels and tissue specificity of gene expression. The effect of these genetic elements can be either activation of transcription or insulation (Chung et al., 1993) of promoters from the effects of enhancers or silencers of nearby genes or heterochromatin. These remote control regions were first discovered for the β-globin gene cluster, where sequences known as *locus control regions* (LCRs) were detected both functionally (Blom van Assendelft et al., 1989; Talbot et al., 1989; Tuan et al., 1989) and because they represented sites of erythroid-specific DNase hypersensitivity (Forrester et al., 1987; Tuan et al., 1985). The globin LCRs, which reside up to 50 kilobases upstream of the genes they activate, consist of four hypersensitive (HS) regions. The HS sites contain clusters of sequences that bind transcription factors, including multiple GATA sites and functionally important paired NFE2 sites (Dillon and Grosveld, 1993; Grosveld et al., 1990).

There are specific DNA segments that act to shield a gene or gene cluster from the activating or repressing influences of surrounding chromosomal DNA. These elements are referred to as *boundary elements* or *insulators* and are being studied intensively in *Drosophila* (e.g., Hagstrom et al., 1996) and in the globin cluster (Chung et al., 1993). The LCR acts as a boundary element on the chromosome and can protect genes from the inhibitory influence of local chromatin. For instance, in transgenic mice that are prepared with constructs containing globin genes flanked by LCRs, there is full activity of the globin gene, independent of its site of integration. The effect of the LCR on gene activity is mediated by increasing the number of cells expressing the gene rather than by increasing the rate of transcription of the gene in each cell (Boyes and Felsenfeld, 1996). In particular, with the full LCR, pericentromeric transgenes are still expressed in all murine fetal liver erythroid cells, while constructs carrying incomplete LCRs are expressed in only a fraction of the cells at any one time (Milot et al., 1996). The latter phenomenon is reminiscent of position effect variegation of periheterochromatic gene expression in *Drosophila.* The chicken insulator from the globin cluster has been narrowed down to a 250 basepair region that contains a DNase hypersensitive site and binding sites for a number of proteins (Chung et al., 1997), but is itself different from a promoter. The mechanism of action of boundary elements and their specificity with respect to enhancers and promoters are important subjects that require further investigation.

Widely Expressed Factors With a Role in Hematopoiesis

HOX Cluster Positional Identity Genes in Hematopoiesis? The study of mutations in *Drosophila* that affect the body plan has led to an emerging understanding of molecular control of development. One particular class of mutations, in which one body segment is converted into another, has yielded a wealth of information. This class, termed *homeotic,* consists of mutations in genes that are clustered in the *Drosophila* genome and exhibit a high degree of similarity to one another. These genes share a common element, the homeobox, which consists of a 183 nucleotide segment encoding a helix-turn-helix protein motif that mediates interaction with DNA, primarily with the sequence ATTA (Lawrence and Morata, 1994). These proteins function as transcription factors to regulate the expression of subordinate genes (largely unidentified) that presumably enact specific features of body plan development (Botas, 1993). In higher eukaryotes, including mammals, homeobox genes are clustered into four groups. A, B, C, and D, each of which is homologous to the single cluster in *Drosophila* (Kappen and Ruddle, 1993; Maconochie et al., 1996; Ruddle et al., 1994).

The analysis of the pattern of HOX gene expression within developing organisms revealed the remarkable feature that the spatial extent of expression in the embryo correlated with the position of the gene on the chromosome in both insects and, with additional complexities, vertebrates (Krumlauf, 1993, 1994): HOX genes located at the 3′ end of the cluster are expressed in domains that extend into the anterior portions of the embryo, while those at the 5′ end are expressed only toward the posterior pole. Cell fate determination seems to be controlled by the 3′-most limit of the HOX gene cluster expressed in the cell.

Through a comparison of similarly positioned genes between groups of vertebrate HOX genes, it was found that homeobox genes often bore more resemblance to genes in the analogous position in a different cluster (known as a *paralog*) than to genes within their own cluster, although not all members of the family are represented in any single cluster (Maconochie et al., 1996). These findings are consistent with the establishment of an ancestral HOX cluster by local gene duplication, followed by duplication of the entire cluster, and this duplication provides some of the stronger evidence for the hypothesis that tetraploidization occurred one or more times during early vertebrate evolution (Holland and Garcia-Fernandez, 1996). In vertebrates a major effect of expression of HOX genes may be to promote the expansion of particular cell populations as well as or instead of specifying the cell fate.

In addition to these "clustered" HOX genes are a number of dispersed, nonclustered genes that are scattered throughout the genome. These are, for the most part, more distantly diverged from the clustered HOX genes. Some of these have been identified at sites of chromosomal breakpoints in leukemias and appear to play important roles in hematopoiesis (Hatano et al., 1991; Kamps et al., 1991).

The binding specificity of HOX proteins is, in general, not high: They all appear to bind to the ATTA motif despite the fact that they have different functions in development and, thus, presumably different target genes. A possible resolution of this dilemma is provided by the findings that, within the cell, HOX proteins likely bind to DNA as heterodimers with other proteins and that one result of this complex formation is acquisition of a much higher degree of binding specificity. For instance, it was recently found that divergent homeobox proteins of the Pbx family can heterodimerize with various members of the clustered HOX family of proteins (Shen et al., 1996) and that in doing so the HOX proteins acquired the ability to bind to a more defined DNA-binding site (Chang et al., 1996). There are three Pbx genes that have been identified, as well as a family of more distantly related genes, called Meis (Nakamura et al., 1996a). Interestingly, two of these genes were identified because of their role in leukemia: Pbx is located at a t(1;19) in human pre-B cell ALL, a translocation that results in the formation of a fusion protein between the helix-loop-helix protein E2A and Pbx (Kamps et al., 1991). Meis1 was identified as a site of retroviral insertion in murine myeloid leukemias (Moskow et al., 1995). Remarkably, in the same tumors, retroviral insertions are also found within the HOX A cluster, resulting in activation of HOXA9 and HOXA7 (Nakamura et al., 1995). This provides genetic evidence for cooperativity between the divergent and the clustered HOX genes, which may have as its basis the need for both heterodimerizing partners to be activated for the particular leukemogenic effect. It is not known precisely which HOX genes are on in ALLs with the t(1;19) translocation or what target genes are critical for Pbx- or Meis-induced leukemogenesis.

The pattern of HOX gene expression in hematopoiesis has been studied fairly extensively and is the subject of an excellent recent review (Lawrence et al., 1996), and it is clear that, in addition to their well-established role in body plan determination in segmented regions of the organism, they play an important role in hematopoiesis. Certain genes of the HOX A, B, and C clusters have been found to be expressed in both hematopoietic cell lines and in primary bone marrow cells. In cell lines with erythroid potential, most of the genes of the HOX B cluster (eight of nine genes) and some of the HOX C cluster are expressed. In myelomonocytic cells, HOXA10 is expressed, and in lymphoid cells certain HOXB genes (B7 and B4) and HOXC4 are expressed. In leukemic cells, HOXA10 is expressed at high levels in myeloid leukemias and may be a marker of diagnostic utility (Lawrence et al., 1996). As mentioned above, two other HOXA genes, HOXA9 and HOXA7, are sites of retroviral insertions in murine myeloid leukemias, and in human leukemias HOXA9 is part of a chimeric protein with NUP98, a nucleoporin protein, encoded by t(7,11) (p15;p15) (Borrow et al., 1996; Nakamura et al., 1996b). In the acute myelomonocytic cell line WEHI3B, there is aberrant activation of HOXB8 via insertion of the retrovirus-like intracisternal A-type particle genome.

In primary human bone marrow cells, the level of HOX genes is such that HOXA > HOXB ≫ HOXC, and HOXD gene transcripts are undetectable (Lawrence et al., 1996). In $CD34^+$ progenitors, both HOXA and HOXB genes are expressed. During commitment to myeloid and erythroid progenitors, genes in the 3′ end of these clusters (e.g., HOXB3) are suppressed, while at later stages, the 5′ genes (e.g., HOXA10) are downregulated (Sauvageau et al., 1994). HOXB3 expression appears to be confined to the very early progenitor, the long-term culture-initiating cell (LTC-IC [Sauvageau et al., 1994]). Another HOXB member, HOXB6, is expressed in murine yolk sac hematopoietic cells. Strikingly, the HOXD cluster of genes does not seem to be expressed at all in hematopoietic lineages.

Functional studies that would support a specific role for the HOX genes in hematopoiesis are limited, yet compelling. First, as mentioned above, a number of HOX genes are implicated in acute leukemias, indicating that their overexpression can lead to a deregulation of control of hematopoietic cell growth and differentiation. Second, overexpression studies indicate a dominant effect of HOXB4 and HOXB8 expression in primary murine hematopoietic progenitor cells: HOXB8 transduction led to enhanced ability to generate IL-3–dependent cell lines and an expansion of progenitors. At low frequency, the mice developed leukemia (Perkins and Cory, 1993). HOXB4 expression in transduced bone marrow cells in repopulated mice led to an increase in the most primitive cells, but in no abnormalities in the peripheral blood and no increased incidence of leukemia (Sauvageau et al., 1995), and overexpression of HOXB4 in embryonic stem cells increased the number of progenitors for mixed myeloid-erythroid colonies and definitive erythroid colonies (Helgason et al., 1996). HOXA10 expression in bone marrow of reconstituted mice led to an increased number of megakaryocytic and primitive blast colonies, as well as an increased progenitor pool. In older animals, this progressed to a CML-like illness, while in secondary recipients of transduced marrow fulminant acute myeloid leukemias developed (Thorsteinsdottir et al., 1997). Although HOXB3 is downregulated earlier than other HOX genes in hematopoiesis, overproduction of HOXB3 surprisingly leads to a increase in myeloid precursors but not an expansion of the stem cell population. Further support for a role of HOX genes in hematopoiesis comes from gene targeting studies: Homozygous mutation of HOXA9 in mice causes a generalized hematopoietic hypoplasia, reduced numbers of lumphocytes and granulocytes in the peripheral blood, small spleen and thymus, and decreased numbers of committed progenitors, but no change in the number of pluripotent progenitors (Lawrence et al., 1995). The combined results of hypoplasia in HOXA9 deficiency and leukemia with HOXA9 overexpression indicate a role for the gene in expansion of hematopoietic cells, but at a point beyond the pluripotent progenitor.

Other Homeobox Genes. Other homeobox genes that may play specific roles in hematopoietic development include the HOX11 gene. Absence of this gene in mice causes asplenia (Dear et al., 1995; Roberts et al., 1994, 1995), while overexpression of an unrearranged HOX11 gene is associated with some T-cell leukemias (Lichty et al., 1995) and forced expression of HOX11 in murine hematopoietic precursor cells led to a high incidence of cell immortalization (Hawley et al., 1994). Curiously, the HOX11 protein may interact directly with protein phosphatases and thus disrupt the G2/M cell cycle checkpoint (Kawabe et al., 1997).

A special subset of homeobox genes, the PAX family (Mansouri et al., 1996) is defined by the presence of a conserved paired box motif. The mammalian genome contains at least nine PAX genes. These genes play a role in specification of a number of differentiation events, such as eye development, vertebral specification, and genesis of neural crest–derived tissues. A particular PAX gene, called MIX-1, has been identified as one that responds with immediate early kinetics to the TGF-β-related intercellular mediators activin and BMP-4 (Mead et al., 1996). The latter protein is a putative ventralizing signal molecule that may be involved in blood cell development (Zhang and Evans, 1996), as injection of BMP-4 cDNA induces ectopic globin formation in developing frog embryos. Dominant negative mutants of MIX-1 inhibit ventralization by BMP-4 (Mead et al., 1996). Recent studies further suggest that MIX proteins may have their activity modified by heterodimer formation *in vitro*. A possible pathway through which BMP-4 can regulate MIX gene transcription is via activation of MAD proteins (Thomsen, 1996), known to be downstream effectors of signaling molecules in the TGF-β family. This provides a skeletal outline of one part of a possible path by which the erythropoietic system is induced to form during embryogenesis. This aspect of the pathway may involve other homeobox genes, since MIX-2, a close homolog of MIX-1, has recently been identified and has a pattern of expression similar to that of MIX-1. Furthermore, additional homeobox genes such as PV.1 (Ault et al., 1966) and Xom (Ladher et al., 1996) are induced by BMP-4 and have ventralizing effects.

The POU homeobox family of proteins (Verrijzer and Van der Vliet, 1993; Wegner et al., 1993) is a subset of homeobox-containing transcription factors that also have a second DNA-binding domain, termed the POU domain, linked to the homeodomain by a flexible linker (Herr and Cleary, 1995). Both domains bind to DNA via the major groove, but do so on opposite faces. The two domains together recognize the octamer sequence 5′ATGCAAAT3′. The POU homeobox genes play numerous roles in development through the control of the transcription of tissue-specific as well as "housekeeping" genes. This family of proteins includes the "octamer" binding factors OCT-1 and OCT-2 that bind to an eight basepair motif present in an immunoglobulin enhancer, as well as other factors that have been studied particularly for their role in nervous system or in embryonic development. An extensive discussion of these proteins is largely out of the scope of the present chapter. We only note that OCT-2 (Latchman, 1996), originally considered to be a factor important for general B-cell development and immunoglobulin expression, may have a more limited role in the regulation of late B-cell-specific genes (Corcoran et al., 1993; Konig et al., 1995). Also, the specific activation via octamer sequences in B cells involves, in addition to OCT-1 or OCT-2, a B-cell-specific coactivator, OCA-B (Luo et al., 1992), that binds DNA in the presence of OCT-2 (Cepek et al., 1996; Gstaiger et al., 1996), making specific contacts within the octamer motif. Interestingly, OCA-B can bind to and act synergistically with CIITA (Kern et al., 1995) to activate class II MHC gene promoters (Fontes et al., 1996).

c-myc and myb: Proliferation vs. Differentiation

Two transcription factors, myc and myb (Graf, 1992), have been implicated in the repression of differentiation and stimulation of proliferation in hematopoietic precursors.

The myb family consists of three members a-myb, b-myb, and c-myb (Introna et al., 1994). c-myb is expressed mostly in hematopoietic tissues, while b-myb is broadly expressed. a-myb expression within the hematopoietic system is largely limited to B cells and related lymphomas and leukemias. All three myb genes bind to the hexanucleotide PyAACG/TG and seem to activate transcription with the relative effectiveness a-myb > c-myb > b-myb. c-myb has been studied to the greatest extent and seems to have both a specific effect in inducing hematopoietic genes and a general effect in blocking hematopoietic differentiation and promoting proliferation of precursor and differentiating cell lines.

The c-myb protooncogene was initially recognized because it was the progenitor of the v-myb oncogene, associated with retrovirally induced avian myeloblastosis. v-myb cooperates with C/EBP to induce the expression of the promyelocyte-specific mim-1 gene in erythroid cells or fibroblasts so that the synergistic action of these two transcription factors can act as a switch to activate myelocyte-specific genes (Mink et al., 1996a). Furthermore, embryonic stem cells transfected with intact, but not mutant, c-myb underwent erythromyeloid differentiation and formed colonies in methylcellulose cultures in the absence of growth factors (Melotti and Calabretta, 1996) and constitutive expression of c-myb transformed avian yolk sac cells to produce cell lines with characteristics of both myeloid and macrophage lineages (Fu and Lipsick, 1997).

The c-myb gene is most abundantly expressed in hematopoietic progenitor cells, and its level falls as the cells mature and differentiate, a pattern of expression that is consistent with a role in promoting cell proliferation and inhibiting cell differentiation. While a number of potential gene targets for myb transcriptional activation have been identified, none of these fully explain the role that Myb has in myelopoiesis. These putative targets include AML1 (Ghozi et al., 1996), c-kit (Chu and Besmer, 1995), GATA-1 (Melotti and Calabretta, 1996), the CD34 gene that encodes a cell surface marker often used to identify early human hematopoietic precursors (Perrotti et al., 1996), and the c-myc gene. However, the block in differentiation of hematopoietic cells caused by myb overexpression seems to occur at a stage earlier than that caused by myc (see below), and thus the myb-induced block is not likely to be a consequence of myb-induced myc expression.

Homozygous knockout of the c-myb gene in mice results in embryonic lethality at day 15 of gestation (ML Mucenski et al., 1991). Yolk sac erythropoiesis is apparently normal. However, the animals become profoundly anemic by embryonic day 15, with a severely reduced number (although not a total absence) of enucleated red cells and a ninefold decrease in intrahepatic CFU-GM progenitor cells. Lymphopoiesis was also reported to be blocked in these mice, although megakaryocytes were generated at wild-type levels in the mutant embryos. The evidence indicates that the defect in these mice is related to an inability to expand the definitive precursor cell populations rather than a complete block to maturation, at least in the erythropoietic lineage. These findings suggest an essential role for myb in the proliferation of early hematopoietic cells (H.H. Lin et al., 1996).

myc is a basic leucine zipper helix-loop-helix transcription factor that has been implicated in proliferation, cell death, differentiation, and oncogenesis. myc transcripts are rapidly induced in a large variety of cell types in the course of mitogenic stimulation, and myc expression is low in quiescent cells, suggesting an important role in cellular proliferation. As with myb, transfection of myc blocks the differentiation of erythroid cells, suggesting an important role in the control of cellular differentiation and/or proliferation. Consistent with this, myc has been implicated in the etiology of several neoplasms: The myc family was originally recognized because a mutated c-myc gene was incorporated, as v-myc, into an avian retrovirus that produced leukemia. In addition, c-myc is deregulated by a number of chromosomal translocations in B-cell neoplasms in humans. Conversely myc is downregulated when terminal differentiation is induced, and forced downregulation of c-myc with antisense oligonucleotides inhibits proliferation and induces differentiation (Liebermann and Hoffman, 1994).

Knockout of the c-myc gene in mice leads to early lethality and widespread defects, suggesting an essential role in development (Davis et al., 1993). Embryonic stem cells homozygous for c-myc null alleles are fully viable and can contribute to most cell lineages in chimeric animals, which shows that c-myc does not play an essential role for proliferation and differentiation within specific organs. This suggests that other proteins, perhaps c-myc family members, carry out essential c-myc functions. In addition to having a role in proliferation and differentiation, expression of c-myc can promote apoptosis on growth factor withdrawal, for example, in 32Dc13 myeloid progenitor cells. Expression of myc can induce apoptosis (Packham and Cleveland, 1995), even in the absence of cell cycle progression (Packham et al., 1996).

The regulation of the levels of c-myc gene product is exceedingly complex (Marcu et al., 1992; Michelotti et al., 1996). Transcription occurs from four promoters, and these may be regulated by ribonucleoprotein binding and by factor binding to single-stranded DNA or DNA in the H configuration (Mirkin and Frank-Kamenetskii, 1994) as well as conventional double-stranded DNA-binding factors, including myb (Cogswell et al., 1993). In addition, regulation is effected at the level of transcript elongation (Roberts and Bentley, 1992). Both c-myc mRNA and protein are quite unstable, with half lives in the range of a half hour or less, providing additional levels at which regulation might occur.

The myc protein interacts with a number of proteins through a variety of dimerization motifs. myc contains both a helix-loop-helix domain, as well as a leucine zipper motif, both of which mediate dimerization. myc protein mediates transcription through dimerization with another helix-loop-helix protein, MAX. Current understanding of myc action is that the myc–MAX heterodimer acts as a transcriptional activator via a CAACG/TG motif, known as an E box binding site. MAX may also heterodimerize with a growing family of other proteins, termed MAD proteins (Hurlin et al., 1994). MAD–MAX heterodimers may act as transcriptional repressors, in some cases interacting physically with mammalian homologs of the yeast sin3 transcriptional repressor (Harper et al., 1996), and these products in turn may interact with yet other repressors such as the N-cor corepressor of transcription factors of the steroid receptor class (Aland et al.).

The list of direct transcriptional targets of myc is growing, although not yet sufficient to account for all the actions of myc. Proposed targets include prothymosin α, the ODC (ornithine) decarboxylase gene and (Ryan and Birnie, 1996), the cell cycle regulatory phosphatase cdc25 (Galaktionov et al., 1996). In addition, c-myc seems to have separable effects on repressing transcription. myc can interact (Roy et al., 1993a) with TFII-I (Roy et al., 1993b), a transcription factor proposed to act at initiator elements, which are defined as DNA sequences adjacent to the transcription initiation site that may promote transcription initiation even in the absence of a TATA box. This interaction of myc and TFII-I may act *in vitro* and *in vivo* to inhibit initiator-dependent transcription, including that of cyclin D1. This action is independent of MAX. Recently, additional actions of myc have been described, including the inhibition of the activity of C/EBP-α and -β transcription factors (Mink et al., 1996b). This inhibition does not seem to involve myc DNA-binding domains, and the mode of action is currently unknown. This is of interest though, as the C/EBP genes have been implicated in induction of various differentiation genes in hematopoietic, hepatic, and adipose cells and is an attractive candidate for at least one of the pathways by which myc can arrest differentiation. myc also interacts with and may modify the action of the YY-1 zinc finger transcription factor (Shrivastava and Calame, 1995), a factor that can either activate or repress transcription, depending on the promoter context. Finally, myc can interact with both the Rb-like protein p107 (Ryan and Birnie, 1996) and the components of the basal transcription apparatus, specifically the TATA-binding protein (TBP) (McEwan et al., 1996) and the large subunit of TFII-F (McEwan et al., 1996). The relative importance of these interactions is not clear.

Nuclear Receptor Family

This protein family consists of a number of zinc finger–containing transcription factors that are related in structure and show similarities in the sequences of their DNA-binding sites. They are characterized functionally by their ability to bind both a small molecule ligand, frequently lipophilic, and a bipartite DNA sequence. Many binding sites consist of direct repeats of the AGGTCA half-site. The half-sites may be separated by between one and five bases, depending on which specific receptor they bind, and may sometimes be arranged as inverted rather than tandem repeats (Glass, 1994). For example, half-sites separated by a single base may bind the RXR receptor; half-sites separated by two bases may serve as a retinoic acid response element; three base separation provides a site for the vitamin D receptor; four bases a site for the thyroid hormone receptor; and five for a retinoic acid receptor (Mangelsdorf and Evans, 1995). The receptors may bind DNA as either homo- or heterodimers and can both positively and negatively regulate transcription. Characteristic members include the receptors for thyroid hormone, vitamin D3, retinoic acid receptors, and so forth.

The steroid receptors proper include the androgen, estrogen, progesterone, and steroid receptors. They function as homodimers that bind to a palindrome with half-site sequence AGAACA (Beato et al., 1996). In addition to their role in activating transcription, steroid receptors may act as reciprocally antagonistic mediators to block AP1, NF-κB, and relA transcriptional activation (Caldenhoven et al., 1995; van der Saag et al., 1996), perhaps accounting for some of their antiinflammatory actions.

The majority of nonsteroid ligand-binding receptors function as heterodimers, commonly in association with a member of the RXR receptor group, receptors that bind specifically to 9-*cis* retinoic acid. Most of these receptors are activated by specific ligands, although there are a number of orphan receptors for which no ligand has yet been identified. Conversely, there are additional natural ligands for the retinoic acid receptors (Achkar et al., 1996). Some members of the family appear to bind tightly to DNA in the absence of ligand, and these may have no natural ligand. In addition to dimerization, various members of this family of factors can associate with cofactors including the SMRT (also termed TRAC) and N-cor genes that function as transcription repressors. They can also associate with one of several coactivator factors (Beato et al., 1995; Mangelsdorf and Evans, 1995), including RIP140 and RIP160, the RING finger protein TIF1, (Beato et al., 1995; Fraser et al., 1997; Mangelsdorf and Evans, 1995); the CBP/p300 proteins involved in interactions with a number of transcription factors (Kamei et al., 1996); and even the TFIIB component of the basal transcription apparatus (Glass, 1994).

Retinoic acid, in particular, is a well-known and extensively studied inducer of differentiation in various developmental systems. All-*trans* retinoic acid can stimulate transcription through RAR–RXR heterodimers binding to direct repeats of the consensus sequence AGGTCA separated by two or five bases. 9-*cis* retinoic acid can stimulate transcription through RXR homodimers binding at consensus sites separated by a single T residue. There are retinoic acid receptor-binding sites in the promoter regions of various homeobox genes, and these could mediate some of the developmental effects of retinoic acid.

The mRNA for the retinoic acid receptor RARα is expressed widely in hematopoietic cells. Retinoic acid has been shown to induce the myelomonocytic leukemia cell line HL60 to differentiate down the neutrophil lineage. On the contrary, retinoic acid can inhibit erythroid colony formation and also inhibit G-CSF–induced myeloid colony growth from human murine and bone marrow (Rusten et al., 1996). Curiously, it may have either no effect or may stimulate IL-3– or GM-CSF–induced GM colony formation. Expression of a dominant-negative form of the retinoic acid receptor in murine progenitors can cause a block to the formation of granulocytic/monocytic lineages that can be overcome with high doses of retinoic acid. Differentiation into other lineages is not blocked (S. Tsai et al., 1994). Transgenic mice expressing a dominant-negative RXR receptor also showed, in at least some cases, a block in myeloid lineage maturation (Sunaga et al., 1997). These lines of evidence suggest that retinoic acid may regulate the choice between proliferation and differentiation as well as the type of differentiation that occurs during hematopoiesis, but more study is clearly needed. Vitamin D_3 also promotes monocytic maturation, and in conjunction with retinoids one of its effects is to alter the levels of RARα in these cells. It also appears to directly upregulate Cip1/p21, a negative regulator of the cell cycle (Liu et al., 1996; Munker et al., 1996). This may cause slowing of the cell cycle, known to occur in vitamin D–induced differentiation.

High-dose estrogen treatment may suppress erythropoiesis, delay maturation of erythroid precursors, and downregulate the expression of a number of erythroid genes in cell culture. Recent investigations indicate that this may be mediated through inhibition of the activating functions of GATA-1 and that the estrogen receptor may interact with GATA-1 in a ligand-dependent manner (Blobel and Orkin, 1996; Blobel et al., 1995). Thyroid hormone may also modulate erythropoiesis, and recently an interaction between the thyroid receptor as well as other nuclear hormone receptors and the erythroid transcription factor NF-E2 has been demonstrated to occur and be mediated by CBP (Cheng et al., 1997).

Glucocorticoids are known to affect erythropoiesis (reviewed by Wessely et al., 1997), enhancing the proliferative capacity of erythroid progenitors and inhibiting globin accumulation in mouse erythroleukemic cell lines. This may occur by inhibitory interactions of the glucocorticoid receptor with GATA-1 (Chang et al., 1993). In chicken erythroid precursors, c-myb is a potential target of the glucocorticoid receptor, and an activated c-myb gene can replace the requirement for the glucocorticoid receptor for self-renewal (Wessely et al., 1997).

The Helix-Loop-Helix E and Id Proteins

The helix-loop-helix family of proteins (Littlewood et al., 1995; Murre et al., 1994) include both broadly expressed E proteins and cell type–specific transcriptional activators and inhibitory Id proteins (Benezra et al., 1990) that lack the DNA-binding motif but retain protein–protein interaction motifs. The factors associate *in vivo* as homo- or heterodimers whose relative abundance may vary from cell type to cell type. Cell type–differentiative events may be controlled by heterodimers formed between tissue-specific and ubiquitous helix-loop-helix proteins and may be negatively regulated by Id proteins that competitively dimerize with the E proteins. The broadly expressed proteins include E12 and E47, which are encoded by the E2A gene, as well as HEB and E2-2, which are encoded by separate genes.

The role of helix-loop-helix transcription factors in B-cell development provides a particularly instructive model. Although no B-cell-specific helix-loop-helix protein has been found, homozygous knockout of the E2A gene prevented B-cell development (Zhuang et al., 1994), and heterozygous knockouts of HEB or E2-2 synergized with a heterozygous E2A knockout to decrease B-cell development (Zhuang et al., 1996). Conversely, constitutive expression of Id genes impaired B-cell development (Wilson et al., 1991). Knockouts of HEB or E2-2 reduced but did not abolish B-cell formation. The E47 homodimer was found in B cells but not in other cell types, and E47 was able to activate immunoglobulin gene rearrangement in pre-T cells. Of incidental note, the dimerization of E2A may require not only specific phosphorylation, but the formation of a covalent disulfide link between subunits (Benezra, 1994). A model has been proposed to explain these results (Zhuang et al., 1996). The model postulates that E2A homodimers serve to activate B-cell differentiation and that this dimerization is antagonized by the Id protein. The role of HEB and E2-2 in this model is to bind Id proteins, releasing E47 for homodimerization.

The broadly expressed helix-loop-helix proteins also may play a specific role in T-cell development. The CD4 upstream enhancer is associated specifically with HEB proteins and an E12-related protein (Zawada and Littman, 1993). Homozygous HEB knockout mice showed reduced levels of expression on thymic cells. Mice doubly heterozygous for E2A and HEB knockouts also showed a reduction in the level of expression of these T-cell surface markers, but E2-2 knockout had no effect on expression of the T-cell-specific genes, in contrast to the effects on B-cell differentiation.

Lim Domain Proteins

The lim domain is a recently defined protein domain involved in specific interactions with other proteins (Gill, 1975; Sanchez-Garcia and Rabbitts,

1993, 1994). The lim domain protein previously termed RBTN2 and now called LMO2 can specifically interact with the potentially leukemogenic basic helix-loop-helix proteins Tal-1, Tal-2, and Lyl-1, even when they are complexed with other helix-loop-helix proteins such as E47. Knockout of RBTN2 was originally reported to produce a defect specifically in erythron development (Warren et al., 1994), but more recent reports suggests these mice have a general defect in hematopoietic development, comparable to that produced by Tal-1 knockout.

Hematopoietic-Specific Factors

Tal-1. Gene knockout studies have implicated a number of genes in the earliest stages of hematopoiesis, including erythropoiesis and lymphopoiesis. Transcription factors whose knockout leads to an absence of both hematopoietic and endothelial cells might function independently in each lineage or be necessary for a common precursor to both lineages. In the absence of further information it is perhaps premature to say that such factors play a direct role in specification of hematopoietic stem cells.

Tal-1 encodes a basic helix-loop-helix protein. The gene is expressed early in hematopoiesis (Begley et al., 1989), during the differentiation of erythroid, megakaryocytic, and mast cell lineages as well as in progenitors of the vascular system (Kallianpur et al., 1994).

The expression of Tal-1 is regulated coordinately with that of GATA-1 in erythroid, megakaryocytic, and mast cell lineages (Mouthon et al., 1993; Green et al., 1992). Of the various factors that have been studied, the evidence implicating Tal-1 and LMO2 in stem cell specification is particularly intriguing. These two proteins interact *in vivo.* Knockout of Tal-1 leads to a complete absence of either yolk sac or intraembryonic hematopoiesis (Porcher et al., 1993; Shivdasani et al., 1995). Embryonic stem knockout cells fail to contribute to hematopoiesis in chimeric animals, and no hematopoietic colony-forming cells can be generated from the mutant embryonic stem cells. Initially knockouts of LMO2 were reported to produce cells that could still form macrophages, but a recent review suggests that this may have to be modified. Tal-1 was also found in individual bipotential and erythroid precursor cells from marrow, but was absent in precursors of monocytes and granulocytes, while forced expression of Tal-1 inhibited the monocytic differentiation of a bipotential precursor cell line (Hoang et al., 1996), indicating that Tal-1 has a lineage-specific role in erythroid development. It is also possible that Tal-1 forms a heterodimer with some factor other than LMO2 and that this heterodimer is essential for stem cell formation (Voronova and Lee, 1994).

AML-1 (CBFα2) and CBFβ. Knockout of other transcription factors also has profound deleterious effects on general hematopoiesis. AML-1 is one of at least three genes whose protein products may form a dimeric complex with the ubiquitously expressed protein CBFβ. CBFβ acts to stabilize the binding of AML-1 to DNA, and the dimer, known as CBF or PEPB2, is an active transcription factor. AML-1 itself is expressed in fetal liver and newborn bone marrow of mice as well as in the thymus and adult T cells. The AML-1 gene initially drew attention because it was involved in translocations in about 15% of *de novo* cases of acute myeloid leukemia (Nucifora and Rowley, 1995), and it has more recently been found to be involved in a translocation in about 25% of B-cell acute lymphocytic leukemia of children. The AML-1 gene is widely expressed in a large number of cell types and tissues, but mice with a homozygous knockout of the AML-1 gene die at about embryonic day 12.5 because of a complete failure of definitive hematopoiesis, including, in addition to an absence of intrahepatic hematopoiesis, a lack of detectable colony-forming cells in the yolk sac (Okuda et al., 1996). Nevertheless, primitive yolk sac erythropoiesis is normal in these mice (Okuda et al., 1996; Sasaki et al., 1996). It would be interesting to know if these mice develop the recently described intraembryonic para-aortic hematopoietic stem cells.

GATA-1. Multiple representations of the sequence "GATA" were recognized early to be a characteristic feature of erythroid-specific promoters such as those for globin, the erythropoietin receptor, and enzymes of the heme biosynthesis pathway, as well as megakaryocyte-specific promoters and at least one mast cell promoter. This led initially to the cloning of the relatively abundant transcription factor GATA-1. Once the structure of GATA-1 was known, several other partially homologous transcription factors were recognized, including GATA-3, which is preferentially expressed in T cells, and GATA-2, expressed *inter alia* in early hematopoietic precursors (Weiss and Orkin, 1995a).

GATA-1 was shown to be capable of mediating specific expression of genes with appropriate sequences in their promoters and could promote both upregulation of genes such as the globin genes and downregulation of genes such as GATA-2 and c-myb that might act to promote proliferation of erythroid precursors rather than differentiation (Briegel et al., 1996). Furthermore, GATA-1 could physically interact with other transcription factors such as SP1 and EKLF and could synergize with these factors *in vivo* to promote transcription (Merika and Orkin, 1995). It was tempting to consider that GATA-1 was a key regulator or erythroid determination, even though it was also expressed in other differentiated lineages, in particular megakaryocytes and mast cells. Initial murine gene knockout studies appeared to support the essential nature of GATA-1 for erythropoiesis, as GATA-1 knockout mice were markedly deficient in erythroid precursors. Further examination, including reseeding of differentiating embryonal stem cells lacking the GATA-1 gene, showed that while mature erythropoiesis was severely impaired, red cell precursors at the level of proerythroblasts were generated (Pevny et al., 1995), but then underwent apoptosis (Weiss and Orkin, 1995b). GATA-2 recognizes binding sites similar to those of GATA-1, and its own promoter contains GATA sites. As mentioned before in the GATA-1–deficient cells, the level of GATA-2 is 50-fold elevated, perhaps due to lack of inhibition by GATA-1. This upregulation has been suggested to be a compensatory mechanism permitting some erythroid differentiation in GATA-1$^-$ cells.

GATA-1 is located on the X chromosome; thus heterozygous females contain a mixture of GATA-1–deficient and normal precursors, predictably producing some impairment of erythropoiesis. GATA-2 knockout heterozygotes show some reduction in hematopoietic precursors. Curiously, mice doubly heterozygous for GATA-1 and GATA-2 knockouts show a profound anemia during development and do not survive.

The regulation of GATA-1 is complex. For example, GATA-1 is phosphorylated *in vivo,* although the functional significance of this phosphorylation is unknown. Also, GATA-1 accumulates in the cytoplasm of avian erythroid progenitor cells, but shifts to the nucleus on differentiation (Briegel et al., 1996). GATA-1 forms incompletely characterized complexes with several other proteins *in vivo* and may be associated with nuclear bodies visible by light (Elefanty et al., 1996). Truncation mutants of GATA-1 that lack the transactivation domain may still function to promote erythropoiesis (Weiss et al., 1997). Recently a protein, FOG (friend of GATA), has been isolated using GATA-1 in a two-hybrid system to isolate interacting proteins (Tsang et al., 1997). FOG, a multitype zinc finger protein, acts as a cofactor for transcription factor GATA-1 in erythroid and megakaryocytic differentiation. Not only does FOG associate physically with GATA-1 but it can act synergistically with GATA-1 to stimulate transcription from promoters containing GATA-binding sites.

GATA-2 and GATA-3. GATA-2 is expressed earlier in embryogenesis than GATA-1 and at least initially is not strictly restricted to hematopoietic lineage cells, but is also found in endothelial cells, megakaryocytes, mast cells, fibroblasts, and embryonic brain and liver cells. The level of GATA-2 declines as GATA-1 levels rise during erythroid maturation. In contrast to the GATA-1–deficient embryos, GATA-2 knockout mice show markedly reduced yolk sac hematopoiesis affecting all lineages (F.Y. Tsai et al., 1994). Although the embryos died by day 10.5 postconception, their blood contained apparently normal nucleated red cells. Hematopoietic colony assays showed that the yolk sac cells could form a small number of mixed colonies, and disaggregated embryoid body cells could differentiate to form primitive and definitive erythroid, macrophage, and mast cell colonies. Transplant of GATA-2 knockout embryonal stem cells into RAG$^-$ blastocysts produced a small number of embryonal stem cell–derived T cells and IgM indicating that GATA-2 is not strictly necessary for lymphocyte differentiation. Overall, however, there was a marked cell-autonomous deficiency in the hematopoietic potential of GATA-2$^-$ cells, and one possible explanation is that GATA-2 is necessary for the proliferation but not the determination or maturation of all hematopoietic lineages. Interestingly, GATA-1 was stated to be at normal levels in the colonies derived from embryoid body cells so that GATA-2 is not necessary to activate GATA-1 expression. In further contrast, knockout of GATA-3 has been reported to leave yolk sac erythropoiesis largely unaffected but produced embryos whose livers contained very few cells that could give rise to multipotential or differentiated colonies *in vitro*. The GATA-3 knockout animals were also anemic, but the animals showed massive internal bleeding by day 11.5 (Pandolfi et al., 1995). GATA-$3^{-/-}$ embryonal stem cells can contribute to the development of the mature erythroid, myelomonocytic, and B-cell lineages in RAG^{-} mice, but not to the formation of thymocytes, mature peripheral T cells, or even the double-negative thymocyte population (Ting et al., 1996).

PU.1. Clearly it is necessary to distinguish between loss of the capacity to generate any cells in a particular lineage and the failure to expand the earliest determined cells or to protect them from apoptosis. In the case of the GATA-1 knockout, the latter case was detected by careful work and because of the advantage of the relatively detailed knowledge of the early erythroid precursor cell types. In the case of stem cells, this analysis is more difficult both because of the small number of cells and because of the lack of information as to whether there are any intermediate cell types between undifferentiated mesenchymal cells and functional hematopoietic stem cells. One must be wary of the possibility that knockouts may indirectly affect cell lineages by blocking the production of cytokines that act transcellularly to promote differentiation or expansion or of cytokine receptors. For example, this could well contribute to some of the effects of PU.1 knockouts. PU.1 is an ets family transcription factor that is expressed in most hematopoietic lineages, but not in T cells. PU.1 knockout animals show a multilineage defect involving B- and T-cell, monocyte, and granulocyte precursors (Scott et al., 1994). Embryonal stem cells lacking the PU.1 gene show early myeloid development, including expression of GM-CSFR and G-CSFR but lack M-CSFR and other markers of terminal myeloid differentiation (Olson et al., 1995). PU.1 is implicated in the regulation of a number of genes in these lineages, including receptors for M, G, and GM-CSFs (Zhang et al., 1996), and defects in expression of these or other receptors hypothetically could contribute to the lineage defects seen in the knockout animals.

NF-E2. NF-E2 is a heterodimeric protein consisting of a small subunit (18 kD) and a large unit (45 kD) both of which have b-zip motifs. MAF proteins form heterodimers with the large subunit. The small subunit of NF-E2 is encoded by the broadly expressed MAFK gene (Igarashi et al., 1995), and the large subunit is encoded by a gene whose expression is largely limited to erythroid, megakaryocytic, mast cells, and myeloid cells (Toki et al., 1996). NF-E2 binds to sequences that include an AP1 site plus one additional residue. These sequences are found in the globin gene promoters and, as mentioned above, play a critical role in the function of globin LCRs. It was therefore a considerable surprise when NF-E2 large subunit knockout mice displayed apparently normal erythropoiesis but were thrombocytopenic. Knockout of the small MAF subunit had no major effect on erythropoiesis or on thrombopoiesis (Kotkow and Orkin, 1996), suggesting that other gene products may substitute for its function. Interestingly, a member of the MAF family has recently been implicated in the specific expression of IL-4 in the T-cell subset Th2 genes (Ho et al., 1996).

PAX-5 (BSAP). PAX genes (see p. 101) play fundamental and remarkably conserved roles in development. In the present context, the PAX-5 gene is of particular interest. PAX-5 is expressed specifically in B-cell development but not in mature plasma cells. It is also expressed in the developing nervous system in the midbrain and in the spinal cord, and in adult testes. Potential binding sites for the PAX gene have been identified in the genes for a number of B-cell proteins, and the gene may also be involved in stimulating immunoglobulin rearrangements (Busslinger and Urbanek, 1995; Michaelson et al., 1996). Knockout of the PAX-5 gene produces runted mice, most of which die within 3 weeks of birth (Urbanek et al., 1994). The animals showed abnormalities of midbrain and cerebellum development, consistent with a role for the PAX-5 gene in regulating the proliferation of certain neurons. The animals entirely lacked conventional or CD5 B cells as well as plasma cells and did not produce circulating immunoglobulins. However, they had relatively normal numbers of large $B220^+$ B-cell precursors, indicating that the block was not in B-cell determination but in the maturation of B-cell precursors. The B cells completely lacked the PAX-5 target CD19 and showed J or D–H rearrangements in their immunoglobulin genes, but were deficient in V–J rearrangement (Nutt et al., 1997).

Lineage-Specific HMG Box Genes. As discussed above, a group of proteins that share the HMG box motif probably act by binding to DNA in the minor groove and causing the bending of DNA, expediting the binding and interaction of other factors at active promoters. In addition to the ubiquitous HMGI(Y) protein, there are several lineage-specific HMG box proteins. The subfamily of these genes is called SOX genes, based on their resemblance to the sex-determining factor SRY. SOX-4 in particular is an HMG box gene that is expressed at several sites in murine embryogenesis but only in B and T cells of adult animals (van de Wetering et al., 1993). Knockout of the SOX-4 gene produces embryonic lethality at day 14 due to impaired formation of the heart. However, when lethally irradiated mice were reconstituted with cells from $SOX\text{-}4^{-/-}$ mice, an extensive but not absolute block in B-cell development was observed at the pro-B-cell stage in which D–J rearrangements start to occur (Schilham et al., 1996). This block has been described as being at a later stage than that in mice lacking E2A or EBF (see below) but is at a similar stage as that seen in mice lacking PAX-5.

Two structurally similar HMG box genes, LEF-1 and TCF, are expressed in T-lymphoid cells of adult mice. Both proteins are also expressed in partly nonoverlapping patterns in several nonhematopoietic tissues during mouse embryogenesis and in the adult testes (Oosterwegel et al., 1993). TCF-1 is expressed in the thymus at 100-fold higher levels than LEF-1, while LEF-1 is also expressed in pre-B cells. These proteins share a virtually identical HMG box and may be derived from a recent gene duplication event, as the chicken has only a single homolog (Gastrop et al., 1992). TCF-1 knockout mice show a severe impairment in T-cell development, with the cells largely blocked at the immature single positive thymocyte stage (Verbeek et al., 1995). The block is leaky, however, and some mature T cells do develop. In contrast, LEF-1 knockout mice have an intact immune system, although they show abnormalities in hair follicle, tooth, mammary gland, and neural development. Interestingly, LEF-1 interacts with β-catenin in a yeast two-hybrid system and forms an intranuclear complex with β-catenin in animal cells (Gastrop et al., 1992), raising the possibility that it functions in cell signaling as well as in transcriptional regulation. Recently, a protein that specifically bends DNA in the embryonic globin gene promoter has been described (Dyer et al., 1996), and presumably "architectural proteins" will ultimately be shown to be part of the complexes assembled on many, if not most, promoters.

EBF. The early B-cell factor (EBF) is a zinc finger transcription factor that is expressed in all stages of the B-cell lineage except for differentiated plasma cells (Hagman et al., 1993). EBF knockout mice have a complete lack of mature B cells, but show expression of the germline mu gene and contain B-cell precursors expressing CD43 and the B220 lineage marker cell surface antigen (Lin and Grosschedl, 1995).

EKLF. EKLF (the erythroid-specific Kruppel-like factor) is a zinc finger–containing transcription factor that was originally identified through differential screening of erythroid-specific clones (Miller and Bieker, 1993). It is expressed specifically in erythroid cells at all stages of development (Southwood et al., 1996) and in lower levels in mast cells. This factor binds the sequence CACCC that lies upstream of the β-globin CAAT sequence and resembles a binding site for the general transcription factor SP1. EKLF is present at about threefold greater abundance in adult relative to fetal erythroid cells. Knockout of the EKLF gene leads to a major defect in production of murine adult β-globin (Nuez et al., 1995; Perkins et al., 1995) and a complete absence of production of human β-globin from a transgene. Interestingly, there is a reciprocal severalfold increase in the level of transgenic γ-globin gene mRNA accumulation in the knockout mice (Perkins et al., 1995; Wijgerde et al., 1996) and a shift towards γ-globin production is seen even in the heterozygous knockout animal. Mutations in the EKLF-binding site that interfere with binding also produce β-thalassemia in humans (Feng et al., 1994), but it is not clear that the change in levels of EKLF during development is sufficient to account for globin switching. There are several other transcription factors that belong to the same family on the basis of homology in the DNA-binding domain, and it is possible that some EKLF-binding sites but remain active in the knockout mouse.

Ikaros. The Ikaros gene encodes several alternatively spliced forms of a zinc finger transcription factor (Molnar and Georgopoulos, 1994). The gene is expressed in both developing B and T cells, particularly in maturing thymocytes (Molnar et al., 1996), and there are high-affinity binding sites for Ikaros in the promoter regions of a number of lymphocyte-specific genes, including CD3 components, RAG-1 recombination mediating protein, terminal nucleotide transferase, the immunoglobulin heavy chain locus, and the immunoglobulin-associated mb-1 gene. Ikaros knockout mice completely lack T or B cells or their progenitors but have normal or supranormal levels of erythroid and myeloid cells (Georgopoulos et al., 1994; Wang et al., 1996).

Egr-1. Egr-1 (Gashler and Sukhatme, 1995) is a widely expressed early response zinc finger gene that was isolated as a clone that became expressed upon induction of macrophage differentiation of the HL60 cell line. HL60 cells constitutively producing Egr-1 could no longer be induced to form myeloid cells, and antisense Egr-1 nucleotides blocked macrophage differentiation of normal or leukemic myeloid cells (Nguyen et al., 1993).

CONCLUSIONS

The reductionist approach of analysis of transcription factors has provided a great deal of knowledge of the role of various factors in differentiation of hematopoietic lineages. Much less is known about the role of transcription factors in initiating differentiation of stem cells or maintaining cells in a pluripotent state. This is due to the lack of sources of large amounts of pure stem cells or their progenitors and the paucity of genes that have been identified to date as specific markers for these early hematopoietic precursors. In addition, the approach of studying factors in depth singly, valuable as it is, does not deal with differentiative phenomena that result from interactions among many or all the factors binding to specific promoters or enhancers, including both factors that directly recognize specific DNA sequences and other proteins that may bind either to single transcription factors or to complexes as they are assembled on regulatory DNA regions.

The former difficulty will presumably be at least partially overcome as newer methods for examining gene expression in smaller numbers of cells as well as better cell purification and propagation methods evolve. The latter difficulty is common to the study of the transcriptional regulation of most genes and will presumably occupy molecular biologists for some years to come.

REFERENCES

Achkar CC, Derguini F, Blumberg B, Langston A, Levin AA, Speck J, Evans RM, Bolado J Jr, Nakanishi K, Buck J, et al. (1996): 4-Oxoretinol, a new natural ligand and transactivator of the retinoic acid receptors. *Proc Natl Acad Sci USA* 93:4879–4884.

Alkema MJ, Bronk M, Verhoeven E, Otte A, van't Veer LJ, Berns A, van Lohuizen M (1997): Identification of Bmi1-interacting proteins as constituents of a multimeric mammalian polycomb complex. *Genes Dev* 11:226–240.

Amon A (1996): Mother and daughter are doing fine: Asymmetric cell division in yeast. *Cell* 84:651–654.

Amrolia PJ, Cunningham JM, Ney P, Nienhuis AW, Jane SM (1995): Identification of two novel regulatory elements within the 5′-untranslated region of the human A gamma-globin gene. *J Biol Chem* 270:12892–12898.

Ashwell JD, King LB, Vacchio MS (1996): Cross-talk between the T cell antigen receptor and the glucocorticoid receptor regulates thymocyte development. *Stem Cells* 14:490–500.

Auerbach R, Huang H, Lu L (1996): Hematopoietic stem cells in the mouse embryonic yolk sac. *Stem Cells* 14:269–280.

Ault KT, Dirksen ML, Jamrich M (1966): Novel homeobox gene pv.1 mediates induction of ventral mesoderm in *Xenopus* embryos. *Proc Natl Acad Sci USA* 93:6415–6420.

Bannister AJ, Kouzarides T (1996): The CBP co-activator is a histone acetyltransferase. *Nature* 384:641–643.

Bartek J, Bartkova J, Lukas J (1996): The retinoblastoma protein pathway and the restriction point. *Curr Opin Cell Biol* 8:805–814.

Beato M, Herrlich P, Schutz G (1995): Steroid hormone receptors: Many actors in search of a plot. *Cell* 83:851–857.

Beato M, Truss M, Chavez S (1996): Control of transcription by steroid hormones. *Ann NY Acad Sci* 784:93–123.

Begley CG, Aplan PD, Denning SM, Haynes BF, Waldmann TA, Kirsch IR (1989): The gene SCL is expressed during early hematopoiesis and encodes a differentiation-related DNA-binding motif. *Proc Natl Acad Sci USA* 86:10128–10132.

Benezra R (1994): An intermolecular disulfide bond stabilizes E2A homodimers and is required for DNA binding at physiological temperatures. *Cell* 79:1057–1067.

Benezra R, Davis RL, Lockshon D, Turner DL, Weintraub H (1990): The protein Id: A negative regulator of helix-loop-helix DNA binding proteins. *Cell* 61:49–59.

Bentley D, Selfridge J, Millar JK, Samuel K, Hole N, Ansell JD, Melton DW (1996): DNA ligase I is required for fetal liver erythropoiesis but is not essential for mammalian cell viability. *Nature Genet* 13:489–491.

Blobel GA, Orkin SH (1996): Estrogen-induced apoptosis by inhibition of the erythroid transcription factor GATA-1. *Mol Cell Biol* 16:1687–1694.

Blobel GA, Sieff CA, Orkin SH (1995): Ligand-dependent repression of the erythroid transcription factor GATA-1 by the estrogen receptor. *Mol Cell Biol* 15:3147–3153.

Blom van Assendelft G, Hanscombe O, Grosveld F, Greaves DR (1989): The beta-globin dominant control region activates homologous and heterologous promoters in a tissue-specific manner. *Cell* 56:969–977.

Bobola N, Jansen RP, Shin TH, Nasmyth K (1996): Asymmetric accumulation of Ash1p in postanaphase nuclei depends on a myosin and restricts yeast mating-type switching to mother cells. *Cell* 84:699–709.

Borden KL, Freemont PS, (1996): The RING finger domain: A recent example of a sequence-structure family. *Curr Opin Struct Biol* 6:395–401.

Borrow J, Shearman A, Stanton V, et al (1996): The t(7;11)(p15;p15) translocation in acute myeloid leukemia fuses the genes for nucleoporin NUP98 and class I homeoprotein HOXA9. *Nature Genet* 12:159–167.

Botas J (1993): Control of morphogenesis and differentiation by HOM/Hox genes. *Curr Opin Cell Biol* 5:1015–1022.

Boyes J, Felsenfeld G (1996): Tissue-specific factors additively increase the probability of the all-or-none formation of a hypersensitive site. *EMBO J* 15:2496–2507.

Briegel K, Bartunek P, Stengl G, Lim KC, Beug H, Engel JD, Zenke M (1996): Regulation and function of transcription factor gata-1 during red blood cell differentiation. *Development* 122:3839–3850.

Brownell JE, Allis CD (1996): Special HATs for special occasions: Linking histone acetylation to chromatin assembly and gene activation. *Curr Opin Genet Dev* 6:176–184.

Brownell JE, Zhou J, Ranalli T, Kobayashi R, Edmondson DG, Roth SY, Allis CD (1996): *Tetrahymena* histone acetyltransferase A: A homolog to yeast Gcn5p linking histone acetylation to gene activation. *Cell* 84:843–851.

Busslinger M, Urbanek P (1995): The role of BSAP (Pax-5) in B-cell development. *Curr Opin Genet Dev* 5:595–601.

Cairns BR, Lorch Y, Li Y, Zhang M, Lacomis L, Erdjument-Bromage H, Tempst P, Du J, Laurent B, Kornberg RD (1996): RSC, an essential, abundant chromatin-remodeling complex. *Cell* 87:1249–1260.

Caldenhoven E, Liden J, Wissink S, Van de Stolpe A, Raaijmakers J, Koenderman L, Okret S, Gustafsson JA, Van der Saag PT (1995): Negative cross-talk between RelA and the glucocorticoid receptor: A possible mechanism for the antiinflammatory action of glucocorticoids. *Mol Endocrinol* 9:401–412.

Calkhoven CF, Ab G (1996): Multiple steps in the regulation of transcription-factor level and activity. *Biochem J* 317:329–342.

Carver-Moore K, Broxmeyer HE, Luoh SM, Cooper S, Peng J, Burstein SA, Moore MW, de Sauvage FJ (1996): Low levels of erythroid and myeloid progenitors in thrombopoietin– and c-mpl–deficient mice. *Blood* 88:803–808.

Cepek KL, Chasman DI, Sharp PA (1996): Sequence-specific DNA binding of the B-cell-specific coactivator OCA-B. *Genes Dev* 10:2079–2088.

Chang CP, Brochieri L, Shen WF, Largman C, Cleary ML (1996): Pbx modulation of Hox homeodomain amino-terminal arms establishes different DNA-binding specificities across the Hox locus. *Mol Cell Biol* 16:1734–1745.

Chang TJ, Scher BM, Waxman S, Scher W (1993): Inhibition of mouse GATA-1 function by the glucocorticoid receptor: Possible mechanism of steroid inhibition of erythroleukemia cell differentiation [published erratum appears in Mol Endocrinol 1993 7(6):786]. *Mol Endocrinol* 7:528–542.

Chen H, Ray-Gallet D, Zhang P, Heterington CJ, Gonzalez DA, Zhang DE, Moreau-Gachelin F, Tenen DG (1995): PU.1 (Spi-1) autoregulates its expression in myeloid cells. *Oncogene* 11:1549–1560.

Chen PL, Riley DJ, Chen-Kiang S, Lee WH (1996): Retinoblastoma protein directly interacts with and activates the transcription factor NF-IL6. *Proc Natl Acad Sci USA* 93:465–469.

Chen PL, Riley DJ, Lee WH (1995): The retinoblastoma protein as a fundamental mediator of growth and differentiation signals. *Crit Rev Eukaryot Gene Exp* 5:79–95.

Chen XD, Turpen JB (1995): Intraembryonic origin of hepatic hematopoiesis in *Xenopus laevis*. *J Immunol* 154:2557–2567.

Chen ZJ, Parent L, Maniatis T (1996): Site-specific phosphorylation of IkappaBalpha by a novel ubiquitination-dependent protein kinase activity. *Cell* 84:853–862.

Cheng XB, Reginato MJ, Andrews NC, Lazar MA (1997): The trancriptional integrator creb-binding protein mediates positive cross talk between nuclear hormone receptors and the hematopoietic bzip protein p45/nf-e2. *Mol Cell Biol* 17:1407–1416.

Choo Y, Sanchez-Garcia I, Klug A (1994): *In vivo* repression by a site-specific DNA-binding protein designed against an oncogenic sequence. *Nature* 372:642–645.

Chu TY, Besmer P, (1995): Characterization of the promoter of the proto-oncogene c-kit. *Proc Natl Sci Counc Repub China B* 19:8–18.

Chung JH, Bell AC, Felsenfeld G (1997): Characterization of the chicken beta-globin insulator. *Proc Natl Acad Sci USA* 94:575–580.

Chung JH, Whiteley M, Felsenfeld G (1993): A 5′ element of the chicken beta-globin domain serves as an insulator in human erythroid cells and protects against position effect in *Drosophila*. *Cell* 74:505–514.

Cogswell JP, Cogswell PC, Kuehl WM, Cuddihy AM, Bender TM, Engelke U, Marcu KB, Ting JP (1993): Mechanism of c-myc regulation by c-Myb in different cell lineages. *Mol Cell Biol* 13:2858–2869.

Cohen KJ, Hanna JS, Prescott JE, Dang CV (1996): Transformation by the Bmi-1 oncoprotein correlates with its subnuclear localization but not its transcriptional suppression activity. *Mol Cell Biol* 16:5527–5535.

Corcoran LM, Karvelas M, Nossal GJ, Ye ZS, Jacks T, Baltimore D (1993): Oct-2, although not required for early B-cell development, is critical for later B-cell maturation and for postnatal survival. *Genes Dev* 7:570–582.

Cress WD, Triezenberg SJ (1991): Critical structural elements of the VP16 transcriptional activation domain. *Science* 251:87–90.

Davis AC Wims M, Spotts GD, Hann SR, Bradley A (1993): A null c-myc mutation causes lethality before 10.5 days of gestation in homozygotes and reduced fertility in heterozygous female mice. *Genes Dev* 7:671–682.

Dear TN, Colledge WH, Carlton MB, Lavenir I, Larson T, Smith AJ, Warren AJ, Evans MJ, Sofroniew MV, Rabbitts TH (1995): The Hox11 gene is essential for cell survival during spleen development. *Development* 121:2909–2915.

de Sauvage FJ, Carver-Moore K, Luoh SM, Ryan A, Dowd M, Eaton DL, Moore MW (1996): Physiological regulation of early and late stages of megakaryocytopoiesis by thrombopoietin. *J Exp Med* 183:651–656.

Dexter TM, Spooncer E (1987): Growth and differentiation in the haemopoietic system. *Annu Rev Cell Biol* 3:423–441.

Dieterlen-Lievre F, Godin I, Pardanaud L (1997): Where do hematopoietic stem cells come from? *Int Arch Allergy Immunol* 112:3–8.

Dillon N, Grosveld F (1993): Transcriptional regulation of multigene loci: Multilevel control. *Trends Genet* 9:134–137.

Donze D, Townes TM, Bieker JJ (1995): Role of erythroid Kruppel-like factor in human gamma- to beta-globin gene switching. *J Biol Chem* 270:1955–1959.

Downing JR, Look AT (1996): MLL fusion genes in the 11q23 acute leukemias. *Cancer Treatment Res* 84:73–92.

Dranoff G, Mulligan RC (1994): Activities of granulocyte-macrophage colony-stimulating factor revealed by gene transfer and gene knockout studies. *Stem Cells* 1:173–182.

Dyer MA, Naidoo R, Hayes RJ, Larson CJ, Verdine GL, Baron MH (1996): A DNA-bending protein interacts with an essential upstream regulatory element of the human embryonic beta-like globin gene. *Mol Cell Biol* 16:829–838.

Elefanty AG, Antoniou M, Custodio N, Carmo-Fonseca M, Grosveld FG (1996): GATA transcription factors associate with a novel class of nuclear bodies in erythroblasts and megakaryocytes. *EMBO J* 15:319–333.

Faibairn L, Cowling G, Reipert B, Dexter T (1993): Suppression of apoptosis allows differentiation and development of a multipotent hemopoietic cell line in the absence of added growth factors. *Cell* 74:823–832.

Feng WC, Southwood CM, Bieker JJ (1994): Analyses of beta-thalassemia mutant DNA interactions with erythroid Kruppel-like factor (EKLF), an erythroid cell-specific transcription factor. *J Biol Chem* 269:1493–1500.

Fidanza V, Melotti P, Yano T, Nakamura T, Bradley A, Canaani E, Calabretta B, Croce CM (1996): Double knockout of the ALL-1 gene blocks hematopoietic differentiation *in vitro. Cancer Res* 56:1179–1183.

Fields S, Song O (1989): A novel genetic system to detect protein–protein interactions. *Nature* 340:245–246.

Fontes JD, Jabrane-Ferrat N, Toth CR, Peterlin BM (1996): Binding and cooperative interactions between two B cell–specific transcriptional coactivators. *J Exp Med* 183:2517–2521.

Forrester WC, Takegawa S, Papayannopoulou T, Stamatoyannopoulos G, Groudine M (1987): Evidence for a locus activation region: The formation of developmentally stable hypersensitive sites in globin-expressing hybrids. *Nucleic Acids Res* 15:10159–10177.

Fraser RA, Rossignol M, Heard DJ, Egly JM, Chanbon (1997): Sug1, a putative transcriptional mediator and subunit of the pa700 proteasome regulatory complex, is a DNA helicase. *J Biol Chem* 272:7122–7126.

Fu SL, Lipsick JS (1997): Constitutive expression of full-length c-myb transforms avian cells characteristic of both the monocytic and granulocytic lineages. *Cell Growth Differ* 8:35–45.

Galaktionov K, Chen X, Beach D (1996): Cdc25 cell-cycle phosphatase as a target of c-myc. *Nature* 382:511–517.

Gashler A, Sukhatme VP (1995): Early growth response protein 1 (Egr-1): Prototype of a zinc-finger family of transcription factors. *Prog Nucleic Acids Res Mol Biol* 50:191–224.

Gastrop J, Hoevenagel R, Young JR, Clevers HC (1992): A common ancestor of the mammalian transcription factors TCF-1 and TCF-1 alpha/LEF-1 expressed in chicken T cells. *Eur J Immunol* 22:1327–1330.

Georgopoulos K, Bigby M, Wang JH Molnar A, Wu P, Winandy S, Sharpe A (1994): The Ikaros gene is required for the development of all lymphoid lineages. *Cell* 79:143–156.

Ghozi MC, Bernstein Y, Negreanu V, Levanon D, (1996): Expression of the human acute myeloid leukemia gene AML1 is regulated by two promoter regions. *Proc Natl Acad Sci USA* 93:1935–1940.

Gill GN (1995): The enigma of LIM domains. *Structure* 3:1285–1289.

Glass CK (1994): Differential recognition of target genes by nuclear receptor monomers, dimers, and heterodimers. *Endocr Rev* 15:391–407.

Godin IE, Garcia-Porrero JA, Coutinho A, Dieterlen-Lievre F, Marcos MA (1993): Para-aortic splanchnopleura from early mouse embryos contains B1a cell progenitors. *Nature* 364:67–70.

Graf T (1992): Myb: A transcriptional activator linking proliferation and differentiation in hematopoietic cells [published erratum appears in Curr Opin Genet Dev 1992 2(3):504]. *Curr Opin Genet Dev* 2:249–255.

Green AR, Lints T, Visvader J, Harvey R, Begley CG (1992): SCL is coexpressed with GATA-1 in hemopoietic cells but is also expressed in developing brain [published erratum appears in Oncogene 1992 7(7):1459]. *Oncogene* 7:653–660.

Grosveld F, Greaves D, Philipsen S, Talbot D, Pruzina S, deBoer E, Hanscombe O, Belhumeur P, Hurst J, Fraser P, et al (1990): The dominant control region of the human beta-globin domain. *Ann NY Acad Sci* 612:152–159.

Gstaiger M, Georgiev O, van Leeuwen H, van der Vliet P, Schaffner W (1996): The B cell coactivator Bob1 shows DNA sequence-dependent complex formation with Oct-1/Oct-2 factors, leading to differential promoter activation. *EMBO J* 15:2781–2790.

Hagman J, Belanger C, Travis A, Turck CW, Grosschedl R (1993): Cloning and functional characterization of early B-cell factor, a regulator of lymphocyte-specific gene expression. *Genes Dev* 7:760–773.

Hagstrom K, Muller M, Schedl P (1996): Fab-7 functions as a chromatin domain boundary to ensure proper segment specification by the *Drosophila* bithorax complex. *Genes Dev* 10:3202–3215.

Hampsey M, Reinberg D (1997): Transcription—why are tafs essential? *Curr Biol* 7:46.

Harper SE, Qiu Y, Sharp PA (1996): Sin3 corepressor function in Myc-induced transcription and transformation. *Proc Natl Acad Sci USA* 93:8536–8540.

Hatano M, Boberts C, Minden M, et al (1991): Deregulation of a homeobox gene, HOX1, by the t(10;14) in T cell leukemia. *Science* 253:79–82.

Hawley RG, Fong AZ, Lu M, Hawley TS (1994): The HOX11 homeobox-containing gene of human leukemia immortalizes murine hematopoietic precursors. *Oncogene* 9:1–12.

Helgason CD, Sauvageau G, Lawrence HJ, Largman C, Humphries RK (1996): Overexpression of HOXB4 enhances the hematopoietic potential of embryonic stem cells differentiated *in vitro. Blood* 87:2740–2749.

Herr W, Cleary MA (1995): The POU domain: Versatility in transcriptional regulation by a flexible two-in-one DNA-binding domain. *Genes Dev* 9:1679–1693.

Ho IC, Hodge MR, Rooney JW, Glimcher LH (1996): The proto-oncogene c-maf is responsible for tissue-specific expression of interleukin-4. *Cell* 85:973–983.

Hoang T, Paradis E, Brady G, Billia F, Nakahara K, Iscove NN, Kirsch IR (1996): Opposing effects of the basic helix-loop-helix transcription factor SCL on erythroid and monocytic differentiation. *Blood* 87:102–111.

Hobert O, Jallal B, Ullrich A (1996): Interaction of Vav with ENX-1, a putative transcriptional regulator of homeobox gene expression. *Mol Cell Biol* 16:3066–3073.

Holland PW, Garcia-Fernandez J (1996): Hox genes and chordate evolution. *Dev Biol* 173:382–395.

Horvitz HR, Herskowitz I (1992): Mechanisms of asymmetric cell division: Two Bs or not two Bs, that is the question. *Cell* 68:237–255.

Hu M, Krause D, Greaves M, Sharkis S, Dexter M, Heyworth C, Enver T (1997): Multilineage gene expression precedes commitment in the hemopoietic system. *Genes Dev* 11:774–785.

Hurlin PJ, Ayer DE, Grandori C, Eisenman RN (1994): The Max transcription factor network: Involvement of Mad in differentiation and an approach to identification of target genes. *Cold Spring Harbor Symp Quant Biol* 59:109–116.

Igarashi K, Itoh K, Motohashi H, Hayashi N, Matuzaki Y, Nakauchi H, Nishizawa M, Yamamoto M (1995): Activity and expression of murine small Maf family protein MafK. *J Biol Chem* 270:7615–7624.

Ihle JN (1996): Janus kinases in cytokine signalling. *Philos Trans R Soc Lond B Biol Sci* 351:159–166.

Introna M, Luchetti M, Castellano M, Arsura M, Golay J (1994): The myb oncogene family of transcription factors: Potent regulators of hematopoietic cell proliferation and differentiation. *Semin Cancer Biol* 5:113–124.

Jacobson MD, Weil M, Raff MC (1997): Programmed cell death in animal development. *Cell* 88:347–354.

Jane SM, Nienhuis AW, Cunningham JM (1995): Hemoglobin switching in man and chicken is mediated by a heteromeric complex between the ubiquitous transcription factor CP2 and a developmentally specific protein [published erratum appears in EMBO J 1995 14(4):854]. *EMBO J* 14:97–105.

Janknecht R, Hunter T (1996): Transcriptional control—Versatile molecular glue. *Curr Biol* 6:951–954.

Jansen RP, Dowzer C, Michaelis C, Galova M, Nasmyth K (1996): Mother cell-specific HO expression in budding yeast depends on the unconventional myosin myo4p and other cytoplasmic proteins. *Cell* 84:687–697.

Joh T, Kagami Y, Yamamoto K, Segawa T, Takizawa J, Takahashi T, Ueda R, Seto M (1996): Identification of MLL and chimeric MLL gene products involved in 11q23 translocation and possible mechanisms of leukemogenesis by MLL truncation. *Oncogene* 13:1945–1953.

Jones R, Wagner J, Celano P, Zicha M, Sharkis S (1990): Separation of pluripotent hematopoietic stem cells from spleen colony-forming cells. *Nature* 347:188.

Kallianpur AR, Jordan JE, Brandt SJ (1994): The SCL/TAL-1 gene is expressed in progenitors of both the hematopoietic and vascular systems during embryogenesis. *Blood* 83:1200–1208.

Kamei Y, Xu L, Heinzel T, Torchia J, Kurokawa R, Gloss B, Lin SC, Heyman RA, Rose DW, Glass CK, Rosenfeld MG (1996): A CBP integrator complex mediates transcriptional activation and AP-1 inhibition by nuclear receptors. *Cell* 85:403–414.

Kamps M, Look T, Baltimore D (1991): The human t(1;19) translocation in pre-B ALL produces multiple nuclear E2A-Pbx1 fusion proteins with differing transforming potentials. *Genes Dev* 5:358–368.

Kappen C, Ruddle FH (1993): Evolution of a regulatory gene family: HOM/HOX genes. *Curr Opin Genet Dev* 3:931–938.

Kawabe T, Muslin AJ, Korsmeyer SJ (1997): HOX11 interacts with protein phosphatases PP2A and PP1 and disrupts a G2/M cell-cycle checkpoint. *Nature* 385:454–458.

Kennison JA (1995): The Polycomb and trithorax group proteins of *Drosophila:* Transregulators of homeotic gene function. *Annu Rev Genet* 29:289–303.

Kern I, Steimle V, Siegrist CA, Mach B (1995): The two novel MHC class II transactivators RFX5 and CIITA both control expression of HLA-DM genes. *Int Immunol* 7:1295–1299.

Kieran MW, Perkins AC, Orkin SH, Zon LI (1996): Thrombopoietin rescues *in vitro* erythroid colony formation from mouse embryos lacking the erythropoietin receptor. *Proc Natl Acad Sci USA* 93:9126–9131.

Knegtel RM, van Tilborg MA, Boelens R, Kaptein R (1995): NMR structural studies on the zinc finger domains of nuclear hormone receptors. *Exs* 73:279–295.

Kollia P, Fibach E, Najjar SM, Schechter AN, Noguchi CT (1996): Modifications of RNA processing modulate the expression of hemoglobin genes. *Proc Natl Acad Sci USA* 93:5693–5698.

Konig H, Pfisterer P, Corcoran LM, Wirth T (1995): Identification of CD36 as the first gene dependent on the B-cell differentiation factor Oct-2. *Genes Dev* 9:1598–1607.

Kopf M, Brombacher F, Hodgkin P, Ramsay A, Milbourn E, Dai W, Ovington K, Ca B, Kohler G, Young I, Matthaei K (1996): IL-5–deficient mice have a developmental defect in $CD5^+$ B-1 cells and lack eosinophilia but have normal antibody and cytotoxic T cell responses. *Immunity* 4:15.

Kotkow KJ, Orkin SH (1996): Complexity of the erythroid transcription factor NF-E2 as revealed by gene targeting of the mouse p18 NF-E2 locus. *Proc Natl Acad Sci USA* 93:3514–3518.

Krause D, et al (1994): *Blood* 84:691.

Krumlauf R (1993): Mouse Hox genetic functions. *Curr Opin Genet Dev* 3:621–625.

Krumlauf R (1994): Hox genes in vertebrate development. *Cell* 78:191–201.

Kuo MH, Brownell JE, Sobel RE, Ranalli TA, Cook RG, Edmondson DG, Roth SY, Allis CD (1996): Transcription-linked acetylation by Gcn5p of histones H3 and H4 at specific lysines. *Nature* 383:269–272.

Ladher R, Mohun TJ, Smith JC, Snape AM (1996): Xom—A *xenopus* homeobox gene that mediates the early effects of bmp-4. *Development* 122:2385–2394.

Laemmli UK, Tjian R (1996): A nuclear traffic jam: Unraveling multicomponent machines and compartments. *Curr Opin Cell Biol* 8:299–302.

Landsberger N, Wolffe AP (1995): Role of chromatin and *Xenopus laevis* heat shock transcription factor in regulation of transcription from the *X. laevis* hsp70 promoter *in vivo*. *Mol Cell Biol* 15:6013–6024.

Latchman DS (1996): The Oct-2 transcription factor. *Int J Biochem Cell Biol* 28:1081–1083.

Lawrence H, Helgason C, Sauvageau G, et al (1995): Mice homozygous for a targeted disruption of the homeobox gene HOXA9 have defects in lymphoid and myeloid hematopoiesis. *Blood* 86:254a.

Lawrence HJ, Sauvageau G, Humphries RK, Largman C (1996): The role of HOX homeobox genes in normal and leukemic hematopoiesis. *Stem Cells* 14:281–291.

Lawrence PA, Morata G (1994): Homeobox genes: Their function in *Drosophila* segmentation and pattern formation. *Cell* 78:181–189.

Lee FS, Hagler J, Chen ZJ, Maniatis T (1997): Activation of the IkappaB alpha kinase complex by MEKK1, a kinase of the JNK pathway. *Cell* 88:213–222.

Lichty BD, Ackland-Snow J, Noble L, Kamel-Reid S, Dube ID (1995): Dysregulation of HOX11 by chromosome translocations in T-cell acute lymphoblastic leukemia: A paradigm for homeobox gene involvement in human cancer. *Leuk Lymphoma* 16:209–215.

Liebermann DA, Hoffman B (1994): Differentiation primary response genes and proto-oncogenes as positive and negative regulators of terminal hematopoietic cell differentiation. *Stem Cells* 12:352–369.

Lieschke G, Grail D, Hodgson G, Metcalf D, Stanley E, Cheers C, Fowler K, Basu S, Zhan Y, Dunn A (1994a): Mice lacking granulocyte colony-stimulating factor have chronic neutropenia, granulocyte and macrophage progenitor cell deficiency, and impaired neutrophil mobilization. *Blood* 84:1737.

Lieschke GJ, Stanley E, Grail D, Hodgson G, Sinickas V, Gall JA, Sinclair RA, Dunn AR (1994b): Mice lacking both macrophage and granulocyte-macrophage colony-stimulating factor have macrophages and coexistent osteopetrosis and severe lung disease. *Blood* 84:27–35.

Lin CS, Lim SK, D'Agati V, Costantini F (1996): Differential effects of an erythropoietin receptor gene disruption on primitive and definitive erythropoiesis. *Genes Dev* 10:154–164.

Lin H, Grosschedl R (1995): Failure of B-cell differentiation in mice lacking the transcription factor EBF. *Nature* 376:263–267.

Lin HH, Sternfeld DC, Shinpock SG, Popp RA, Mucenski ML (1996): Functional analysis of the c-myb proto-oncogene. *Curr Top Microbiol Immunol* 211:79–87.

Littlewood TD, Evan GI (1995): Transcription factors 2: Helix-loop-helix. *Protein Profile* 2:621–702.

Liu M, Lee MH, Cohen M, Bommakanti M, Freedman LP (1996): Transcriptional activation of the Cdk inhibitor p21 by vitamin D3 leads to the induced differentiation of the myelomonocytic cell line U937. *Genes Dev* 10:142–153.

Lovering R, Hanson IM, Borden KL, Martin S, O'Reilly NJ, Evan GI, Rahman D, Pappin DJ, Trowsdale J, Freemont PS (1993): Identification and preliminary characterization of a protein motif related to the zinc finger. *Proc Natl Acad Sci USA* 90:2112–9116.

Lumelsky NL, Forget BG (1991): Negative regulation of globin gene expression during megakaryocytic differentiation of a human erythroleukemic cell line. *Mol Cell Biol* 11:3528–3536.

Luo Y, Fujii H, Gerster T, Roeder RG (1992): A novel B cell–derived coactivator potentiates the activation of immunoglobulin promoters by octamer-binding transcription factors. *Cell* 71:231–241.

Mach B, Steimle V, Martinez-Soria E, Reith W (1996): Regulation of MHC class II genes: Lessons from a disease. *Annu Rev Immunol* 14:301–331.

Maconochie M, Nonchev S, Morrison A, Krumlauf R (1996): Parologous hox genes—Function and regulation *Annu Rev Genet* 30:529–556.

Mangelsdorf DJ, Evans RM (1995): The RXR heterodimers and orphan receptors. *Cell.* 83:841–850.

Mansouri A, Hallonet M, Gruss P (1996): Pax genes and their role in cell differentiation and development. *Curr Opin Cell Biol* 8:851–857.

Marcu KB, Bossone SA, Patel AJ (1992): Myc function and regulation. *Annu Rev Biochem* 61:809–860.

Mbangkollo D, Burnett R, McCabe N, Thirman M, Gill H, Yu H, Rowley JD, Diaz MO (1995): The human MLL gene: Nucleotide sequence, homology to the *Drosophila* trx zinc-finger domain, and alternative splicing. *DNA Cell Biol* 14:475–483.

McEwan IJ, Dahlman-Wright K, Ford J, Wright AP (1996): Functional interaction of the c-Myc transactivation domain with the TATA binding protein: Evidence for an induced fit model of transactivation domain folding. *Biochemistry* 35:9584–9593.

Mead PE, Brivanlou IH, Kelley CM, Zon LI (1996): BMP-4–responsive regulation of dorsal-ventral patterning by the homeobox protein Mix.1. *Nature* 382:357–360.

Medvinsky AL, Samoylina NL, Muller AM, Dzierzak EA (1993): An early pre-liver intraembryonic source of CFU-S in the developing mouse. *Nature* 364:64–67.

Melotti P, Calabretta B (1996): Induction of hematopoietic commitment and erythromyeloid differentiation in embryonal stem cells constitutively expressing c-myb. *Blood* 87:2221–2234.

Merika M, Orkin SH (1995): Function synergy and physical interactions of the erythroid transcription factor GATA-1 with the Kruppel family proteins Sp1 and EKLF. *Mol Cell Biol* 15:2437–2447.

Metcalf D (1989): The molecular control of cell division, differentiation, commitment, and maturation in haemopoietic cells. *Nature* (*Lond*) 339:27–30.

Metcalf D, Robb L, Dunn AR, Mifsud S, Di Rago L (1996): Role of granulocyte-macrophage colony-stimulating factor and granulocyte colony-stimulating factor in the development of an acute neutrophil inflammatory response in mice. *Blood* 88:3755–3764.

Michaelson JS, Singh M, Birshtein BK (1996): B cell lineage-specific activator protein (BSAP). A player at multiple stages of B cell development. *J Immunol* 156:2349–2351.

Michelotti GA, Michelotti EF, Pullner A, Duncan RC, Eick D, Levens D (1996): Multiple single-stranded cis elements are associated with activated chromatin of the human c-myc gene *in vivo. Mol Cell Biol* 16:2656–2669.

Miller IJ, Bieker JJ (1993): A novel, erythroid cell–specific murine transcription factor that binds to the CACCC element and is related to the Kruppel family of nuclear proteins. *Mol Cell Biol* 13:2776–2786.

Milot E, Strouboulis J, Trimborn T, Wijgerde M, de Boer E, Langeveld A, Tan-Un K, Vergeer W, Yannoutsos N, Grosveld F, Fraser P (1996): Heterochromatin effects on the frequency and duration of LCR-mediated gene transcription. *Cell* 87:105–114.

Mink S, Kerber U, Klempnauer KH (1996a): Interaction of C/EBPbeta and v-Myb is required for synergistic activation of the mim-1 gene. *Mol Cell Biol* 16:1316–1325.

Mink S, Mutschler B, Weiskirchen R, Bister K, Klempnauer KH (1996b): A novel function for Myc: Inhibition of C/EBP-dependent gene activation. *Proc Natl Acad Sci USA* 93:6635–6640.

Mirkin SM, Frank-Kamenetskii MD (1994): H-DNA and related structures. *Annu Rev Biophys Biomol Struct* 23:541–576.

Mizzen CA, Yang XJ, Kokubo T, Brownell JE, Bannister AJ, Owen-Hughes T, Workman J, Wang L, Berger SL, Kouzarides T, Nakatani Y, Allis CD (1996): The TAF(II)250 subunit of TFIID has histone acetyltransferase activity. *Cell* 87:1261–1270.

Molnar A, Georgopoulos K (1994): The Ikaros gene encodes a family of functionally diverse zinc finger DNA-binding proteins. *Mol Cell Biol* 14:8292–8303.

Molnar A, Wu P, Largespada DA, Vortkamp A, Scherer S, Copeland NG, Jenkins NA, Bruns G, Georgopoulos K (1996): The Ikaros gene encodes a family of lymphocyte-restricted zinc finger DNA binding proteins, highly conserved in human and mouse. *J Immunol* 156:585–592.

Morrison SJ, Shah NM, Anderson DJ (1997): Regulatory mechanisms in stem cell biology. Cell 88:287–298.

Moskow JJ, Bullrich F, Huebner K, Daar IO, Buchberg AM (1995): Meis1, a PBX1-related homeobox gene involved in myeloid leukemia in BXH-2 mice. *Mol Cell Bio* 15:5434–5443.

Mouthon MA, Bernard O, Mitjavila MT, Romeo PH, Vainchenker W, Mathieu-Mahul D (1993): Expression of tal-1 and GATA-binding proteins during human hematopoiesis. Blood 81:647–655.

Mucenski ML, McLain K, Kier AB, Swerdlow SH, Schreiner CM, Miller TA, Pietryga DW, Scott WJ, Potter SS (1991): A functional c-myb gene is required for normal fetal hepatic hematopoiesis. *Cell* 65:677–689.

Muller AM, Medvinsky A, Strouboulis J, Grosveld F, Dzierzak E (1994): Development of hematopoietic stem cell activity in the mouse embryo. *Immunity* 1:291–301.

Muller R (1995): Transcriptional regulation during the mammalian cell cycle. *Trends Genet* 11:173–178.

Munker R, Kobayashi T, Elstner E, Norman AW, Uskokovic M, Zhang W, Andreeff M, Koeffler HP (1996): A new series of vitamin D analogs is highly active for clonal inhibition, differentiation, and induction of WAF1 in myeloid leukemia. *Blood* 88:2201–2209.

Murre C, Bain G, van Dijk MA, Engel I, Furnari BA, Massari ME, Matthews JR, Quong MW, Rivera RR, Stuiver MH (1994): Structure and function of helix-loop-helix proteins. *Biochim Biophys Acta* 1218:129–135.

Murrell AM, Bockamp EO, Gottgens B, Chan YS, Cross MA, Heyworth CM, Green AR (1995): Discordant regulation of SCL/TAL-1 mRNA and protein during erythroid differentiation. *Oncogene* 11:131–139.

Nakamura T, Jenkins NA, Copeland NG (1996a): Identification of a new family of Pbx-related homeobox genes. *Oncogene* 13:2235–2242.

Nakamura T, Largaespada D, Lee M, et al (1996b): Fusion of the nucleoporin gene NUP98 to HOXA9 by the chromosome translocation t(7;11)(p15;p15) in human myeloid leukemia. *Nature Genet* 12:154–158.

Nakamura T, Largaespada D, Shaughnessy J, Jenkins N, Copeland N (1995): Cooperative activation of Hoxa and PBX-1–related genes in murine myeloid leukemias. *Nature Genet* 12:149–153.

Nelson HC (1995): Structure and function of DNA-binding proteins. *Curr Opin Genet Dev* 5:180–189.

Newman AP, White JG, Sternberg PW (1996): Morphogenesis of the *C. elegans* hermaphrodite uterus. *Development* 122:3617–3626.

Nguyen HQ, Hoffman-Liebermann B, Liebermann DA (1993): The zinc finger transcription factor Egr-1 is essential for and restricts differentiation along the macrophage lineage. *Cell* 72:197–209.

Nishinakamura R, Miyajima A, Mee P, Tybulewicz V, Murray R (1996): Hematopoiesis in mice lacking the entire granulocyte-macrophage colony-stimulating factor/interleukin-3/interleukin-5 functions. *Blood* 88:2458–2464.

Nucifora G, Rowley JD (1995): AML1 and the 8;21 and 3;21 translocations in acute and chronic myeloid leukemia. *Blood* 86:1–14.

Nuez B, Michalovich D, Bygrave A, Ploemacher R, Grosveld F (1995): Defective haematopoiesis in fetal liver resulting from inactivation of the EKLF gene. *Nature* 375:316–318.

Nutt SL, Urbanek P, Rolink A, Busslinger M (1997): Essential functions of pax5 (bsap) in pro-B cell development—Difference between fetal and adult B lymphopoiesis and reduced V-to-DJ recombination at the IGH locus. *Genes Dev* 11:476–491.

Ogryzko VV, Schiltz RL, Russanova V, Howard BH, Nakatani Y (1996): The transcriptional coactivators p300 and CBP are histone acetyltransferases. *Cell* 87:953–959.

Okuda T, van Deursen J, Hiebert SW, Grosveld G, Downing JR (1996): AML1, the target of multiple chromosomal translocations in human leukemia, is essential for normal fetal liver hematopoiesis. *Cell* 84:321–330.

Olson MC, Scott EW, Hack AA, Su GH, Tenen DG, Singh H, Simon MC (1995): PU.1 is not essential for early myeloid gene expression but is required for terminal myeloid differentiation. *Immunity* 3:703–714.

Oosterwegel M, van de Wetering M, Timmerman J, Kruisbeek A, Destree O, Meijlink F, Clevers H (1993): Differential expression of the HMG box factors TCF-1 and LEF-1 during murine embryogenesis. *Development* 118:439–448.

Orlando V, Paro R (1995): Chromatin multiprotein complexes involved in the maintenance of transcription patterns. *Curr Opin Genet Dev* 5:174–179.

Orphanides G, Lagrange T, Reinberg D (1996): The general transcription factors of RNA polymerase II. *Genes Dev* 10:2657–2683.

Osawa M, Hanada K, Hamada H, Nakauchi H (1996): Long-term lymphohematopoietic reconstitution by a single CD34-low/negative hematopoietic cell. *Science* 273:242–245.

Packham G, Cleveland JL (1995): c-Myc and apoptosis. *Biochim Biophys Acta* 1242:11–28.

Packham G, Porter CW, Cleveland JL (1996): c-Myc induces apoptosis and cell cycle progression by separable, yet overlapping, pathways. *Oncogene* 13:461–469.

Pandolfi PP, Roth ME, Karis A, Leonard MW, Dzierzak E, Grosveld FG, Engel JD, Lindenbaum MH (1995): Targeted disruption of the GATA3 gene causes severe abnormalities in the nervous system and in fetal liver haematopoiesis *Nature Genet* 11:40–44.

Park J, Yaseen NR, Hogan PG, Rao A, Sharma S (1995): Phosphorylation of the transcription factor NFATp inhibits its DNA binding activity in cyclosporin A–treated human B and T cells. *J Biol Chem* 270:20653–20659.

Parthun MR, Widom J, Gottschling DE (1996): The major cytoplasmic histone acetyltransferase in yeast: Links to chromatin replication and histone metabolism. *Cell* 87:85–94.

Pearce JJ, Singh PB, Gaunt SJ (1992): The mouse has a Plycomb-like chromobox gene. *Development* 114:921–929.

Perkins A, Cory S (1993): Conditional immortalization of mouse myelomonocytic, megakaryocytic and mast cell progenitors by the *Hox-2.4* homeobox gene. *EMBO J* 12:3835–3846.

Perkins AC, Gaensler K, Orkin SH (1996): Silencing of human fetal globin expression is impaired in the absence of the adult beta-globin gene activator protein eklf. *Proc Natl Acad Sci USA* 12267–12271.

Perkins AC, Sharpe AH, Orkin SH (1995): Lethal beta-thalassaemia in mice lacking the erythroid CACCC-transcription factor EKLF. *Nature* 375:318–322.

Perrotti D, Bellon T, Trotta R, Martinez R, Calabretta B (1996): A cell proliferation–dependent multiprotein complex nc-3a positively regulates the cd34 promoter via a tcattt-containing element. *Blood* 88:3336–3348.

Perrotti D, Melotti P, Skorski T, Casella I, Peschle C, Calabretta B (1995): Overexpression of the zinc finger protein MZF1inhibits hematopoietic development from embryonic stem cells: Correlation with negative regulation of CD34 and c-myb promoter activity. *Mol Cell Biol* 15:6075–6087.

Pevny L, Lin CS, D'Agati V, Simon MC, Orkin SH, Costantini F (1995): Development of hematopoietic cells lacking transcription factor GATA-1. *Development* 121: 163–172.

Pieler T, Bellefroid E (1994): Perspectives on zinc finger protein function and evolution—An update. *Mol Biol Rep* 20:1–8.

Pirrotta V, Chan CS, McCabe D, Qian S (1995): Distinct parasegmental and imaginal enhancers and the establishment of the expression pattern of the Ubx gene. *Genetics* 141:1439–1450.

Porcher C, Swat W, Rockwell K, Fujiwara Y, Alt FW, Orkin SH (1996): The T cell leukemia oncoprotein SCL/tal-1 is essential for development of all hematopoietic lineages. *Cell* 86:47–57.

Quivy JP, Becker PB (1996): The architecture of the heat-inducible *Drosophila* hsp27 promoter in nuclei. *J Mol Biol* 256:249–263.

Ramsfjell V, Borge OJ, Veiby OP, Cardier J, Murphy MJ, Lyman SD, Lok S, Jacobsen S (1996): Thrombopoietin, but not erythropoietin, directly stimulates multilineage growth of primitive murine bone marrow progenitor cells in synergy with early acting cytokines—distinct interactions with the lignads for c-kit and flt3. *Blood* 88:4481–4492.

Razin A, Shemer R (1995): DNA methylation in early development. *Hum Mol Genet* 1751–1755.

Roberts CW, Shutter JR, Korsmeyer SJ (1994): Hox11 controls the genesis of the spleen. *Nature* 368:747–749.

Roberts CW, Sonder AM, Lumsden A, Korsmeyer SJ, (1995): Development expression of Hox11 and specification of splenic cell fate. *Am J Pathol* 146:1089–1101.

Roberts S, Bentley DL (1992): Distinct modes of transcription read through or terminate at the c-myc attenuator. *EMBO J* 11:1085–1093.

Roy AL, Carruthers C, Gutjahr T, Roeder RG (1993a): Direct role for Myc in transcription initiation mediated by interactions with TFII-I. *Nature* 365:359–361.

Roy AL, Malik S, Meisterenst M, Roeder RG (1993b): An alternative pathway for transcription initiation involving TFII-I. *Nature* 365:355–359.

Rubnitz JE, Behm FG, Downing JR (1996): 11q23 rearrangements in acute leukemia. *Leukemia* 10:74–82.

Ruddle FH, Bartels JL, Bentley KL, Kappen C, Murtha MT, Pendleton JW (1994): Evolution of Hox genes. *Annu Rev Genet* 28:423–442.

Ruff VA, Leach KL (1995): Direct demonstration of NFATp dephosphorylation and nuclear localization in activated HT-2 cells using a specific NFATp polyclonal antibody. *J Biol Chem* 270:22602–22607.

Russell JE, Liebhaber SA (1996): The stability of human beta-globin mRNA is dependent on structural determinants positioned within its 3′ untranslated region. *Blood* 87:5314–5323.

Rusten LS, Dybedal I, Blomhoff HK, Blomhoff R, Smeland EB, Jacobsen SE (1996): the RAR-RXR as well as the RXR-RXR pathway is involved in signaling growth inhibition of human CD34$^+$ erythroid progenitor cells. *Blood* 87:1728–1736.

Ryan KM, Birnie GD (1996): Myc oncogenes: The enigmatic family. *Biochem J* 314:713–721.

Sanchez I, Dynlacht BD (1996): Transcriptional control of the cell cycle. *Curr Opin Cell Biol* 8:318–324.

Sanchez-Garcia I, Rabbitts TH (1993): LIM domain proteins in leukaemia and development. *Semin Cancer Biol* 4:349–358.

Sanchez-Garcia I, Rabbitts TH (1994): The LIM domain: A new structural motif found in zinc-finger-like proteins. *Trends Genet* 10:315–320.

Sasaki K, Yagi H, Bronson RT, Tominaga K, Matsunashi T, Deguchi K, Tani Y, Kishimoto, T, Komori T (1996): Absence of fetal liver hematopoiesis in mice deficient in transcriptional coactivator core binding factor beta. *Proc Natl Acad Sci USA* 93:12359–12363.

Saurin AJ, Borden KL, Boddy MN, Freemont PS (1996): Does this have a familiar RING? *Trends Biochem Sci* 21:208–214.

Sauvageau G, Lansdorp PM, Eaves CJ, Hogge DE, Dragowska WH, Reid DS, Largman C, Lawrence HJ, Humphries RK, (1994): Differential expression of homeobox genes in functionally distinct CD34$^+$ subpopulations of human bone marrow cells. *Proc Natl Acad Sci USA* 91:12223–12227.

Sauvageau G, Thorsteinsdottir U, Eaves CJ, Lawrence HJ, Largman C, Lansdorp PM, Humphries RK (1995): Overexpression of HOXB4 in hematopoietic cells causes the selective expansion of more primitive populations *in vitro* and *in vivo*. *Genes Dev* 9:1753–1765.

Sawada S, Littman DR (1993): A heterodimer of HEB and an E12-related protein interacts with the CD4 enhancer and regulates its activity in T-cell lines. *Mol Cell Biol* 13:5620–5628.

Schilham MW, Oosterwegel MA, Moerer P, Ya J, de boer PA, van de Wetering M, Verbeek S, Lamers WH, Kruisbeek AM, Cumano A, Clevers H (1996): Defects in cardiac outflow tract formation and pro-B-lymphocyte expansion in mice lacking Sox-4. *Nature* 380:711–714.

Scott EW, Simon MC, Anastasi J, Singh H (1994): Requirement of transcription factor PU.1 in the development of multiple hematopoietic lineages. *Science* 265:1573–1577.

Sellers WR, Novitch BG, Miyake S, Heitl A, Otterson GA, Kaye FJ, Lassar AB, Kaelin WG Jr (1998): Stable binding to E2F is not required for the retinoblastoma protein to activate transcription, promote differentiation, and suppress tumor cell growth. *Genes and Dev* 12(1):95–106.

Shaw KT, Ho AM, Raghavan A, Kim J, Jain J, Park J, Sharma S, Rao A, Hogan PG (1995): Immunosuppressive drugs prevent a rapid dephosphorylation of transcription factor NFAT1 in stimulated immune cells. *Proc Natl Sci USA* 92:11205–11209.

Shen WF, Chang CP, Rozenfeld S, Sauvageau G, Humphries RK, Lu M, Lawrence HJ, Cleary ML, Largman C (1996): Hox homeodomain proteins exhibit selective complex stabilities with Pbx and DNA. *Nucleic Acids Res* 24:898–906.

Shivdasani RA, Mayer EL, Orkin SH (1995): Absence of blood formation in mice lacking the T-cell leukaemia oncoprotein tal-1/SCL. *Nature* 373:432–434.

Shrivastava A, Calame K (1995): Association with c-Myc: An alternated mechanism for c-Myc function. *Curr Top Microbiol Immunol* 194:273–282.

Sil A, Herskowitz I, (1996): Identification of asymmetrically localized determinant, Ash1p, required for lineage-specific transcription of the yeast HO gene. *Cell* 84:711–722.

Smith LT, Hohaus S, Gonzalez DA, Dziennis SE, Tenen DG (1996): PU.1 (Spi-1) and C/EBP alpha regulate the granulocyte colony-stimulating factor receptor promoter in myeloid cells. *Blood* 88:1234–1247.

Southwood CM, Downs KM, Bieker JJ (1996): Erythroid Kruppel-like factor exhibits an early and sequentially localized pattern of expression during mammalian erythroid ontogeny. *Dev Dyn* 206:248–259.

Spangrude GJ, Heimfeld S, Weissman IL (1988): Purification and characterisation of mouse hematopoietic stem cells. *Science* 241:58–62.

Sun XH, Copeland NG, Jenkins NA, Baltimore D (1991): Id proteins Id1 and Id2 selectively inhibit DNA binding by one class of helix-loop-helix proteins. *Mol Cell Biol* 11:5603–5611.

Sunaga S, Maki K, Lagasse E, Blanco JC, Ozato K, Miyazaki J, Ikuta K (1997): Myeloid differentiation is impaired in transgenic mice with targeted expression of a dominant negative form of retinoid X receptor beta. *Br J Haematol* 96:19–30.

Svaren J, Horz W (1997): Transcription factors vs. nucleosomes-regulation of the pho5 promoter in yeast. *Trends Biochem Sci* 22:93–97.

Tagawa M, Sakamoto T, Shigemoto K, Matsubara H, Tamura Y, Ito T, Nakamura I, Okitsu A, Imai K, Taniguchi M (1990): Expression of novel DNA-binding protein with zinc finger structure in various tumor cells. *J Biol Chem* 265:20021–20026.

Talbot D, Collis P, Antoniou M, Vidal M, Grosveld F, Greaves DR (1989): A dominant control region from the human beta-globin locus conferring integration site-independent gene expression. *Nature* 338:352–355.

Taunton J, Hassig CA, Schreiber SL (1996): A mammalian histone deacetylase related to the yeast transcriptional regulator Rpd3p. *Science* 272:408–411.

Thomsen GH (1996): *Xenopus* mothers against decapentaplegic is an embryonic ventralizing agent that acts downstream of the BMP-2/4 receptor. *Development* 122:2359–2366.

Thorsteinsdottir U, Sauvageau G, Hough MR, Dragowska W, Lansdorp PM, Lawrence HJ, Largman C, Humphries RK (1997): Overexpression of hoxa10 in murine hematopoietic cells perturbs both myeloid and lymphoid differentiation and leads to acute myeloid leukemia. *Mol Cell Biol* 17:495–505.

Ting CN, Olson MC, Barton KP, Leiden JM (1996): Transcription factor GATA-3 is required for development of the T-cell lineage. *Nature* 384:474–478.

Toki T, Itoh J, Arai K, Kitazawa J, Yokoyama M, Igarashi K, Yamamoto M, Ito E (1996): Abundant expression of erythroid transcription factor P45 NF-E2 mRNA in human peripheral granulocytes. *Biochem Biophys Res Commun* 219:760–765.

Triezenberg SJ (1995): Structure and function of transcriptional activation domains. *Curr Opin Genet Dev* 5:190–196.

Tsai FY, Keller G, Kuo FC, Weiss M, Chen J, Rosenblatt M, Alt FW, Orkin SH (1994): An early haematopoietic defect in mice lacking the transcription factor GATA-2. *Nature* 371:221–226.

Tsai S, Bartelmez S, Sitnicka E, Collins S (1994): Lymphohematopoietic progenitors immortalized by a retroviral vector harboring a dominant-negative retinoic acid receptor can recapitulate lymphoid, myeloid, and erythroid development. *Genes Dev* 8:2831–2841.

Tsang A, Visvader J, Turner C, Fujiwara Y, Yu C, Weiss M, Crossley M, Orkin S (1997): FOG, a multitype zinc finger protein, acts as a cofactor for transcription factor GATA-1 in erythroid and megakaryocytic differentiation. *Cell* 90:109–119.

Tuan D, Solomon W, Li Q, London IM, (1985): The "beta-like-globin" gene domain in human erythroid cells. *Proc Natl Acad Sci USA* 82:6384–6388.

Tuan DY, Solomon WB, London IM, Lee DP (1989): An erythroid-specific, developmental-stage-independent enhancer far upstream of the human "beta-like globin" genes. *Proc Natl Acad Sci USA* 86:2554–2558.

Urbanek P, Wang ZQ, Fetka I, Wagner EF, Busslinger M (1994): Complete block of early B cell differentiation and altered patterning of the posterior midbrain in mice lacking Pax5/BSAP. *Cell* 79:901–912.

van de Wetering M, Oosterwegel M, van Norren K, Clevers H (1993): Sox-4, an Sry-like HMG box protein, is a transcriptional activator in lymphocytes. *EMBO J* 12:3847–3854.

van der Saag PT, Caldenhoven E, van de Stolpe A (1996): Molecular mechanisms of steroid action: A novel type of cross-talk between glucocorticoids and NF-kappa B transcription factors. *Eur Respir J Suppl* 22:146s–153s.

Vanderlugt N, Alkema M, Berns A, Deschamps J (1996): The polycomb-group homolog bmi-1 is a regulator of murine hox gene expression. *Mech Dev* 58:1–2

van Lohuizen M, Verbeek S, Scheijen B, Wientjens E, van der Gulden H, Berns A (1991): Identification of cooperating oncogenes in Eμ-myc transgenic mice by provirus tagging. *Cell* 65:737–752.

Varga-Weisz PD, Blank TA, Becker PB (1995): Energy-dependent chromatin accessibility and nucleosome mobility in a cell-free system. *EMBO J* 14:2209–2216.

Verbeek S, Izon D, Hofhuis F, Robanus-Maandag E, te Riele H, van de Wetering M, Oosterwegel M, Wilson A, MacDonald HR, Clevers H (1995): An HMG-box-containing T-cell factor required for thymocyte differentiation. *Nature* 374:70–74.

Verreault A, Kaufman PD, Kobayashi R, Stillman B (1996): Nucleosome assembly by a complex of CAF-1 and acetylated histones H3/H4. *Cell* 87:95–104.

Verrijzer CP, Van der Vliet PC (1993): POU domain transcription factors. *Biochim Biophys Acta* 1173:1–21.

Voronova AF, Lee F (1994): The E2A and tal-1 helix-loop-helix proteins associate *in vivo* and are modulated by Id proteins during interleukin 6-induced myeloid differentiation. *Proc Natl Sci USA* 91:5952–5952.

Wang JH, Nichogiannopoulou A, Wu L, Sun L, Sharpe AH, Bigby M, Georgopoulos K (1996): Selective defects in the development of the fetal and adult lymphoid system in mice with an Ikaros null mutation. *Immunity* 5:537–549.

Wang L, Mizzen C, Ying C, Candau R, Barley N, Brownell J, Allis CD, Berger SL (1997): Histone acetyltransferase activity is conserved between yeast and human GCN5 and is required for complementation of growth and transcriptional activation. *Mol Cell Biol* 17:519–527.

Wang X, Kiledjian M, Weiss IM, Liebhaber SA (1995): Detection and characterization of a 3′ untranslated region ribonucleoprotein complex associated with human alpha-globin mRNA stability [published erratum appears in *Mol Cell Biol* 1995 15:2331]. *Mol Cell Biol* 15:1769–1777.

Warren AJ, Colledge WH, Carlton MB, Evans MJ, Smith AJ, Rabbitts TH (1994): The oncogenic cysteine-rich LIM domain protein rbtn2 is essential for erythroid development. *Cell* 78:45–57.

Wegner M, Drolet DW, Rosenfeld MG (1993): POU-domain proteins: Structure and function of developmental regulators. *Curr Opin Cell Biol* 5:488–498.

Weintraub H (1993): The MyoD family and myogenesis: Redundancy, networks, and thresholds. *Cell* 75:1241–1244.

Weiss A, Keshet I, Razin A, Cedar H (1996): DNA demethylation *in vitro:* Involvement of RNA. *Cell* 86:709–718.

Weiss IM, Liebhaber SA (1994): Erythroid cell-specific determinants of alpha-globin mRNA stability. *Mol Cell Biol* 14:8123–8132.

Weiss MJ, Keller G, Orkin SH (1994): Novel insights into erythroid development revealed through *in vitro* differentiation of GATA-1 embryonic stem cells. *Genes Dev* 8:1184–1197.

Weiss MJ, Orkin SH (1995a): GATA transcription factors: Key regulators of hematopoiesis. *Exp Hematol* 23:99–107.

Weiss MJ, Orkin SH (1995b): Transcription factor GATA-1 permits survival and maturation of erythroid precursors by preventing apoptosis. *Proc Natl Acad Sci USA* 92:9623–9627.

Weiss MJ, Yu CN, Orkin SH (1997): Erythroid-cell–specific properties of transcription factor gata-1 revealed by phenotypic rescue of a gene-targeted cell line. *Mol Cell Biol* 17:1642–1651.

Wessely O, Deiner EM, Beug H, Vonlindern M (1997): The glucocorticoid receptor is a key regulator of the decision between self-renewal and differentiation in erythroid progenitors. *EMBO J* 16:267–280.

Whetton A, Dexter T (1989): Myeloid haematopoietic growth factors. *Biochem Biophys Acta* 989:111–132.

Wijgerde M, Gribnau J, Trimborn T, Nuez B, Philipsen S, Grosveld F, Fraser P (1996): The role of EKLF in human beta-globin gene competition. *Genes Dev* 10:2894–2902.

Wilson RB, Kiledjian M, Shen CP, Benezra R, Zwollo P, Dymecki SM, Desiderio SV, Kadesch T (1991): Repression of immunoglobulin enhancers by the helix-loop-helix protein Id: Implications for B-lymphoid–cell development. *Mol Cell Biol* 11:6185–6191.

Wolffe AP (1996): Histone deacetylase: A regulator of transcription. *Science* 272: 371–372.

Wu H, Liu X, Jaenisch R, Lodish HF (1995): Generation of committed erythroid BFU-E and CFU-E progenitors does not require erythropoietin or the erythropoietin receptor. *Cell* 83:59–67.

Yu BD, Hess JL, Horning SE, Brown GA, Korsmeyer SJ (1995): Altered Hox expression and segemental identity in Mll-mutant mice. *Nature* 378:505–508.

Zawel L, Reinberg D (1995): Common themes in assembly and function of eukaryotic transcription complexes. *Annu Rev Biochem* 64:533–561.

Zhang C, Evans T (1996): BMP-like signals are required after the midblastula transition for blood cell development. *Dev Genet* 18:267–278.

Zhang DE, Hohaus S, Voso MT, Chen HM, Smith LT, Hetherington CJ, Tenen DG (1996): Function of PU.1 (Spi-1), C/EBP, and AML1 in early myelopoiesis: Regulation of multiple myeloid CSF receptor promoters. *Curr Top Microbiol Immunol* 211:137–147.

Zhuang Y, Cheng P, Weintraub H (1996): B-lymphocyte development is regulated by the combined dosage of three basic helix-loop-helix genes, E2A, E2-2, and HEB. *Mol Cell Biol* 16:2898–2905.

Zhuang Y, Soriano P, Weintraub H (1994): The helix-loop-helix gene E2A is required for B cell formation. *Cell* 79:875–884.

Zilberman Y, Yefenof E, Oron E, Dorogin A, Guy R (1996): T cell receptor–independent apoptosis of thymocyte clones induced by a thymic epithelial cell line is mediated by steroids. *Cell Immunol* 170:78–84.

Zink D, Paro R (1995): *Drosophila* polycomb-group regulated chromatin inhibits the accessibility of a trans-activator to its target DNA. *EMBO J* 14:5660–5671.

Zon LI (1995): Developmental biology of hematopoiesis. *Blood* 86:2876–2891.

5

HEMATOPOIETIC STEM CELLS: PROLIFERATION, PURIFICATION, AND CLINICAL APPLICATIONS

RUTH PETTENGELL AND MALCOLM A. S. MOORE

Department of Hemotology, St. George's Hospital Medical School, London, United Kingdom (R.P.), James Ewing Laboratory of Developmental Hematopoiesis, Memorial Sloan–Kettering Cancer Center, New York, NY 10021 (M.A.S.M.)

INTRODUCTION

Primitive hematopoietic cells, including self-renewing pluripotent stem cells and nonself-renewing, lineage-committed progenitor cells, are present within the $CD34^+$ cell population of marrow and blood. Their detection in the circulation led to clinical studies in which cytokine-mobilized peripheral blood or umbilical cord blood was used as a substitute for bone marrow in autologous and, more recently, allogeneic transplantation. Because the $CD34^+$ cell population is heterogeneous, there has been interest in exploiting different subsets for different applications. A mixture of primitive pluripotent and lineage-committed progenitors is required for autologous hematopoietic reconstitution after myeloablative therapy, but umbilical cord blood may be preferred for allogeneic transplantation because of its low immunogenicity. The most primitive stem cells will be required for corrective gene therapy. Here we review progress in the identification, mobilization, isolation, and culture of different hematopoietic progenitor subgroups and discuss their potential applications.

Stem Cell Biology and Gene Therapy, Edited by Peter J. Quesenberry, Gary S. Stein, Bernard G. Forget, and Sherman M. Weissman
ISBN 0-471-14656-0

STEM CELL ASSAYS

Although lineage-committed progenitors are readily detected by *in vitro* cloning assays, there is no method that positively identifies the most primitive totipotent hematopoietic stem cells, which are generally quiescent and resistant to 5-fluorouracil (5-FU) or 4-hydroperoxy-cyclophosphamide (4HC). *In vivo* assays using severe combined immunodeficiency (SCID) mice and human cytokine support have been developed to measure long-term repopulating capacity (and secondary transfer potential) of human hematopoietic cells (Lapidot et al., 1992) and have already led to clinical experiments in SCID patients (Wengler et al., 1996). An alternative *in vivo* assay system uses irradiated human fetal bone xenografts (Kyoizumi et al., 1992). Human stem cells can also be assayed by transplantation into fetal lambs with subsequent measurement of long-term hematopoietic chimerism (Srour et al., 1993; 1996). Although cumbersome and at best semiquantitative, only *in vivo* assays can confirm that pluripotent stem cells are present in a test cell population.

In vitro stem cell assays are based on assessing the clonal expansion of cells in the presence of cytokine combinations, marrow stromal support, or both (Moore, 1995). Limiting dilution methods can be used to obtain quantitative estimates of stem cell numbers. For example, the Delta assay involves expansion of limiting numbers of cells resistant to 5-FU or 4-HC (Schneider et al., 1994) or selected populations of $CD34^+$ or $CD34^+$, lin^- cells in suspension culture with multiple cytokines (Shapiro et al., 1994). At weekly intervals, granulocyte-macrophage colony-forming units are assayed and cells recultured at the same starting cell concentration in fresh medium and cytokines until there is no further cellular expansion. Cumulative expansion of progenitors and total cells provides an index of the proliferative potential of the input population and, when carried out at limiting dilutions on marrow stroma, can be used to quantitate long-term culture-initiating cells (LTC-IC) (Pettengell et al., 1994; Sutherland et al., 1990). An alternative method assesses the ability of LTC-IC to give rise to "cobblestone areas" on irradiated bone marrow stroma and measures the number of GM-CFC produced by such cells (Pettengell et al., 1994).

STEM CELL PHENOTYPES

Both stem cells and lineage-committed hematopoietic progenitor cells (HPC) express the CD34 antigen (Krause DS et al., 1996). $CD34^+$ cells constitute 1%–5% of cells in adult bone marrow and 5%–10% in fetal bone marrow. Mice constitutively lacking the CD34 antigen have reduced numbers of hematopoietic precursors, which have poor colony-forming activity in bone marrow and spleen and reduced proliferation in response to hematopoietic growth factors, suggesting that CD34 plays an important role in the formation of progenitor cells in both fetal and adult hematopoiesis (Cheng et al., 1996). Primitive stem cells, however, lack the differentiation antigens that are present on lineage-committed progenitors and are thus $CD38^-$, $CD45^{lo}$, $CD71^{lo}$ (Craig et al., 1993). HLA-DR is absent or is expressed at low levels on adult stem cells but is present on fetal and neonatal hematopoietic stem cells (Lansdorp et al., 1993). Thy-1 antigen is present on all human fetal and neonatal hematopoietic cells but is only expressed on a proportion of lineage-committed progenitors in the adult (Craig et al., 1993; Mayani and Lansdorp, 1994; Watt and Visser, 1992). The MDR1 gene encodes a P-glycoprotein that functions as an efflux pump in the cell membrane. MDR1 is strongly expressed in hematopoietic stem cells and confers on them the ability to exclude (among others) the mitochondrial-binding dye rhodamine 123 (Chaudhary and Roninson, 1991). The receptors kit and flk-2, which have intrinsic tyrosine kinase activity, are expressed on both stem cells and progenitors, but some kit^+ cells with stem cell function can be flk-2^- (Zeigler et al., 1994). AC133 is a recently described antigen with a novel 5-transmembrane molecule. The function of the antigen is unknown (Miraglia S et al., 1997). The AC133 antibody selects a subset of $CD34^+$ cells ($CD34^{bright}$) which contains both short and long term repopulating cells. It therefore offers an alternative to the CD34 antigen for cell selection (Yin AH et al., 1997). Thus the closest approximation to a hematopoietic stem cell is given by the phenotype shown in Table 1.

A high level of $CD34^+$ cells in the hematopoietic product does not necessarily correlate with high numbers of cells with stem cell phenotype (Baumann et al., 1996; Weaver et al., 1996). This implies that it can be misleading to measure a single progenitor cell subpopulation, such as GM colony-forming cells (CFC) or $CD34^+$ cells, when assessing the capacity of a drug to mobilize HPC. For example, we have shown that a much weaker correlation exists between $CD34^+$ cells and pre-CFU in an assay in which *in vitro* 4HC-purged $CD34^+$ cells were cultured for 7 days with cytokines and the capacity to generate secondary CFU was used as an indirect measure of "pre-CFU" (Schneider et al., 1994). This lack of correlation was particularly evident in heavily pretreated breast cancer patients.

TABLE 1 Phenotypes of Adult Human Hematopoietic Stem and Progenitor Cells

Stem Cells	Progenitor Cells
$CD34^+$	$CD34^+$
$AC133^+$	$AC133^+$
lin^-	$CD33^+$, $CD54^+$, $CD7^+$, $CD19^+$, $CD24^+$ (3%–30%) $CD9^+$, $CD18^+$, $CD29^+$, $CD31^+$, $CD38^+$, $CD44^+$
CD45-RA^{lo} (>70%)	$CD45^+$
Thy-1^+	Thy-1^+ (5%–25%)
HLA-DR^-	HLA-DR^+
c-kit^+	kit^+ (70%–80%)
flk-2^+	flk-2^+ (20%–50%)
$MDR1^{hi}$	$MDR1^{lo}$
$Rhodamine^{dull}$	$Rhodamine^{bright}$

STROMAL REQUIREMENTS FOR HEMATOPOIESIS

The bone marrow stroma provides a structural framework and a responsive physiological environment that are both essential for hematopoiesis. The extracellular components play an important role. Fibronectin promotes cell proliferation by selective adhesion processes, whereas glycosaminoglycans specifically retain and deliver growth factors. Fibrinogen and its D fragment potentiate the effects of IL-3 on early hematopoietic cells (Zhou et al., 1993). Direct contact between the HPC and stroma is needed for the survival of human repopulating cells. For maintenance of progenitors and some LTC-IC subsets, separation by a semipermeable membrane is feasible (Verfaillie, 1992, 1993). Stromal cells comprise a variety of cell types, including fibroblasts, endothelial cells, and adipocytes (Deryugina and Muller-Sieburg, 1993). The stromal cells secrete diverse positive and negative growth factors, both constitutively and in response to cytokines such as IL-1 and tumor necrosis factor (TNF). Distinct stromal cell lines can be isolated that support the growth of early and late hematopoietic progenitors, and it is postulated that these form niches within bone marrow where the microenvironment supports different cell lineages (Wineman et al., 1996).

SOURCES OF HEMATOPOIETIC PROGENITOR CELLS

Aspirated bone marrow was initially used for hematopoietic reconstitution following myeloablative therapy. This has the advantage of supplying the full range of hematopoietic and stromal cells. Attempts to improve the yield of specific hematopoietic subpopulations by priming bone marrow donors with 5-FU (Stewart et al., 1993) or cytokines have met with limited success (Naparstek et al., 1992; Ratajczak et al., 1994). For autologous transplantation, mobilized HPC collected by apheresis have almost completely replaced bone marrow and are increasingly used in allogeneic transplantation (Beelen DW et al., 1997). Comparative studies have shown that mobilized HPC lead to earlier recovery of neutrophil and platelet counts than bone marrow transplantation, reducing hospital costs and stays (Beyer et al., 1995; Schmitz et al., 1996; Beelen DW et al., 1997).

Mobilized HPC

A wide variety of cytokines and cytotoxics, alone and in combination, can mobilize HPC from the marrow to the peripheral blood. The optimum HPC-mobilizing regimen will depend on the proposed application. For example, cytotoxic chemotherapy will not be acceptable to mobilize HPC in normal donors for allogeneic transplantation. A variety of other proteins also mobilize HPC. Administration of antibodies to the adhesion factor VLA-4 results in rapid mobilization (within 30 minutes) of progenitor cells (Papayannopoulou and Nakamoto, 1993; Craddock et al., 1997). Different proportions of CFCs and more primitive cells are mobilized by different stimuli and with different time courses (Verbik et al., 1995), making selective mobilization a possibility.

Cytokines can be used alone to mobilize HPC. Paradoxically, lineage-specific cytokines such as granulocyte colony-stimulating factor (G-CSF) and GM-CSF are more potent stimuli to HPC mobilization than early-acting factors such as interleukin (IL)-3 flt-3/flk-2 ligand and the kit ligand, but the combination of early- and late-acting factors gives highest yields (Weaver et al., 1996; Brasel K et al., 1997). G-CSF treatment increases the numbers of circulating clonogenic cells 6–58-fold and CD34$^+$ cells 4–62-fold (Pettengell and Testa, 1995). Maximal HPC numbers are observed after 4–6 days of cytokine administration. Importantly, both primitive and lineage-committed progenitors can be mobilized in this way. The addition of kit ligand to G-CSF appears to have a role in heavily pretreated patients (Shapiro et al., 1997). The flk-2/flt-3 ligand synergizes with many of the same growth factors as kit ligand (Rusten et al., 1996). It does not activate mast cells and so has a better side effect profile than kit ligand (Lebsack ME et al., 1997). It has no effect on erythropoiesis but can stimulate B, T, and dendritic cells (Jacobsen et al., 1996; McKenna et al., 1995; Lebsack ME et al., 1997). Thrombopoietin alone stimulates modest increases in GM-CFC (maximum sevenfold) and megakaryocyte (M)-CFC (maximum fourfold) and acts additively with chemotherapy and G-CSF to mobilize them (58- and 66-fold, respectively) (Basser et al., 1996).

A number of cytotoxic drug combinations have been used to mobilize HPC (Pettengell and Testa, 1995) based on popular therapeutic regimens. Comparisons between them are difficult because the degree of HPC mobilization varies widely with pretreatment, disease extent, and pathology, as well as the mobilizing chemotherapy used. Even within a single treatment center there can be considerable variability. This is not surprising in light of the laboratory and assay variables and the high interpatient variability (up to 100-fold) in numbers of progenitors mobilized. Many, but not all, cytotoxic drugs stimulate HPC release. In general, it seems that drugs such as melphalan, busulphan, and platinum, which induce severe thrombocytopenia, are most likely to be toxic to hematopoietic stem cells and ineffective mobilizers. As with cytokines, the mechanism of HPC mobilization remains unknown. No relationship has yet emerged with the mode of antitumor action, class of drug, or degree of myelosuppression induced. The alkylating agent cyclophosphamide is the best documented and most widely used drug for HPC mobilization. Many studies suggest that the combination of chemotherapy and cytokines stimulates more early HPC into the circulation than either alone, particularly in heavily pretreated patients (Mohle et al., 1994; Pettengell et al., 1993b; To et al., 1994). Following the licensing of filgrastim and lenograstim for HPC mobilization, G-CSF alone and cyclophosphamide/G-CSF have become the most popular regimens.

The proliferative potential of mobilized HPC is reduced in heavily pretreated patients. Mobilized CD34$^+$ cells from a previously untreated patient achieve a 100–200-fold increase in progenitors in the Delta assay, peaking by 3 weeks and subsequently declining to baseline levels by 4–5 weeks. In heavily pretreated patients with ovarian cancer, expansion may be normal for 1–2 weeks and then fall to baseline by 3 weeks, suggesting a deficit of early precursors that generate secondary progenitors at later stages of the culture (Fig. 1). These early precursors may include true self-renewing stem cells (LTC-IC). Similar differences have been observed in patients receiving treatment for germ cell tumors and lymphoma (Shapiro et al., 1994, 1997).

Bone marrow or HPC collected by leukapheresis can be either immediately infused into the recipient or stored by cryopreservation for future use. The use of HPC in whole blood stored for up to 48 hours at 4°C is a recent development that has extended the use of HPC to support multicyclic dose-intensive chemotherapy (Pettengell et al., 1995b).

Umbilical Cord Blood

The use of either related or unrelated donor umbilical cord blood stem cells for allogeneic transplantation is firstly to try to reduce transplant related

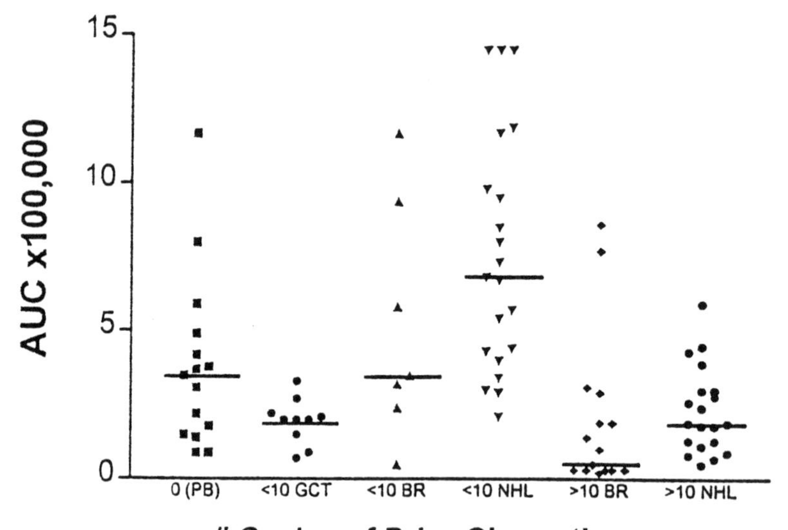

FIGURE 1. Area under the curve (AUC) of cumulative progenitor cell expansions in Delta assays of CD34$^+$ cells isolated from mobilized peripheral blood (PB) from patients with untreated ovarian cancer (0) or from moderately (<10 cycles of prior chemotherapy) or heavily (>10 cycles) pretreated patients with germ cell tumor (GCT), stage 4 breast cancer (BR), or non-Hodgkin's lymphoma (NHL). Each point is an assay of CD34 cells harvested from a single apheresis.

complications and secondly to augment the pool of potential donors (Cairo MS and Wagner JE, 1997). Volume for volume, human umbilical cord blood is at least as rich a source of hematopoietic progenitor cells as bone marrow (Broxmeyer et al., 1992; DiGiusto et al., 1996; Hows et al., 1992). The numbers of LTC-IC and 9M-CFC per Mononuclear cells in human umbilical cord blood are 2.5 and 2 times, respectively, higher than in leukapheresis product (Hirao et al., 1994). The proliferative potential of LTC-IC from umbilical cord blood exceeds that of adult bone marrow (Hao et al., 1995; Mayani and Lansdorp, 1995), compensating in part for the lower number of cells that can be obtained from a single donor compared with a conventional bone marrow harvest. Indeed, sustained hematopoietic engraftment after myeloablation has been obtained with as few as 2×10^4 LTC-IC from umbilical cord blood (Wagner et al., 1996). Among 50 reported umbilical cord blood transplants, the median number of GM-CFC infused was 1.9×10^4/kg (range $<0.1–25.6 \times 10^4$/kg). The recipients ranged in age from 1.3 to 47.8 years and had a median weight of 20 kg (7.5–98.8 kg). Engraftment of donor cells was seen in 39 of 44 evaluable transplants, with neutrophil recovery to 0.5×10^9/L at a median of 22 days and platelet recovery to 50×10^9/L at a median of 49 days. There was no correlation between nucleated cell count or CFC content of the graft and time to neutrophil recovery or probability of engraftment (Wagner et al., 1995, 1996). These data indicate that umbilical cord blood provides sufficient transplantable hematopoietic stem cells for children with HLA-identical or single HLA antigen-disparate sibling donors, but whether this will prove adequate for two- or three-HLA antigen–disparate sibling donors and adults remains to be determined. Recently, successful engraftment has been reported in three adult patients of >50 kg transplanted with umbilical cord blood (Kurtzberg et al., 1996; Laporte et al., 1996).

In reported umbilical cord blood transplants, the incidence of graft-versus-host disease is low. One of the unanswered questions about umbilical cord blood transplantation is whether the reduced immunogenicity of the graft (compared with bone marrow from an unrelated donor) will also result in less graft-versus-leukemia effect (Van Zant et al., 1994). A recent report suggested that in umbilical cord blood a graft-versus-leukemia effect was mediated by natural killer and lymphokine-activated killer cells independent of T-cell-mediated graft-versus-host disease (Harris, 1995).

Fetal Hematopoietic Progenitors

The liver is a major site of hematopoiesis in the human fetus from weeks 5 to 7 of gestation onward. Hematopoiesis in fetal bone marrow becomes important from about week 16 of gestation. Fetal hematopoietic progenitors have a greater growth potential than those in umbilical cord blood, adult bone marrow, or leukapheresis product (Lansdorp et al., 1993; Shapiro et al., 1994). Such cells are unlikely to become routinely available for clinical use, but could be targets for corrective gene therapy of inherited disorders (Humeau L et al., 1997).

STEM CELL SEPARATION DEVICES

Several methods have been developed for the clinical purification of hematopoietic cell subsets and have predominantly been used to obtain CD34$^+$ cell-rich fractions. They can be used for the separation of cells from bone marrow, leukapheresis product, umbilical cord blood, or whole blood. Mononuclear cells for separation are first obtained by density gradient centrifugation. These methods vary in the yield (total number of cells obtained), the purity (the proportion of CD34$^+$ cells in the product), and the recovery (the proportion of the total number of CD34$^+$ cells retained (McNiece I et al., 1997; Williams SF et al., 1996; Watts MJ et al., 1997)). Where the percentage of CD34$^+$ cells in the starting product is low ($<0.5\%$), the final purity is also lower ($<80\%$) (Flasshove et al., 1995a,b). The different systems use monoclonal antibodies directed against different CD34$^+$ epitopes. These epitopes are characterized by different sensitivities to enzymatic degradation with neuraminidase and a glycoprotease from *Pasteurella haemolytica.* The Cellpro and Isolex systems use antibodies directed against the class I epitope, Miltenyi against the class II epitope and Dynal beads against class III. It is possible that expression of different epitopes of the CD34$^+$ molecule could reflect different stages of maturation in the cells. The different methods in use are compared in Table 2.

Fluorescence-Activated Cell Sorting

Fluorescence-activated cell sorting (FACS; Systemix) using two parameters, CD34$^+$ and Thy-1$^+$, can be used to isolate essentially pure populations of

TABLE 2 Comparison of Different Methods of Hematopoietic Stem Cell Separation

	Yield	Purity	Recovery	T-Cell Depletion	Tumor Cell Log Depletion/kg
FACS	Low	+++	+	5 Log	10^2–10^3
Biotin-avidin (Ceprate)	High	++	+++	3 Log	10^5
Panning	Medium	+	+		
Immunomagnetic					
Isolex 300 (Dynal)	Medium–high	+++	+++	4 Log	10^4–10^5
Miltenyi	Medium	+++	+++	5 Log	10^4–10^5

+ low, ++ medium, +++ high.

cells expressing antigens that are recognized by fluorescent-labeled antibodies. However, the procedure is time consuming and expensive, limiting the number of cells available for transplant. The system can be used to sort a wide range of expressed antigens singularly or concurrently (Murray et al., 1995) and offers greater purging efficiency with respect to tumor cells or T cells. Systemix has developed FACS machines that allow processing on a scale suitable for clinical use.

Biotin-Avidin Affinity Column

In this process cells are incubated with a biotinylated antibody to $CD34^+$ (antibody 12.8) and then passed through an avidin column. $CD34^+$ cells are retained within the column and then released by agitation (Ceptrate, System, Cell Pro Inc., Bothell, WA). Cell recovery is high, but purity ranges from 50% to 85%. To date this system has been the most used clinically.

Panning

Cells are incubated in polystyrene flasks coated with soybean agglutinin. This agglutinates red blood cells, B and T lymphocytes, monocytes, and stromal cells, resulting in a 1.5–5-fold enrichment of $CD34^+$ cells. The nonadherent cells are passed into a second flask coated with CD34 antibody (ICH3) from which $CD34^+$ cells are then removed by agitation. Recovery and purity are low.

Immunomagnetic Beads

Following incubation with murine anti-CD34 antibody, the cells are mixed with paramagnetic beads that are coated with antimurine antibodies in solution and passed through a column in a magnetic field. The cells are removed from the Miltenyi column by pressure. The Miltenyi beads average 0.1 μm in diameter and may be infused and degraded *in vivo*. In contrast, Dynal beads are 1–2 μm in diameter must be removed by chymopapain or $PR34^+$ (a competitive binding peptide (Baxter Inc., Santa Clara, CA; Isolex system) or by DETACHaBEAD (Dynal AS, Oslo, Norway). These systems give excellent purity with reasonable yields and recovery, and with the $PR34^+$-releasing agent the phenotype of the cells is unchanged.

ADVANTAGES OF STEM CELL PURIFICATION

In pediatric patients, $CD34^+$ cell selection is favored because it reduces the volume and DMSO content of the reinfused hematopoietic product, reducing the risk of adverse effects. Similarly, CD34 purification is favored in *ex vivo* expansion systems because reducing the volume leads to lower costs for media, growth factors, and so forth. Furthermore, in conventional static *ex vivo* culture systems the degree of progenitor expansion is closely correlated with the $CD34^+$ enrichment of the input cell population. For retroviral gene therapy, a purified population of stem cells is preferred to target transduction to the most primitive cells. Transduction efficiency is low because stem cells divide rarely or because they lack receptors for the retroviral vector. Immunocytochemical analysis of BudR incorporation indicates that a very low percentage of $CD34^+$ cells in umbilical cord blood or in mobilized HPC are in cell cycle. Exposure to two or more cytokines initiated entry into cell cycle and proliferation in 24 hours with a maximum of 37%–54% of the cells in S phase by 72 hours (Flasshove et al., 1995a,b; Traycoff et al., 1994).

Available methods of $CD34^+$ cell selection result in a 10^3–10^6-fold depletion of T cells but also deplete antigen-presenting cells such as dendritic cells. Although this may be desirable for allogeneic transplantation or gene therapy applications, it is not needed for autologous transplantation except in autoimmune disease. In circumstances where T-cell depletion is required, $CD34^+$ cell selection is not sufficient (Broxmeyer and Carow, 1993). Additional methods of T-cell depletion include elutriation and the Campath-1H antibody, further selection or depletion using a cocktail of antibodies or *ex vivo* culture using myeloid culture conditions.

There is considerable concern about possible tumor contamination of HPC collected by apheresis in cancer patients. Direct evidence that reinfused tumor cells contribute to relapse was furnished by the gene-marking studies of Brenner et al. (1993). They transfected the neomycin resistance gene into bone marrow cells before autologous transplantation and demonstrated incorporation of marked cells in the tumor population at relapse. Since most solid tumors do not express CD34, the selection of $CD34^+$ cells has been used to reduce tumor cell contamination of hematopoietic products used for autologous transplantation for patients with these tumors. $CD34^+$ cell selection can reduce tumor cell contamination by a factor of 10–10^4 using immunomagnetic beads or biotin-avidin columns (Farley TJ, 1997; Shpall et al., 1994) or by 10^3 using panning (Lebkowski et al., 1992). Further tumor cell purging requires additional treatment of the hematopoietic product with chemotherapy of monoclonal antibodies directed against tumor antigens or *ex vivo* culture using myeloid culture conditions.

Shpall et al. (1996) have shown that positive selection of $CD34^+$ HPC from breast cancer patients resulted in a marked reduction in the number of tumor cells reinfused after high dose chemotherapy. However, the relapse rate was no different. Vogel et al. (1996) assessed blood progenitor cell products from stage 3 breast cancer patients expanded in K36EIL-1 for 14–21 days and did not detect tumor cells. Future studies will determine the extent to which $CD34^+$ HPC contain residual tumor cells and their clinical significance (Rizzoli and Carlo-Stella, 1995).

EX VIVO HPC EXPANSION

Ex vivo expansion is a technique using nutrients and cytokines (added or supplied by a nutrient stromal layer) to favor the growth of hematopoietic progenitors *in vitro*. Such cells can then be used for a variety of clinical applications, including HPC transplantation, immunotherapy, and gene therapy. Proliferation of primitive cells with self-renewal capacity is necessary for the sustained production of HPC and differentiated cells. Many different systems have been developed to expand hematopoietic progenitor cell numbers *ex vivo* using defined media containing combinations of growth factors. In some, cell culture occurs entirely in liquid phase (using gas-permeable bags), while others use solid matrices (Cellco) or stromal-based systems (Aastrom Biosciences Inc., Ann Arbor, MI). In some, periodic media changes are required (gas-permeable bags), whereas others use continuous media perfusion (Aastrom, Cellco). In a system using gas-permeable bags or 100 ml culture flasks, without changing the medium, 15–40-fold expansion of myeloid progenitors was observed in 7–14 days, without LTC-IC depletion, using a cocktail of five cytokines and 1%–2% of autologous plasma (Shapiro et al., 1994). Similarly, using stirred suspension cultures supplemented with IL-3 and kit ligand, a sevenfold increase in LTC-IC and a 22-fold increase in committed progenitors was seen in 4 weeks (Shapiro et al., 1994). In direct comparisons, the stromal-based system yielded fivefold more LTC-IC than the stroma-free cultures (Koller et al., 1993a,b). In umbilical cord blood, expansion of LTC-IC was enhanced on stroma. Direct contact between stroma and HPC cells however, may not be necessary, as higher recoveries of LTC-IC have been reported from noncontact cultures (Verfaillie, 1992, 1993). Negative regulatory influences, in particular transforming growth factor-β, can influence the proliferation of stem cells in the adherent layer (Cashman et al., 1990). Ongoing studies have four general goals: to further reduce the period of pancytopenia following myeloablative chemotherapy, to reduce the number of HPC required for transplantation, to expand selected subsets of HPC, and to reduce malignant cell contamination of HPC.

Reduction of the Period of Pancytopenia Following Myeloablative Chemotherapy

The first goal is the expansion of the lineage-committed progenitor cell population to abolish pancytopenia following myeloablative chemotherapy. This depends on the beliefs that relatively mature cells are responsible for early engraftment and that, if they can be expanded to sufficiently high numbers, the period of pancytopenia can be progressively reduced. For the expansion of committed myeloid progenitors, the optimum combination of cytokines includes early-acting cytokines such as flk-2/flt-3 ligand, kit ligand, IL-3, IL-1, IL-6, and late-acting factors G- or GM-CSF (Haylock et al., 1992; Shapiro et al., 1994). In one study, allogeneic transplantation of bone marrow cultured with IL-3 and GM-CSF led to a reduction in cytopenia by 4 days. This approach is of particular importance in umbilical cord blood transplantation where total cell dose is limited, and even when high nucleated cell numbers are given haemopoietic engraftment is slow (Cairo MS and Wagner JE, 1997). Clinical trials are in progress to test whether *ex vivo* expansion of HPC will result in faster recovery following transplantation compared with unprocessed cells.

Reduction in the Number of HPC Required for Transplantation

By reducing the number and length of HPC collection procedures, one apheresis could suffice for repeated transplantations or one umbilical cord for several procedures. Recent studies have shown that significant expansion of HPC is possible in *ex vivo* cultures of mononuclear cells or purified $CD34^+$ cells (Traycoff et al., 1995). Using a perfusion bioreactor with continuous media replacement and 20% oxygen, starting with mononuclear cells that established a stroma, a 15–20-fold increase in lineage-committed progenitors and a 3.4–9.8-fold increase in LTC-IC were obtained in 14 days. Extrapolation from these figures suggests that 10–15 ml of unseparated marrow could be used to initiate cultures in a 350–750 ml bioreactor to generate sufficient HPC for clinical marrow reconstitution.

Controversy continues over whether umbilical cord blood HPC should be maintained or expanded in bioreactors. In one report, umbilical cord blood mononuclear cells and LTC-IC expanded in a bioreactor in the presence of irradiated bone marrow stroma, IL-3, IL-6, and kit ligand yielded sufficient cells for transplantation from an inoculum of $3–4 \times 10^8$ mononuclear cells (10–15 ml) (Wagner et al., 1996). Thus, the infusion of *ex vivo* manipulated and expanded primitive and lineage-committed HPC from umbilical cord blood appears to be feasible and safe, but clinical benefit is still unproven.

Expansion of Selected HPC Subsets

The expansion of selected subsets of HPC could be used to provide "stem cells" for gene therapy, or various lineage-committed progenitors, such as megakaryocytes, for the correction of thrombocytopenia or lymphocytes for adoptive immunotherapy (Young and Inaba, 1996; Miller JS et al., 1994). Production of lineage-restricted progenitors and differentiated populations can be manipulated by varying the cytokine combinations. For example, IL-5 favors the production of eosinophils, M-CSF favors monocytes and macrophages, and erythropoietin favors erythrocyte growth. Megakaryocyte-lineage cells expressing CD61 increased 1,500–2,000-fold after $CD34^+$ cells were cultured for 7 days with IL-3, IL-6, kit ligand, G-CSF, and erythropoietin (Schneider et al., 1994). Thrombopoietin alone or in combination with kit ligand is even more effective at promoting differentiation into the megakaryocyte pathway (Broudy et al., 1995).

Dendritic cells are potent antigen-presenting cells that can be used to elicit antigen-specific immune responses. They can be generated from bone marrow, peripheral blood, or umbilical cord blood CD34$^+$ cells cultured with GM-CSF and TNF-α. The addition of kit ligand results in a further 100–1,000-fold expansion of dendritic cell numbers by days 14 and 21, respectively (Szabolcs et al., 1995). Addition of IL-4 to the cytokine cocktail interferes with the number of monocytes, reducing overall cell yield but increasing the percentage of CD1a$^+$ cells (Strunk et al., 1996), which strongly stimulate a mixed lymphocyte reaction; this is a test of the capacity of the cells to stimulate allogeneic T lymphocytes (Rosenzwajg et al., 1996).

Culture conditions that favor myeloid cell expansion lead to the production of fewer lymphoid cells. This results in T-cell depletion, which is desirable for allogeneic transplantation. Functional NK cells with a CD56$^+$, CD3$^-$ phenotype can be generated from CD34$^+$, CD33$^-$ marrow after 2–5 weeks of culture with IL-2 (Miller et al., 1994). Verbik et al. (1995) show that the frequency of cytotoxic effector cells inducing lymphokine-activated killer cell precursors and lymphocytes are significantly higher in early cultures, whereas more GM-CFU are found in later cultures. Dendritic cells generated *in vitro* culture can prime naive T cells and process antigens for MHC class II presentation (Caux et al., 1995; Garrigan et al., 1996).

The presence of stem cells in expanded populations is critical if *ex vivo* HPC expansion is to be used for transplantation. An increase in the number of cells exhibiting a stem cell phenotype can be documented, but not all such cells are capable of extensive self-renewal. Of concern, surface antigen expression can be modulated by culture conditions without reflecting changes in proliferative potential. The source of input HPC greatly influences the fate of stem cells in culture (Pawliuk et al., 1996). CD34$^+$ cells from fetal liver or umbilical cord blood can be expanded up to 20-fold in 7–14 days, whereas adult bone marrow or peripheral blood CD34$^+$ cells show at best two- to threefold expansion. These findings have been confirmed by fluorescent tracking dye studies, which suggest that primitive cells in bone marrow and peripheral blood persist (up to 80%) but do not enter into the cell cycle (Traycoff et al., 1995; Young et al., 1996). It is possible that adult stem cells are incapable of self-replication in the presence of soluble growth factors and need membrane-associated forms or perhaps additional stromal-derived factors. flk-2/flt-3 ligand alone or in combination with kit ligand may offer better support for the expansion of LTC-IC, particularly at high concentrations (Piacibello et al., 1997).

While it has been demonstrated that CD34$^+$ cells can be expanded in culture (Henschler et al., 1994; Tjonnfjord et al., 1994; Traycoff et al., 1995), controversy continues over whether LTC-IC can be expanded and to what extent LTC-IC represent true "stem cells" (Sutherland et al., 1994; Traycoff et al., 1995). There appears to be a functional hierarchy within the LTC-IC population, with the most quiescent population being the closest to the human repopulating stem cell (Hao et al., 1995). In cell cultures, proliferative senescence eventually supervenes. This is associated with reducing chromosome telomere length, explained by the incomplete replication of the 3′ termini of the chromosome by DNA polymerase after each cell division (Hayflick and Moorehead, 1961; Vaziri et al., 1993, 1994). In cultures of fetal, neonatal, and adult human hematopoietic cells, 40–50 telomere base pairs are lost during cell division (Vaziri et al., 1994). In normal life, peripheral leukocyte telomeres shorten by approximately 9 base pairs per year. Senescence is seen after 30–75 population doublings (Hayflick limit) (Hayflick and Moorehead, 1961). In the setting of clinical HPC transplantation, stem cell senescence could limit the usefulness of expansion procedures. If the hematopoietic stem cells have proliferated extensively prior to transplantation and few are engrafted, then they may reach their Hayflick limit, leading to delayed marrow failure. $CD34^+$ cells grown in liquid culture upregulate telomerase but continue to lose approximately 0.4 kilobase pair each week of culture. The rate of telomere loss is slower during the first 2 weeks in culture, when telomerase activity is highest. Telomere loss accelerates during weeks 3 and 4 of culture, when telomerase levels become undetectable. Therefore, telomerase activity in hematopoietic cells reduces but does not prevent telomere shortening during proliferation Telomerase also fails to prevent telomere shortening in vivo. Young recipients of allogeneic bone marrow transplants have shorter telomeres than their donors, reflecting the increased proliferative demand on stem cells at the time of transplantation and engraftment (Wynn R et al., 1998). A further study has shown that the extent of the reduction of the telomere correlated inversely with the number of nucleated cells infused (Notaro R et al., 1997). Telomere length analysis showed that fetal liver and umbilical cord blood have longer telomeres than adult peripheral blood or bone marrow (Engelhardt et al., 1997). It is estimated that umbilical cord blood cells can go through 33 more population doublings than adult cells before reaching senescence. This increased replicative potential combined with their greater expansion potential would support the value of this stem cell source for allogeneic transplantation.

These data suggest a "repression, expansion, and cell cycle" model for telomerase regulation in hematopoietic cells in which telomerase is repressed in quiescent stem cells ($CD34^+$ $CD38^-$); is activated upon cell proliferation, expansion, cell cycle entry, and progression into the progenitor compartment ($CD34^+$, $CD38^+$); and is repressed again upon further differentiation ($CD34^-$ (Holt et al., 1996; Zhu et al., 1996).

Reduction of Malignant Cell Contamination of HPC

The fourth goal of *ex vivo* HPC expansion is to eliminate malignant cells in culture by favoring the differential growth of HPC over tumor cells. This goal remains aspirational and has been challenged by the finding that cytokine regimens that mobilize HPC can also mobilize solid tumor cells, which suggests that tumor growth may also be promoted in the conditions of *ex vivo* expansion (Brugger et al., 1994). The presence of receptors for hematopoietic growth factors on some tumor cells suggests that direct stimulation could occur. It is encouraging to note, however, that both normal and malignant lymphoid cells rapidly disappear under culture conditions optimal for myeloid expansion. Widmer et al. (1995) used a competitive polymerase chain reaction titration assay for t(14:18) in NHL to assess residual lymphoma cells before and after $CD34^+$ cell selection and *ex vivo* expansion. From an initial tumor load of 10–4,000 lymphoma cells/10^6 mononuclear cells, they achieved a 1–4 log depletion with $CD34^+$ selection. The final purity of the selected $CD34^+$ population was 88% (76%–94%). Following expansion in KL, IL-1β, IL-3, and IL-6 for 7–14 days, two of nine samples were negative for the translocation. To date, *ex vivo* growth of normal or malignant human breast epithelial cells with cytokines has not shown any stimulatory effect on the breast cells (Emerman and Eaves, 1994; Vogel et al., 1996).

Differential cytoadhesion of myeloid leukemic cells to stroma also provides a selection procedure for retention of normal stem cells in culture (Dexter and Chang, 1994; Udomsakdi et al., 1992). Autologous transplantation of such cultured marrow produced a substantial degree of Ph chromosome negative reconstitution in patients with chronic myeloid leukemia (Barnett et al., 1994). The use of all-*trans* retinoic acid (ATRA) in acute promyelocytic leukemia (PML) or 1,25$(OH)_2$ vitamin D_3 and analogs in acute myeloid leukemia is generally considered to be effective by inducing differentiation of the leukemic cells. 1,25$(OH)_2D_3$ appears to act by blocking leukemic stem cell self-renewal with or without inducing differentiation. Vitamin D preferentially reduces the clonogenic potential of a variety of cell lines and leukemic states relative to normal bone marrow while preserving the early hematopoietic cells (Pettengell et al., 1995a).

CLINICAL APPLICATIONS OF HEMATOPOIETIC PROGENITORS

High-dose (myeloablative) therapy has been used in a variety of solid and liquid tumors with the aim of eradicating malignant cells. Hematopoietic progenitor cell transplantation is used to restore normal hematopoiesis. This requires the presence in the graft of lineage-committed progenitors to effect early engraftment and primitive progenitors to effect long-term reconstitution. Transplantation of $CD34^+$ cells from bone marrow or peripheral blood has demonstrated that both primitive and lineage-committed progenitor populations reside in the $CD34^+$ fraction (Shpall et al., 1994). Karyotyping of recipients after sex-mismatched allogeneic bone marrow transplantation in humans has proved that donor cells are responsible for long-term engraftment. Retroviral gene-marking techniques have now been used in children undergoing autologous HPC transplantation to show that cells of both myeloid and lymphoid lineages were derived from a common precursor and that such cells are capable of sustained hematopoietic reconstitution over 18 months (Brenner et al., 1993). Sex-mismatched allogeneic transplantation using mobilized HPC is expected to confirm that functional blood stem cells are also collected from the peripheral blood at apheresis.

Undoubtedly, the number of HPC reinfused following myeloablative therapy determines the time to hematopoietic recovery (Bensinger et al., 1995; Pawliuk et al., 1996). The minimum number of HPC required to effect hematopoietic reconstitution after myeloablation has not been determined. This is partly due to the incomplete information available on the composition of the hematopoietic product obtained after various HPC mobilizing regimens. Higher cell numbers are required for allogeneic than autologous transplantation. Various medical centers have established threshold doses appropriate to their own patient population, HPC mobilizing, myeloablative, and post-transplant treatment regimens (Dreger et al., 1995; Sutherland et al., 1994), but it is difficult to compare these. For autologous transplantation in most patients with solid tumors, a minimum of 2×10^6 $CD34^+$ cells/kg or 10^5 GM-CFC cells/kg are recommended (Mavroudis et al., 1996; Weaver et al., 1995). Caution is required, however, because high levels of $CD34^+$ cells do not always indicate high stem cell numbers (Mohle et al., 1994).

Myeloablative Therapy and HPC Transplantation

Myeloablative therapy was first routinely used with bone marrow transplantation to eliminate minimal residual disease in patients with chemosensitive leukemias. Its use has now been extended to other chemosensitive tumors, but in most of these (with the exception of relapsed lymphomas) it remains experimental. Rapid and sustained hematopoietic recovery occurs after autologous blood progenitor cells transplantation, with a shorter period of thrombocytopenia, thus improving the risk/benefit ratio of the procedure compared with bone marrow transplantation (Beyer et al., 1995; Pettengell et al., 1993a,b; Schmitz et al., 1996; Sheridan et al., 1992). Other advantages include the avoidance of a general anesthetic for bone marrow harvesting and the ability to offer high dose therapy to patients with poor marrow reserve after previous chemoradiotherapy or with bone marrow involvement by tumor. For these reasons, blood progenitor cells have replaced autologous bone marrow as a source of HPC for transplantation in many medical centers, and high-dose treatments are now being offered to a wider range of patients. There are, however, no prospective randomized trials showing improved disease outcome in patients treated with blood progenitor cell transplantation over bone marrow transplantation. One study with non-Hodgkin's lymphoma showed a significantly better survival for a poor prognosis group of patients with bone marrow involvement receiving HPC transplantation compared with a better prognosis group receiving bone marrow transplantation (Vose et al., 1993). Whether this is due to a lower likelihood of occult tumor cells, a greater number of cytotoxic effector cells, or a different and advantageous pattern of immunological recovery is not yet known. The issue of whether dose-intensive chemotherapy with HPC support is more effective as induction or consolidation therapy is not resolved. The rationale for early treatment is to minimize tumor resistance and ensure adequate HPC yields. However, the increased risk of tumor contamination must be weighed against this.

Immune reconstitution after both autologous and allogeneic transplantation is faster following BPCT than BMT. Whether this contributes to a lower incidence of infectious complications is as yet unknown (Ottinger et al., 1996). The balance of T-cell depletion versus graft function and graft versus leukaemia (GVL) also needs careful study. An increased rate of lymphoid reconstitution has been reported following blood progenitor cell transplantation in humans and may be important, given the probability that part of the antitumor effect of allogeneic bone marrow transplantation results from a graft-versus-tumor effect (Roberts et al., 1993; Scheid et al., 1995). The ratio of T cells to progenitors and the distribution of T-cell subsets differs between apheresis and bone marrow harvests (Galy et al., 1994). Autologous blood progenitor cell rescue results in earlier immunological reconstitution than bone marrow transplantation, with apparently complete recovery within 6 months. Whether this will result in a graft-versus-tumor effect and enhanced disease-free survival, as suggested by the Nebraska group (Vose et al., 1993), is yet to be determined. The potential to manipulate the graft, either *in vitro* or *in vivo,* with immunomodulatory agents may also prove beneficial.

Allogeneic HPC Transplantation

Early fears that the large number of T lymphocytes in the HPC product would increase the incidence and severity of graft-versus-host disease in allogeneic HPC transplantation have not been sustained. Allogeneic HPC are well tolerated and may engraft faster than bone marrow (Korbling et al., 1995; Schmitz et al., 1995). *Ex vivo* manipulation of accessory cells is also easier using leukapheresis product. However, the cellular composition of the graft is extremely important to hematopoietic recovery and for graft-versus-tumor effect. Therefore, immunologically tailored grafts will need to be extensively evaluated both *in vitro* and *in vivo.* The use of human umbilical cord blood for allogeneic transplantation is an exciting research area (see above). The use of donor HPC avoids concerns about the quantity and quality of prior therapy and contaminating tumor cells when autologous bone marrow or HPC are used.

Multiple High-Dose Treatments with Transplantation

Multiple cycles of myeloablative treatment, each supported by autologous HPC transplantation, offers the opportunity to maximize cytotoxic dose intensity for solid tumors in which a single high-dose treatment is likely to be inadequate. Phase I and II studies of dose escalation by high-dose sequential therapy with repeated transplantation show that this approach is possible, with manageable hematological toxicity, but with significant nonmyeloid toxicity

(Gianni et al., 1997). Such an approach carries substantial morbidity and mortality, in addition to requiring prolonged hospitalization. Most medical centers using this approach collect enough HPC for all the transplantation procedures before embarking on the first myeloablative treatment. This has the advantage of ensuring that the hematopoietic cells are not damaged by the treatment, but damage to the marrow stroma may still delay re-engraftment in later cycles. Furthermore, careful screening and purging will be needed to minimize the risk of tumor contamination in the HPC product. It remains to be determined whether multiple high-dose cycles will prove better than a single cycle of myeloablative treatment.

Multiple Subablative Treatments with HPC Support

Hematopoietic growth factors have been used to improve the delivery of the planned dose intensity, but their capacity to increase cytotoxic dose intensity is limited (Woll et al., 1995). There is therefore interest in using HPC to support increased dose intensity of conventional chemotherapy (Crown et al., 1993; Shea et al., 1992; Tepler et al., 1993). In most published studies, the same strategy has been adopted as that used for multiple transplants—that of using aliquots of HPC collected before starting the dose-intensive treatment. An alternative strategy is to collect and reinfuse HPC at each treatment cycle (Pettengell et al., 1995b). This has the potential advantage of reducing the risk of malignant cell contamination, because the blood will be purged of malignant cells *in vivo* by each chemotherapy cycle.

CONCLUSION

Increasing interest in the use of stem cells for gene therapy has focused attention on hematopoietic stem cells—the most accessible human stem cell population. In this chapter, we outlined methods of identifying, mobilizing, and culturing different hematopoietic progenitor cell populations. To date, most workers have concentrated on obtaining a mixture of primitive and lineage-committed progenitors to optimize engraftment following myeloablative therapy. It now appears that cytokines and other agents can be used to modulate hematopoietic cell populations *in vivo* and manipulate them *in vitro*. Thus diverse subpopulations can be exploited for different applications, including immunotherapy and protective and corrective gene therapy.

REFERENCES

Barnett MJ, Eaves CJ, Phillips GL, Gascoyne RD, Hogge DE, Horsman DE, Humphries RK, Klingemann HG, Lansdorp PM, Nantel SH, et al. (1994): Autografting with cultured marrow in chronic myeloid leukemia: Results of a pilot study. *Blood* 84:724–732.

Basser R, Rasko J, Clarke K, Green M, Cebon J, Grigg A, Berndt M, Zalcberg J, Marty J, Menchaca D, Tomita D, Fox R, Begley G (1996): Pegylated megakaryocyte growth and development factor (peg-rhumgdf) enhances the mobilization of peripheral-blood progenitor cells (pbpc) by chemotherapy and filgrastim. *Blood* 88:2554.

Baumann I, Swindell R, Van Hoeff MEHM, Dexter TM, de Wynter E, Lange C, Luft T, Howell A, Testa N (1996): Mobilisation kinetics of primitive haemopoietic cells following G-CSF with or without chemotherapy for advanced breast cancer. *Ann Oncol* 7:1051–1057.

Beelen DW, Ottinger HD, Elmaagacli A, Scheulen B, Basu O, Kremens B, Havers W, Grosse-Wilde H, Schaefer UW (1997): Transplantation of filgrastim mobilized peripheral blood stem cells from HLA-identical sibling or alternative family donors in patients with hematologic malignancies: A prospective comparison on clinical outcome, immune reconstitution, and hematopoietic chimerism. *Blood* 90:4725–4735.

Bensinger W, Appelbaum F, Rowley S, Storb R, Sanders J, Lilleby K, Gooley T, Demirer T, Schiffman K, Weaver C, et al. (1995): Factors that influence collection and engraftment of autologous peripheral-blood stem cells. *J Clin Oncol* 13:2547–2555.

Beyer J, Schwella N, Zingsem J, Strohscheer I, Schwaner I, Oettle H, Serke S, Huhn D, Stieger W (1995): Hematopoietic rescue after high-dose chemotherapy using autologous peripheral-blood progenitor cells or bone marrow: A randomized comparison. *J Clin Oncol* 13:1328–1335.

Brasel K, McKenna HJ, Charrier K, Morrissey PJ, Williams DE, Lyman SD (1997): Fl+-3 ligand synergizes with granulocyte-macrophage colony-stimulating factor or granulocyte colony-stimulating factor to mobilize hematopoietic progenitor cells into the peripheral blood of mice. *Blood* 90:3781–3788.

Brenner MK, Rill DR, Holladay MS, Heslop HE, Moen RC, Buschle M, Krance RA, Santana VM, Anderson WF, Ihle JN (1993): Gene marking to determine whether autologous marrow infusion restores long-term haemopoiesis in cancer patients. *Lancet* 342:1134–1137.

Brenner MK, Rill DR, Moen RC, Krance RA, Mirro J, Jr, Anderson WF, Ihld JN (1993): Gene-marking to trace origin of relapse after autologous bone-marrow transplantation. *Lancet* 341:85–86.

Broudy VC, Lin NL, Kaushansky K (1995): Thrombopoietin (c-mpl ligand) acts synergistically with erythropoietin, stem cell factor, and interleukin-11 to enhance murine megakaryocyte colony growth and increases megakaryocyte ploidy *in vitro*. *Blood* 85:1719–1726.

Broxmeyer HE, Carow CE (1993): Characterization of cord blood stem/progenitor cells. *J Hematother* 2:197–199.

Broxmeyer HE, Hangoc G, Cooper S, Ribeiro RC, Graves V, Yoder M, Wagner J, Vadhan-Raj S, Benninger L, Rubinstein P, et al. (1992): Growth characteristics and expansion of human umbilical cord blood and estimation of its potential for transplantation in adults. *Proc Nat Acad Sci USA* 89:4109–4113.

Brugger W, Bross KJ, Glatt M, Weber F, Mertelsmann R, Kanz L (1994): Mobilization of tumor cells and hematopoietic progenitor cells into peripheral blood of patients with solid tumors. *Blood* 83:636–640.

Cairo MS, Wagner JE (1997): Placental and/or Umbilical cord blood: An alternative source of hematopoietic stem cells for transplantation. *Blood* 90:4665–4678.

Cashman JD, Eaves AC, Raines EW, Ross R, Eaves CJ (1990): Mechanisms that regulate the cell cycle status of very primitive hematopoietic cells in long-term human marrow cultures. I. Stimulatory role of a variety of mesenchymal cell activators and inhibitory role of TGF-beta. *Blood* 75:96–101.

Caux C, Massacrier C, Dezutter-Dambuyant C, Vanbervliet B, Jacquet C, Schmitt D, Banchereau J (1995): Human dendritic Langerhans cells generated *in vitro* from $CD34^+$ progenitors can prime naive $CD4^+$ T cells and process soluble antigen. *J Immunol* 155:5427–5435.

Chaudhary PM, Roninson IB (1991): Expression and activity of P-glycoprotein, a multidrug efflux pump, in human hematopoietic stem cells. *Cell* 66:85–94.

Cheng J, Baumhueter S, Cacalano G, Carver-Moore K, Thibodeaux H, Thomas R, Broxmeyer HE, Cooper S, Hague N, Moore M, Lasky LA (1996): Hematopoietic defects in mice lacking the sialomucin CD34. *Blood* 87:479–490.

Craddock CF, Nakamoto B, Andrew RG, Priestley GV, Papayannopoulou T (1997): Antibodies to VLA-4 integrin mobilize long-term repopulating cells and augment cytokine-induced mobilization in primates and mice. *Blood* 90:4779–4788.

Craig W, Kay R, Cutler RL, Lansdorp PM (1993): Expression of Thy-1 on human hematopoietic progenitor cells. *J Exp Med* 177:1331–1342.

Crown J, Kritz A, Vahdat L, Reich L, Moore M, Hamilton N, Schneider J, Harrison M, Gilewski T, Hudis C, et al (1993): Rapid administration of multiple cycles of high-dose myelosuppressive chemotherapy in patients with metastatic breast cancer. *J Clin Oncol* 11:1144–1149.

Deryugina EI, Muller-Sieburg CE (1993): Stromal cells in long-term cultures: Keys to the elucidation of hematopoietic development? *Crit Rev Immunol* 13:115–150.

Dexter TM, Chang J (1994): New strategies for the treatment of chronic myeloid leukemia. *Blood* 84:673–675.

DiGiusto DL, Lee R, Moon J, Moss K, O'Toole T, Voytovich A, Webster D, Mule JJ (1996): Hematopoietic potential of cryopreserved and *ex vivo* manipulated umbilical cord blood progenitor cells evaluated *in vitro* and *in vivo*. *Blood* 87:1261–1271.

Dreger P, Kloss M, Petersen B, Haferlach T, Loffler H, Loeffler M, Schmitz N (1995): Autologous progenitor cell transplantation: Prior exposure to stem cell–toxic drugs determines yield and engraftment of peripheral blood progenitor cell but not of bone marrow grafts. *Blood* 86:3970–3978.

Emerman JT, Eaves CJ (1994): Lack of effect of hematopoietic growth factors on human breast epithelial cell growth in serum-free primary culture. *Bone Marrow Transplant* 13:285–291.

Engelhardt M, Kumar R, Albanell J, Pettengell R, Han W, Moore MAS (1997): Telomerase regulation, cell cycle and telomere stability in primitive hematopoietic cells. *Blood* 90:182–193.

Farley TJ, Ahmed T, Fitzgerald M, Preti RA (1997): Optimization of $CD34^+$ cell selection using immunomagnetic beads: Implications for use in cryopreserved peripheral blood stem cell collections. *Journal of Hematotherapy* 6:53–60.

Flasshove M, Banerjee D, Bertino JR, Moore MA (1995a): Increased resistance to methotrexate in human hematopoietic cells after gene transfer of the Ser31 DHFR mutant. *Leukemia* 9(suppl 1):S34-7.

Flasshove M, Banerjee D, Mineishi S, Li MX, Bertino JR, Moore MA (1995b): *Ex vivo* expansion and selection of human CD34$^+$ peripheral blood progenitor cells after introduction of a mutated dihydrofolate reductase cDNA via retroviral gene transfer. *Blood* 85:566–574.

Galy AH, Webb S, Cen D, Murray LJ, Condino J, Negrin RS, Chen BP (1994): Generation of T cells from cytokine-mobilized peripheral blood and adult bone marrow CD34$^+$ cells. *Blood* 84:104–110.

Garrigan K, Moronirawson P, Mcmurray C, Hermans I, Abernethy N, Watson J, Ronchese F (1996): Functional comparison of spleen dendritic cells and dendritic cells cultured *in vitro* from bone-marrow precursors. *Blood* 88:3508–3512.

Gianni AM, Bregni M, Siena S, et al. (1997): High-dose chemotherapy and autologous bone marrow transplantation compared with MACOP-B in aggressive B-cell lymphoma. *N Engl J Med* 336:1290–1297.

Hao QL, Shah AJ, Thiemann FT, Smogorzewska EM, Crooks GM (1995): A functional comparison of CD34$^+$ CD38$^-$ cells in cord blood and bone marrow. *Blood* 86:3745–3753.

Harris DT (1995): *In vitro* and *in vivo* assessment of the graft-versus-leukemia activity of cord blood. *Bone Marrow Transplant* 15:17–23.

Hayflick L, Moorehead PS (1961): The serial cultivation of human diploid cell strains. *Exp Cell Res* 25:585–621.

Haylock DN, To LB, Dowse TL, Juttner CA, Simmons PJ (1992): *Ex vivo* expansion and maturation of peripheral blood CD34$^+$ cells into the myeloid lineage. *Blood* 80:1405–1412.

Henschler R, Brugger W, Luft T, Frey T, Mertelsmann R, Kanz L (1994): Maintenance of transplantation potential in *ex vivo* expanded CD34($^+$)-selected human peripheral blood progenitor cells. *Blood* 84:2898–9203.

Hirao A, Kawano Y, Takaue Y, Suzue T, Abe T, Sato J, Saito S, Okamoto Y, Makimoto A, Kawahito M, et al. (1994): Engraftment potential of peripheral and cord blood stem cells evaluated by a long-term culture system. *Exp Hematol* 22:521–526.

Holt SE, Shay JW, Wright WE (1996): Refining the telomere–telomerase hypothesis of aging and cancer. *Nature Biotechnol* 14:836–839.

Hows JM, Bradley BA, Marsh JC, Luft T, Coutinho L, Testa NG, Dexter TM (1992): Growth of human umbilical-cord blood in longterm haemopoietic cultures. *Lancet* 340:73–76.

Humeau L, Chabannon C, Firpo MT, Mannoni P, Bagnis C, Roncarolo M-G, Namikawa R (1997): Successful reconstitution of human hematopoiesis in the SCID-Hu mouse by genetically modified, highly enriched progenitors isolated from fetal liver. *Blood* 90:3496–3506.

Jacobsen SE, Veiby OP, Myklebust J, Okkenhaug C, Lyman SD (1996): Ability of flt3 ligand to stimulate the *in vitro* growth of primitive murine hematopoietic progenitors is potently and directly inhibited by transforming growth factor-beta and tumor necrosis factor-alpha. *Blood* 87:5016–5026.

Krause DS, Fackler MJ, Civin CI, May WS (1996): CD34: Structure, biology and clinical utility. *Blood* 87:1–3.

Koller MR, Bender JG, Miller WM, Papoutsakis ET (1993a): Expansion of primitive human hematopoietic progenitors in a perfusion bioreactor system with IL-3, IL-6, and stem cell factor. *Bio/Technology* 11:358–363.

Koller MR, Emerson SG, Palsson BO (1993b): Large-scale expansion of human stem and progenitor cells from bone marrow mononuclear cells in continuous perfusion cultures. *Blood* 82:378–384.

Korbling M, Huh YO, Durett A, Mirza N, Miller P, Engel H, Anderlini P, van Besien K, Andreeff M, Przepiorka D, et al. (1995): Allogeneic blood stem cell transplantation: Peripheralization and yield of donor-derived primitive hematopoietic progenitor cells ($CD34^+$ Thy-1 dim) and lymphoid subsets, and possible predictors of engraftment and graft-versus-host disease. *Blood* 86:2842–2848.

Kurtzberg J, Laughlin M, Graham ML, Smith C, Olson JF, Halperin EC, Ciocci G, Carrier C, Stevens CE, Rubinstein P (1996): Placental blood as a source of hematopoietic stem cells from transplantation into unrelated recipients. *N Engl J Med* 335:157–166.

Kyoizumi S, Baum CM, Kaneshima H, McCune JM, Yee EJ, Namikawa R (1992): Implantation and maintenance of functional human bone marrow in SCID-hu mice. *Blood* 79:1704–1711.

Lansdorp PM, Dragowska W, Mayani H (1993): Ontogeny-related changes in proliferative potential of human hematopoietic cells. *J Exp Med* 178:787–791.

Lapidot T, Pflumio F, Doedens M, Murdoch B, Williams DE, Dick JE (1992): Cytokine stimulation of multilineage hematopoiesis from immature human cells engrafted in SCID mice. *Science* 255:1137–1141.

Laporte JP, Gorin NC, Rubinstein P, Lesage S, Portnoi MF, Barbu V, Lopez M, Douay L, Najman A (1996): Cord-blood transplantation from an unrelated donor in an adult with chronic myelogenous leukemia. *N Engl J Med* 335:167–170.

Lebkowski JS, Schain LR, Okrongly D, Levinsky R, Harvey MJ, Okarma TB (1992): Rapid isolation of human CD34 hematopoietic stem cells—Purging of human tumor cells. *Transplantation* 53:1011–1019.

Lebsack MG, Hoek JA, Maraskovsky E, McKenna HJ (1997): Flt3-ligand induces stem and dendritic cell mobilization in healthy volunteers. Proceedings of ISHAGE 3rd International Meeting p 49.

Mavroudis D, Read E, Cottlerfox M, Couriel D, Molldrem J, Carter C, Yu M, Dunbar C, Barrett J (1996): Cd34(+) cell dose predicts survival, posttransplant morbidity, and rate of hematologic recovery after allogeneic marrow transplants for hematologic malignancies. *Blood* 88:3223–3229.

Mayani H, Lansdorp PM (1994): Thy-1 expression is linked to functional properties of primitive hematopoietic progenitor cells from human umbilical cord blood. *Blood* 83:2410–2417.

Mayani H, Lansdorp PM (1995): Proliferation of individual hematopoietic progenitors purified from umbilical cord blood. *Exp Hematol* 23:1453–1462.

McKenna HJ, de Vries P, Brasel K, Lyman SD, Williams DE (1995): Effect of flt3 ligand on the *ex vivo* expansion of human $CD34^+$ hematopoietic progenitor cells. *Blood* 86:3413–3420.

McNiece I, Briddell R, Stoney G, Kern B, Zilm K, Recktenwald D, Miltenyi S (1997): Large-scale isolation of $CD34^+$ cells using the Amgen cell selection device results in high levels of purity and recovery. *Journal of Hematotherapy* 6:5–11.

Miller JS, Klingsporn S, Lund J, Perry EH, Verfaillie C, McGlave P (1994): Large scale *ex vivo* expansion and activation of human natural killer cells for autologous therapy. *Bone Marrow Transplant* 14:555–562.

Miraglia S, Godfrey W, Yin AH, Atkins K, Warnke R, Holden JT, Bray RA, Waller EK, Buck DW (1997): A novel five-transmembrane hematopoietic stem cell antigen: Isolation, characterization, and molecular cloning. *Blood* 90:5013–5021.

Mohle R, Pforsich M, Fruehauf S, Witt B, Kramer A, Haas R (1994): Filgastim postchemotherapy mobilizes more $CD34^+$ cells with a different antigenic profile compared with use during steady-state hematopoiesis [published erratum appears in Bone Marrow Transplant 1995 15(4):655]. *Bone Marrow Transplant* 14:827–832.

Moore MAS (1995): Hematopoietic reconstruction—New approaches. *Clin Cancer Res* 1:3–9.

Murray L, Chen B, Galy A, Chen S, Tushinski R, Uchida N, Negrin R, Tricot G, Jagannath S, Vesole D, et al (1995): Enrichment of human hematopoietic stem cell activity in the $CD34^+Thy^-1^+Lin^-$ subpopulation from mobilized peripheral blood. *Blood* 85:368–378.

Naparstek E, Hardan Y, Ben-Shahar M, Nagler A, Or R, Mumcuoglu M, Weiss L, Samuel, S, Slavin S (1992): Enhanced marrow recovery by short preincubation of marrow allografts with human recombinant interleukin-3 and granulocyte-macrophage colony-stimulating factor. *Blood* 80:1673–1678.

Notaro R, Cimmino A, Tabarini D, Rotoli B, Luzzatto L (1997): *In vivo* dynamics of human hematopoietic stem cells. *Proc Natl Acad Sci USA* 94:13782–85.

Ottinger HD, Beelen DW, Scheulen B, Schaefer UW, Grossewilde H (1996): Improved immune reconstitution after allotransplantation of peripheral-blood stem-cells instead of bone-marrow. *Blood* 88:2775–2779.

Papayannopoulou T, Nakamoto B (1993): Peripheralization of hemopoietic progenitors in primates treated with anti-VLA4 integrin. *Proc Natl Acad Sci USA* 90:9374–9378.

Pawliuk R, Eaves C, Humphries RK (1996): Evidence of both ontogeny and transplant dose-regulated expansion of hematopoietic stem-cells *in vivo*. *Blood* 88:2852–2858.

Pettengell R, Luft T, Henschler R, Hows JM, Dexter TM, Ryer D, Testa NG (1994): Direct comparison by limiting dilution analysis of long-term culture-initiating cells in human bone marrow, umbilical cord blood, and blood stem cells. *Blood* 84:3653–3659.

Pettengell R, Morgenstern GR, Woll PJ, Chang J, Rowlands M, Young R, Radford JA, Scarffe JH, Testa NG, Crowther D (1993a): Peripheral blood progenitor cell transplantation in lymphoma and leukemia using a single apheresis. *Blood* 82:3770–3777.

Pettengell R, Shido K, Kanz L, Berman E, Moore MAS (1995a): Preferential expansion of hematopoietic versus acute myeloid-leukemia cells *ex vivo*. *Blood* 86:568.

Pettengell R, Testa NG (1995): Biology of blood progenitor cells used in transplantation. *Int J Hematol* 61:1–15.

Pettengell R, Testa NG, Swindell R, Crowther D, Dexter TM (1993b): Transplantation potential of hematopoietic cells released into the circulation during routine chemotherapy for non-Hodgkin's lymphoma. *Blood* 82:2239–2248.

Pettengell R, Woll PJ, Thatcher N, Dexter TM, Testa NG (1995b): Multicyclic, dose-intensive chemotherapy supported by sequential reinfusion of hematopoietic progenitors in whole blood. *J Clin Oncol* 13:148–156.

Piacibello W, Sanavio F, Garetto L, Severino A, Bergandi D, Ferrario J, Fagioli F, Berger M, Aglietta M (1997): Extensive amplification and self-renewal of human primitive haematopoietic stem cells from cord blood. *Blood* 89:2644–2653.

Ratajczak MZ, Ratajczak J, Kregenow DA, Gewirtz AM (1994): Growth factor stimulation of cryopreserved CD34+ bone marrow cells intended for transplant: An *in vitro* study to determine optimal timing of exposure to early acting cytokines. *Stem Cells* 12:599–603.

Rizzoli V, Carlo-Stella C (1995): Stem cell purging: An intriguing dilemma. *Exp Hematol* 23:296–302.

Roberts MM, To LB, Gillis D, Mundy J, Rawling C, Ng K, Juttner CA (1993): Immune reconstitution following peripheral blood stem cell transplantation, autologous bone marrow transplantation and allogeneic bone marrow transplantation. *Bone Marrow Transplant* 12:469–475.

Rosenzwajg M, Canque B, Gluckman JC (1996): Human dendritic cell-differentiation pathway from cd34(+) hematopoietic precursor cells. *Blood* 87:535–544.

Rusten LS, Lyman SD, Veiby OP, Jacobsen SE (1996): The FLT3 ligand is a direct and potent stimulator of the growth of primitive and committed human CD34+ bone marrow progenitor cells *in vitro. Blood* 87:1317–1325.

Scheid C, Pettengell R, Ghielmini M, Radford JA, Morgenstern GR, Stern PL, Crowther D (1995): Time-course of the recovery of cellular immune function after high-dose chemotherapy and peripheral blood progenitor cell transplantation for high-grade non-Hodgkin's lymphoma. *Bone Marrow Transplant* 15:901–906.

Schmitz N, Dreger P, Suttorp M, Rohwedder EB, Haferlach T, Loffler H, Hunter A, Russell NH (1995): Primary transplantation of allogeneic peripheral blood progenitor cells mobilized by filgrastim (granulocyte colony-stimulating factor). *Blood* 85:1666–1672.

Schmitz N, Linch DC, Dreger P, Goldstone AH, Boogaerts MA, Ferrant A, Demuynck MH, Link H, Zander A, Barge A, et al. (1996): Randomised trial of filgrastim-mobilised peripheral blood progenitor cell trasnplantation versus autologous bone-marrow transplantation in lymphoma patients. *Lancet* 347:353–357.

Schneider JG, Crown JP, Wasserheit C, Kritz A, Wong G, Reich L, Norton L, Moore MA (1994): Factors affecting the mobilization of primitive and committed hematopoietic progenitors into the peripheral blood of cancer patients. *Bone Marrow Transplant* 14:877–884.

Shapiro F, Yao TJ, Moskowitz C, Reich L, Weust DL, Heimfeld S, McNiece I, Gabrilove JL, Nimer S, Moore MAS (1997): The effects of prior chemotherapy on the *in vitro* proliferative potential of stem cell factor plus filgrastim-mobilized CD34-positive progenitor cells. Submitted. *Clinical Cancer Research* 3:1571–1578.

Shapiro F, Yao TJ, Raptis G, Reich L, Norton L, Moore MA (1994): Optimization of conditions for *ex vivo* expansion of CD34+ cells from patients with stage IV breast cancer. *Blood* 84:3567–3574.

Shea TC, Mason JR, Storniolo AM, Newton B, Breslin M, Mullen M, Ward DM, Miller L, Christian M, Taetle R (1992): Sequential cycles of high-dose carboplatin administered with recombinant human granulocyte-macrophage colony-stimulating factor and repeated infusions of autologous peripheral-blood progenitor cells: A novel and effective method for delivering multiple courses of dose-intensive therapy. *J Clin Oncol* 10:464–473.

Sheridan WP, Begley CG, Juttner CA, Szer J, To LB, Maher D, McGrath KM, Morstyn G, Fox RM (1992): Effect of peripheral-blood progenitor cells mobilised by filgrastim (G-CSF) on platelet recovery after high-dose chemotherapy. *Lancet* 339:640–644.

Shpall EJ, Gee AP, Hogan C, Cagnoni P, Gehling U, Hami L, Franklin W, Bearman SI, Ross M, Jones RB (1996): Bone marrow metastases. *Hematol Oncol Clin North Am* 10:321–343.

Shpall EJ, Jones RB, Bearman SI, Franklin WA, Archer PG, Curiel T, Bitter M, Claman HN, Stemmer SM, Purdy M, et al. (1994): Transplantation of enriched CD34-positive autologous marrow into breast cancer patients following high-dose chemotherapy: Influence of CD34-positive peripheral-blood progenitors and growth factors on engraftment. *J Clin Oncol* 12:28–36.

Srour EF, Bregni M, Traycoff CM, Ero BA, Kosak ST, Hoffman R, Siena S, Gianni AM (1996): Long-term hematopoietic culture-initiating cells are more abundant in mobilized peripheral-blood grafts than in bone-marrow but have a more limited *ex vivo* expansion potential. *Blood Cells Mol Dis* 22:68–81.

Srour EF, Zanjani ED, Cornetta K, Traycoff CM, Flake AW, Hedrick M, Brandt JE, Leemhuis T, Hoffman R (1993): Persistence of human multilineage, self-renewing lymphohematopoietic stem cells in chimeric sheep. *Blood* 82:3333–3342.

Stewart FM, Crittenden RB, Lowry PA, Pearson-White S, Quesenberry PJ (1993): Long-term engraftment of normal and post-5-fluorouracil murine marrow into normal nonmyeloablated mice. *Blood* 81:2566–2571.

Strunk D, Rappersberger K, Egger C, Strobl H, Kromer E, Elbe A, Maurer D, Stingl G (1996): Generation of human dendritic cells/Langerhans cells from circulating $CD34^+$ hematopoietic progenitor cells. *Blood* 87:1292–1302.

Sutherland HJ, Eaves CJ, Lansdorp PM, Phillips GL, Hogge DE (1994): Kinetics of committed and primitive blood progenitor mobilization after chemotherapy and growth factor treatment and their use in autotransplants. *Blood* 83:3808–3814.

Sutherland HJ, Lansdorp PM, Henkelman DH, Eaves AC, Eaves CJ (1990): Functional characterization of individual human hematopoietic stem cells cultured at limiting dilution on supportive marrow stromal layers. *Proc Nat Acad Sci USA* 87:3584–3588.

Szabolcs P, Feller ED, Moore MA, Young JW (1995): Progenitor recruitment and *in vitro* expansion of immunostimulatory dendritic cells from human $CD34^+$ bone marrow cells by c-kit–ligand, GM-CSF, and TNF alpha. *Adv Exp Med Biol* 378:17–20.

Tepler I, Cannistra SA, Frei E, 3d, Gonin R, Anderson KC, Demetri G, Niloff J, Goodman H, Muntz H, Muto M, et al. (1993): Use of peripheral-blood progenitor cells abrogates the myelotoxicity of repetitive outpatient high-dose carboplatin and cyclophosphamide chemotherapy. *J Clin Oncol* 11:1583–1591.

Tjonnfjord GE, Steen R, Evensen SA, Thorsby E, Egeland T (1994): Characterization of $CD34^+$ peripheral blood cells from healthy adults mobilized by recombinant human granulocyte colony-stimulating factor. *Blood* 84:2795–2801.

To LB, Haylock DN, Dowse T, Simmons PJ, Trimboli S, Ashman LK, Juttner CA (1994): A comparative study of the phenotype and proliferative capacity of peripheral blood (PB) $CD34^+$ cells mobilized by four different protocols and those of steady-phase PB and bone marrow $CD34^+$ cells. *Blood* 84:2930–2939.

Traycoff CM, Abboud MR, Laver J, Clapp DW, Srour EF (1994): Rapid exit from G0/G1 phases of cell cycle in response to stem cell factor confers on umbilical cord blood $CD34^+$ cells an enhanced *ex vivo* expansion potential. *Exp Hematol* 22:1264–1272.

Traycoff CM, Kosak ST, Grigsby S, Srour EF (1995): Evaluation of *ex vivo* expansion potential of cord blood and bone marrow hematopoietic progenitor cells using cell tracking and limiting dilution analysis. *Blood* 85:2059–2068.

Udomsakdi C, Eaves CJ, Swolin B, Reid DS, Barnett MJ, Eaves AC (1992): Rapid decline of chronic myeloid leukemic cells in long-term culture due to a defect at the leukemic stem cell level. *Proc Nat Acad Sci USA* 89:6192–6196.

Van Zant G, Rummel SA, Koller MR, Larson DB, Drubachevsky I, Palsson M, Emerson SG (1994): Expansion in bioreactors of human progenitor populations from cord blood and mobilized peripheral blood. *Blood Cells* 20:482–491.

Vaziri H, Dragowska W, Allsopp RC, Thomas TE, Harley CB, Lansdorp PM (1994): Evidence for a mitotic clock in human hematopoietic stem cells: Loss of telomeric DNA with age. *Proc Nat Acad Sci USA* 91:9857–9860.

Vaziri H, Schachter F, Uchida I, Wei L, Zhu X, Effros R, Cohen D, Harley CB (1993): Loss of telomeric DNA during aging of normal and trisomy 21 human lymphocytes. *Am J Hum Genet* 52:661–667.

Verbik DJ, Jackson JD, Pirruccello SJ, Patil KD, Kessinger A, Joshi SS (1995): Functional and phenotypic characterization of human peripheral-blood stem-cell harvests—A comparative analysis of cells from consecutive collections. *Blood* 85:1964–1970.

Verfaillie CM (1992): Direct contact between human primitive hematopoietic progenitors and bone marrow stroma is not required for long-term *in vitro* hematopoiesis. *Blood* 79:2821–2826.

Verfaillie CM (1993): Soluble factor(s) produced by human bone marrow stroma increase cytokine-induced proliferation and maturation of primitive hematopoietic progenitors while preventing their terminal differentiation. *Blood* 82:2045–2053.

Vogel W, Behringer D, Scheding S, Kanz L, Brugger W (1996): *Ex vivo* expansion of cd34($^+$) peripheral-blood progenitor cells—Implications for the expansion of contaminating epithelial tumor cells. *Blood* 88:2707–2713.

Vose JM, Anderson JR, Kessinger A, Bierman PJ, Coccia P, Reed EC, Gordon B, Armitage JO (1993): High-dose chemotherapy and autologous hematopoietic stem-cell transplantation for aggressive non-Hodgkin's lymphoma. *J Clin Oncol* 11:1846–1851.

Wagner JE, Kernan NA, Steinbuch M, Broxmeyer HE, Gluckman E (1995): Allogeneic sibling umbilical-cord-blood transplantation in children with malignant and non-malignant disease. *Lancet* 346:214–219.

Wagner JE Rosenthal J, Sweetman R, Shu XO, Davies SM, Ramsay NK, McGlave PB, Sender L, Cairo MS (1996): Successful transplantation of HLA-matched and HLA-mismatched umbilical cord blood from unrelated donors: Analysis of engraftment and acute graft-versus-host disease. *Blood* 88:795–802.

Watt SM, Visser JW (1992): Recent advances in the growth and isolation of primitive human haemopoietic progenitor cells. *Cell Proliferation* 25:263–297.

Watts MJ, Sullivan AM, Ings SJ, Leverett D, Peniket AJ, Perry AR, Williams CD, Devereux S, Goldstone AH, Linch DC (1997): Evaluation of clinical scale CD34$^+$ cell purification: experience of 71 immunoaffinity column procedures. *Bone Marrow Transplantation* 20:157–162.

Weaver A, Ryder D, Crowther D, Dexter TM, Testa NG (1996): Increased numbers of long-term culture-initiating cells in the apheresis product of patients randomized to receive increasing doses of stem-cell factor administered in combination with chemotherapy and a standard-dose of granulocyte-colony-stimulating factor. *Blood* 88:3323–3328.

Weaver CH, Hazelton B, Birch R, Palmer P, Allen C, Schwartzberg L, West W (1995): An analysis of engraftment kinetics as a function of the CD34 content of peripheral blood progenitor cell collections in 692 patients after the administration of myeloablative chemotherapy. *Blood* 86:3961–3969.

Wengler GS, Lanfranchi A, Frusca T, Verardi R, Neva A, Brugnoni D, Giliani S, Fiorini M, Mella P, Guandalini F, Mazzolari E, Pecorelli S, Notarangelo LD, Porta F, Ugazio AG (1996): In-utero transplantation of parental cd34 hematopoietic progenitor cells in a patient with X-linked severe combined immunodeficiency (scidxi). *Lancet* 348:1484–1487.

Widmer L, Pichert G, Jost LM, Stahel RA (1995): Autologous transplantation of cd34$^+$-selected and immunomagnetically purged hematopoietic progenitor cells mediates rapid hematological recovery after high-dose chemotherapy in patients with follicular lymphoma. *Blood* 86:9145.

Williams SF, Lee WJ, Bender JG, Zimmerman T, Swinney P, Blake M, Carreon J, Schilling M, Smith S, Williams DE, Oldham F, Van Epps D (1996): Selection and expansion of peripheral blood CD34$^+$ cells in autologous stem cell transplantation for breast cancer. *Blood* 87:1687–1691.

Wineman J, Moore K, Lemischka I, Muller-Sieburg C (1996): Functional heterogeneity of the hematopoietic microenvironment: Rare stromal elements maintain long-term repopulating stem cells. *Blood* 87:4082–4090.

Woll PJ, Hodgetts J, Lomax L, Bildet F, Cour-Chabernaud V, Thatcher N (1995): Can cytotoxic dosed-intensity be increased by using granulocyte colony-stimulating factor? A randomized controlled trial of lenograstim in small-cell lung cancer. *J Clin Oncol* 13:652–659.

Wynn RF, Cross MA, Hatton C, Will AM, Lashford LS, Dexter TM, Testa NG (1998): Accelerated telomere shortening in young recipients of allogeneic bone-marrow transplants. *Lancet* 351:178–181.

Yin AH, Miraglia S, Zanjani ED, Almeida-Porada G, Ogawa M, Leary AG, Olwaus J, Kearney J, Buck DW (1997): AC133, a novel marker for human hematopoietic stem and progenitor cells. *Blood* 90:5002–5012.

Young JC, Varma A, DiGiusto D, Backer MP (1996): Retention of quiescent hematopoietic cells with high proliferative potential during *ex vivo* stem cell culture. *Blood* 87:545–556.

Young JW, Inaba K (1996): Dendritic cells as adjuvants for class-i major histocompatibility complex-restricted antitumor immunity. *J Exp Med* 183:7–11.

Zeigler FC, Bennett BD, Jordan CT, Spencer SD, Baumhueter S, Carroll KJ, Hooley J, Bauer K, Matthews W (1994): Cellular and molecular characterization of the role of the flk-2/flt-3 receptor tyrosine kinase in hematopoietic stem cells. *Blood* 84:2422–2430.

Zhou YQ, Levesque JP, Hatzfeld A, Cardoso AA, Li ML, Sansilvestri P, Hatzfeld J (1993): Fibrinogen potentiates the effect of interleukin-3 on early human hematopoietic progenitors. *Blood* 82:800–806.

Zhu X, Kumar R, Mandal M, Sharma N, Sharma HW, Dhingra U, Sokoloski JA, Hsiao R, Narayanan R (1996): Cell cycle–dependent modulation of telomerase activity in tumor cells. *Proc Nat Acad Sci USA* 93:6091–6095.

6

DELIVERY SYSTEMS FOR GENE THERAPY: THE ADENOVIRUS

THOMAS SHENK

Howard Hughes Medical Institute, Department of Molecular Biology, Princeton University, Princeton, NJ 08544-1014

INTRODUCTION

Adenoviruses were first isolated in 1953 from patients with acute respiratory illness (Rowe et al., 1953; Hilleman and Werner, 1954). Subsequently, members of the adenovirus family have been identified that infect a wide range of mammalian and avian hosts, and 50 human serotypes have been distinguished on the basis of cross-neutralization studies. Soon after their discovery, it became clear that adenoviruses generally are not responsible for the common cold. They are responsible for only a minor portion of acute respiratory morbidity in both children and adults. Besides respiratory disease, some adenovirus serotypes cause conjunctivitis, gastroenteritis, and a variety of other less common syndromes.

Adenoviruses became the subjects of intense study after human adenovirus type 12 (Ad12) was shown to induce malignant tumors in newborn hamsters (Trentin et al., 1962). This finding, together with the fact that adenoviruses contain a DNA genome, led them to be classified as DNA tumor viruses. Extensive surveys failed to reveal an association between adenovirus and human malignancy, but the investigation of adenovirus-coded oncoproteins has provided key insights into mechanisms controlling cellular proliferation

Stem Cell Biology and Gene Therapy, Edited by Peter J. Quesenberry, Gary S. Stein, Bernard G. Forget, and Sherman M. Weissman
ISBN 0-471-14656-0

and has revealed key points at which this regulatory machinery is sensitive to perturbations that lead to cancer.

Adenovirus-based vectors for the delivery of therapeutic genes are now under development. The group C viruses, adenovirus type 2 and type 5 (Ad2 and Ad5), are the most intensively studied serotypes at the molecular level, and vector applications have focused almost exclusively on them. The closely related Ad2 and Ad5 are attractive vector candidates. Their structural and genetic organizations, are well as their gene functions and interactions with host cells, are relatively well understood (reviewed by Shenk, 1996). Procedures have been developed that permit the facile manipulation of the viral genome to produce derivatives that lack undesirable viral genes and carry nonviral genes. Adenoviruses and adenovirus vectors can be propagated relatively easily to produce high-titer stocks. Adenoviruses will express their genes in cells that are not actively growing and dividing, suggesting that they might be especially useful for *in vivo* vector applications where the majority of target cells will not be growing. Most importantly, administration of these viruses to humans appears to be safe. Most children exhibit immunological evidence of infection with group C adenoviruses at an early age without serious morbidity, and adenovirus vaccines have been used extensively in the military with good results (reviewed in Horwitz, 1996).

ADENOVIRUS BIOLOGY

Horwitz (1996) and Shenk (1996) provide detailed recent overviews of adenovirus biology. All adenoviruses contain a linear double-stranded genome of about 36,000 base pairs. The DNA is associated with four basic virus-coded proteins to form a core structure that is encapsidated in a protein shell comprised of seven virus-coded polypeptides. The icosahedral virions are about 70 nm in diameter, and they are distinguished by the presence of projections, comprised of three copies of the fiber polypeptide, from each vertex of the icosahedron.

With one exception (the avian CELO virus; Chiocca et al., 1996), all adenovirus genomes that have been examined exhibit the same organization, that is, the *cis*-acting elements and the genes encoding specific functions are located at the same position on the chromosome (Fig. 1A). The genome includes three known *cis*-acting sequences that comprise several hundred base pairs, and these must be maintained for its propagation as a virus. Two are origins of DNA replication that are located at the ends of the viral DNA, and the third is a packaging sequence that is found near the left end of the chromosome. The packaging sequence must be located within several hundred base pairs of the end of the chromosome to direct the interaction of the viral DNA with its encapsidating proteins (Hearing et al., 1987). The genome includes seven units transcribed by RNA polymerase II and one or two units, depending on the serotype, transcribed by RNA polymerase III. Transcripts from the

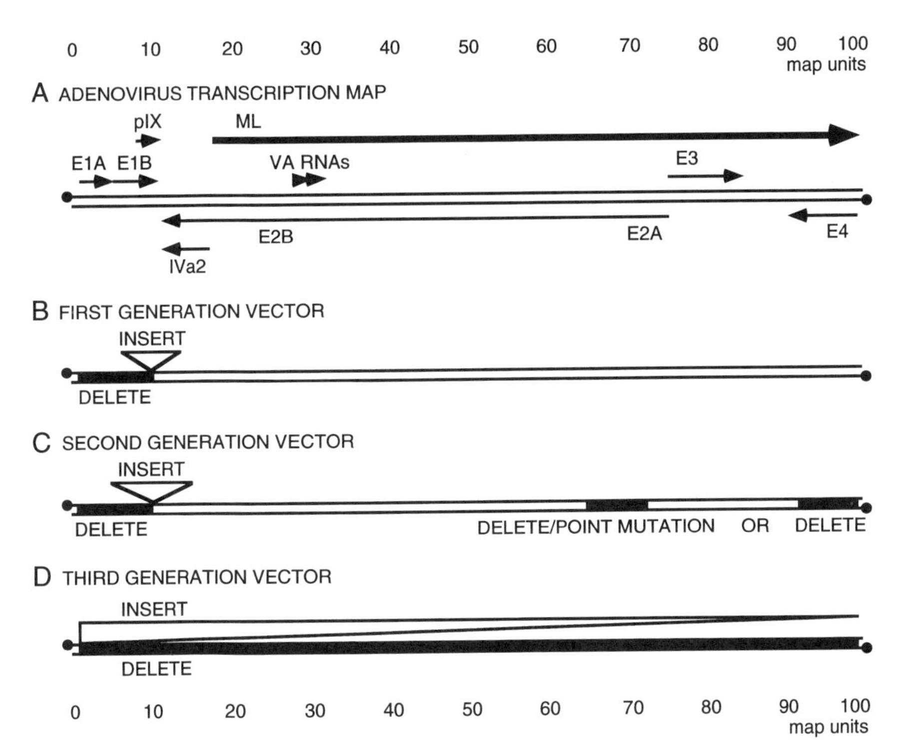

FIGURE 1. Diagram of the adenovirus transcription map and current adenovirus vectors. **A:** The adenovirus map is presented in its conventional orientation. The viral DNA is represented by two parallel lines, and the closed circles at the ends of the DNA strands represent the viral terminal protein that is covalently attached to 5′ ends. Transcription units are designated by arrows whose directions indicate the direction of transcription. Units transcribed by RNA polymerase II: early region 1A (E1A), early region 1B (E1B), early region 2A (E2A), early region 2B (E2B), early region 3 (E3), early region 4 (E4), intermediate region encoding protein IX (pIX), intermediate region encoding protein IVa2 (IVa2), major late unit (ML). Units transcribed by polymerase III: virus-associated RNAs (VA RNAs). First (**B**), second (**C**), and third (**D**) generation vectors are shown with deletions of viral sequences (delete) and insertions of transgene sequences (insert) indicated.

protein-coding units are processed to give rise to multiple mRNAs by differential splicing and the use of multiple poly(A) addition sites. As a result, the seven transcription units give rise to about 35 known polypeptides.

Lytic Replication in Human Cells

Adenoviruses bind to cells in a two-step process. First, the terminal knob of the viral fiber protein interacts with an unknown cellular receptor, and then the penton base protein interacts with members of the integrin family of cell surface proteins. The fiber projects from the penton base that sits at each vertex of the particle. Both interactions are necessary for successful adsorption and internalization. Adsorbed virus is internalized through receptor-mediated endocytosis, and then, as the pH drops in endosomes, the virus escapes to the cytosol very efficiently through a process that is not yet understood. The disassembly of the virion begins as soon as the virus is internalized. Partially disassembled virions have been visualized at the nuclear pores of infected cells, presumably delivering their DNA to the nucleus. When the viral DNA reaches the nucleus, it associates with the nuclear matrix through a viral polypeptide that is covalently attached to each 5′ end of the chromosome, termed the *terminal protein* (Schaack et al., 1990), and then the program of viral gene expression is initiated.

As for many DNA viruses, adenovirus gene expression is divided by convention into two phases, early and late, that are separated by the onset of viral DNA replication. The first early transcription unit to become active is early region 1A (E1A). E1A proteins are promiscuous transcriptional activating proteins, and they help to activate expression of all the remaining adenovirus transcription units. E1A proteins activate transcription by binding to cellular transcriptional regulatory proteins and modulating their activity. In addition to activating the expression of viral genes, the E1A proteins induce quiescent cells to enter the S phase of the cell cycle. This would be expected to create an environment optimally conducive to viral replication. The key to understanding how E1A proteins manipulate cell cycle regulation came from the observation that a defined set of cellular proteins can be coimmunoprecipitated with antibodies to E1A (Yee and Branton, 1985; Harlow et al., 1986). A 105 kD coprecipitating protein was the first to be identified (Whyte et al., 1988). It is the retinoblastoma tumor suppressor protein, pRB, that normally binds to the cellular E2F transcription factor and inhibits its activity. During normal cell growth, pRB becomes hyperphosphorylated late in the G1 phase of the cell cycle, and this modification causes it to dissociate from E2F, which can then activate the transcription of a series of S phase–specific cellular genes. E1A binds to the same domain on pRB as does E2F, and, when E1A accumulates in the infected cells, it dissociates pRB from E2F, freeing E2F to activate expression of S phase–specific genes (reviewed by Nevins, 1992).

As E1A disrupts the normal regulatory activity of pRB and the other cellular regulatory proteins to which it binds, the p53 tumor suppressor protein is stabilized (Lowe and Ruley, 1993), and its level rises. The elevated level of p53 can antagonize viral replication by blocking cell cycle progression and inducing apoptosis, but two polypeptides encoded by the early region 1B (E1B) transcription unit block these antiviral activities. The E1B–55kD protein binds directly to p53 and interferes with its transcriptional activation function that is needed to block cell cycle progression (Sarnow et al., 1982; Yew et al., 1994). The E1B–19kD protein is related in its sequence to the cellular bcl-2 family of proteins, and, like bcl-2, it blocks the induction of apoptosis by p53 and a variety of other inducers (Rao et al., 1992).

The early region 2 (E2) transcription unit encodes three proteins that function directly in adenovirus DNA replication: DNA polymerase, single-stranded DNA-binding protein, and terminal protein. The terminal protein is covalently attached to a molecule of dGTP, and it serves as a primer, preserving the integrity of the viral termini through multiple rounds of replication. It remains attached to progeny molecules, serving to mediate their attachment to the nuclear matrix, as mentioned above. These three viral proteins function in concert with several cellular proteins to replicate the viral chromosome that appears to proceed in two stages. First, synthesis is initiated at one end of the linear duplex and proceeds toward the other end, generating a new duplex plus a displaced single strand. In the second stage, the single strand circularizes through its self-complementary termini, which contain the *cis*-acting replication origins, to produce a panhandle molecule with the same structure at its ends as at the termini of normal double-stranded viral DNA. When replication is initiated at the end of the panhandle, a second progeny duplex molecule is formed (Lechner and Kelly, 1977).

With the onset of viral DNA replication, adenovirus late genes are expressed efficiently. Most late coding regions are organized into a single large transcription unit known as the major late (ML) unit. Expression of this unit is controlled by the ML promoter, which is weakly active early after infection and highly active late after infection. Its activation is controlled in part by a cascade of adenovirus gene products: the E1A protein accumulates and activates the IVa2 promoter, and then the IVa2 gene product accumulates and binds downstream of the ML transcriptional start site, helping to activate the ML promoter (Tribouley et al., 1994). The primary transcript generated from the ML unit is 29,000 nucleotides long (Evans et al., 1977), and it is processed to generate about 18 different late mRNAs. During the late phase of the viral replication cycle, the cytoplasmic accumulation of cellular mRNAs is inhibited (Beltz and Flint, 1979). The block is mediated by a complex of the E1B–55kD and E4–34kD proteins, and the same complex facilitates the cytoplasmic accumulation of viral mRNAs late after infection (reviewed by Ornelles and Shenk, 1991).

The replication of viral DNA coupled with the synthesis of virion proteins encoded primarily by the ML unit sets the stage for the assembly of progeny virions (reviewed by Hasson et al., 1992). An empty capsid is formed, and then a viral DNA molecule is inserted in a polar fashion, beginning with the left end defined by the conventional adenovirus map. DNA–capsid recognition is mediated by the packaging sequence, but the proteins that bind to this element and mediate the recognition have not yet been identified. A virus-coded cysteine proteinase is essential for the assembly process; it cleaves at least four virion polypeptides to produce a mature viral particle.

Adenovirus can persist in its human host for years after a primary infection. This persistence is undoubtedly facilitated by a series of adenovirus gene products that are known to antagonize host antiviral responses. The E1A proteins and VA RNAs inhibit the response of infected cells to interferon (IF)-α and -β. E1A functions at least in part by blocking the activity of ISGF3, a cellular transcription factor that activates IF-responsive genes (Reich et al., 1988). The two small VA RNAs, which are transcribed by RNA polymerase III, bind to the IF-inducible, double-stranded RNA-dependent protein kinase (PKR), blocking its activation and preventing the phosphorylation of eIF-2α, a translational initiation factor (reviewed by Mathews and Shenk, 1991). In the absence of the VA RNAs, the phosphorylation of eIF-2α by PKR in response to IF leads to the sequestration of eIF-2B in nonfunctional complexes with phosphorylated eIF-2α. This causes cellular protein synthesis to cease, inhibiting viral replication and spread. The other adenovirus gene products known to antagonize host antiviral defenses are encoded by the early region 3 (E3) transcription unit (reviewed by Gooding, 1992). The E3–19kD glycoprotein resides in the membrane of the endoplasmic reticulum and binds to the peptide-binding domain of MHC class I antigens, retaining class I antigen in the endoplasmic reticulum and causing its concentration on the cell surface to decrease. Since cytotoxic T lymphocytes (CTLs) recognize infected cells through foreign antigens displayed on the cell surface in the context of class I antigen, the E3–19kD interaction with class I antigen inhibits recognition of adenovirus-infected cells. A protective role for the E3–19kD protein can be inferred from experiments in cotton rats where pulmonary infections produced by a mutant virus unable to express the E3–19kD protein generate markedly increased inflammation in comparison to infections with wild-type virus (Ginsberg et al., 1989). Finally, both the E3–14.7kD polypeptide and the complex of E3–14.5kD and E3–10.4kD polypeptides prevent cytolysis by tumor necrosis factor (TNF)-α, but their mechanism of action is not understood.

Transformation of Rodent Cells

All human adenoviruses that have been tested can transform cultured rodent cells, and some human serotypes, e.g., Ad12, are able to directly induce sarcomatous tumors in rodents (reviewed by Shenk, 1996). It is also possible to transform rodent cells, which are not fully permissive for replication of human adenoviruses, by infection or by transfection with plasmids encoding the E1A and E1B genes. It is easy to rationalize the transforming activity of the E1A oncoproteins, given their ability to modulate cell cycle regulation by inhibiting the activity of the pRB family of tumor suppressor proteins. E1A alone, however, only rarely induces transformed cells, because, as discussed above, E1A stabilizes p53. It is necessary to inhibit p53 function in order to generate transformed cells, and this is accomplished by the E1B oncoproteins. Thus E1A and E1B cooperate to transform cells, inactivating the function of two tumor suppressor proteins that control cell cycle progression.

Recently, an Ad5 early region 4 (E4)–coded protein has been shown to have oncogenic potential. The E4–34kD protein can bind to p53 and interfere with its ability to activate transcription (Dobner et al., 1996). Given its ability to antagonize p53 function, it is perhaps not surprising that E4–34kD can substitute for E1B and cooperate with E1A to transform cells (Moore and Shenk, 1996). Furthermore, E4–34kD increases the efficiency with which E1A and E1B transform cells, and it enhances the tumorigenicity of transformants.

ADENOVIRUS VECTORS

A variety of earlier reviews summarize the use of adenoviruses as vectors for gene therapy (Goff and Shenk, 1993; Kozarsky and Wilson, 1993; Berns and Giraud, 1995; Kremer and Perricaudet, 1995; Shenk, 1995; Horwitz, 1996).

First-Generation Vectors

The first adenovirus vectors were designed to lack the E1A and E1B genes, and in some cases they also lacked portions of the E3 unit (Fig. 1B). It made good sense to remove E1A and E1B for several reasons. First, safety considerations dictated that these genes should be removed since they encoded products that function as oncoproteins in rodent cells and pervert cell cycle control in human cells. Second, earlier genetic analysis predicted that their removal would inhibit lytic growth of the vector in the human host, a desirable property for a vector that must deliver a therapeutic gene without killing the targeted cell. Third, it is easy to propagate adenoviruses lacking E1A and E1B using 293 cells (Graham et al., 1977). These are adenovirus-transformed human embryonic kidney cells that express the E1A and E1B proteins and efficiently complement viruses lacking one or both of the E1 genes. The E3 gene was deleted in some first-generation vectors to provide extra space for nonviral genes. This region, as discussed above, encodes products that mitigate the host response to infection, and the unit is not essential for growth of the virus in cultured cells. As a result, $E1A^-$, $E1B^-$, $E3^-$ viruses grow as well as wild-type adenovirus on 293 cells. Since the virus can encapsidate a DNA that is about 5% larger than the wild-type genome (Shenk, unpublished observations), the $E1A^-$, $E1B^-$, $E3^-$ virus can accommodate an insert of as much as ~9 kbp.

First-generation vectors generally were found to sponsor only short-term expression of transgenes. Longer term expression was achieved when the vectors were introduced into severe combined immunodeficient (SCID) mice or athymic nude mice that are unable to mount a class I–restricted CTL response (Tripathy et al., 1994; Yang et al., 1994; reviewed by Kremer and Perricaudet, 1995), consistent with earlier work showing that the pathogenic response to wild-type adenovirus is significantly more reduced in nude than in immunocompetent mice (Ginsberg et al., 1991). These and other observations (reviewed by Engelhardt et al., 1994; Shenk, 1995) led to the view that cells transduced with the first-generation vectors were being killed by host antiviral defenses, possibly as a result of residual, low level expression of viral genes. Thus, inflammation and its associated cell killing limited the duration of transgene expression mediated by E1A$^-$, E1B$^-$ vectors.

Second-Generation Vectors

The concept behind the second generation of adenovirus vectors was to mutate a third gene in addition to E1A and E1B that is essential for viral replication (Fig. 1C). Presumably this third deficiency would inhibit residual viral gene expression and improve the performance of the vector. The initial second-generation vector (Engelhardt et al., 1994) carried a temperature-sensitive mutation, ts125, in the E2-coded single-stranded DNA-binding protein. This mutation was known to block viral DNA replication at the nonpermissive temperature (van der Vliet et al., 1975). The E1A$^-$, E1B$^-$, E2A$^-$ vector was propagated at its permissive temperature in 293 cells, and its performance appeared to be markedly better than first-generation vectors when tested in a mouse liver model. It induced less CTL infiltration, and expression of its marker gene (lacZ) continued for >70 days. This result supported the view that the first-generation E1A$^-$, E1B$^-$ vectors induce an inflammatory response leading to the death of transduced cells because they continue to express viral proteins at a low level. The second-generation E1A$^-$, E1B$^-$, E2A$^-$ vector minimized this inflammatory response and allowed longer term expression of β-galactosidase.

Subsequently, cell lines were generated that could complement E2A deletions (Gorziglia et al., 1996; Zhou et al., 1996), and an E1A$^-$, E1B$^-$, E2A$^-$ variant carrying a deletion in the E2A gene was confirmed to be substantially more crippled for residual gene expression than an E1A$^-$, E1B$^-$ virus. A vector with a deletion is obviously preferable to a variant with a point mutation, since the deletion will not revert and it provides additional space for incorporation of a large transgene. Cell lines have also been reported that complement mutations in the adenovirus E2B-coded proteins, the terminal protein (Schaack et al., 1995), and the DNA polymerase (Amalfitano et al., 1996), providing the opportunity to generate vectors lacking any one of the critical E2 gene products that function directly in adenovirus DNA replication.

The E4 region was also mutated to produce a second-generation vector. Mutations in this gene inhibit viral gene expression and DNA replication (Halbert et al., 1985; Weinberg and Ketner, 1986); and, in addition to further crippling the virus, deletion of E4 should enhance the safety of the vector given the ability of the E4–34kD protein to bind to p53 and cooperate with E1A to transform rodent cells (Dobner et al., 1996; Moore et al., 1996). The E4–34kD protein is the only E4 product needed for efficient growth of adenovirus in cultured cells (Halbert et al., 1985), and Weinberg and Ketner (1983) had previously produced an E4-expressing monkey cell line, W162 cells. Given this precedent, several groups readily produced 293 cells that express the E4–34kD protein and demonstrated that they support the growth of E1A$^-$, E1B$^-$, E4$^-$ vectors (Wang et al., 1995; Brough et al., 1996; Gao et al., 1996; Yeh et al., 1996). As was the case for the E1A$^-$, E1B$^-$, E2A$^-$ vectors, the E1A$^-$, E1B$^-$, E4$^-$ vectors sponsored longer term expression of transgenes than E1A$^-$, E1B$^-$ vectors (Gao et al., 1996), but second-generation vectors have nevertheless induced an immune response. As discussed later in this chapter, the residual inflammation appears to result primarily from expression of a foreign transgene.

Third-Generation Vectors

The ultimate adenovirus vector should contain only the minimal *cis*-acting DNA elements needed for replication and packaging of the vector DNA (Fig. 1D). Since this vector contains no viral genes, it should completely avoid problems associated with residual low level expression of viral genes. The first step in this direction was to show that it is possible to complement with wild-type virus the growth of a vector lacking a 7.2 kbp segment of the E2B and ML transcription units and then partially separate the vector from helper by equilibrium density centrifugation (Mitani et al., 1995). Subsequently, a helper virus was used to complement the growth of a vector containing only the terminal *cis*-acting replication origins and packaging element, and again the helper and vector were separated on the basis of differences in their densities (Fisher et al., 1996; Kochanek et al., 1996). However, this approach is unlikely to produce useful preparations of vectors for therapeutic applications. The vector and helper can recombine, yields of the vector are relatively low, and purification of the vector from the helper is cumbersome and incomplete.

A promising genetic trick has been developed to exclude most of the helper virus from stocks of vector that contain only the terminal *cis*-acting sequences plus transgene plus nonadenovirus stuffer DNA (Parks et al., 1996; Hardy et al., 1997). The trick is to utilize the Cre–loxP recombination system from bacteriophage P1. Cre encodes a recombinase that mediates highly efficient recombination between lox DNA elements. When recombination occurs between lox sites residing in the same orientation on the same DNA molecule, the sequence between lox sites is deleted. To exploit this system, lox sites were inserted on either side of the packaging element of an E1A$^-$, E1B$^-$ helper virus. When the helper viruses were grown in standard 293 cells, the lox sites did not recombine; but when the viruses were grown on 293 cells expressing the Cre recombinase, the packaging element was excised, and, as a result, the helpers were not encapsidated. When a mixture of a vector plus helper with lox sites was propagated on Cre-expressing 293 cells, the viral yield consisted mainly of vector. Residual E1A$^-$, E1B$^-$ helper in the vector stocks was partially removed by equilibrium density centrifugation. The biological properties of vectors produced using this system have not yet been reported.

Finally, the Cre–loxP system has been used to delete a large segment of DNA (~25 kbp) from the vector as it was propagated in Cre-expressing 293 cells (Lieber et al., 1996); no separate helper virus was added. Using this approach, a vector with a 9 kbp genome carrying a human α_1-antitrypsin expression cassette was generated, and equilibrium density centrifugation removed residual vector containing full-length genome to produce 9 kbp genome vector preparations that were >99% pure. It is intriguing that it was possible to generate vector particles with such small genomes, and it is of interest to learn whether these particles are as stable as normal virions. The 9 kbp genome vector transduced cultured cells as efficiently as the parental, undeleted vector, indicating that its infectivity was not crippled. Curiously, expression from the vector was transient when introduced into cultured cells or tested *in vivo* in the mouse liver. It could be stabilized by co-administration of an E1A$^-$ vector, suggesting that limited viral gene expression is needed for long-term expression. The authors speculated that the instability of the vector DNA might result from a requirement for transient or persistent replication of the vector DNA for its maintenance in the transduced cells.

Clemens et al. (1996) produced an adenovirus vector containing only the terminal adenovirus *cis*-acting sequences plus a full-length dystrophin cDNA. Intramuscular injection of the vector into 6-day-old dystrophic (mdx) mice resulted in expression of the transgene; but, consistent with the results obtained for the 9 kbp vector (Lieber et al., 1996), expression from this vector was transient. Whatever the explanation for their instability, these vectors lacking most or all viral genes appear to have revealed a previously unsuspected requirement for adenovirus persistence.

Host Immune Response

It seems clear that much of the problem with limited term expression from adenovirus vectors results from a host immune response. This problem first became evident when longer term expression was achieved in immunologically deficient mice (Tripathy et al., 1994; Yang et al., 1994). The immune response causes the destruction of transduced cells. The destruction is rapid and complete with first-generation E1A$^-$, E1B$^-$ vectors, and it is much reduced with second-generation E1A$^-$, E1B$^-$, E2A$^-$ or E1A$^-$, E1B$^-$, E4$^-$ vectors. The improved performance of the second-generation vectors almost certainly results from minimal viral gene expression, since they are much more severely crippled than first-generation vectors.

What is responsible for the residual immune response to the second-generation vectors? Two recent reports indicate that the expressed transgene can induce inflammation. Tripathy et al. (1996) compared erythropoietin expression in mice receiving an intramuscular dose of a first-generation E1A$^-$, E1B$^-$, E3$^-$ vector carrying either the murine or human gene. These two proteins are 79% identical at the amino acid level. Animals receiving the vector expressing the human protein exhibited an elevated hematocrit that peaked after 14 days, whereas animals receiving the same vector expressing the murine protein displayed an elevated hematocrit throughout the 112 day duration of the experiment. Long-term expression of murine erythropoietin was observed in five different mouse strains. The "foreign" human protein elicited both humoral and cellular immune responses, and it must be the response to the therapeutic protein and not to adenovirus proteins that led to the rapid cessation of expression. In a similar vein, Gao et al. (1996) generated an $E1A^-$, $E1B^-$, $E4^-$ vector carrying a lacZ transgene. β-Galactosidase expression was diminished by a factor of 10 after 60 days in the liver of mice instilled with the vector by tail vein injection. When transgenic ROSA-26 mice that constitutively express β-galactosidase and see the bacterial enzyme as "self" received the same vector, the level of β-galactosidase expression dropped by a factor of less than two during the 60 day period. Both of these reports demonstrate that expression of a transgene product that is recognized as foreign by the recipient host results in the rapid diminution of expression, presumably as a result of the destruction of transduced cells. Although additional work is needed to generalize these results, they are highly encouraging and suggest that second-generation adenoviruses have the potential to sponsor long-term expression of transgenes that are seen as "self" in animal and human hosts.

Additional approaches are being evaluated to control the immune destruction of cells transduced by adenovirus vectors. Continuous immunosuppression of animals with cyclosporin A (Dai et al., 1995; Fang et al., 1995) or FK506 (Vilquin et al., 1995) led to long-term expression of transgenes from first-generation $E1A^-$, $E1B^-$ vectors. These results probably serve more as a confirmation of the nature of the immunological problem than as a solution, since it seems unlikely that prolonged immunosuppression will be acceptable as a co-therapy in the vast majority of gene transfer applications. However, immunosuppression with FK506 for 1 month (Lochmüller et al., 1996) or transient treatment with antibodies that interfere with T-cell activation at the time of vector application (Kay et al., 1995; Yang et al., 1996a) also facilitated long-term expression in murine models from adenovirus vectors. These treatments, as well as transient administration of IL-12 (Yang et al., 1995b), also attenuated the production of neutralizing antibodies. Treatments that transiently antagonize cellular and humoral immune responses hold great promise not only to promote long-term expression of transgenes but also to facilitate readministration of the vector.

Another approach under development that has potential to overcome the immunological barrier to readministration of adenovirus vectors is the use of vectors derived from different human adenovirus serotypes (Kass-Eisler et al., 1996). If a set of different adenovirus vectors were available that are not cross-neutralized, they could be used sequentially to readminister the same transgene.

Targeting

Work has begun to redirect binding of the adenovirus fiber protein (Michael et al., 1995; Krasnykh et al., 1996) with the goal of achieving tissue-specific targeting similar to that being developed for retrovirus vectors (e.g., Kasahara et al., 1994; Somia et al., 1995). Bispecific antibodies binding to an epitope in the adenovirus penton base and a cell surface integrin also show promise for directing adenovirus adsorption (Wickham et al., 1995, 1996). If the specificity of the vector–cell interaction can be modified and controlled, the utility of adenovirus as an *in vivo* vector will be dramatically enhanced. In addition to facilitating the transduction of cells that adenovirus generally does not efficiently enter, targeting should reduce the delivery of transgenes to untargeted tissue, allowing the administration of lower doses and avoiding undesirable side effects.

Applications

Adenovirus vectors have perhaps been most thoroughly explored as vectors for treatment of cystic fibrosis. Initially, it seemed likely that this might be an ideal application for adenovirus vectors, given the apparent predilection of the virus to infect the respiratory tract. However, studies using $E1A^-$, $E1B^-$ vectors documented inflammation and short-term expression of transgenes such as lacZ. The destruction of transduced cells in these studies results from the combined action of CTLs specific for the transgene and viral antigens (Yang et al., 1995a, 1996b). Presumably, second- and third-generation vectors will exhibit improved performance if they express transgenes that are not seen as neoantigens in animal or human hosts (Gao et al., 1996; Tripathy et al., 1996). However, a major obstacle to the treatment of cystic fibrosis will likely remain, since it appears that adenovirus does not efficiently infect columnar epithelial cells due to a paucity of the cellular $\alpha_v\beta_5$ integrin receptor for the virus in this cell type (Goldman and Wilson, 1995). Perhaps this problem can be solved by redirecting the vector to another receptor system that is plentiful in the target epithelial cells of the lung.

Irrespective of immune responses to adenovirus vectors, it is likely that they will find a useful niche in therapies that require only short-term expression and in vaccination protocols where long-term expression of the therapeutic gene is not required and limited inflammation might be useful. The virus is safe as a vehicle for vaccination; indeed the U.S. military has immunized recruits against several adenoviruses for years, and a variety of animal studies suggest that adenovirus vectors induce both humoral and cellular immunity (reviewed by Randrianarison-Jewtoukoff and Perricaudet, 1995).

Many potential cancer therapies would likely require only short-term expression of a transduced gene product such as a prodrug. For example, adenovirus vectors can transfer the herpesvirus thymidine kinase gene to tumors *in vivo* (Chen et al., 1994; Tong et al., 1996), sensitizing tumor cells to the antiherpetic drug ganciclovir. The nucleoside analog is phosphorylated and activated by the herpes thymidine kinase but not by cellular kinases, and when activated it acts as a chain terminator in cells synthesizing DNA, killing tumor cells. Furthermore, when the thymidine kinase gene is introduced into tumors by direct injection of a vector, cells that do not receive vector are subsequently killed by administration of ganciclovir, presumably because phosphorylated drug moves to adjacent cells by gap junction transfer.

Will adenovirus vectors be useful in applications targeting hematopoietic stem cells? Mitani et al. (1994) reported that human bone marrow cells can be infected with adenovirus, but they did not ascertain whether primitive cell types were infected. More recently, Neering et al. (1996) transfected human bone marrow with an adenovirus vector expressing a membrane-associated alkaline phosphatase marker protein that can be monitored by flow cytometry. Marker expression was detected on ~45% of the primitive CD34$^+$ population, as well as on more primitive CD34$^+$, CD38$^-$ cells, indicating that a significant portion of bone marrow cells with a primitive phenotype can be transduced by an adenovirus vector. Thus, adenovirus vectors may prove useful for expression of therapeutic proteins in at least some classes of hematopoietic stem cells.

At first consideration, adenoviruses seem almost uniquely suited as vectors for gene therapy. They are relatively well understood, they express their genes in cells that are not actively growing, they are easily manipulated and produced in large quantities, they can potentially accommodate large transgenes, and they appear to be safe. In spite of these advantageous features, adenovirus vectors were initially disappointing in that they have failed to sponsor long-term expression of transgenes. However, newer vectors show promise for longer term expression, and it seems likely that further improvements in the performance of adenovirus vectors will come with a better understanding of the immune responses that they elicit in their human hosts.

ACKNOWLEDGMENTS

I thank Steve Hardy for helpful discussions about the state of adenovirus vectorology. I am an American Cancer Society Professor and an Investigator of the Howard Hughes Medical Institute.

REFERENCES

Amalfitano A, Begy CR, Chamberlain JS (1996): Improved adenovirus packaging cell lines to support the growth of replication-defective gene-delivery vectors. *Proc Natl Acad Sci USA* 93:3352–3356.

Beltz GA, Flint SJ (1979): Inhibition of HeLa cell protein synthesis during adenovirus infection. *J Mol Biol* 131:353–373.

Berns KI, Giraud C (1995): Adenovirus and adeno-associated virus as vectors for gene therapy. *Ann NY Acad Sci* 772:95–104.

Brough DE, Lizonova A, Hsu C, Kulesa VA, Kovesdi I (1996): A gene transfer vector-cell line system for complete functional complementation of adenovirus early regions E1 and E4. *J Virol* 70:6497–6501.

Chen S-H, Shine HD, Goodman JC, Grossman RG, Woo SL (1994): Gene therapy for brain tumors: Regression of experimental gliomas by adenovirus-mediated gene transfer *in vivo. Proc Natl Acad Sci USA* 91:3054–3057.

Chiocca S, Kurzbauer R, Schaffner G, Baker A, Mautner V, Cotten M (1996): The complete DNA sequence and genomic organization of the avian adenovirus CELO. *J Virol* 70:2939–2949.

Clemens PR, Kochanek S, Sunada Y, Chan S, Chen H-H, Campbell KP, Caskey CT (1996): *In vivo* muscle gene transfer of full length dystrophin with an adenoviral vector that lacks all viral genes. *Gene Ther* 3:965–972.

Dai Y, Schwarz EM, Gu D, Zhang W-W, Sarvetnick N, Verma I (1995): Cellular and humoral immune responses to adenoviral vectors containing factor IX gene: Tolerization of Factor IX and vector antigens allows for long-term expression. *Proc Natl Acad Sci USA* 92:1401–1405.

Dobner T, Horikoshi N, Rubenwolf S, Shenk T (1996): Blockage by adenovirus E4orf6 of transcriptional activation by the p53 tumor suppressor. *Science* 272:1470–1473.

Engelhardt JF, Ye X, Doranx B, Wilson JM (1994): Ablation of E2A in recombinant adenoviruses improves persistence and decreases inflammatory response in mouse liver. *Proc Natl Acad Sci USA* 91:6196–6200.

Evans RM, Fraser N, Ziff E, Weber J, Wilson M, Darnell JE (1977): The initiation sites for RNA transcription in Ad2 DNA. *Cell* 12:733–739.

Fang B, Eisensmith RC, Wang H, Kay MA, Cross RE, Landen CN, Gordon G, Bellinger DA, Read MS, Hu PC, Brinkhaus KM, Woo SLC (1995): Gene therapy for hemophilia B: Host immunosuppression prolongs the therapeutic effect of adenovirus-mediated factor IX expression. *Hum Gene Ther* 6:1039–1044.

Fisher KJ, Choi H, Burda J, Chen S-J, Wilson JM (1996): Recombinant adenovirus deleted of all viral genes for gene therapy of cystic fibrosis. *Virology* 217:11–22.

Gao G-P, Yang Y, Wilson JM (1996): Biology of adenovirus vectors with E1 and E4 deletions for liver-directed gene therapy. *J Virol* 70:8934–8943.

Ginsberg HS, Lundholm-Beauchamp U, Horswood RL, Pernis B, Wold WSM, Chanock RM, Prince GA (1989): Role of early region 3 (E3) in pathogenesis of adenovirus disease. *Proc Natl Acad Sci USA* 86:3823–3827.

Ginsberg HS, Moldawer LL, Sehgal PB, Redington M, Kilian PL, Chanock RM, Prince GA (1991): A mouse model for investigating the molecular pathogenesis of adenovirus pneumonia. *Proc Natl Acad Sci USA* 88:1651–1655.

Goff SP, Shenk T (1993): Sleeping with the enemy: Viruses as gene therapy vectors. *Curr Opin Genet Dev* 3:71–73.

Goldman MJ, Wilson JM (1995): Expression of alpha v beta 5 integrin is necessary for efficient adenovirus-mediated gene transfer in the human airway. *J Virol* 69:5951-5958.

Gooding LR (1992): Virus proteins that counteract host immune defenses. *Cell* 71:5–7.

Gorziglia MI, Kadan MJ, Yei S, Lim J, Lee GM, Luthra R, Trapnell BC (1996): Elimination of both E1 and E2a from adenovirus vectors further improves prospects for *in vivo* human gene therapy. *J Virol* 70:4173–4178.

Graham FL, Smiley J, Russell WC, Nairu R (1977): Characterization of a human cell line transformed by DNA from human adenovirus type 5. *J Gen Virol* 36:59–72.

Halbert DN, Cutt JR, Shenk T (1985): Adenovirus early region 4 encodes functions required for efficient DNA replication, late gene expression and host cell shutoff. *J Virol* 56:250–257.

Hardy S, Kitamura M, Harris-Stansil T, Dai Y, Phipps ML (1997): Construction of adenovirus vectors using cre–lox recombination. *J Virol* 71:1842–1849.

Harlow E, Whyte P, Franza BR, Schley C (1986): Association of adenovirus early region 1A proteins with cellular polypeptides. *Mol Cell Biol* 6:1579–1589.

Hasson TB, Ornelles DA, Shenk T (1992): Adenovirus L1 52- and 55-kDa proteins are present within assembling virions and colocalize with nuclear structures distinct from replication centers. *J Virol* 66:6133–6142.

Hearing P, Samulski R, Wishart W, Shenk T (1987): Identification of a repeated sequence element required for efficient encapsidation of the adenovirus type 5 chromosome. *J Virol* 61:2555–2558.

Hilleman MR, Werner JH (1954): Recovery of new agents from patients with acute respiratory illness. *Proc Soc Exp Biol Med* 85:183–188.

Horwitz MS (1996): Adenoviruses. In Fields, BN, Knipe DM, Howley, PM (eds): Fields Virology. Philadelphia: Lippincott-Raven, pp 2149–2172.

Kasahara N, Dozy A, Kan Y (1994): Tissue-specific targeting of retroviral vectors through ligand-receptor interactions. *Science* 266:1373–1376.

Kass-Eisler A, Leinwand L, Gall J, Bloom B, Falck-Pedersen E (1996): Circumventing the immune response to adenovirus-mediated gene therapy. *Gene Ther* 3:54–162.

Kay MA, Holterman A-X, Meuse L, Gown A, Ochs HD, Linsley PS, Wilson CB (1995): Long-term hepatic adenovirus-mediated gene expression in mice following CTLA4Ig administration. *Nature Genet* 11:191–197.

Kochanek S, Clemens PR, Mitani K, Chen H-H, Chan S, Caskey CT (1996): A new adenoviral vector: Replacement of all viral coding sequences with 28 kb of DNA independently expressing both full length dystrophin and β-galactosidase. *Proc Natl Acad Sci USA* 93:5731–5736.

Kozarsky K, Wilson JM (1993): Gene therapy: Adenovirus vectors. *Curr Opin Genet Dev* 3:499–503.

Krasnykh VN, Mikheeva GV, Douglas JT, Curiel DT (1996): Generation of recombinant adenovirus vectors with modified fibers for altering viral tropism. *J Virol* 70:6839–6846.

Kremer EJ, Perricaudet M (1995): Adenovirus and adeno-associated virus mediated gene transfer. *Br Med Bull* 51:31–44.

Lechner RL, and Kelly TJ Jr (1977): The structure of replicating adenovirus 2 DNA molecules. *Cell* 12:1007–1020.

Lieber A, He C-Y, Kirillova I, Kay M (1996): Recombinant adenoviruses with large deletions generated by cre-mediated excision exhibit different biological properties compared with first generation vectors *in vivo* and *in vitro*. *J Virol* 70:8944–8960.

Lochmüller H, Petrof BJ, Pari G, Larochelle N, Dodelet V, Wang Q, Allen C, Prescott S, Massie B, Nalbantoglu J, Karpati G (1996): Transient immunosuppression by FK506 permits a sustained high-level dystrophin expression after adenovirus-mediated dystrophin minigene transfer to skeletal muscles of adult dystrophic (mdx) mice. *Gene Ther* 3:706–716.

Lowe SW, Ruley HE (1993): Stabilization of the p53 tumor suppressor is induced by adenovirus 5 E1A and accompanies apoptosis. *Genes Dev* 7:535–545.

Mathews MB, Shenk T (1991): Adenovirus virus-associated RNA and translational control. *J Virol* 65:5657–5662.

Michael SI, Hong JS, Curiel DT, Engler JA (1995): Addition of a short peptide ligand to the adenovirus fiber protein. *Gene Ther* 2:660–668.

Mitani K, Graham FL, Caskey CT (1994): Transduction of human bone marrow by adenovirus vector. *Hum Gene Ther* 5:941–948.

Mitani K, Graham FL, Caskey CT, Kochanek S (1995): Rescue, propagation and partial purification of a helper virus-dependent adenovirus vector. *Proc Natl Acad Sci USA* 92:3854–3858.

Moore M, Horikoshi N, Shenk T (1996): Oncogenic potential of the adenovirus E4orf6 protein. *Proc Natl Acad Sci USA* 93:11295–11301.

Neering SJ, Hardy SF, Minamoto D, Spratt SK, Jordan CT (1996): Transduction of primitive human hematopoietic cells with recombinant adenovirus vectors. *Blood* 88:1147–1155.

Nevins JR (1992): E2F: A link between the Rb tumor suppressor protein and viral oncoproteins. *Science* 258:424–429.

Ornelles D, Shenk T (1991): Localization of the adenovirus E1B-55kDa protein during lytic infection: Association with nuclear viral inclusions requires the E4-34kDa protein. *J Virol* 65:424–439.

Parks RJ, Chen L, Anton M, Sankar U, Rudnicki MA, Graham F (1996): A helper-dependent adenovirus vector system: Removal of helper virus by cre-mediated excision of the viral packaging signal. *Proc Natl Acad Sci USA* 93:13565–13570.

Randrianarison-Jewtoukoff V, Perricaudet M (1995): Recombinant adenoviruses as vaccines. *Biologicals* 23:145–157.

Rao L, Debbas M, Sabbatini P, Hockenbery D, Korsmeyer S, White E (1992): The adenovirus E1A proteins induce apoptosis which is blocked by the E1B 19kDa and Bcl-2 proteins. *Proc Natl Acad Sci USA* 89:7742–7746.

Reich N, Pine R, Levy D, Darnell JE Jr (1988): Transcription of interferon-stimulated genes is induced by adenovirus particles but is suppressed by E1A gene products. *J Virol* 62:114–119.

Rowe WP, Huebner RJ, Gilmore LK, Parrott RH, Ward TG (1953): Isolation of a cytopathogenic agent from human adenoids undergoing spontaneous degeneration in tissue culture. *Proc Soc Exp Biol Med* 84:570–573.

Sarnow P, Ho Y-S, Williams J, Levine AJ (1982): Adenovirus E1b-58kd tumor antigen and SV40 large tumor antigen are physically associated with the same 54 kd cellular protein in transformed cells. *Cell* 28:387–394.

Schaack J, Guo X, Ho WY-W, Karlok M, Chen C, Ornelles D (1995): Adenovirus type 5 precursor terminal protein-expressing 293 and HeLa cell lines. *J Virol* 69:4079–4085.

Schaack J, Ho WY-W, Freimuth P, Shenk T (1990): Adenovirus terminal protein mediates both nuclear matrix association and efficient transcription of adenovirus DNA. *Genes Dev* 4:1197–1208.

Shenk T (1995): Group C adenoviruses as vectors for gene therapy. In Kaplitt M, Loewy A (eds): Viral Vectors. San Diego: Academic Press, pp 43–54.

Shenk T (1996): Adenoviridae: The viruses and their replication. In Fields BN, Knipe DM, Howley PM (eds): Fields Virology. Philadelphia: Lippincott-Raven, pp 2111–2148.

Somia N, Zoppe M, Verma I (1995): Generation of targeted retroviral vectors by using single-chain variable fragment: An approach to *in vivo* gene delivery. *Proc Natl Acad Sci USA* 92:7570–7574.

Tong X-W, Block A, Chen S-H, Contant CF, Agoulnik I, Blankenburg K, Kaufman RH, Woo SL, Kieback DG (1996): In vivo gene therapy of ovarian cancer by adenovirus-mediated thymidine kinase gene transduction and ganciclovir administration. *Gynec Oncol* 61:175–179.

Tripathy SK, Black HB, Goldwasser E, Leiden J (1996): Immune responses to transgene-encoded proteins limit the stability of gene expression after injection of replication-defective adenovirus vectors. *Nature Med* 2:545–550.

Trentin JJ, Yabe Y, Taylor G (1962): The quest for human cancer viruses. *Science* 137:835–849.

Tribouley C, Lutz P, Staub A, Kedinger C (1994): The product of the adenovirus intermediate gene IVa2 is a transcriptional activator of the major late promoter. *J Virol* 68:4450–4457.

Tripathy SK, Goldwasser E, Lu M-M, Barr E, Leiden J (1994): Stable delivery of physiologic levels of recombinant erythropoietin to the systemic circulation by intramuscular injection of replication-defective adenovirus. *Proc Natl Acad Sci USA* 91:11557–11561.

van der Vliet PC, Levine AJ, Ensinger MJ, Ginsberg HS (1975): Thermolabile DNA-binding protein from cells infected with a temperature-sensitive mutant of adenovirus defective in viral DNA synthesis. *J Virol* 15:348–354.

Vilquin JT, Guerette B, Kinoshita I, Roy B, Goulet M, Gravel C, Roy R, Tremblay JP (1995): FK506 immunosuppression to control the immune reactions triggered by first generation adenovirus-mediated gene transfer. *Hum Gene Ther* 6:1391–1401.

Wang Q, Jia X-C, Finer M (1995): A packaging cell line for propagation of recombinant adenovirus vectors containing two lethal gene-region deletions. *Gene Ther* 2:775–783.

Weinberg DH, Ketner G (1983): A cell line that supports the growth of a defective early region 4 deletion mutant of human adenovirus type 2. *Proc Natl Acad Sci USA* 80:5383–5386.

Weinberg DH, Ketner G (1986): Adenoviral early region 4 is required for efficient viral DNA replication and for late gene expression. *J Virol* 57:833–838.

Whyte P, Buchkovich KJ, Horowitz JM, Friend SH, Raybuck M, Weinberg RA, Harlow E (1988): Association between an oncogene and an antioncogene: The adenovirus E1A proteins bind to the retinoblastoma gene product. *Nature* 334:124–129.

Wickham TJ, Carrion ME, Kovesdi I (1995): Targeting of adenovirus penton to new receptors through replacement of its RGD motif with other receptor-specific peptide motifs. *Gene Ther* 2:750–756.

Wickham TJ, Segal DM, Roelvink PW, Carrion ME, Lizonova A, Lee GM, Kovesdi I (1996): Targeted adenovirus gene transfer to endothelial and smooth muscle cells by using bispecific antibodies. *J Virol* 70:6831–6838.

Yang Y, Li Q, Ertl HCJ, Wilson JM (1995a): Cellular and humoral immune responses to viral antigens create barriers to lung-directed gene therapy with recombinant adenoviruses. *J Virol* 69:2004–2015.

Yang Y, Nunes F, Berensci K, Furth E, Gonczol E (1994): Cellular immunity to viral antigens limits E1-deleted adenoviruses for gene therapy. *Proc Natl Acad Sci USA* 91:4407–4411.

Yang Y, Su Q, Grewal IS, Schilz R, Flavell RA, Wilson JM (1996a): Transient subversion of CD40 ligand function diminishes immune responses to adenovirus vectors in mouse liver and lung tissues. *J Virol* 70:6370–6377.

Yang Y, Su Q, Wilson JM (1996b): Role of viral antigens in destructive cellular responses to adenovirus vector-transduced cells in mouse lungs. *J Virol* 70:7209–7212.

Yang Y, Trinchieri G, Wilson JM (1995b): Recombinant IL-12 prevents formation of blocking IgA antibodies to recombinant adenovirus and allows repeated gene therapy to mouse lung. *Nature Med* 1:890–893.

Yee S, Branton P (1985): Detection of cellular proteins associated with human adenovirus type 5 early region 1A polypeptides. *Virology* 147:142–153.

Yeh P, Dedieu J-F, Orsini C, Vigne E, Denefle P, Perricaudet M (1996): Efficient dual complementation of adenovirus E1 and E4 regions from a 293-derived cell line expressing a minimal E4 functional unit. *J Virol* 70:559–565.

Yew PR, Liu X, Berk AJ (1994): Adenovirus E1B protein tethers a transcriptional repression domain to p53. *Genes Dev* 8:190–202.

Zhou H, O'Neal W, Morral N, Beaudet AL (1996): Development of a complementing cell line and a system for construction of adenovirus vectors with E1 and E2a deleted. *J Virol* 70:7030–7038.

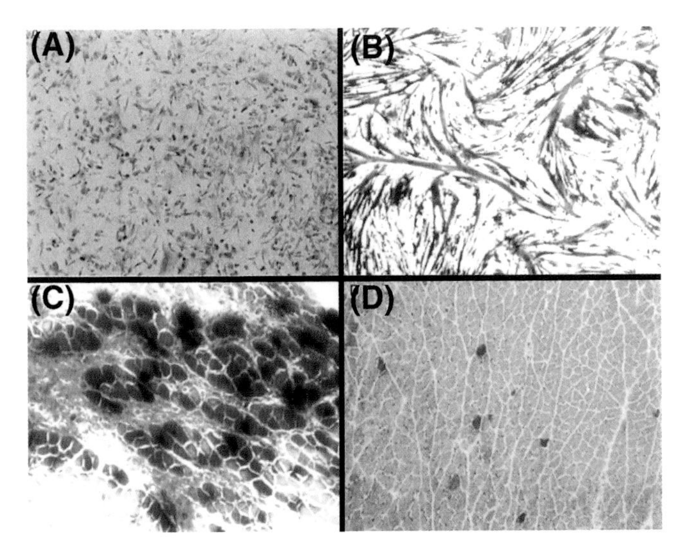

CHAPTER 7, FIGURE 2. HSV-1 vector-mediated β-galactosidase transgene expression in myogenic cells. L6 mouse myoblasts (**A**) and differentiated myotubes (**B**) were infected with replication-defective SHZ.1 vector at an MOI of 1.0 and stained with X-gal to detect lacZ expressing cells. Both myoblasts and differentiated myotubes are highly transduced by the HSV-1 vector. Newborn (**C**) and adult (**D**) mouse gastrocnemius muscle were injected with 2.5×10^6 pfu of SHZ.1 vector and the reporter gene expression examined at 3 days postinjection. A significant number of fibers are transduced in newborn but not in adult muscle.

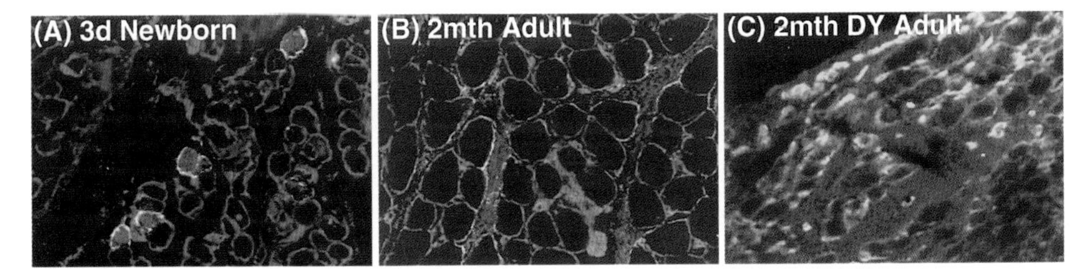

CHAPTER 7, FIGURE 3. Role of the basal lamina in limiting HSV-1 infection in mature myofibers. Normal newborn (**A**), normal adult (**B**), and adult merosin-deficient dy/dy mouse gastrocnemius (**C**) was infected with 2.5×10^6 pfu of the SHZ.1 vector, and collagen IV (fluorescein; green) as well as HSV-1 infiltration (Cy3; red) were co-localized by immunofluorescence. In newborn animals, we observed a high number of muscle fibers infected by the HSV-1 vector since both immunofluorescent markers are localized in the same myofibers (**A**). In adult muscle fibers (**B**), we observed that HSV-1 remains outside the muscle fibers, since the red fluoresence signal does not co-localize with the green immunofluorescence. However, in the dy/dy adult mice (**C**), HSV-1 vector penetrates the mature myofibers since both signals co-localize to the inside of mature myofibers as observed in newborn muscle.

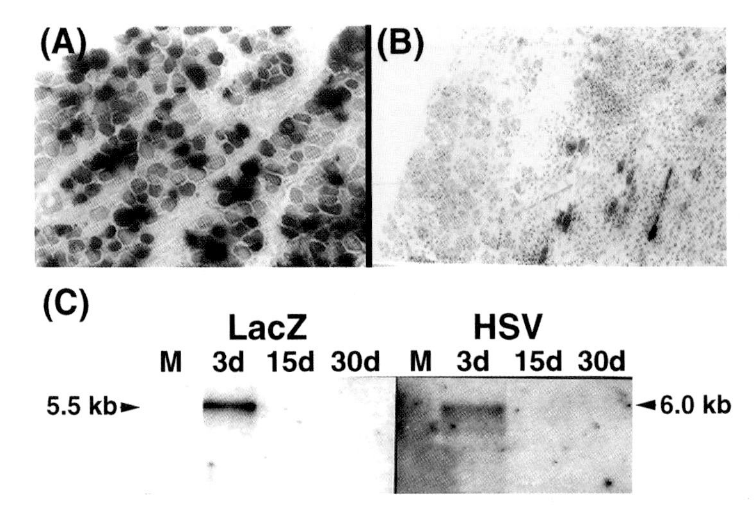

CHAPTER 7, FIGURE 4. Barriers to long-term HSV-1 vector-mediated β-galactosidase expression in muscle. SHZ.1 vector 1 x 107 pfu was injected into newborn gastrocnemius and ß-galctosidase expression was examined by X-gal staining at 3 (**A**) and 6 (**B**) days postinoculation. The large number of β-gal-positive muscle fibers in newborn (**A**) present at 3 days post inoculation disappear by day 6 (**B**). Similar results were detected in adult muscle, even though the transduction frequency in adults at 3 days is greatly reduced compared with newborn (see Fig. 2C,D). Southern blot analysis for the persistence of the HSV-1 vector-mediated transgene DNA following intramuscular injection of HSV-1 (SHZ-1) vector were performed (**C**). DNA 8 μg extracted out of the injected muscle was digested with BamHI, transferred to nytran membranes, and hybridized with two different probes (lacZ and HSV). The lacZ probe hybridized to a 5.5 kb band in extracted muscle DNA at 3 days but did not hybridize to muscle DNA from 15 to 30 day animals. The HSV probe hybridized also with a 6 kb band only in muscle DNA at 3 days, demonstrating that the vector did not persist.

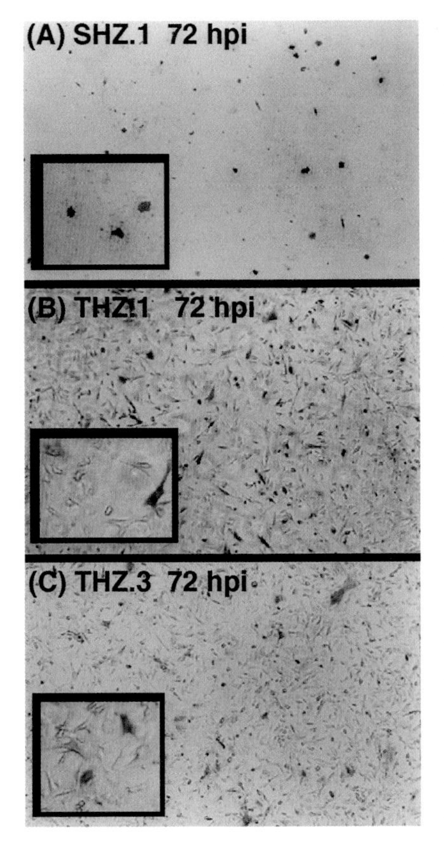

CHAPTER 7, FIGURE 5. HSV-1-mediated cytotoxicity and transgene expression in myoblasts *in vitro*. G8 mouse myoblasts were infected with various HSV-1 mutant vectors at an MOI of 5. At various times postinfection, viable cell counts were obtained by trypan blue exclusion assay, cell monolayers were fixed and stained with X-gal, or the level of β-galactosidase was determined by ONPG assay. Few SHZ.1 (**A**) infected myoblasts survive or express β-galactosidase by 72 hours postinfection. However, both THZ.1 (**B**) and THZ.3 (**C**) infected myoblasts survive and continue to express the reporter gene at 72 hours postinfection. The inserts represent higher magnifications in which individual cells can be clearly seen. For SHZ,1 (**A**), the majority of the cell monolayer has been destroyed. The levels of cytotoxicity for the three vector constructs versus mock infected myoblasts is depicted over the various time points (**D**). As additional toxic genes are deleted from the virus vector, the level of toxicity is significantly decreased and approaches that seen in mock infected G8 myoblasts.

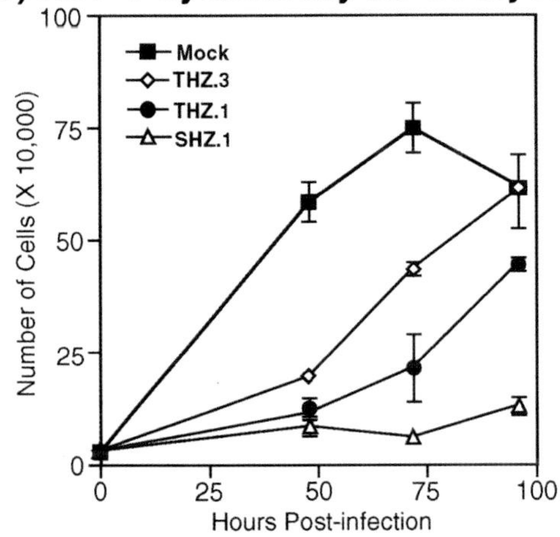

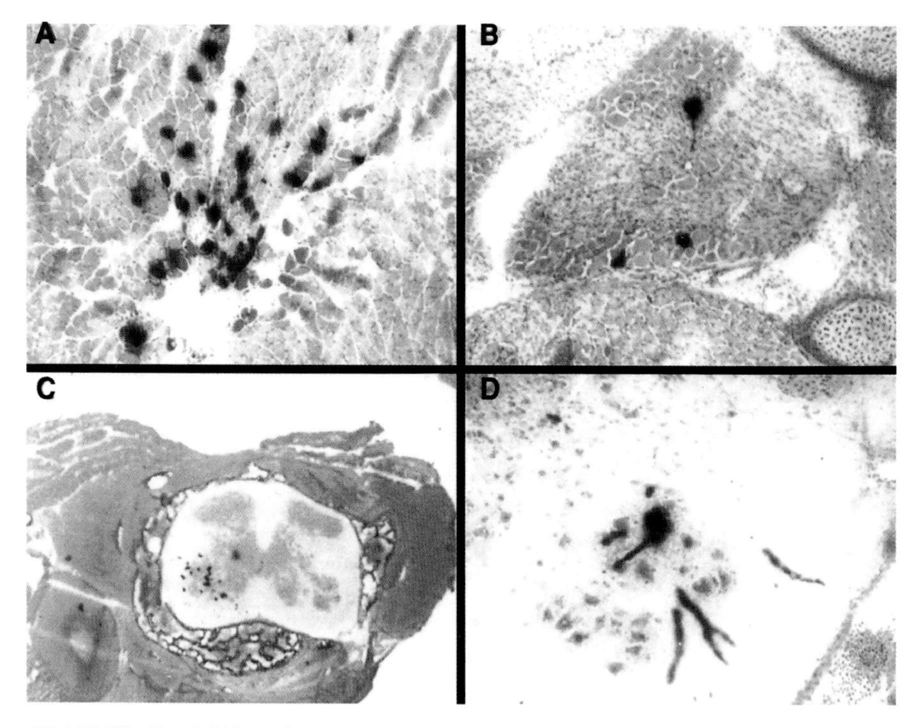

CHAPTER 7, FIGURE 6. HSV-1 vector-mediated gene transfer to dorsal root ganglia and spinal cord from direct inoculation of mouse gastrocnemius muscle. Newborn mice were infected with 1×10^7 pfu of SHZ.1 intramuscularly into the gastrocnemius muscle. At 3 days postinfection, the animals were sacrificed, and the gastrocnemius, dorsal root ganglia, and spinal cord were removed by microdissection. The tissues were frozen, sectioned, and stained with X-gal to detect lacZ gene activity. Intact muscle fibers expressing the reporter gene (**A**) were readily detectable at 3 days. The vector spread by retrograde axonal transport from muscle to the dorsal root ganglia that innervate this region of the gastrocnemius as evidenced by expression of the reporter gene in the peripheral nerves (**B**). The vector was also capable of spreading to the ventral horn of the spinal cord (**C,D**) which possesses projections to the site of inoculation.

7

GENE TRANSFER TO MUSCLE AND SPINAL CORD USING HERPES SIMPLEX VIRUS–BASED VECTORS

JOHNNY HUARD, WILLIAM F. GOINS, GIRIDHAR R. AKKARAJU, DAVID KRISKY, TOM OLIGINO, PEGGY MARCONI, CHARLES S. DAY, AND JOSEPH C. GLORIOSO

Department of Molecular Genetics and Biochemistry (W.F.G., G.R.A., D.K., T.O., P.M., J.C.G.) and Department of Orthopedic Surgery, Musculoskeletal Research Center, Children's Hospital of Pittsburgh (J.H., C.S.D.), University of Pittsburgh School of Medicine, Pittsburgh, PA 15261

INTRODUCTION

Gene therapy is a novel form of molecular medicine that can have a major impact on human health care in the future. A variety of inheritable muscular and neuromuscular disorders exist that may be amenable to treatment by gene therapy. Duchenne muscular dystrophy (DMD) is an X-linked recessive muscle-wasting disease resulting from the absence of the cytoskeletal protein dystrophin within the sarcolemma of myofibers. The biochemical and pathophysiological events that lead to the eventual severe loss of muscle fibers is not well understood, but relates to the general instability of the defective muscle fiber membranes. Over time these DMD patients succumb to congestive cardiac and respiratory failure. Because no treatment exists for this and other fatal muscle disorders, several approaches have been developed to rescue the diseased muscle tissue.

Stem Cell Biology and Gene Therapy, Edited by Peter J. Quesenberry, Gary S. Stein, Bernard G. Forget, and Sherman M. Weissman
ISBN 0-471-14656-0

Even though the current techniques for gene transfer to muscle are aimed primarily at delivering a normal copy of a gene (such as dystrophin for DMD) that is missing or defective in afflicted individuals in sufficient quantities to be therapeutically efficient for ameliorating muscle diseases, this approach can also be used to deliver trophic factors to promote the recovery of muscle following injury. Muscle injury can occur by a variety of mechanisms, ranging from ischemia to direct and indirect muscle trauma, including lacerations, contusions as well as complete and partial muscle tears. The recovery of injured muscle depends on several factors, such as muscle regeneration, revascularization, and reinnervation. Recovery can be improved with gene therapy procedures that deliver substances that promote the recovery of the injured muscle and may also prevent complications associated with muscle healing and contractures. Two different approaches have been employed to deliver genes to muscle: cell therapy based on myoblast transplantation (MT) and gene therapy (GT) using viral and nonviral vectors.

CELL THERAPY

MT consists of the implantation of myoblast precursors (satellite cells) into injured muscle to create a reservoir of normal myoblasts that can fuse and deliver genes to defective muscle tissue. This approach has been extensively used to deliver the dystrophin gene to muscle for treating DMD (Law et al., 1991, 1992; Gussoni et al., 1992; Karpati and Worton, 1992; Huard et al., 1992a,b; Karpati et al., 1993; Tremblay et al., 1993). However, several obstacles have limited the success of this technique, such as inefficient spread and poor survival of the injected myoblasts, as well as immunological problems associated with the recognition of the donor myoblasts (Alamedine et al., 1989; Karpati et al., 1989; Partridge et al., 1989; Morgan et al., 1990; Huard et al., 1991, 1994b,c; Partridge, 1991; Vilquin et al., 1993; Kinoshita et al., 1994).

Immune rejection of transplants remains one of the major limitations associated with MT. To combat this problem *ex vivo* approaches using autologous myoblast transfer have been employed. Autologous myoblast transfer (AMT) represents a form of somatic cell gene transfer in which primary myoblasts removed from a patient and expanded in cell culture are transduced with viral vectors carrying a transgene and subsequently injected back into muscle from the same host. Isogenic myoblasts transduced with adenovirus and retrovirus vectors fused with fibers within the injected muscle and expressed the transgene (Salvatori et al., 1993; Huard et al., 1994a). This technique therefore permits the introduction of myoblasts capable of expressing the transgene into muscle without apparent immune rejection normally associated with injected myoblasts (Salvatori et al., 1993; Huard et al., 1994a).

In addition to use as a gene delivery vehicle for muscle, the myoblast-mediated *ex vivo* gene transfer approach has been applied to gene delivery for (1) transfer and expression of factor IX for hemophilia B (Dai et al., 1992),

(2) systemic delivery of human growth hormone for growth retardation (Dhawan et al., 1991), (3) gene delivery of human adenosine deaminase for adenosine deaminase immunodeficient syndrome (Lynch et al., 1992), (4) gene transfer of human proinsulin for diabetes (Simonsen et al., 1996), and (5) transfer and expression of tyrosine hydroxylase for Parkinson's disease (Jiao et al., 1993).

GENE THERAPY

Direct gene transfer using different vector systems represents an alternative therapy approach to cell-mediated therapy for treating muscle disease. Plasmid DNA, liposomes, and viral vectors (adenovirus, adeno-associated virus, retrovirus, and herpes simplex virus) have been used to transfer foreign genes into muscle cells. The strengths and weaknesses of these vector systems are discussed below.

Direct transfection of naked DNA into muscle cells *in vitro,* as well as intramuscular injection of naked DNA or complexed DNA (liposomes) carrying reporter genes *in vivo,* has been found to be very inefficient (Acsadi et al., 1991) for transfering genes to muscle cells. A strength of this approach is that transgene expression has been observed to persist over long periods (at least 1.5 years) (Wolff et al., 1992), suggesting reduced immunogenicity to the transduced muscle fibers even though humoral and cellular immunity have been detected following direct DNA injection (Katsumi et al., 1994). Therefore, while these systems may circumvent immune rejection, they are unlikely to be useful for gene replacement therapies.

Retroviral vectors have been employed for a variety of *ex vivo* GT applications (Naviaux and Verma, 1992). Their use in *in vivo* approaches for muscle disease is limited due to the difficulties in producing high titer virus stocks and by the fact that they cannot infect and express the transgene in nondividing cells, although this problem may be overcome by the use of lentiviral vectors (Naldini et al., 1996). The packaging potential further limits the application of these vectors for treatment of DMD. Current retroviral vectors have been used to introduce the Becker dystrophin minigene cassette into dystrophic myoblasts *in vitro,* but these vectors are unable to infect nondividing cells such as differentiated myotubes and muscle fibers since they require dividing cells for integration and expression (Dunckley et al., 1992). However, an intermediate level of infection has been observed in regenerating muscle with retrovirus due to transduction of activated satellite cells (Dunckley et al., 1992; Dunckley et al., 1993).

Adeno-associated virus (AAV) suffers from the same production and packaging limitations as retroviruses (Muzyczka, 1992; Flotte and Carter, 1995); however, AAV vectors are capable of infecting both dividing and nondividing cells (Muzyczka, 1992; Flotte et al., 1994; Kaplitt et al., 1994). Recently, AAV vectors were used to transduce newborn and adult muscle fibers (Xiao et al.,

1996) *in vivo.* Although transgene expression persisted to 1.5 years in the absence of a cellular immune response to the vector, the ability to accommodate only 5 kb of foreign DNA will restrict the use of these vectors for the treatment of DMD.

Adenovirus (AV) based vectors possess two distinct advantages over both retroviral and AAV vectors. First, high titer preparations of recombinant virus (10^{11}–10^{13} virus particles/ml) are readily generated, and, second, these vectors infect a wide variety of cell types, including quiescent nondividing cells. AV can efficiently infect myoblasts in cell culture, yet low-level transduction has been observed in differentiated myotubes (Ascadi et al., 1994A). Adenoviral vectors have been employed to successfully infect and deliver foreign reporter genes in newborn muscle fibers following intramuscular inoculation (Quantin et al., 1992; Ragot et al., 1993; Vincent et al., 1993; Acsadi et al., 1994). Some limitations still remain concerning the use of the AV as a gene delivery vector to muscle, such as differential transducibility throughout muscle maturation, immunological response induced by AV transduction, and restricted packaging capacity of the AV genome (Acsadi et al., 1994B; Yang et al., 1994) although newly generated vectors have been engineered with increased carrying capacity (Kochanek et al., 1996).

The vector systems described above all allow for the transduction of muscle cells. However, the optimal vector has yet to be devised. Based on these findings, we have examined the feasibility of using herpes simplex virus (HSV) based vectors for gene transfer to muscle. The virus naturally establishes a latent or "quiescent" infection in cells of the nervous system, a state in which viral genomes persist for the life of the host without integrating into host DNA or altering host cell metabolism. Whereas the natural biology of HSV makes it well suited for gene transfer to neurons, the virus is capable of infecting a wide variety of postmitotic terminally differentiated cells and tissues. Although latency is limited to neurons for wild-type virus, replication-defective recombinants that are unable to replicate *in vivo* may persist as defective viral genomes in muscle cells.

In this chapter, we summarize current revelant aspects of HSV molecular biology and the status of HSV-mediated gene transfer to muscle and muscle-associated neural tissues. We include experimental findings related to the efficiency of HSV-1 gene transfer to muscle cells *in vitro* and *in vivo,* the hurdles associated with gene transfer to muscle, and the targeting of spinal cord neurons following intramuscular inoculation of the viral vector.

THE MOLECULAR BIOLOGY OF HSV-1

The HSV particle is 110 nm in diameter (Roizman and Furlong, 1974) and composed of an icosahedral-shaped nucleocapsid surrounded by a protein matrix, the tegument, which in turn is encompassed by a glycolipid-containing envelope. The envelope contains at least 11 virus-encoded glycoproteins that play an essential role in the adsorption and penetration of the host cell (Spear, 1993). The adsorption process involves initial binding to heparan sulfate moieties on the cell surface primarily by glycoproteins B (gB) and C (gC) (Wudunn and Spear, 1989; Herold et al., 1991; Shieh and Spear, 1994), followed by an apparently higher affinity binding to a second receptor recognized by glycoprotein D (gD) (Ligas and Johnson, 1988; Johnson et al., 1990; Montgomery et al., 1996). The virus penetrates the cell by fusion of the virus envelope with the cell surface membrane and requires the presence of gD, gB, and the gH–gL complex (Cai et al., 1987; Desai et al., 1988; Ligas and Johnson, 1988; Hutchinson et al., 1992; Roop et al., 1993). The de-enveloped particle enters the cytoplasm, where it is guided along microtubules to the nucleus. The viral DNA subsequently enters the nucleus to begin the productive replication cycle.

The viral genome is a linear double-stranded DNA molecule 152 kb in length and contains two unique segments (U_L and U_S) each flanked by inverted repeat components. Of the 81 known genes (Roizman and Sears, 1996), 38 are essential for production of infectious virus particles in cell culture while the remaining 43 genes are not essential for replication *in vitro* but contribute to the virus life cycle *in vivo*. Nonessential genes can be individually deleted from the viral genome without preventing virus replication under permissive tissue culture conditions used for culturing virus. Some deletion mutants grow less vigorously than wild-type virus, and the removal of multiple genes can significantly impair replication. In general, the accessory genes contribute to the virus host range, increase cell-to-cell spread, increase pathogenesis, help the virus-infected cell elude immune surveillance, and increase virus growth in nondividing cells *in vivo*.

Following entry into the cell nucleus, the viral genome circularizes, and the cascade of viral gene expression (Honess and Roizman, 1974) is initiated by binding of a viral tegument protein, VP16, in combination with two cellular transcription factors Oct-1 and HCF, to TAATGARAT enhancer sequences located in the promoters of the five immediate early (IE) genes (Gerster and Roeder, 1988; Preston et al., 1988; O'Hare and Goding, 1988; Katan et al., 1990; Kristie and Sharp, 1993; Werstuck and Capone, 1993; Wilson et al., 1993). These genes encode infected cell polypeptides (ICP) 0, 4, 22, 27, and 47, named according to their molecular sizes in order of appearance in SDS-PAGE gels. A sixth gene, ICP6, is expressed both as an IE and an early (E) function since its promoter contains the VP16 responsive element and is also transactivated by ICP0 (Desai et al., 1993). The IE gene products ICP4 and ICP27 are essential (Sacks et al., 1985; DeLuca and Schaffer, 1985), while the other IE gene products are accessory functions. ICP0, ICP4, and ICP27 enhance expression of E and late (L) genes (Preston, 1979; Dixon and Schaffer, 1980; Watson and Clements, 1980; Sacks et al., 1985; DeLuca and Schaffer, 1985). In addition, L gene expression requires viral DNA synthesis (Holland et al., 1980; Mavromara-Nazos and Roizman, 1987). The IE genes are the only viral genes that can be expressed in the absence of viral protein synthesis.

In addition to transcriptional regulation functions, ICP27 affects the splicing, polyadenylation, and stability of mRNA (Smith et al., 1992; Brown et al., 1995; Sandri-Goldin et al., 1995; McGregor et al., 1996; Sandri-Goldin and Hibbard, 1996); ICP22 may aid the usurping of cellular RNA polymerase by phosphorylation (Rice et al., 1994); and ICP47 inhibits MHC class I antigen presentation (York et al., 1994; Fruh et al., 1995; Hill et al., 1995). The E genes are expressed in response to IE gene induction and are largely products that carry out viral DNA synthesis. Nine of these gene products, including the viral DNA polymerase and origin-binding protein, are essential for viral genome replication (Challberg, 1986). The genome is thought to be replicated by a rolling circle mechanism forming head-to-tail concatemers (Jacob et al., 1979; Skaliter et al., 1996). During DNA replication, the U_L and U_S components can invert by homologous recombination events involving the inverted repeat sequences forming four possible isomers all of which appear to be infectious (Jacob et al., 1979; Mocarski and Roizman, 1982; Davison and Wilkie, 1983). The L genes encode mainly structural proteins that assemble into capsids in a well-ordered manner and assist in head-full packaging of the viral genome that stabilizes the nucleocapsid (Frenkel et al., 1976; Deiss et al., 1986). The tegument assembles around the mature capsids prior to budding of the virion through the nuclear membrane where the virus acquires its envelope. The lytic cycle almost always results in cell lysis with the possible exception of replication in sensory neurons during the establishment of latency or following viral reactivation.

HSV-1 VECTOR-MEDIATED GENE DELIVERY TO MUSCLE

First-generation replication-defective HSV-1 vectors deleted for the essential IE gene ICP4 were employed in the initial studies of reporter gene transfer and expression in myogenic cells in culture and following direct inoculation of mouse skeletal muscle (Huard et al., 1995). The recombinant vector SHZ.1 was engineered to contain the *Escherichia coli* lacZ reporter gene under the transcriptional control of the strong human cytomegalovirus (HCMV) IE promoter (Fig. 1). We observed that the replication-defective vector was capable of infecting both myoblasts (Fig. 2A) and differentiated myotubes (Fig. 2B) in cell culture. Because the efficiency of transduction was similar (Huard et al., 1995, 1996, 1997), it was apparent that heparan sulfate, the cellular ligand responsible for the initial attachment interaction for HSV-1, was preserved throughout muscle differentiation *in vitro*. We have also shown that the same vector (SHZ.1) can infect and express a foreign reporter gene in a significant number of muscle fibers in newborn muscle (Fig. 2C) and some fibers in adult muscle (Fig. 2D). The inability to efficiently infect adult muscle fibers has also been observed with adenoviral vectors and remains one of the major limitations associated with gene transfer to muscle. Even though the first-generation replication-defective HSV-1 vectors were able to efficiently deliver and express a reporter gene in myogenic cells, a number of hurdles remain to be overcome before HSV-1 can be used successfully for gene delivery to skeletal muscle.

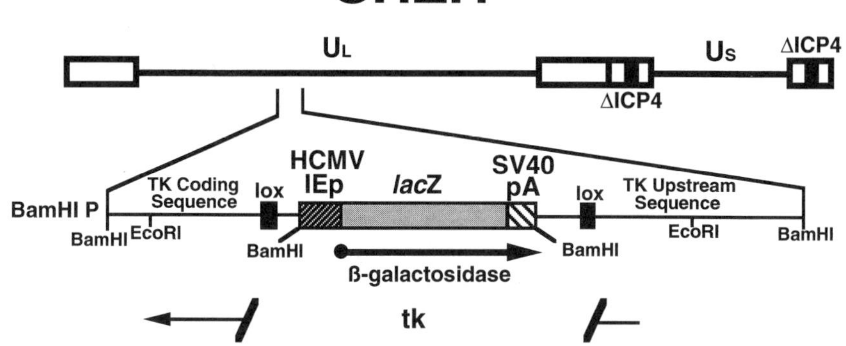

FIGURE 1. Replication-defective HSV-1 vector SHZ.1. SHZ.1 contains an expression cassette with the HCMV IE promoter driving lacZ in the thymidine kinase (tk) locus of an $ICP4^-$ virus (Mester et al., 1995; Rasty et al., 1995).

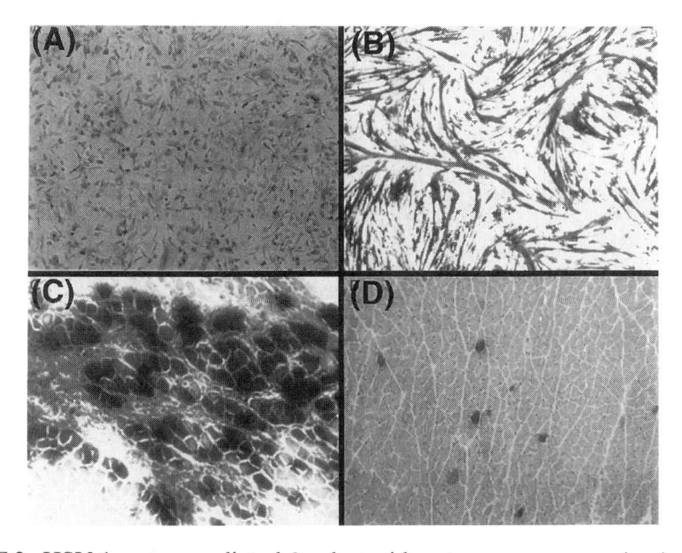

FIGURE 2. HSV-1 vector-mediated β-galactosidase transgene expression in myogenic cells. L6 mouse myoblasts (**A**) and differentiated myotubes (**B**) were infected with replication-defective SHZ.1 vector at an MOI of 1.0 and stained with X-gal to detect lacZ expressing cells. Both myoblasts and differentiated myotubes are highly transduced by the HSV-1 vector. Newborn (**C**) and adult (**D**) mouse gastrocnemius muscle were injected with 2.5×10^6 pfu of SHZ.1 vector and the reporter gene expression examined at 3 days postinjection. A significant number of fibers are transduced in newborn but not in adult muscle. Figure also appears in color section.

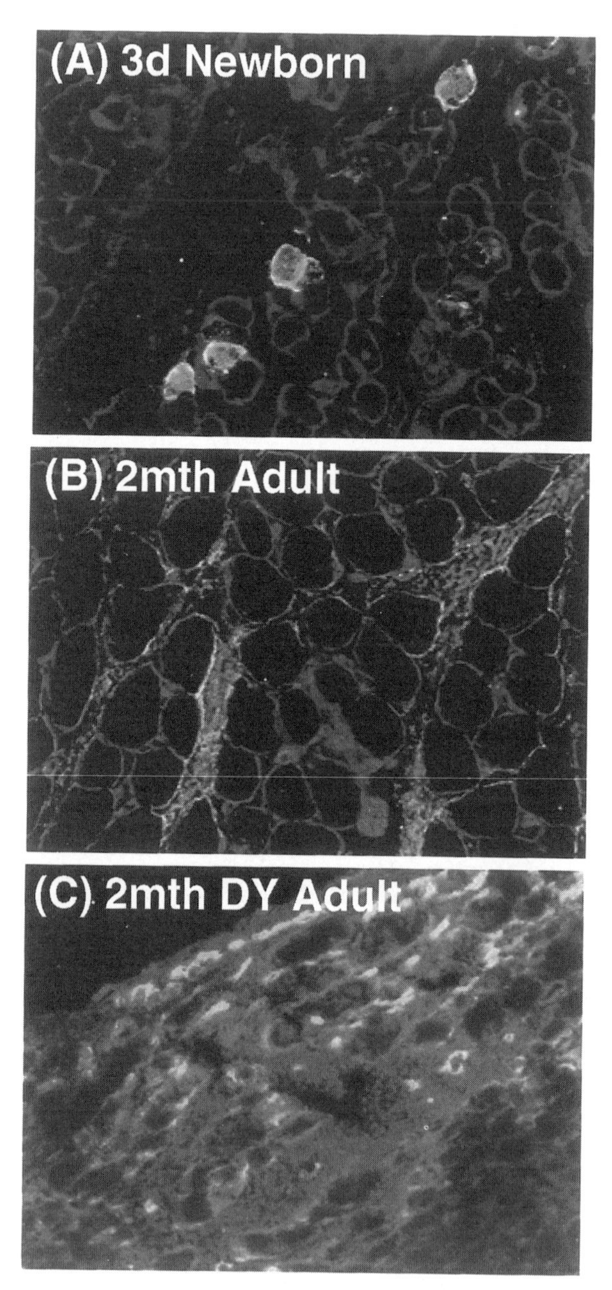

FIGURE 3. Role of the basal lamina in limiting HSV-1 infection in mature myofibers. Normal newborn (**A**), normal adult (**B**), and adult merosin-deficient dy/dy mouse gastrocnemius (**C**) was infected with 2.5×10^6 pfu of the SHZ.1 vector, and collagen IV (fluorescein; green) as well as HSV-1 infiltration (Cy3; red) were co-localized by immunofluoresence. In newborn animals, we observed a high number of muscle fibers

IMPEDIMENTS TO HSV-1 GENE TRANSFER TO MUSCLE

Differential Transducibility Throughout Muscle Fiber Maturation

Low-level transduction of mature muscle by HSV-1 may be the result of several factors such as the changes in immunocompetency between newborn and adult animals, the loss of attachment sites for HSV-1 on the surface of mature myofibers, or the maturation of the basal lamina that can act as a barrier to viral entry in mature myofibers.

To examine the role of the immune system in the differential transduction observed throughout muscle development, the transduction efficiency of the replication-defective recombinant vector (SHZ.1) was compared following intramuscular injection of adult immunodeficient and immunocompetent mice. As was the case with immunocompetent adult mice, we observed that HSV-1 poorly transduced adult muscle fibers of both nude and severe combined immunodeficient mice in contrast to the efficient transduction observed with newborn animals (Huard et al., 1995, 1996, 1997), indicating that the poor level of HSV-1 transduction in adult muscle is not directly related to immune-mediated clearance of infectious virus or infected cells.

We next examined the possibility that the resistance of adult muscle fibers to HSV infection was due to a loss in HSV receptors during the maturation process. We observed that HSV-1 is capable of efficiently infecting myoblasts, myotubes, and isolated myofibers from 14-day-old mice, thus elminating the possibility that adult myofibers lacked viral receptors. Experiments using a specific monoclonal antibody against the initial viral receptor, the heparan sulfate proteoglycan, confirmed that the receptor was uniformly distributed on newborn and adult muscle fibers (Huard et al., 1996).

A third possible impediment to infection of adult muscle could involve the gradual maturation of the basal lamina, which could provide a physical barrier to HSV-1 infection (Huard et al., 1996). Immunohistochemical analyses of vector injected tissue using antibodies to collagen IV, a major component of the basal lamina, and HSV-1 viral antigens in co-localization studies demonstrated that HSV-1 cannot penetrate the adult muscle basal lamina (Fig. 3B), which contrasts with observations in newborn muscle fibers (Fig. 3A). This conclusion was further supported by the finding that HSV-1 vectors displayed an intermediate level of transduction of mature dy/dy mouse muscle fibers (Fig. 3C) from merosin-deficient mice defective for normal basal lamina formation.

(*continued*) infected by the HSV-1 vector since both immunofluorescent markers are localized in the same myofibers (A). In adult muscle fibers (B), we observed that HSV-1 remains outside the muscle fibers, since the red fluoresence signal does not co-localize with the green immunofluorescence. However, in the dy/dy adult mice (C), HSV-1 vector penetrates the mature myofibers since both signals co-localize to the inside of mature myofibers as observed in newborn muscle. Figure also appears in color section.

Experiments are in progress to develop methods to permeabilize the basal lamina with fenestrating agents to allow HSV-1 penetration and subsequent transduction of normal mature myofibers.

Transient HSV Vector-Mediated Transgene Expression in Myofibers

Persistence of transgene expression mediated by HSV-1 vectors was found to be very limited in the injected muscle. Transgene expression in transduced myofibers present at 3 days postinjection in newborn and adult muscle disappeared by 15 days (Fig. 4A,B). The disappearance of transgene expression may have resulted from shut-off of the HCMV IE promoter, cytotoxic effects of the virus vector, or immune rejection of the transduced muscle fibers. Since the viral genome could not be detected in injected muscle at 15 days postinjection by Southern blot analysis using lacZ and HSV-1 DNA probes (Fig. 4C), the transient nature of transgene expression cannot be related to

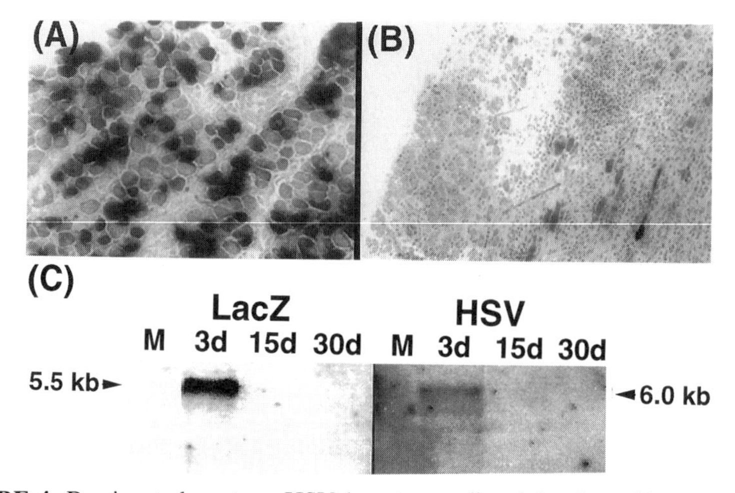

FIGURE 4. Barriers to long-term HSV-1 vector-mediated β-galactosidase expression in muscle. SHZ.1 vector 1×10^7 pfu was injected into newborn gastrocnemius and β-galactosidase expression was examined by X-gal staining at 3 (**A**) and 6 (**B**) days postinoculation. The large number of β-gal–positive muscle fibers in newborn (**A**) present at 3 days post inoculation disappear by day 6 (**B**). Similar results were detected in adult muscle, even though the transduction frequency in adults at 3 days is greatly reduced compared with newborn (see Fig. 2C,D). Southern blot analysis for the persistence of the HSV-1 vector-mediated transgene DNA following intramuscular injection of HSV-1 (SHZ.1) vector was performed (**C**). DNA 8 μg extracted out of the injected muscle was digested with BamHI, transferred to nytran membranes, and hybridized with two different probes (lacZ and HSV). The lacZ probe hybridized to a 5.5 kb band in extracted muscle DNA at 3 days but did not hybridize to muscle DNA from 15 and 30 day animals. The HSV probe hybridized also with a 6 kb band only in muscle DNA at 3 days, demonstrating that the vector did not persist. Figure also appears in color section.

the HCMV promoter shut-off. Moreover, histological staining of the HSV-1–injected muscle revealed some areas free of muscle fibers, suggesting that clearance of virus-infected cells is due to either toxicity or immune rejection of the infected muscle fibers (Huard et al., 1997).

It is presumed that HSV vector-associated cytotoxicity observed using ICP4-deficient viruses (d120) results from the overexpression (Dixon and Schaffer, 1980; DeLuca et al., 1985; DeLuca and Schaffer, 1985) of the other IE gene products (ICP0, ICP22, ICP27, and ICP47) since UV irradiation almost entirely reduces virus toxicity in myogenic (Huard et al., 1997) and other (Leiden et al., 1980) cells in culture. Deletion of any of these products alone or deletion of the UL41 gene, which encodes the virion host shut-off function, did not reduce toxicity (Johnson et al., 1992). However, ICP4, ICP0, ICP27, and ICP22 have all been shown to be toxic in stable transfection assays (Johnson et al., 1994), supporting the hypothesis that deletion of these genes in combination may be required to reduce vector-associated toxicity to that observed with UV-irradiated virus. To date, we have deleted all five targeted cytotoxic genes (ICP4, ICP0, ICP22, ICP27, and UL41) in various combinations (Marconi et al., 1996). Two multiple deletion viruses, THZ.1 ($ICP4^-/22^-/27^-$) and THZ.3 ($ICP4^-/22^-/27^-/UL41^-$), display markedly reduced toxicity in Vero cells in culture (Krisky and Glorioso, unpublished data).

We evaluated toxicity and gene expression from these new multiple gene deletion vectors in myogenic cells in culture. As observed with the Vero cell studies, the first-generation single ICP4 gene deletion vector SHZ.1 was very cytotoxic, resulting in 80% of the myoblasts (G8) being killed within 5 days (Fig. 5A). Both of the multiple IE gene deletion vectors (THZ.1 and THZ.3) were found to be significantly less cytotoxic for G8 myoblasts (Fig. 5B,C, respectively) than SHZ.1 (Fig. 5A). At an MOI of 5, both vectors left the majority of myoblasts unharmed, whereas the first-generation vector (SHZ.1) killed these cells (Fig. 5D). Moreover, transgene (lacZ) expression persisted to at least 96 h postinfection with the THZ.1 and THZ.3 vectors, in contrast to only 24 h with the SHZ.1 vector. These findings are encouraging, and the use of these vectors for gene transfer to murine muscle *in vivo* is currently being explored.

Although UV irradiation resulted in increased myoblast survival *in vitro,* transduced muscle fibers were still eliminated following *in vivo* injection of UV-inactivated virus, suggesting that a second mechanism for vector elimination was acting *in vivo.* Our observation that transgene expression persisted for greater periods in severe combined immunodeficient mice clearly showed that an important impediment to HSV-mediated gene transfer and transgene expression in muscle *in vivo* was the induction of immune mechanisms resulting in rejection of transduced cells. The number of transduced muscle fibers in immunodeficient mice did not significantly decrease between 2 and 6 days postinjection, in contrast to the rapid elimination of transduced cells in immunocompetent mice (Huard et al., 1997). These observations indicate that methods for curtailing immune rejection of transduced cells will be required to

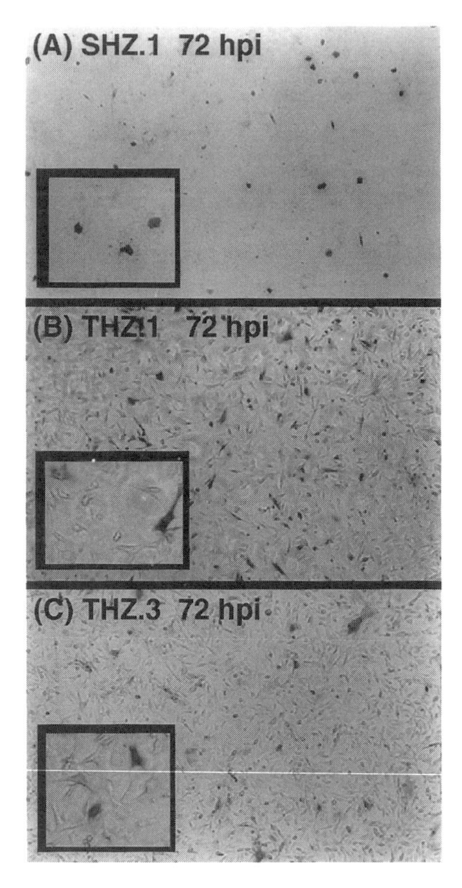
(A) SHZ.1 72 hpi
(B) THZ.1 72 hpi
(C) THZ.3 72 hpi

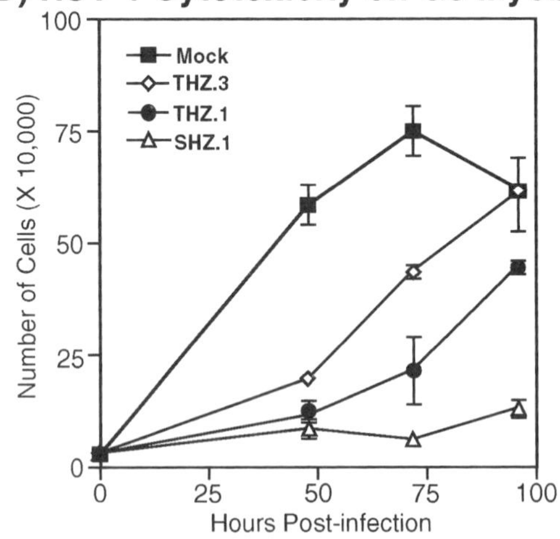
(D) HSV-1 Cytotoxicity on G8 Myoblast
Number of Cells (X 10,000)
Hours Post-infection
Mock
THZ.3
THZ.1
SHZ.1
0
25
50
75
100

achieve persistent gene expression in immunocompetent hosts. Together, these data demonstrate that, in addition to the barrier imposed by the natural basal lamina, both vector toxicity and vector-induced immune responses are significant obstacles to long-term transgene expression mediated by HSV-1 vectors in muscle.

TARGETING SPINAL CORD NEURONS FOLLOWING INTRAMUSCULAR INOCULATION OF HSV-1

Neuromuscular degenerative diseases affecting motor neurons of the spinal cord, including amyotrophic lateral sclerosis and spinal muscular atrophy, result in progressive muscle atrophy leading to paralysis and eventual death of the afflicted individual. These intractable disorders are complicated by the lack of an effective method to deliver the therapeutic gene product since systemic administration or direct injection of spinal parenchyma fails to target motor neurons specifically (Barinaga, 1994; Dittrich et al., 1994; Lisovoski et al., 1994; Yan et al., 1994). One potential approach to achieve specific gene delivery to spinal cord neurons involves intramuscular inoculation of the vector, in which the vector can reach afferent neurons innervating the injected muscle by retrograde axonal transport. Several vectors, including plasmid DNA (Shanek et al., 1993), adenovirus (Lisovoski et al., 1994; Finiels et al., 1995) and HSV (Keir et al., 1995; Levatte et al., 1995), have been used to achieve specific targeting and gene expression into the spinal cord by peripheral intramuscular inoculation of the vectors. We have initiated studies using HSV-1 as a gene delivery vector to the spinal cord following intramuscular injection of replication-defective and -competent recombinant HSV-1 vectors inoculated unilaterally into the gastrocnemius muscle. Expression of the β-galactosidase reporter gene was examined 3 days postinjection in both muscle

FIGURE 5. HSV-1–mediated cytotoxicity and transgene expression in myoblasts *in vitro*. G8 mouse myoblasts were infected with various HSV-1 mutant vectors at an MOI of 5. At various times postinfection, viable cell counts were obtained by trypan blue exclusion assay, cell monolayers were fixed and stained with X-gal, or the level of β-galactosidase was determined by ONPG assay. Few SHZ.1 (**A**) infected myoblasts survive or express β-galactosidase by 72 hours postinfection. However, both THZ.1 (**B**) and THZ.3 (**C**) infected myoblasts survive and continue to express the reporter gene at 72 hours postinfection. The inserts represent higher magnifications in which individual cells can be clearly seen. For SHZ.1 (**A**), the majority of the cell monolayer has been destroyed. The levels of cytotoxicity for the three vector constructs versus mock infected myoblasts is depicted over the various time points (**D**). As additional toxic genes are deleted from the virus vector, the level of toxicity is significantly decreased and approaches that seen in mock infected G8 myoblasts. Figure also appears in color section.

and spinal cord to determine whether this approach may be used to target specific populations of spinal cord neurons.

Replication-competent KHZ.1 (Mester et al., 1995; Rasty et al., 1995) vector (2.7×10^6 pfu) was injected into the gastrocnemius muscle of newborn mice (P5), and the muscle as well as the spinal cord were cryostat sectioned and stained for the presence of the reporter gene expression at 3 days postinfection. A significant number of transduced muscle fibers were detected in the injected muscle with replication-competent HSV-1, as well as neurons of the dorsal root ganglia and the ventral horn of the spinal cord, which project to the injection site. Direct inoculation of replication-defective SHZ.1 vector (1×10^7 pfu) into gastrocnemius of 28-day-old mice resulted in transduction of muscle fibers (Fig. 6A). Some motorneurons (Fig. 6C,D) as well as dorsal

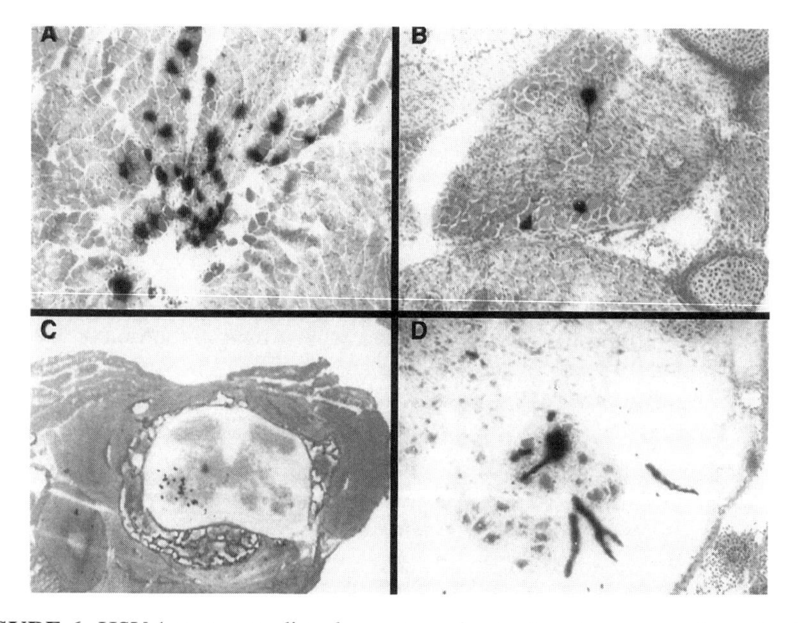

FIGURE 6. HSV-1 vector-mediated gene transfer to dorsal root ganglia and spinal cord from direct inoculation of mouse gastrocnemius muscle. Newborn mice were infected with 1×10^7 pfu of SHZ.1 intramuscularly into the gastrocnemius muscle. At 3 days postinfection, the animals were sacrificed, and the gastrocnemius, dorsal root ganglia, and spinal cord were removed by microdissection. The tissues were frozen, sectioned, and stained with X-gal to detect lacZ gene activity. Intact muscle fibers expressing the reporter gene (**A**) were readily detectable at 3 days. The vector spread by retrograde axonal transport from muscle to the dorsal root ganglia that innervate this region of the gastrocnemius as evidenced by expression of the reporter gene in the peripheral nerves (**B**). The vector was also capable of spreading to the ventral horn of the spinal cord (**C,D**), which possesses projections to the site of inoculation. Figure also appears in color section.

root ganglia neurons (Fig. 6B) were also transduced by the vector at 3 days postinjection.

These experiments suggest that both replication-competent and -defective HSV-1 vectors are capable of retrograde transport into the spinal cord following inoculation into skeletal muscle. Infection and reporter gene expression observed in spinal cord following retrograde transport of the replication-defective vector was less abundant yet more specific than that observed with the replication-competent virus. We have employed this system to deliver and express nerve growth factor from the vector in dorsal root ganglia and spinal cord (Goins, Huard, and Glorioso, unpublished data). Long-term expression of neurotrophic factors mediated by HSV-1 vectors in spinal cord, as well as the biological effects of this expression, remain to be determined.

SUMMARY

HSV possesses many biological features that make it attractive for gene delivery to muscle. The ability to generate replication-defective recombinants deleted for multiple genes that may be maintained in postmitotic cells for extended periods and are capable of incorporating large amounts of foreign DNA support the use of these vectors to treat muscle-specific disorders such as DMD. We have recently engineered recombinants that express either the full-length dystrophin cDNA or the Becker minigene cassette in dystrophin-deficient myoblasts and myotubes from mdx mice in culture (Akkaraju and Glorioso, unpublished data), demonstrating the ability of HSV vectors to handle large segments of foreign DNA. We have demonstrated that replication-defective HSV-1 vectors can efficiently transduce both myoblasts and differentiated myotubes in culture as well as immature myofibers following direct injection of skeletal muscle. We have shown that the basal lamina affects the ability of the vector to efficiently penetrate and transduce adult muscle fibers, suggesting that fenestrating agents may be required to increase the transduction efficiency of HSV-1 in mature muscle. The loss of transgene expression over time suggests that vector toxicity and/or vector-induced immune responses are responsible for the transient nature of reporter gene expression as supported by our experiments using immunodeficient animals. The construction of new replication-defective vectors deleted for multiple toxic functions has demonstrated that virus vector-induced toxicity was one factor responsible for the loss of transduced muscle fibers even in immunoincompetent mice. The full potential of these new vectors in reducing toxicity and clearance of transduced fibers while modifying transgene persistence *in vivo* remains to be determined. The ability to introduce therapeutic genes into spinal cord neurons following retrograde transport of HSV-1 vectors from muscle may be important in neurobiological investigations and suggests possible therapeutic applications for GT for heritable diseases affecting spinal cord.

REFERENCES

Acsadi G, Dickson G, Love DR, Jani A, Walsh FS, Gurusinghe A, Wolff JA, Davies KE, et al. (1991): Human dystrophin expression in mdx mice after intramuscular injection of DNA constructs. *Nature* 352:815–818.

Ascadi G, Jani A, Huard J, Blaschuk K, Massie B, Holland P, Lochmuller H, Karpati G (1994A): Cultured human myoblasts and myotubes show markedly different transducibility by replication-defective adenovirus recombinants. *Gene Ther* 1:338–340.

Acsadi G, Jani A, Massie B, Simoneau M, Holland P, Blaschuk K, Karpati G (1994B): A differential efficiency of adenovirus-mediated *in vivo* gene transfer into skeletal muscle cells of different maturity. *Hum Mol Genet* 3:579–584.

Alamedine HS, Dehaupas M, Fardeau M (1989): Regeneration of skeletal muscle fiber from autologous satellite cells multiplied *in vitro. Muscle Nerve* 12:544–555.

Barinaga M (1994): Neurotrophic factors enter the clinic. *Science* 264:772–773.

Brown CR, Nakamura MS, Mosca JD, Hayward GS, Straus ST, Perera LP (1995): Herpes simplex virus *trans*-regulatory protein ICP27 stabilizes and binds of 3′ ends to labile mRNA. *J Virol* 69:7187–7195.

Cai W, Person S, Warner S, Zhou J, Glorioso J (1987): Linker-insertion nonsense and restriction-site deletion mutations of the gB glycoprotein gene of herpes simplex virus type 1. *J Virol* 61:714–721.

Challberg M (1986): A method of identifying the viral genes required for herpesvirus DNA replication. *Proc Natl Acad Sci USA* 83:9094–9098.

Dai Y, Roman M, Naviaux RK, Verrma IM (1992): Gene therapy via primary myoblasts, long term expression of factor IX protein following transplantation in vivo. *Proc Natl Acad Sci USA* 89:10892–10895.

Davison A, Wilkie N (1983): Inversion of the two segments of the herpes simplex virus genome in intertypic recombinants. *J Gen Virol* 64:1–18.

Deiss L, Chou J, Frenkel N (1986): Functional domains within the *a* sequence involved in the cleavage-packaging of herpes simplex virus DNA. *J Virol* 57:605–618.

DeLuca NA, McCarthy AM, Schaffer PA (1985): Isolation and characterization of deletion mutants of herpes simplex virus type 1 in the gene encoding immediate-early regulatory protein ICP4. *J Virol* 56:558–570.

DeLuca NA, Schaffer PA (1985): Activation of immediate-early, early, and late promoters by temperature-sensitive and wild-type forms of herpes simplex virus type 1 protein ICP4. *Mol Cell Biol* 5:1997–2008.

Desai P, Ramakrishnan R, Lin ZW, Ozak B, Glorioso JC, Levine M (1993): The RR1 gene or herpes simplex virus type 1 is uniquely transactivated by ICP0 during infection. *J Virol* 67:6125–6135.

Desai P, Schaffer P, Minson A (1988): Excretion of non-infectious virus particles lacking glycoprotein H by a temperature-sensitive mutant of herpes-simplex virus type 1: Evidence that gH is essential for virion infectivity. *J Gen Virol* 69:1147–1156.

Dhawan J, Pan LC, Pavlath GK, Travis MA, Lanctot AM, Blau HM (1991): Systemic delivery of human growth hormone by injection of genetically engineered myoblasts. *Science* 254:1509–1512.

Dittrich F, Thoenen H, Sendtner M (1994): Ciliary neurotrophic factor: Pharmacokinetics and acute-phase response in rat. *Ann Neurol* 35:151–163.

Dixon RAF, Schaffer PA (1980): Fine-structure mapping and functional analysis of temperature-sensitive mutants in the gene encoding the herpes simplex virus type 1 immediate early protein VP175. *J Virol* 36:189–203.

Dunckley M, Love D, Davies K, Walsh F, Morris G, Discon G (1992): Retroviral-mediated transfer of a dystrophin minigene into mdx mouse myoblasts *in vitro*. *FEBS Lett* 2:128–134.

Dunckley MG, Wells DS, Walsh FS, Dickson G (1993): Direct retroviral-mediated transfer of a dystrophin minigene into male mouse muscle *in vivo*. *Hum Mol Genet* 2:717–723.

Finiels F, Ribotta MGY, Barkats M, Sanolyk ML, Robert JJ, Privat A, Revah F, and Mattet J (1995): A specific and efficient gene transfer strategy offers new potentialities for the treatment of motor neuron disease. *Neuro Report* 7:373–378.

Flotte TR, Afione SA, Zeitlin PL (1994): Adeno-associated virus vector gene expression occurs in nondividing cells in the absence of vector DNA integration. *Am J Respir Cell Mol Biol* 11:517–521.

Flotte TR, Carter BJ (1995): Adeno-associated virus vectors for gene therapy. *Gene Ther* 2:357–362.

Frenkel N, Locker H, Batterson W, Hayward G, Roizman B (1976): Anatomy of herpes simplex DNA. VI. Defective DNA originates from the S component. *J Virol* 20:527–531.

Fruh K, Ahn K, Djaballah H, Sempe P, van Endert PM, Tampe R, Petersen PA, Yang Y (1995): A viral inhibitor of peptide transporters for antigen presentation. *Nature* 375:415–418.

Gerster T, Roeder R (1988): A herpesvirus trans-activating protein interacts with transcription factor OTF-1 and other cellular proteins. *Proc Natl Acad Sci USA* 85:6347 6351.

Gussoni E, Pavlath PK, Lanctot AM, Sharma K, Miller RG, Steinamn L, Blau HM (1992): Normal dystrophin transcripts detected in DMD patients after myoblast transplantation. *Nature* 356:435–438.

Herold B, WuDunn D, Soltys N, Spear P (1991): Glycoprotein C of herpes simplex virus type 1 plays a principal role in the adsorption of virus to cells and in infectivity. *J Virol* 65:1090–1098.

Hill A, Jugovic P, York I, Russ G, Bennink J, Yewdell J, Ploegh H, Johnson D (1995): Herpes simplex virus turns off the TAP to evade host immunity. *Nature* 375:411–415.

Holland LE, Anderson KP, Shipman C, Wagner EK (1980): Viral DNA synthesis is required for efficient expression of specific herpes simplex virus type 1 mRNA. *Virology* 101:10–24.

Honess R, Roizman B (1974): Regulation of herpes simplex virus macromolecular synthesis. I. Cascade regulation of the synthesis of three groups of viral proteins. *J Virol* 14:8–19.

Huard J, Acsadi G, Jani A, Massie B, Karpati G (1994a): Gene transfer into skeletal muscles by isogenic myoblasts. *Hum Gene Ther* 5:949–958.

Huard J, Akkaraju G, Watkins SC, Pike-Cavalcoli M, Glorioso JC (1997): LacZ gene transfer to skeletal muscle using a replication defective herpes simplex virus type 1 vector. *Hum Gene Ther* 8:439–452.

Huard J, Bouchard JP, Roy R, Malouin F, Dansereau G, Labrecque C, Albert N, Richards CL, Lemieuex B, Tremblay JP (1992a): Human myoblast transplantation: Preliminary results of four cases. *Muscle Nerve* 15:550–560.

Huard J, Feero WG, Watkins SC, Hoffman EP, Rosenblatt DJ, Glorioso JC (1996): The basal lamina is a physical barrier to HSV mediated gene delivery to mature muscle fibers. *J Virol* 70:8117–8123.

Huard J, Goins WF, Glorioso JC (1995): Herpes simplex virus type 1 vector mediated gene transfer to muscle. *Gene Ther* 2:385–392.

Huard J, Guerette B, Verreault S, Tremblay G, Roy R, Lille S, Tremblay JP (1994b): Human myoblast transplantation in immunodeficient and immunosuppressed mice: Evidence of rejection. *Muscle Nerve* 17:224–234.

Huard J, Labrecque C, Dansereau G, Robitaille L, Tremblay JP (1991): Dystrophin expression in myotubes formed by the fusion of normal and dystrophic myoblasts. *Muscle Nerve* 14:178–182.

Huard J, Roy R, Bouchard J, Maloin F, Richard C, Tremblay J (1992b): Human myoblast transplantation between immunohistocompatible donors and recipients produces immune reactions. *Transplant Proc J* 24:3049–3051.

Huard J, Verreault S, Roy R, Tremblay M, Tremblay JP (1994c): High efficiency muscle regeneration following human myoblast clone transplantation in SCID mice. *J Clin Invest* 93:586–599.

Hutchinson L, Brown H, Wargent V, Davis-Poynter N, Primorac S, Goldsmith K, Minson A, Johnson D (1992): A novel herpes simplex virus glycoprotein, gL, forms a complex with glycoprotein H (gH) and affects normal folding and surface expression of gH. *J Virol* 6:2240–2250.

Jacob R, Morse L, Roizman B (1979): Anatomy of herpes simplex virus DNA XII. Accumulation of head-to-tail concatemers in nuclei of infected cells and their role in the generation of the four isomeric arrangements of viral DNA. *J Virol* 29:448–457.

Jiao S, Gurevich V, Wolff J (1993): Long-term correction of rat model of Parkinson's disease by gene therapy. *Nature* 362:450–453.

Johnson D, Burke R, Gregory T (1990): Soluble forms of herpes simplex virus glycoprotein D bind to limited number of cell surface receptors and inhibit virus entry into cells. *J Virol* 64:2569–2576.

Johnson P, Miyanohara A, Levine F, Cahill T, Friedmann T (1992): Cytotoxicity of a replication-defective mutant herpes simplex virus type 1. *J Virol* 66:2952–2965.

Johnson P, Wang M, Friedman T (1994): Improved cell survival of the reduction of immediate-early gene expression in replication-defective mutants of herpes simplex virus type 1 but not by mutation of the viron host shutoff function. *J Virol* 68:6347–6362.

Kaplitt M, Leone P, Samulski R, Xiao X, Pfaff D, O'Malley K, During M (1994): Long-term gene expression and phenotype correction using adeno-associated virus vectors in the mammalian brain. *Nature Genet* 8:148–154.

Karpati G, Ajdukovic D, Arnold D, Gledhill RB, Guttman R, Holland P, Koch PA (1993): Myoblast transfer in Duchenne muscular dystrophy. *Ann Neurol* 34:8–17.

Karpati G, Pouliot Y, Zubrzycka-Gaarn EE, Carpenter S, Ray PN, Worton RG, Holland P (1989): Dystrophin is expressed in mdx skeletal muscle fibers after normal myoblast implantation. *Am J Pathol* 135:27–32.

Karpati G, Worton RG (1992): Myoblast transfer in DMD: Problems and interpretation of efficiency. *Muscle Nerve* 15:1209.

Katan M, Haigh A, Verrijzer C, Vliet PVD, O'Hare P (1990): Characterization of a cellular factor which interacts functionally with Oct-1 in the assembly of a multicomponent transcription complex. *Nucleic Acids Res* 18:6871–6880.

Katsumi A, Emi N, Abe A, Hasegawa Y, Ito M, Saito H (1994): Humoral and cellular immunity to an encoded protein induced by direct DNA injection. *Hum Gene Ther* 5:1335–1339.

Keir SD, Mitchell WJ, Feldman LT, Martin JR (1995): Targeting and gene expression in spinal cord motor neurons following inoculation of an HSV-1 vector. *J Neurovirol* 1:259–267.

Kinoshita I, Vilquin JT, Guerette B, Asselin I, Roy R, Tremblay JP (1994): Very efficient myoblast allotransplantation in mice under FK506 immunosuppression. *Muscle Nerve* 17:1407–1415.

Kochanek S, Clemens PR, Mitani K, Chen HH, Chan S, Caskey CT (1996): A new adenoviral vector: Replacement of all viral coding sequences with 28 kb of DNA independently expressing both full-length dystrophin and beta-galactosidase. *Proc Natl Acad Sci USA* 93:5731–5736.

Kristie T, Sharp P (1993): Purification of the cellular C1 factor required for the stable recognition of the Oct-1 homeodomain by herpes simplex virus α-*trans*-induction factor (VP16). *J Biol Chem* 268:6525–6534.

Law PK, Goodwin TG, Fang Q, Chen M, Li HJ, Florendo JA, Kirby DS, Bertorini T, Herrod H, Golden G (1991): Pioneering development of myoblast transfer therapy. In Angelini C, Darrieli GA, Fontanen D (eds): *Muscular Dystrophy Research.* New York: Elsevier, pp 109–116.

Law PK, Goodwin TG, Fang Q, Duggirala V, Larkin C, Florendo JA, Kirby DS, Deering MB, Li HG, Chen M, et al. (1992): Feasibility, safety, and efficacy of myoblast transfer therapy on Duchenne muscular dystrophy boys. *Cell Transplant* 1:235–244.

Leiden J, Frenkel N, Rapp F (1980): Identification of the herpes simplex virus DNA sequences present in six herpes simplex virus thymidine kinase–transformed mouse cell lines. *J Virol* 33:272–285.

Levatte MA, Weaver LC, York IA, Johnson DC, Dekaban GA (1995): Delivery of a foreign gene to sympathetic preganglionic neurons using recombinant herpes simplex virus. *Neuroscience* 66:737–750.

Ligas M, Johnson D (1988): A herpes simplex virus mutant in which glycoprotein D sequences are replaced by β-galactosidase sequences binds to but is unable to penetrate into cells. *J Virol* 62:1486–1494.

Lisovoski F, Cadusseau J, Akli S, Caillaud C, Vigne E, Poenaru L, Stratford-Perricaudet L, Perricaudet M, Kahn A, Peschanski M (1994): *In vivo* transfer of a marker gene to study motorneuronal development. *Neuroreport* 5:1069–1072.

Lynch CM, Clowes MM, Osbornes WRA, Clowes AW, Miller AD (1992): Long term expression of human adenosine deaminase in vascular smooth muscle cells of rats. *Proc Natl Acad Sci USA* 89:1138–1142.

Marconi P, Krisky D, Oligino T, Poliani PL, Ramakrishnan R, Goins WF, Fink DJ, Glorioso JC (1996): Replication-defective HSV vectors for gene transfer *in vivo*. *Proc Natl Acad Sci USA* 93:11319–11320.

Mavromara-Nazos P, Roizman B (1987): Activation of herpes simplex virus 1 $\gamma 2$ genes by viral DNA replication. *Virology* 161:593–598.

McGregor F, Phelan A, Dunlop J, Clements J (1996): Regulation of herpes simplex virus poly(A) site usage and the action of immediate-early protein IE63 in the early-late switch. *J Virol* 70:1931–1940.

Mester JC, Pitha P, Glorioso JC (1995): Anti-viral activity of herpes simplex virus vectors expressing alpha-interferon. *Gene Ther* 3:187–196.

Mocarski E, Roizman B (1982): Structure and role of the herpes simplex virus DNA termini in inversion, circularization and generation of virion DNA. *Cell* 31:89–97.

Montgomery RI, Warner MS, Lum BJ, Spear PG (1996): Herpes simplex virus-1 entry into cells mediated by a novel member of the TNF/NGF receptor family. *Cell* 87:427–436.

Morgan JE, Hoffman EP, Partridge TA (1990): Normal myogenic cells from newborn mice restore normal histology to degenerating muscle of the mdx mouse. *J Cell Biol* 111:2437–2449.

Muzyczka N (1992): Use of adeno-associated virus as a general transduction vector for mammalian cells. *Curr Top Microbiol Immunol* 158:97–129.

Naldini L, Blomer U, Gallay P, Ory D, Mulligan R, Gage FH, Verma IM, Trono D (1996): *In vivo* gene delivery and stable transduction of nondividing cells by a lentiviral vector. *Science* 272:263–267.

Naviaux RK, Verma IM (1992): Retroviral vectors for persistent expression *in vivo*. *Curr Opin Biotechnol* 3:540–547.

O'Hare P, Goding C (1988): Herpes simplex virus regulatory elements and the immunoglobulin octamer domain bind a common factor and are both targets for virion transactivation. *Cell* 52:435–445.

Partridge TA (1991): Myoblast transfer: A potential therapy for inherited myopathies. *Muscle Nerve* 14:197–212.

Partridge TA, Morgan JE, Coulton GR, Hoffman EP, Kunkel LM (1989): Conversion of mdx myofibers from dystrophin negative to positive by injection of normal myoblasts. *Nature* 337:176–179.

Preston C (1979): Control of herpes simplex virus type 1 mRNA synthesis in cells infected with wild-type virus or the temperature-sensitive mutant tsK. *J Virol* 29:275–284.

Preston C, Frame M, Campbell M (1988): A complex formed between cell components and an HSV structural polypeptide binds to a viral immediate early gene regulatory DNA sequence. *Cell* 52:425–434.

Quantin B, Perricaudet L, Tajbakhsh S, Mandel J-L (1992): Adenovirus as an expression vector in muscle cells *in vivo*. *Proc Nat Acad Sci USA* 89:2581–2584.

Ragot T, Vincent N, Chafey P, Gilgenkrantz H, Couton D, Cartaud J, Briand P, Kaplan J-C, Perricaudet M, Kahn A (1993): Efficient adenovirus-mediated transfer of a human minidystrophin gene to skeletal muscle of mdx mice. *Nature* 321:647–650.

Rasty S, Goins WF, Glorioso JC (1995): Site-Specific Integration of Multigenic Shuttle Plasmids Into the Herpes Simplex Virus Type 1 (HSV-1) Genome Using a Cell-Free Cre-lox Recombination System. New York: Academic Press.

Rice S, Long M, Lam V, Spencer C (1994): RNA polymerase II is aberrantly phosphorylated and localized to viral replication compartments following herpes simplex virus infection. *J Virol* 68:988–1001.

Roizman B, Furlong D (1974): The Replication of Herpesviruses. New York: Plenum Press.

Roizman B, Sears A (1996): Herpes simplex viruses and their replication. In Fields BN, Knipe DM, Howley PM (eds): *Fields Virology.* Philadelphia: Lippincott-Raven, pp 2231–2295.

Roop C, Hutchinson L, Johnson D (1993): A mutant herpes simplex virus type 1 unable to express glycoprotein 1 cannot enter cells, and its particles lack glycoprotein H. *J Virol* 67:2285–2297.

Sacks W, Greene C, Aschman D, Schaffer P (1985): Herpes simplex virus type 1 ICP27 is essential regulatory protein. *J Virol* 55:796–805.

Salvatori G, Ferrari G, Mezzogiorno A, Servidei S, Coletta M, Tonali P, Giavazzi R, Cossu G, Mavilio F (1993): Retroviral vector–mediated gene transfer into human primary myogenic cells leads to expression in muscle fibers *in vivo. Hum Gen Ther* 4:713–723.

Sandri-Goldin R, Hibbard M (1996): The herpes simplex virus type 1 regulatory protein ICP27 coimmunoprecipitates with anti-sm antiserum, and the C terminus appears to be required for this interaction. *J Virol* 70:108–118.

Sandri-Goldin R, Hibbard M, Hardwicke M (1995): The C-terminal repressor region of herpes simplex virus type 1 ICP27 is required for the redistribution of small nuclear ribonucleoprotein particles and splicing factor SC25; however, these alterations are not sufficient to inhibit host cell splicing. *J Virol* 69:6063–6076.

Shanek Z, Seharaseyon J, Mendell JR, Burghes AHM (1993): Gene delivery to spinal motor neurons. *Brain Res* 606:126–129.

Shieh M-T, Spear P (1994): Herpes virus–induced cell fusion that is dependent on cell surface heparan sulfate on soluble heparin. *J Virol* 68:1224–1228.

Simonsen GD, Groskreutz DJ, Gorman CM, Macdonald, MJ (1996): Synthesis and processing of genetically modified human proinsulin by rat myoblast primary culture. *Hum Gene Ther* 7:71–78.

Skaliter R, Makhov A, Griffith J, Lehman I (1996): Rolling circle DNA replication by extracts of herpes simplex virus type 1-infected human cells. *J Virol* 70:1132–1136.

Smith IL, Hardwicke MA, Sandri-Goldin RM (1992): Evidence that the herpes simplex virus immediate early protein ICP27 acts post-transcriptionally during infection to regulate gene expression. *Virology* 186:74–86.

Spear P (1993): Membrane fusion induced by herpes simplex virus. In Bentz J (eds): *Viral Fusion Mechanisms.* CRC Press, Boca Raton, FL: pp 201–232.

Tremblay JP, Malouin F, Roy R, Huard J, Bouchard JP, Satoh A, Richards CL (1993): Results of a blind clinical study of myoblast transplantation without immunosuppressive treatment in young boys with Duchenne muscular dystrophy. *Cell Transplant* 2:99–112.

Vilquin JT, Wagner E, Kinoshita I, Roy R, Tremblay JP (1993): Successful histocompatible myoblast transplantation in dystrophin-deficient mdx mouse despite the production of antibodies against dystrophin. *J Cell Biol* 131:975–988.

Vincent N, Ragot T, Gilgenkrantz H, Couton D, Chafey P, Gregoire A, Briand P, Kaplan J-C, Kahn A, Perricaudet M (1993): Long-term correction of mouse dystrophic degeneration by adenovirus-mediated transfer of a minidystrophin gene. *Nature Genet* 5:130–134.

Watson R, Clements J (1980): A herpes simplex virus type 1 function continuously required for early and late virus RNA synthesis. *Nature* 285:329–330.

Werstuck G, Capone J (1993): An unusual cellular factor potentiates protein–DNA complex assembly Oct-1 and Vmw65. *J Biol Chem* 268:1272–1278.

Wilson A, LaMarco K, Peterson M, Herr W (1993): The VP16 accessory protein HCF is a family of polypeptides processed from a large precursor protein. *Cell* 74:115–125.

Wolff JA, Ludtke JJ, Acsadi G, Williams P, Jani A (1992): Long-term persistence of plasmid DNA and foreign gene expression in mouse muscle. *Hum Mol Genet* 1:363–369.

Wudunn D, Spear P (1989): Initial interaction of herpes simplex virus with cells is binding to heparan sulfate. *J Virol* 63:52–58.

Xiao X, Li J, Samulski RJ (1996): Efficient long-term gene transfer into muscle tissue of immunocompetent mice by adeno-associated virus vector. *J Virol* 70:8098–8108.

Yan Q, Matheson C, Lopez OT, Miller JA (1994): The biological responses of axotomized adult motorneurons to brain-derived neurotrophic factor. *J Neurosci* 14:5281–5291.

Yang Y, Nunes FA, Berencis K, Gonczol E, Englehardt JF, Wilson JM (1994): Inactivation of E2a in recombinant adenoviruses improves the prospect for gene therapy in cystic fibrosis. *Nature Genet* 7:362–369.

York I, Roo C, Andrews D, Riddell S, Graham F, Johnson D (1994): A cytosolic herpes simplex virus protein inhibits antigen presentation to $CD8^+$ T lymphocytes. *Cell* 77:525–535.

8

HERPES VIRUS VECTORS

XANDRA O. BREAKEFIELD, PETER PECHAN,
KAREN JOHNSTON, AND DAVID JACOBY
Molecular Neurogenetics Unit, Neurology Department, Massachusetts General Hospital and Harvard Medical School, Charlestown, MA 02129

OVERVIEW

Virus-derived vectors provide an efficient means of gene delivery into dividing and nondividing cells in culture and *in vivo.* Vectors derived from herpes simplex virus type 1 (HSV-1) have some particular advantages as compared with other vectors. HSV-1 is a large (152 kb) double-stranded DNA virus with a broad tropism (Roizman and Sears, 1996). It is a pathogen in humans causing skin lesions during productive infection and able to assume a latent state in sensory neurons for very long periods (years). The virions can deliver genes with very high efficiency into most mammalian cell types. Virions enter cells by direct fusion with the cell membrane and are transported to the cell nucleus by microtubule-mediated transport. Viral DNA is extruded in through the nuclear pore in postmitotic nuclei or has direct access to the nuclear matrix in mitotic cells. In some postmitotic cells the virus DNA is maintained in latency as an episomal element with minimal transcriptional activity and as such is "benign" to the cell (Stevens, 1989). Wild-type virus can reactivate from this latent state, resulting in virus replication, but a number of different mutations in the virus genome can prevent reactivation while still allowing

Stem Cell Biology and Gene Therapy, Edited by Peter J. Quesenberry, Gary S. Stein, Bernard G. Forget, and Sherman M. Weissman
ISBN 0-471-14656-0

establishment of latency. HSV-derived vectors thus appear ideally suited for gene delivery and expression in neurons and other cell types.

Two types of vectors have been developed from this virus, termed *recombinant virus vectors* and *amplicon vectors,* which together provide the high versatility of this delivery system (Breakefield and DeLuca, 1991; Leib and Olivo, 1993; Glorioso et al., 1995). Recombinant virus vectors incorporate foreign DNA within the virus genome through deletion of portions of the virus genome and replacement with transgenes by homologous recombination. Homologous recombination occurs between viral DNA and plasmids bearing the foreign gene sequences flanked by viral sequences homologous to the target site in the viral genome (Smiley et al., 1981; Mocarski et al., 1980). By disrupting virus genes needed for viral functions, these vectors can be rendered relatively nontoxic and replication defective. HSV-1 vectors have a large transgene capacity, at least 30 kb (Roizman and Jenkins, 1985), and can be generated at high titers. These advantages, combined with the stability of the virions, allows effective gene delivery in the context of the nervous system where only small volumes of vehicle can be injected without causing damage to neuronal structures. These vectors retain the ability to enter a latent, stable state in some postmitotic cells and can mediate transgene expression for months. Amplicon vectors consist of plasmid DNA bearing the transgene(s) and two noncoding elements of the virus, a DNA origin of replication (ori_S) and a virion packaging signal (pac), which allow the DNA to be replicated and packaged into virions in the context of viral functions provided by helper virus (Spaete and Frenkel, 1982; Geller and Breakefield, 1988). These vectors have a more limited gene capacity, about 15 kb, but are very easy to generate and can be produced free of helper virus (Fraefel et al., 1996). Although they do not enter latency *per se,* they can be retained for months as transcriptionally active elements in nuclei of postmitotic cells. They allow a means to package plasmid DNA, free of viral-coding sequences, into highly infectious virion particles.

HSV-derived vectors can be used for gene delivery to a wide range of cell types, although they seem especially well suited for delivery to neurons compared with other vectors. They can deliver genes to postmitotic cells in culture with very high efficiency such that in a culture of rodent sympathetic or sensory neurons most cells will express the transgene at a multiplicity of infection (MOI) of 1, or one infectious particle per cell. For wild-type virus, the ratio of infectious virions to noninfectious virions is about 1 : 10 (Browne et al., 1996), a ratio much higher than that seen for some other viruses used as vectors, such as adenovirus, which has a ratio of 1 : ≥100 (Weitzman et al., 1995). Noninfectious particles can be toxic in their own right due to functions inherent in the virion itself. For example, in HSV-1 the VP16 protein contained within the virion can cause untoward transcriptional activation or repression of host cell genes when present in high copy number. As with other recombinant virus vectors, viral proteins expressed even by replication-incompetent vectors can be toxic to cells. Extensive efforts have been undertaken to delete toxic genes from HSV constructs that are used for recombinant virus vectors and as helper virus for amplicon vectors. Deletion of one or more of the immediate early viral genes yields vectors that have no apparent toxicity at low MOI (less than 1) but that continue to manifest some toxicity at MOI of 10 (Wu et al., 1996; Johnson et al., 1994). Interestingly, toxicity appears to be less in neurons than in other cell types in culture, possibly reflecting the tendency of the virus to enter latency in the former. Probably the real forte of HSV-1 vectors lies in their capacity for efficient delivery and long-term gene expression in neurons *in vivo*. This capacity lies in their ability to be transported extended distances by rapid retrograde transport from nerve terminals to cell nuclei and to enter a state of latency in which episomal viral DNA retains transcriptional activity. HSV-1 vectors have also proven to be highly effective for treatment of experimental brain tumors in that certain mutations can render the virus replication competent in dividing tumor cells, while being replication defective in nondividing normal cells (Martuza et al., 1991; Breakefield et al., 1995).

LIFE CYCLE OF THE VIRUS

The life cycle of HSV-1 has been extensively characterized (Roizman and Sears, 1996) and is summarized here. HSV-1 is a large double-stranded DNA enveloped virus. The envelope consists of lipid bilayer containing at least 10 viral membrane glycoproteins (Fig. 1) (Steven and Spear, 1996). The envelope surrounds an interior isosahedral capsid composed of seven capsid proteins and containing the viral genome and the viral core proteins. The capsid is surrounded by an amorphous proteinaceous matrix called the tegument. The 152 kb viral genome has been completely sequenced and codes for at least 74 genes arranged within unique long (U_L) and short (U_S) sequences and flanking repeat elements, a, b, and c (Fig. 2A). Infection proceeds by a staged process, beginning with binding of glycoproteins C and B in the envelope to heparan sulfate ligands on the cell surface (Fig. 3). This relatively nonspecific binding is followed by binding of a viral glycoprotein, as yet unidentified, to a cell surface receptor, which is a member of the low affinity nerve growth factor/tumor necrosis factor alpha family (Montgomery et al., 1996). This triggers a direct fusion process between glycoprotein D, and possibly other glycoproteins in the viral envelope, with the cell membrane, thereby releasing the capsid and associated tegument proteins into the cytoplasm. Tegument proteins and/or the capsid itself associates with the dynein complex that mediates ATP-dependent, vectorial transport to the cell nucleus along microtubules (Sodeik et al., 1996; Topp et al., 1994). At the nuclear membrane the capsid opens, extruding viral DNA and associated viral proteins into the nucleus through a nuclear pore. Once the viral DNA reaches the nucleus, the lytic cycle of HSV is initiated in which the genome is expressed in a coordinately regulated temporal pattern. Initially, two of the tegument proteins, VP16 and

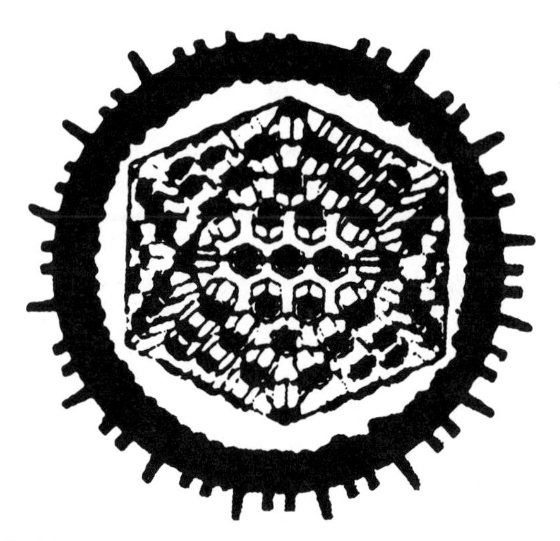

FIGURE 1. HSV virion. Schematic cross section of the herpes virion. Outer layer is the envelope with glycorproteins (spikes) projecting from the surface and an irregular inner perimeter representing association with tegument proteins, which reside in the space between the envelope and capsid. Capsid itself is isosadeltahedron with twofold symmetry. (Reproduced from Roizman 1996, with permission of the publisher.)

the virion host shut-off protein (vhs) are associated with activation of viral gene expression (Stern et al., 1989) and the disruption of host cellular RNA and protein synthesis (Kwong and Frenkel, 1989; Oroskar and Read, 1989), respectively.

Viral genes are classified into three groups depending on their order of expression during productive viral infection (Fig. 4). The immediate early (IE) genes (alpha), encoding the protein ICP4, ICP22, ICP27, ICP0, and ICP47, have complex functions that include transcriptional activation of viral genes, alteration of RNA splicing and prevention of viral antigen presentation. The early (E) genes (beta) include those coding for enzymes utilized for viral DNA synthesis, such as thymidine kinase, UTPase, and ribonucleotide reductase. Viral DNA replication appears to proceed at discrete loci along the inner nuclear membrane as a rolling circle DNA concatenate utilizing three origins of replication. After DNA synthesis, the late (L) genes (gamma) are expressed, which include the viral structural proteins. Capsids are assembled within the nucleus and mature as viral DNA is packaged within them. This is achieved by "stuffing" the capsid with viral DNA to capacity and cleavage of DNA at a packaging sequence (pac). The mechanism by which the mature virion acquires the tegument and envelope remain controversial. Capsids appear to bud through the nuclear membrane and to transit through the Golgi (Avitabile et al., 1995). In some cells, enveloped virions within vesicles are present in the cytoplasm and appear to fuse with the cell membrane, resulting in release of enveloped particles (Lycke et al., 1988). However, in neurons, naked capsids have been noted passing in an anterograde manner, apparently picking up the virion envelope from patches on the cell membrane containing viral glycoproteins as they bud out of the cell (Penfold et al., 1994).

In neurons and possibly other postmitotic cells, HSV-1 can enter a latent state. During latency, the viral genome follows the same initial route of entry into the cell nucleus, but immediate early genes and later viral gene expression is repressed. The viral DNA becomes circular and highly condensed in a nucleosomal structure, which exists separate from the cellular genome in an episomal state (Dressler et al., 1987; Deshmane and Fraser, 1989). A neuron can harbor many copies of the HSV genome, up to thousands per cell, in this state (Sawtell and Thompson, 1992; Ramakrishnan et al., 1994). In latency, there is little transription of the virus genome, with the exception of two latency-associated transcripts (LATs) (Stevens et al., 1987). These are spliced, nonpolyadenylated RNAs found in the cell nucleus of unknown function. It is not clear whether all cells that harbor the virus in latency express LATs. Moreover, the relationship between LATs and latency is unresolved, but deletions of this region reduce the ability of the virus to enter latency and to reactivate from it (Sawtell and Thompson, 1992). The "decision" of the virus to enter latency rather than to mount a productive infection probably depends on a number of viral and cellular factors, including cell type, MOI, genotype of infecting virus, transcriptional state of the cell, and levels of modulatory factors. Low MOI appears to favor latency, which probably correlates at least in part with lower amounts of VP16 entering with fewer virus particles. VP16 binds the cellular transcription factor Oct-1 and activates transcription of immediate early viral genes. In general immediate early gene transcription is repressed when the virus enters latency. A mechanism of latency has been proposed in which the isoforms of Oct-2 that are expressed in sensory neurons can prevent VP16-mediated activation of immediate early viral genes (Lillycrop et al., 1994). In this model, Oct-2 preferentially bind the HSV-1 VP16, thus forming a transcriptionally inactive complex. Several mutations in the HSV-1 genome can prevent viral propagation while still allowing entrance into latency, including mutations in the immediate early genes, ICP4, ICP22, and ICP27, although, interestingly, expression of ICP0 seems to promote latency (Smith et al., 1996). Sensory neurons are the preferred site of latency *in vivo,* but HSV-1 can enter a latent state in central nervous system neurons and possibly in glia and other nonmitotic cell types; presumably the latent genome would be lost if cells underwent mitosis.

Similarly it is not clear how the virus undergoes reactivation. Clearly, the transition from latency to productive infection can be triggered by exposure of cells harboring latent virus to stress, including heat, certain hormones, and stimulation, which would also alter the transcriptional state of the cells (Stevens, 1989). Activation of the ICP4 gene would be sufficient to start the cascade of productive infection. It remains possible that neurons produce limited amounts of virus during reactivation, which are not sufficient to kill the neurons themselves, but which then go on to produce lytic infection in

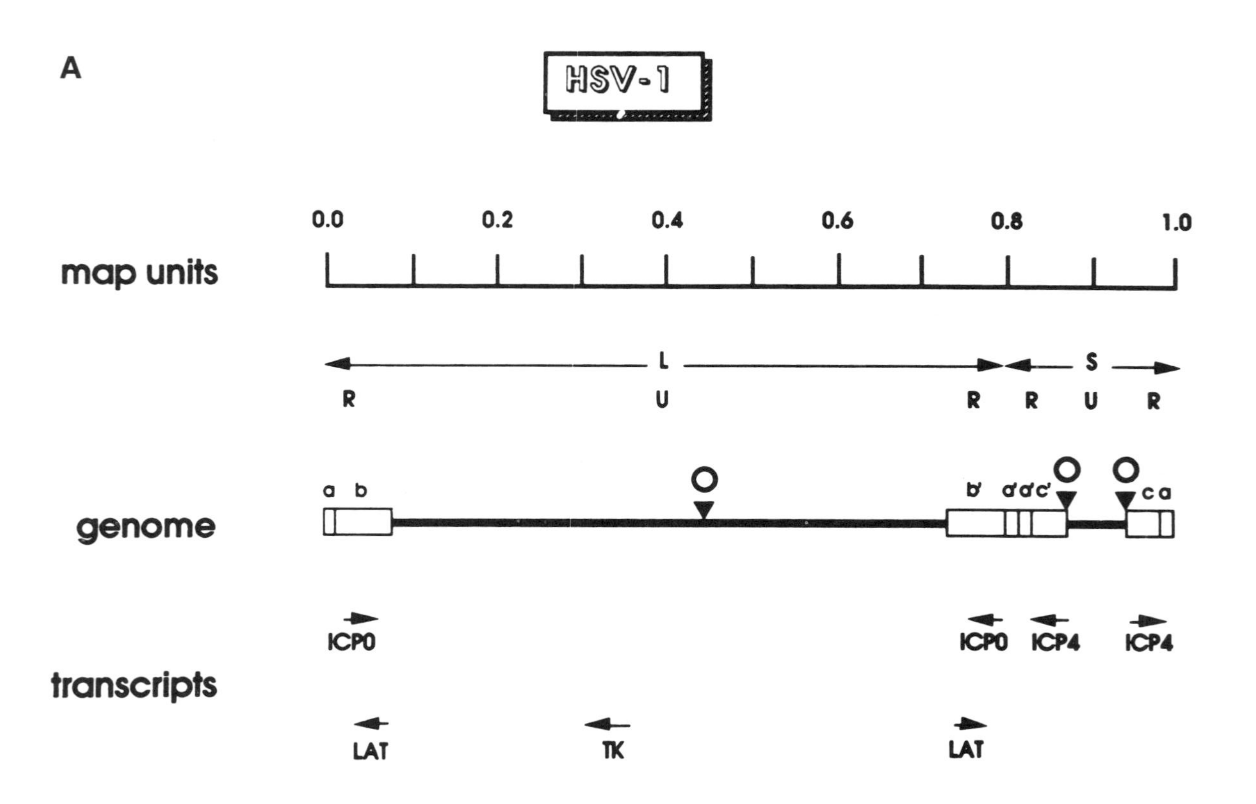
A
HSV-1
map units
0.0
0.2
0.4
0.6
0.8
1.0
L
S
R
U
R
R
U
R
genome
a b
b' a' a' c'
c a
transcripts
ICP0
ICP0
ICP4
ICP4
LAT
TK
LAT

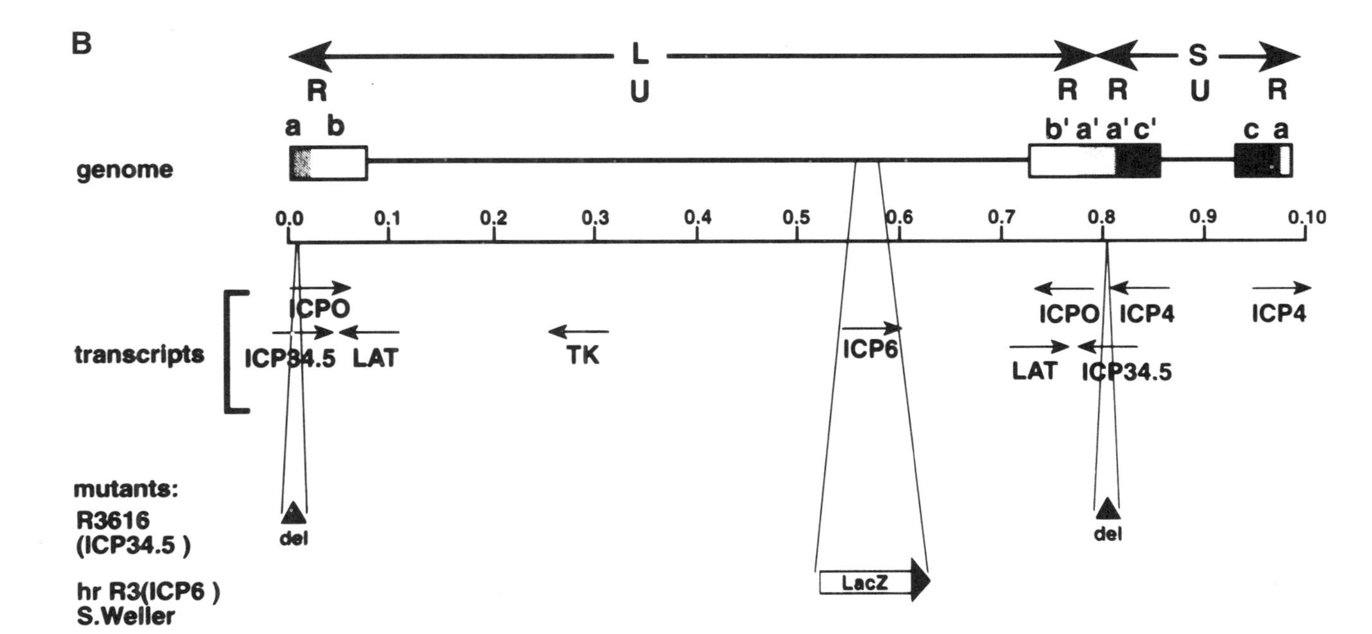

FIGURE 2. HSV-1 genome and recombinant virus vector. **A:** Virus genome. The 152 kb viral genome consists of two unique (U) regions, long (L) and short (S); with repeat elements (R) ab and b′a′ flanking U_L and a′c′ and ca flanking U_S. The positions of the origins of DNA replication (O), and some of the approximately viral genes that have been manipulated in vectors are indicated. **B:** Recombinant virus vector. As an example, the vector MGH1 was derived by homologous recombination of a plasmid bearing the lacZ gene within the ICP6 (ribonucleotide reductase) gene (Goldstein and Weller, 1988) into a virus-bearing mutation in the gamma 34.5 gene (R3616) (Chou et al., 1990) by Dr. Christof Kramm, Ms. Maureen Chase, and Dr. Antonio Chiocca (MGH). (This construct is similar to that described by Mineta et al. [1994]). Mutation in ICP6 renders the vector replication conditional in dividing cells; mutation in gamma 34.5 reduces neurovirulence, and the lacZ gene allows detection of viral infection. (Reproduced from Breakefield et al., 1996, with permission of the publisher.)

FIGURE 3. Virus entry into cells. Enveloped virions bind to cells by interaction between glycoproteins C and B and heparin sulfate on the cell surface. Tighter binding and fusion of the envelope to the plasma membrane is achieved by interaction with other glycoproteins on the envelope and a membrane receptor, which is a member of the low-affinity nerve growth/tumor necrosis factor-α family (Montgomery et al., 1996). The capsid with associated tegument proteins is thought to bind to a dynein complex and be transported along microtubules to the nuclear membrane. The capsid then opens, releasing DNA into the nucleus through the nuclear pore.

surrounding cells following release from the neurons. In sensory neurons, virus is preferentially released at the same nerve terminals that served as the initial site of infection, thus causing recurrent local infections. Systemic spread of the virus is limited by a strong immune response to it; most humans have circulating, neutralizing antibodies to HSV-1 (Baringer and Swoveland, 1973).

RECOMBINANT VIRUS VECTORS

Genetic manipulations of HSV-1 have been carried out extensively by herpes virologists (Roizman and Jenkins, 1985; Glorioso et al., 1995). The viral genome has been completely sequenced and most of the genes identified (McGeoch et al., 1988). Recently the viral genome was cloned into overlapping cosmid clones, allowing direct manipulation of the viral genome (Cunningham and Davison, 1993). Thus the technology to engineer the genome is well developed, and many mutant forms of the virus exist that can be used as backbones for the construction of vectors (or as helper virus for amplicons). In creating vectors for gene delivery to normal cells, several parameters are important—reduced toxicity, expanded transgene capacity, efficiency of vector generation, and promoter fidelity.

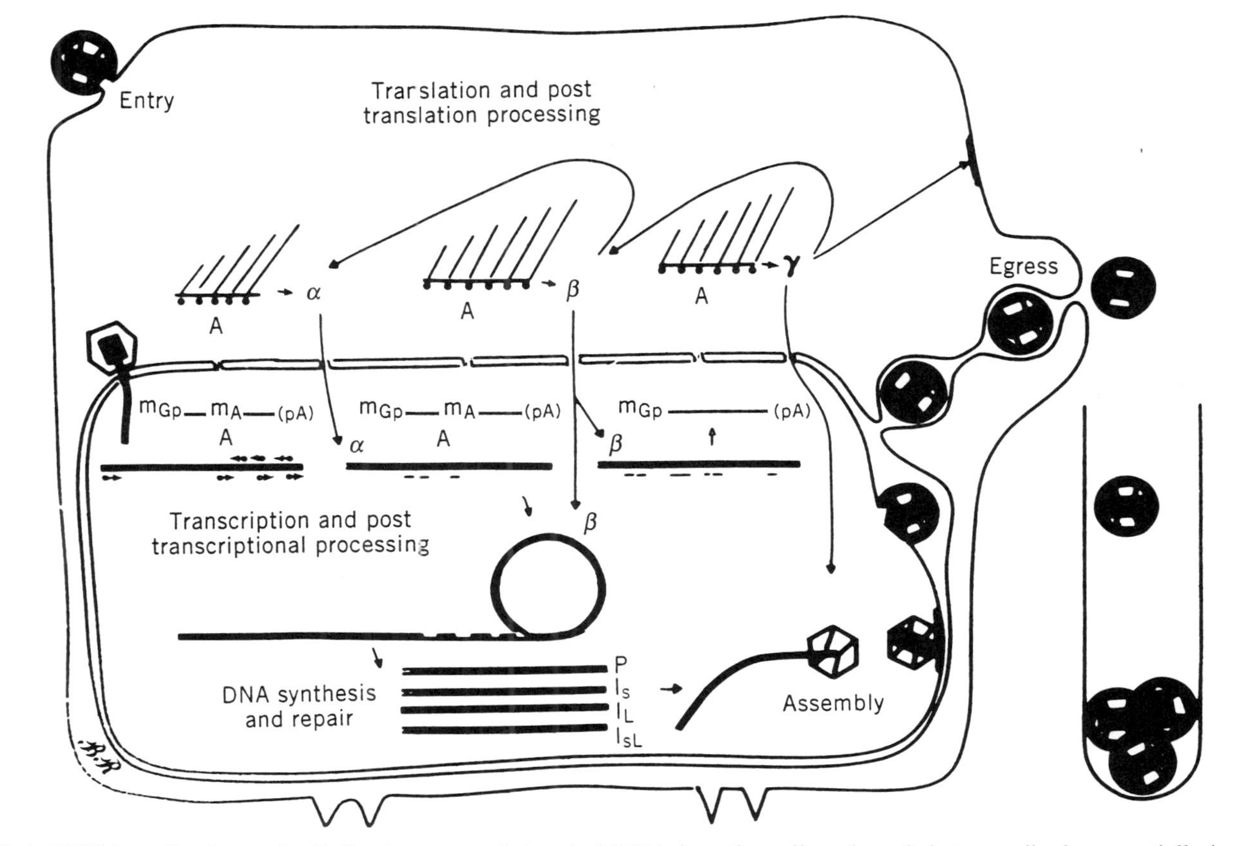

FIGURE 4. HSV-1 replicative cycle. Following entry of the viral DNA into the cell nucleus, it is transcribed sequentially in a cascade of immediate early (α), early (β) and late (γ) genes. Viral proteins are translated in the cytoplasm. Viral DNA is replicated and packaged into capsids in the nucleus. Capsids bud out through the nuclear membrane, acquiring the envelope during transit. Enveloped virions are released at the cell membrane. (Reproduced from Roizman and Batterson, 1985, with permission of the publisher.)

Recombination of a transgene into the HSV-1 genome leads to insertional disruption at that site. Sites for insertion fall into two classes, essential and nonessential loci. These terms refer to the ability of the mutant virus to propagate in permissive cells in culture, typically Vero or baby hamster kidney (BHK) cells. Essential loci are required for virus production and cannot be replaced by normal cell functions. Thus mutations in essential genes can only be propagated in cell lines in which the missing function is complemented by constitutive expression of the corresponding viral gene(s). Although it requires more effort to obtain mutations in essential genes, as transfected host cell lines must also be generated, in fact disruption of these genes can also reduce the toxicity of the virus and render it replication defective *in vivo* (see below). Nonessential loci encode functions that are not essential for virus propagation in culture, but may be critical during infection *in vivo.*

There are over 40 nonessential loci, including genes for thymidine kinase (TK), ribonucleotide reductase (RR), LATs, ICP47 (which blocks viral antigen presentation by infected cells) (York et al., 1994), and some of the envelope glycoproteins. In some regions of the viral genome these nonessential genes occur in tandem, allowing up to 10 kb of transgene insertion in at least four sites (Glorioso et al., 1995). In fact the whole U_S region (approximately 30 kb) contains only one essential gene (glycoprotein D) and one nonessential gene (UL22; DeLuca, personal communication) needed for high viral titers. By moving these gene(s) into the U_L region, a large "cargo" space can be provided for transgenes (Rasty and Glorioso, 1996). Where possible, it is ideal to delete entire viral genes when inserting a transgene, both to increase the transgene capacity and to reduce the chance of recombination with other viral elements and generation of "wild-type" virus.

Toxicity of HSV-1 is conferred by many viral functions, and productive viral infection leads to cell lysis. Vectors can be classified with respect to their level of replication competence. Fully replication-competent virus can be used as vectors when inoculated at low MOI through an appropriate route. For example, LAT^- vectors inoculated at the cornea or footpad of mice result in expression of the transgene in some neurons in the central nervous system and sensory ganglia, respectively (Wolfe et al., 1992; Carpenter and Stevens, 1996). Such vectors inoculated directly into the mouse brain would be lethal at high doses due to encephalitis. Replication-compromised vectors have reduced kinetics of infection, but may still exhibit cytopathic effect. For example, an $ICP0^-$ virus causes a slow degeneration in the rat brain over months after injection at high titers (Huang et al., 1992). Mutants are termed replication conditional if they can propagate efficiently in some cells, typically permissive cells in culture, but not in other cells, such as neural cells. These would include mutants with reduced neurovirulence (or ability to propagate in the nervous system), such as those lacking gamma 34.5 (Chou and Roizman, 1992) and UL5 (Bloom and Stevens, 1994). HSV-1 mutants lacking DNA synthetic enzymes, such as RR (Goldstein and Weller, 1988), TK (Coen et al., 1989), and UTPase (Pyles et al., 1992) are replication conditional in that they replicate well in dividing cells and poorly in nondividing cells. Rapidly dividing cells can complement the missing function of these viral mutants due to the upregulation of DNA synthetic enzymes; whereas postmitotic or quiescent cells, such as found in the central nervous system, have low levels of these enzymes. This type of replication-conditional mutant is effective in gene therapy for experimental brain tumors, as they propagate in and thereby kill tumor cells while largely sparing neurons (Martuza et al., 1991; Breakefield et al., 1995).

Many of the viral genes are toxic when expressed in cells so that even replication-defective vectors can be toxic. Toxicity is usually assessed in cultured cells and is a component of, but not equivalent to, neurovirulence and neuroinvasiveness, which reflect the dose needed to cause fatal encephalitis or the tendency to be transmitted along neuronal pathways, respectively, *in vivo*. Toxicity if affected by the MOI and the cell type. A number of mutations have been shown to reduce toxicity, including mutations in genes for the immediate early genes ICP4, ICP22, and ICP27 (Wu et al., 1996). In fact, virus with mutations in all three genes markedly reduced toxicity even at MOI 10, although changes in cellular DNA synthesis and loss of cell viability are still noted after a few days. Other vector backbones include mutations in genes for vhs, VP16, and ICP0. The latter protein, however, also appears to have a role in "steering" the virus into latency (Smith et al., 1996) and upregulating expression of mammalian and viral promoters (Cai and Schaffer, 1989; Everett, 1987). Interestingly, it appears that neurons may be less sensitive to the toxic effects of HSV-1 virus than other cell types and less susceptible *in vivo* than in culture. For example, differentiated PC12 cells survive infection with a virus, mutated in ICP4 and vhs, much better than BHK cells do (Johnson et al., 1994). However, it is difficult in such comparisons to determine whether lower toxicity is due to reduced infectibility, tendency of the virus to enter latency, or other cellular resistance factors. It may also be possible to increase the resistance of cells to virus infection. Thus it has been reported that NGF (Wilcox et al., 1990), bcl2 (Wang et al., 1995), certain isozymes of Oct-2 (Jacoby et al., 1995; Lillycrop et al., 1994), interferon-α (Chatterjee and Burns, 1990), and inhibitors of DNA synthesis (Nichol et al., 1996) can reduce the propagation and/or toxic effects of HSV-1 infection in some cells. In general, cells from mice, humans, and nonhuman primates appear to be more susceptible than those from rats and nonmammalian species, with some strains, e.g., *Aotus* monkeys, being highly susceptible and younger animals being more susceptible than adult animals.

Recombinant virus vectors are usually generated by recombining transgenes into the target HSV-1 locus cloned in a plasmid vector such that the transgene is flanked by at least 300 bp of HSV-1 sequence from the target locus (Johnson and Friedmann, 1994; Lowenstein and Enquist, 1996). The cloned transgene cassette is then transfected into permissive cells in culture and then the cells are either infected with HSV-1 or transfected with infectious HSV-1 DNA (Fig. 5). Recombination occurs during propagation of the virus with recombinants representing about 1/1,000 of the viral progeny. Recombinant plaques

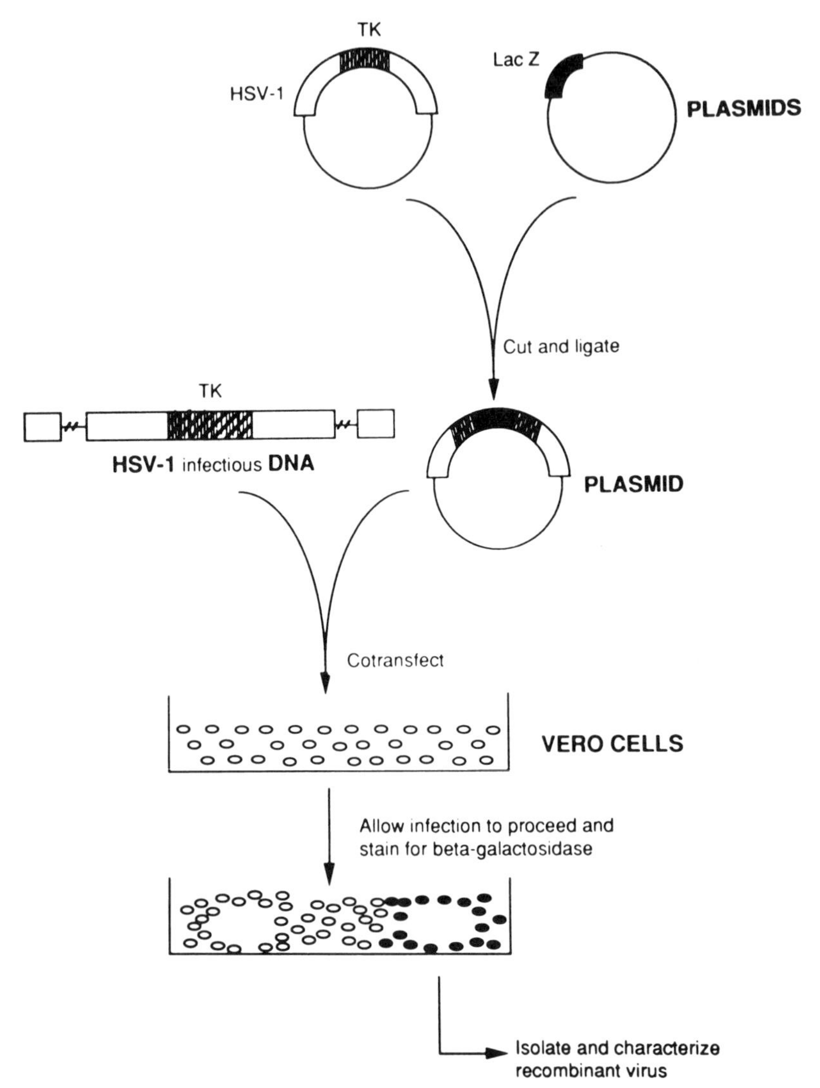

FIGURE 5. Generation of recombinant virus vector. In this scheme, a DNA fragment containing the *E. coli* β-galactosidase gene (lacZ) is cloned into a plasmid such that it disrupts the HSV TK gene and is under control of that viral promoter. Plasmid DNA is co-transfected into Vero cells with infectious wild-type HSV-1 DNA. During virus replication homologous recombination can occur between the intact TK gene in the viral genome and the disrupted TK gene in the plasmid. A few days later virus is harvested and replated onto Vero cells in the presence of acyclovir to select for TK^- virus. Plaques are stained histochemically for β-galactosidase to identify those bearing the lacZ gene. These recombinant virus vector plaques are picked, expanded, and plaque purified several times. (Reproduced from Andersen and Breakefield, 1995, with permision of the publisher.)

can then be selected by a number of means: loss or restoration of TK activity (Roizman and Jenkins, 1985), rescue of an essential viral gene linked to a transgene ("marker rescue," Desai et al., 1994), loss or gain of a marker gene such as lacZ or green fluorescent protein (e.g., Goldstein and Weller, 1988; Ho and Mocarski, 1988; Chase et al., in preparation), or *in situ* hybridization using a probe for the inserted transgene (Homa et al., 1986; Dobson et al., 1989). For loci that are present in duplicate copies in the HSV genome, e.g., ICP4 and LAT, it is necessary to go through two cycles of selection/identification to ensure that both loci are altered, as if only one is altered the vector will be unstable. With the availability of a set of cosmids spanning the HSV genome it is possible to insert the transgene into the viral locus directly by restriction digestion and ligation and then to co-transfect that cosmid with the complementary set of cosmids into permissive cells, in which case all viral progeny will be recombinant. The insertion of a lox sequence both into a cloned fragment of the HSV genome and into the transgene allows recombination to take place directly *in vitro* in the presence of Cre recombinase (Gage et al., 1992). Recombinant HSV genomes can be packaged into the HSV-1 virion if they are within 8 kb of the normal 152 kb genome. All vectors should be checked for integrity of the HSV-1 genome by Southern blot analysis. Titers of recombinant virus vectors are typically in the range of 10^9–10^{11} transducing units per milliliter.

Experimental applications of recombinant virus vectors fall into two groups, gene delivery to neurons and other normal cell types and gene therapy for tumors. HSV-1 vectors are well suited for gene delivery to neurons (as reviewed by Glorioso et al., 1995; Leib and Olivo, 1993) because they can enter a benign, stable, episomal state in many neuronal types and can be efficiently delivered through application at nerve terminals, with rapid retrograde transport to the cell nuclei. Most of the work to date has been directed toward reducing residual toxicity and regulating transgene expression using the lacZ gene as a marker of gene delivery (Dobson et al., 1990). Vectors appear to have lowest toxicity when inoculated at the cornea or footpad for delivery to sensory neurons, although replication-competent vectors can propagate in these neurons and be transported transsynaptically back to the brain, causing encephalitis. Direct inoculation into the brain parenchyma even of replication-defective vectors can cause focal toxicity due to the high MOI at the site of injection but achieves gene delivery to hundreds of neural cells in the immediate area and those projecting to that region (Fig. 6). Gene delivery over a wider range can be achieved by inoculation of larger volumes of vector stock into the ventricles (Kramm et al., 1996a,b) or into the carotid artery with temporary disruption of the blood–brain barrier (Muldoon et al., 1995). Recombinant virus vectors have also been used to deliver genes to hepatocytes (Miyanohara et al., 1992), muscle cells (Huard and Glorioso, 1995), and keratinocytes (Visalli et al., 1996) and can probably be used for a large number of nondividing cell types.

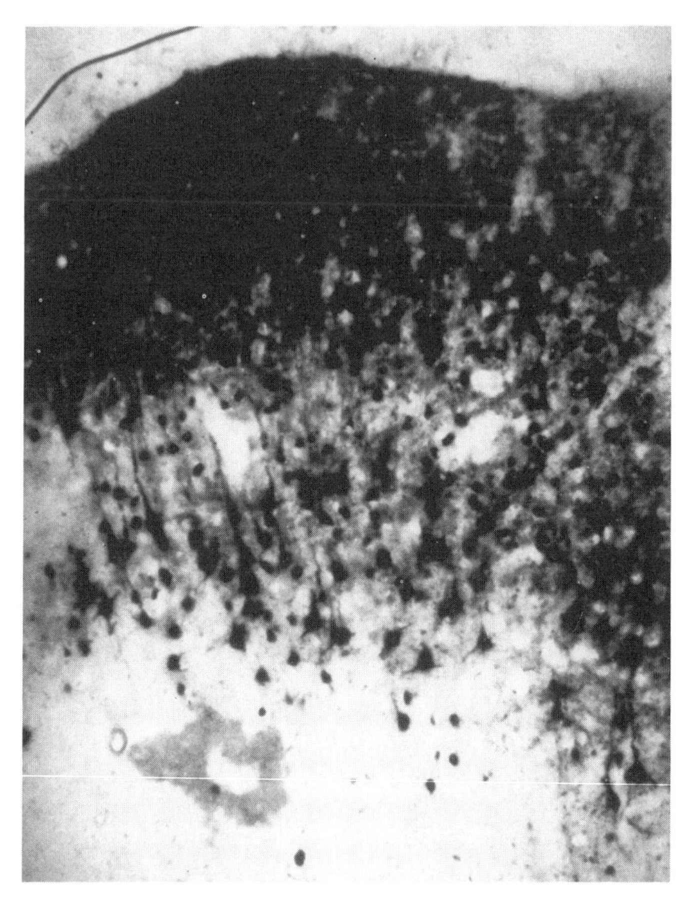

FIGURE 6. Transduction of neurons in brain in the recombinant virus vector. A replication-compromised HSV-1 vector bearing the lacZ gene (ICP0$^-$) (Cai and Schaffer, 1989) was stereotactically injected (2×10^5 pfu in 2 μl) into the frontal cortex of adult rat brain. Three days later animals were sacrificed, and sections of the brain were stained for β-galactosidase. This section shows extensive staining of neurons in the cortex. (Reproduced from Chiocca et al., 1990, with permission of the publisher.)

The most problematic aspect of gene delivery to neurons has been control of transgene expression with respect to stability, cell specificity, and regulation. Even though HSV-1 DNA can remain in the latent state with LAT expression in some neurons for years, transgene expression is usually extinguished within weeks after infection. Several groups have achieved transgene expression for months in sensory and even CNS neurons by placing them under modified LAT promoters (Carpenter and Stevens, 1996; Chen et al., 1995; Lokensgard et al., 1994). The integrity of cell-specific promoters is usually lost in the context of the HSV genome, probably due to the high density of overriding viral information and the nucleosomal structure of the latent DNA. For example, neither the neuronal-specific enolase (Andersen et al., 1992) nor the neurofilament (Carpenter and Stevens, 1996) promoters confers stable neuronal-specific expression *in vivo*. However, recently Rabkin et al. (1996) achieved acute liver-specific transgene expression using the albumin enhancer/promoter placed in the TK locus. Also hormone-responsive transgene expression has been achieved by including five yeast GAL4 transcriptional elements upstream of the cytomegalovirus (CMV) promoter in the TK locus (Oligino et al., 1996). Regulation is achieved by incorporation of a tripartite transactivator gene comprising the GAL4 DNA-binding domain, the transactivation domain of VP16, and a mutated hormone-binding domain of the progesterone receptor into the gC locus of the virus. Transgene expression can be induced in culture (30-fold) and in the brain by administration of a hormone analog. To date very little work has gone into exploration of the ability of these vectors to alter cell physiology, but vectors have been constructed encoding NGF to prevent neuronal degeneration (Carpenter and Stevens, 1996; Goins et al., 1996), neuropeptides to block pain transmission (Davar et al., 1996), and β-glucuronidase to mediate enzyme replacement in the lysosomal deficiency state (Wolfe et al., 1992).

Use of recombinant HSV vectors for treatment of brain tumors in experimental animals has proceeded quickly, as features of vector toxicity and transient transgene expression are not as critical as in delivery to neurons. The strategy has been to use vectors that are capable of propagation in tumor cells but that have reduced neurovirulence for normal brain. Mutants used for this purpose include disruptions of TK (Martuza et al., 1991) or RR (Mineta et al., 1994; Kaplitt et al., 1994; Boviatsis et al., 1994), which render the virus selectively replication competent in dividing cells (for review, see Breakefield et al., 1995; Hunter et al., 1995). In addition, several groups have shown that virus defective in gamma 34.5 (Chambers et al., 1995) or in both gamma 34.5 and RR (Mineta et al., 1994) (Fig. 2) can be effective in killing tumor cells, while being nontoxic to normal brain. Vectors can be inoculated directly into the tumor mass, through the cerebrospinal fluid, or across the brain–tumor barrier. The latter is "weaker" than the blood–brain barrier, and the virus preferentially enters tumors after intracarotid inoculation combined with osmotic (Nilaver et al., 1995) or pharmacological (Rainov et al., 1996) opening of the barrier. The therapeutic efficacy of these vectors probably also involves enhancement of the immune response to tumor cells via a strong immune response to the virus. This effect is enhanced in rat tumor models in which the gliomas used, e.g., C6 and 9L, are antigenic and the host is naive to the virus, thus allowing efficient infection and spread of the virus before the host mounts an immune response to it. The therapeutic efficacy of the RR-deficient vectors is enhanced by treatment of animals with ganciclovir, which is converted by viral TK into a nucleotide analog that disrupts DNA synthesis (Boviatsis et al., 1994; Kramm et al., 1995). Incorporation of other prodrug activating enzymes and cytokines into HSV-1 vectors should further increase their antineoplastic capacity (Chiocca et al., 1994; Kramm et al., 1995). The effectiveness of these vectors in experimental brain tumor models has compelled the evaluation of these vectors for potential clinical use in humans (for discussion, see Breakefield et al., 1995).

AMPLICON VECTORS

Amplicon vectors were derived from the observation that defective HSV virions contained repeated units of the origin of replication, ori_S, and the packaging signal, pac (Spaete and Frenkel, 1982; Kwong and Frenkel, 1995). The term *amplicon* refers to the fact that a DNA sequence can be amplified in a head-to-tail arrangement in defective, concatemeric virus genomes and packaged into HSV-1 virions. Amplicon vectors thus contain two noncoding sequences of HSV, ori_S and pac, which allow replication in eukaryotic cells infected with HSV-1 and packaging in HSV-1 virions, as well as an *Escherichia coli* origin of DNA replication and an antibiotic resistance gene to allow plasmid expansion and selection in bacteria (Fig. 7). Ori_S and pac are both contained within the a repeat element flanking U_S (Fig. 2A). Although they are each only a few hundred bp in length, they contain within them elements that affect gene transcription (TAATGARAT box and SP1 sites in ori_S; Stern et al., 1989) and promote recombination (repeat sequence in pac; Umene, 1993; Dutch et al., 1994) and can confound generation and use of vectors. In some vectors, ori_S and pac sequences are contained in 1–2 kb fragments that seem to promote the efficiency of replication and packaging, but also introduce other viral elements, including IE4 and IE3 promoters, which are activated by VP16 (Geller and Breakefield, 1988). Amplicons appear to be more stable when $<$15 kb in size, allowing for transgene inserts of 10–12 kb, but can theoretically be much larger in size, as the virion will package up to 152 kb of DNA in concatenated units. Amplicons to date have been constructed with two transgenes (New and Rabkin, 1996) and could hold more. The primary advantages of these vectors are their ease of construction, lack of viral genes, and the fact that packaging in the HSV-1 virion allows very efficient transgene delivery to both dividing and nondividing cells.

Packaging of the amplicon DNA involves expression of numerous helper functions by the virus. To generate amplicon vectors, amplicon DNA is first transfected into cells and then cells are infected with a helper virus or transfected with infectious HSV-1 DNA (Fig. 8A). Typically the helper virus is replication defective, e.g., $ICP4^-$, and complementing cells transfected with the corresponding essential gene are used to generate amplicon stocks. Stocks are passaged several times and then assessed for titers of amplicon vectors (transgene expression on nontransfected cells), helper virus (plaque formation on transfected cells), and recombinant wild-type virus (plaque formation on nontransfected cells). Titers of amplicon vectors are in the range of 10^4–10^6 transducing units per milliliter with 1–10 times as much helper virus. Several techniques have been used to increase the ratio of amplicon : helper virus,

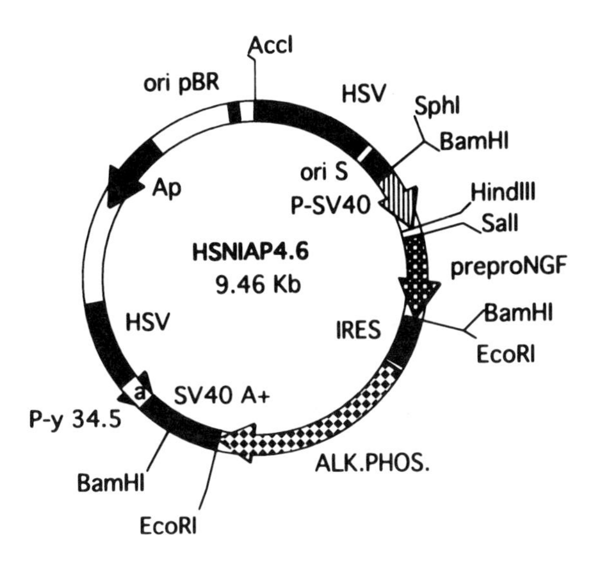

Plasmid name: HSNIAP4.6
Plasmid size: 9.46 kb
Constructed by: Peter Pechan, Ph.D., Breakefield Lab

FIGURE 7. Amplicon plasmid. This 9.46 kb plasmid contains three kinds of genetic elements: (1) sequences that allow propagation in *E. coli*—the ampillicin resistance gene (Ap) and the *E. coli* origin of DNA replication (ori pBr); (2) sequences that support propagation and packaging of plasmid DNA by a helper HSV-1 virus—an HSV-1 origin of DNA replication (ori_S) and a packaging site (a); and (3) a transcription unit including the SV40 promoter and the mouse prepro-NGF cDNA, an IRES translational read-through, the human alkaline phosphatase (alk phos) marker gene, and the SV40 polyadenylation site (SV40 A^+). (Construct provided by Dr. Peter Pechan.)

including coupling of their propagation by including a gene in the amplicon that is necessary for propagation of the replication-deficient helper virus, such as ICP4 (Pechan et al., 1996), UL42 (Berthomme et al., 1996), or TK (in the setting of cells deficient in TK; Zhang et al., 1996). Methods for packaging these vectors with helper virus have been described (Ho, 1994; Wu et al., 1995; Lowenstein and Enquist, 1996).

Recently a method has been described for packaging helper virus-free amplicons (Fraefel et al., 1996). Helper virus functions are provided by the overlapping cosmid set that covers the virus genome (Cunningham and Davi-

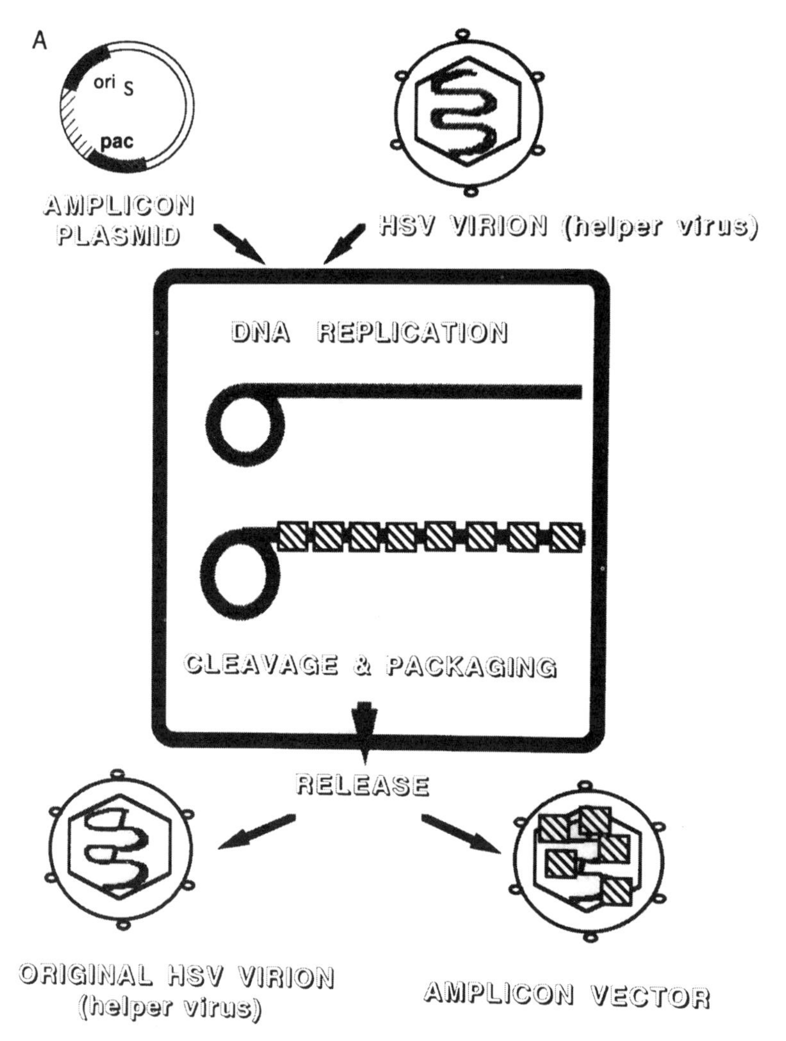

FIGURE 8. Generation of amplicon vector stocks. **A:** Amplicon vector and helper virus. Amplicon plasmid DNA is transfected into permissive cells in culture (typically cells are stably transfected with an essential HSV-1 gene), and cells are subsequently infected with a helper virus (typically one mutated in the same essential gene). Amplicon DNA is replicated as a concatenate and packaged into virus particles (typically 10–15 copies of the amplicon plasmid per virion) in parallel with the helper virus. Cells release both amplicon vectors and helper virus (Modified from Leib and Olivo, 1993.) **B:** Helper-free amplicon vectors. Amplicon plasmids are co-transfected into Vero cells with an overlapping set of cosmids (Cunningham and Davison, 1993) that cover the HSV-1 genome and have been mutated in pac sequences such that the recombinant wild-type HSV-1 genome cannot be packed into virions (Fraefel et al., 1996). Amplicon DNA is replicated and packaged into virus particles, and only amplicon vectors are released from cells.

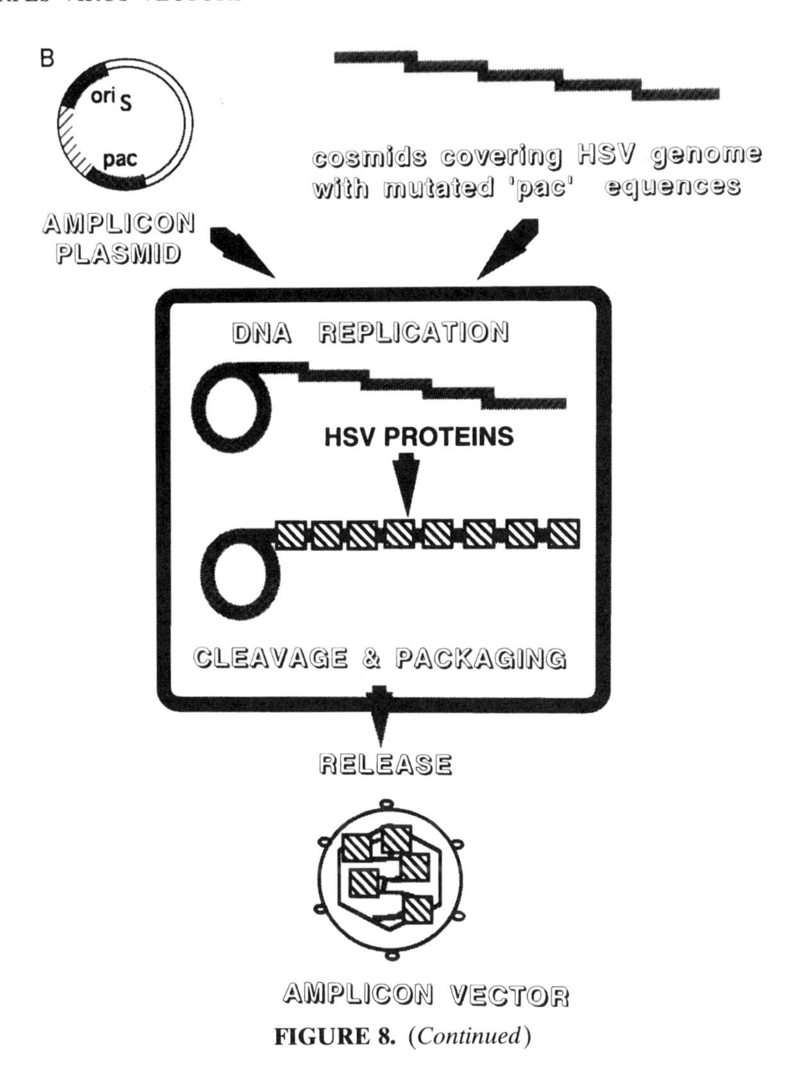

FIGURE 8. (*Continued*)

son, 1993) with mutations introduced into the pac signals present in the repeat units. This cosmid set and the amplicon plasmid are co-transfected into cells, resulting in replication and packaging of amplicon vectors, but not helper virus (Fig. 8B). Titers of up to 10^8 amplicon vectors/ml can be generated in this way. This provides a "virtual" synthetic vector system that contains only plasmid DNA and no viral genes packaged in highly infectious HSV-1 virions. Toxic functions of VP16 and vhs proteins associated with the virion itself remain, but at low MOI would not be present in sufficient amounts or for long enough to cause disruption of host cell metabolism.

Amplicon vectors introduce their DNA into the cells nucleus, as does HSV-1. The state and fate of amplicon DNA in the cell nucleus has not been evaluated. Based on the stability of transgene expression, it appears that the DNA is maintained in a functional state in the nucleus for weeks to months in nondividing cells, such as neurons (Geller, 1993), and that is appears to be lost in dividing cells (Johnston et al., 1997; Boothman et al., 1989). Two amplicon hybrid vectors have been developed recently to help control the fate of the amplicon DNA after it enters the cell nucleus. Vos and coworkers (Wang and Vos, 1996) have inserted the EBNA1 gene and origin of DNA replication, ori_P, from the Epstein-Barr virus genome into the amplicons to allow replication of the amplicon DNA in synchrony with host cell DNA synthesis. EBNA1 in combination with cellular factors allows initiation of episomal DNA replication at ori_P, and at the same time ori_P has a retention function that mediates binding of episomal DNA to chromosomal DNA, thereby facilitating the distribution of amplicon DNA to daughter cells at mitosis. Our group (Johnston et al., 1997) has incorporated adeno-associated virus (AAV) elements into the amplicon in order to permit amplification/integration of transgene sequences in the nucleus. In these constructs the transgene cassette is flanked by the AAV inverted terminal repeat (ITR) elements, and the AAV rep gene is included under its own promoter outside the ITRs. The ITRs potentially allow replication (amplification) of the transgene cassette in the cell nucleus, and the ITRs in the presence of rep isozymes may also mediate integration of the transgene into the cellular genome, in particular in the AAVS1 locus on human chromosome 19q (Kotin et al., 1992, 1996). These hybrid vectors illustrate the potential of amplicons to incorporate multiple viral elements, as well as transgenes, in a DNA construct that can be efficiently delivered to the cell nucleus via HSV-1 virions.

Amplicons also provide a plasmid DNA platform for regulation of transgene expression. HSV-1 elements used for replication and packaging do contain some potential enhancer/silencer activities that must be assessed to determine their effect on promoter expression, and there will undoubtedly also be position effects within the amplicon, but these can be carefully assessed and controlled by reduction in the size of the HSV-1 elements and site-directed mutagenesis of interfering sequences. In the first amplicon used to express lacZ in neurons, the immediate early IE3 promoter was used to drive transgene expression (Geller and Breakefield, 1988). This gives very robust expression within a few days after infection due to the presence of VP16 contained in the virions. Expression drops to a lower level within a few days after infection, but can be restimulated, even months later, by infection with replication-defective HSV (Starr et al., 1996). Many other viral and mammalian promoters have been used effectively in these vectors, including those for cytomegalovirus (Ho et al., 1993), tyrosine hydroxylase (Oh et al., 1996), JC virus, and SV40 (Pechan et al., 1996). Inducible transgene expression in the context of helper virus infection has been achieved using the glucocorticoid responsive element (Lu and Federoff, 1995) and the tetracycline response operon and transactivator developed Gossen and Bujard (1992) (Jacoby et al., 1995).

Amplicon vectors have been used effectively to change cellular morphology and physiology in culture and *in vivo* over periods of weeks to months. As examples, in culture an amplicon vector bearing GAP43 changed the morphology of PC12 cells (Neve et al., 1991), and one bearing the low-affinity NGF receptor changed the NGF responsiveness of cells (Battleman et al., 1993). Craig et al. (1995) have used an amplicon bearing a membrane-targeted protein, CD8, in cultured hippocampal neurons and observed preferential addition of CD8 at axonal growth cones. Amplicon vectors can also be used to explore molecular mechanisms of pathogenesis, for example, by expression of the calcium-binding protein parvalbumin in the presence of NMDA (Hartley et al., 1996) or expression of the GluR6 kainate subtype of the glutamate receptor (Bergold et al., 1993). Many different gene delivery models have been explored *in vivo*—including delivery of the glucose transporter (Ho et al., 1993) and bcl-2 (Lawrence et al., 1996) to promote central nervous system neuronal survival; bcl-2 (Linnik et al., 1995) and vascular endothelial growth factor (VEGF) (Mesri et al., 1995) to stimulated angiogenesis in models of cardiac ischemia; NGF to help sensory neurons survive axotomy (Federoff et al., 1992); and BDNF to potentiate neurite extension from damaged spiral ganglia neurons in the ear (Geschwind et al., 1997). The therapeutic effectiveness of these vectors has been demonstrated in a rat model of Parkinson's disease, where dopaminergic neurons in the substantia nigra are damaged with 6-hydroxydopamine (During et al., 1994). Amplicon-mediated delivery of the tyrosine hydroxylase gene under its own promoter was permissive for behavioral recovery in an apomorphine-induced rotation paradigm. Overall amplicon vectors provide a very promising means to bring genes into neurons and other cell types in culture and *in vivo* for experimental evaluation of neuronal functions and for therapeutic intervention.

FUTURE DIRECTIONS

It can be anticipated that HSV-1–derived vectors will have a major role in the upcoming development of vectors for gene delivery in three ways. First, the amplicon vectors packaged in a helper-free system offer a "virtual" synthetic vector for efficient, nontoxic gene delivery with very few modifications. Studies need to be undertaken to reduce the toxicity of elements in the virion itself, VP16, vhs, and possible other tegument proteins; to define the minimal, enhancer-free ori_S and pac sequences that will allow efficient replication and packaging in a helper-free system, mediate reliable transgene expression, and reduce recombinations and rearrangements; and to introduce other DNA elements that will allow the amplicon sequences to be maintained in both dividing and nondividing cells. Second, the virus itself has a number of novel functions that could be utilized to increase transduction efficiency of truly synthetic vectors. These include envelope glycoproteins that mediate fusion between lipid bilayer membranes at neutral pH; tegument and/or capsid pro-

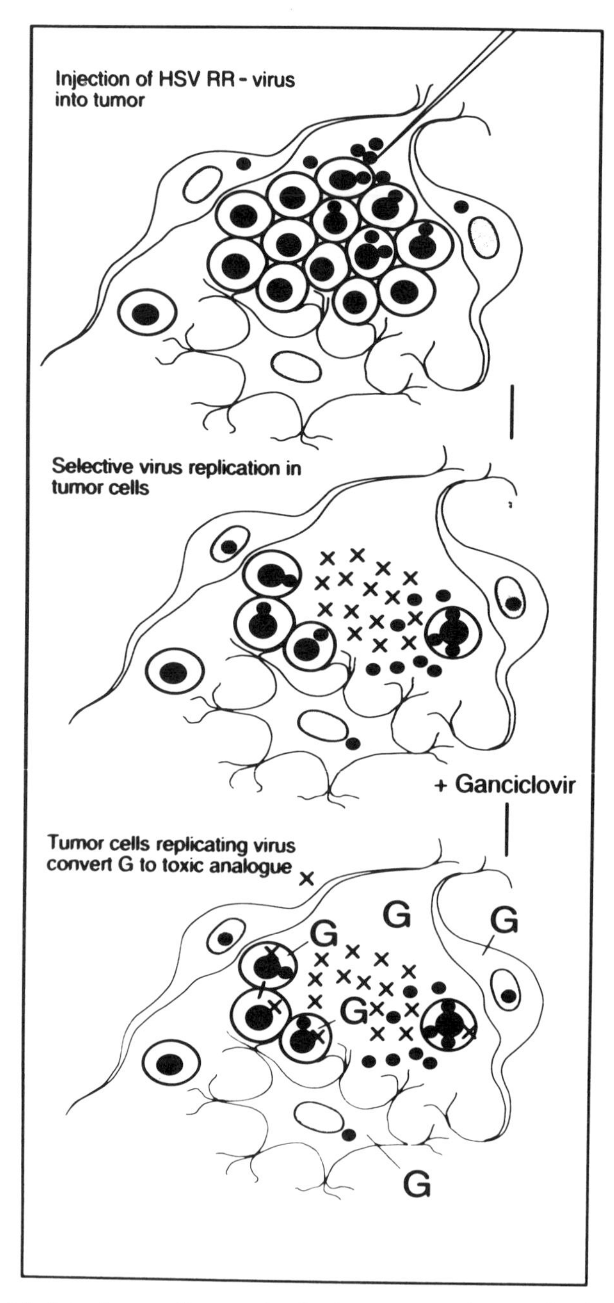

FIGURE 9. Model of gene therapy for brain tumors using replication-conditional HSV-1 vector. The replication-conditional RR^- HSV-1 vector is inoculated directly into the tumor mass. Dividing tumor cells replicate the virus and are killed in the process. Nondividing neural cells do not replicate the virus, and viral DNA enters latency or is eventually degraded in normal cells. After allowing virus replication to

teins that allow rapid retrograde and anterograde transport along microtubules; tegument and/or capsid proteins that facilitate entry of DNA into postmitotic nuclei; and viral proteins, such as VP22, that are transferred efficiently from infected cells into the nuclei of noninfected cells, possibly through an actin-mediated mechanism (Elliott and O'Hare, 1996). Third, HSV-derived vectors provided an efficient backbone for construction of vectors designed to increase transgene capacity and efficiency of transgene expression. Both recombinant and amplicon vectors allow efficient gene delivery to nondividing cells in culture and *in vivo;* this is especially useful in the case of neurons for which other means of gene delivery are much less efficient. For safe delivery to neurons it will be important to further reduce the residual toxicity of the virus by making additional mutations in the genome of the helper virus or recombinant helper virus vectors. Efforts are also ongoing in a number of laboratories to simplify construction and packaging methods, to modify the viral envelope for cell-specific targeting of gene delivery, and to allow stable transgene expression in the latent state. The major goal of recombinant virus vectors remains to create a nontoxic vector that will achieve stable transgene expression in neurons via an episomal viral genome.

Replication-conditional recombinant HSV vectors are highly effective vehicles for gene delivery to brain tumors (Martuza et al., 1991; Breakefield et al., 1995). Such vectors can be made less toxic to normal brain by mutations in neurovirulence genes and can be made more effective at therapy by including therapeutic genes for prodrug activation, immune enhancement, and anti-angiogenesis. It should also be possible to combine amplicon and recombinant virus vectors to expand the therapeutic capacity. In the "piggyback" system, by using a recombinant helper virus vector with mutations in both IE3 and ICP6 (RR) and an amplicon vector bearing IE3, it would be possible to retain division-specific propagation, at the same time introducing another level of safety to the vector, since propagation depends on co-infection with both vectors and only dividing cells would support replication (Pechan et al., 1996).

HSV-1 vectors appears to be poised for clinical trials. These will include therapy for life-threatening brain tumors and metastatic tumors using replication-conditional or replication-defective vectors that bear antineoplastic genes, such as TK (Fig. 9) Administration can be by direct injection into tumors or through the vasculature or cerebrospinal fluid or, in the case of

---

proceed for a therapeutic time window, animals are treated systematically with ganciclovir (G). In cells actively propagating the vector, levels of HSV TK are very high.HSV TK converts ganciclovir to toxic nucleotide analogs that can be transferred across gap junctions to uninfected neighboring cells. These toxic analogs are incorporated into replicating genomic DNA, thereby disrupting it and leading to cell death. Ganciclovir also blocks further virus replication. (Reproduced from Breakefield et al., 1996, with permission of the publisher.)

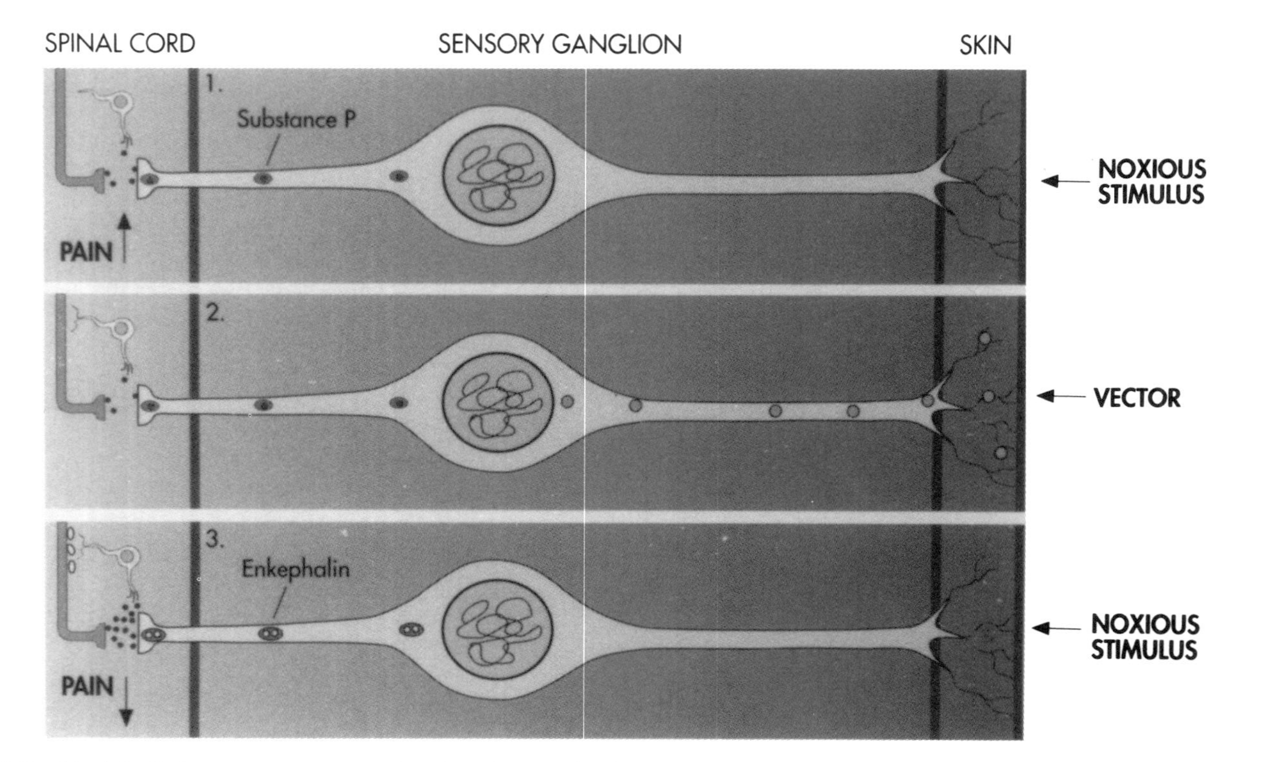

FIGURE 10. Model of therapy for peripheral pain using replication-defective HSV-1 vector. Sensory neurons project from the skin (**right**) to the spinal cord (**left**). Pain transmission is mediated in response to noxious stimuli by release of substance P and other transmitters in the spinal cord to the nerve endings of projection neurons that transfer signals to the brain. Vectors encoding the proenkephalin cDNA are applied at nerve terminals on the skin and taken by retrograde transport back to the cell nucleus. The vector enters a state of latency in the nucleus and expresses the proenkephalin precursor protein. This precursor is processed to the transmitter enkephalin within secretory vesicles, along with substance P. Vesicles are transported to nerve terminals in the spinal cord such that release of substance P is now accompanied by release of enkephalin. Enkephalin acts through interneurons in the spinal cord to downregulate signaling by projection neurons to the brain, thereby "muting" the pain signal.

cytokines, by peripheral vaccination with irradiated tumor cell transduced with the vector. Other therapeutic applications appear more problematic. It seems hard to justify use of a potentially toxic vector with limited transgene expression as a therapy for a neurodegenerative condition in which the patient is trying to maintain as much neurological function as possible for as long as possible and where introduction of the vector itself into the brain requires an invasive neurosurgical procedure. In addition, the modes of delivery and relatively low titers of some of the vectors reduces the number of neurons that can be transduced. The most feasible neuronal application to date would be for peripheral inoculation of sensory or motor neurons, for example to promote neurite regeneration, to block pain transmission, or to treat movement disorders. For example, both the vitreous of the eye and the cerebrospinal fluid provide fluid compartments where a substantial amount of vector could be injected and "bathe" neurons at a constant MOI. In these compartments vectors could be used to deliver trophic factors that would promote neuronal survival over weeks to months following injury. Alternatively, vectors could be injected at sites of peripheral nerve injury to promote neurite extension (Breakefield, 1994). Treatment of peripheral pain could be effected by inoculation of vectors onto cutaneous receptors of sensory neurons to alter the transmitter phenotype of these neurons such that they would release analgesic peptides into the spinal cord along with their normal transmitters. The therapeutic peptides would serve to downregulate normal pain transmission (Fig. 10). A similar strategy could also be used to block transmission of motor neuron signaling to skeletal muscles involved in dystonic posturing. Currently botulinum toxin is applied to these nerve terminals, which results in a dying back of neurites innervating muscle (Tsui et al., 1995). Use of an HSV-1 vector to block synthesis of acetylcholine at neuromuscular junctions would reduce muscle contractions without damaging motor neurons. As with other virus vectors, extensive preclinical testing must be done to ensure that the vectors are as safe as possible and do not provide a public health hazard. However, new and original HSV vectors are being designed to minimize such risks and should offer effective alternatives to current modes of therapy.

REFERENCES

Andersen JK, Breakefield XO (1995): Gene delivery to neurons of the adult mammalian nervous system using herpes and adenovirus vectors. In Chang PL (ed): Somatic Gene Therapy. London: CRC Press, pp 135–160.

Andersen JK, Garber DA, Meaney CA, Breakefield XO (1992): Gene transfer into mammalian central nervous system using herpes virus vectors: Extended expression of bacterial lacZ in neurons using the neuron-specific enolase promoter. *Hum Gene Ther* 3:487–499.

Avitabile E, De Gaeta S, Torrisi MR, Ward PL, Roizman B, Campadelli-Fiume G (1995): Redistribution of microtubules and Golgi apparatus in herpes simplex virus infected cells and their role in viral exocytosis. *J Virol* 69:7472–7482.

Baringer JR, Swoveland P (1973): Recovery of herpes-simplex virus from human trigeminal ganglions. *N Engl J Med* 188:648–650.

Battleman DS, Geller AI, Chao MV (1993): HSV-1 vector-mediated gene transfer of the human nerve growth factor p75hNGFR defines high-affinity NGF binding. *J Neurosci* 13:941–951.

Bergold PJ, Casaccia-Bonnefil P, Zeng XL, Federoff HJ (1993): Transsynaptic neuronal loss induced in hippocampal slice cultures by a herpes simplex virus vector expressing the GluR6 subunit of the kainate receptor. *Proc Natl Acad Sci USA* 90:6165–6169.

Berthomme H, Fournel S, Epstein AL (1996): Increased transcomplementation properties of plasmids carrying HSV-1 origin of the replication and packaging signals. *Virology* 216:437–443.

Bloom DC, Stevens JG (1994): Neuron-specific restriction of a herpes simplex virus recombinant maps to the UL5 gene. *J Virol* 68:3761–3772.

Boothman DA, Geller AI, Pardee AB (1989): Expression of the *E. coli* lacZ gene from a defective HSV-1 vector in various human normal, cancer-prone and tumor cells. *FEBS Lett* 258:159–162.

Boviatsis EJ, Parks S, Sena-Esteves M, Kramme C, Chase M, Efird JT, Wei M, Breakefield XO, Chiocca EA (1994): Long-term survival of rats harboring brain neoplasms treated with ganciclovir and a herpes simplex virus vector that retains an intact thymidine kinase gene. *Cancer Res* 54:5745–5751.

Breakefield XO (1994): Gene therapy for spinal cord injury? *ISRT Res Digest* 6:3.

Breakefield XO, DeLuca NA (1991): Herpes simplex virus for gene delivery to neurons. *New Biol* 3:203–218.

Breakefield XO, Kramm CM, Chiocca EA, Pechan PA (1995): Herpes simplex virus vectors for tumor therapy. In Sobel RE, Scanlon KJ (eds): The Internet Book of Gene Therapy: Cancer Gene Therapeutics. Appleton and Lange, pp 41–56.

Breakefield XO, Pechan P, Johnston K, Herrlinger U, Jacoby D, Chase M, Borghesani P, Dunn R, Smith F, Chiocca E (1996): Novel herpes virus amplicon vectors for therapy of experimental brain tumors. 21st Herpes Virus Workshop, DeKalb, Illinois, Abstract 476.

Browne H, Bell S, Minson T, Wilson DW (1996): An endoplasmic reticulum-retained herpes simplex virus glycoprotein H is absent from secreted virions: Evidence for reenvelopment during egress. *J. Virol* 70:4311–4316.

Cai W, Schaffer PA (1989): Herpes simplex virus type 1 ICP0 plays a critical role in the *de novo* synthesis of infectious virus following transfection of viral DNA. *J Virol* 63:4579–4589.

Carpenter DE, Stevens JG (1996): Long-term expression of a foreign gene from a unique position in the latent herpes simplex virus genome. *Hum Gene Ther* 7:1447–1454.

Chambers R, Gillespie GY, Soroceanu L, Andrearsky S, Catterjee S, Chou J, Roizman B, Whitley RJ (1995): Comparison of genetically engineered herpes simplex viruses for the treatment of brain tumors in a *scid* mouse model of human malignant glioma. *Proc Natl Acad Sci USA* 92:1411–1415.

Chatterjee S, Burns P (1990): Expression of herpes simplex virus type 1 glycoproteins in interferon-treated human neuroblastoma cells. *J Virol* 64:5209–5213.

Chen X, Schmidt MC, Goins WF, Glorioso JC (1995): Two herpes simplex virus type 1 latency-active promoters differ in their contributions to latency-associated transcript expression during lytic and latent infections. *J Virol* 69:7899–7908.

Chiocca EA, Andersen J, Takamiya Y, Martuza R, Breakefield XO (1994): Virus-mediated genetic treatment of rodent gliomas: In Wolff JA (ed): Gene Therapeutics. Boston: Birkhauser, pp 245–262.

Chiocca EA, Choi BB, Cai W, DeLuca NA, Schaffer PA, DiFiglia M, Breakefield XO, Martuza RL (1990): Transfer and expression of the lacZ gene in rat brain neurons mediated by herpes simplex virus mutants. *New Biol* 2:739–746.

Chou J, Kern ER, Whitley RJ, Roizman B (1990): Mapping of herpes virus-I neurovirulence to gamma 34.5, a gene nonessential for growth in culture. *Science* 250:1262–1266.

Chou J, Roizman B (1992): The gamma 34.5 gene of herpes simplex virus 1 precludes neuroblastoma cells from triggering total shutoff of protein synthesis characteristics of programmed cell death in neuronal cells. *Proc Natl Acad Sci USA* 89:3266–3270.

Coen DM, Kosz-Vnenchak M, Jacobson JG, Leib DA, Bogard CL, Schaffer PA, Tyler KL, Knipe DM (1989): Thymidine kinase–negative herpes simplex virus mutants establish latency in mouse trigeminal ganglia but do not reactivate. *Proc Natl Acad Sci USA* 86:4736–4740.

Craig AM, Wyborski RJ, Banker G (1995): Preferential addition of newly synthesized membrane protein at axonal growth cones. *Nature* 375:592–594.

Cunningham C, Davison AJ (1993): A cosmid-based system for constructing mutants of herpes simplex virus type 1. *Virology* 197:116–124.

Davar G, Bebrin WR, Day R, Breakefield XO (1995): Gene delivery to mouse sensory neurons with herpes simplex virus: a model for postherpetic neuralgia and its treatment? In: *Neurology* 45(12 Suppl 8):S69.

Desai P, Homa FL, Person S, Glorioso JC (1994): A genetic selection method for transfer of HSV-1 glycoprotein B mutations from the plasmid of the viral genome. *Virology* 204:312–322.

Deshman SL, Fraser NW (1989): During latency, herpes simplex virus type 1 DNA is associated with nucleosomes in a chromatin structure. *J Virol* 63:943–947.

Dobson AT, Margolis TP, Sedarati F, Stevens JG, Feldman T (1990): A latent, non-pathogenic HSV-1–derived vector stably expresses beta-galactosidase in mouse neurons. *Neuron* 5:353–360.

Dobson AT, Sederati F, Devi-Rao G, Flanagan WM, Farrell MJ, Stevens JG, Wagner EK, Feldman LT (1989): Identification of the latency-associated transcript promoter by expression of rabbit beta-globin mRNA in mouse sensory nerve ganglia latently infected with a recombinant herpes simplex virus. *J Virol* 63:3844–3851.

Dressler GR, Rock DL, Fraser NW (1987): Latent herpes simplex virus type 1 DNA is not extensively methylated *in vivo*. *J Gen Virol* 68:1761–1765.

During MJ, Naegele JR, O'Malley KL, Geller AI (1994): Long-term behavioral recovery in parkinsonian rats by an HSV vector expressing tyrosine hydroxylase. *Science* 266:1399–1403.

Dutch RE, Zemelman BC, Lehman IR (1994): Herpes simplex virus type 1 recombination: The U_c-DR1 region is required for high-level a-sequence–mediated recombination. *J Virol* 68:3733–3741.

Elliott G, O'Hare P (1996): The HSV tegument protein VP22 as a novel vector for gene peptide delivery. 21st Herpes Virus Workshop, DeKalb, Illinois, Abstract 464.

Everett RD (1987): The regulation of transcription of viral and cellular genes by herpes virus immediate-early gene products. *Anticancer Res* 7:589–604.

Federoff HJ, Geschwind MD, Geller AI, Kessler JA (1992): Expression of nerve growth factor *in vivo* from a defective herpes simplex virus 1 vector prevents effects of axotomy on sympathetic ganglia. *Proc Natl Acad Sci USA* 89:1636–1640.

Fraefel C, Song S, Lim F, Lang P, Yu L, Wang Y, Wild P, Geller AI (1996): Helper virus–free transfer of herpes simplex virus type 1 plasmid vectors into neural cells. *J Virol* 70:7190–7197.

Gage PJ, Sauer B, Levine M, et al (1992): A cell-free recombination system for site-specific integration of multigenic shuttle plasmids into the herpes simplex virus type 1 genome. *J Virol* 66:5509–5515.

Geller AI (1993): Herpes viruses: Expression of genes in postmitotic brain cells. *Curr Opin Genet Dev* 3:81–88.

Geller AI, Breakefield XO (1988): A defective HSV-1 vector expresses *Escherichia coli* beta-galactosidase in cultured peripheral neurons. *Science* 241:1667–1669.

Geschwind MD, Hartnick CJ, Liu W, Amat J, Van De Water TR, Federoff HJ (1996): Defective HSV-1 vector expressing BDNF in auditory ganglia elicits neurite outgrowth: Model for treatment of neuron loss following cochlear degeneration. *Hum Gene Ther* 7:173–182.

Glorioso JC, Bender MA, Goins WF, Fink DJ, DeLuca N (1995): Herpes simplex virus as a gene-delivery vector for the central nervous system. In Kaplitt MG, Loewy AD (eds): Viral Vectors. New York: Academic Press, pp 1–23.

Goins WF, Huard J, Pike-Cavalcoli M, DeKosky S, Fink DJ, Glorioso JC (1996): HSV vector-mediated nerve growth factor expression in the mouse peripheral nervous system and spinal chord. 21st Herpes Virus Workshop, DeKalb, Illinois, Abstract 473.

Goldstein DJ, Weller SK (1988): Herpes simplex virus type 1–induced ribonucleotide reductase activity is dispensable for virus growth and DNA synthesis; isolation and characterization of an ICP6 lacZ insertion mutant. *J. Virol* 62:196–205.

Gossen M, Bujard H (1992): Tight control of gene expression in mammalian cells by tetracycline-responsive promoters. *Proc Natl Acad Sci USA* 89:5547–5551.

Hartley DM, Neve RL, Bryan J, Ullrey DB, Bak S-Y, Lang P, Geller AI (1996): Expression of the calcium-binding protein, parvalbumin, in cultured cortical neurons using a HSV-1 vector system enhances NMDA. *Mol Brain Res* 40:285–296.

Ho DY (1994): Amplicon-based herpes simplex virus vectors. *Methods Cell Biol* 43:191–210.

Ho DY, Mocarski ED (1988): Beta-galactosidase as a marker in the peripheral and neural tissues of the herpes simplex virus–infected mouse. *Virology* 167:279–283.

Ho DY, Mocarski ES, Sapolsky RM (1993): Altering central nervous system physiology with a defective herpes simplex virus vector expressing the glucose transporter gene. *Proc Natl Acad Sci USA* 90:3655–3659.

Homa FL, Otal TM, Glorioso JC, et al. (1986): Transcriptional control signals of herpes simplex virus type 1 late (gamma 2) gene lie within bases −34 to +124 relative to the 5′ terminus of the mRNA. *Mol Cell Biol* 6:3652–3666.

Huang Q, Vonsattel JP, Schaffer PA, Martuza RL, Breakefield XO, DiFiglia M (1992): Introduction of a foreign gene (*Escherichia coli* lacZ) into rat neostriatal neurons using herpes simplex virus mutants: A light and electron microscopic study. *Exp Neurol* 115:303–316.

Huard J, Glorioso JC (1995): Herpes simplex virus type 1 vector mediated gene transfer to muscle. *Gene Ther* 2:385–392.

Hunter W, Rabkin S, Martuza R (1995): Brain tumor therapy using genetically engineered replication-competent virus. In Kaplitt MG, and Lowey AD, (eds): Viral Vectors. New York: Academic Press, pp. 259–274.

Jacoby DR, Breakefield XO, Latchman DS (1995): Construction of an HSV-1 amplicon with an inducible Oct-2 operon to modulate cytopathic effect. In Conf Gene Ther CNS, Philadelphia, PA, Abstract 1.31.

Johnson PA, Friedmann T (1994): Replication-defective recombinant herpes simplex virus vectors. *Methods Cell Biol* 43:211–230.

Johnson PA, Wang MJ, Friedmann T (1994): Improved cell survival by the reduction of immediate-early gene expression in replication-defective mutants of herpes simplex virus type 1 but not by mutation of the virion host shutoff function. *J Virol* 68:6347–6362.

Johnston KM, Jacoby D, Pechan P, Fraefel C, Borghesani P, Schuback D, Dunn R, Smith F, Breakefield XO (1997): HSV-1 amplicon/AAV hybrid vectors extend transgene expression in human glioma cells. *Human Gene Ther* 8:359–370.

Kaplitt MG, Tjuvajev JG, Leib DA, Berk J, Pettigrew DKl, Posner JB, Pfaff DW, Rabkin SD, Blasberg RG (1994): Mutant herpes simplex virus induced regression of tumors growing in immunocompetent rats. *J Neuro-Oncol* 19:137–147.

Kotin RM, Linden RM, Berns KI (1992): Characterization of a preferred site on human chromosome 19q for integration of adeno-associated virus DNA by non-homologous recombination. *EMBO J* 11:5071–5078.

Kotin RM, Linden RM, Berns KI (1996): Characterization of a preferred site on human chromosome 19q for integration of adeno-associated virus DNA by non-homologous recombination. *EMBO J* 11:5071–5078.

Kramm CM, Rainov NG, Sena-Esteves M, Barnett FH, Chase M, Herrlinger U, Pechan PA, Chiocca EA, Breakefield XO (1996a): Long-term survival in a rodent model of disseminated brain tumors by combined intrathecal delivery of herpes vectors and ganciclovir treatment. *Hum Gene Ther* 7:1989–1994.

Kramm CM, Rainov NG, Sena-Esteves M, Chase M, Pechan PA, Chiocca EA, Breakefield XO (1996b): Herpes vector-mediated delivery of thymidine kinase gene to disseminated CNS tumors. *Hum Gene Ther* 7:291–300.

Kramm CM, Sena-Esteves M. Barnett FH, Rainov NG, Schuback DE, Yu JS, Pechan PA, Paulus W, Chiocca EA, Breakefield XO: Gene Therapy for Brain Tumors. *Brain Pathol* 1995 Oct;5(4):345–381.

Kwong AD, Frenkel N (1989): The herpes simplex virus virion host shut off function. *J Virol* 63:4834–4839.

Kwong AD, Frenkel N (1995): Biology of herpes simplex virus (HSV) defective viruses and development of the amplicon system. In Kaplitt MG, Loewy AD (eds): Viral Vectors. New York: Academic Press, pp 25–42.

Lawrence MS, Ho DY, Sun GH, Steinberg GK, Sapolsky RM (1996): Overexpression of bcl-2 with herpes simplex virus vectors protects CNS neurons against neurological insults *in vitro* and *in vivo*. *J Neurosci* 16:486–496.

Leib DA, Olivo PD (1993): Gene delivery to neurons: Is herpes simplex virus the right tool for the job? BioEssays 15:547–554.

Lillycrop KA, Howard MK, Estridge J, Latchman DS (1994): Inhibition of herpes simplex virus infection by ecotopic expression of neuronal splice variants of the Oct-2 transcription factor. *Nucleic Acids Res* 22:815–820.

Linnik MD, Zahos P, Geschwind MD, Federoff HJ (1995): Expression of bcl-2 from a defective herpes simplex virus-1 vector limits neuronal death in focal cerebral ischemia. *Stroke* 26:1670–1674.

Lokensgard JR, Bloom DC, Dobson AT, Feldman LT (1994): Long-term promoter activity during herpes simplex virus latency. *J Virol* 68:7148–7158.

Lowenstein PR, Enquist LW (1996): Protocols for Gene Transfer in Neuroscience. New York: John Wiley & Sons.

Lu B, Federoff HJ (1995): Herpes simplex virus type 1 amplicon vectors with glucocorticoid-inducible gene expression. *Hum Gene Ther* 6:419–428.

Lycke E, Hamark B, Johansson M, Krotochwil A, Lycke J, Svennerholm B (1988): Herpes simplex virus infection of the human sensory neuron: An electron microscopy study. *Arch Virol* 101:87–104.

Martuza RL, Malick A, Markert JM, Ruffner KL, Coen DM (1991): Experimental therapy of human glioma by means of a genetically engineered virus mutant. *Science* 252:854–856.

McGeoch DJ, Dalrymple MA, Davison AJ, Dolan A, Frame MC, McNab D, Perry LJ, Scott JE, Taylor P (1988): The complete DNA sequence of the long unique region in the genome of herpes simplex virus type 1. *J Gen Virol* 69:1531–1574.

Mesri EA, Federoff HJ, Brownier M (1995): Expression of vascular endothelial growth factor from a defective herpes simplex virus type 1 amplicon vector induces angiogenesis in mice. *Circ Res* 76: 161–167.

Mineta T, Rabkin SD, Martuza RL (1994): Treatment of malignant gliomas using ganciclovir-hypersensitive, ribonucleotide reductase–deficient herpes simplex viral mutant. *Cancer Res* 54:3963–3966.

Miyanohara A, Johnson PA, Elam RI, et al (1992): Direct gene transfer to the liver with herpes simplex virus type 1 vectors: Transient production of physiologically relevant levels of circulating factor IX. *New Biol* 4:238–246.

Mocarski ES, Post LE, Roizman B (1980): Molecular engineering of the herpes simplex virus genome: Insertion of second L–S junction into the genome causes additional genome inversions. *Cell* 22:243–255.

Montgomery RL, Warner MS, Lum BJ, and Spear PG (1996): Herpes simplex virus-1 entry into cells mediated by a novel member of the TNF/NGF receptor family. *Cell* 87:427–436.

Muldoon LL, Nilaver G, Kroll RA, Pagel MA, Breakefield XO, Chiocca EA, Davidson BL, Weissleder R, Neuwelt EA (1995): Comparison of intracerebral inoculation and osmotic blood–brain barrier disruption for delivery of adenovirus, herpesvirus and iron oxide particles to normal rat brain. *Am J Pathol* 147:1840–1851.

Neve RL, Ivins KJ, Benowitz LI, During MJ, Geller AI (1991): Molecular analysis of the function of the neuronal growth-associated protein GAP-43 by genetic intervention. *Mol Neurobiol* 5:131–141.

New KC, Rabkin SD (1996): Co-expression of two gene products in the CNS using double-cassette defective herpes simplex virus vectors. *Mol Brain Res* 37:317–323.

Nichol PF, Chang JY, Johnson EMJ, Olivo PD (1996): Herpes simplex virus gene expression in neurons: Viral DNA synthesis is a critical regulatory event in the branch point between the lytic and latent pathways. *J Virol* 70:5476–5486.

Nilaver G, Muldoon U, Kroll RA, Pagel MA, Breakefield XO, Davidson BL, Neuwelt EA (1995): Delivery of herpes virus and adenovirus to nude rat intracerebral tumors following osmotic blood–brain barrier disruption. *Proc Natl Acad Sci USA* 92:9829–9833.

Oh YJ, Moffat M, Wong S, Ullrey D, Geller AI, O'Malley KL (1996): A herpes simplex virus-1 vector containing the rat tyrosine hydroxylase promoter directs cell type-specific expression of beta-galactosidase in cultured rat peripheral neurons. *Mol Brain Res* 35:227–236.

Oligino R, Poliani PL, Wang Y, Tsai S, O'Malley BW, Fink DJ, Glorioso JC (1996): Regulatable transgene expression in brain using a herpes simplex virus vector containing a steroid responsive inducible promoter. 21st Herpes Virus Workshop, DeKalb, Illinois, Abstract 461.

Oroskar AA, Read GS (1989): Control of mRNA stability by the virion host shutoff function of herpes simplex virus. *J Virol* 63:1897–1906.

Pechan PA, Fotaki M, Thompson RL, Dunn R, Chase M, Chiocca EA, Breakefield XO (1996): A novel "piggyback" packaging system for herpes simplex virus amplicon vectors. *Hum Gene Ther* 7:2003–2013.

Penfold ME, Armati P, Cunningham AL (1994): Axonal transport of herpes simplex virions to epidermal cells: Evidence for a specialized mode of virus transport and assembly. *Proc Natl Acad Sci USA* 91:6529–6533.

Pyles RB, Sawtell NM, Thompson RL (1992): Herpes simplex virus type 1 dUTPase mutants are attenuated for neurovirulence, neuroinvasiveness, and reactivation from latency. *J Virol* 66:607–671.

Rabkin SD, Martuza RL, Miyatake S-I (1996): Transcriptional targeting of herpes simplex virus for cell-specific replication. 21st Herpes Virus Workshop, DeKalb, Illinois, Abstract 462.

Rainov NG, Kramm CM, Aboody-Guterman K, Chase M, Ueki K, Louis D, Harsh G, Chiocca EA, Breakefield XO (1996): Retrovirus-mediated gene therapy of experimental brain neoplasms using the herpes simplex virus-thymidine kinase/ganciclovir paradigm. *Cancer Gene Ther* 3:99–106.

Ramakrishnan R, Levine M, Fink DJ (1994): PCR-based analysis of herpes simplex virus type 1 latency in the rat trigeminal ganglion established with a ribonucleotide reductase–deficient mutant. *J Virol* 68:7083–7091.

Rasty S, Glorioso JC (1996): Deletion of the S component inverted repeat sequence c′ and the nonessential genes U_S1–U_S5 from the herpes simplex virus type 1 genome substantially impairs productive viral infection in cell culture and pathogenesis in rat CNS. 21st Herpes Virus Workshop, DeKalb, Illinois, Abstract 485.

Roizman B (1996): Herpesviridae. In Fields BM, Knipe DM, Howley PM et al (eds): Fields Virology. Philadelphia: Lippincott-Raven Publishers, pp 2221–2230.

Roizman B, Batterson W (1985): Herpes viruses and their replication. In Fields, BN (ed): Virology. New York: Raven Press, pp 497–526.

Roizman B, Jenkins FJ (1985): Genetic engineering of novel genomes of large DNA viruses. *Science* 229:1208–1214

Roizman B, Sears AE (1996): Herpes simplex viruses and their replication. In Fields BN, Knipe DM, Howley PM (eds): Fields Virology. Philadelphia. Lippincott-Raven Publishers, pp 2231–2296.

Sawtell NM, Thompson RL (1992): Herpes simplex virus type 1 latency-associated transcription unit promotes anatomical site-dependent establishment and reactivation from latency. *J Virol* 66:2157–2169.

Smiley JR, Fong BS, Leung WC (1981): Construction of a double-joined herpes simplex viral DNA molecule: Inverted repeats are required for segment inversions, and direct repeats promote deletions. *Virology* 113:345–362.

Smith RL, Mysofski D, Everett RD, Wilcox CL (1996): The HSV-1 immediate early protein ICP0 is necessary for the efficient establishment of the latent infection. 21st Herpes Virus Workshop, DeKalb, Illinois, Abstract 271.

Sodeik B, Ebersold MW, Helenius A Microtuble-mediated transport of incoming herpes simplex virus 1 capsids to the nucleus. In: *J Cell Biol* (1997) 136(5):1007–21.

Spaete R, Frenkel N (1982); The herpes virus amplicon: A new eucaryotic defective–virus cloning-amplifying vector. *Cell* 30:295–304.

Starr PA, Lim F, Grant FD, Trask L, Lang P, Yu L. Geller AI (1996): Long-term persistance of defective HSV-1 vectors in the rat brain is demonstrated by reactivation of vector gene expression. *Gene Ther* 3:615–623.

Stern S, Tanaka M, Herr W (1989): The Oct-1 homeo domain directs formation of a multiprotein–DNA complex with the HSV transactivator VP16. *Nature* 341:624–630.

Steven AC, Spear PG (1996): Herpesvirus Capsid Assembly and Envelopment. New York: Oxford University Press (in press).

Stevens JG (1989): Human herpesviruses: A consideration of the latent state. *Microbiol Rev* 53:318–332.

Stevens JG, Wagner EK, Devi-Rao GB, Cook ML, Feldman LT (1987): RNA complementary to a herpesvirus alpha gene mRNA is prominent in latently infected neurons. *Science* 235:1056–1059.

Topp KS, Meade LB, LaVail JH (1994): Microtubule polarity in the peripheral processes of trigeminal ganglion cells: Relevance for the retrograde transport of herpes simplex virus. *J Neurosci* 14:318–325.

Tsui JK, Hayward M, Mak EK, Schulzer M (1995): Botulinum toxin type B in the treatment of cervial dystonia: A pilot study. *Neurology* 45:2109–2110.

Umene K (1993): Herpes simplex virus type 1 variant *a* sequence generated by recombination and breakage of the *a* sequence in defined regions, including the one involved in recombination. *J Virol* 67:5685–5691.

Visalli RJ, Courtney RJ, Meyer C: Infection and replication of herpes simplex virus type 1 in an organotypic epithelial culture system. In: *Virology* (1997) 230(2):236–43.

Wang MJ, Friedman T, Johnson PA (1995): Differentiation of PC12 cells by infection with an HSV-1 vector expression nerve growth factor. *Gene Ther* 2:323–335.

Wang S, Vos J-M (1996): A hybrid infectious vector based on Epstein-Barr virus and herpes simplex virus type 1 for gene transfer into human cells *in vitro* and *in vivo*. *J Virol* 70:8422–8430.

Weitzman MD, Wilson JM, Eck SL (1995): Adenovirus vectors in cancer gene therapy. In Sobol RE, Scanlon KJ (eds): The Internet Book of Gene Therapy. Norwalk, CT: Appleton and Lange, p 17.

Wilcox CL, Smith RL, Freed CR, Johnson EMJ (1990): Nerve growth factor-dependence of herpes simplex virus latency in peripheral sympathetic and sensory neurons *in vitro*. *J Neurosci* 10:1268–1275.

Wolfe JH, Deshmane SL, Fraser NW (1992): Herpesvirus vector gene transfer and expression of beta-glucuronidase in the central nervous system of MPS VII mice. *Nature Gene* 1:379–384.

Wu N, Watkins SC, Schaffer PA, DeLuca NA (1996): Prolonged gene expression and cell survival after infection by a herpes simplex virus mutant defective in the immediate-early genes encoding ICP4, ICP27, and ICP22. *J Virol* 70:6358–6369.

Wu X, Leduc Y, Cynader M, Trufaro F (1995): Examination of conditions affecting the efficiency of HSV-1 amplicon packaging. *J Virol Methods* 52:219–229.

York IA, Roop C, Andrews DW, Riddell SR, Graham FL, Johnson DC (1994): A cytosolic herpes simplex virus protein inhibits antigen presentation to $CD8^+$ T lymphocytes. *Cell* 77:525–535.

Zhang X, Efstathious S, Inglis S (1996): A selection for packaging HSV amplicon into viral particles and the subsequent gene deliver. 21st Herpes Virus Workshop, DeKalb, Illinois, Abstract.

9

DELIVERY SYSTEMS FOR GENE THERAPY: ADENO-ASSOCIATED VIRUS

Gabriele Kroner-Lux, Christopher E. Walsh, and Richard Jude Samulski
Gene Therapy Center, University of North Carolina, Chapel Hill, NC 27599

INTRODUCTION

Gene transfer into hematopoietic cells for the purpose of gene therapy is a relatively new and promising field. Blood and bone marrow cells are easily accessible and can be manipulated *ex vivo* before reinfusion into a recipient patient. This feature makes hematopoietic gene transfer an attractive system to test a variety of vectors and cell subtypes for gene therapy. The purpose of this chapter is to provide an overview of recombinant adeno-associated virus (rAAV) vectors and the current "state of the art" of their potential utility in gene therapy of hematopoietic disorders. We briefly review AAV biology and genetics and suggest more general reviews about AAV (Flotte and Carter, 1995; Rolling and Samulski, 1995; Skulimowski and Samulski 1995).

The basic requirements that need to be fulfilled for gene therapy as treatment of hematopoietic disorders are the following:

1. The transgene must be integrated into the host chromosome of the target stem cells, thereby achieving transmission to all progeny cells to obtain a long-lasting effect.

Stem Cell Biology and Gene Therapy, Edited by Peter J. Quesenberry, Gary S. Stein, Bernard G. Forget, and Sherman M. Weissman
ISBN 0-471-14656-0

2. The transgene should be stably expressed, and the efficiency of transfer must be high enough to affect the disease phenotype.
3. The treatment needs to be without risk for the patient, which implies that integration should take place in a safe chromosomal location to avoid insertional mutagenesis or dysregulation of growth control genes.

AAV VECTOR BIOLOGY

AAV is a single-stranded DNA virus with a genome size of 4.7 kb. The viral genome is flanked by palindromic terminal repeats (TRs) that are necessary for encapsidation, replication, and integration of the viral genome. AAV-based vectors are nonpathogenic, possess an extremely wide host and tissue range, integrate stablely into cellular DNA, and transduce both proliferating and nonproliferating cells (Flotte et al., 1993, 1994; Kaplitt et al., 1994, 1996; Lebkowski et al., 1988; McCown et al., 1996; Muzyczka, 1992; Xiao et al., 1996). Although a high percentage of the population is seropositive for AAV antibodies (Flotte and Carter, 1995; Muzyczka, 1992), the virus has not been associated with human disease. For a productive AAV infection, coinfection with a helper virus, either adenovirus (Ad) or herpes virus, is required (Berns and Bohenzky, 1987; Buller et al., 1981). In the absence of helper virus, AAV integrates in a stable fashion into a specific site on chromosome 19 (Kotin et al., 1990; Samulski et al., 1991). Although this feature of site-specific integration is attractive for gene therapy approaches, current rAAV vectors do not retain targeting capability. Recent analyses of the molecular fate of rAAV-transduced genes suggest that the vector integrates randomly or persists as high molecular size episomal concatemers. These vector structures can achieve persistent (>1.5 years) gene expression in this form (Xiao et al., 1996) and have achieved one of the major goals of gene therapy: *the ability to obtain long-term expression.*

Cell Tropism of AAV

Another feature of parvovirus vectors that lends them to testing efficient gene delivery is cell tropism. AAV infects a broad range of host cell types and has been isolated from primate as well as from nonprimate species. The viral receptor is unknown. A variety of both primary and transformed cell types originating from different tissues (canine, murine, bovine) have been demonstrated to be permissive for infection (Berns, 1996; Cukor et al., 1984). This promiscuous infectivity suggests that some common membrane protein may be the receptor, such as the MHC class I protein identified for Ad (Hong et al., 1997), which would help explain these findings. However, a few cell types resistant to AAV infection like the megakaryocytic leukemia cell lines M07e and MB-02 (Ponnazhagan et al., 1996) and erythroid lineages (UT-7) (Mizukami et al., 1996) suggest that it may be possible to identify the AAV receptor with the use of these cell types.

While there has been evidence for AAV infection of human peripheral blood cells (Grossman et al., 1992), another human parvovirus, B19, shows a specific tropism for human erythroid progenitor cells. This cell specificity is mediated by a specific globoside receptor, the erythrocyte P-antigen receptor (Brown et al., 1993). B19 replication requires erythroid cell-specific factors (Liu et al., 1992), causing erythrocyte destruction during a productive infection. This is an important step in B19 pathogenesis exemplified by nonerythroid cells that express the globoside receptor but are nonpermissive for B19 replication (Brown et al., 1994; Rouger et al., 1987). While B19 is pathogenic for humans and in its wild-type state not suitable for gene therapy, the B19 parvovirus can package 6 kb of DNA, a full 1 kb more than the rAAV. Recent efforts using B19/AAV chimeras suggest that AAV viral tropism can be manipulated and may suggest that an increase in vector packaging capacity may be obtained (Rabinowitz and Samulski, personal communication).

Production of Recombinant Virus

A rate-limiting step in the use of rAAV vectors has been the ability to generate high titer stocks. To package transgenes into AAV capsids, different packaging systems have been developed. Typically a transient transfection system using plasmid substrates is implied. The initial system established used an infectious plasmid (psub201, SSV9) incorporating the AAV genome flanked between the two palindromic terminal repeats. This construct was modified to allow for removal of the AAV genome and insertion of foreign genes and their regulatory elements (Laughlin et al., 1983; Samulski et al., 1982, 1987). The vector plasmid contains only the foreign transgene flanked by the 145 bp inverted TRs, the only *cis* element required for DNA replication, packaging, and integration (Muzyczka, 1991; Samulski et al., 1989). With all of the viral genes removed, the rAAV vector has a strict packaging limit of 5 kb (typically up to 108% of wild-type AAV genome size). To generate infectious rAAV vectors, a second (helper) plasmid encodes the AAV genes required for productive infection. This molecule is devoid of the *cis*-acting TRs and homologous sequences and therefore cannot replicate or generate wild-type recombinants. Co-transfection of the helper and vector plasmids in Ad-infected cells (helper virus) is required to complete the packaging system and generate rAAV stocks. To eliminate wild-type Ad either heat treatment or physical separation by CsCl is performed. Several of the first experiments with rAAV in the field of hematopoietic cells used crude heat-inactivated cell lysates containing rAAV. Virus titers ranged from 10^3 to 10^5 infectious particles per milliliter. Improvements with co-transfection, modifications of the helper plasmid (Li et al., 1997), and purification methods now routinely produce titers of 10^7–10^9 particles/ml. This system, although improved, has two major disadvantages: One is the requirement of an adenoviral coinfection, and the other is the inconvenience and inconsistency of the co-transfection procedure.

While it is possible to purify rAAV from the contaminating adenoviral helper virus by CsCl centrifugation and heat inactivation of the virus preparation, it would be advantageous to generate rAAV stocks that are totally devoid of adenovirus. A genetic approach to generate rAAV without the need for Ad helper virus infection has been developed (Xiao, Ferrari, and Samulski, unpublished results). This method requires the transfection of three plasmid constructs: the recombinant plasmid harboring the transgene with AAV TRs, the helper plasmid harboring the AAV replication and packaging genes, and a third plasmid that contains the essential adenoviral helper genes. With this procedure it is possible to generate high titer rAAV stocks that are completely free of Ad. With this method, titers of 10^9–10^{10} rAAV particles ml can be produced (Xiao, Ferrari, and Samulski, unpublished results). While these various approaches have continued to improve rAAV production, they still imply transient procedures, which may be problematic when large-scale production is required.

To circumvent the necessity of co-transfection of the different vector and helper plasmids, there have also been efforts to generate producer cell lines. Cell lines that contain the rep gene under the control of inducible promoters (Hölscher et al., 1994) allow the production of recombinant virus but only in low yield. A HeLa cell line containing an integrated AAV genome without the TRs has been generated in an effort to establish a packaging cell line (Vincent et al., 1990). After transfection of the recombinant vector, these producer cells supply the AAV genes in *trans.* Although attractive, this approach yielded low rAAV titers of 10^3–10^4 per milliliter and suggests that improvements in this strategy are warranted. Along these lines, new cell lines that incorporated the vector plasmid into the cellular genome have been established. In this case, only the helper plasmid needs to be transfected, which results in a simplified procedure with higher vector yield (Flotte et al., 1995). While attractive, the need for transient transfection still impedes efficient large-scale vector production.

Recently an AAV-packaging cell line was reported (Clark et al., 1995) that may have resolved some of the earlier problems. The authors generated HeLa cells that harbored both AAV coding and AAV vector sequences in an integrated tandem array. After infection with adenovirus, the cell line produced rAAV stocks of more than 10^9 transducing units per milliliter. This cell line generates comparable yields of rAAV when compared with the transfection method. This is the first practical rAAV producer cell line reported, and it holds promise for the generation of an optimum packaging line in the future.

Determination of rAAV Titers

The wide range of rAAV titers that has been reported (from 10^5/ml to 10^{11}/ml) reflects not only the different vectors and production protocols but also how titers are measured and defined. Basically, rAAV titers can be expressed in three different ways:

1. *particles per milliliter,* which reflects the number of DNase-resistant virions containing the rAAV genome without any statement about their infectivity or functionality. Usually the unpacked genomes in an rAAV vector stock are digested with DNase, and then the viral particles are disrupted, thereby releasing the DNA genomes and quantitating them by DNA dot blot, Southern blot, or semiquantitative polymerase chain reaction (PCR).
2. *infectious units* (or replication units), which are determined by infectious center assay (or replication center assay). For this assay rAAV is co-infected into cells with sufficient amounts of wild-type AAV and Ad to provide the necessary helper functions for the rAAV vector DNA to amplify within the cells. The cells are transferred on a nylon membrane and probed with radiolabeled rAAV-specific DNA. Each dot on an autoradiograph represents one rAAV-infected cell and therefor 1 infectious unit.
3. *transducing units,* which are determined by functional assays for the rAAV transgenes like blue cells produced by rAAV–lacZ virus or immunochemical-positive cells resulting from specific transgene expression. The virus titers estimated by this method are the most informative titers but also depend on the choice of transgene, promoter, and cell types.

Since AAV does not form plaques, the plaque assay cannot be used to determine viral titers. The infectious center assay should generate titers with the least vector-to-vector variation, but it is also the most laborious procedure. In summary, for a given rAAV preparation the dot blot will usually generate the highest titer, followed by the infectious center assay, and finally the transducing unit titer. The ratio of these three titering methods is roughly in the range of 1,000 (particles) : 10 (infectious units) : 1 (transducing unit) although the actual ratio may vary between different preparations and various vector constructs. Taken together, it is important to pay attention to the rAAV titers used when interpreting rAAV delivery results.

Status of the rAAV Genome: Integrated versus Episomal

Integration of wild-type AAV is part of the viral life cycle that is one of the advantages of rAAV vectors as gene therapy vectors (Hermonat and Muzyczka, 1984; Tratschin et al., 1985). *In vitro* wild-type AAV shows site-specific integration into human chromosome 19 (Kotin et al., 1990; Samulski et al., 1991). For targeted integration of wild-type AAV, the rep gene products are necessary (Kotin et al., 1990; Samulski, 1993). This information was first deduced from rAAV latent cell lines that lost chromosome 19 targeting (Walsh et al., 1992). These initial observations have been extended by studies that demonstrate that in the presence of rep gene products, plasmids containing AAV inverted TR sequences can target chromosome 19 in a manner similar to wild-type AAV (Balague et al., 1997; Shelling and Smith, 1994; Xiao, 1996).

Although well documented in tissue culture cells, integration has not been well characterized in *in vivo* settings. While wild-type AAV-2 DNA has been detected by PCR in peripheral blood leukocytes in 2 of 55 healthy blood donors and in 2 of 16 hemophilic patients, integration of the virus was not examined in that study (Grossman et al., 1992). The methodology by which the rAAV genome is assayed for integration has been problematic. Usually total DNA is isolated and characterized by Southern blot with or without single-cut restriction digestion. In undigested lanes the vector DNA should be associated with high molecular weight DNA. However it is usually difficult or impossible to rule out the existence of large concatemeric episomes or minichromosomes containing vector/chromosomal DNA (Xiao et al., 1996). For this reason, other procedures such as FISH are required to validate the integrated state. Since rAAV is only 5 kb, this makes the use of this technique difficult (FISH probe limitation is 3–5 kb). Demonstration of rAAV integration in hematopoietic cells has been achieved and is described below.

Recently an rAAV–lacZ vector has been reported to be maintained in nondividing tissue culture cells as functional double-stranded DNA episomes (Flotte et al., 1994). Another report revealed that an rAAV–CFTR vector has persisted as episomal DNA *in vivo* in monkey lungs for more than 3 month. The vectors were detected in low molecular weight DNA fractions but also in the genomic fraction of the transduced monkey lung (Afione et al., 1996). These results demonstrate the need for further analysis of the molecular fate of rAAV genomes *in vivo*.

An attractive explanation for episomal persistence is that the AAV inverted TRs may also function as an origin of replication in a Rep-independent manner. AAV inverted TRs have been observed to act as origin of replication mediating circular plasmid DNA replication in genotoxic reagent stressed cells (Yalkinoglu et al., 1991). An alternative explanation is that the inverted TRs may simply stabilize the episomal vector DNA from degradation in nondividing cells, while the vector may proceed to integrate in rapidly dividing cells. Data that support this suggestion have shown that single-stranded vector episomes were also detected in growth-inhibited cells for nearly 2 weeks postinfection. However, after releasing the growth block of these cells the single-stranded episomes were converted to double-stranded integrated DNA molecules (Russell et al., 1994). Therefore, rAAV integration may be a slow process, especially in nondividing or slowly growing cells, where the vector DNA may persist as an episome and convert into duplex DNA, which may be the substrate required for integration into the host chromosome. It is clear that future studies in this area will not only yield answers to these possible explanations, but also delineate the expectations and the limitations of this vector in dividing versus nondividing cells.

AAV-MEDIATED TRANSDUCTION OF MARKER GENES IN HEMATOPOIETIC CELLS AND ANALYSIS OF INTEGRATION

Early experiments to establish the feasibility of rAAV vectors used selectable marker genes such as the *Escherichia coli* β-galactosidase and the neomycin resistance genes. Cells that express the transgene can easily be detected by blue staining (β-galactosidase) or G418 resistance (neomycin). Initial studies employed K562 cells. This erythroleukemia cell line responds to hemin stimulation by differentiation and maturation into erythroid cells with concomitant increases in levels of γ-globin gene transcription and globin protein production (Rutherford et al., 1979). The cell line is therefore a suitable model system to represent gene transfer in an erythroid-specific lineage.

With rAAV that expresses the *E. coli* β-galactosidase gene (AAV β-gal), the K562 cell line and primary hematopoietic progenitor cells were used to assess transduction (Goodman et al., 1994). Transduction efficiency of the infected K562 cells was detected directly by β-gal staining, following infection with AAV β-gal. Stable expression of the transduced gene for 5 days could be shown. About 2%–3% of the cells were positive following exposure to the virus at an multiplicity of infection (MOI) of 1. The rAAV preparation that was used for this assay was crude heat-inactivated cell extract, and the titer was estimated by immunostaining for capsid proteins in infected cells or by quantitative Southern blot analysis at 10^4–10^6 infectious particles per milliliter.

Target cells for the treatment of hematopoietic diseases are the stem cells. Stem cells are defined by their ability to self-renew and repopulate the bone marrow after experimental marrow ablation and transplantation. These cells can be isolated from human bone marrow, mobilized peripheral blood, and umbilical cord blood. Both primitive and committed cells express the CD34 molecule on their membrane surface. Subsets of $CD34^+$ cells have been identified based on the pattern of antigen expression. Cells having the phenotype $CD34^+$, $CD33^-$, $CD38^-$, HLA-DR^-, and $Thy1^+$ are thought to include the repopulating stem cells (Craig et al., 1993; Lansdorp et al., 1992).

The $CD34^+$ cells of humans can be used to address the same question as above, namely, whether rAAV vectors can infect these cells and express rAAV transgenes. Using normal human bone marrow, cells were immunoselected for $CD34^+$ cells, incubated in suspension culture with growth factors, and infected at an MOI between 1 and 10 over a period of 3 days. From 60% to 70% of the $CD34^+$ cells stained positive for β-gal 16 h after transfection. To quantitate rAAV transduction frequency, a DNA PCR assay was used to assess the copy number of the vector relative to the cellular genome. A cell line that was previously determined to carry a single integrated copy of the AAV β-gal vector served to calibrate the system, setting it up in a way that it was sensitive to the vector copy number over a broad range of genomic DNA concentrations. The copy number of the vector relative to the cellular genome was determined using primers specific for the cellular β-actin gene to produce a standard of comparison. Infections of $CD34^+$-enriched cells were performed at an MOI of approximately 1 and plated under methylcellulose for C-Colony-forming unit-cell analysis in the absence of selection. Individual colonies were isolated 12–14 days postinfection, and 60% of the colonies yielded PCR signals indicating the presence of the lacZ sequence. A copy number of 1 or 2 vector genomes per cell was determined by PCR.

Similar experiments were performed with $CD34^+$ progenitor cells isolated from rhesus monkeys and again showed a rAAV–β gal transduction frequency of 66%. Half of the positive clones yielded lacZ-specific PCR signals. A copy number of 1–2 rAAV genomes per cell was assessed. The molecular status of the rAAV genome in the transduced cells was left unanswered as the DNA PCR results could not distinguish between episomal or integrated virus genomes. These studies give no definitive answer to the molecular state of the rAAV genome in transduced human and primate $CD34^+$ cells.

If rAAV can transduce hematopoietic cells, is transduction influenced by cell prestimulation with cytokines? While stem cells are noncycling or slow cycling, the treatment with cytokines would stimulate cycling and perhaps transduction/integration. A study to determine if cytokine prestimulation is required (Zhou et al., 1994) used the neomycin resistance gene as a marker to determine transduction. $CD34^+$-enriched cells from human cord blood were infected with the rAAV-neomycin virions. The neomycin gene was expressed under the control of either the herpes virus thymidine kinase (TK) or the human parvovirus B19 p6 promoter as well as an upstream erythroid cell-specific enhancer (HS-2) from the locus control region of the human β-globin gene cluster. After infecting the $CD34^+$ progenitor cells with an MOI of 1, the cells were placed under G418 selection. While 12% of the mock-infected granulocyte-macrophage colony-forming unit (GM-CFU) colonies survived, the numbers increased to 25% with rAAV–TK–Neo, 38% with HS-2–TK–Neo, 39% with B19–Neo, and 27% with HS-2–B19–Neo. For the erythroid burst-forming unit (E-BFU) colonies similar results were observed, as 15% of the mock-infected cultures survived in G418 selection and 37%–49% of the various rAAV-infected cultures were resistant to G418. Using a DNA PCR assay, the neo gene could be detected in the primitive as well as in the more mature hematopoietic progenitor cells infected with the recombinant AAV–Neo virus. In mock-transfected cells no Neo DNA product could be amplified. However, the DNA PCR used to detect transduced cells cannot distinguish between vector DNA contamination versus intracellular viral genomes whether integrated or episomal.

After determining these baselines, the authors next tried to prestimulate the cord blood cells with various growth factors before transduction with the recombinant virus. Recombinant human interleukin-3 (rhIL-3), recombinant human GM colony-stimulating factor (rhGM-CSF), and recombinant human steel factor (rhSCF) were used. Neither GM-CFU nor E-BFU was measurably affected when the cells were prestimulated with growth factors for 48 h. These results were reproduced with IL-3, IL-6, and GM-CSF to prestimulate cord blood-derived $CD34^+$ cells (Fisher-Adams et al., 1996). Again, prestimulation with cytokines did not affect rAAV transduction efficiency.

In contrast to using retroviral vectors where at least one round of cell division seems to be necessary before random integration of a transduced gene occurs (Miller et al., 1990), these experiments suggest that AAV can express transgenes in slow or noncycling human hematopoietic progenitor cells from cord blood in the absence of growth factor stimulation. It is unclear if the transgene sequences are expressed from integrated or episomal rAAV and when the integration of the virus into the genome takes place. To test if rAAV vectors require active cell division for transduction, cells were induced into a nonproliferating state by treatment with cell cycle inhibitors (Podsakoff et al., 1994b). At the time of growth arrest cells were infected with rAAV/β-gal. No difference in transduction was observed when cell cycle–arrested cells were compared with actively cycling cells. Integrated viral genomes were detected by Southern hybridization in the cycle-arrested cells. The data suggest that rAAV transduction and integration occur in the absence of cell cycling. However, this experimental approach does not rule out the possibility that the rAAV genome can persist in nondividing cells in an episomal form and integrate after the cell cycle block is released.

Studies have shown that wild-type AAV integrates into a specific region on chromosome 19 in the absence of helper virus in many cell lines (Kotin et al., 1990; Samulski, 1993). It is of interest if the virus behaves in the same way in hematopoietic cells. To determine if wild-type AAV integrates in a site-specific manner in primary hematopoietic progenitors, genomic DNA isolated from $CD34^+$ cells infected with wild-type virus at a high multiplicity (MOI of 100–1,000) was analyzed by PCR (Goodman et al., 1994). A set of nested primers specific for the vector sequence and the chromosome 19 integration locus amplified junction sequences between the proviral and cellular DNA sequences. PCR products generated with these primers indicate site-specific integration at chromosome 19. Pools of infected cells derived from 40–80 individual colonies were then tested for the presence of chromosome 19–AAV junction sequences. Each pool yielded PCR signals consistent with the presence of AAV–chromosome 19 junctions. The wild-type AAV showed targeted integration in hematopoietic progenitor cells with a frequency of 14% (4 of 27 pools). Because the first AAV-specific primer of the nested set annealed to a sequence unique to the right-hand end of the viral genome, this assay could detect AAV integration in only one orientation, thereby probably underestimating the frequency of targeted integration.

A comparison of rAAV transduction and the integration status of rAAV has been performed in K562 cells. Using a rAAV vector that carries a truncated rat nerve growth factor receptor (tNGFR) under the control of the Moloney murine leukemia virus long terminal repeat, gene transduction in K562 cells was followed periodically by flow cytometry. Using an MOI of 1.3 infectious units, 26%–38% of cells expressed tNGFR on their surface shortly after transduction (7 days), but after 1 month this rate dropped to 3%. If a much higher MOI was used (130 infectious units per cell), 90% of cells expressed tNGFR after 7 days, and but after 1 month only 62% of the cells were still expressing the transgene and this number further declined in the next 3 month. Integration analysis showed that overall integration was inefficient in the absence of selective pressure. At an MOI of 1.3, about 2% of the cells showed integration; at an MOI of 130, about 49% of the cells have integrated the transgene. Similar results were obtained analyzing tNGFR expression in human $CD34^+$ cells. Again, the levels of expression correlate with the virus input. The integration status was not assessed.

In an effort to correlate rAAV transduction and integration frequency in $CD34^+$ cells from human bone marrow or umbilical cord blood, $CD34^+$ cells from bone marrow of four donors were infected with rAAV vectors at an MOI of 2 for one day and then analyzed by Southern blot (Fisher-Adams et al., 1996). In each case, the probes detected vector-specific bands in sizes that were larger than the single vector, suggesting chromosomal integration of the vector. However, such bands could also derive from extrachromosomal vector sequences in rearranged forms. Another analysis was conducted using umbilical cord blood cells that were infected at an MOI of 0.2. The cells were cultivated without selective pressure, and 2 weeks after infection FISH analysis was performed. The transgene sequences were present in a minimum of 3% of cord blood cells 7 days after infection. The signal could be detected on both chromatids, suggesting that the integrated vector replicated during cell division. Previous studies using FISH analysis of wild-type AAV integration on chromosome 19 also detected signals on both chromatids (Kotin et al., 1990; Samulski et al., 1991). However, the larger number of $CD34^+$ cells expressing the transgene (50%) compared with the fraction showing integrated vector sequences suggests that the majority of transduced cells must carry double-stranded episomal vector genomes. A concern using this method of analysis is the very low number of metaphase spreads that show a signal for the vector probe. In addition, the small probe employed was at the limit of detection (3.6 kb), and due to random integration a control probe labeling the second allele or the chromosome could not be used.

The experiments mentioned above show convincingly that hematopoietic progenitor cells can be transduced by rAAV. Wild-type AAV integrates in hematopoietic cells in a site-specific locus in chromosome 19. rAAV integration occurs in hematopoietic cell lines but stable, long-lasting expression can also be achieved from episomes. In primary human $CD34^+$ cells only one report suggests integration into the genome. Clearly more work has to be done to confirm these results. If the low transduction frequencies could be overcome, rAAV vectors could be very promising for targeting progenitor cells with therapeutic genes.

AAV- MEDIATED TRANSDUCTION OF THERAPEUTIC GENES IN HEMATOPOIETIC CELLS

While experiments with marker genes suggested that hematopoietic cells can be infected with rAAV and a foreign gene expressed therein, these experiments give no information about the level and stability of expression of a therapeutic gene. After showing that rAAV can stably transduce and express marker genes for short time periods of time in hematopoietic progenitor cells, the next step was to investigate the regulated expression of therapeutic genes. The initial experiments dealt with globin gene expression. Enhancer/promoter elements for the globin gene cluster fall into two general groups: the proximal promoter elements associated with each gene of the globin cluster and the locus control region (LCR). The LCR contains several DNase hypersensitive sites (HS), termed HS-1 to HS-5. A 225 bp fragment of HS-2 normally situated approximately 10 kb upstream from the β-like globin gene cluster is sufficient to confer high level, cell type-specific hemin inducibility upon the globin promoters (Philipsen et al., 1990).

The HS-2 element incorporated into a rAAV vector linked with the γ-globin gene (Walsh et al., 1992) was first to be investigated. The crude cell lysate virus preparation was heat inactivated. K562 cell infection with the vector resulted in stable integrated clones following neomycin selection. Each clone tested consistently maintained one to two unrearranged copies per cell. Pools and individual clones were tested for the transcription of the transduced Aγ-globin sequences with and without hemin induction. Following hemin induction, comparable levels of γ-mRNA transcripts of the endogenous gene and the transgene were detected. The two genes could be distinguished because the transgene harbors a 6 bp deletion in the 5′-untranslated region. Upon analysis of individual clones using quantitative reverse transcriptase PCR and RNase protection assays, the transgenes of seven individual clones were determined to be transcribed to approximately 40%–50% of the levels of the endogenous genes after correction for copy number. Following hemin induction, the level of transcripts from the transduced genes rose to approximately 85% of the induced levels of the endogenous genes. This demonstrated that the AAV Aγ-globin constructs were capable of directing the expression of transgenes in amounts comparable to those of the endogenous genes. The levels of uninduced transcription of the endogenous gene and the transgene were similar.

The ability of an AAV-based vector to stably transduce a population of cells and to respond to cell type-specific stimuli makes it a powerful tool for the study of the behavior of transgenes and associated regulatory elements. As AAV can only carry a limited size of foreign genetic material, the precise understanding of these regulatory elements will allow inclusion of only the necessary regulatory elements. A precise investigation of the regulatory elements of the globin genes was performed (Miller et al., 1993) using mutational analysis to investigate the role of putative HS-2 *cis*-acting sequences. Mutation of the HS-2 NF-E2 and the GATA-1 transcription factor binding sites were incorporated into HS2/γ-globin rAAV vectors. Following rAAV infection of K562 cells, in the absence of hemin induction, mutation of the NF-E2 binding site did not alter the basal level of transgene expression. However, no induction of γ-globin transcription could be observed after adding hemin. A different pattern of vector expression could be observed by mutating the GATA-1 site. Here, the hemin induction lead to comparable levels of γ-globin mRNA, but a higher degree of variability in the level of induced and uninduced expression from the GATA-1 mutants was observed. This suggested that although GATA-1 was not essential for either basal or induced expression, the ability to bind this factor mitigated the effects of the site of integration, which was probably the only difference between these cell clones.

These experiments demonstrate two important results about AAV-derived vectors. First, the exogenous Aγ-globin gene under the control of its core promoter and a small segment of the globin LCR was inducible with hemin and expressed the globin message. Thus, within the context of the AAV provirus, the behavior of the transgene was indistinguishably from that of the endogenous gene in these cells. Second, the expression of the transgene was independent of its chromosomal position as long as the proviral construct contained the appropriate sequences (HS-2). This is an important observation given the fact that AAV integration is probably random if the viral Rep proteins are not present in the cell (Russell et al., 1994; Walsh et al., 1992).

One approach for the treatment of β-thalassemia or sickle cell anemia requires the transfer of a human β-globin gene into a patient's hematopoietic stem/progenitor cells. High-level regulated β-globin expression requires the presence of introns, a minimal β-globin promoter, and sequences located 50–60 kb upstream of the human β-globin gene designated the LCR. An HS-2 regulated β-globin expression rAAV vector in K562 and KB cells (Zhou et al., 1996) was examined. With Northern blot as well as RNase protection assays, a regulated expression of the human β-globin gene was observed in K562 cells, with expression of the transduced human β-globin in 10%–20% of the infected K562 cells. Similar experiments used a rAAV vector containing the HS-4, HS-3, and HS-2 LCR nano-elements linked to a human β-globin gene (Einerhand et al., 1995) and transduced 293 cells and murine erythroleukemia (MEL) cells. After selection of positive clones with G418, Southern blot analysis detected one integrated provirus per clone. Nuclease S1 protection analysis showed low preinduction levels of human β-globin. Upon erythroid induction of the transduced MEL cells, variable levels of β-globin RNA expression were observed, the average being about 50% of the level of the endogenous murine β^{maj} expression. However, 50% of the transduced clones contained a rearranged β-globin expression cassette. An interesting finding in this study was the variable expression amongst clones, suggesting that the LCR region elements used were not sufficient for position-independent expression. Taken together, these experiments demonstrate that rAAV is able to transduce transgenes into hematopoietic progenitor cells and that these genes exhibit almost the same regulated expression as the endogenous genes. However, more research has to be done to delineate the minimal combination of LCR elements necessary to preserve position-independent expression.

Another possible approach to gene therapy for β-thalassamia is the suppression of α-globin production, since the accumulation of free α-globin reduces the lifespan of red blood cells in these patients. rAAV/antisense α-globin vectors (Ponnazhagan et al., 1994) were used to infect K562 cells. Stable neomycin-selected clones were tested for antisense mRNA transcription. Northern blot analysis of total cellular RNA showed that while a thymidine kinase promoter construct had no effect on the α-globin RNA level, an SV40 (simian virus 40) promoter-driven construct reduced α-globin expression by 29% and an α-globin promoter-driven construct reduced the α-globin gene expression by 91% on the transcriptional level.

A different approach to gene therapy for malignant hematopoietic disorders has been investigated through the expression of specific cytokines in Cos-1 cells. A rAAV vector expressing human granulocyte-macrophage colony-stimulating factor (GM-CSF) under the control of the SV40 early promoter was used for infection of Cos-1 cells (Lou et al., 1995). The cytokine was released into the supernatant at about 67–118 U/ml and had biological activity on a GM-CSF–dependent human megakaryocytic leukemia cell line. Recombinant GM-CSF has been used in clinical trials for several hematopoietic malignant disorders. The generation of GM-CSF secreting cells will be useful for a wide range of hematopoietic disorders. However, as GM-CSF has shown toxic effects in animal models after long-term exposure to high amounts of the factor, it is of crucial importance to achieve regulated expression of the GM-CSF gene *in vivo*. Therefore, the *cis*-controlling elements of the cytokine have to be determined, and the safety and efficacy in animal models need to be established.

Another important goal is the phenotypic correction of a disease by expression of a transferred gene. This question has been addressed using Fanconi's anemia (FA) as a disease model (Liu et al., 1994; Walsh et al., 1994b). The disease is an autosomal recessive disorder that appears to stem from deficiencies in DNA repair. There are five known FA complementation groups (A, B, C, D, E), and the genes for group A (FA-A) and group C (FA-C) have been cloned (Lo Ten Foe et al., 1996; Strahdee et al., 1992). The genes for both complementation groups A and C can correct the phenotypic defect when transfected into FA cells in culture.

A rAAV vector was constructed that expressed the FA-C cDNA under the control of the RSV promoter and also included a neo marker gene (Walsh et al., 1994b). Lymphoblasts from FA-C patients were immortalized by Epstein-Barr virus transformation and then infected with the rAAV vector and selected for neo expression. Phenotypic expression of the FA-C gene was determined by cell resistance to the DNA clastogen mitomycin C. Molecular confirmation of rAAV transduction of FA-C lymphoblasts was performed by Southern blot analysis and reverse transcriptase PCR. In transduced cells, transgene mRNA expression levels were similar to endogenous gene expression. Using immunoprecipitation of radiolabeled transduced cell extracts, the level of protein expression was comparable with levels in normal cells. The hallmark of this disease is the hypersensitivity of FA cells to DNA cross-linking agents. Transduction of the FA gene in hematopoietic stem/progenitor cells should confer a growth advantage for these cells when grown in the presence DNA-damaging agents. This was demonstrated following the transduction of $CD34^+$ cells from an FA-C patient. In a short-term colony-forming assay, viral-infected cells yielded four times more colonies than mock-infected cells. Surprisingly, reverse transcriptase PCR analysis yielded a positive signal for only 60% of the cells. The integration status of the transduced cells was not documented.

AAV-MEDIATED TRANSDUCTION *IN VIVO*

While experiments in cell culture provide important information about the feasibility of a vector system, they give no information about how the vector will behave *in vivo.* Questions regarding long-term expression, level of expression, and side effects like toxicity can only be addressed in animal models. As we have already described, the majority of rAAV experiments have utilized a variety of vector constructs in established cell lines in culture. However, there are few reports about the use of rAAV vectors in primary mouse bone marrow cells and in mouse animal models.

Demonstration that primary mouse bone marrow cells can be infected by rAAV was performed using a rAAV vector carrying a neomycin gene as the reporter gene (Zhou et al., 1993). Recombinant virions were used to infect unfractionated murine bone marrow cells. With short-term colony cultures, colony growth was assayed with G418 selection, and stable integration of the constructs could be demonstrated on Southern blots by hybridization with a neo probe.

The expression of human α- and β-globin in MEL cells and mouse bone marrow cells has been addressed (Ohi and Kim, 1996). With a rAAV vector, the globin genes were expressed under the wild-type AAV p40 promoter. In the case of the β-globin construct the enhancers HS-1 and HS-2 were included. Messenger RNA and protein of both globin chains could be detected, with levels of expression similar to that of the endogenous mouse β-globin.

The murine model can also be useful for analysis of the fate of human transduced cells. The ability to engraft human hematopoietic cells in various immunodeficient mouse strains provides an interesting human/mouse *in vivo* model system. A rAAV vector encoding the gene for FA-C was used to determine "gene marking" of human umbilical cord blood progenitor cells *in vivo* using CB17 scid/scid mice. (Walsh et al., 1994a). Following transduction, vector-specific sequences could be identified by DNA PCR 11 weeks post-transplantation in three of four mice.

Another assay for stem cell function is long-term multilineage hematopoietic reconstitution. rAAV transduced murine bone marrow cell engraftment was performed in lethally irradiated mice, and evidence of transgene expression was observed for more than 6 months in peripheral blood mononuclear cells (Podsakoff et al., 1994a). Vector-specific sequences were detected in marrow, thymus, spleen, peripheral blood, and brain 6 and 7.5 months post-transplantation. In both of these studies no information on lineage marking or transgene expression was reported.

AAV-mediated gene transfer of the human γ-globin gene was also tested by serial bone marrow transplantation in mice (Ponnazhagan et al., 1997). The $^{A}\gamma$-globin gene was expressed under the control of the human β-globin promoter and the HS-2 enhancer. Recombinant AAV containing the transgene was CsCl purified and heated to eliminate contaminating adenovirus. Low-density bone marrow cells from mice were infected at MOIs of 1 and 10 and transplanted into six lethally irradiated mice. The level of engraftment could be measured because donor cells showed a specific enzyme phenotype (glucose phosphate isomerase) that was different from the enzyme phenotype of the recipient mice. Reconstitution of the bone marrow was complete at 4 months after infusion. Six months post transplantation the mice were sacrificed, and cells from bone marrow, spleen, and thymus were analyzed for gene marking and transgene expression by PCR. The γ-globin sequences could be detected in bone marrow and, with the exception of one animal, also in spleen, with a transduction efficiency of about 7%. Expression analysis using reverse transcriptase PCR could only show expression in bone marrow cells. The level of expression was low, reaching only 4%–6% of the level of expression of the endogenous mouse β^{maj}-globin gene. A portion of the bone marrow obtained 6 months after transplantation was used to engraft secondary recipients. These were analyzed 3 months post-transplantation by again using reverse transcriptase PCR to determine the expression of the transgene. Only one of the six animals showed the presence of the transduced human γ-globin sequences in bone marrow and spleen cells. The level of expression of the transduced γ-globin gene was determined by reverse transcriptase PCR to be 0.4% compared with the endogenous mouse β^{maj}-globin gene. However, the presence of the vector DNA sequences for 6 months in primary recipients and for an additional 3 months in secondary recipients suggests that a progenitor cell may have been transduced. It is not clear which cells have been transduced with this vector, and the fate of the vector genome has not been addressed.

CURRENT PROBLEMS

Several issues concerning gene delivery and expression in hematopoietic cells are still unresolved. First, efficiency of gene transfer must be increased in order to achieve a therapeutic effect in the hematopoietic system. A possible solution to this problem could be *in vivo* selection and expansion of transduced cells. An example of such an approach utilized constructs containing the multidrug resistance gene encoding the P-glycoprotein gene product. In the mouse model, three groups have shown that retroviral transfer of the multidrug resistance gene (MDR-1) into mouse bone marrow cells allows *in vivo* selection or chemoprotection after engraftment (Hanania and Deisseroth, 1994; Podda et al., 1992; Sorrentino et al., 1992). After a single dose of taxol, mice engrafted with transduced cells showed dramatic and stable increases in the number of circulating cells containing the MDR-1 provirus.

Second, the mechanism of rAAV transduction is still unresolved. Several DNA-damaging agents, UV and γ-irradiation, increase transduction efficiencies 100–1,000-fold (Ferrari et al., 1996; Fisher et al., 1996). In addition, cells from different patients show tremendous heterogeneity in transduction events, even if the same virus preparation is used (McKeon and Samulski, 1996). In this case, the difference in MOI required to transduce cells suggests that a rate-limiting step either at viral entry and uptake (receptor variation) or at virus uncoating and gene expression is at play.

Third, rAAV vector production is still unsatisfactory. Currently the most commonly used procedure relies on production of helper-free rAAV vectors using transient transfections. As illustrated by the new AAV packaging cell lines, these technical problems should be overcome soon.

Fourth, the size constraint of AAV vectors can be problematic. It is unlikely that the AAV viral vector will ever overcome this limitation; however, preliminary results using AAV integrating plasmids suggest that nonviral delivery may be an alternative approach for solving this problem. The evolution of such novel strategies may be the template for second-generation AAV vectors.

CONCLUSIONS

AAV vectors provide a powerful tool for the transduction of therapeutic genes into hematopoietic cells. They are able to infect non- or slow-dividing cells. With rAAV vectors, hematopoietic progenitor cells can be transduced, resulting in potentially therapeutic expression levels of transgenes. Phenotypic correction of FA and early experiments with bone marrow reconstitution in mice appear promising. To date, the question of rAAV integration in hematopoietic stem cells remains unanswered. However, long-term expression of transgenes using rAAV has been demonstrated over 1 year in rat brain (McCown et al., 1996; McCown, personal communication), 1.5 years in mouse muscle (Xiao et al., 1996), and 9 months in a mouse hematopoietic serial transplantation models (Fisher-Adams et al., 1996), making AAV an attractive vector for gene therapy of hematopoietic diseases. However, only from more animal studies can we prove that AAV can be useful for gene therapy of human hematopoietic disorders.

REFERENCES

Afione SA, Conrad CK, Kearns WG, Chundru S, Adams R, Reynolds T, Guggino WB, Cutting GR, Carter BJ, Flotte TR (1996). *In vivo* model of adeno-associated virus vector persistence and rescue. *J Virol* 70:3235–3241.

Balague C, Kalla M, Zhang WW (1997). Adeno-associated virus Rep78 protein and terminal repeats inhance integration of DNA sequences into the cellular genome. *J Virol,* 71:3299–3306.

Berns KI (1996). Parvorviridae and their replication. In Fields BN, Knipe DM, Howely PME (eds): Fields Virology. Philadelphia: Lippincott-Raven, pp 2173–2197.

Berns KI, Bohenzky RA (1987). Adeno-associated viruses: An update. *Adv Virus Res* 32:243–306.

Brown KE, Anderson SM, Young NS (1993). Erythrocyte P antigen: Cellular receptor for B19 parvovirus. *Science,* 262:114–117.

Brown KE, Young NS, Liu JM (1994). Molecular, cellular and clinical aspects of parvovirus B19 infection. *Crit Rev Oncol Hematol* 16:1–31.

Buller RM, Janik JE, Sebring ED, Rose JA (1981). Herpes simplex virus types 1 and 2 completely help adenovirus-associated virus replication. *J Virol* 40:241–247.

Clark KR, Voulgaropoulou F, Fraley DM, Johnson PR (1995). Cell lines for the production of recombinant adeno-associated virus. *Hum Gene Ther* 6:1329–41.

Craig W, Kay R, Cutler R, Lansdorp P (1993). Expression of Thy-1 on human hematopoietic progenitor cells." *J Exp Med* 177:1331.

Cukor G, Blacklow NR, Hoggan D, Berns KI (1984). Biology of adeno-associated virus. Berns KI (ed): The Parvoviruses. New York: Plenum Press, pp 33–66.

Einerhand MPW, Antoniou M, Zolotukhin S, Muzyczka N, Berns KI, Grosveld F, Valerio D (1995). Regulated high level human β-globin gene expression in erythroid cells following recombinant adeno-associated virus-mediated gene transfer. *Gene Ther* 2:336–343.

Ferrari FK, Samulski T, Shenk T, Samulski RJ (1996). Second-strand synthesis is a rate limiting step for efficient transduction by recombinant adeno-associated virus vectors. *J Virol* 70:3227–3234.

Fisher KJ, Gao GP, Weitzman MD, DeMatteo R, Burda JF,Wilson JM (1996). Transduction with recombinant adeno-associated virus for gene therapy is limited by leading-strand synthesis. *J Virol* 70:520–532.

Fisher-Adams G, Wong KK, Podsakoff G, Forman SJ, Chatterjee S (1996). Integration of adeno-associated virus vectors in $CD34^+$ human hematopoietic progenitor cells after transduction. *Blood* 88:492–504.

Flotte TR, Afione SA, Conrad C, McGrath SA, Solow R, Oka H, Zeitlin PL, Guggino WB, Carter BJ (1993). Stable *in vivo* expression of the cystic fibrosis transmembrane conductance regulator with an adeno-associated virus vector. *Proc Natl Acad Sci USA* 90:10613–10617.

Flotte TR, Afione SA, Zeitlin PL (1994). Adeno-associated virus vector gene expression occurs in nondividing cells in the absence of vector DNA integration. *Am J Respir Cell Mol Biol* 11:517–521.

Flotte TR, Barraza-Ortiz X, Solow R, Afione SA, Carter BJ, Guggino WB (1995). An improved system for packaging recombinant adeno-associated virus vectors capable of *in vivo* transduction. *Gene Ther* 2:29–37.

Flotte TR, Carter BJ (1995). Adeno-associated virus vectors for gene therapy. *Gene Ther* 2:357–362.

Goodman S, Xiao X, Donahue RE, Moulton A, Miller J, Walsh C, Young NS, Samulski RJ, Nienhuis AW (1994). Recombinant adeno-associated virus-mediated gene transfer into hematopoietic progenitor cells. *Blood* 84:1492–1500.

Grossman Z, Mendelson E, Brok-Simoni F, Mileguir F, Leitner Y, Rechavi G, Ramot B (1992). Detection of adeno-associated virus type 2 in human peripheral blood cells. *J Gen Virol* 73:961–966.

Hanania EG, Deisseroth AB (1994). Serial transplantation shows that early hematopoietic precursor cells are transduced by MDR-1 retroviral vector in a mouse gene therapy model. *Cancer Gene Ther* 1:21–25.

Hermonat PL, Muzyczka N (1984). Use of adeno-associated virus as a mammalian DNA cloning vector: Transduction of neomycin resistance into mammalian tissue culture cells. *Proc Natl Acad Sci USA,* 81:6466–6470.

Hölscher C, Horer M, Kleinschmidt JA, Zentgraf H, Burkle A, Heilbronn R (1994). Cell lines inducibly expressing the adeno-associated virus (AAV) rep gene: Requirements for productive replication of rep-negative AAV mutants. *J Virol* 68:7169–7177.

Hong SS, Karayan L, Tournier J, Curiel DT, Boulanger PA (1997). Adenovirus type 5 fiber knob binds to MHC class I a2 domain at the surface of human epithelial and B lymphoblast cells. *EMBO J* 16:2294–2306.

Kaplitt MG, Leone P, Samulski RJ, Xiao X, Pfaff DW, O'Malley KL, During MJ (1994). Long-term gene expression and phenotypic correction using adeno-associated virus vectors in the mammalian brain. *Nature Genet* 8:148–154.

Kaplitt MG, Xiao X, Samulski RJ, Ojamaa K, Klein IL, Makimura H, Kaplitt MJ, Strumpf RK, Diethrich EB (1996). Long-term gene transfer in porcine myocardium after coronary infusion of an adeno-associated virus vector. *Ann Thorac Surg* 62:1669–1676.

Kotin RM, Siniscalco M, Samulski RJ, Zhu XD, Hunter L, Laughlin CA, McLaughlin S, Muzyczka N, Rocchi M, Berns KI (1990). Site-specific integration by adeno-associated virus. *Proc Natl Acad Sci USA,* 87:2211–2215.

Lansdorp P, Schmitt C, Sutherland H, et al. (1992). Hematopoietic stem cell characterization. *Prog Clin Biol Res* 377:475.

Laughlin CA, Tratschin J-D, Coon H, Carter BJ (1983). Cloning of infectious adeno-associated virus genomes in bacterial plasmids. *Gene* 23:65–73.

Lebkowski JS, McNally MM, Okarma TB, Lerch LB (1988). Adeno-associated virus: A vector system for efficient introduction of DNA into a variety of mammalian cell types. *Mol Cell Biol* 8:3988–3996.

Li J, Samulski RJ, Xiao X (1997). Role for highly regulated rep gene expression in adeno-associated virus vector production. *J Virol* 71:5236–5243.

Liu JM, Buchwald M, Walsh CE, Young NS (1994). Fanconi anemia and novel strategies for therapy. *Blood* 84:3995–4007.

Liu J, Green S, Shimada T, Young NS (1992). A block in full-length transcript maturation in cells nonpermissive for B19 parvovirus. *J Virol* 66:4686–4692.

Lo Ten Foe J, Rooimans M, Bonoyan-Collins L, Alon N, Wijker M, Parker L, Lightfoot J, Carreau M, Callen D, Savoia A, Cheng N, van Berkel C, Strunk M, Gille J, Pals G, Kruyt F, Pronk J, Arwet F, Buchwald M, Joenje H (1996). Expression cloning of a cDNA for the major Fanconi anaemia gene, FAA. *Nature Genet* 14:320.

Lou F, Zhou SZ, Cooper S, Munshi NC, Boswell H, Broxmeyer HE, Srivastava A (1995). Adeno-associated virus 2–mediated gene transfer and functional expression of the human granulocyte-macrophage colony stimulating factor. *Exp Hematol* 23:1261–1267.

McCown TJ, Xiao X, Li J, Breeze GR, Samulski RJ (1996). Differential and persistent expression patterns of CNS gene transfer by an adeno-associated virus (AAV) vector. *Brain Res* 713:99–107.

McKeon C, Samulski RJ (1996). NIDDK Workshop on AAV vectors: Gene transfer into quiescent cells. *Hum Gene Ther* 7:1615–1619.

Miller DG, Adam MA, Miller AD (1990). Gene transfer by retrovirus vectors occurs only in cells that are actively replicating at the time of infection. *Mol Cell Biol* 10:4239–4242.

Miller JL, Walsh CE, Ney PA, Samulski RJ, Nienhuis AW (1993). Single-copy transduction and expression of human gamma-globin in K562 erythroleukemia cells using recombinant adeno-associated virus vectors: The effect of mutations in NF-E2 and GATA-1 binding motifs within the hypersensitivity site 2 enhancer. *Blood* 82:1900–1906.

Mizukami H, Young N, Brown K (1996). Adeno-associated virus type 2 binds to a 150 kilodalton cell membrane glycoprotein. *Virology* 217:124–130.

Muzyczka N (1991). *In vitro* replication of adeno-associated virus DNA. *Semin Virol* 2:281–290.

Muzyczka N (1992). Use of adeno-associated virus as a general transduction vector for mammalian cells. *Curr Top Microbiol Immunol* 158:97–129.

Ohi S, Kim BC (1996). Synthesis of human globin polypeptides mediated by recombinant adeno-associated virus vectors. *J Pharmaceut Sci* 85:274–281.

Philipsen S, Talbot D, Fraser P, Grosveld F (1990). The beta-globin dominant control region: Hypersensitive site 2. *EMBO J* 9:2159–2167.

Podda S, Ward M, Himelstein A, Richardson C, Flor-Weiss E, Smith L, Gottesmann M, Pastan I, Bank A (1992). Transfer and expression of the human multiple drug resistance gene into live mice. *Proc Natl Acad Sci USA* 89:9676–9680.

Podsakoff G, Shaughnessy EA, Lu D, Wong KK, Chatterjee S (1994a). Long term *in vivo* reconstitution with murine marrow cells transduced with an adeno-associated virus vector. *Blood* 84(Suppl):256a.

Podsakoff G, Wong KK, Chatterjee S (1994b). Efficient gene transfer into non-dividing cells by adeno-associated virus-based vectors. *J Virol* 68:5656–5666.

Ponnazhagan S, Nallari ML, Srivastava A (1994). Suppression of human alpha-globin gene expression mediated by the recombinant adeno-associated virus 2-based antisense vectors. *J Exp Med* 179:733–738.

Ponnazhagan S, Wang X-S, Woody MJ, Luo F, Kang LY, Nallari M, Munshi NC, Zhou SZ, Srivastava A (1996). Differential expression in human cells from the p6 promoter of human parvovirus B19 following plasmid transfection and recombinant adeno-associated virus 2 (AAV) infection: Human megakaryocytic leukemia cells are non-permissive for AAV infection. *J Gen Virol* 77(pt 6):1111–1122.

Ponnazhagan S, Yoder M, Srivastava A (1997). Adeno-associated virus type 2–mediated transduction of murine hematopoietic cells with long-term repopulating ability and sustained expression of a human globin gene *in vivo. J Virol* 71:3098–3104.

Rolling F, Samulski RJ (1995). AAV as a viral vector for human gene therapy. Generation of recombinant virus. *Mol Biotechnol* 3:9–15.

Rouger P, Gane P, Salmon C (1987). Tissue distribution of H, Lewis and P antigens as shown by a panel of 18 monoclonal antibodies. *Rev Fr Trans Immunohematol* 30:699–708.

Russell DW, Miller AD, Alexander IE (1994). Adeno-associated virus vectors preferentially transduce cells in S phase. *Proc Natl Acad Sci USA* 91:8915–8919.

Rutherford TR, Clegg JB, Weatherall DJ (1979). K562 human leukemic cells synthesize embryonic haemoglobin in response to haemin. *Nature* 280:164–165.

Samulski RJ (1993). Adeno-associated virus: Integration at a specific chromosomal locus. *Curr Opin Genet Dev* 3:74–80.

Samulski RJ, Berns KI, Tan M, Muzyczka N (1982). Cloning of adeno-associated virus into pBR322: Rescue of intact virus from the recombinant plasmid in human cells. *Proc Natl Acad Sci USA* 79:2077–2081.

Samulski RJ, Chang L-S, Shenk T (1987). A recombinant plasmid from which an infectious adeno-associated virus genome can be excised *in vitro* and its use to study viral replication. *J Virol* 61:3096–3101.

Samulski RJ, Chang L-S, Shenk T (1989). Helper-free stocks of recombinant adeno-associated viruses: Normal integration does not require viral gene expression. *J Virol* 63:3822–3828.

Samulski RJ, Zhu X, Xiao X, Brook JD, Housman DE, Epstein N, Hunter LA (1991). Targeted integration of adeno-associated virus (AAV) into human chromosome 19 [published erratum appears in *EMBO J* 1992 11, 1228]. *EBO J* 10:3941–3950.

Shelling A, Smith MG (1994). Targeted integration of transfected and infected adeno-associated virus vectors containing the neomycin resistance gene. *Gene Ther* 1:165–169.

Skulimowski AW, Samulski RJ (1995). Adeno-associated virus: Intergrating vectors for human gene therapy. In Adolph KW (ed): *Methods in Molecular Genetics.* New York: Academic Press, pp 3–12.

Sorrentino BP, Brandt SJ, Bodine D, et al. (1992). Selection of drug-resistant bone marrow cells *in vivo* after retroviral transfer of human MDR-I. *Science* 257:99–103.

Strahdee CA, Gavish H, Shannon WR, Buchwald M (1992). Cloning of cDNAs for Fanconi's anemia by functional complementation. *Nature* 356:763–776.

Tratschin JD, Miller IL, Smith MG, Carter BJ (1985). Adeno-associated virus vector for high-frequency integration, expression, and rescue of genes in mammalian cells. *Mol Cell Biol* 5:3251–3260.

Vincent KA, Moore GK, Haigwood NL (1990). Replication and packaging of HIV envelope genes in a novel adeno-associated virus vector system. *Vaccine* 90:353–359.

Walsh CE, Liu IM, Wang S, Xiao X, Hashmi NJ, Zwerdling T, Agarwal R (1994a). *In vivo* gene transfer with a novel adeno-associated virus vector to human hematopoietic cells engrafted in SCID-hu mice. *Blood* 84:256a.

Walsh CE, Liu JM, Xiao X, Young NS, Nienhuis AW, Samulski RJ (1992). Regulated high level expression of a human gamma-globin gene introduced into erythroid cells by an adeno-associated virus vector. *Proc Natl Acad Sci USA* 89:7257–7261.

Walsh CE, Nienhuis AW, Samulski RJ, Brown MG, Miller JL, Young NS, Liu JM (1994b). Phenotypic correction of Fanconi anemia in human hematopoietic cells with a recombinant adeno-associated virus vector [see comments]. *J Clin Invest* 94:1440–1448.

Xiao W (1996). Characterization of cis- and trans-elements essential for the targeted integration of rAAV plasmid vectors. Ph.D. Thesis, University of North Carolina at Chapel Hill.

Xiao X, Li, J, Samulski RJ (1996). Efficient long-term gene transfer into muscle tissue of immunocompetent mice by adeno-associated virus vector. *J Virol* 70:8098–8108.

Yalkinoglu AO, Zentgraf H, Hubscher U (1991). Origin of adeno-associated virus DNA replication is a target of carcinogen-inducible DNA amplification. *J Virol* 65:3175–3184.

Zhou SZ, Broxmeyer HE, Cooper S, Harrington MA, Srivastava A(1993). Adeno-associated virus 2–mediated gene transfer in murine hematopoietic progenitor cells. *Exp Hematol* 21:928–933.

Zhou SZ, Cooper S, Kang LY, Ruggieri L, Heimfeld S, Srivastava A, Broxmeyer HE (1994). Adeno-associated virus 2–mediated high efficiency gene transfer into immature and mature subsets of hematopoietic progenitor cells in human umbilical cord blood. *J Exp Med* 179:1867–1875.

Zhou S Z, Li Q, Stamatoyannopoulos G, Srivastava A (1996). Adeno-associated virus 2–mediated transduction and erythroid cell-specific expression of a human β-globin gene. *Gene Ther* 3:223–229.

10

DELIVERY SYSTEMS FOR GENE THERAPY: ADENO-ASSOCIATED VIRUS 2

ARUN SRIVASTAVA

Department of Medicine, Microbiology and Immunology, Division of Hematology/ Oncology, Walther Oncology Center, Indiana University School of Medicine and Walther Cancer Institute, Indianapolis, IN 46202-5120

INTRODUCTION

Adeno-associated virus 2 (AAV) is a single-stranded DNA-containing human parvovirus (Siegl et al., 1985; Berns and Bohenzky, 1987; Berns, 1990). It is becoming increasingly clear that the AAV-based vectors may prove to be potentially useful for human gene therapy. The basis for this optimism is twofold. First, between 80% and 90% of the human population has been exposed to AAV, and yet no symptoms or pathology have thus far been attributed to AAV infection (Blacklow, 1988). Second, the wild-type (wt) AAV has been documented to establish a stable, latent infection in human cells where the viral genome has been shown to integrate into the chromosomal DNA in a site-specific manner (Kotin et al., 1990, 1991, 1992; Kotin and Berns, 1989; Samulski et al., 1991). Thus, the potential nonpathogenic nature of AAV vectors coupled with stable integration of the proviral genome may be of significant benefit compared with other viral vectors that are based on retroviruses and adenoviruses, which have already been employed in a number of clinical trials (Rosenberg et al., 1990; Zabner et al., 1993).

Stem Cell Biology and Gene Therapy, Edited by Peter J. Quesenberry, Gary S. Stein, Bernard G. Forget, and Sherman M. Weissman
ISBN 0-471-14656-0

Despite the initial success with retroviral vectors (Grossman et al., 1994; Blaese et al., 1995; Bordignon et al., 1995), their safety in long-term applications may be of some concern since in nonhuman primate studies these vectors have been reported to cause T-cell lymphoma (Donahue et al., 1992). Similarly, the efficacy of adenoviral vectors in gene therapy of cystic fibrosis has been questioned (Knowles et al., 1995). It is of note that retroviruses as well as most of the DNA-containing viruses are the etiological agents of, or are intimately associated with, malignant disorders (Weiss et al., 1984; Tooze, 1981). Parvoviruses, on the other hand, constitute the only group of viruses that have thus far not been known to cause any malignant disease (Cotmore and Tattersall, 1987; Pattison, 1988).

Interestingly, AAV has been shown to possess antitumor properties (Mayor et al., 1973; Cukor et al., 1975; Ostrove et al., 1981), which may be an added desirable feature of the AAV-based vector system. Several reviews on AAV vectors have recently been published (Muzyczka, 1992; Carter, 1993; Xiao et al., 1993; Samulski, 1993; Srivastava, 1994; Kotin, 1994; Flotte and Carter, 1995; Berns and Linden, 1995; Srivastava et al., 1996). This chapter attempts to highlight the salient features of AAV in general and outline a number of advantages of AAV vectors for their potential use in human gene therapy.

LIFE-CYCLE OF AAV

The AAV genome is a single-stranded DNA consisting of 4,680 nucleotides (Srivastava et al., 1983). The genomic organization of AAV is schematically represented in Figure 1. The viral genome is flanked by inverted terminal repeats (ITRs) that are 145 nucleotides in length, are palindromic, and form T-shaped hairpin structures (Lusby et al., 1980). The left half of the AAV genome contains the rep genes that encode four distinct nonstructural Rep proteins that are required for viral DNA replication, and the right half of the genome contains the cap genes that encode three viral capsid (Cap) proteins (Srivastava et al., 1983). The AAV genome contains three distinct promoters at map units 5 (p5), 19 (p19), and 40 (p40), respectively (Muzyczka, 1992). Expression of the viral rep gene is under the control of the p5 and p19

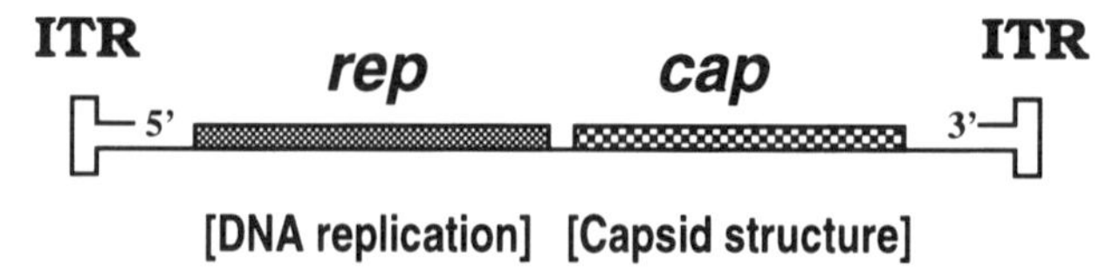

FIGURE 1. Schematic representation of the organization of the AAV genome. The viral rep and cap gene sequences are indicated as stippled and checkered boxes, respectively. The ITRs are depicted as T-shaped hairpin structures.

promoters, whereas expression of the viral cap gene is controlled by the p40 promoter.

The life cycle of AAV is schematically depicted in Figure 2. AAV requires co-infection with a helper virus, such as adenovirus, herpesvirus, or vaccinia virus (Berns, 1990), or conditions of genotoxic stress (Yacobson et al., 1987; Yalkinoglu et al., 1988) for its optimal replication. In the absence of a helper virus, however, the AAV genome integrates into the host cell chromosomal DNA (Cheung et al., 1980; Berns and Linden, 1995). Upon superinfection with a helper virus, the proviral genome undergoes rescue, followed by replication much the same way as during a lytic infection (Berns and Hauswirth, 1979; Berns and Linden, 1995). Detailed accounts of these pathways in the AAV life cycle have previously been published (Berns and Bohenzky, 1987; Berns, 1990; Berns and Linden, 1995).

Virus–Host Cell Interaction

Because AAV possesses a broad host range that transcends the species barrier (Muzyczka, 1992), it has been suspected that AAV infection is either nonspecific or mediated by a putative receptor that is present ubiquitously. Ponnazhagan et al. (1994b, 1996d) have addressed this issue more directly. In their studies, a number of different human cell lines were transduced with a recombinant AAV vector containing the firefly luciferase (Luc) reporter gene under the control of the thymidine kinase (TK) gene promoter (vTKp-Luc). It was noted that whereas significant expression of Luc gene occurred from the TK promoter in a human erythroleukemia cell line (K562), no expression from

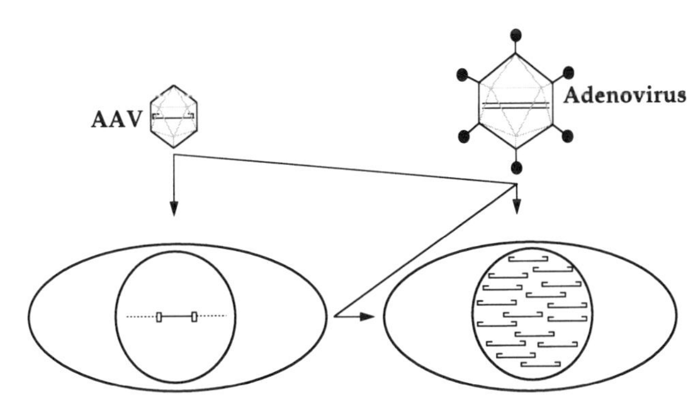

FIGURE 2. Life cycle of AAV. In the presence of a helper virus, the wt AAV undergoes a lytic infection, but in the absence of a helper virus, the wt AAV establishes a latent infection and the proviral genome integrates into the chromosomal DNA in a site-specific manner. Following superinfection with a helper virus, the proviral AAV genome undergoes rescue and replication, and a lytic cycle ensues.

this promoter occurred in two different human megakaryocytic leukemia cell lines (MB-02 and M07e) under identical conditions. This observation was initially attributed to the lack of strength of the TK promoter. However, when these cells were transduced with the same reporter gene under the control of the cytomegalovirus (CMV) promoter (vCMVp-Luc), again a significantly high level of Luc activity was detected in K562 cells, but no detectable activity was observed in MB-02 or M07e cells. These results are shown in Figure 3. Similar results were obtained when the bacterial β-galactosidase (lacZ) gene was used as a reporter. Taken together, these results strongly suggested, although did not prove, that the human megakaryocytic leukemia cells lacked the putative receptor for AAV and, as a result, were nonpermissive for AAV infection.

Since it remained possible that expression of the transduced Luc or lacZ genes were below the detection limit, Ponnazhagan et al. (1996) next examined whether wt AAV could undergo replication in these cells in the presence of a helper virus, such as adenovirus (Ad2), since previous studies have established abundant replication of viral replicative DNA intermediates in permissive human cells, such as HeLa, in the presence of Ad2 that is easily detectable

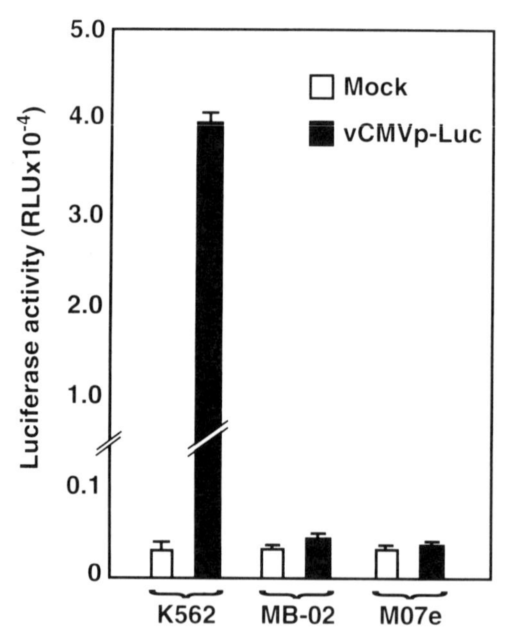

FIGURE 3. Comparative analysis of Luc activity in K562, MB-02, and M07e cells transduced with the recombinant vCMVp-Luc virions. Equivalent numbers of K562, MB-02, and M07e cells were either mock infected (open bars) or infected with the recombinant vCMVp-Luc virions (closed bars) and analyzed for Luc activity 48 h postinfection as previously described (Ponnazhagan et al., 1996).

(Nahreini and Srivastava, 1989; Samulski et al., 1989). When equivalent numbers of HeLa, K562, MB-02, and M07e cells were mock infected or infected with wt AAV in the presence or absence of Ad2 under identical conditions, it was evident that whereas HeLa and K562 cells allowed high and low levels of replication, respectively, of AAV DNA, but only in the presence of Ad2, no viral replicative DNA intermediates were detected either in MB-02 or M07e cells. These results are shown in Figure 4A. Although K562 cells are known to be susceptible to infection by recombinant AAV (Walsh et al., 1992; Ponnazhagan et al., 1994a; Zhou et al., 1996), the low level replication of AAV in these cells, and lack of AAV DNA replication in MB-02 and M07e cells, could also be due to lack of infectivity of these cells by Ad2.

This possibility was examined by reprobing the Southern blot shown in Figure 4A with an Ad2-specific DNA probe. The results shown in Figure 4B indicate abundant replication of Ad2 DNA in HeLa cells followed by that in K562 cells. Because low-level replication of Ad2 DNA could also be detected in MB-02 and M07e cells, these results further suggest that lack of Ad2 infection is insufficient to account for the absence of AAV DNA replicative DNA intermediates in these cells.

Additional experiments further corroborated this contention. For example, no G418-resistant clones of MB-02 and M07e cells could be obtained following AAV-mediated transduction with a selectable gene such as neo^R. In addition, infection of these cells with wt AAV followed by serial passage failed to reveal the presence of integrated proviral AAV genomes on Southern blots (Ponnazhagan et al., 1996). Thus, this is the first instance that a human cell type has been identified that cannot be infected by AAV. These results further suggest that the human megakaryocytic leukemia cells are resistant to infection by AAV and that AAV infection of human cells is receptor-mediated. More recently, an as yet unidentified host membrane protein has tentatively been implicated to mediate successful infection of human cells by AAV (Srivastava et al., 1995; Mizukami et al., 1996; C. Mah and A. Srivastava, unpublished results). The availability of this putative receptor, and the possibility that this receptor could be introduced into the potential target cells, should facilitate further studies on the virus–host cell interactions. The identification and characterization of the putative receptor for AAV also has important implications in the potential use of AAV-based vectors in human gene therapy.

Lytic Versus Latent Infection

Following infection of a permissive cell by AAV, at least one round of viral DNA synthesis must precede prior to initiation of viral gene expression because of the single-stranded nature of the viral genome. The palindromic nature of the viral ITRs facilitates their use as primers for viral DNA replication, which utilize the host cell DNA polymerases. In addition to the viral ITR sequences that serve as the origin of DNA replication (Lusby et al., 1980; Srivastava, 1987), the viral Rep proteins also play a crucial role in replication and resolution of the

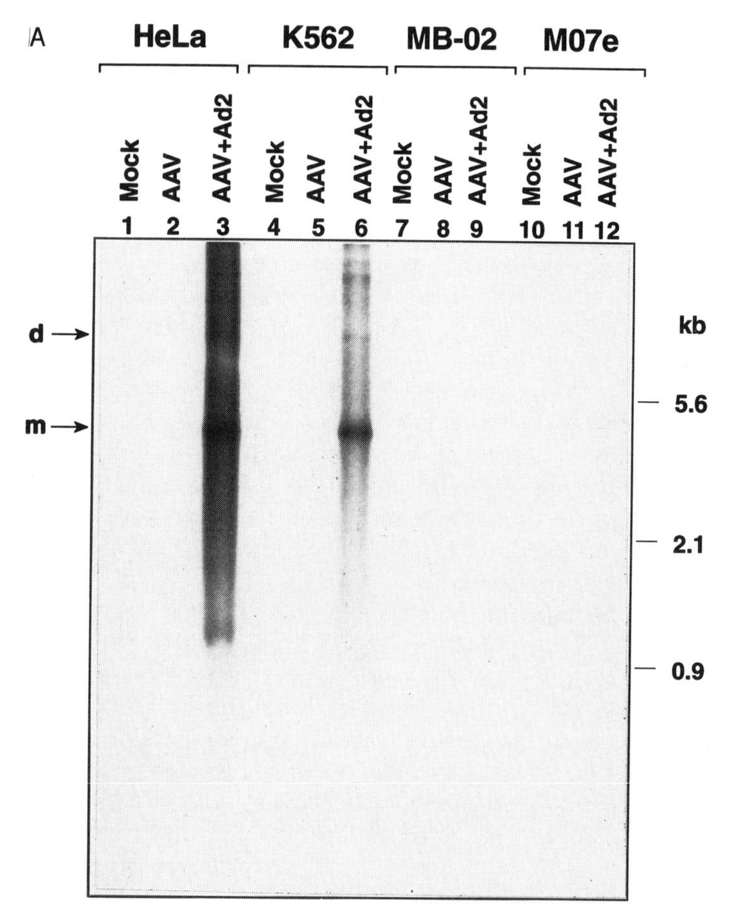

FIGURE 4. Southern blot analysis of replication of the wt AAV in HeLa, K562, MB-02, and M07e cells. **A:** Equivalent numbers of cells were either mock-infected (lanes 1, 4, 7, and 10), infected with AAV alone (lanes 2, 5, 8, and 11), or co-infected with AAV + Ad2 (lanes 3, 6, 9, and 12). Seventy-two hours postinfection, low M_r DNA samples were isolated by the method described by Hirt (1967) and analyzed on Southern blots (Southern, 1975) using a ^{32}P-labeled AAV-specific DNA probe. The AAV monomeric (m) and the dimeric (d) forms of the DNA replicative intermediates are indicated. **B:** Same as in A except that the filter was probed with a ^{32}P-labeled Ad2 DNA probe as previously described (Ponnazhagan et al., 1996).

AAV genome (Muzyczka, 1992). Expression of the major Rep proteins occurs from the p5 promoter (Hermonat et al., 1984). In addition, the Rep proteins are required for successful rescue and DNA replication of the cloned wt AAV genome in human cells in the presence of a helper virus (Samulski et al., 1982, 1983; Nahreini and Srivastava, 1989, 1992) since AAV genomes with mutations in the rep gene are defective for viral DNA replication (Hermonat et al., 1984; Tratschin et al., 1984; Owens and Carter, 1992).

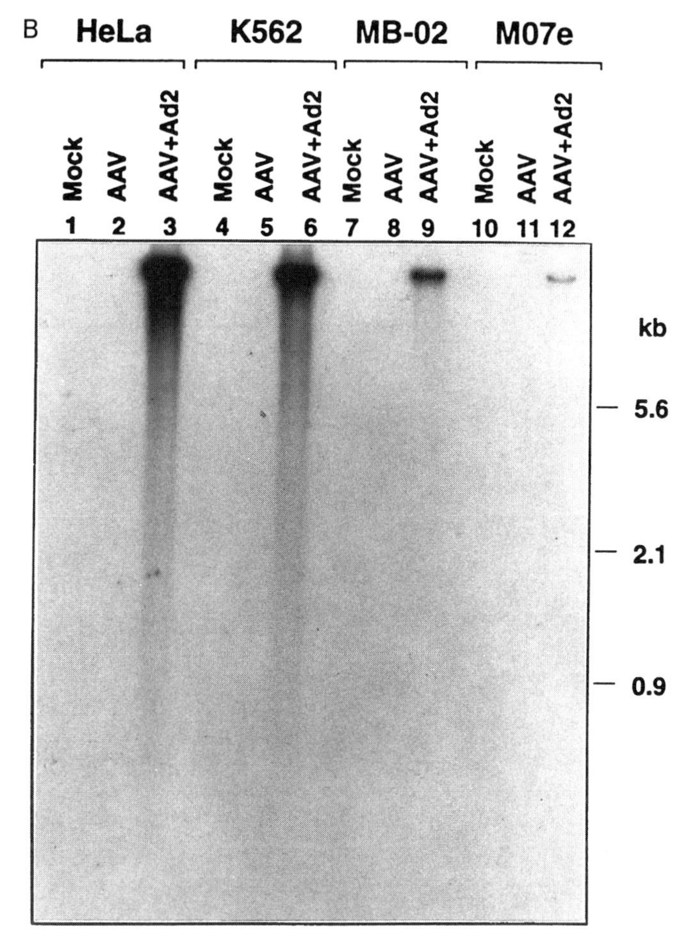

FIGURE 4. (*Continued*).

The original model for replication of linear, single-stranded DNA suggested by Cavalier-Smith (1974) and all available data on AAV DNA replication share remarkable similarities (Srivastava, 1987). Two of the viral Rep proteins (Rep78 and Rep68) are site-specific and strand-specific endonucleases that specifically bind to and cleave at the terminal resolution site (trs) within the AAV ITRs (Ashktorab and Srivastava, 1989; Im and Muzyczka, 1989, 1990, 1992; Snyder et al., 1990). The terminal 125 nucleotides form a palindrome that can fold back on itself to form a T-shaped hairpin (HP) structure. The terminal hairpin is used as a primer for initiation of viral DNA replication (Lusby et al., 1980; Berns and Bohenzky, 1987; Srivastava, 1987; Muzyczka, 1992). Previous *in vivo* and *in vitro* studies have demonstrated that the intact ITRs are required in *cis* for AAV DNA replication as well as for rescue or excision from prokaryotic plasmids (Samulski et al., 1982, 1983; Senapathy et

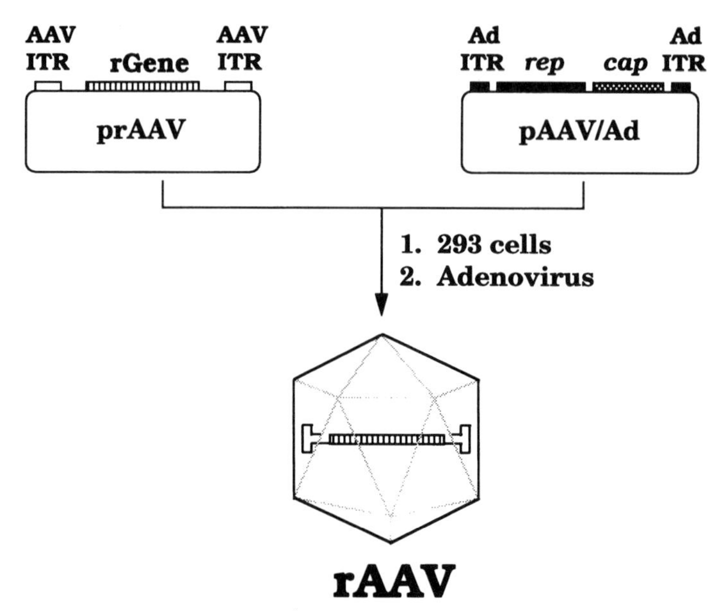

FIGURE 5. Strategy for generating recombinant AAV. Recombinant AAV vectors are generated by co-transfecting a plasmid containing the gene of interest between the AAV ITRs (open boxes) and a plasmid containing the wt AAV coding sequences between the adenovirus ITRs (closed boxes) in adenovirus-infected human cells. Recombinant AAV can be physically separated from the wt adenovirus by centrifugation on CsCl density gradients (Wang et al., 1995b).

al., 1984; Gottlieb and Muzyczka, 1988; Hong et al., 1992, 1994; Ni et al., 1994). The ITR also contains a stretch of 20 nucleotides, designated the D sequence, that is not involved in the hairpin formation. Recently, Wang et al. (1995a, 1996b) provided experimental evidence to suggest that the D sequence may be the packaging signal for AAV. Furthermore, although previous studies have documented that the AAV Rep proteins specifically interact with the AAV ITRs that are present in a cruciform structure (Ashktorab and Srivastava, 1989; Im and Muzyczka, 1989) and catalyze replication and resolution of the viral genome (Im and Muzyczka, 1990, 1992), the existence of hitherto unknown cellular proteins that specifically interact with the D sequence has also been demonstrated (Wang et al., 1996b; Qing et al., 1997, 1998). It is highly likely that these cellular proteins are specifically recruited by AAV to ensure efficient replication and encapsidation of the viral genomes (Wang et al., 1996b; Qing et al., 1997, 1998). Several early gene products of adenovirus have been implicated in the regulation of AAV gene expression (Berns and Bohenzky, 1987). The p5 promoter also contains two binding sites for the host cell transcription factor YY1, and the adenovirus E1A protein has been shown to relieve the YY1-mediated negative regulation of expression from the p5 promoter (Shi et al., 1991).

Following viral DNA replication and Cap protein synthesis, progeny virions are assembled in infected cell nuclei. Although this poorly understood process is believed to be spontaneous, recent studies from two laboratories have begun to examine the AAV assembly in some detail (Hölscher et al., 1994, 1995; Wistuba et al., 1995; Prasad and Trempe, 1995). In addition, Wang et al. (1996b) have identified the packaging signal for the AAV genome, and studies by Kube et al. (1997) have revealed that the viral Rep proteins are also encapsidated in mature progeny virions. Because productive replication of the helper virus leads to a lytic response, the role of AAV in this process is not entirely clear (Berns and Linden, 1995).

In the absence of a helper virus, the wt AAV genome becomes integrated with the host cell chromosomal DNA in a site-specific manner (Kotin et al., 1990, 1991; Samulski et al., 1991). Little expression from any of the AAV promoters occurs in the absence of a helper virus (Berns and Linden, 1995). The precise underlying molecular mechanism by which the remarkable site specificity of integration is accomplished remains unknown, but it is becoming clear that AAV Rep proteins and specific sequences in the viral ITR are pivotal (Kotin et al., 1992; Weitzman et al., 1994).

RECOMBINANT AAV VECTORS

The nonpathogenic nature of AAV has prompted a number of investigators to develop AAV as an alternative to the more commonly used retrovirus- and adenovirus-based vectors. However, the general concept that recombinant AAV vectors are also nonpathogenic as well as potentially capable of site-specific delivery of therapeutic genes has not been rigorously examined. Furthermore, one of the major limitations in utilizing AAV vectors for human gene therapy has been the relative difficulties in generating large quantities of high titer vector stocks. It is also well known that during a natural infection with the wt AAV, defective-interfering (DI) particles are generated (Hauswirth and Berns, 1979). It has been estimated that the defective:infectious particle ratio for the wt AAV ranges between 20 and 200 (Muzyczka, 1992), but this ratio for the recombinant AAV vectors has been reported to vary widely depending on the methodology employed in different laboratories. A brief account of these issues follows.

Strategies of Construction

The general strategy to construct recombinant AAV vectors is outlined in Figure 5. Briefly, a gene of interest molecularly cloned between the two AAV ITRs (Hermonat and Muzyczka, 1984; Samulski et al., 1987; Nahreini et al., 1993; Srivastava et al., 1989) in a recombinant bacterial plasmid (prAAV) is co-transfected into adenovirus-infected human 293 cells with a helper plasmid (pAAV/Ad) that contains the entire AAV coding sequences but lacks the

AAV ITRs, flanked by the adenovirus ITRs (Samulski et al., 1989). The AAV genome in the helper plasmid provides all the viral gene products in *trans,* but is itself unable to undergo DNA rescue and replication. The Rep proteins mediate the rescue and replication of the recombinant AAV genome followed by encapsidation into recombinant AAV progeny virions. In early studies, AAV DNA sequences containing insertions that were too large to be packaged into progeny virions were used as helper plasmids (Hermonat and Muzyczka, 1984). However, because of extensive DNA sequence homology in the ITR sequences between the recombinant and the helper plasmids, this led to generation of the wt AAV (McLaughlin et al., 1988).

Samulski et al. (1987, 1989) devised a strategy to circumvent this problem by removing the homologous sequences between the recombinant and the helper plasmids, but it appears that low levels of contamination with AAV, which resembles the wt, in some preparations of highly purified recombinant AAV stocks can still be detected (Flotte et al., 1995; Kube et al., 1997). Recent studies suggest that non-homologous recombination between the recombinant and the helper plasmids leads to the generation of contaminating AAV that is replication competent in the presence of co-infection with adenovirus (Wang et al., 1998). Alternative strategies to generate recombinant AAV stocks that are completely free of wt AAV have also been developed (Wang et al., 1998).

High Titer Stocks and DI Particles

A number of investigators have attempted to circumvent the potential problem of relatively low titers of recombinant AAV by using a variety of means. For example, Ponnazhagan et al. (1995a) have described several strategies that include the elimination of the need for two-plasmid co-transfection by inserting the AAV rep and cap gene sequences in the same recombinant AAV plasmid, development of helper plasmids that undergo rescue and replication but are unable to get packaged, and development of AAV rep$^+$ cell lines that produce high levels of Rep proteins following adenovirus infection that mediate rescue and replication of the recombinant AAV genomes. Kube et al. (1997) have described additional refinements in transfection protocols and subsequent purification of progeny virions on sucrose cushion and CsCl density gradients that have facilitated the production of recombinant viral titers ranging between 10^{11} and 10^{12} particles/ml. Maxwell et al. (1995) have also been able to increase the yield of recombinant AAV by using NB324K cells and dl309 mutant adenovirus. An improved method of obtaining tiers in the range of 10^{11} ml has also been described by Flotte et al. (1995) that utilizes the HIV LTR promoter to drive the expression of the AAV rep gene and cell lines that contain rescuable recombinant AAV genomes. Similarly, Chiorini et al. (1995) have described a method utilizing an SV40 replicon to amplify the AAV structural genes in cos cells that yields 60-fold higher viral titers than a nonreplicating helper plasmid, and Mamounas et al. (1995) have utilized the adenovirus–polylysine DNA complex system to obtain an increase of viral titers by two orders of magnitude over the conventional method.

The development of a number of different AAV rescue and packaging cell lines that can be induced to express the viral proteins have recently been reported by several investigators (Hölscher et al., 1994, 1995; Q. Yang et al., 1994; Ponnazhagan et al., 1995a; Luhovy et al., 1996; Tamayose et al., 1996). The development of a "semipackaging" (Luhovy et al., 1996) and an AAV packaging cell line (Clark et al., 1995; Tamayose et al., 1996) that yield transducing viral titers ranging between 10^6 and 10^9/ml has been reported as well. Further refinements of these systems may yield viral titers exceeding those generated by the conventional two-plasmid co-transfection method (Samulski et al., 1989). Finally, a novel means to generate high titer (10^{11}–10^{12} particles/plate) recombinant AAV in the absence of co-infection by adenovirus has been developed by Colosi et al. (1995). Thus, it appears that in due course the availability of high titer, clinical-grade recombinant AAV will no longer be a limitation.

Although the defective:infectious ratio for the wt AAV ranges between 20 and 200 (Muzyczka, 1992), it is intriguing that the same ratio for the recombinant AAV has been reported to be as high as 10^6 (Russell et al., 1995). The molecular basis for this apparent discrepancy has been examined by Wang et al. (1996a), who have carried out rescue and replication of the wt AAV genome from recombinant plasmids in adenovirus-infected human cells because, in contrast to infection with wt AAV, rescue of the recombinant AAV genome must first occur from a plasmid. In recent *in vitro* studies, the purified Rep68 protein has been shown to bind not only to the ITR but also to some linear DNA sequences such as the A sequence and the AAVp5, AAVp19, and AAVp40 promoter sequences in the viral genome (McCarty et al., 1994a,b). In addition to Rep binding to the Rep-binding site (RBS) in AAVp5 promoter, which is believed to be involved in Rep gene expression (Kÿostio et al., 1995), and AAV DNA integration (Giraud et al., 1995), Rep binding followed by cleavage of these sequences in adenovirus-infected human cells has also been detected (Wang et al., 1996a; Wang and Srivastava, 1997). Thus, it is clear that additional putative "trs-like" sequences in the AAV genome exist other than that present within the ITRs that are utilized in the Rep-mediated cleavage of the viral genome during a natural AAV infection. These results provide further evidence that the "trs-like" site near the AAV promoter sequences also constitutes a binding site for the AAV Rep proteins *in vivo* since Rep-mediated cleavage could be abolished following deletion of the RBS (Wang et al., 1996a; Wang and Srivastava, 1997) previously shown to bind the AAV Rep proteins *in vitro* (McCarty et al., 1994a,b). Interestingly, Wang et al. (1996b) have also documented that wt AAV genomes containing only one ITR sequence can be successfully packaged into progeny virions. Taken together, these studies are beginning to provide some clues as to how the naturally occurring AAV DI particles might be generated and why only 1 out of 20–200 wt AAV particles is infectious. Thus, by deliberately mutating the putative RBS in the recombinant AAV genomes, without altering the transgene coding sequences, it may be feasible to significantly diminish the production of DI particles.

AAV-MEDIATED TRANSDUCTION AND GENE EXPRESSION

A number of investigators have utilized the AAV-based vector system to successfully transduce and obtain short- and long-term expression of a variety of reporter genes *in vitro* as well as *in vivo* (Muzyczka, 1992; Srivastava, 1994; Kotin, 1994; Flotte and Carter, 1995; Berns and Linden, 1995). A number of potentially therapeutic gene sequences that have been transduced using this vector system are listed in Table 1. However, whether AAV is capable of transducing nondividing cells has been a subject of some debate (Podsakoff

TABLE 1 AAV-Mediated Transduction of Potentially Therapeutic Genes

Genes	References
Antisense α-globin	Ponnazhagan et al. (1994a)
Antisense HIV	Chatterjee et al. (1992)
Arylsulfatase A	Wei et al. (1994)
B7-2[a]	Chiorini et al. (1995)
β-Globin	Dixit et al. (1991), Einerhand et al. (1995), Zhou et al. (1996)
CFTR[b]	Egan et al. (1992), Flotte et al. (1993a,b)
FAC-C[c]	Walsh et al. (1993)
Flt-3[d]	Broxmeyer et al. (1996)
FR[e]	Sun et al. (1995)
γ-Globin	Walsh et al. (1992), Miller et al. (1994), Ponnazhagan et al. (1997c)
Glucocerebrosidase	Wei et al. (1994)
GM-CSF[f]	Luo et al. (1995)
IL-2	Philip et al. (1994)
MDR-1[g]	Shaughnessy et al. (1995)
NADPH-oxidase	Thrasher et al. (1995)
Neuropeptide Y	Wu et al. (1994)
TH[h]	Kaplitt et al. (1994)
TK	Su et al. (1996)

[a] T-cell co-stimulatory protein.
[b] Cystic fibrosis trans-membrane conductance regulator.
[c] Fanconi's anemia–complementation group C.
[d] Ligand for flt-3/flk-2 tyrosine kinase receptor.
[e] Folate receptor.
[f] Granulocyte-macrophage colony-stimulating factor.
[g] Multidrug resistance.
[h] Tyrosine hydroxylase.

et al., 1994b; Russell et al., 1994; Alexander et al., 1994; Kaplitt et al., 1994). Whereas it is clear that DNA-damaging agents can greatly increase AAV-mediated transduction (Alexander et al., 1994, 1996), it appears that the lack of efficient transduction may also be due to the rate-limiting second-strand DNA synthesis step of the recombinant AAV genome (Fisher et al., 1996; Ferrari et al., 1996). Thus, it is becoming clear that the efficiency of AAV-mediated transduction may be cell type specific (Ponnazhagan et al., 1996b,d; Qing et al., 1997, 1998). Further studies are warranted to examine this issue. Also, since one of the ultimate goals of successful gene therapy is to be able to deliver a therapeutic gene of interest directly into the body, studies on the fate of AAV vectors *in vivo* are needed as well.

Direct Injection *In Vivo*

Although a number of studies have reported successful AAV-mediated transduction and expression of therapeutic genes *in vitro* (Chatterjee et al., 1992; Walsh et al., 1992, 1994; Flotte et al., 1993b; Ponnazhagan et al., 1994a; 1997b, 1997c; Miller et al., 1994; Einerhand et al., 1995; Luo et al., 1995; Zhou et al., 1996), few studies have examined the safety and efficacy of the AAV vectors *in vivo* (Flotte et al., 1993a; Kaplitt et al., 1994). Using a murine model system *in vivo,* Ponnazhagan et al. (1997b) have systematically evaluated the fate of the AAV vectors following direct intravenous (i.v.) injection of AAV. Two different highly purified recombinant AAV, the cytomegalovirus (CMV) promoter-driven β-galactosidase gene (vCMVp-lacZ) and the human β-globin promoter-driven human $^{A}\gamma$-globin gene containing the DNase hypersensitive-site 2 (HS-2) enhancer element (Tuan et al., 1989) from the locus control region (LCR) from the human β-globin gene cluster (vHS2-βp-$^{A}\gamma$-globin), were utilized. Approximately 1×10^{10} viral particles of vCMVp-lacZ were injected i.v. into the tail vein of C57BL/6 mice. These animals were sacrificed at various times postinjection, and equivalent amounts of tissues from various organs were examined for the presence of recombinant viral genome by polymerase chain reaction (PCR) amplification using the lacZ-specific primer-pair followed by Southern blot analysis.

The results shown in Figure 6 document that the recombinant AAV genomes were detected predominantly in the liver tissues up to 1 week postinjection in each group of animals, suggesting that AAV may possess *in vivo* organ tropism for liver. These results were corroborated by injecting the recombinant vHS2-βp-$^{A}\gamma$-globin virions under identical conditions and by examining tissues from various organs 7 weeks postinjection using the β-globin promoter-$^{A}\gamma$-globin gene-specific primer-pair as described above. The results shown in Figure 7 further document that liver cells accumulate the bulk of the recombinant AAV followed by lung and heart. The extent of liver-specific human globin gene delivery was determined by semiquantitative PCR analysis (Goodman et al., 1994) using the mouse β-actin gene sequences as a control. Approximately 4% of liver cells contained the transduced human globin gene 7 weeks

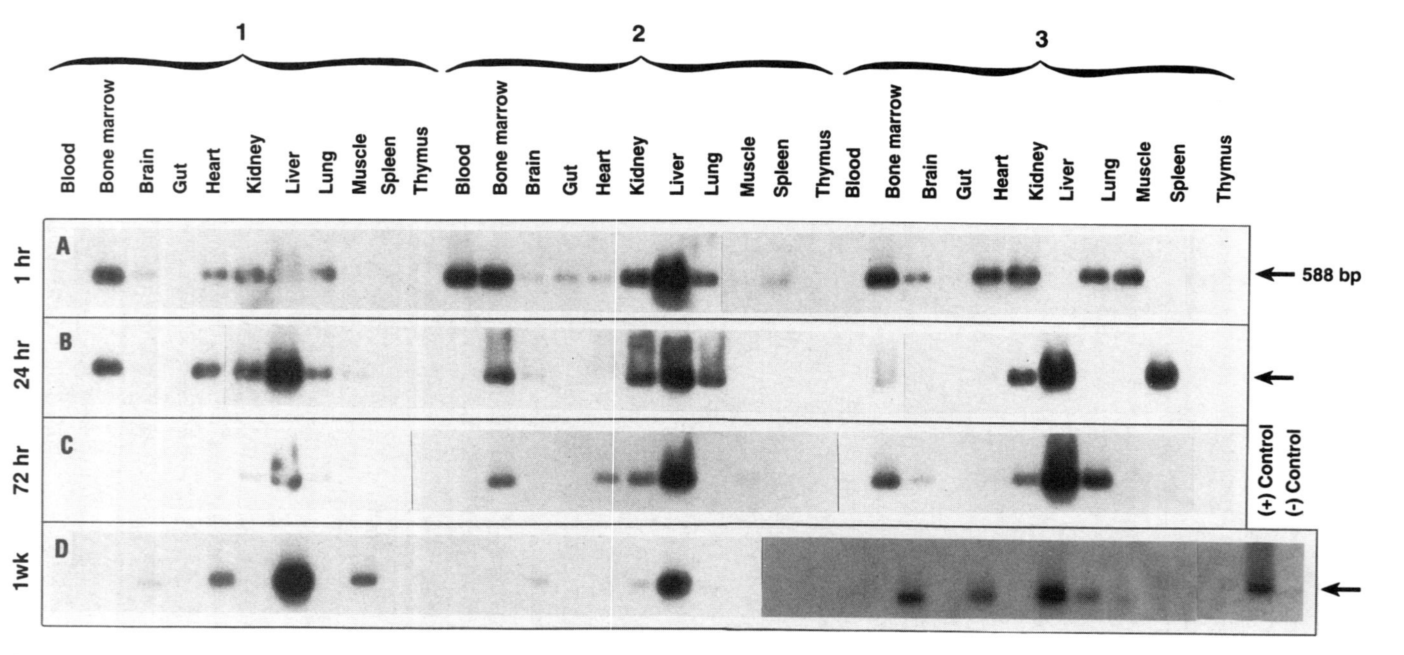

FIGURE 6. Southern blot analysis of PCR-amplified DNA fragments of the recombinant vCMVp-lacZ genome in various tissues of mice. Approximately 1×10^{10} particles of the recombinant vCMVp-lacZ virus were injected into the tail vein of 8-week-old C57BL/6 mice. Three animals per group were sacrificed at indicates times postinjection (1 h, 24 h, 72 h, 1 week; **A–D**) Individual tissues and organs were obtained and rinsed extensively with phosphate-buffered saline, and equivalent amounts were used in a 35-cycle PCR amplification reaction using the lacZ-specific primer pair (5′-GATGAGCGTGGTGGTTATG-3′, 5′-TACAGCGCGTCGTGATTAG-3′). Plasmids pCMVp-lacZ (Ponnazhagan et al., 1997b) and pUC19 (Sambrook et al., 1989) were used as positive and negative controls, respectively. The PCR products were analyzed on Southern blots using a lacZ-specific ^{32}P-labeled DNA probe as previously described (Ponnazhagan et al., 1997b). Arrows indicate the 588 bp lacZ-specific DNA fragment.

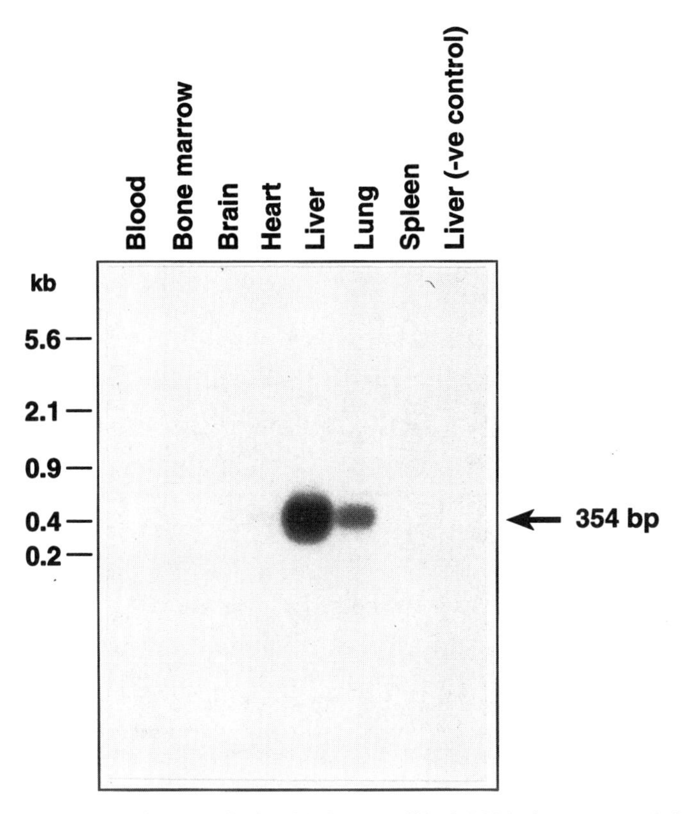

FIGURE 7. Southern blot analysis of PCR-amplified DNA fragments of the recombinant vHS2-βp-$^{A}\gamma$-globin genome in various tissues of mice. Approximately 1×10^{10} particles of the recombinant vHS2-βp-$^{A}\gamma$-globin virus were injected i.v. as described in the legend to Fig. 6. Seven weeks postinjection, the indicated organs were obtained and analyzed for the presence of the recombinant viral genome using the human β-globin promoter (5′-GATGGTATGGGGCCAAGAGA-3′)-specific and $^{A}\gamma$-globin gene (5′-GGGTTTCTCCTCCAGCATCT-3′)-specific oligonucleotide primer-pair. Liver tissues obtained from a mock-injected mouse was also included as a negative control. The arrow indicates the 354 bp human γ-globin–specific DNA fragment (Ponnazhagan et al., 1997b).

postinjection (Ponnazhagan et al., 1997b). The transcription potential of the lacZ gene delivered by direct injection of the recombinant AAV was also determined. Whereas no expression was observed in liver cells from mock-injected animals, lacZ gene expression was readily detected in hepatocytes in liver from animals injected with vCMVp-lacZ (Ponnazhagan et al., 1997b). Interestingly, however, a cytotoxic T lymphocyte (CTL) response against β-galactosidase expressed from recombinant AAV in these mice 1 week postinjection could not be detected. These results are shown in Figure 8. It remains possible, however, that the level of β-galactosidase expression was insufficient

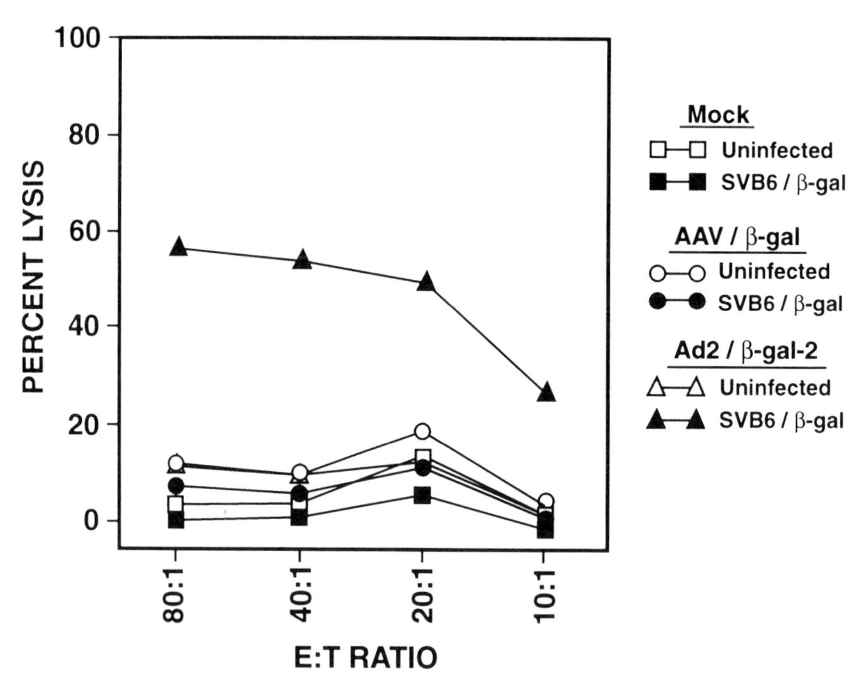

FIGURE 8. Cytotoxic T lymphocyte assays. Single cell suspensions from spleens obtained from three mock-injected, or three vCMVp-lacZ-injected mice at 7 days postinjection were re-stimulated *in vitro* with syngeneic SV40-transformed C57BL/6 fibroblasts (SVB6KHA fibroblasts) infected with a β-galactosidase–encoding adenovirus vector (AD2/β-gal-2) to stimulate expansion of CTLs specific for β-galactosidase. Spleen cells were cultured for 6 days at 37°C in 5% CO_2 and were assessed for cytolytic activity against uninfected SVB6KHA fibroblasts (background lysis) or β-galactosidase–expressing SVB6KHA/β-galactosidase fibroblasts transfected with a retrovirus vector encoding β-galactosidase (β-galactosidase–specific lysis). All targets were treated with recombinant mouse interferon-γ for approximately 24 h (to maximize MHC class I expression and antigen presentation) and labeled with ^{51}Chromium overnight before the CTL assay. Effector cells were plated in triplicate with the target cells at various effector:target (E : T) ratios and incubated for 5 h at 37°C in 5% CO_2. A 100 μl aliquot of cell-free supernatant was collected from each well and counted in a gammacounter. The percent specific lysis was calculated as follows:

$$\% \text{ Lysis} = \frac{(\text{cpm in test sample}) - (\text{spontaneous cpm})}{(\text{total cpm}) - (\text{spontaneous cpm})} \times 100$$

Spontaneous chromium release was assessed by incubating target cells alone in medium, and the total amount of chromium incorporated by the targets was determined by adding 1% Triton X-100 to the target cells. The SVB6KHA/β-galactosidase targets were confirmed to express β-gal by X-galactosidase staining. In a separate CTL assay performed concurrently with effector cells from mice instilled with an adenovirus vector encoding β-galactosidase, using the same reagents and target cells, significant lysis of SVB6KHA/β-galactosidase fibroblasts was observed, indicating that the conditions of the assay were adequate to detect β-galactosidase–specific lysis (Ponnazhagan et al., 1997b).

to induce a CTL response since *Escherichia coli* β-galactosidase can serve as a CTL target in mice (Gavin et al., 1993) and is immunogenic when expressed at high levels from a recombinant adenovirus vector (Fig. 8) (Van Ginkel et al., 1995).

Since AAV infection involves a hitherto unknown host cell receptor (Ponnazhagan et al., 1996), it is perhaps not surprising that organ tropism of AAV was evident, suggesting that murine liver cells may express the highest number of the putative receptor. Whether human liver cells also show preferential transducibility remains to be determined. If so, AAV-mediated liver-specific gene delivery may be feasible. It is intriguing, however, that the presence of the recombinant viral genome in thymus tissue of any of the animals was not detected, the significance of which is currently unknown. Further studies with additional recombinant viral vectors are needed to establish organ-tropism of AAV as well as to determine whether the proviral genome undergoes stable integration into the host chromosomal DNA or remains episomal.

Ex Vivo Transduction

The efficiency of AAV-mediated transduction of murine hemopoietic progenitor cells *ex vivo* and the potential of expression of the transduced gene *in vivo* following transplantation into recipient mice have also been evaluated (Ponnazhagan et al., 1996b). Highly purified murine hematopoietic progenitor cells (Sca-1^+, lin^-) obtained from C57BL/6 mice were either mock-infected or infected with the recombinant vCMVp-lacZ virions. Approximately 250 cells were transplanted into lethally-irradiated syngeneic recipients. Twelve days post-transplantation, the animals were sacrificed. Spleens from these animals were obtained and enumerated for colony formation (Yoder et al., 1993). Individual spleen colonies were examined for the presence of the transgene by PCR analysis as described above. The results shown in Figure 9 for 10 such colonies indicated the presence of transduced gene sequences only in mice transplanted with the vCMVp-lacZ–infected cells. Equivalent numbers of cells from spleen colonies obtained from these mice were also examined for expression of the transduced lacZ gene by fluorescence-activated cell-sorting (FACS). These results shown in Figure 10 for 10 individual colonies clearly indicated expression of the transgene only in vCMVp-lacZ–infected cells.

The nonpathogenic nature of AAV vectors was further evidenced by the fact that AAV infection had no effect on the spleen colony formation by murine hemopoietic progenitor cells. Although abundant short-term expression of the transgene was evident following both direct injection and *ex vivo* transduction and transplantation, it remains unclear whether the recombinant AAV genome was stably integrated into the host chromosomal DNA. Although further studies are needed to address this issue, Ponnazhagan et al. (1997c), using a serial bone marrow transplant model, have obtained evidence to indicate the potential of long-term *in vivo* expression of the AAV-mediated

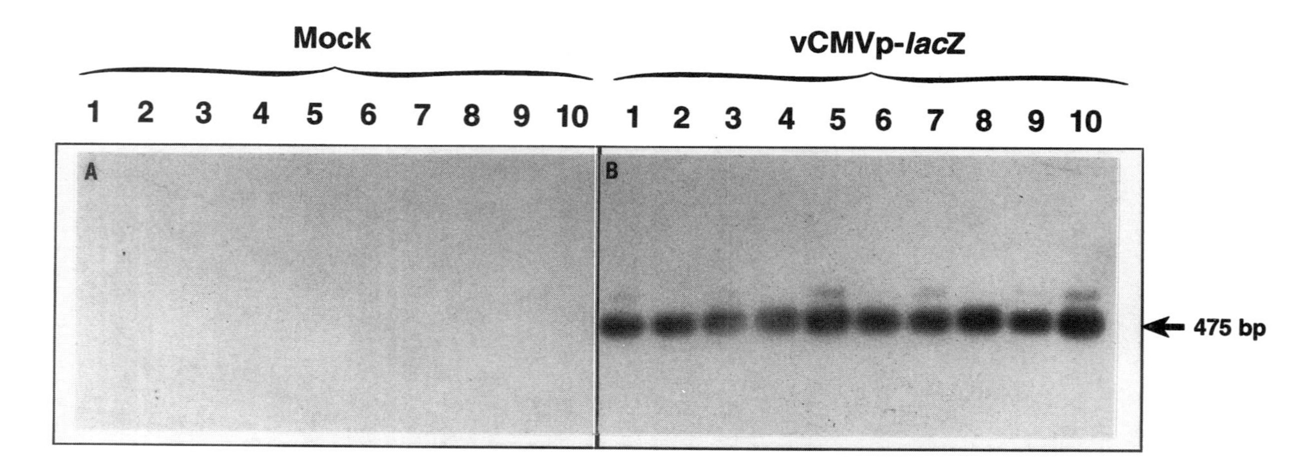

FIGURE 9. Southern blot analysis of PCR-amplified DNA fragments of the recombinant vCMVp-lacZ genome in spleen colonies from mice transplanted with primitive hematopoietic progenitor cells. Sca-1$^+$, lin$^-$ cells enriched in murine hematopoietic stem and progenitor cells were obtained from low-density bone marrow mononuclear cells from adult C57BL/6 mice using flow cytometry as previously described (Yoder et al., 1993). Equivalent numbers of these cells were either mock-infected or infected with the vCMVp-lacZ and injected into the tail vein of lethally-irradiated recipient mice (250 cells/mouse). Twelve days post-transplantation, mice were sacrificed, spleen colonies were enumerated, and individual colonies from several mice were obtained. Ten colonies each from mice transplanted either with mock-infected cells (**A**) or with the vCMVp-lacZ–infected cells (**B**) were analyzed by PCR followed by Southern blotting. The arrow indicates the 475 bp DNA fragment generated by PCR using the CMV promoter- and lacZ-specific primer-pair (Ponnazhagan et al., 1997).

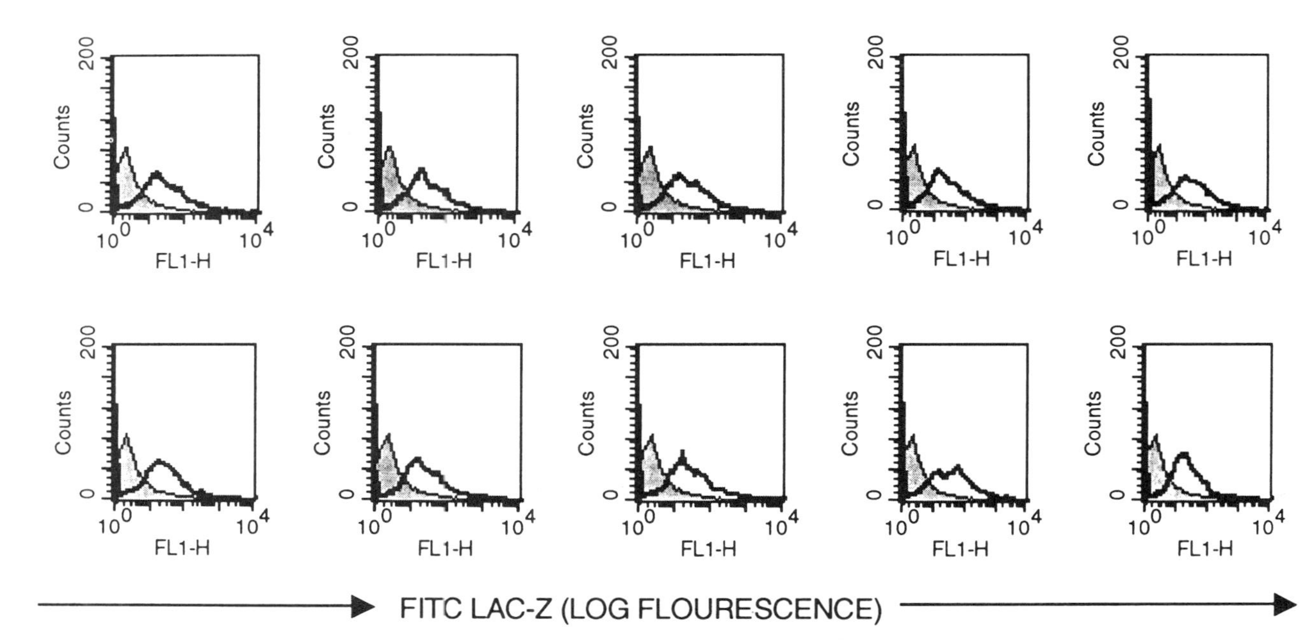

FIGURE 10. Flow cytometric analysis of production of the β-galactosidase protein in spleen cell colonies. Equivalent numbers of cells from 10 colonies each from mice transplanted either with mock-infected cells (shaded) or with the vCMVp-lacZ–infected cells (clear) were incubated with Imagene green C_{12}FDG β-galactosidase substrate (Molecular Probes Inc., Eugene, OR) for 30 min at 37°C. Following centrifugation, the cells were analyzed using a Becton-Dickinson Fluorescence-Activated Cell Scanner as previously described (Clapp et al., 1995). Counts refer to the number of events per minute (Ponnazhagan et al., 1997b).

transduced human globin gene, implying stable integration of the recombinant viral genome in murine hemopoietic stem cells.

Based on these results, it can be concluded that highly purified recombinant AAV vectors directly injected i.v. into mice predominantly target the liver. The transgene expression occurs in hepatocytes without a detectable CTL response. These results raise the possibility that liver-specific gene delivery may be feasible, which, in turn, may prove useful for the potential gene therapy of a variety of liver diseases such as familial hypercholesterolemia, α_1-antitrypsin deficiency, phenylketonuria, and hemophilia B (Grossman et al., 1992). Recombinant AAV vectors can also be successfully used for high-efficiency transduction of murine hematopoietic stem cells *ex vivo* without any cytotoxic effect, and these cells can be transplanted in mice, leading to the generation of progeny cells that not only retain the recombinant proviral genome, but also allow expression of the transduced gene. Since no adverse effects are evident following either direct viral injection or reconstitution with AAV-infected cells, these studies strongly suggest that AAV vectors may prove to be a safe and effective alternative to the more commonly used retrovirus- and adenovirus-based vectors. Further studies with a nonhuman primate model system are warranted to establish the safety and efficacy of the AAV-based vector system prior to its potential use in human gene therapy.

Finally, although previous studies by Kotin et al. (1990, 1991, 1992), Kotin and Berns (1989), and Samulski et al. (1991) have established that the wt AAV genome integrates site specifically on human chromosome 19q13.3-qter at a very high frequency in established human cell lines, the site specificity of integration of recombinant AAV vectors has not been documented (Xiao et al., 1993; Goodman et al., 1994; Kearns et al., 1996; Ponnazhagan et al., 1997a). Previous studies with the wt AAV have also revealed that integrated proviral genomes are almost always present in linear, tandem repeats (Giraud et al., 1994, 1995). However, in an *in vivo* rhesus macaque model system, the recombinant AAV vector sequences have been shown to persist in an episomal form for up to 34 months (Afione et al., 1996). Interestingly, in murine bone marrow serial transplant model systems, sustained presence of, and expression from, the recombinant proviral genomes in progeny hematopoietic cells approximately 1 year post-transplantation have been documented (Podsakoff et al., 1994a; Ponnazhagan et al., 1995b, 1997c). These studies strongly suggest that successful transduction and, by inference, stable integration of the proviral genome into chromosomal DNA of pluripotent hematopoietic stem cells does indeed occur. These observations underscore the need to critically evaluate the recombinant AAV–diploid cell interaction since primary cells are the most likely targets for AAV-mediated gene transfer in a gene therapy setting. Further studies are therefore warranted to systematically investigate the molecular interaction of recombinant AAV with human cells in general and diploid cells in particular (Srivastava, 1994; Zhou et al., 1993, 1994).

ADVANTAGES AND DISADVANTAGES OF AAV VECTORS

It is becoming increasingly clear that the more commonly used retrovirus- and adenovirus-based vectors may be associated with certain complications. For example, retroviruses are associated with neoplastic events (Donahue et al., 1992) and adenoviruses induce a CTL response (Y. Yang et al., 1994). Thus, the search is on for an alternative vector that is neither pathogenic nor immunogenic (Hodgson, 1995; Miller and Vile, 1995). In contrast to retroviruses and all other DNA-containing viruses, parvoviruses remain the only group of viruses that have to date not been associated with any malignant disease and in fact have been shown to possess antitumor properties. Specifically, the nonpathogenic nature of AAV makes this virus extremely attractive for the purpose of developing vectors for human gene therapy. The following brief account lends credence to the argument that the advantages of the AAV-based vector system for human gene therapy might outnumber its disadvantages.

Host Range and Safety and Efficacy Issues

AAV has thus far been shown to possess an extremely broad host range that transcends species barriers (Muzyczka, 1992), with the possible exception of human megakaryocytic leukemia cells (Ponnazhagan et al., 1996). Additional desirable features of AAV vectors include their ability to transduce nondividing cells as well as stable and potentially nonrandom integration of the proviral genome in human cells. Although vector mobilization remains a finite possibility, it is less likely since simultaneous infections by wt AAV and adenovirus would be required. Rescue of the integrated proviral genome has been shown to occur in only approximately 30% of the transduced cells, at least *in vitro* (McLaughlin et al., 1988). Superinfection immunity by AAV, on the other hand, has not been observed, at least *ex vivo* (Lebkowski et al., 1988), suggesting that multiple transductions might be feasible. Since AAV is of human origin, it is conceivable that, in contrast to murine retroviral vectors, AAV-based vectors might be physiologically more relevant for gene therapy in humans.

The possibility of an adverse immune response to AAV vectors *in vivo* appears to be greatly diminished (Ponnazhagan et al., 1997b), in contrast to adenovirus-based vectors, since none of the wt AAV gene sequences is present in most of the recombinant AAV vectors. However, because 80%–90% of the human population is seropositive for AAV, successful transduction of human cells *in vivo* would be a formidable, if not insurmountable, challenge. These problems notwithstanding, recent success with animal models *in vivo* documenting long-term expression of AAV-mediated transduced genes without significant pathological consequences (Flotte et al., 1993a; 1993b; Kaplitt et al., 1994; Chatterjee et al., 1995; Broxmeyer et al., 1995; Ponnazhagan et al.,

1997b, 1997c) strongly suggests that the AAV-based vector system may prove to be safe and effective in human gene therapy.

Large-Scale Production and Size Limitation

Because of the small size of AAV, the main limitation of AAV vectors is the size of a DNA sequence that can be packaged into mature virions. However, the use of cDNA copies of most, if not all, therapeutic genes may circumvent this problem. The problem of large-scale, commercial production of high titer recombinant AAV vectors does not now appear to be as unsurmountable as was initially perceived. Although contamination with low-levels of wt AAV in some of the recombinant vectors has been detected, it remains to be determined whether this would be detrimental in a gene therapy setting. Finally, the heat stability, resistance to lipid solvents, and stability between pH 3.0 and 9.0 should facilitate worldwide application of the AAV-based vector system for human gene therapy.

SUMMARY

From the foregoing discussion, it is evident that the AAV-based vector system has the potential to be a safe and effective alternative to the more commonly used retrovirus- and adenovirus-based vectors in human gene therapy. However, it is also clear that a number of questions with regard to the basic biology of AAV still need to be resolved. Although it is clear that extensive studies on AAV vectors are currently being pursued in a number of laboratories, our own efforts have focused on the development of recombinant AAV vectors that not only mediate stable transduction but also allow tissue-specific expression of transduced genes (Srivastava et al., 1996). It is anticipated that the AAV-based vector system will prove useful in the potential gene therapy of a variety of human diseases in general and human hemoglobinopathies in particular.

FUTURE PROSPECTS

Further molecular analyses of the AAV–primary human cell interactions should provide new and useful information that will undoubtedly contribute to the further development of the AAV-based vector system. Detailed studies on the molecular mechanisms underlying viral assembly, viral entry into target cells, and stability of integration and long-term, regulated expression of transduced genes will also be critical in realizing the full potential of this vector system. Finally, a better understanding of the host immunological responses to AAV vectors will be crucial in achieving the safe and efficacious application of this unique vector system in gene therapy of a variety of clinical disorders in humans, both genetic and acquired, in hopefully the not too distant future.

ACKNOWLEDGMENTS

I thank Dr. Kenneth I. Berns for his encouragement and support. I also thank Dr. Richard J. Samulski for his kind gift of recombinant plasmids and the past and present members of my laboratory for helpful discussions.

The research in the author's laboratory was supported in part by grants from the National Institutes of Health (Public Health Service grants AI-26323, HL-48342, HL-53586, and DK-49218, Centers of Excellence in Molecular Hematology), Phi Beta Psi Sorority, and an Established Investigator Award from the American Heart Association.

REFERENCES

Afione SA, Conrad CK, Kearns, WG, Chunduru S, Adams R, Reynolds TC, Guggino WG, Cutting GR, Carter BJ, Flotte TR (1996): *In vivo* model of adeno-associated virus vector persistence and rescue. *J Virol* 70:3235–3241.

Alexander IA, Russell DW, Miller AD (1994): DNA-damaging agents greatly increase the transduction of nondividing cells by adeno-associated virus vectors. *J Virol* 68:8282–8287.

Alexander IA, Russell DW, Spence AM, Miller AD (1996): Effects of gamma irradiation on the transduction of dividing and nondividing cells in brain and muscle of rats by adeno-associated virus vectors. *Hum Gene Ther* 7:841–850.

Ashktorab H, Srivastava A (1989): Identification of nuclear proteins that specifically interact with adeno-associated virus type 2 inverted terminal repeat hairpin DNA. *J Virol* 63:3034–3039.

Berns KI (1990): Parvovirus replication. *Microbiol Rev* 54:316–329.

Berns KI, Bohenzky RA (1987): Adeno-associated viruses: An update. *Adv Virus Res* 32:243–306.

Berns KI, Hauswirth WW (1979): Adeno-associated viruses. *Adv Virus Res* 25:407–449.

Berns KI, Linden RM (1995): The cryptic life style of adeno-associated virus. *BioEssays* 17:237–245.

Blacklow NR (1988): Parvoviruses and Human Disease. Boca Raton: CRC Press.

Blaese RM, Culver KW, Miller AD, Carter CS, Fleisher T, Clerici M, Shearer G, Chang L, Chiang Y, Tolstoshev P, Greenblatt JJ, Rosenberg SA, Klein H, Berger M, Mullen CA, Ramsey WJ, Muul L, Morgan RA, Anderson WF (1995): T lymphocyte–directed gene therapy for ADA$^-$ SCID: Initial trial results after 4 years. *Science* 270:475–480.

Bordignon C, Notarangelo LG, Nobili N, Ferrari G, Casorati G, Panina P, Mazzolari E, Maggioni D, Rossi C, Servida P, Ugazio AG, Mavilio F (1995): Gene therapy in peripheral blood lymphocytes and bone marrow for ADA⁻ immunodeficient patients. *Science* 270:470–475.

Broxmeyer HE, Cooper S, Etienne-Julan M, Bruan S, Lu L, Lyman SD, Srivastava A (1996): Cord blood hemopoietic stem/progenitor cell growth and transduction using adeno-associated and retroviral vectors. Proceedings of the Colloquium on Foetal and Neonatal Hemopoiesis and the Mechanisms of Bone Marrow Failure. Montrouge, France, *John Libbey Eurotext, Ltd.* 235:109–114.

Broxmeyer HE, Cooper S, Etienne-Julan M, Wang X-S, Ponnazhagan S, Braun S, Lu L, Srivastava A (1995): Cord blood transplantation and the potential for gene therapy: Gene transduction using a recombinant adeno-associated viral vector. *Ann NY Acad Sci* 770:105–115.

Carter BJ (1993): Adeno-associated virus vectors. *Curr Opin Biotech* 3:533–539.

Cavalier-Smith T (1974): Palindromic base sequences and replication of eukaryotic chromosome ends. *Nature* 350:467–470.

Chatterjee S, Johnson PR, Wong KK (1992): Dual-target inhibition of HIV-1 by means of an adeno-associated virus antisense vector. *Science* 258:1485–1488.

Chatterjee S, Lu D, Podsakoff G, Wong KK (1995): Strategies for efficient gene transfer into hemopoietic cells: The use of adeno-associated virus vectors in gene therapy. *Ann NY Acad Sci* 770:79–90.

Cheung AKM, Hoggan MD, Hauswirth WW, Berns KI (1980): Integration of the adeno-associated virus genome into cellular DNA latently infected human Detroit 6 cells. *J Virol* 33:739–748.

Chiorini JA, Wendtner CM, Urcelay E, Safer B, Hallek M, Kotin RM (1995): High-efficiency transfer of the T cell co-stimulatory molecule B7-2 to lymphoid cells using high-titer recombinant adeno-associated virus vectors. *Hum Gene Ther* 6:1531–1541.

Clapp DW, Freie B, Srour E, Yoder MC, Fortney K, Gerson SL (1995): Myeloproliferative sarcoma virus directed expression of β-galactosidase following retroviral transduction of murine hemopoietic cells. *Exp Hematol* 23:630–638.

Clark KR, Voulgaropoulou F, Fraley DM, Johnson PR (1995): *Hum Gene Ther* 6:1329–1341.

Colosi P, Elliger S, Elliger C, Kurtzman G (1995): AAV vectors can be efficiently produced without helper virus. *Blood* 10:627a.

Cotmore SF, Tattersall PJ (1987): The autonomously replicating parvoviruses of vertebrates. *Adv Virus Res* 33:91–174.

Cukor G, Blacklow NR, Kibrick S, Swan IC (1975): Effect of adeno-associated virus on cancer expression by herpesvirus-transformed hamster cells. *J Natl Cancer Inst* 55:957–959.

Dixit M, Webb MS, Smart WC, Ohi S (1991): Construction and expression of a recombinant adeno-associated virus that harbors a human β-globin-encoding cDNA. *Gene* 104:253–257.

Donahue RE, Kessler SW, Bodine D, McDonagh K, Dunbar C, Goodman S, Agricola B, Byrne E, Raffeld M, Moen R, Bacher J, Zsebo KM, Nienhuis AW (1992): Helper virus induced T cell lymphoma in non-human primates after retroviral mediated gene transfer. *J Exp Med* 176:1125–1135.

Egan M, Flotte TR, Afione S, Solow R, Zeitlin PL, Carter BJ, Guggino WG (1992): Defective regulation of outwardly rectifying Cl^- channels by protein kinase A corrected by insertion of CFTR. *Nature* 358:581–584.

Einerhand MPW, Antoniou M, Zolotukhin S, Muzyczka N, Berns KI, Grosveld F, Valerio D (1995): Regulated high-level human beta-globin gene expression in erythroid cells following recombinant adeno-associated virus-mediated gene transfer. *Gene Ther* 2:336–343.

Ferrari FK, Samulski T, Shenk T, Samulski RJ (1996): Second-strand synthesis is a rate-limiting step for efficient transduction by recombinant adeno-associated virus vectors. *J Virol* 70:3227–3234.

Fisher KJ, Gao G-P, Weitzman MD, DeMatteo R, Burda JF, Wilson JM (1996): Transduction with recombinant adeno-associated virus for gene therapy is limited by leading-strand synthesis. *J Virol* 70:520–532.

Flotte TR, Afione SA, Conrad C, McGrath SA, Solow R, Oka H, Zeitlin P, Guggino WB, Carter BJ (1993a): Stable *in vivo* expression of the cystic fibrosis transmembrane conductance regulator with an adeno-associated virus vector. *Proc Natl Acad Sci USA* 90:10613–10617.

Flotte TR, Afione SA, Solow R, Drumm ML, Markakis D, Guggino WB, Zeitlin P, Carter BJ (1993b): Expression of the cystic fibrosis transmembrane conductance regulator from a novel adeno-associated virus promoter. *J Biol Chem* 268:3781–3790.

Flotte TR, Barraza-Ortiz X, Solow R, Afione SA, Carter BJ, Guggino WB (1995): An improved system for packaging recombinant adeno-associated virus vectors capable of *in vivo* transduction. *Gene Ther* 2:29–37.

Flotte TR, Carter BJ (1995): Adeno-associated virus vectors for gene therapy. *Gene Ther* 2:357–362.

Gavin MA, Gilbert MJ, Riddell SR, Greenberg PD, Bevan MJ (1993): Alkali hydrolysis of recombinant proteins allows for the rapid identification of class I MHC-restricted CTL epitopes. *J Immunol* 151:3971–3980.

Giraud C, Winocour E, Berns KI (1994): Site-specific integration by adeno-associated virus is directed by a cellular DNA sequence. *Proc Natl Acad Sci USA* 91:10039–10043.

Giraud C, Winocour E, Berns KI (1995): Recombinant junctions formed by site-specific integration of adeno-associated virus into an episome. *J Virol* 69:6917–6924.

Goodman S, Xiao X, Donahue RE, Moulton A, Miller J, Walsh C, Young NS, Samulski RJ, Nienhuis AW (1994): Recombinant adeno-associated virus mediated gene transfer into hemopoietic progenitor cells. *Blood* 84:1492–1500.

Gottlieb J, Muzyczka N (1988): *In vitro* excision of adeno-associated virus DNA from recombinant plasmids: Isolation of an enzyme fraction from HeLa cells that cleaves DNA at poly(G) sequences. *Mol Cell Biol* 6:2513–2522.

Grossman M, Raper SE, Kozarsky K, Stein EA, Engelhardt JF, Muller D, Lupien PJ, Wilson JM (1994): Successful *ex vivo* gene therapy directed to liver in a patient with familial hypercholesterolaemia. *Nature Genet* 6:335–341.

Grossman M, Raper SE, Wilson JM (1992): Transplantation of genetically modified autologous hepatocytes into non-human primates: Feasibility and short-term toxicity. *Hum Gene Ther* 3:501–510.

Hauswirth WW, Berns KI (1979): Adeno-associated virus DNA replication: Non unit length molecules. *Virology* 93:57–68.

Hermonat PL, Labow MA, Wright R, Berns KI, Muzyczka N (1984): Genetics of adeno-associated virus: Isolation and preliminary characterization of adeno-associated virus type 2 mutants. *J Virol* 51:329–339.

Hermonat PL, Muzyczka N (1984): Use of adeno-associated virus as a mammalian DNA cloning vector. Transduction of neomycin resistance into mammalian tissue culture cells. *Proc Natl Acad Sci USA* 81:6466–6470.

Hirt B (1967): Selective extraction of polyoma DNA from infected mouse cultures. *J Mol Biol* 26:365–369.

Hodgson CP (1995): The vector void in gene therapy. *Bio/Technol* 13:222–225.

Hölscher C, Hörer M, Kleinschmidt JA, Zentgraf H. Bürkle A, Heilbronn R (1994): Cell lines inducibly expressing the adeno-associated virus (AAV) rep gene: Requirements for productive replication of rep-negative AAV mutants. *J Virol* 68:7169–7177.

Hölscher C, Kleinschmidt JA, Bürkle A (1995): High-level expression of adeno-associated virus (AAV) Rep78 protein is sufficient for infectious-particle formation by a rep-negative AAV mutant. *J Virol* 69:6880–6885.

Hong G, Ward P, Berns KI (1992): *In vitro* replication of adeno-associated virus DNA. *Proc Natl Acad Sci USA* 89:4673–4677.

Hong G, Ward P, Berns KI (1994): Intermediates of adeno-associated virus DNA replication *in vitro*. *J Virol* 68:2011–2015.

Im D-S, Muzyczka N (1989): Factors that bind to adeno-associated virus terminal repeats. *J Virol* 63:3095–3104.

Im D-S, Muzyczka N (1990): The AAV origin binding protein Rep68 is an ATP-dependent site-specific endonuclease with DNA helicase activity. *Cell* 61:447–457.

Im D-S, Muzyczka, N (1992): Partial purification of adeno-associated virus Rep78, Rep52, and Rep40 proteins and their biochemical characterization. *J Virol* 66:1119–1128.

Kaplitt MG, Leone P, Samulski RJ, Xiao X, Pfaff DW, O'Malley KL, During MJ (1994): Long-term gene expression and phenotypic correction using adeno-associated virus vectors in the mammalian brain. *Nature Genet* 8:148–153.

Kearns WG, Afione SA, Flumer SB, Pang MG, Erikson D, Egan M, Landrum M, Flotte TR, Cutting GR (1996): Recombinant adeno-associated virus (AAV-CFTR) vectros do not integrate in a site-specific fashion in an immortalized epithelial cell line. *Gene Ther* 3:748–755.

Knowles MR, Hohneker KW, Zhou Z, Olsen JC, Noah TL, Hy P-C, Leigh MW, Engelhardt JF, Edwards LJ, Jones KR, Grossman M, Wilson JM, Johnson LG, Boucher RC (1995): A controlled study of adenoviral vector-mediated gene transfer in the nasal epithelium of patients with cystic fibrosis. *N Engl J Med* 333:823–831.

Kotin RM (1994): Prospects for the use of adeno-associated virus as a vector for human gene therapy. *Hum Gene Ther* 5:793–801.

Kotin RM, Berns KI (1989): Organization of adeno-associated virus DNA in latently infected Detroit 6 cells. *Virology* 170:460–467.

Kotin RM, Linden RM, Berns KI (1992): Characterization of a preferred site on human chromosome 19q for integration of adeno-associated virus DNA by non-homologous recombination. *EMBO J* 11:5071–5078.

Kotin RM, Menninger JC, Ward DC, Berns KI (1991): Mapping and direct visualization of a region-specific viral DNA integration site on chromosome 19q13-qter. *Genomics* 10:831–834.

Kotin RM, Siniscalco M, Samulski RJ, Zhu X, Hunter L, Laughlin CA, McLaughlin S, Muzyczka N, Rocchi M, Berns KI (1990): Site-specific integration by adeno-associated virus. *Proc Natl Acad Sci USA* 87:2211–2215.

Kube DM, Ponnazhagan S, Srivastava A (1997): Encapsidation of adeno-associated virus 2 Rep proteins in wild-type and recombinant progeny virions: Rep-mediated growth inhibition of primary human cells. *J Virol* 71:7364–7371.

Kÿostio SRM, Wonderling RS, Owens RA (1995): Negative regulation of the adeno-associated virus (AAV) p5 promoter involves both the p5 rep binding site and the consensus ATP-binding motif of the AAV rep68 protein. *J Virol* 69:6787–6796.

Lebkowski JS, McNally MM, Okarma TB, Lerch LB (1988): Adeno-associated virus: A vector system for efficient introduction of DNA into a variety of mammalian cell types. *Mol Cell Biol* 8:3988–3996.

Luhovy M, McCune S, Dong J-Y, Prchal JF, Townes TM, Prchal JT (1996): Stable transduction of recombinant adeno-associated virus into hematopoietic stem cells from normal and sickle cell patients. *Biol Blood Marrow Transplant* 2:24–30.

Luo F, Zhou SZ, Cooper S, Munshi NC, Boswell HS, Broxmeyer HE, Srivastava A (1995): Adeno-associated virus 2–mediated gene transfer and functional expression of the human granulocyte-macrophage colony-stimulating factor. *Exp Hematol* 23:1261–1267.

Lusby E, Fife KH, Berns KI (1980): Nucleotide sequence of the inverted terminal repetition in adeno-associated virus DNA. *J Virol* 34:402–409.

Mamounas M, Leavitt M, Yu M, Wong-Staal F (1995): Increased titer of recombinant AAV vectors by gene transfer with adenovirus coupled with DNA–polylysine complexes. *Gene Ther* 2:429–432.

Maxwell F, Harrison G, Maxwell IH (1995): Improved methods for production of recombinant AAV and determination of infectious titer. VIth Parvovirus Workshop, Montpellier, France, p 72.

Mayor HD, Houlditch GS, Mumford DM (1973): Influence of adeno-associated satellite virus on adenovirus-induced tumors in hamsters. *Nature New Biol* 241:44–46.

McCarty DM, Pereira DJ, Zolotukhin I, Zhou X, Ryan JH, Muzyczka N (1994a): Identification of linear DNA sequences that specifically bind the adeno-associated virus rep protein. *J Virol* 68:4988–4997.

McCarty DM, Ryan JH, Zolotukhin I, Zhou X, Muzyczka N (1994b): Interaction of the adeno-associated virus rep protein with a sequence within the palindrome of the viral terminal repeat. *J Virol* 68:4998–5006.

McLaughlin SK, Collis P, Hermonat PL, Muzyczka N (1988): Adeno-associated virus general transduction vectors: Analysis of proviral structures. *J Virol* 62:1963–1973.

Miller JL, Donahue RE, Sellers SE, Samulski RJ, Young NS, Nienhuis AW (1994): Recombinant adeno-associated virus (rAAV)-mediated expression of human γ-globin gene in human progenitor-derived erythroid cells. *Proc Natl Acad Sci USA* 91:10183–10187.

Miller N, Vile R (1995): Targeted vectors for gene therapy. *FASEB J* 9:190–199.

Mizukami H, Young NS, Brown KE (1995): Adeno-associated virus type 2 binds to a 150 kilodalton glycoprotein. *Virology* 217:124–130.

Muzyczka N (1992): Use of adeno-associated virus as a general transduction vector for mammalian cells. *Curr Top Microbiol Immunol* 158:97–129.

Nahreini P, Srivastava A (1989): Rescue and replication of the adeno-associated virus 2 genome in mortal and immortal human cells. *Intervirology* 30:74–85.

Nahreini P, Srivastava A (1992): Rescue of the adeno-associated virus 2 genome correlates with alterations in DNA-modifying enzymes in human cells. *Intervirology* 33:109–115.

Nahreini P, Woody MJ, Zhou SZ, Srivastava A (1993): Versatile adeno-associated virus 2–based vectors for constructing recombinant virions. *Gene* 124:257–262.

Ni T-H, Zhou X-H, McCarty DM, Zolotukhin I, Muzyczka N (1994): *In vitro* replication of adeno-associated virus DNA. *J Virol* 68:1128–1138.

Ostrove JM, Duckworth DH, Berns KI (1981): Inhibition of adenovirus-transformed cell oncogenicity by adeno-associated virus. *Virology* 113:521–533.

Owens RA, Carter BJ (1992): *In vitro* resolution of adeno-associated virus DNA hairpin termini by wild-type Rep protein is inhibited by a dominant-negative mutant of Rep. *J Virol* 66:1236–1240.

Pattison JR (1988): Parvovirus and Human Disease. Boca Raton: CRC Press.

Philip R, Brunette E, Kilinski L, Murugesh D, McNally MA, Ucar K, Rosenblatt J, Okarma TB, Lebkowski JS (1994): Efficient and sustained gene expression in primary T lymphocytes and primary and cultured tumor cells mediated by adeno-associated virus plasmid DNA complexed to cationic liposomes. *Mol Cell Biol* 14:2411–2418.

Podsakoff G, Shaughnessy EA, Lu D, Wong KK, Chatterjee S (1994a): Long term *in vivo* reconstitution with murine marrow cells transduced with an adeno-associated virus vector. *Blood* 84:256a.

Podsakoff G, Wong KK, Chatterjee S (1994b): Stable and efficient gene transfer into non-dividing cells by adeno-associated virus (AAV)-based vectors. *J Virol* 68:5656–5666.

Ponnazhagan S, Erikson D, Kearns WG, Zhou SZ, Nahreini P, Wang X-S, Srivastava A (1997a): Lack of site-specific integration of the recombinant adeno-associated virus genomes in human cells. *Hum Gene Ther* 8:275–284.

Ponnazhagan S, Mukherjee P, Yoder MC, Wang X-S, Zhou SZ, Kaplan J, Wadsworth S, Srivastava A (1997b): Adeno-associated mediated gene transfer *in vivo:* Organ tropism and expression of transduced sequences in mice. *Gene* 190:203–210.

Ponnazhagan S, Nallari ML, Srivastava A (1994a): Suppression of human α-globin gene expression mediated by the recombinant adeno-associated virus 2–based antisense vectors. *J Exp Med* 179:733–738.

Ponnazhagan S, Yoder MC, Srivastava A (1997c): Adeno-associated virus 2–mediated transduction of murine hemopoietic cells with long-term repopulating ability and sustained expression of a human globin gene *in vivo. J Virol* 71:3098–3104.

Ponnazhagan S, Wang X-S, Kang LY, Woody MJ, Nallari ML, Munshi NC, Zhou SZ, Srivastava A (1994b): Transduction of human hematopoietic cells by the adeno-associated virus 2 vectors is receptor-mediated. *Blood* 84:742a.

Ponnazhagan S, Wang X-S, Srivastava A (1995a): Alternative strategies for generating recombinant AAV vectors. VIth Parvovirus Workshop, Montpellier, France, p 71.

Ponnazhagan S, Wang X-S, Srivastava A, Yoder MC (1995b): Adeno-associated virus 2–mediated gene transfer and expression in murine hemopoietic progenitor cells *in vivo. Blood* 86:240a.

Ponnazhagan S, Wang X-S, Woody MJ, Luo F, Kang LY, Nallari ML, Munshi NC, Zhou SZ, Srivastava A (1996): Differential expression in human cells from the p6 promoter of human parvovirus B19 following plasmid transfection and recombinant adeno-associated virus 2 (AAV) infection: Human megakaryocytic leukaemia cells are non-permissive for AAV infection. *J Gen Virol* 77:1111–1122.

Prasad K-MR, Trempe JP (1995): The adeno-associated virus Rep78 protein is covalently linked to viral DNA in a preformed virion. *Virology* 214:360–370.

Qing KY, Wang X-S, Kube DM, Ponnazhagan S, Bajpai A, Srivastava A (1997): Role of tyrosine phosphorylation of a cellular protein in adeno-associated virus 2-mediated transgene expression. *Proc Natl Acad Sci USA* 94:10879–10884.

Qing KY, Khuntirat B, Mah C, Kube DM, Wang X-S, Ponnazhagan S, Zhou SZ, Dwarki VJ, Yoder MC, Srivastava A (1998): Adeno-associated virus type 2-mediated gene transfer: Correlation of tyrosine phosphorylation of the cellular single-stranded D sequence-binding protein with transgene expression in human cells *in vitro* and murine tissues *in vivo*. *J Virol* 72:1593–1599.

Rosenberg SA, Abersold P, Cornetta K, Kasid A, Morgan RA, Moen R, Karson EM, Lotze MT, Yang JC, Topalian SL, Merino MJ, Culver K, Miller AD, Blaese RM, Anderson WF (1990): Gene transfer into humans—Immunotherapy of patients with advanced melanoma, using tumor-infiltrating lymphocytes modified by retroviral gene transduction. *N Engl J Med* 323:570–577.

Russell DW, Alexander IE, Miller AD (1995): DNA synthesis and topoisomerase inhibitors increase transduction by adeno-associated virus vectors. *Proc Natl Acad Sci USA* 92:5719–5723.

Russell DW, Miller AD, Alexander IE (1994): Adeno-associated virus vectors preferentially transduce cells in S phase. *Proc Natl Acad Sci USA* 91:8915–8919.

Sambrook J, Fritsch EF, Maniatis T (1989): Molecular Cloning: A Laboratory Manual. Cold Spring Harbor, NY: Cold Spring Harbor Laboratory Press, pp 1.53–1.110.

Samulski RJ (1993): Adeno-associated virus: Integration at a specific chromosomal locus. *Curr Opin Genet Dev* 3:74–80.

Samulski RJ, Berns KI, Tan M, Muzyczka N (1982): Cloning of adeno-associated virus into pBR322. Rescue of intact virus from the recombinant plasmid in human cells. *Proc Natl Acad Sci USA* 79:2077–2081.

Samulski RJ, Chang L-S, Shenk T (1987): A recombinant plasmid from which an infectious adeno-associated virus genome can be excised *in vitro* and its use to study viral replication. *J Virol* 61:3096–3101.

Samulski RJ, Chang L-S, Shenk T (1989): Helper-free stocks of recombinant adeno-associated viruses: Normal integration does not require viral gene expression. *J Virol* 63:3822–3828.

Samulski RJ, Srivastava A, Berns KI, Muzyczka N (1983): Rescue of adeno-associated virus from recombinant plasmids: Gene correction within the terminal repeats of AAV. *Cell* 33:135–143.

Samulski RJ, Zhu X, Xiao X, Brook J, Houseman DE, Epstein N, Hunter LA (1991): Targeted integration of adeno-associated virus (AAV) into human chromosome 19. *EMBO J* 10:3941–3950.

Senapathy P, Tratschin J-D, Carter BJ (1984): Replication of adeno-associated virus DNA. Complementation of naturally occurring rep$^-$ mutants by a wild-type genome or an ori$^-$ mutant and correction of terminal palindrome deletions. *J Mol Biol* 179:1–20.

Shaughnessy E, Wong KK, Podsakoff G, Kane S, Chatterjee S (1995): Gene transfer of MDR-1 into primary human CD34$^+$ enriched marrow cells mediated by an adeno-associated virus–based vector. *Blood* 86:244a.

Shi Y, Seto E, Chang L-S, Shenk T (1991): Transcriptional repression by YY1, a human GL1-Kruppel–related protein, and relief of repression by adenovirus E1A protein. *Cell* 67:377–388.

Siegl G, Bates RC, Berns KI, Carter BJ, Kelly DC, Kurstak E, Tattersall P (1985): Characteristics and taxonomy of parvoviridae. *Intervirology* 23:61–73.

Snyder RO, Samulski RJ, Muzyczka N (1990): *In vitro* resolution of covalently joined AAV chromosome ends. *Cell* 60:105–113.

Southern EM (1975): Detection of specific sequences among DNA fragments separated by gel electrophoresis. *J Mol Biol* 98:503–517.

Srivastava A (1987): Replication of the adeno-associated virus DNA termini *in vitro. Intervirology* 27:138–147.

Srivastava A (1994): Parvovirus-based vectors for human gene therapy. *Blood Cells* 20:531–538.

Srivastava A, Lusby EW, Berns KI (1983): Nucleotide sequence and organization of the adeno-associated virus 2 genome. *J Virol* 45:555–564.

Srivastava A, Wang X-S, Ponnazhagan S, Zhou SZ, Yoder MC (1995): Parvovirus-based vectors for human gene therapy. VIth Parvovirus Workshop, Montpellier, France, p 29.

Srivastava A, Wang X-S, Ponnazhagan S, Zhou SZ, Yoder MC (1996): Adeno-associated virus 2–mediated transduction and erythroid lineage-specific expression in human hemopoietic progenitor cells. *Curr Top Microbiol Immunol* 218:93–117.

Srivastava CH, Samulski RJ, Lu L, Larsen SH, Srivastava A (1989): Construction of a recombinant human parvovirus B19: Adeno-associated virus 2 (AAV) DNA inverted terminal repeats are functional in an AAV-B19 hybrid virus. *Proc Natl Acad Sci USA* 86:8078–8082.

Su H, Chang JC, Xu SM, Kan YW (1996): Selective killing of AFP-positive hepatocellular carcinoma cells by adeno-associated virus transfer of the herpes simplex virus thymidine kinase gene. *Hum Gene Ther* 7:463–470.

Sun X-L, Murphy BR, Li QJ, Gullapalli S, Mackins J, Jayaram HN, Srivastava A, Antony AC (1995): Transduction of folate receptor cDNA into cervical carcinoma cells using recombinant adeno-associated virions delays cell proliferation *in vivo. J Clin Invest* 96:1535–1547.

Tamayose K, Hirai Y, Shimada T (1996): A new strategy for large-scale preparation of high-titer recombinant adeno-associated virus vectors by using sulfonated cellulose column chromatography. *Hum Gene Ther* 7:507–513.

Thrasher AJ, de Alwis M, Casimir CM, Kinnon C, Page K, Lebkowski J, Segal SW, Levinsky RJ (1995). Functional reconstitution of the NADPH-oxidase by adeno-associated virus gene transfer. *Blood* 86:761–765.

Tooze J (1981): DNA Tumor Viruses. Cold Spring Harbor, NY: Cold Spring Harbor Laboratory Press.

Tratschin J-D, Miller IL, Carter BJ (1984): Genetic analysis of adeno-associated virus: Properties of deletion mutants constructed *in vitro* and evidence for an adeno-associated virus replication function. *J Virol* 51:611–619.

Tuan DYH, Solomon WB, London IM, Lee DP (1989): An erythroid specific development stage-independent enhancer far upstream of the human "β-like globin" genes. *Proc Natl Acad Sci USA* 86:2554–2559.

Van Ginkel FW, Liu C, Simecka JW, Dong J-Y, Greenway T, Frizzell RA, Kiyono H, McGhee JR, Pascaul DW (1995): Intratracheal gene delivery with adenoviral vector induces elevated systemic IgG and mucosal IgA antibodies to adenovirus and β-galactosidase. *Hum Gene Ther* 6:895–903.

Walsh CE, Liu JM, Miller JL, Nienhuis AW, Samulski RJ (1993): Gene therapy for human hemoglobinopathies. *Proc Soc Exp Biol Med* 204:289–300.

Walsh CE, Liu JM, Xiao X, Young NS, Nienhuis AW, Samulski RJ (1992): Regulated high level expression of a human γ-globin gene introduced into erythroid cells by an adeno-associated virus vector. *Proc Natl Acad Sci USA* 89:7257–7261.

Walsh CE, Nienhuis AW, Samulski RJ, Brown MG, Miller JL, Young NS, Liu JM (1994): Phenotypic correction of Fanconi anemia in human hematopoietic cells with a recombinant adeno-associated virus vector. *J Clin Invest* 94:1440–1448.

Wang X-S, Khuntirat B, Qing KY, Ponnazhagan S, Kube DM, Zhou SZ, Dwarki VJ, Srivastava A (1998): Characterization of wild-type adeno-associated virus type 2-like' particles generated during recombinant viral vector production and strategies for their elimination. *J Virol* 72:in press.

Wang X-S, Ponnazhagan S, Srivastava A (1995a): Rescue and replication signals of the adeno-associated virus 2 genome. *J Mol Biol* 250:573–580.

Wang X-S, Ponnazhagan S, Srivastava A (1996b): Rescue and replication of adeno-associated virus type 2 genome as well as vector DNA sequences from recombinant plasmids containing deletions in the viral inverted terminal repeats: Selective encapsidation of the viral genomes in progeny virions. *J Virol* 70:1668–1677.

Wang X-S, Srivastava A (1997): A novel terminal resolution-like site in the adeno-associated virus 2 genome. *J Virol* 71:1140–1146.

Wang X-S, Yoder MC, Zhou SZ, Srivastava A (1995b): Parvovirus B19 promoter at map unit 6 confers replication competence and erythroid specificity to adeno-associated virus 2 in primary human hemopoietic progenitor cells. *Proc Natl Acad Sci USA* 92:12416–12420.

Wei JF, Wei FS, Samulski RJ, Barranger JA (1994): Expression of the human glucocerebrosidase and arylsulfatase A genes in murine and patient primary fibroblasts transduced by an adeno-associated virus vector. *Gene Ther* 1:261–268.

Weiss RN, Teich N, Varmus HE, Coffin JM (1984): RNA Tumor Viruses. Cold Spring Harbor, NY: Cold Spring Harbor Laboratory Press.

Weitzman MD, Kÿostio SRM, Kotin RM, Owens RA (1994): Adeno-associated virus (AAV) Rep proteins mediate complex formation between AAV DNA and its integration site in human DNA. *Proc Natl Acad Sci USA* 91:5808–5812.

Wistuba A, Weger S, Kern A, Kleinschmidt JA (1995): Intermediates of adeno-associated virus type 2 assembly: Identification of soluble complexes containing Rep and Cap proteins. *J Virol* 69:5311–5319.

Wu P, Ziska D, Bonell MA, Grouzmann E, Millard WJ, Meyer EM (1994): Differential neuropeptide Y gene expression in post-mitotic versus dividing neuroblastoma cells driven by an adeno-associated virus vector. *Brain Res Mol Brain Res* 24:27–33.

Xiao X, deVlaminck W, Monahan J (1993): Adeno-associated virus (AAV) vectors for gene transfer. *Adv Drug Del Rev* 12:201–215.

Yacobson B, Koch T, Winocour E (1987): Replication of adeno-associated virus in synchronized cells without the addition of a helper virus. *J Virol* 61:972–981.

Yalkinoglu AO, Heilbronn R, Burkle A, Schlehofer JR, zur Hausen H (1988): DNA amplification of adeno-associated virus as a response to cellular genotoxic stress. *Cancer Res* 48:3123–3125.

Yang Q, Kadam A, Trempe JP (1994): Characterization of cell lines that inducibly express the adeno-associated virus Rep proteins. *J Virol* 68:4847–4856.

Yang Y, Ertl HC, Wilson JM (1994): MHC class 1–restricted cytotoxic T lymphocytes to viral antigens destroy hepatocytes in mice infected with E1-deleted recombinant adenoviruses. *Immunology* 1:433–442.

Yoder MC, Du X-X, Williams DA (1993): High proliferative potential colony-forming cell heterogeneity identified using counterflow centrifugal elutriation. *Blood* 82:385–391.

Zabner J, Couture LA, Gregory RJ, Graham SM, Smith AE, Welsh MJ (1993): Adenovirus-mediated gene transfer transiently corrects the chloride transport defect in nasal epithelia of patients with cystic fibrosis. *Cell* 75:207–216.

Zhou SZ, Broxmeyer HE, Cooper S, Harrington MA, Srivastava A (1993): Adeno-associated virus 2–mediated gene transfer in murine hemopoietic progenitor cells. *Exp Hematol* 21:928–933.

Zhou SZ, Cooper S, Kang LY, Ruggieri L, Heimfeld S, Srivastava A, Broxmeyer HE (1994): Adeno-associated virus 2–mediated high efficiency gene transfer into immature and mature subsets of hemopoietic progenitor cells in human umbilical cord blood. *J Exp Med* 179:1867–1875.

Zhou SZ, Li Q, Stamatoyannopoulos G, Srivastava A (1996): Adeno-associated virus 2–mediated transduction and erythroid cell-specific expression of a human β-globin gene. *Gene Ther* 3:223–229.

11

RIBOZYME GENE THERAPY TARGETING STEM CELLS FOR HUMAN IMMUNODEFICIENCY VIRUS INFECTION

ANTHONY D. HO, PING LAW, XINQIANG LI, AND FLOSSIE WONG-STAAL

Department of Medicine (A.D.H., P.L.) and Department of Biology (X.L., F.W.S.), University of California, LaJolla, CA 92093

INTRODUCTION

Most infectious diseases are self-limiting or are responsive to antimicrobial agents. Infection with the human immunodeficiency virus (HIV), however, is chronic, usually lasting for more than a decade until the patient finally succumbs to the disease. Recently, promising advances have been made on the clinical front for treating HIV-infected patients. The use of new protease inhibitors, particularly in conjunction with other antiviral drugs such as reverse transcriptase inhibitors, have demonstrated greater and more sustained antiviral effects than previously seen (Stephenson, 1996). However, the chronicity of HIV infection and the probable need for lifelong therapy call for the development of more long-term treatment strategies.

Gene therapy, although still in its infancy, promises to be an exciting new treatment regimen. Its scope has greatly expanded beyond its original conception as a gene replacement strategy for hereditary disorders and now embodies

Stem Cell Biology and Gene Therapy, Edited by Peter J. Quesenberry, Gary S. Stein, Bernard G. Forget, and Sherman M. Weissman
ISBN 0-471-14656-0

disease targets that include chronic infectious diseases, cancer, and degenerative disorders. For HIV infection, many different approaches using gene transfer to either limit virus replication or stimulate host immunity are in preclinical development, and a number of clinical investigations are currently underway (for recent reviews, see Yu et al., 1994; Poeschla and Wong-Staal, 1995–96). The antiviral approach has also been referred to as *intracellular immunization* (Baltimore, 1988), as the goal is to render cells genetically resistant to infection. This chapter is not meant to provide an exhaustive and comprehensive review of the entire field, but focuses on the use of hemopoietic stem cells as vehicles for antiviral gene therapy for HIV infection.

HIV INFECTION AND HEMATOPOIESIS

The acquired immunodeficiency syndrome (AIDS) is one of the major threats to public health on a global scale. Other than immunodeficiencies, ineffective hematopoiesis is commonly observed in patients with AIDS. The mechanisms for the hematological abnormalities are not yet defined but may include accelerated destruction of cells due to infections, specific autoimmunity, or direct action of HIV on hematopoietic precursors, their progeny cells, or bone marrow accessory cells such as stromal cells and T lymphocytes. As most of the cells targeted by HIV-1 infection, e.g., T-helper lymphocytes, monocytes, and microglia cells, are all derived from hematopoietic stem cells, immunization of the latter by means of genetic manipulations might eventually confer protection against HIV-1 invasion and restore immune function. Gene therapy targeting hematopoietic stem cells, therefore, may pre-empt infection of the mature target cells by the virus.

LIFE CYCLE OF HIV: STRATEGIES FOR GENE THERAPY

HIV-1 is a complex retrovirus that infects target cells by binding to its cell surface receptor, the CD4 molecule. By interacting with the latter and at least one other cell surface molecule, a conformational change occurs in the envelope glycoproteins to reveal the hydrophobic sequence of the transmembrane envelope protein. This leads to a fusion of the viral envelope with the cell membrane and delivery of the viral capsid into the cytoplasm of the cell. Within the capsid structure, the diploid RNA genome is reverse transcribed by the viral reverse transcriptase (RT) to an RNA–DNA heterodimer; subsequently the RNA is degraded and a DNA duplex synthesized. The double-stranded DNA genome is integrated into the host cell chromosomes where it might become latent or transcriptionally active. Transcription is mediated through promoter and enhancer sequences in the long terminal repeat (LTR) and initially gives rise to RNAs that are multiply spliced to form the codes for the regulatory proteins tat, rev, and nef. The tat protein is a transcriptional transactivator that interacts with an LTR encoded RNA stem-loop structure (TAR) in the nascent transcript. The rev protein interacts with a *cis*-acting stem-loop RNA structure embedded in the env coding sequence (the rev response element [RRE]), which is removed in the multiply spliced RNA. This interaction apparently shunts the RRE-containing RNA (unspliced or singly spliced viral mRNA) from the splicing to the nuclear export pathway and allows its utilization in the cytoplasm for protein synthesis. The singly spliced mRNA codes for the envelope glycoprotein and the unspliced message for the gag and pol proteins. Envelope glycoproteins have a signal sequence that leads to translocation of the polypeptide into the endoplasmic reticulum lumen where it is processed and glycosylated. The envelope glycoprotein is then transported to the cell membranes, which are assembling the unglycosylated gag and gag/pol polyproteins. In association with the envelope glycoproteins, gag and gag/pol assemble at the membrane, and the virus buds from the cell. After budding, autocatalytic cleavage of the gag and gag/pol polyprotein occurs, leading to production of mature and infectious progeny viruses.

The above-described life cycle shows many vulnerable steps that could be disrupted by genetic modifications. In particular, processes specific to the virus are most amenable to manipulations: entry into the cell, reverse transcription, integration, virus budding, and maturation. The tat and rev RNA–protein interactions and the packaging of the RNA genome are additional virus-specific functions distinct from normal cellular processes. Current antiviral gene therapy strategies include the use of receptor decoys (e.g., CD4, gp120), intracellular expression of RNA decoys (monomeric and polymeric TAR, RRE decoys, packaging site decoys), transdominant proteins (e.g., transdominant RT, tat, rev, gag, env), antisense RNA, and ribozymes. Recent phase I studies of various gene therapy trials have revealed that expression of foreign proteins in the genetically modified cells is likely to efficiently induce immune clearance of these cells (Riddell et al., 1996). Since persistence of the modified cells is key to the concept of "intracellular immunization," these observations indicate that therapeutic genes that express noncodogenic RNA would be preferable over those that express foreign viral proteins (e.g., transdominant mutant viral proteins), which are immunogenic. Therefore, we only expand on the several approaches that utilize RNA as inhibitory molecules.

RNA DECOYS

This strategy seeks to sequester viral regulatory proteins through overexpression of their cognate RNAs, e.g., TAR and RRE. The HIV-1 TAR (Sullenger et al., 1990) and RRE (Smith et al., 1993) sequences have been introduced into both LTRs of a retroviral vector with an internal pol III ($tRNA^{met}$) promoter driving the expression. CEM SS cells transduced with these constructs were specifically inhibited in HIV expression and replication. One potential concern for the decoy strategy is that cellular factors that bind TAR or RRE will also be sequestered. In the case of RRE, the regions binding to rev and to cellular factors are apparently distinct (Xu et al., 1997; Lee et al., 1992), and a minimal version of the RNA was efficient in inhibiting rev function and viral replication (Lee et al., 1992). In contrast, conservation of the loop of TAR which is responsible for binding cellular factor but not tat, is essential for the antiviral effect (Lee et al., 1992). One way of minimizing the potential toxicity of TAR is to render its expression tat inducible. Therefore, only HIV-infected cells expressing tat will express the TAR decoy, which provides feedback inhibition for viral transcription (Lisziewicz et al., 1993). It should be noted that RNA decoys are only effective after virus infection and integration, but would not be able to prevent the establishment of infection.

ANTISENSE RNA

Antisense RNA or DNA utilizes the specificity of Watson-Crick base-pairing to interfere with gene expression in a sequence-specific fashion. Initial studies employing antisense RNA to inhibit HIV-1 replication showed limited efficacy with pol II–promoted transcripts complementary to tat, rev, vpu (Rhodes and James, 1990, 1991), gag (Sczakiel and Pawlita, 1991), and the primer binding site (Joshi et al., 1991). Significantly greater inhibition was observed using a recombinant adeno-associated virus (AAV) vector to deliver an antisense sequence complementary to a 63 bp sequence in the HIV-1 LTR, which is present at both 5′ and 3′ ends of all HIV-1 transcripts. Transduction of a human T- cell line with the antisense vector resulted in a three log reduction of virus titer after challenge.

RIBOZYMES

Ribozymes are RNA molecules that possess both antisense and sequence-specific RNA cleaving properties. Because of their catalytic properties, ribozymes can theoretically be effective at much lower concentrations. The greater constraint inherent in the rules for substrate selection should also minimize potential toxicity due to nonspecific gene inhibition. Sarver et al. (1990) first described the use of a hammerhead ribozyme as an anti-HIV agent. Since then, several groups have reported the design and functional study of hammerhead ribozymes targeting other HIV sequences including vif (Lorentzen et al., 1991), integrase (Sioud and Drlica, 1991), and the leader sequence (+133, NL43 [Weerasinghe et al., 1991]), all reporting various delays or reduction in virus expression. We have designed a hairpin ribozyme that cleaves HIV RNA in the 5′-leader sequence (positions +111/112 from the cap site in HXB2). Human T-cell lines and primary T cells transduced with retroviral vectors containing this hairpin ribozyme showed resistance to the challenge of HIV virus for a prolonged period of time. A clinical trial using $CD4^+$ cells as targets is now under way. As lymphocytes have a limited life span, hematopoietic stem cells represent an alternative and perhaps an ideal target for human gene therapy.

HEMATOPOIETIC STEM CELLS AS VEHICLES

Pluripotent hematopoietic stem cells are precursor cells from which all mature blood cells, including the T and B lymphocytes, are derived. In contrast to other somatic cells such as monocytes/macrophages, hepatocytes, and muscle cells, stem cells are capable of self-renewal as well as differentiation into all hematopoietic lineages throughout adult life. Successful insertion of anti-HIV ribozyme genes into the true pluripotent stem cells may protect all the progeny cells, including the monocytes and T cells from infection by HIV. The long-term goal of gene therapy for AIDS patients is, therefore, to use autologous or allogeneic grafts of stem cells equipped with anti-HIV ribozyme genes to reconstitute the immune function of HIV-infected individuals. The hypothesis is that stem cells equipped with ribozyme genes will give rise to progeny cells, especially lymphocytes and monocytes, that carry the anti-HIV transgene and are hence protected as they mature and become prime targets for HIV infection. This protection might provide them with survival benefit versus progeny cells derived from untreated stem cells. Autologous or allogeneic transplantation with genetically engineered stem cells might represent an ultimate cure for patients with AIDS.

PLURIPOTENT STEM CELLS VERSUS LINEAGE-COMMITTED PROGENITOR CELLS

Hematopoietic stem/progenitor cells have been demonstrated to be associated with the surface antigenic marker CD34 (Civin et al., 1987). At steady state, about 1%–2% of the cells from the bone marrow and <0.1% (almost undetectable) of mononuclear cells (MNC) from the peripheral blood are positive for CD34 (Terstappen et al., 1991; Bender et al., 1991). By means of multicolor, multidimensional flow cytometry, it has been shown that the $CD34^+$ cells can be subcategorized into early, or pluripotent, stem cells and lineage-committed progenitors (Terstappen et al., 1991; Bender et al., 1991). The pluripotent stem cells are characterized by $CD34^+$, $CD38^-$ and can be further subclassified according to the presence or absence of HLA-DR (Terstappen et al., 1991), Thy-1 (Lansdorp and Dragowska, 1992), or staining with rhodamine, and so forth (Wagner, 1993). The lineage-committed subsets are characterized by $CD34^+$, $CD38^+$, and the co-expression of one of other leukocyte surface antigens depending on the lineage commitment, e.g., CD10, CD19 for B lymphoid cells; CD7, CD10 for T lymphoid cells; CD33, CD14, CD15 for myeloid cells; and CD71 for erythroid cells (Terstappen et al., 1991).

It is not yet definitively known which $CD34^+$ subset represents the population with the highest self-renewal or multilineage potential. Possible candidates include $CD34^+$ $CD38^-$, $HLA\text{-}DR^+$ (Terstappen et al., 1991; Huang and Terstappen, 1994), $CD34^+$, $Thy1^+$ and negative for all mature lineages (lin^-), or $CD34^+$, $CD45RA^{lo}$, $CD71^{lo}$, and so forth, probably depending on the assay system as well as on the source of the cells (Davis et al., 1995; Mayani and Lansdorp, 1995; Verfaille and Miller, 1995; Mayani et al., 1995; Traycoff et al., 1995). Applying a single cell culture technique, Huang and Terstappen (1994) recently provided evidence that among the $CD34^+$ populations derived from fetal liver or fetal marrow, the subset characterized by $CD34^+$, $CD38^-$, $HLA\text{-}DR^+$ gives rise to both myeloid and lymphoid precursors. This observation has been confirmed in our laboratory by "index sorting." Previous single cell sorting studies were performed using flow cytometers with an automated cell deposition unit (ACDU). The cells were deposited into single wells according to preset sort gates. An index sorting unit is a device that links the ACDU (coordinates of the wells) to the sorted events (list-mode data of the cells), and the information is stored in a computer file. Therefore, index sorting allows the assignment of list-mode data of each cell to the location of the cell in the microtiter plates. The technology enables us to (1) retain list-mode data of sorted cells, (2) identify any duplication of cells in a single well during sorting, and (3) correlate the functional capacity (growth pattern and replating potential) to the assigned markers and the level of surface markers on the cell surface. Using this technique, single-sorted cells derived from cord blood or fetal tissue that were $CD34^+$, $CD38^-$, $HLA\text{-}DR^+$ have been shown to have the highest replating potential, giving rise to blast and cluster colonies even up to fourth-generation cells upon repetitive replating (S. Huang et al., unpublished results).

SOURCES OF STEM CELLS

For clinical purposes such as for transplantation, hematopoietic stem cells can be derived from a healthy donor, i.e., syngeneic, from an identical twin; allogeneic, from a matched related or unrelated donor; or (autologous), from the patient. The conventional source of hematopoietic stem cells for clinical use has been the bone marrow. In the past decade, mobilized peripheral blood progenitor cells have been increasingly used in lieu of bone marrow for allogeneic as well as autologous transplantation. A major advantage is the accelerated hematopoietic recovery achieved by blood-derived versus marrow-derived stem cells (Weerasinghe et al., 1991; Kessinger and Armitage, 1991; Körbling et al., 1986; Kessinger et al., 1986; Sheridan et al., 1992; Bensinger et al., 1993; Ho et al., 1993).

Essential for a successful engraftment is the collection and transplantation of an adequate amount of both pluripotent stem cells and lineage-committed progenitors. At steady state, the concentration of stem/progenitor cells in circulating blood is less than one tenth of that in bone marrow. Chemotherapy, with or without the addition of hematopoietic growth factors, has been used successfully to enhance and mobilize stem/progenitor cells into peripheral blood for patients with cancer (Sheridan et al., 1992; Bensinger et al., 1993; Ho et al., 1993; Pettengell et al., 1993; Peters et al., 1993). Granulocyte colony-stimulating factor (G-CSF) has been used extensively for mobilization of stem/progenitor cells from normal individuals for allogenic transplantation (Grigg et al., 1995; Bensinger et al., 1995; Körbling et al., 1995; Schmitz et al., 1995; Goldman, 1995; Russell et al., 1993). We have demonstrated that injection of G-CSF (10 μg/kg body weight) for 4 days is as effective as the combination of G- and granulocyte-macrophage (GM)-CSF (at 5 μg/kg body weight of each) for the same time period (Lane et al., 1995). Moreover, concurrent administration of G- and GM-CSF (5 μg/kg/day of each) appears to be as effective as sequential administration of GM-CSF (10 μg/kg/day for 3 days) followed by G-CSF (10 mg/kg/day for 2 days).

Within the overall objective of identifying and isolating the optimal graft for transplantation and gene therapy, we have compared the immunophenotypes and colony formation capacities of the $CD34^+$ cells mobilized by different growth factors in normal subjects (Ho et al., 1996). Using three-color and five-dimensional flow cytometry to compare the pluripotent $CD34^+$ subsets, we have found evidence that the combination of G-CSF and GM-CSF stimulates a significantly higher proportion of pluripotent CD34 subsets than G-CSF alone, whereas GM-CSF by itself is not very efficient in mobilizing an adequate number of $CD34^+$ cells. The combination of G- and GM-CSF might therefore offer the advantage of stimulating both quantitatively an adequate number of total $CD34^+$ cells and a higher proportion of the rare pluripotent stem/progenitor cells into the peripheral blood. The profile of $CD34^+$ subsets mobilized by the combination of G- and GM-CSF approaches that of fetal cord blood, which is known to contain a higher proportion of primitive, pluripotent stem/progenitor cells (Ho et al., 1996). Such pluripotent progenitors might be of special relevance for gene therapy.

Other sources of hematopoietic stem cells for clinical use, such as PUCB and fetal liver tissue, are being studied and their roles in human gene therapy explored (Broxmeyer et al., 1992; Wagner, 1993; Gluckman et al., 1992; Rubinstein et al., 1993; Rice et al., 1994; Roncarolo et al., 1991). Present evidence indicates that there are significant differences in the proportion of pluripotent stem cells versus lineage-committed progenitors among fetal tissue, umbilical cord blood, adult bone marrow, and peripheral blood. Such differences are of relevance for gene therapy, as long-lasting effects can only be expected from transduction of pluripotent stem cells with self-renewal capacity. In terms of collecting large quantities of stem cells, mobilized peripheral blood is the easiest approach among all sources.

INFECTIBILITY OF HUMAN HEMATOPOIETIC STEM CELLS BY HIV

Conflicting results have been reported regarding the infectibility of human hematopoietic stem cells. It is not definitively known if and which of the pluripotent stem cell or committed progenitor cell compartments are susceptible to HIV infection (Busch M et al., 1986; Davis BR et al., 1991; Molina JM et al., 1990; Stanley SK et al., 1992; Von Laer D et al., 1990). The reason for such discrepancies might be caused by the different methods for separation of stem/progenitor cells, which selects the cells based on the antigenic marker CD34. As mentioned above, the population defined by the antigen CD34 is very heterogeneous and contains only a very small fraction of pluripotent stem cells among mainly lineage-committed progenitors (Ho et al., 1996). Stanley et al. (1992) reported that $CD34^+$ cells from 14%–37% of HIV-1–infected patients are infected, with up to 0.2% of $CD34^+$ cells positive by polymerase chain reaction (PCR) or after co-culturing; this percentage exceeded the percentage of peripheral blood lymphocytes that were positive in the same patients. Others have reported virus-specific RNA in megakaryocytes and granulocytes (Busch et al., 1986) and DNA in PCR studies of hematopoietic colonies (Kaczmarski et al., 1992). Davis et al. (1991) reached the opposite conclusion while using PCR to assay for proviral DNA in GM colony-forming units (CFU) derived from bone marrow aspirates from HIV-seropositive donors. Similarly, von Laer et al. (1990) and Molina et al. (1990) did not detect HIV-infected cells by PCR in CD34-enriched bone marrow cells or in GM-CFU or erythroid blast-forming units (E-BFU) from HIV-infected subjects. *In situ* hybridization and immunohistochemistry studies on single colonies derived *in vitro* from bone marrow of three AIDS patients also failed to detect HIV RNA or protein (Ganser et al., 1990). Most recently, De Luca et al. (1993) found that $CD34^+$-enriched cells, GM-CFU colonies, and purified $CD34^+$ cells after 3 weeks of liquid culture from six HIV-infected patients were negative by PCR. However, viral sequences were always detected in bone marrow mononuclear cells from the same patients. In this study, HIV-1 proviral copies decreased with increasing enrichment for CD34; the investigators concluded that at most $1/10^4$ $CD34^+$ cells are infected *in vivo*. It seems indisputable from all of these studies that, even if present, HIV-infected progenitor cells comprise a very small percentage of the hematopoietic stem cell population.

Whether purified $CD34^+$ bone marrow cells from healthy uninfected donors are infected *in vitro* by HIV is also controversial. Whereas viral replication was detected after approximately 1 month of continuous culture of $CD34^+$ cells from an uninfected bone marrow sample following exposure to HIV-1, all the cells with detectable virus expressed a $CD4^+$, $CD34^-$ phenotype with esterase staining indicative of mature monocytes. Further complicating the interpretation of this and other studies (Folks et al., 1988; Kitano et al., 1991), has been determining whether the true pluripotent $CD34^+$ cells were infected in these experiments or whether the contaminating mature monocytes (with HIV-1 infection) subsequently infected the culture-induced progeny monocytes derived from the $CD34^+$ cells. A third possibility is that a very small subset of truly pluripotent $CD34^+$ cells expresses the CD4 marker.

We have recently identified HIV-1 genes and $CD34^+$ markers within the same cell. In one infant born of an HIV-1 positive mother, we have studied the enriched $CD34^+$ cells obtained from umbilical cord blood at the time of delivery by double staining techniques and flow cytometry (Li et al., 1996). We demonstrated that 0.5%–1.0% of the mononuclear cells stained positively with both phycoerythrin-conjugated anti-CD34 and FITC-linked anti-HIV-1 p24. This was subsequently confirmed by positive PCR-amplified products from colonies derived from $CD34^+$ cells from the same cord blood sample using HIV-1–specific primers. This infant became HIV-1 positive at 3 months after birth. Thus our data indicate that a small percentage of the $CD34^+$ cells can be infected by HIV-1. However, the majority of the $CD34^+$ cells are not infective and do not appear to be functionally or phenotypically compromised (see below). These should still be appropriate targets for gene therapy.

MOBILIZATION OF STEM CELLS FROM AIDS PATIENTS

There is no definite information concerning the feasibility of mobilizing $CD34^+$ cells into the peripheral circulation of HIV-1–infected patients. Potentially, HIV-1 infection and antiretroviral therapy has an impact on hematopoiesis and might influence the response to G-CSF or GM-CSF. Moreover, it is possible, although unlikely, that administration of GM-CSF or G-CSF might also induce an activation of HIV-1 in a clinical setting.

Experience with GM-CSF and G-CSF in HIV-1–infected patients has been largely limited to the use of these agents to increase circulating neutrophils during HIV-related pancytopenia, including primary bone marrow failure due to HIV infection, pancytopenia due to granulomatous infections, and (most commonly) neutropenia induced by antiretroviral (AZT) or antiherpetic (ganciclovir) antiviral agents. In addition, considerable experience exists concerning the use of these agents for support of neutrophils during chemotherapy for AIDS-related malignancies, especially lymphoma.

Information concerning the potential of these cytokines to induce HIV expression *in vivo* is limited. While GM-CSF has been shown to be capable of accelerating or increasing the amount of HIV produced by infected monocyte-macrophages, T cells, and $CD34^+$, $CD4^-$ progenitor cells *in vitro* (Koyangi et al., 1988; Perno et al., 1992), experience with clinical use of GM-CSF has not demonstrated prominent virological effects, as assessed by determination of plasma HIV-1 gag p24 antigen by EIA and by semiquantitative virus culture (Krown et al., 1992; Kaplan et al., 1991; Lafevillade et al., 1992; Davison et al., 1994). One study did show an average increase of 243% in p24 antigen concentrations in non-Hodgkin's lymphoma patients undergoing chemotherapy after the second cycle (Kaplan et al., 1991). One recent study employed PCR measurement of proviral DNA in peripheral blood mononuclear cells under GM-CSF administration has not detected any effect (Kimura et al., 1990).

In vitro studies have shown a lack of effect of G-CSF on viral production in *in vitro* culture systems (Koyangi et al., 1988; Perno et al., 1992) and in neutropenic pediatric patients and adults undergoing zidovudine therapy or therapy for AIDS-related malignancy (Kimura et al., 1990; Sloand et al., 1993) as assessed by p24 EIA or semiquantitative culture. G-CSF has also been used in AIDS patients with bacterial sepsis, aphthous ulcers, and refractory fungal infection (Manders et al., 1995; Vecchiarelli et al., 1995) as adjunctive immunotherapy. No increase in viremia was attributable to G-CSF injections. However, the effects of high doses of G-CSF in non-neutropenic, asymptomatic HIV-seropositive patients, who might be assumed to be at risk for treatment-induced activation of virus expression through indirect mechanisms, have not been determined. Likewise, it is not clear how well such subjects will tolerate the side effects commonly associated with G-CSF. Studies are currently underway to evaluate the use of G-CSF to mobilize hematopoietic stem/progenitor cells from AIDS patients.

GENE THERAPY USING HEMATOPOIETIC STEM CELLS

Despite some encouraging advances made in the identification and enrichment of hematopoietic stem cells and in vector development, there remain two major obstacles to the transfection of hematopoietic stem cells (Friedmann, 1989). The first is a low efficiency of gene expression after transduction (Benveniste and Reshef, 1990; Lu et al., 1993). Similar to the situation in other primate and human somatic cells, only a small amount of foreign DNA can be integrated into human stem cells. The second problem is that the transgene has a short-lived expression even after successful transfection.

The Effect of Stroma on Adhesion Molecules

Some studies have shown that the extent of gene transfer into marrow-derived hematopoietic progenitor cells is increased by using the cytokine combination of interleukin (IL)-3, IL-6, and stem cell factor (SCF) (Lu et al., 1993, 1994). Others have shown that the presence of irradiated marrow stroma cells during transduction of hematopoietic progenitor cells increases gene transfer and eliminates the need to use cytokines (Nolta et al., 1995). There are other reports that have shown relatively efficient gene transfer into peripheral blood stem/progenitor cells in the absence of stroma (Lu et al., 1994). Nolta et al. (1995) have demonstrated that the optimal transduction conditions differed for marrow and peripheral blood-derived stem/progenitor cells. A stromal underlayer appeared to be essential for efficient retroviral-mediated transduction of lineage-committed progenitors as well as for early, primitive stem cells from bone marrow. In contrast, stem/progenitor cells from G-CSF–mobilized peripheral blood were effectively transduced without stroma if cytokines were present. It has been suggested that the presence of bone marrow stroma possessed dual benefits in that it increased gene transfer efficiency and is essential for survival of long-lived human hematopoietic progenitors.

Several mechanisms have been proposed by which irradiated stroma can increase the extent of retroviral-mediated gene transfer into human hematopoietic progenitor cells (Nolta et al., 1995). A co-localization effect may exist by which retroviruses adhere to elements of the stroma matrix. This effect might facilitate the entry of the virions into hematopoietic cells attached to fibronectin or other adhesion molecules on the stroma layer. Moritz et al. (1994) showed that flasks coated with the fibronectin fragment increased the extent of gene transfer into hematopoietic progenitor cells as efficiently as stroma. Another mechanism may be the result of direct stimulation by attachment of stem/progenitor to the stroma layer. Integrins can transmit activation signals via tyrosine phosphorylation in response to attachment to matrix molecules. This activation might be able to induce quiescent stem/progenitor cells into cell cycle. Another possibility is the production of cytokines (sequential, combination, or yet to be identified factors) by the stroma layer that induce cycling of quiescent hematopoietic progenitors, allowing retroviral integration.

Induction of Stem/Progenitor Cells Into Cell Cycle

One of the major limitations in the use of retroviral vectors is that their integration occurs only in replicating cells (Miller et al., 1992). As hematopoietic stem/progenitor cells are predominantly in a resting state and little is known about the replication and cell cycle kinetics of the pluripotent stem cells *in vivo,* various combinations of cytokines have been used empirically to stimulate the stem/progenitor cells into cycle. The cells are exposed to cytokines and then to retroviral vectors with the hope that gene transfer will ensue (Hughes et al., 1989; Gregni et al., 1992; Cassel et al., 1993). We have examined systematically the impact of various cytokines on the cell cycle of $CD34^+$ cells derived from peripheral blood from normal subjects after mobilization using G-CSF or GM-CSF. The $CD34^+$ cells enriched by the immunomagnetic procedures using the Baxter Isolex System were incubated with or without SCF, either alone or in combination with IL-1 (100 ng/ml), IL-3 (500 U/ml), IL-6 (500 U/ml), IL-11 (50 ng/ml), G-CSF (50 ng/ml), GM-CSF (50 ng/ml), or flk-2/flt-3 ligand (100 ng/ml). The cell cycle kinetics of three consecutive cell divisions were studied by labeling with 5 mM bromodeoxyuridine, HO33258, and/or ethidium bromide staining (Agrawal et al., 1996). The best cytokine combinations for inducing $CD34^+$ cells into mitosis after 48 h were determined to be SCF/IL-1, SCF/IL-3, SCF/G-CSF, and SCF/GM-SCF. Without any exogenous cytokines, the cells remained in the G1 noncycling state. (Clear differences were observed in the abilities of different cytokines to induct $CD34^+$ cells into cell cycle when aliquots of cytokine-treated samples were analyzed after 24, 48, or 72 h of incubation.) By 72 h, the majority of cells had cycled once, and some were in the third cycle of cell division in all cytokine combinations. However, even under the most favorable stimulating conditions, 20%–30% of cells remained quiescent and never entered the cell cycle.

We then analyzed the cell cycle of $CD34^+$, $CD38^-$ subsets using 7-aminoactinomycin-D (7-AAD) staining of the stimulated cells. The results suggest that, under the conditions of the assay, most of the $CD34^+$, $CD38^-$ cells derived from mobilized blood do not undergo cell division within 72 h (Agrawal et al., 1996). Other studies have confirmed that the pluripotent stem cells with the highest self-renewal capacity remain initially quiescent under cytokine stimulation (Young et al., 1996; Nolta et al., 1995).

A few clinical trials have suggested that successful transfection of pluripotent stem cells can be achieved using retroviral vectors. The transgenes can be detected in the bone marrow and peripheral blood cells for a long period (>1 year). However, the proportion of transfected progeny cells and the levels of gene expression were low (Brenner et al., 1993; Dunbar et al., 1995). Although the clinical results were encouraging, the technical hurdle of inducing early stem/progenitor cells into cell cycle remained unresolved.

Transduction of Stem/Progenitor Cells Using the Hairpin Ribozyme

We have previously shown that macrophage progeny cells derived from cord blood $CD34^+$ cells transduced with ribozyme vectors were resistant to HIV replication (Li et al., 1996). We have initiated studies to determine whether transfection of PUCB $CD34^+$ stem cells from HIV-exposed infants would compromise progenitor colony formation. The cells were transfected with cell-free recombinant retroviral vectors expressing a single ribozyme against the U5 region of HIV: MJT, MJV, and a double ribozyme vector MY-2 containing an additional ribozyme targeting pol region. No difference in either the clonogenic activity or the transduction efficiency by retroviruses (70%–90%) was observed between $CD34^+$ cells derived from normal donors and those from HIV-exposed donors.

Li et al. (1996) recently studied the use of enriched $CD34^+$ cells from PUCB of newborns of HIV-infected mothers for gene therapy. $CD34^+$ cells enriched from HIV-exposed infants were stimulated (IL-3 at 500 U/ml, IL-6 at 500 U/ml, and SCF at 25 ng/ml) for 24 h and transduced by LNL, MJV, or MY-2 viral supernatants. The transduced cells were plated in flasks coated with an irradiated marrow stromal layer after 6 days. After culturing for 20+ days, supernatant cells were collected and subsequently cultured for colony formation. Individual colonies were removed, and DNA PCR was performed using neo markers as primer. Transduction efficiencies were as high as described for a normal UCB sample. Moreover, when the progeny macrophages of the transduced $CD34^+$ cells were challenged with HIV-pal (a macrophage-tropic strain) or with the HIV isolate of the mother, the cells were protected for up to a period of 56 days (Li et al., 1996). Thus our studies have demonstrated the efficiency and safety of this strategy.

PROSPECTS

For HIV gene therapy to be effective, it should be implemented at a very early stage after HIV infection before extensive damage to lymphoid organs such as the thymus. As most of the antiviral drugs also cause severe myelosuppression, use of high-dose antiviral treatment with a combination of anti-HIV agents will be similar to the use of high-dose chemotherapy and stem cell transplant for patients with cancer. Stem cells will be harvested and, if successfully transduced, can be a durable cure for patients with HIV infection. The ability to harness pluripotent hematopoietic stem cells with self-renewal capacity may represent one solution to circumvent the finite life span of genetically modified mature lymphocytes. Studies are underway to assess the capacity of genetically modified pluripotent hematopoietic stem cells to undergo multilineage development in HIV-infected individuals.

ACKNOWLEDGMENTS

This work was supported by National Institutes of Health grants R01 DK49619-01, R01 DK49619-02, 1 U 19 AI36612-02, and 1 U 19 AI36612-03; by CRCC University of California Systemwide training grant 5-0873; by grants from the American Cancer Society, the Pete Lopiccola Memorial Foundation, the UCSD Cancer Center Foundation, and the UCSD Center for AIDS Research.

REFERENCES

Agrawal YP, Agrawal RS, Sinclair AM, Young D, Maruyama M, Levine F, Ho AD (1996): Cell cycle kinetics and VSV-G pseudotyped retrovirus mediated gene transfer in blood-derived CD34+ cells. *Exp Hematol* 24:738–747.

Baltimore D (1988): Gene therapy: Intracellular immunization. *Nature* 335:395–396.

Bender JG, Universagt KL, Walker DE, Lee W, Van Epps DE, Smith DH, Stewart CC, To LB (1991): Identification and comparison of CD34-positive cells and their subpopulations from normal peripheral blood and bone marrow using multicolor flow cytometry. *Blood* 77:2591–2596.

Bensinger W, Singer J, Appelbaum F, et al. (1993): Autologous transplantation with peripheral blood mononuclear cells collected after administration of recombinant granulocyte stimulating factor. *Blood* 81:3158–3163.

Bensinger WI, Weaver CH, Appelbaum FR, Rowley S, Demirer T, Sanders J, Storb R, Buckner CD (1995): Transplantation of allogeneic peripheral blood stem cells mobilized by recombinant human granulocyte colony-stimulating factor. *Blood* 85:1655.

Benveniste N, Reshef L (1990): Direct introduction of genes into rats and expression of the genes. *Proc Natl Acad Sci USA* 83:275–276.

Brenner MK, Rill DR, Holladay MS, Heslop HE, Moen RC, Buschle M, Krance RA, Santana VM, Anderson WF, Ihle JN (1993): Gene marking to determine whether autologous marrow infusion restores long-term haemopoiesis in cancer patients. *Lancet* 342:1134–1137.

Broxmeyer HE, Hangoc G, Cooper S, Ribeiro RC, Graves V, Yoder, M, Wagner J, Vadhan-Raj S, Bennminger L, Rubinstein P, Broun ER (1992): Growth characteristics and expansion of human umbilical cord blood and estimation of its potential for transplantation in adults. *Proc Natl Acad Sci USA* 89:4109.

Busch M, Beckstead J, Gantz D, Vyas G (1986): Detection of HIV infection of myeloid precursors in bone marrow samples from AIDS patients. *Blood* 68:122a.

Cassel A, Cottler-Fox M, Doren S, Dunbar CE (1993): Retroviral mediated gene transfer into CD34-enriched human peripheral blood stem cells. *Exp Hematol* 21:585–591.

Civin I, Banquerigo ML, Strauss LC, Loken MR (1987): Antigenic analysis of hematopoiesis. IV. Characterization of MY10-positive progenitor cells in normal human bone marrow. *Exp Hematol* 15:10–17.

Davis TA, Robinson DH, Lee KP, Kessler SW (1995): Porcine brain microvascular endothelial cells support the *in vitro* expansion of human primitive hematopoietic bone marrow progenitor cells with a high replating potential: Requirement for cell-to-cell interact and colony-stimulating factors. *Blood* 85:1751–1761.

Davis BR, Schwartz DH, Marx JC, et al. (1991): Absent or rare human immunodeficiency virus infection of bone marrow stem/progenitor cells *in vivo*. *J Virol* 65:1985–1990.

Davison FD, Kaczmarski RS, Pozniak A, Mufti GJ, Sutherland S (1994): Quantification of HIV by PCR in monocytes and lymphocytes in patients receiving antiviral treatment and low dose recombinant human granulocyte macrophage colony stimulating factor. *J Clin Pathol* 47:855–857.

De Luca A, Teofill L, Antinori A, et al. (1993): Haemopoietic $CD34^+$ progenitor cells are not infected by HIV-1 *in vivo* but show impaired clonogenesis. *Br J Haematol* 85:20–24.

Dunbar CE, Cottler-Fox M, O'Shaugnessy JA, Doren S, Carter C, Berenson R, Brown S, Moen RC, Greenblatt J, Stewart FM, Leitman SF, Wilson WH, Cowan K, Young NS, Nienhuis AW (1995): Retrovirally marked CD34-enriched peripheral blood and bone marrow cells contribute to long-term engraftment after autologous transplantation. *Blood* 85:3048–3057.

Friedmann T (1989): Progress toward human gene therapy. *Science* 244:1275–1281.

Folks TM, Kessler SW, Orenstein JM, Justement JS, Jaffe E, Fauci AS (1988): Infection and replication of HIV-1 in purified progenitor cells of normal human bone marrow. *Science* 242:919–922.

Ganser A, Ottman OC, von Briesen H, Volkers B, Rubsamen-Waigmann H, Hoelzer D (1990): Changes in the haematopoietic progenitor cell compartment in the acquired immunodeficiency syndrome. *Res Virol* 141:185.

Gluckman E, Thierry D, Lesage S, et al. (1992): Clinical applications of stem cell transfusion from cord blood. *Transfusion Sci* 13:415.

Goldman (1995): Peripheral blood stem cells for allografting. *Blood* 85:1413.

Gregni M, Magni M, Siena S, DiNicola M, Bonadonna G, Gianni AM (1992): Human peripheral blood hematopoietic progenitors are optimal targets of retroviral-medicated gene transfer. *Blood* 80:1418.

Grigg AP, Roberts AW, Raunow H, Houghton S, Layton JE, Boyd AW, McGrath KM, Maher D (1995): Optimizing dose and scheduling of Filgrastim (granulocyte colony-stimulating factor) for mobilization and collection of peripheral blood progenitor cells in normal volunteers. *Blood* 86:4437–4445.

Ho AD, Mason J, Corringham RET (1993): Hematopoietic progenitor cells: Sources, applications, and expansion. *Oncology* 7:17–20.

Ho AD, Young D, Maruyama M, et al. (1996): Pluripotent and lineage-committed $CD34^+$ subsets in leukapheresis products mobilized by G-CSF, GM-CSF versus a combination of both. *Exp Hematol* 24:1460–1468.

Huang S, Terstappen LW (1994): Lymphoid and myeloid differentiation of single human $CD34^+$, $HLA\text{-}DR^+$, $CD38^-$ hematopoietic stem cells. *Blood* 83:1515–1526.

Hughes PFD, Eaves LJ, Hogge DE, Humphries EK (1989): High-efficiency gene transfer to human hematopoietic cells maintained in long-term culture. *Blood* 74:1915–1922.

Joshi S, Van Brunschot A, Asad S, van der Elst I, Read SE, Bernstein A (1991): Inhibition of human immunodeficiency virus type 1 multiplication by antisense and sense RNA expression. *J Virol* 65:5524–5530.

Kaczmarski RS, Davison F, Blair E, Sutherland S, Moxham J, McManus T, Mufti GJ (1992): Detection of HIV in haemopoietic progenitors. *Br J Haematol* 82:764–769.

Kaplan LD, Kahn JO, Crowe S, Northfelt D, Neville P, Grossberg H, Abrams DI, Tracey J, Mills J, Volberding PA (1991): Clinical and virologic effects of recombinant human granulocyte-macrophage colony-stimulating factor in patients receiving chemotherapy for human immunodeficiency virus-associated non-Hodgkin's lymphoma: Results of a randomized trial. *J Clin Oncol* 9:929–940.

Kessinger A, Armitage JO (1991): The evolving role of autologous peripheral stem cell transplantation following high-dose therapy for malignancies. *Blood* 77:211–213.

Kessinger A, Armitage JO, Landmark JD, et al. (1986): Reconstitution of human hematopoietic function with autologous cryopreserved circulating stem cells. *Exp Hematol* 114:192–196.

Kimura S, Matsuda J, Ikematsu S, Miyazono K, Ito A, Nakahata T, Minamitani M, Shimada K, Shiokawa Y, Takaku F (1990): Efficacy of recombinant human granulocyte colony-stimulating factor on neutropenia in patients with AIDS. *AIDS* 12:1251–1255.

Kitano K, Abboud CN, Ryan DH, Quan SG, Baldwin GC, Golde DW (1991): Macrophage-active colony-stimulating factors enhance human immunodeficiency virus type 1 infection in bone marrow stem cells. *Blood* 77:1699–1705.

Körbling M, Dörken B, Ho AD, et al. (1986): Autologous transplantation of blood-derived hemopoietic stem cells after myeloablative therapy in a patient with Burkitt's lymphoma. *Blood* 67:529–532.

Körbling M, Przepiorka D, Huh YO, Engel H, van Besien K, Giralt S, Andersson B, Kleine HD, Seong D, Deisseroth AB, Andreeff M, Champlin R (1995): Allogeneic blood stem cell transplantation for refractory leukemia and lymphoma: Potential advantage of blood over marrow allografts. *Blood* 85:1659.

Koyangi Y, O'Brien WA, Zhao JQ, Golde DW, Gassen JC, Chen IS (1988): Cytokines alter production of HIV-1 from primary mononuclear phagocytes. *Science* 241:1673–1675.

Krown SE, Paredes J, Bundow D, Polsky B, Gold JW, Flomenberg N (1992): Interferon-alpha, zidovudine, and granulocyte-macrophage colony-stimulating factor: A phase I AIDS Clinical Trials Group study in patients with Kaposi's sarcoma associated with AIDS. *J Clin Oncol* 10:1344–1351.

Lafeuillade A, Tamalet C, Pellegrino P, Tourres C, Vignoli C, Quilichini R (1992): HIV-1 viraemia during granulocyte-macrophage colony stimulating factor therapy. *AIDS* 6:1550–1551.

Lane TA, Law P, Maruyama M, Young D, Burgess J, Mullen M, Mealiffe M, Terstappen LWMM, Hardwick A, Moubayed M, Oldham F, Corringham RET, Ho AD (1995): Harvesting and enrichment of hematopoietic progenitor cells mobilized into the peripheral blood of normal donors by granulocyte macrophage-colony stimulating factor (GM-CSF) or G-CSF: Potential role in allogeneic marrow transplantation. *Blood* 85:275–282.

Lansdorp PM, Dragowska W (1992): Long-term erythropoiesis from constant numbers of $CD34^+$ cells in serum-free cultures initiated with highly purified progenitor cells from human bone marrow. *J Exp Med* 175:1501–1509.

Lee TC, Sullenger BA, Gallardo HF, Ungers GE, Gilboa E (1992): Overexpression of RRE-derived sequences inhibits HIV-1 replication in CEM cells. *New Biologist* 4:66–74.

Li X, Gervaix A, Kang D, Peterson S, Law P, Spector SA, Ho AD, Wong-Staal F (1998): Gene therapy targeting cord blood derived $CD34^+$ cells from HIV-exposed infants: Preclinical studies. *Gene Therapy* 5:233–239.

Lisziewicz J, Sun D, Smythe J, et al. (1993): Inhibition of human immunodeficiency virus type 1 replication by regulate expression of a polymeric Tat activation response RNA decoy as a strategy for gene therapy in AIDS. *Proc Natl Acad Sci USA* 90:8000–8004.

Lorentzen EU, Wieland U, Kuhn JE, Braun RW (1991): *In vitro* cleavage of HIV-1 vif RNA by a synthetic ribozyme. *Virus Genes* 5:17–23.

Lu M, Maruyama M, Zhang N, et al. (1994): High efficiency retroviral-mediated gene transduction into $CD34^+$ cells purified from peripheral blood. *Hum Gene Ther* 5:203–208.

Lu L, Xiao M, Clapp DW, Li Z-H, Broxmeyer HE (1993): High efficiency retroviral mediated gene transduction into single isolated immature and replatable $CD34^+$ hematopoietic stem/progenitor cells from human umbilical cord blood. *J Exp Med* 178:2089–2096.

Manders SM, Kostman JR, Mendez L, Russin VL (1995): Thalidomide-resistant HIV-associated aphthae successfully treated with granulocyte colony-stimulating factor. *J Am Acad Dermatol* 33(2 Pt 2):380–382.

Mayani H, Lansdorp PM (1995): Proliferation of individual hematopoietic progenitors purified from umbilical cord blood. *Exp Hematol* 23:1453–1462.

Mayani H, Little MT, Dragowska W, Thornbury G, Lansdorp PM (1995): Differential effects of the hematopoietic inhibitors MIP-1α, TGF-β, and TNF-α on cytokine-induced proliferation of subpopulations of CD34+ cells purified from cord blood and fetal liver. *Exp Hematol* 23:422–427.

Miller DG, Adam MA, Miller AD (1992): Gene transfer by retrovirus vectors occurs only in cells that are actively replicating at the time of infection. *Mol Cell Biol* 10:4239–4242.

Molina JM, Scadden DT, Sakaguchi M, et al (1990): Lack of evidence for infection of or effect on growth of hematopoietic progenitor cells after *in vivo* or *in vitro* exposure to human immunodeficiency virus. *Blood* 86:2476–2482.

Moritz T, Patel VP, Williams DA (1994): Bone marrow extracellular matrix molecules improve gene transfer into human hematopoietic cells via retroviral vectors. *J Clin Invest* 93:1451–1457.

Nolta JA, Smogorzewska EM, Kohn DB (1995): Analysis of optimal conditions for retroviral-mediated transduction of primitive human hematopoietic cells. *Blood* 86:101–110.

Perno CF, Cooney DA, Gao WY, Hao Z, Johns DG, Foli A, Hartman NR, Calio R, Broder S, Yarchoan R (1992): Effects of bone marrow stimulatory cytokines on human immunodeficiency virus replication and the antiviral activity of dideoxynucleosides in cultures of monocyte/macrophages. *Blood* 80:995–1003.

Peters WP, Rosner G, Ross M, Vredenburgh J, Eisenberg B, Gilbert C, Kurtzberg J (1993): Comparative effects of granulocyte-macrophage colony-stimulating factor (GM-CSF) and granulocyte colony-stimulating factor (G-CSF) on priming peripheral blood progenitor cells for use with autologous bone marrow after high-dose chemotherapy. *Blood* 81:1709–1719.

Pettengell R, Testa NG, Swindell R, Crowther D, Dexter MT (1993): Transplantation potential of hematopoietic cells released into the circulation during routine chemotherapy for non-Hodgkin's lymphoma. *Blood* 82:2239–2248.

Poeschla E, Wong-Staal F (1995–96): Gene therapy and HIV disease. *AIDS Clin Rev* 3–45.

Rhodes A, James W (1990): Inhibition of human immunodeficiency virus replication in cell culture by endogenously synthesized antisense RNA. *J Gen Virol* 71:1965–1974.

Rhodes A, James W (1991): Inhibition of heterologous strains of HIV by antisense RNA. *AIDS* 5:145–151.

Rice HE, Emani VR, Skarsgard ER, Knazek RA, Zanjani ED, Flake AW, Harrison MR (1994): Human fetal liver hematopoietic cell expansion with a novel bioreactor system. *Transplant Proc* 26:3338–3339.

Riddell SR, Elliott M, Lewinsohn DA, Gilbert MJ, Wilson L, Manley SA, Lupton SD, Overell RW, Reynolds TC, Corey L, et al (1996): T-cell mediated rejection of gene-modified HIV-specific cytotoxic T lymphocytes in HIV-infected patients. *Nature Med* 2:216–223.

Roncarolo MG, Bacchetta R, Bigler M, Touraine JL, de Vries JE, Spits H (1991): A SCID patient reconstituted with HLA-incompatible fetal stem cells as a model for studying transplantation tolerance. *Blood Cells* 17:391–402.

Rubinstein P, Rosenfield RE, Adamson JW, Stevens CS (1993): Stored placental blood for unrelated bone marrow reconstitution. *Blood* 81:1679.

Russell NH, Hunter A, Rogers S, Hanley J, Anderson D (1993): Peripheral blood stem cells as an alternative to marrow for allogeneic transplantation. *Lancet* 341:1482.

Sarver N, Cantin EM, Chang PS, et al (1990): Ribozymes as potential anti-HIV-1 therapeutic agents. *Science* 247:1222–1225.

Schmitz N, Dreger P, Suttorp M, Rohwedder EBK, Haferlach T, Löffler H, Hunter A, Russell NH (1995): Primary transplantation of allogeneic peripheral blood progenitor cells mobilized by filgrastim (granulocyte colony-stimulating factor). *Blood* 85:1666.

Sczakiel G, Pawlita M (1991): Inhibition of human immunodeficiency virus type 1 replication in human T cells stably expressing antisense RNA. *J Virol* 65:468–472.

Sheridan WP, Begley CG, Juttner CA, et al (1992): Effect of peripheral-blood progenitor cells mobilized by filgrastim (G-CSF) on platelet recovery after high-dose chemotherapy. *Lancet* 339:640–644.

Sioud M, Drlica K (1991): Prevention of human immunodeficiency virus type 1 integrase expression in *Escherichia coli* by a ribozyme *Proc Natl Acad Sci USA* 88:7303–7307.

Sloand E, Kumar PN, Pierce PF (1993): Chemotherapy for patients with pulmonary Kaposi's sarcoma: Benefit of filgrastim (G-CSF) in supporting dose administration. *S Med J* 86:1219–1224.

Smith C, Lee SW, Sullenger BA, Gallardo HF, Ungers GE, Gilboa E (1993): Intracellular immunization against HIV using RNA decoys. The 1993 UCLA/UCI AIDS Symposium: Gene Therapy Approaches to Treatment of HIV Infection. *Conference Program,* p 25.

Stanley SK, Kessler SW, Justement JS, Schnittman SM, Greenhouse JJ, Brown CC, Musongela L, Musey K, Kapita B, Fauci AS (1992): CD34^{+} bone marrow cells are infected with HIV in a subset of seropositive individuals. *J Immunol* 149:689–697.

Stephenson J (1996): New anti-HIV drugs and treatment strategies buoy AIDS researchers. *JAMA* 275:579–580.

Sullenger BA, Gallardo HF, Ungers GE, Gilboa E (1990): Overexpression of TAR sequences renders cells resistant to human immunodeficiency virus replication. *Cell* 63:601–608.

Terstappen LWMM, Huang S, Safford M, Lansdorp PM, Loken MR (1991): Sequential generations of hematopoietic colonies derived from single nonlineage-committed CD34^{+}/CD38^{-} progenitor cells. *Blood* 77:1218–1227.

Traycoff CM, Kosak ST, Grigsby S, Srour EF (1995): Evaluation of *ex vivo* expansion potential of cord blood and bone marrow hematopoietic progenitor cells using cell tracking and limiting dilution analysis. *Blood* 85:2059–2068.

Vecchiarelli A, Monari C, Baldelli F, Pietrella D, Retini C, Tascini C, Francisci D, Bistoni F (1995): Beneficial effect of recombinant human granulocyte colony-stimulating factor on fungicidal activity of polymorphonuclear leukocytes from patients with AIDS. *J Infect Dis* 171:1448–1454.

Verfaillie CM, Miller JS (1995): A novel single-cell proliferation assay shows that long-term culture-initiating cell (LTC-IC) maintenance over time results from the extensive proliferation of a small fraction of LTC-IC. *Blood* 86:2137–2145.

von Laer D, Hufert FT, Fenner TE, Schwander S, Dietrich M, Schmitz H, Kern P (1990): CD34^{+} hematopoietic progenitor cells are not a major reservoir of the human immunodeficiency virus. *Blood* 76:1281–1286.

Wagner JR (1993): Umbilical cord blood stem cell transplantation. *Am J Pediatr Hematol Oncol* 15:169–174.

Weerasinghe M, Liem SE, Asad S, Read SE, Joshi S (1991): Resistance to human immunodeficiency virus type 1 (HIV-1) infection in human $CD4^+$ lymphocyte-derived cell lines conferred by using retroviral vectors expressing an HIV-1 RNA-specific ribozyme. *J Virol* 65:5531–5534.

Xu Y, Reddy TR, Fischer WH, Wong-Staal F (1996): A novel HnRNP specifically interacts with HIV-1 RRE RNA. *J Biomed Sci* 3:82–91.

Young JC, Varma A, DiGiusto D, Backer MP (1996): Retention of quiescent hematopoietic cells with high proliferative potential during *ex vivo* stem cell culture. *Blood* 87:545–556.

Yu M, Poeschla E, Wong-Staal F (1994): Progress towards gene therapy for HIV infection. *Gene Ther* 1:13–26.

12

ELEMENTS OF DNA VACCINE DESIGN

MICHAEL J. CAULFIELD AND MARGARET A. LIU

Department of Virus and Cell Biology, Merck Research Laboratories, West Point, PA 19454

Overview

The concept of immunization using nucleic acid vectors encoding antigens rather than purified protein vaccines has numerous theoretical advantages over the use of conventional subunit vaccines. The central question is, however, does this novel methodology work? The answer is clearly yes—the original work has been reproduced in many laboratories and extended to many different preclinical disease models. Do DNA vaccines work as well as conventional vaccines? The answer to this question may depend on the particular immune response necessary for protection. If cell-mediated immune responses or a more diverse array of antibody isotypes are desired, then DNA vaccines may have the advantage over certain conventional vaccines that induce antibody responses sometimes containing undesirable (e.g., IgE) isotypes. Clearly, however, immunization using DNA (or RNA) vectors will enhance vaccine development by facilitating the discovery of new antigens and by allowing the testing of modified or hybrid proteins without the painstaking process of expression and purification of each antigen—a process that may require the development of new purification schemes for each new protein modification. In contrast, the molecular biology necessary to construct plasmid DNA expres-

Stem Cell Biology and Gene Therapy, Edited by Peter J. Quesenberry, Gary S. Stein, Bernard G. Forget, and Sherman M. Weissman
ISBN 0-471-14656-0

sion vectors is usually straightforward, and purification of plasmid DNA is the same regardless of the inserted gene. The field of DNA vaccines has been reviewed extensively in recent articles from the perspective of disease targets and efficacy, particularly in animal models (Donnelly et al., 1997; Manickan et al., 1997b; Shiver et al., 1996c; Ulmer et al., 1996b). Therefore, the present chapter focuses on strategies for the design of vaccines that exploit the advantages of the DNA vaccine approach.

Antigen Expression

Because bacterial plasmids did not evolve to vector foreign genes into mammalian cells, additional elements must be incorporated to allow expression of the target gene. An expression vector must contain, at a minimum, a promoter and a terminator. The promoter (which is usually of viral origin) is required to enlist nuclear-binding factors from the host cell to allow initiation of transcription. A terminator is required to provide the genetic information necessary for the addition of a poly(A) tail, which is needed for exit from the nucleus to the cytoplasm where translation into protein occurs. Once in the cytoplasm, the mRNA can be translated into protein. Since the fate of a protein depends on its sequence, the potential exists to alter the coding sequence or its regulatory elements using standard recombinant DNA technology to target the molecule along different intracellular pathways to effect secretion, cell membrane expression, or degradation. In this way, the resulting immune response may be redirected from, e.g., a humoral response to a cell–mediated response.

Immunogenicity of DNA Vaccines

Following the demonstration by Wolff et al. (1990) that intramuscular (i.m.) injection of naked DNA could result in expression of reporter genes in transfected myocytes, Tang et al. (1992) showed that immunization using DNA attached to gold beads delivered via a "gene gun" was effective at inducing antibody responses. Ulmer et al. (1993) demonstrated that i.m. injection of plasmid DNA encoding influenza nucleoprotein (NP) resulted in the generation of NP-specific cytotoxic T lymphocytes (CTLs) and cross-strain protection from death and disease after a live virus challenge. Since then DNA vaccines have been used to induce immune responses to a wide variety of microbial antigens as well as tumor-associated antigens, as summarized in Table 1.

The ultimate goal of immunization is to induce protective immunity. This entails eliciting the appropriate type of immune response. For most diseases, neutralizing antibodies are desired for protection. However, for some diseases (such as tuberculosis), a cell-mediated immune response is required for protection (Lowrie et al., 1995). For development of a therapeutic vaccine, a cell-mediated response may be required.

TABLE 1 Disease Targets and Selected Experimental Models for DNA Vaccines

Disease	Antigen(s)	Challenge Model	Reference
Bovine herpes	gIV	Cattle	Babiuk et al. (1995)
Colon carcinoma	Carcinoembryonic antigen	Mouse	Conry et al. (1994, 1995)
Hepatitis B	HBs, preS, core	—	Davis et al. (1996), Michel et al. (1995), Schirmbeck et al. (1995)
Hepatitis C	Core antigen	—	Geissler et al. (1997), Lagging et al. (1995)
Herpes simplex 1/2	env gB/gD	Mouse, guinea pig	Bourne et al. (1996), Manickan et al. (1995), McClements et al. (1996)
HIV	env, gag, nef, pol, rev	SHIV-rhesus monkey	Asakura et al. (1996), Boyer et al. (1996), Fuller et al. (1996), Lekutis et al. (1997), Liu et al. (1996), Okuda et al. (1995), Shiver et al. (1995, 1996a,b), Wahren et al. (1995), Wang et al. (1993, 1995), Yasutomi et al. (1996)
Influenza	HA, NP	Mouse, ferret	Donnelly et al. (1995), Fynan et al. (1993a), Montgomery et al. (1993), Robinson et al. (1993), Ulmer et al. (1993, 1995)
Leishmaniasis	gp63	Mouse	Xu and Liew (1995)
Lyme disease	OspA	Mouse	Simon et al. (1996)
Lymphoma leukemia	Idiotype	Mouse	Stevenson et al. (1995), Syrengelas et al. (1996)
Malaria	Circumsporozoite	Mouse	Doolan et al. (1996), Sedegah et al. (1994)
Melanoma	MAGE	Mouse	Bueler and Mulligan (1996)
Papilloma	L1 capsid protein	Rabbit	Donnelly et al. (1996)
Rabies	Glycoprotein	—	Ertl et al. (1995), Xiang et al. (1995)
Schistosomiasis	Paramyosin	—	Yang et al. (1995)
Streptococcal pneumonia	PspA	Mouse	McDaniel et al. (1997)
Tuberculosis	Antigen 85, hsp65	Mouse	Huygen et al. (1996), Lowrie et al. (1994), Tascon et al. (1996)

SHIV; simian HIV; gIV; bovine herpes virus envelope glycoprotein IV; HSV; herpes simplex virus; gB; HSV envelope glycoprotein B; gD; HSV envelope glycoprotein D; HA; influenza hemagglutinin; PspA; pneumococcal surface protein A; MAGE: melanoma antigen gene.

DELIVERY

Route of Injection

For naked DNA injected by syringe, the route of injection is important in optimizing antibody responses. For influenza vaccines, the i.m. route is clearly superior to intravenous (i.v.) or intraperitoneal (i.p) injection (Donnelly et al., 1994). Intradermal injection has been shown to induce potent antibody and CTL responses (Raz et al., 1994, 1996) but may be less effective in the induction of helper T-cell responses (Shiver et al., 1995, 1996b) or protection from live virus challenge (Donnelly et al., 1994). Delivery of DNA on gold particles using a "gene gun" has been shown to induce potent immune responses with much lower amounts of DNA compared with i.m. injection (Fynan et al., 1993b; Herrmann et al., 1996; Tang et al., 1992); however, the response to gene gun immunization is qualitatively different from that obtained with i.m. injection of DNA. In a comparative study, i.m. inoculation was shown to induce Th1-like responses as defined by elevated IgG2a levels, production of interferon-γ (IFN-γ), CTL activity, and lack of interleukin (IL)-4 production. By contrast, the gene gun was reported to induce a mixed Th1/Th2 response consisting of IFN-γ and CTL activity characteristic of Th1 responses as well as elevated IgG1 antibodies and IL-4 production characteristic of a Th2 response (Pertmer et al., 1996). In a subsequent study, Feltquate et al. (1997) reported that gene gun administration of an influenza hemagglutinin (HA) expressing plasmid led to a predominantly Th2 response, whereas i.m. injection of the vaccine resulted in a Th1 response profile.

Delivery Vehicles

A number of strategies have been used to enhance the potency and stability of DNA vaccines. Delivery vehicles range from inert liposome vesicles designed to protect DNA from degradation to formulations that include immunostimulating substances. Studies on the mechanism of DNA delivery with cationic lipids support the concept that the DNA–lipid complexes fuse with cell membranes, thereby facilitating entry of intact DNA into the cytoplasm. Most of the widely available cationic lipids, including Lipofectin and Lipofectamine, are reported to have very poor ability to enhance DNA expression in the lung over that of naked DNA (Felgner et al., 1995). The intracellular fate of injected DNA-containing liposomes remains largely unknown. Presumably, the DNA–lipid complexes are taken up into phagolysosomes from which a fraction of the DNA escapes into the cytoplasm before being degraded. Although various compounds have been evaluated for their ability to aid *in vitro* transfection of cells by DNA, it is not clear that any correlation exists between the *in vitro* utility of a delivery vehicle and its usefulness for generating or enhancing immune responses *in vivo*.

Intracellular Fate of Plasmid DNA

Injection of naked plasmid DNA into muscles has been shown to result in uptake and expression of the encoded gene product. How naked DNA enters cells is not known in detail; however, there are reports that specific receptors on the cell membrane of human leukocytes mediate binding and internalization of DNA (Bennett et al., 1985, 1986). Whether naked DNA is taken up into phagosomes or whether it directly enters the cytoplasm is unknown, but it is likely that most plasmid DNA is degraded (Bennett et al., 1985) to mono- or oligonucleotides within the lysosome and that only a fraction of intact DNA escapes to the cytoplasm. The oligonucleotides resulting from degradation of the plasmid DNA may not be inert, as it has been demonstrated that certain DNA motifs containing CpG dinucleotides have potent immunostimulatory properties (Krieg et al., 1995; Yamamoto et al., 1992). This issue is discussed in detail below.

Migration of DNA to the Nucleus

In a recent paper, Dean (1997) showed that pBR322 plasmid DNA was not imported into the nucleus of mammalian cells, whereas plasmids containing the SV40 origin of replication and eukaryotic promoter sequences were imported at the same rate as the SV40 genome. By contrast, pBR plasmids containing the SV40 T antigen, polyadenylation signal, or pBR322–SV40 hybrid plasmids lacking early and late promoters as well as the origin of SV40 remained in the cytoplasm. Since the SV40 promoter region contains consensus binding sites for eukaryotic transcription factors such as Oct-1 and NF-κB (Jones et al., 1989), these results suggest that viral/eukaryotic promoter sequences inserted into DNA vaccines serve two purposes: (1) to target the DNA for import to the nucleus and (2) to enable efficient transcription of mRNA once the plasmid reaches the nucleus. Whereas nuclear transport of proteins is reported to take only 30 min (Adam and Gerace, 1991), nuclear import of plasmid DNA requires 6–8 h (Dean, 1997).

GENE AND PROTEIN EXPRESSION

Transcription

Promoter Effects. Once in the nucleus, the plasmid DNA must compete with host DNA for transcription factors and other DNA-binding proteins enabling the transcription and processing of DNA into mRNA. Early studies used strong promoters from potentially oncogenic viruses such as Rous sarcoma virus (RSV) (Gorman et al., 1982) or SV40 (Moreau et al., 1981). Subsequent studies were done to evaluate various promoters from more acceptable sources for relative strength as measured by expression of reporter genes *in vitro* and *in vivo.* Most laboratories currently utilize the immediate early promoter from human cytomegalovirus (CMV) (Boshart et al., 1985). This promoter (with or without intron A) has been used in numerous expression vectors and DNA vaccines as reviewed elsewhere (Donnelly et al., 1997; Ulmer et al., 1996b). An interesting exception to the general finding that eukaryote/viral promoters are superior to prokaryotic promoters is the observation of Simon et al. (1996) that plasmid DNA encoding the outer surface lipoprotein A (OspA) of *Borrelia burgdorferi* under the control of its own bacterial promoter induced OspA-specific antibody and T-cell responses. These investigators identified a 140 bp region immediately upstream from the OspA open reading frame (ORF) that was essential for transcription. Deletion of this region eliminated expression and immunogenicity of a plasmid in which OspA was situated next to the human elongation factor 1α (EF-1α) promoter.

Enhancers. As the name implies, enhancer elements increase expression of genes by enhancing (but not substituting for) the activity of the promoter. Enhancer elements can be upstream or downstream from the ORF of the gene of interest. For DNA vaccines, enhancer elements have not been studied systematically, and their presence in most DNA vaccines is serendipitous.

Transcriptional Transactivation. Genes encoding transcriptional transactivators have not been studied systematically in DNA vaccines. The HBX gene, a known transactivator located about 500 bp downstream from the gene encoding hepatitis B envelope (HBs) protein has been included in the DNA vaccines studied by Davis et al. (1993); however, vectors lacking the HBX gene have not been directly compared with those containing this transactivator. Therefore, the contribution of the HBX gene to the immunogenicity of the HBs gene remains unknown. Most transactivator genes studied are of viral origin, and some are associated with oncogenesis. Thus, addition of such genes to DNA vaccines intended for mass immunization might be problematic. Nevertheless, systematic investigation of transactivating genes that increase expression and immunogenicity of target antigens may be a fertile area for future investigation. Studies on nonviral transactivators such as the tetR element used in so called maximum expression and regulated vectors (MERVs) that have been reported to increase expression by ~10-fold (Liang et al., 1996) may prove to be beneficial for certain applications of DNA vaccination.

Termination/Poly(A) Signal Site. Transcription of mRNA in mammalian cells is accompanied by the addition of a poly(A) tail required for exit of the message from the nucleus to the cytoplasm where protein synthesis takes place. The poly(A) tail is added 11–30 nucleotides downstream from a conserved sequence (AAUAAA) in the 3′ end of the mRNA transcript. Many DNA vaccines use the bovine growth hormone (BGH) terminator sequence (Montgomery et al., 1993), whereas others use endogenous terminators that are downstream from the ORF of the gene of interest (e.g., the 3′-untranslated region [UTR] of the HBs transcript contains an endogenous terminator that can replace the BGH terminator and allow expression in human myoblast [RD] cells [M. Caulfield, X. Liu, and S. Wang, unpublished observations]). In other studies, modifications to the polyadenylation and transcriptional termination sequences have been shown to influence reporter gene expression (Hartikka et al., 1996).

Message Stability. The stability of mRNAs varies from minutes to hours, depending on the sequence of the untranslated region. Long-lived messages (e.g., hemoglobin) contain a ~2 kb UTR and have a half life of ~10 h. By contrast, other genes (especially cytokine genes) contain AT-rich sequences such as ATTTA that are associated with rapid turnover of the message (Kruys et al., 1989; Shaw and Kamen, 1986). Rajagopalan et al. (1995) used a gene gun to transfect peripheral blood mononuclear cells with expression vectors encoding granulocyte-macrophage colony-stimulating factor (GM-CSF) containing wild-type or mutated instability elements. They found that a single base change in the AT-rich UTR (from ATTTA to ATGTA) resulted in mRNA expression that was increased ~ 5-fold from a half life of 20 min to a half life of 95 min. Based on these observations, DNA vaccines should avoid such "instability sequences" in order to prolong expression.

Protein Synthesis

Initiation Site Optimization. Genes derived from viruses and bacteria may not contain a protein synthesis initiation site optimized for use in mammalian cells. The Kozak consensus sequence, GCCGCC(A/G)CCAUGG, defines what is thought to be the optimal context for initiation of protein synthesis in mammalian cells (Kozak, 1987). Many investigators have altered the sequence preceding the initiation codon (AUG) to incorporate this sequence. Again, as for other sequence alterations used in DNA vaccines, systematic comparisons of vectors using or lacking the Kozak sequence have not been studied.

Protein Termination. In mammalian cells translation is initiated at an AUG codon, and termination occurs at UGA codons. Occasionally, "read through" occurs and protein synthesis proceeds until a second stop codon is reached. This can result in synthesis of an oversized aberrant protein that may fold improperly and may be targeted for degradation. To prevent read through, some investigators have inserted double stop codons (e.g., UAGUGA) to ensure proper termination of protein synthesis (Yellen-Shaw and Eisenlohr, 1997). The potential benefit of this practice has not been examined for DNA vaccines.

Expression Versus Immunogenicity

Generally, DNA vaccines are tested for expression of target genes after transient transfection in mammalian cells *in vitro* prior to immunogenicity testing *in vivo.* There are no reported exceptions to the rule that immunogenic DNA vaccines are expressed after *in vitro* transfection into myoblast or other cell types. However, there is less evidence demonstrating that the level of expression *in vitro* correlates directly with immunogenicity *in vivo.* Constructs that are more potent *in vivo* have generally been shown to express better *in vitro;* however, the converse may not be true, and systematic studies of this issue have not been reported.

Correlation between expression *in vivo* and immunogenicity has been more difficult to study. Expression of reporter genes such as β-galactosidase and luciferase indicate that long-lasting expression can occur in mice injected with certain vectors (Wolff et al., 1992). This implies that the immune response to these reporter genes did not eliminate the antigen-expressing cells. By contrast, Davis et al. (1993) have reported that HBs antigen can be detected *in vivo* following injection of an HBs DNA vaccine and that the antigen becomes undetectable at the time anti-HBs antibodies arise. The long-term persistence of immune responses in certain species following DNA inoculation may be reflective of species differences more than an indicator of persistent antigen expression.

ANTIGEN PROCESSING

The issue of antigen presentation after injection of plasmid DNA is of special interest because the antigen can potentially be expressed in multiple cells of the host, i.e., the vaccinee. The pathway of antigen processing appears to depend on whether the antigen expressed by a DNA vaccine is secreted, expressed as a membrane protein, or rapidly degraded into peptides that are then bound to MHC class I molecules. Secreted proteins would be expected to enter the class II processing pathway, whereas membrane-bound or internally processed peptides would be expected to enter the class I pathway.

Targeting Protein to Proteosomes/TAP

An antigen lacking an N-terminal signal peptide would be synthesized in the cytoplasm and subsequently degraded into peptide fragments in proteosomes. Peptides are then transported via the transporter associated with antigen processing (TAP) complex (Monaco et al., 1990; Spies et al., 1992) to the endoplasmic reticulum (ER) where they bind to MHC class I molecules. Indeed, there is now evidence that rapid intracellular degradation of human immunodeficiency virus (HIV) proteins enhances CTL responses *in vivo* (Tobery and Siliciano, 1997) following DNA vaccination.

Targeting Protein to ER for Secretion

The addition of a signal peptide to a protein otherwise destined for the cytoplasm can allow it to be secreted, increasing its availability to professional antigen-presenting cells (APCs). Some antigens used for DNA vaccines contain endogenous signal peptides (McClements et al., 1996; Stevenson et al., 1995; Syrengelas et al., 1996). For those that do not, exogenous signal peptide-encoding sequences (e.g., the adenovirus E3 leader sequence [Ciernik et al., 1996]) or the tissue plasminogen activator (tPA) can be added to the 5′ end of the target gene (Chapman et al., 1991). The signal peptide is cleaved upon entry into the ER and thus is not part of the mature protein, although there is a report of antibody production to a signal peptide following DNA vaccination (Krasemann et al., 1996). During transport across the ER membrane, the protein may associate with chaperone proteins that assist in folding the antigen protein into its native structure. Subsequent modification of the protein includes the addition of sugar residues at *N* or *O* positions containing the corresponding amino acid motifs. *N*-linked glycosylation begins in the ER and is completed upon transfer to the Golgi, whereas *O*-linked glycosylation occurs mainly in the Golgi. Secretion of the mature protein occurs when the Golgi body fuses with the cell membrane, thereby allowing escape of its contents. The state of glycosylation may affect the processing of a protein (e.g., by protecting certain residues from proteolytic degradation and thereby altering the T-cell response to the "protected" peptide). For example, the addition of *N*-linked oligosaccharides to *Mycobacterium tuberculosis* antigen 85 during expression of this bacterial antigen in eukaryotic cells following DNA vaccination appears to interfere with processing and/or presentation of this antigen for induction of T-cell responses (A. M. Yawman et al., unpublished observation).

Membrane-Targeted Antigens

Proteins containing a transmembrane domain (usually near the C terminus for type I proteins) remain membrane bound upon fusion of the Golgi to the cell membrane. Antigens expressed on the cell membrane may present conformational epitopes to B cells. Alternatively, membrane-bound proteins can re-enter the cell through endocytosis, whereupon the antigen can be degraded to peptides for presentation via the MHC class I pathway as described above.

The plasticity of the DNA vaccine approach may allow the development of vaccines that simultaneously enhance both humoral and cell-mediated immunity. For example, a vaccine could contain a combination of DNA vectors expressing the same antigen with or without a leader or transmembrane sequence. Whether such an approach is beneficial or detrimental for the induction of protective immunity remains to be determined.

Requirement for Professional APCs

Although many cell types can take up and express plasmid DNA, recent experiments indicate that bone marrow–derived "professional" APCs are required for generation of a CTL response (Corr et al., 1996; Fu et al., 1997).

To determine whether professional APCs are required for generation of class I–restricted CTL following DNA immunization, parent → F_1 bone marrow chimeras were examined to determine the MHC restriction pattern of CTL from immunized recipient mice. In our studies, BALB/c × C57BL/6 F_1 hybrid mice were irradiated and reconstituted with T cell–depleted bone marrow from either parent strain. Recipient mice were held at least 3 months to allow regeneration of donor bone marrow cells and sufficient exhaustion of residual host-derived APCs (confirmed by FACs analysis). Mice were then immunized with plasmid DNA encoding influenza virus NP. CTLs recovered from chimeric mice immunized with NP DNA were found to be restricted to the haplotype of the bone marrow donor rather than to both parental haplotypes, as would be the case if the myocyte presented antigen directly to the T cell.

These results indicate that bone marrow–derived APCs are required for the generation of CTLs after DNA immunization, suggesting that antigen produced in transfected cells is transferred to professional APCs. This would easily explain the response to secreted proteins; however, it does not adequately explain how CTL responses are generated against nonsecreted proteins. In this case one would have to invoke "cross priming" (Bevan, 1976; Huang et al., 1994) by antigen released from dead cells or a novel phenomenon of transfer of MHC class I–peptide complexes from antigen-expressing cells (e.g., myocytes) to macrophages or dendritic cells. An alternative explanation is that bone marrow–derived cells such as macrophages or dendritic cells may be directly transfected with plasmid DNA even after i.m. injection. In support of this possibility, it has been reported that dendritic cells transfected *in vitro* with a herpes simplex virus (HSV) DNA vaccine induced an enhanced immune response to HSV upon transfer into naive mice (Manickan et al., 1997a). Finally, Condon et al. (1996) have reported that gene gun administration of a DNA vaccine resulted in direct transfection of cutaneous dendritic cells that were shown to localize to draining lymph nodes, where they were presumed to initiate an immune response to the antigen.

Transfer of Antigen to APC

To determine whether an antigen expressed in a nonprofessional APC could be transferred to professional APCs, a myoblast cell line permanently transfected with an NP vector was transplanted into syngeneic or semiallogeneic mice to look for induction of CTLs. Ulmer et al. (1996a) found that transplantation of NP-transfected myoblasts into syngeneic mice led to the generation of NP-specific antibodies, CTLs, and cross-strain protective immunity against a lethal challenge with influenza virus. Furthermore, transplantation of NP-expressing myoblasts (H-2^k) intraperitoneally into F_1 hybrid mice (H-2^d × H-2^k) elicited NP CTL restricted by the MHC haplotype of both parental strains. These results indicate that NP expression by muscle cells after transplantation was sufficient to generate protective cell-mediated immunity and that induction of the CTL response was mediated at least in part by transfer of antigen from the transplanted muscle cells to a host cell. Finally, the strategy of using parent → F_1 bone marrow chimeras was applied to the transplantation model to confirm the requirement for professional APCs in the generation of CTLs after immunization with NP-expressing myoblasts (Fu et al., 1997) and provided further evidence that transfer of antigen in some form from myoblasts to APCs occurs.

TARGETING CELL-MEDIATED VERSUS HUMORAL RESPONSES

Conventional subunit vaccines consisting of purified proteins formulated with aluminum adjuvants induce humoral (antibody) rather than cell-mediated immune responses, whereas vaccination with live virus vaccines induces both antibody and cell-mediated immunity. DNA vaccines tend to mimic live-virus vaccines in that they induce both antibody and cell-mediated immune responses. This property of DNA vaccines is associated with the induction of IFN-γ and IL-2, cytokines associated with the Th1 population of helper T cells (Mosmann et al., 1986). The ability of DNA vaccines to induce cell-mediated immunity may account for the demonstrated activity of these vaccines to induce protective immunity upon challenge with infectious agents, as summarized in Table 1 and discussed in detail elsewhere (Donnelly et al., 1997; Manickan et al., 1997b; Ulmer et al., 1996b).

Immunoenhancing Agents for DNA Vaccines

Although naked DNA has been shown to be immunogenic in a variety of experimental models (Table 1), several approaches have been taken to enhance the immune response to DNA vaccines, including (1) co-administration of DNA vaccines with cytokine proteins, (2) co-administration of DNA vaccines with expression vectors encoding cytokine genes, and (3) incorporating immunostimulatory sequences into DNA vaccine vectors.

Cytokines. Irvine et al. (1996) used an experimental murine tumor, CT26, expressing the model tumor-associated antigen, β-galactosidase (β-gal) to investigate cytokine enhancement of DNA immunization. A plasmid expressing β-gal administered via a "gene gun" induced β-gal-specific antibody and cytolytic responses. DNA immunization alone had little or no impact on the growth of established lung metastases; however, a significant reduction in the number of established metastases was observed when human rIL-2, mouse rIL-6, human rIL-7, or mouse rIL-12 was given after DNA inoculation. rIL-12 was reported to be the most effective adjuvant for this DNA vaccine.

Cytokine Genes. Studies in experimental animals have shown that it is possible to manipulate the immune response to a plasmid-encoded viral antigen by co-inoculation with plasmids expressing cytokines. Xiang and Ertl (1995) showed that co-inoculation of a plasmid expressing the glycoprotein of rabies virus with a vector expressing mouse GM-CSF enhanced the helper B- and T-cell activity to rabies virus, whereas co-inoculation with a plasmid expressing IFN-γ resulted in a decrease in the immune response to the viral antigen (Xiang et al., 1995; Xiang and Ertl, 1995). For HIV, Kim et al. (1997) found that co-delivery of IL-12 expression vectors with DNA vaccines for HIV-1 in mice resulted in splenomegaly, as well as a shift in the specific immune responses induced. Co-delivery of IL-12 genes resulted in the reduction of specific antibody response, whereas co-injection of GM-CSF genes resulted in the enhancement of HIV-specific antibody responses. A dramatic increase in HIV-specific CTL responses was reported in animals co-immunized with the HIV-1 DNA vaccine and IL-12 genes.

Immunostimulatory Sequences in Plasmid DNA (CpG Motifs). Bacterial DNA has been shown to have immunostimulatory properties, including the ability to induce natural killer (NK) cell activity and the induction of cytokines such as IFN-α/β and IFN-γ (Yamamoto et al., 1988). By contrast, mammalian DNA is not mitogenic (Pisetsky, 1996). The active bacterial DNA was found to contain palindromic hexameric sequences such as GACGTC, AGCGCT, or AACGTT (Tokunaga et al., 1992; Yamamoto et al., 1992). More recently, Krieg et al. (1995) have shown that oligonucleotides containing similar motifs to those identified by Yamamoto et al. (1992) (e.g., GACGTC and AACGTT) are mitogenic for mouse B lymphocytes. They further demonstrated that methylation of the cytosine base within the CpG motif resulted in a loss of activity, and they have attributed the lack of mitogenicity of mammalian DNA to the observation that this cytosine base is heavily methylated.

Recently, CpG-containing sequence motifs have been reported to enhance the immunogenicity of a model DNA vaccine encoding β-gal (Sato et al., 1996). In these studies, plasmids containing the ampicillin resistance gene (ampR) induced a stronger antibody and CTL response to β-gal than did plasmids containing the kanamycin-resistance gene (kanR) instead of the ampR gene when injected intradermally. Analysis of the ampR gene revealed that it contained two repeats of 5′-AACGTT-3′ that were reported to have immunostimulatory activity (Yamamoto et al., 1992). By contrast, the kanR gene had no such sequence. Furthermore, insertion of 5′-AACGTT-3′ sequence motifs into the kanR gene within the DNA vaccine resulted in an increase in both antibody and CTL activities. The strategy of incorporating immunostimulatory motifs into DNA vaccines may not be a universal solution, however, since replacement of the kanR gene with the ampR gene did not improve the immune response to a highly immunogenic DNA vaccine encoding influenza virus NP (R. Deck et al., unpublished observation).

TOLERANCE VERSUS IMMUNITY

Tolerance to self-antigens generally occurs during fetal development at which time immature T cells encounter peptides associated with self-MHC in the thymus and are deleted (Fowlkes and Pardoll, 1989). For some antigens the period of tolerance induction extends beyond birth, e.g., mice can be rendered tolerant to alloantigens by injection of allogeneic cells during the neonatal time period (Billingham et al., 1956). For certain polysaccharide antigens, tolerance appears to develop because the B-cell repertoire is incomplete (Chang et al., 1991; Fernandez and Moller, 1978). Recently, Mor et al. (1996) compared a DNA vaccine with the corresponding protein antigen for the induction of neonatal tolerance in mice. They injected 2–5-day-old mice with a DNA vaccine encoding the malaria circumsporozoite (CS) antigen and then re-immunized them at 7 weeks of age with either the DNA vaccine or the CS protein. Three weeks later, sera were tested against a panel of synthetic peptides spanning the length of the *P. yoelii* CS protein. Mice injected neonatally and boosted with the DNA vaccine did not develop an antibody response to any CS peptide in their panel, whereas adult mice injected once or twice with the DNA vaccine developed antibodies to CS peptides 3, 19, 21, and 25. Interestingly, mice immunized with the CS protein in complete Freund's adjuvant developed a different pattern of antibody reactivity. These mice responded vigorously to peptides 9, 14, 15, and 25 but not to peptides 3 and 21. Thus, immunization with the CS DNA vaccine appears to select a different repertoire of B cells than does immunization with the corresponding protein antigen.

Regarding the issue of neonatal tolerance induction, Mor et al. (1996) reported that mice injected neonatally with the DNA vaccine were unresponsive to the DNA vaccine as adults, whereas mice boosted with the CS protein in complete Freund's adjuvant developed antibodies predominantly to CS peptides 14 and 15. They interpret these results to suggest that neonatal immunization with a DNA vaccine induces tolerance; however, the data indicate only that immunization with the DNA vaccine induces unresponsiveness to a second injection of the DNA vaccine. The mice were clearly not tolerant to the CS antigen per se because they responded well to a booster injection of CS protein. Other investigators have found that neonatal injection of DNA vaccines leads to immunity rather than tolerance (Bot et al., 1996; Prince et al., 1996; Wang et al., 1997). For example, Bot et al. (1996) demonstrated that neonatal mice were immunized rather than tolerized by the injection of an influenza virus NP DNA vaccine, and Wang et al. (1997) showed that mice inoculated within 24 h after birth with a plasmid vector expressing the rabies virus glycoprotein developed antibodies as well as helper T cells to the rabies virus glycoprotein—a response that was indistinguishable from that of adult mice.

SUMMARY

This review was intended to outline the elements of DNA vaccine design as they relate to the induction of immunity with the purpose of identifying areas for future research. Although the basic design of plasmid vaccines appears to be set to include a viral promoter (usually CMV), the gene of interest, and a terminator (often from BGH), there may be room to improve the various elements of a DNA vaccine to enhance expression and immunogenicity. The search for better promoters and enhancers will continue, and the observation that sequences from the target gene may influence expression and immunogenicity (as is the case with OspA) indicates the potential need to match target genes with compatible promoters and other regulatory elements. Improving immunogenicity of DNA vaccines by the inclusion or co-administration of cytokine genes or immunostimulatory DNA sequences is a fertile area for further study, especially for therapy of cancer in which there may be synergy between DNA vaccines and cytokines. It is also clear that DNA vaccine delivery can be improved whether the issue is one of increasing the dose that can be delivered via the gene gun or optimizing lipid-based delivery vehicles.

A major consideration in vaccine design is the nature of the immune response that is desired. This is not trivial for some diseases (e.g., HIV) in which early efforts at vaccination were directed toward the induction of antibody responses to env protein, whereas current efforts are directed toward maximizing CTL responses. Although DNA vaccines (due to endogenous expression of target antigens) tend to generate CTL responses, it has been shown that such responses can be enhanced by changing the target gene sequence to effect more rapid degradation of the encoded antigen (Tobery and Siliciano, 1997). Whether such changes will result in a better vaccine is the question. Regardless, the use of plasmid DNA as immunogen will surely help define the vaccines of the future.

ACKNOWLEDGMENTS

We thank Drs. William McClements, John Shiver, and Jeffrey Ulmer for helpful comments and suggestions regarding the manuscript and for sharing unpublished results.

REFERENCES

Adam SA, Gerace L (1991): Cytosolic proteins that specifically bind nuclear location signals are receptors for nuclear import. *Cell* 66:837–847.

Asakura Y, Hamajima K, Fukushima J, Mohri H, Okubo T, Okuda K (1996): Induction of HIV-1 Nef-specific cytotoxic T lymphocytes by Nef-expressing DNA vaccine. *Am J Hematol* 53:116–7.

Babiuk LA, Lewis PJ, Cox G, van Drunen Littel-van den Hurk S, Baca-Estrada M, Tikoo SK (1995): DNA immunization with bovine herpesvirus-1 genes. *Ann NY Acad Sci* 772:47–63.

Bennett R, Gabor G, Merritt M (1985): DNA binding to human leukocytes. Evidence for a receptor-mediated association, internalization, and degradation of DNA. *J Clin Invest* 76:2182–2190.

Bennett R, Peller J, Merritt M (1986): Defective DNA-receptor function in systemic lupus erythematosus and related diseases: Evidence for an autoantibody influencing cell physiology. *Lancet* 1:186–188.

Bevan MJ (1976): Minor H antigens introduced on H-2 different stimulating cells cross-react at the cytotoxic T cell level during *in vivo* priming. *J Immunol* 117:2233–2238.

Billingham RL, Brent L, Medawar P (1956): Quantitative studies on tissue transplantation immunity. *Proc R Soc Lond* 239:44–45.

Boshart M, Weber F, Jahn G, Dorsch-Hasler K, Fleckenstein B, Schaffner W (1985): A very strong enhancer is located upstream of an immediate early gene of human cytomegalovirus. *Cell* 41:521–530.

Bot A, Bot S, Garcia-Sastre A, Bona C (1996): DNA immunization of newborn mice with a plasmid-expressing nucleoprotein of influenza virus. *Viral Immunology* 9:207–210.

Bourne N, Stanberry LR, Bernstein DI, Lew D (1996): DNA immunization against experimental genital herpes simplex virus infection. *J Infect Dis* 173:800–7.

Boyer JD, Wang B, Ugen KE, Agadjanyan M, Javadian A, Frost P, Dang K, Carrano RA, Ciccarelli R, Coney L, Williams WV, Weiner DB (1996): *In vivo* protective anti-HIV immune responses in non-human primates through DNA immunization. *J Med Primatol* 25:242–250.

Bueler H, Mulligan RC (1996): Induction of antigen-specific tumor immunity by genetic and cellular vaccines against MAGE: enhanced tumor protection by coexpression of granulocyte-macrophage colony-stimulating factor and B7-1. *Mol Med* 2:545–555.

Chang T, Capraro E, Kleinman RE, Abbas AK (1991): Anergy in immature B lymphocytes. *J Immunol* 147:750–756.

Chapman BS, Thayer RM, Vincent KA, Haigwood NL (1991): Effect of intron A from human cytomegalovirus (Towne) immediate early gene on heterologous expression in mammalian cells. *Nucleic Acids Res* 19:3979–3986.

Ciernik IF, Berzofsky JA, Carbone DP (1996): Induction of cytotoxic T lymphocytes and antitumor immunity with DNA vaccines expressing single T cell epitopes. *J Immunol* 156:2369–2375.

Condon C, Watkins SC, Celluzzi CM, Thompson K, Falo LD (1996): DNA-based immunization by *in vivo* transfection of dendritic cells. *Nature Med* 2:1122–1128.

Conry RM, Lo Buglio AF, Kantor J, Schlom J, Loechel F, Moore SE, Sumerel LA, Barlow DL, Abrams S, Curiel DT (1994): Immune response to a carcinoembryonic antigen polynucleotide vaccine. *Cancer Res* 54:1164–1168.

Conry RM, Lo Buglio AF, Loechel F, Moore SE, Sumerel LA, Barlow DL. Curiel DT (1995): A carcinoembryonic antigen polynucleotide vaccine has *in vivo* antitumor activity. *Gene Ther* 2:59–65.

Corr M, Lee DJ, Carson DA, Tighe H (1996): Gene vaccination with naked plasmid DNA: Mechanism of CTL priming. *J Exp Med* 184:1555–1560.

Davis HL, McCluskie MJ, Gerin JL, Purcell RH (1996): DNA vaccine for hepatitis B: Evidence for immunogenicity in chimpanzees and comparison with other vaccines. *Proc Natl Acad Sci USA* 93:7213–7218.

Davis HL, Michel ML, Whalen RG (1993): DNA-based immunization induces continuous secretion of hepatitis B surface antigen and high levels of circulating antibody. *Hum Mol Genet* 2:1847–1851.

Dean DA (1997): Import of plasmid DNA into the nucleus is sequence specific. *Exp Cell Res* 230:239–302.

Donnelly JJ, Friedman A, Martinez D, Montgomery DL, Shiver JW, Motzel SL, Ulmer JB, Liu MA (1995): Preclinical efficacy of a prototype DNA vaccine: Enhanced protection against antigenic drift in influenza virus. *Nature Med* 1:583–587.

Donnelly JJ, Martinez D, Jansen KU, Ellis RW, Montgomery DL, Liu MA (1996): Protection against papillomavirus with a polynucleotide vaccine. *J Infect Dis* 173:314–320.

Donnelly JJ, Ulmer JB, Liu MA (1994): Immunization with DNA. *J Immunol Methods* 176:145–152.

Donnelly JJ, Ulmer JB, Shiver JW, Liu MA (1997): DNA vaccines. *Annu Rev Immunol* 15:617–648.

Doolan DL, Sedegah M, Hedstrom RC, Hobart P, Charoenvit Y, Hoffman SL (1996): Circumventing genetic restriction of protection against malaria with multigene DNA immunization: $CD8^+$ cell-, interferon gamma-, and nitric oxide-dependent immunity. *J Exp Med* 183:1739–1746.

Ertl HC, Verma P, He Z, Xiang ZQ (1995): Plasmid vectors as anti-viral vaccines. *Ann NY Acad Sci* 772:77–87.

Felgner PL, Tsai YJ, Sukhu L, Wheeler CJ, Manthorpe M, Marshall J, Cheng SH (1995): Improved cationic lipid formulations for *in vivo* gene therapy. *Ann NY Acad Sci* 772:126–139.

Feltquate DM, Heaney S, Webster R, Robinson H (1997): Different T helper cell types and antibody isotypes generated by saline and gene gun DNA immunization. *J Immunol* 158:2278–2284.

Fernandez C, Moller G (1978): Immunological unresponsiveness to native dextran B512 in young animals of dextran high responder strains is due to lack of Ig-receptor expression. *J Exp Med* 147:645–655.

Fowlkes BJ, Pardoll DM (1989): Molecular and cellular events of T cell development. *Adv Immunol* 44:207–264.

Fu T-M, Ulmer JB, Caulfield MJ, Deck RR, Friedman A, Wang S, Liu X, Donnelly JJ, Liu MA (1997): Priming of cytotoxic T lymphocytes by DNA vaccines: Requirement for professional antigen presenting cells and evidence for antigen transfer from myocytes. *Mol Med* 3:362–371.

Fuller DH, Murphey-Corb M, Clements J, Barnett S, Haynes JR (1996): Induction of immunodeficiency virus–specific immune responses in rhesus monkeys following gene gun-mediated DNA vaccination. *J Med Primatol* 25:236–241.

Fynan EF, Robinson HL, Webster RG (1993a): Use of DNA encoding influenza hemagglutinin as an avian influenza vaccine. *DNA Cell Biol* 12:785–789.

Fynan EF, Webster RG, Fuller DH, Haynes JR, Santoro JC, Robinson HL (1993b): DNA vaccines: Protective immunizations by parenteral, mucosal, and gene-gun inoculations. *Proc Natl Acad Sci USA* 90:11478–11482.

Geissler M, Gesien A, Tokushige K, Wands JR (1997): Enhancement of cellular and humoral immune responses to hepatitis C virus core protein using DNA-based vaccines augmented with cytokine-expressing plasmids. *J Immunol* 158:1231–1237.

Gorman CM, Merlino GT, Willingham MC, Pastan I, Howard BH (1982): The Rous sarcoma virus long terminal repeat is a strong promoter when introduced into a variety of eukaryotic cells by DNA-mediated transfection. *Proc Natl Acad Sci USA* 79:6777–6781.

Hartikka J, Sawdey M, Cornefert-Jensen F, Margalith M, Barnhart K, Nolasco M, Vahlsing H, Meek J, Marquet M, Hobart P, Norman J, Manthorpe M (1996): An improved plasmid DNA expression vector for direct injection into skeletal muscle. *Hum Gene Ther* 7:1205–1217.

Herrmann JE, Chen SC, Fynan EF, Santoro JC, Greenberg HB, Wang S, Robinson HL (1996): Protection against rotavirus infections by DNA vaccination. *J Infect Dis* 174:S93–S97.

Huang AYC, Golumbek P, Ahmadzadeh M, Jaffee E, Pardoll D, Levitsky H (1994): Role of bone marrow–derived cells in presenting MHC class I–restricted tumor antigens. *Science* 264:961–965.

Huygen K, Content J, Denis O, Montgomery DL, Yawman AM, Deck RR, De Witt CM, Orme IM, Baldwin S, D'Souza C, Drowart A, Lozes E, Vandenbussche P, Van Vooren JP, Liu MA, Ulmer JB (1996): Immunogenicity and protective efficacy of a tuberculosis DNA vaccine. *Nature Med* 2:893–898.

Irvine KR, Rao JB, Rosenberg SA, Restifo NP (1996): Cytokine enhancement of DNA immunization leads to effective treatment of established pulmonary metastases. *J Immunol* 156:238–245.

Jones N, Rigby PW, Ziff EB (1989): Trans-acting protein factors and the regulation of eukaryotic transcription: Lessons from studies on DNA tumor viruses. *Genes Dev* 2:267–281.

Kim JJ, Ayyavoo V, Bagarazzi ML, Chattergoon MA, Dang K, Wang B, Boyer JD, Weiner DB (1997): *In vivo* engineering of a cellular immune response by coadministration of IL-12 expression vector with a DNA immunogen. *J Immunol* 158:816–826.

Kozak M (1987): At least six nucleotides preceding the AUG initiator codon enhance translation in mammalian cells. *J Mol Biol* 196:947–950.

Krasemann S, Groschup M, Hunsmann G, Bodemer W (1996): Induction of antibodies against human prion proteins (PrP) by DNA-mediated immunization of PrP0/0 mice. *J Immunol Methods* 199:109–118.

Krieg AM, Yi AK, Matson S, Waldschmidt TJ, Bishop GA, Teasdale R, Koretzky GA, Klinman DM (1995): CpG motifs in bacterial DNA trigger direct B-cell activation. *Nature* 374:546–549.

Kruys V, Marinx O, Shaw G, Deschamps J, Huez G (1989): Translational blockade imposed by cytokine-derived AU-rich sequences. *Science* 245:852.

Lagging LM, Meyer K, Hoft D, Houghton M, Belshe RB, Ray R (1995): Immune responses to plasmid DNA encoding the hepatitis C virus core protein. *J Virol* 69:5859–5863.

Lekutis C, Shiver JW, Liu MA, Letvin NL (1997): An HIV-1 env DNA vaccine administered to rhesus monkeys elicits MHC class II–restricted $CD4^+$ T helper cells that secrete IFN-γ and TNF-α. *J Immunol* 158:4471–4477.

Liang X, Hartikka J, Sukhu L, Manthorpe M, Hobart P (1996): Novel, high expressing and antibiotic-controlled plasmid vectors designed for use in gene therapy. *Gene Ther* 3:350–356.

Liu MA, Yasutomi Y, Davies ME, Perry HC, Freed DC, Letvin NL, Shiver JW (1996): Vaccination of mice and nonhuman primates using HIV-gene–containing DNA. *Antibiot Chemother* 48:100–104.

Lowrie DB, Tascon RE, Colston MJ, Silva CL (1994): Towards a DNA vaccine against tuberculosis. *Vaccine* 12:1537–1540.

Lowrie DB, Tascon RE, Silva CL (1995): Vaccination against tuberculosis. *Int Arch Allergy Immunol* 108:309–312.

Manickan E, Kanangat S, Rouse RJD, Yu Z, Rouse BT (1997a): Enhancement of immune respone to naked vaccine by immunization with transfected dendritic cells. *J Leukocyte Biol* 61:125–132.

Manickan E, Karem KL, Rouse BT (1997b): DNA vaccines—A modern gimmick or a boon to vaccinology? *Crit Rev Immunol* 17:139–154.

Manickan E, Rouse RJD, Yu Z, Wire WS, Rouse BT (1995): Genetic immunization against herpes simplex virus. Protection is mediated by $CD4^+$ T lymphocytes. *J Immunol* 155:259–265.

McClements WL, Armstrong ME, Keys RD, Liu MA (1996): Immunization with DNA vaccines encoding glycoprotein D or glycoprotein B, alone or in combination, induces protective immunity in animal models of herpes simplex virus-2 disease. *Proc Natl Acad Sci USA* 93:11414–11420.

McDaniel LS, Loechel F, Benedict C, Greenway T, Briles DE, Conry RM, Curiel DT (1997): Immunization with a plasmid expressing pneumococcal surface protein A (PspA) can elicit protection against fatal infection with *Streptococcus pneumoniae. Gene Ther* 4:375–377.

Michel ML, Davis HL, Schleef M, Mancini M, Tiollais P, Whalen RG (1995): DNA-mediated immunization to the hepatitis B surface antigen in mice: Aspects of the humoral response mimic hepatitis B viral infection in humans. *Proc Natl Acad Sci USA* 92:5307–5311.

Monaco JJ, Cho S, Attaya M (1990): Transport protein genes in the murine MHC: Possible implications for antigen processing. *Science* 250:1723–1726.

Montgomery DL, Shiver JW, Leander KR, Perry HC, Friedman A, Martinez D, Ulmer JB, Donnelly JJ, Liu MA (1993): Heterologous and homologous protection against influenza A by DNA vaccination: Optimization of DNA vectors. *DNA Cell Biol* 12:777–783.

Mor G, Yamshchikov G, Sedegah M, Takeno M, Wang R, Houghten RA, Hoffman S, Klinman DM (1996): Induction of neonatal tolerance by plasmid DNA vaccination of mice. *J Clin Invest* 98:2700–2705.

Moreau P, Hen R, Wasylyk B, Everett R, Gaub MP, Chambon P (1981): The SV40 72 base pair repeat has a striking effect on gene expression both in SV40 and other chimeric recombinants. *Nucleic Acids Res* 9:6047–6068.

Mosmann TR, Cherwinski H, Bond MW, Gieldin MA, Coffman RL (1986): Two types of murine helper T cell clone. I. Definition according to profiles of lymphokine activities and secreted proteins. *J Immunol* 136:2348–2357.

Okuda K, Bukawa H, Hamajima K, Kawamoto S, Sekigawa K, Yamada Y, Tanaka S, Ishii N, Aoki I, Nakamura M, Yamamoto MH, Cullen BR, Fukushima J (1995): Induction of potent humoral and cell-mediated immune responses following direct injection of DNA encoding the HIV type 1 *env* and *rev* gene products. *AIDS Res Hum Retroviruses* 11:933–943.

Pertmer TM, Roberts TR, Haynes JR (1996): Influenza virus nucleoprotein-specific immunoglobulin G subclass and cytokine responses elicited by DNA vaccination are dependent on the route of vector DNA delivery. *J Virol* 70:6119–6125.

Pisetsky DS (1996): The immunologic properties of DNA. *J Immunol* 156:421–423.

Prince AM, Whalen R, Brotman B (1997): Successful DNA-based immunization of newborn chimpanzees. (Abstract) *Vaccine* 15:916–919.

Rajagopalan LE, Burkholder JK, Turner J, Culp J, Yang N-S, Malter JS (1995): Granulocyte-macrophage coloney-stimulating factor mRNA stabilization enhances transgenic expression in normal cells and tissues. *Blood* 86:2551–2558.

Raz E, Carson DA, Parker SE, Parr TB, Abai AM, Aichinger G, Gromkowski SH, Singh M, Lew D, Yankauckas MA, et al. (1994): Intradermal gene immunization: The possible role of DNA uptake in the induction of cellular immunity to viruses. *Proc Natl Acad Sci USA* 91:9519–9523.

Raz E, Tighe H, Sato Y, Corr M, Dudler JA, Roman M, Swain SL, Spiegelberg HL, Carson DA (1996): Preferential induction of a Th1 immune response and inhibition of specific IgE antibody formation by plasmid DNA immunization. *Proc Natl Acad Sci USA* 93:5141–5145.

Robinson HL, Hunt LA, Webster RG (1993): Protection against a lethal influenza virus challenge by immunization with a haemagglutinin-expressing plasmid DNA. *Vaccine* 11:957–960.

Sato Y, Roman M, Tighe H, Lee D, Corr M, Nguyen MD, Silverman GJ, Lotz M, Carson DA, Raz E (1996): Immunostimulatory DNA sequences necessary for effective intradermal gene immunization. *Science* 273:352–354.

Schirmbeck R, Bohm W, Ando K, Chisari FV, Reiman J (1995): Nucleic acid vaccination primes hepatitis B virus surface antigen-specific cytotoxic T lymphocytes in nonresponder mice. *J Virol* 69:5929–5934.

Sedegah M, Hedstrom R, Hobart P, Hoffman SL (1994): Protection against malaria by immunization with plasmid DNA encoding circumsporozoite protein. *Proc Natl Acad Sci USA* 91:9866–9870.

Shaw G, Kamen R (1986): A conserved AU sequence from the 3′ untranslated region of GM-CSF m-RNA mediates selective mRNA degradation. *Cell* 46:659.

Shiver JW, Davies M-E, Perry HC, Freed DC, Liu MA (1996a): Humoral and cellular immunities elicited by HIV-1 DNA vaccination. *J Pharmacol Sci* 85:1317–1324.

Shiver JW, Perry HC, Davies ME, Freed DC, Liu MA (1995): Cytotoxic T lymphocyte and helper T cell responses following HIV polynucleotide vaccination. *Ann NY Acad Sci* 772:198–208.

Shiver JW, Ulmer JB, Donnelly JJ, Liu MA (1996b): Humoral and cellular immunities elicited by DNA vaccines: Application to the human immunodeficiency virus and influenza. *Adv Drug Delivery Rev* 21:19–31.

Shiver JW, Ulmer JB, Donnelly JJ, Liu MA (1996c): Naked DNA vaccination. In Kaufmann SHE (ed): *Concepts in Vaccine Development.* Berlin: Walter de Gruyter, pp 423–436.

Simon MM, Gern L, Hauser P, Zhong W, Nielsen PJ, Kramer MD, Brenner C, Wallich R (1996): Protective immunization with plasmid DNA containing the outer surface lipoprotein A gene of *Borrelia burgdorferi* is independent of an eukaryotic promoter. *Eur J Immunol* 26:2831–2840.

Spies T, Cerundolo V, Colonna M, Cresswell P, Townsend A (1992): Presentation of viral antigen by MHC class I molecules is dependent of a putative peptide transporter heterodimer. *Nature* 355:644–646.

Stevenson FK, Zhu D, King CA, Ashworth LJ, Kumar S, Hawkins RE (1995): Idiotypic DNA vaccines against B-cell lymphoma. *Immunol Rev* 145:211–228.

Syrengelas AD, Chen TT, Levy R (1996): DNA immunization induces protective immunity against B-cell lymphoma. *Nat Med* 2:1038–1041.

Tang DC, De Vit M, Johnston SA (1992): Genetic immunization is a simple method for eliciting an immune response. *Nature* 356:152–154.

Tascon RE, Colston MJ, Ragno S, Stavropoulos E, Gregory D, Lowrie DB (1996): Vaccination against tuberculosis by DNA injection. *Nat Med* 2:888–892.

Tobery TW, Siliciano RF (1997): Targeting of HIV-1 antigens for rapid intracellular degradation enhances cytotoxic T lymphocyte (CTL) recognition and the induction of *de novo* CTL responses *in vivo* after immunization. *J Exp Med* 185:909–920.

Tokunaga T, Yano O, Kuramoto E, Kimura Y, Yamamoto T, Kataoka T, Yamamoto S (1992): Synthetic oligonucleotides with particular base sequences from the cDNA encoding proteins of *Mycobacterium bovis* BCG induce interferons and activate natural killer cells. *Microbiol Immunol* 36:55–66.

Ulmer JB, Deck RR, DeWitt C, Donnelly J, Liu M (1996a): Generation of MHC class I–restricted cytotoxic T lymphocytes by expression of a viral protein in muscle cells: Antigen presentation by non-muscle cells. *Immunology* 8959–67.

Ulmer JB, Donnelly JJ, Deck RR, De Witt CM, Liu MA (1995): Immunization against viral proteins with naked DNA. *Ann NY Acad Sci* 772:117–125.

Ulmer JB, Donnelly JJ, Liu MA (1996b): Toward the development of DNA vaccines. *Curr Opin Biotechnol* 7:653–658.

Ulmer JB, Donnelly JJ, Parker SE, Rhodes GH, Felgner PL, Dwarki VJ, Gromkowski SH, Deck RR, De Witt CM, Friedman A, et al. (1993): Heterologous protection against influenza by injection of DNA encoding a viral protein. *Science* 259:1745–1749.

Wahren B, Hinkula J, Stahle EL, Borrebaeck CA, Schwartz S, Wigzell H (1995): Nucleic acid vaccination with HIV regulatory genes. *Ann NY Acad Sci* 772:278–281.

Wang B, Boyer J, Srikantan V, Ugen K, Agadjanian M, Merva M, Gilbert L, Dang K, McCallus D, Moelling K, et al. (1995): DNA inoculation induces cross clade anti-HIV-1 responses. *Ann NY Acad Sci* 772:186–197.

Wang B, Ugen KE, Srikantan V, Agadjanyan MG, Dang K, Refaeli Y, Sato AI, Boyer J, Williams WV, Weiner DB (1993): Gene inoculation generates immune responses against human immunodeficiency virus type 1. *Proc Natl Acad Sci USA* 90:4156–4160.

Wang Y, Xiang Z, Pasquini S, Ertl H (1997): Immune response to neonatal genetic immunization. *Virology* 228:278–284.

Wolff J, Malone RW, Williams P, Chong W, Ascadi G, Jani A, Felgner P (1990): Direct transfer into mouse muscle *in vivo*. *Science* 247:1465–1468.

Wolff JA, Ludtke JJ, Williams P, Jani A (1992): Long-term presistence of plasmid DNA and foreign gene expression in mouse muscle. *Hum Mol Genet* 1:363–369.

Xiang Z, Ertl HC (1995): Manipulation of the immune response to a plasmid-encoded viral antigen by coinoculation with plasmids expressing cytokines. *Immunity* 2:129–135.

Xiang ZQ, Spitalnik SL, Cheng J, Erikson J, Wojczyk B, Ertl HC (1995): Immune responses to nucleic acid vaccines to rabies virus. *Virology* 209:569–579.

Xu D, Liew FY (1995): Protection against leishmaniasis by injection of DNA encoding a major surface glycoprotein, gp63, of *L. major. Immunology* 84:173–176.

Yamamoto S, Kuramoto E, Shimada S, Tokunaga T (1988): *In vitro* augmentation of natural killer cell activity of interferon α/β and -γ with deoxyribonucleic acid fraction from *Mycobacterium bovis* BCG. *Jpn J Cancer Res* (*Gann*) 79:866–873.

Yamamoto S, Yamamoto T, Kataoka T, Kuramoto E, Yano O, Tokunaga T (1992): Unique palindromic sequences in synthetic oligonucleotides are required to induce INF and augment INF-mediated natural killer activity. *J Immunol* 148:4072–4076.

Yang W, Waine GJ, McManus DP (1995): Antibodies to *Schistosoma japonicum* (Asian bloodfluke) paramyosin induced by nucleic acid vaccination. *Biochem Biophys Res Commun* 212:1029–1039.

Yasutomi Y, Robinson HL, Lu S, Mustafa F, Lekutis C, Arthos J, Mullins JI, Voss G, Manson K, Wyand M, Letvin NL (1996): Simian immunodeficiency virus-specific cytotoxic T-lymphocyte induction through DNA vaccination of rhesus monkeys. *J Virol* 70:678–681.

Yellen-Shaw AJ, Eisenlohr LC (1997): Regulation of class I–restricted epitope processing by local or distal flanking sequence. *J Immunol* 158:1727–1733.

13

DEVELOPMENT OF GENE THERAPY FOR GAUCHER DISEASE

J.A. Barranger, E.O. Rice, J. Dunigan, M. Eljanne, N. Takiyama, M. Nimgaonkar, J. Mierski, M. Beeler, A. Kemp, J. Lancia, S. Lucot, S. Schierer-Fochler, J. Mannion-Henderson, T. Mohney, W. Swaney, A. Bahnson, V. Bansal, and E. Ball

University of Pittsburgh Medical Center, Pittsburgh, PA 15261

INTRODUCTION

Gaucher Disease

The lysosomal storage disorders are caused by mutations in the genes coding enzymes responsible for catabolic reactions. These degradative processes are required to dispose of the debris of cell turnover. Failure to breakdown these materials results in their accumulation in the lysosome and subsequently cell dysfunction and cell death. The pathobiology of the disorders should dictate the target cell for gene transfer in the development of gene therapy. However, because the pathogenesis of these and most diseases is only partially defined, a variety of gene transfer approaches may need to be studied before a successful therapy is discovered (Table 1).

Gaucher disease (GD) is the most common lysosomal storage disorder. In the Ashkenazic Jewish population, the incidence of the disease is 1 in 450, with approximately 1 person in 10 carrying the gene for the disease (Zimran

Stem Cell Biology and Gene Therapy, Edited by Peter J. Quesenberry, Gary S. Stein, Bernard G. Forget, and Sherman M. Weissman
ISBN 0-471-14656-0

TABLE 1 Gene Transfer Strategies

Ex vivo gene transfer to transplantable cells *involved* in the *pathogenesis*
Distribution *in vivo:* Widely or locally
Examples: Bone marrow, hepatocytes, oligodendrocytes
Ex vivo gene transfer to transplantable cells capable of secreting a *therapeutic protein*
Distribution *in vivo:* Usually locally (miniorgan)
Examples: Fibroblasts, myoblasts, bone marrow
In vivo targeted gene transfer to a single organ in which genetic correction has *broad* therapeutic effects on *multiple* organs
Distribution *in vivo:* Single organ
Example: Liver
In vivo direct gene transfer locally to correct a major manifestation of the disease
Distribution *in vivo:* Local
Examples: Brain, vascular endothelium

et al., 1991). In the non-Ashkenazic Jewish population, the carrier frequency is approximately 1 in 100, with 1 person in 40,000 affected.

Persons with GD are deficient in glucocerebrosidase (GC), a specialized lysosomal enzyme that hydrolyzes glucosylceramide to glucose and ceramide. As a result of the deficiency, glucosylceramide accumulates in the lysosomes of reticuloendothelial cells to produce the characteristic Gaucher cell. These cells are large, lipid-filled macrophages that are particularly abundant in the red pulp of the spleen, sinusoids of the liver, sinusoids and medullary portion of the lymph nodes, and bone marrow. Gaucher cells are histiocytes in the spleen, Kupffer cells in the liver, macrophages in the bone marrow, and periadventitial cells in the Virchow-Robin space of the brain. These distinctive cells are 20–100 μm in size and have an eccentric nucleus and a "wrinkled tissue paper" appearance useful for distinguishing GD from other lysosomal disorders.

There are three types of GD (Barranger and Ginns, 1989). Each is the result of GC deficiency and is inherited in an autosomal recessive manner. The principal difference among the types is the presence and progression of neurological complications. Type 1 GD is the most common and has a chronic, non-neuronopathic course. The age of onset and severity of symptoms vary widely. It is this type of GD that has an increased incidence in the Ashkenazic Jewish population. Type 2 GD is the acute neuronopathic or infantile form. The average age of onset is 3 months, and neurological complications are usually apparent by 6 months of age. This type is fatal at an average of 9 months and has no ethnic predilection. Type 3 GD is panethnic, but there is a genetic isolate of type 3 disease in the population of northern Sweden. Patients with type 3 usually present as children and have slowly progressive neurodegenerative disease. The first neurological sign is typically oculomotor apraxia, an eye movement disorder.

GD is caused by mutations in the gene encoding GC (Ginns et al., 1984). The most common genetic mutation causing type 1 disease is a single base pair substitution in codon 370 (Barranger and Ginns, 1989). It accounts for approximately 70% of the mutant alleles in Ashkenazic Jewish patients with GD. The lysosomal enzyme associated with this mutation has reduced activity but is not diminished in concentration in the lysosome (Furbish et al., 1981). Most patients with the more severe types of GD (types 2 and 3) who present with neurological complications carry at least one allele with a single base substitution in codon 444. This mutation results in an unstable enzyme with little or no enzymatic activity in the lysosome. More than 50 mutations in the GC gene are known, most being private alleles occurring in a single kindred.

Diagnosis

The diagnosis of GD should be considered in any case of unexplained splenomegaly with or without a bleeding diathesis or other disease manifestations in the liver or skeleton. The diagnosis should also be considered in infants with hepatosplenomegaly and a neurodegenerative course. The presence of Gaucher cells in bone marrow aspirates is highly suggestive of the disease; however, bone marrow examination should not be used as the principle diagnostic tool because enzyme assays are readily available. Definitive diagnosis of GD requires the demonstration of GC deficiency, which is apparent in all tissues and cells, including leukocytes, cultured skin fibroblasts, amniocytes, and chorionic villi.

Enzyme activity less than 30% of normal is diagnostic of the disease state. Enzyme activity in excess of 30% but less than normal is consistent with persons who are heterozygous or carriers of a GD mutation. Carrier detection via this method is unreliable, however, due to a high rate of false-negative results. There is overlap between normal and carrier enzyme levels so that about 20% of obligate carriers fall within the normal range. Molecular diagnosis using a polymerase chain reaction (PCR)–based system and oligonucleotide probes specific for the most common alleles detects carriers accurately in kindreds in which the mutations are known.

Treatment

The recent success of enzyme replacement therapy has simplified the management of patients with type 1 GD. Before enzyme replacement therapy became available, severe anemia and thrombocytopenia were treated with splenectomy. Some clinicians maintain that splenectomy contributes to the progression of GD. Enzyme replacement therapy may make splenectomy unnecessary and resolve the controversy over the contribution of splenectomy to the progression of the disease.

Skeletal problems are treated by limitation of activity, prostheses, analgesics, and surgical intervention. Surgical decompression of an acutely infarcted area of the bone may alleviate discomfort, but bone surgery is hazardous for GD patients due to the risks of hemorrhage and infection.

In severely affected patients with liver failure, orthotopic liver transplantation and bone marrow transplantation (BMT) are life-saving options. However, liver transplantation is not a systemic therapy for GD and additional therapies must be considered. Given that macrophages are derived from the bone marrow and are the only storage cells in which glucosylceramide accumulates in GD, successful BMT logically has been curative for this genetic disease. BMT has been reported in several patients with GD, who had resolution of enzyme deficiency in circulating white blood cells, regression of organomegaly, and an improvement in general health. However, the risk of allogenic transplantation does not justify the procedure in patients with mild disease. Moreover, transplant-associated risks increase with the severity of the disease and the age of the patient. The advent of enzyme replacement therapy obviously limits the number of patients who will be considered for BMT.

With the isolation and purification of GC, replacement of the deficient enzyme in persons with GD became a possibility (Barranger et al., 1989). While initial biochemical results were encouraging, further studies revealed that GC in its native form was not effectively delivered to the storage macrophages in which glucosylceramide accumulates. Further research established that if the oligosaccharide side chains of the enzyme were degraded to expose mannose residues, the enzyme will bind to a mannose-specific receptor on the macrophage plasma membrane and be endocytosed (Furbish et al., 1981). Studies in animals have shown more than a 10-fold increase in uptake of the modified (mannose-terminated) enzyme by Gaucher cells compared with uptake of native enzyme by the same cells. Periodic infusion of mannose-terminated GC in patients with GD is effective in reaching the target macrophages and reversing disease manifestations. Patients experience an increase in hemoglobin concentration, an increase in platelet count, a reduction in the size of the liver and spleen, and a gradual improvement in bone manifestations.

Enzyme replacement therapy represents the first therapeutic breakthrough for the treatment of persons with lysosomal storage disease (Barranger et al., 1989; Barton et al., 1991). The pharmacology of mannose-terminated GC is complicated, and multiple variables need to be clarified to fine tune and enhance the success of enzyme replacement therapy. The recommended dosage is 60 U/kg body weight every 2 weeks (Barton et al., 1991). The minimum effective dose and the optimal dosage frequency are only beginning to be determined (Barton et al., 1991; Fallet et al., 1992; Beutler et al., 1991). Additional clinical experience and research using cell cultures and transgenic animal models will provide important tools to further evaluate this therapeutic approach.

The cloning of cDNA for the GC gene made it possible to consider treating GD with somatic cell gene therapy (Ginns et al., 1984). Initially it was shown that GC deficiency can be corrected by transferring the GC gene into cultured fibroblasts of persons with GD (Choudary et al., 1996). Early studies demonstrated that simplified retroviral vectors carrying the human gene can result in sustained expression of GC in the hematopoietic stem cells of mice (Ohashi et al., 1992; Nolta et al., 1990).

DEVELOPMENT OF GENE TRANSFER AS A POTENTIAL THERAPY FOR GD: CONSTRUCTION OF MFG-GC

Initial results with N2-derived vectors demonstrated efficient transduction into murine hematopoietic stem cells in long-term bone marrow cultures (Nolta et al., 1990) and in irradiated syngeneic recipients of transduced bone marrow (Ohashi et al., 1992). Expression of the transgene was observed *in vitro,* but only minimally *in vivo* in spleen colonies derived from recipients' transplanted bone marrow transduced with the N2-SV-GC vector. In contrast, sustained long-term expression was achieved in animals transplanted with bone marrow transduced with the MFG retroviral vector. The features of the MFG-GC vector are that the GC cDNA was transcribed by the retroviral long terminal repeat (LTR) and the start codon of the GC cDNA was placed at the start codon of the deleted envelope protein gene. No internal promoter or dominant selectable marker is included in the construct (Ohashi et al., 1992).

To clone into the MFG vector, we created an NcoI site in the position of the start codon (ATG) of the GC cDNA using PCR. The sense primer (5′-CCACCATGGCTGGCAGCCTC-3′) was made with a one base pair mismatch to create the NcoI site. The antisense primer (5′-GTGTACTCTCATAGCGGCTG-3′) was located down stream of a HindIII site. The product was cut with NcoI and HindIII, and a 60 bp fragment was isolated using 4% NuSieve agarose gel. On the 3′ side of the GC cDNA, an EcoRI fragment of GC cDNA was isolated, and the terminus was filled in with dNTPs by the Klenow fragment to create a blunt end. A BclI linker was ligated in the usual manner and digested with HindIII and BclI. The 1.7 kb HindIII/BclI fragment was isolated from 1% low melting point agarose gel. The 60 bp NcoI/HindIII fragment and 1.7 kb HindIII/BclI fragment were ligated to NcoI/BamH1 site in the MFG vector. The sequence of this construct revealed no PCR errors or cloning artifacts.

Murine Bone Marrow Transplant Model

The MFG-GC vector was cotransfected with pSV2neo into the ecotropic ΨCRE helper line. Following selection in G418 and screening of clones of the supernatants by gene transfer into 3T3 targets, approximately half of 28 G418-resistant clones expressed viral titers in the range of 10^6 integrating copies per milliliter, and five clones expressed titers 5–10-fold higher. The titering method was based on a linear correlation observed between enzyme activity in 3T3 targets with Southern blot hybridization intensity calibrated upon cell lines of known copy number (Bahnson et al., 1994).

The highest titer ecotropic producer, ΨCRE4, was used to transduce marrow from 5-fluoruoracil (5-FU)–treated donor C57BL6/J-GPI[a] mice using a 2 day prestimulation followed by 2 day co-culture with irradiated producer cells in the presence of interleukin-3 (IL-3), IL-6, and stem cell factor (SCF), essentially as outlined by Bodine and coworkers (1986). Irradiated syngeneic recipient mice carried the alternative GPI-1[b] isoenzyme, which permitted estimation of the degree of engraftment of donor cells at various time points after transplantation based on electrophoresis of peripheral blood cell extracts.

The efficiency of transduction and expression of the GC gene in bone marrow cells was estimated by analyzing individual colonies harvested from the spleens of lethally irradiated mice at 12 days after BMT. Eighty-six colonies from the spleens of mice given MFG-GC–transduced cells and 13 colonies from mice reconstituted with marrow infected with N2-SV-GC were analyzed for comparison. Colonies from lethally irradiated mice transplanted with normal syngeneic marrow were used as controls. Southern blot hybridization revealed that essentially all of the spleen colonies from either group of experimental recipients contained integrated vector at similar copy numbers, but colonies from MFG-GC mice gave enzymatic activities that ranged from two- to fivefold above the controls, whereas the colonies from N2-SV-GC mice were not greater than controls.

By 2 months after BMT in mice given 10^6 bone marrow cells, circulating white blood cells were >90% donor type as assessed by GPI and remained >90% until sacrifice. In addition, the majority of circulating white blood cells were positive for the human gene product by immunocytochemical analysis. Animals were sacrificed between 4 and 7 months after BMT. Southern blots were positive for the human gene in tissues from the liver, lung, spleen, thymus, lymph nodes, and bone marrow. The enzymatic activity of hematopoietic tissues from mice reconstituted with MFG-GC–infected marrow exceeded the activity of control tissues an average of threefold in spleen and sixfold in bone marrow. By comparison, the hematopoietic tissues from mice given N2-SV-GC–infected BM cells showed little or no increase above control levels of activity. The data accumulated on nonhematopoietic tissues (liver, lung) were also informative. These tissues normally are supplied with bone marrow–derived cells on a continuing basis. Under normal physiological circumstances, bone marrow–derived cells in these tissues are primarily macrophages. In liver, tissue macrophages (Kupffer cells) constitute approximately 15% of the organ. If all of the liver macrophages were replaced in MFG-GC–reconstituted animals by the progeny of transduced hematopoietic stem cells derived from the bone marrow, the copy number of the vector in the liver should be about 0.15/cell. The results of Southern blot analyses demonstrated that the MFG-GC vector resulted in a copy number in liver of approximately 0.1/cell. This result is consistent with a high transduction efficiency of hematopoietic stem cells by the MFG-GC vector and the ability of hematopoietic stem cells to repopulate the macrophage lineage.

As a further measure of the ability of the MFG-GC vector to transduce self-renewing pluripotent hematopoietic stem cell, we performed secondary BMTs using bone marrow collected from three long-term reconstituted mice. All secondary spleen colonies from animals sacrificed at 12 days were positive for the human GC gene by Southern blot analyses (n = 27), and the enzymatic activities were two to seven times higher than that of control spleen colony-forming units (S-CFU). Long-term secondary transplants were maintained for more than 12 months. Assay of peripheral blood leukocytes (PBL's) for GC activity revealed that cells have 4–10 times the background amount of enzyme.

To investigate vector expression in macrophages, bone marrow from sacrificed recipients was expanded in the presence of macrophage colony-stimulating factor (M-CSF), resulting in nearly pure colonies of macrophages that were actively phagocytic as evidenced by the ability to take up latex beads. The enzymatic activity in these cells was on average fourfold greater than that of control macrophages. Western blots of macrophage extracts showed an intense band of human GC protein not present in control cells cultured from normal mouse bone marrow. The copy number in these cells was approximately one to two per cell by Southern blot analyses. Immunochemical staining of these cultured macrophages for the human gene product revealed that most of the cells exhibited positive staining, whereas there was no staining in control cultures. These results demonstrated conclusively that self-renewing pluripotent stem cells were transduced, that the transgene was expressed at high levels for the life of the mice, and that macrophages, which are the differentiated cells of therapeutic potential for gene therapy, exhibited robust expression of the GC enzyme at low vector copy numbers.

Generation and Characterization of Amphotropic Producer Lines

For preclinical studies and for clinical application, an amphotropic producer cell line was required. Cell surface envelope protein interferes with cross-infection of cells of the same pseudotype, but supernatant from an ecotropic producer can be used to cross-infect an amphotropic packaging cell line to generate stable amphotropic producers and *vice versa.* In this case, the proviral vector DNA integrates without carriage of plasmid sequences, and transduction efficiency may be high enough to clone candidate producers without selection. The high titer amphotropic MFG-GC producer cc-2 was thus generated by cross-infection of ΨCRIP packaging cells with supernatant from ΨCRE4. Among 12 isolated clones, 5 yielded significant titer, and one clone, cc-2, expressed a titer of about 2×10^7 integrating copies/ml.

The MFG-GC vector was subsequently modified by insertion of a SacII linker immediately downstream of the gag start site, yielding a frame shift to prevent expression of partial gag sequences (incorporated into MFG to retain packaging efficiency). Following this change, a set of promising producers was obtained using a two-step process employing the BOSC 23 transient packaging line (Pear et al., 1993) to produce an ecotropic vector-containing supernatant that was then used to cross-infect ΨCRIP packaging cells. The subsequent titer of a selected ΨCRIP producer was highly variable on 3T3 targets in contrast with human TF-1 cell targets, which remained readily transducible with supernatants from the same producer. Southern blot analysis revealed high copy number in the 3T3 targets, indicating efficient transduction without GC enzyme expression in these targets. More definitively, among a recent set of 15 G418-resistant ΨCRIP clones co-transfected with the modified MFG-GC vector and pSV2neo, none raised 3T3 target activity as much as 200 U/mg above background, whereas 8 clones yielded elevations of over 500 U/mg in human TF-1 cell targets. These findings reveal unexpected specificity of expression among slightly different forms of the MFG vector (manuscript in preparation) and imply that potentially useful producer clones may have been overlooked because 3T3 target activity was implicitly relied on for titering.

Transduction of $CD34^+$-Enriched Cells

High titer vector production and efficient expression of the MFG-GC vector facilitated effective supernatant transduction of enriched populations of $CD34^+$ cells obtained from normal cord blood, normal bone marrow, peripheral blood granulocyte (G)-CSF–primed leukemia patients, and GD patient bone marrow collected during surgical procedures for knee replacement. Enzyme activity in transduced cells was compared with normal and abnormal (Gaucher) enzyme activity levels in nontransduced cells, showing that a potentially therapeutic elevation in GC was readily achieved in transduced cells using this experimental protocol and high titer cc-2 supernatants.

In the initial protocol. $CD34^+$-enriched cells were prestimulated in medium containing the cytokines IL-3, IL-6, and SCF, followed by four or five daily exposures to fresh vector-containing supernatants. During the course of these initial studies, experiments indicated that a prestimulation period of 1 day and reduced infection numbers over a shorter time period were equally effective (Bahnson et al., 1994; Nimgaonkar et al., 1994). The rapid appearance of myeloid differentiation antigens on the $CD34^+$ cells over time in culture stressed the importance of minimizing the *ex vivo* period as much as possible (Nimgaonkar et al., 1994).

Transduction was directly analyzed by Southern blot analysis of SstI digests of genomic DNA from expanded transduced $CD34^+$ cells. Vector proviral DNA hybridization intensity in transduced cells was compared with controls consisting of human DNA quantitatively spiked with DNA from a murine fibroblast clone carrying a known vector copy number. Results showed that low copy numbers of the MFG-GC vector (<1 copy/cell) resulted in GC expression levels that more than compensated for the deficiency of this enzyme in GC patient hematopoietic cells in culture. Supporting evidence for normal enzyme expression was provided by immunocytochemical staining, which indicated that at least 20% of the expanded cells expressed the transgene and displayed a staining intensity equal to or greater than that of normal bone marrow cells (Bahnson et al., 1994).

Centrifugal Enhancement of Transduction

Despite the encouraging results described above, variability and reports of lower than expected transduction efficiencies when research procedures were applied in clinical practice (Dunbar et al., 1993; Kohn et al., 1993) spurred us to investigate centrifugation as a method to gain additional advantage over the transduction process (R. W. Atchison, personal communication). In part this effort is based on the assumption that higher observed transduction in progenitor cells will be found to correlate with a higher probability of transduction in the ultimate target, the engrafting pluripotent stem cells. This assumption is necessary at the present time, regardless of the method, since there are no proven assays for this target cell population other than transplantation in human patients.

Initial experiments comparing centrifugal with noncentrifugal transduction of cord blood $CD34^+$-enriched cells demonstrated the potential for improved transduction. In one of the two experiments comparing both methods, PCR analysis of granulocyte-macrophage (GM) CFU indicated a transduction efficiency of 95% using centrifugation in comparison with no PCR-positive colonies detected in control cultures of cells transduced in the same experiment at 1g. A transduction efficiency of 17 to 20% in the long-term culture-initiating cells (LTC-IC) from centrifuged samples was revealed by PCR analysis of GM-CFU taken at 4, 5, and 6 weeks in this experiment. Although these results are among the best reported for retroviral transduction of hematopoietic cells, lower transduction of LTC-IC in comparison to GM-CFU points to and reinforces the need for continuing improvement in transduction efficiency.

Additional experiments with cultured human hematopoietic cells (TF-1) indicated that the enhancement effect of centrifugation is directly related to centrifugal force up to 10,000g and to the time of centrifugation. On the other hand, the effect was inversely related to cell number in a given container, presumably reflecting a requirement for surface area exposure to suspended virus (Bahnson et al., 1995). These variables provide opportunities for further improvement, but for practical application at the present time we have adopted a procedure using blood bags containing up to 5×10^7 cells per bag centrifuged at 2,400g twice for 2 hour periods with a midpoint change of supernatant. A major advantage of the centrifugation protocol is that it reduces to a minimum the time the cells must be kept in culture. This may be particularly important in light of studies by Quesenberry and coworkers showing rapid reduction in engraftment potential with increasing time in culture (Peters et al., 1996).

TRANSDUCTION OF $CD34^+$ CELLS

Preclinical studies demonstrated efficient transduction in $CD34^+$ cord blood (CB) cells as measured by high enzyme activity and positive PCR signals for the GC transgene in GM-CFU and clonogenic cells arising from LTC-IC cultures in transduced fractions (Mannion-Henderson et al., 1995).

Human umbilical CB cells were collected following normal deliveries and enriched for the CD34$^+$ fraction using Ceprate columns (Cellpro, Inc., Bothell, WA). The CD34-enriched fraction was prestimulated for up to 24 hours with 10 ng/ml each of IL-3, IL-6, and stem cell factor (PeproTech) in long-term bone marrow culture media at a concentration of 2×10^5 cells/ml. These cells were transduced with a retroviral vector containing the normal human GC cDNA three times over the course of 24 h at a concentration of 10^5/ml, using a protocol based on a centrifugation-enhanced technique developed by Bahnson, et al. (1995). Nontransduced controls were obtained as fractions from each CB sample. Data from these experiments demonstrate an average transduction efficiency in the CD34$^+$-enriched fraction of 50% as measured by PCR for the integrated GC cDNA in CFU-GM colonies. PCR of GM-CFU harvested from LTC-IC cultures at 6 weeks also indicates transfer of the cDNA to early progenitor cells. Measurements of enzyme activity comparing transduced and nontransduced fractions at 4 or 6 days post-transduction indicate an average enzyme increase of six-fold over nontransduced background levels.

The next question that needed to be addressed was the stability of the transgene and the ability to sustain expression of a functional enzyme over time. Human cord blood cells would not be appropriate for this assessment, since the composition and stability of the CD34$^+$ cell population grown in bulk culture over time is not well defined. The human cell line TF-1, a factor-dependent human erythroleukemia cell line (American Type Culture Collection) (Kitamura et al., 1989), was chosen to address this question of stability. TF-1 cultures provide readily available material in which to optimize a large-scale infection that would be necessary in a clinical trial. These cells were transduced at a concentration of 10^5/ml, using the a retroviral vector containing the GC cDNA and maintained in culture for 44 days. With a successful transduction efficiency at a cell concentration of 10^5/ml, the question of increasing cell concentration against efficiency was addressed. Cells from this experiment remained in culture for 35 days. Although a relationship was observed with increasing cell number to decreasing transduction efficiency, both experiments show enzymatic activities in transduced TF-1 cultures that remained at least 25-fold above the activity of nontransduced cells throughout a 6 week culture period.

The results from this study were used in establishing conditions appropriate for transducing GD patient cells in a clinical trial.

PRECLINICAL STUDIES OF CD34 CELLS FROM GD PATIENTS

Transplantation of CD34$^+$ cells is advantageous for several reasons. First, the number of CD34$^+$ cells required for transplantation is considerably smaller than a whole bone marrow transplant, consequently reducing the side effects associated with cell transplantation as well those associated with infusion of the toxic cryopreservant dimethylsulfoxide. Second, the amount of viral supernatant and cytokines required for transduction and prestimulation is significantly reduced.

We have shown previously that a significant number of $CD34^+$ cells can be collected in patients with GD that are adequate for conventional transplantation in the host (Nimgaonkar et al., 1995). These data are reviewed here.

Patients

Three patients with GD were entered into this Institutional Review Board-approved study. Patient 1 (JH), aged 48 years, was diagnosed in 1968 as having type 1 GD. He was started on enzyme replacement therapy (ERT) in September 1993 and is at present on Ceredase 30 U/kg every 2 weeks. Patient 2 (RH), aged 35 years, was diagnosed in 1965 with type 1 GD and has been maintained on Ceredase 45 U/kg every 2 weeks. Patient 3 (IM), aged 49 years, was diagnosed in 1954 with type 1 GD. This patient has been maintained on Ceredase since June 1992. Her present dose is 30 U/kg every 2 weeks. All three patients have responded to enzyme therapy as evidenced by a reduction in organ size and an increase in hematological indicies. Each continues to experience skeletal complications of the disease. The different doses of ERT for these individuals reflect the efforts to reduce the dose from 60 U/kg every 2 weeks to a lower maintenance dose. Patient characteristics are summarized in Table 2. All patients gave written informed consent.

Mobilization Regimen and Leukapheresis

Patients 1 and 2 received G-CSF (Neupogen [Amgen]) 5 mg/kg/day subcutaneously for 10 days. Patient 3 received G-CSF at a dose of 10 mg/kg/day for 10 days. Pre G-CSF laboratory evaluation for each patient involved GC activity, genotype, complete blood count and differential, platelets, uric acid (UA), alkaline phosphatase, lactate dehydrogenase (LDH), and flow cytometric analysis for $CD34^+$ cells. Daily evaluation while the patient was on G-CSF involved

TABLE 2 Characteristics of Patients

	Age (years)					Bone	
					Present	Marrow	GC
	At	At	Disease		Therapy	Storage	Activity
Patient	Diagnosis	Present	Type	Genotype	(Ceredase)	(nmol/h/mg)	
1 (JH)	21	48	1	N370S/ N370S	60 U/kg	+++	1
2 (RH)	5	35	1	N370S/?	45 U/kg	+++	2
3 (IM)	6	49	1	L444P/?	30 U/kg	+++	2

complete blood count, platelet count and differential, UA, LDH, alkaline phosphatase, and FACS analysis for $CD34^+$ cell number.

Leukapheresis was started in the three patients on day 5. A total of five leukaphereses were done in each patient using the Cobe Apheresis Spectra device. The total volume of the leukapheresis produce was recorded, and the samples were removed for cell counts and further analysis. All cell counts were done on the Coulter Counter ZM to determine the total number of white blood cells in the leukapheresis product. The leukapheresis product was washed with RPMI 1640 on the Cobe 2991. The final volume of the washed cell preparation was approximately 150 ml.

Mobilization of $CD34^+$ Cells

Following leukapheresis, the total number of white blood cells in the apheresis products averaged 1.6×10^{11} (range 1.0×10^{11}–2.5×10^{11}) or 2.1×10^9/kg.

Enrichment Procedure for $CD34^+$ Cells

Enrichment for $CD34^+$ cells was done using the Ceprate column. The washed leukapheresis product was incubated with the biotin-labeled anti-CD34 antibody and 0.1% HSA (25%) for 25 min at room temperature with mixing for 15 min. Following incubation the cells were washed on the Cobe 2991 and resuspended in phosphate-buffered saline to a volume of 300 ml. Samples were removed for further analysis at this stage, which included cell counts, flow cytometry for CD34 and subset analysis, and viability and clonogenic assays. The cells were processed on the Ceprate SC instrument. The enriched and depleted fractions were removed, and aliquots were removed for further analysis. Sterility testing was also performed on the enriched fraction. The enriched fraction was centrifuged at 1,200 rpm for 8 min and the supernatant removed. The enriched $CD34^+$ cells were cryopreserved in Media 199 with dimethylsulfoxide plus 20% autologous plasma using a clinically approved controlled rate freezing protocol.

The number of white cells in the enriched fractions averaged 6.3×10^8 (range 4.5×10^8–7.9×10^8) or 9.6×10^6/kg. Using the clinical Ceprate column, enrichments averaging 195-fold (range 4–625-fold) were observed. These data demonstrate up to a sixfold increase in the percentage of $CD34^+$ cells in the peripheral blood of the three patients respectively.

The recovery varied from a mean of 25.6% (range 4.9%–48.4%), 61.4% (range 22.4%–168.3%), and 36.9% (range 25%–59%) in the three patients, respectively. The total number of $CD34^+$ cells collected was 1.2, 3.5, and 2.1×10^6 cells/kg, respectively.

Flow Cytometric Analysis for $CD34^+$ Cells

Sample Preparation. The cell concentration was adjusted to 10×10^6 cells/ml in phosphate-buffered saline. Then 100 μl of the cell suspension was incubated with the labeled monoclonal antibodies. Dual color staining was done using the following monoclonal antibodies: CD34 (FITC) combined with either CD38 (phycoerythrin) (PE), HLA-DR (PE), Thy-1 (PE), or CD33 (PE). Relevent isotype controls were used. A 100 μl aliquot of stained cells was mixed with an equal volume of PBS containing 1% human albumin and 0.1% sodium azide and incubated for 15 min at room temperature in the dark. The tubes were then centrifuged for 5 min at 250g and the supernatant decanted. Two milliliters of FACS lysing solution was added to each tube and the cells incubated for an additional 10 min. The tubes were then centrifuged again for 5 min at 250g, the supernatant decanted, and the pellet resuspended in 500 μl of 1% paraformaldehyde. The suspensions were stored at 4°C in the dark and submitted to flow cytometric analysis in their respective staining solutions.

Flow Cytometry. Samples were acquired on a FACScan flow cytometer (Becton Dickinson, San Jose, CA.) equipped with a 15 mW, air-cooled, 488 nm argon-ion laser. Fluorescence data were displayed on a four-decade log scale. Analysis of the bivariate data was performed with LYSYS II software (Becton Dickinson).

Analysis. To obtain total $CD34^+$ cell numbers, samples were analyzed based on SSC and CD34 fluorescence intensity within a live cell gate. Gates to identify $CD34^+$ cells were set using the relevant negative isotype control. The $CD34^+$ cells were gated, and the expression of the second antibody was assessed on the $CD34^+$ cells using a dot plot of FL1 versus FL2. Twenty thousand events were acquired for each sample. To improve the accuracy of subpopulation analysis in those samples in which the $CD34^+$ cell count was low, at least 500 $CD34^+$ events were collected.

CD34 Subset Analysis. We analyzed the percentage of $CD34^+$ cells that coexpressed Thy-1, CD38, HLA-DR, and CD33. The mean percentage of $CD34^+$ cells in peripheral blood that co-expressed Thy-1 was 71.6%, 28.9%, and 44.4% in the three patients, respectively. We noted that there was a trend for the percentage of $CD34^+$, $Thy\text{-}1^+$ cells to diminish with enrichment. The reason for this reduction is uncertain and is under investigation. Thy-1 has been shown to be an early hematopoietic marker (Murray et al., 1995). Impairment of long-term engraftment by the diminution of these cells is unlikely, since $CD34^+$ cells using the CellPro column have been shown to engraft as well as bone marrow (Berenson et al., 1991). Therefore, the significance of this observation remains unknown. The percentages of other CD34 subsets remained unaltered with manipulation.

Transduction of Human Hematopoietic Progenitors

We have previously demonstrated high transduction efficiencies using a centrifugation-promoted infection protocol (Bahnson et al., 1995). This method involves centrifugation of small numbers of cells in tubes, which is not feasible in the clinical setting. We have therefore developed a method using blood collection bags for the centrifugation of large numbers of cells applicable to the clinical trial. The patients' CD34-enriched cells that were frozen were thawed rapidly while mixing constantly. The cells were washed immediately in long-term bone marrow culture media, counted, and resuspended at a concentration of 2×10^5 cells/ml. Pre-stimulation of the cells was performed using the cytokines IL-3, IL-6, and SCF at concentrations of 10 ng/ml for 24 h. The cell concentration was maintained at 2×10^5/ml throughout the pre-stimulation procedure. Following this, the cells were resuspended at 1×10^7–10^8 cells in 50 ml of long-term bone marrow culture media. The cytokine concentration was maintained at 10 ng/ml, and protamine sulfate was added to achieve a concentration of 4 mg/ml. A 60 ml syringe was used to inject the cell suspension into a 150 ml capacity blood collection bag. Then 50 ml of viral supernatant was injected into the same blood collection bag. The air pocket was removed, the bag sealed, and the excess tubing removed. The blood bags were then centrifuged at 2,400 g at 24°C for 2 hours. The bags were removed from the centrifuge and the contents transferred to a 50 ml conical tube. The bag was rinsed with 50 ml of long-term bone marrow culture media to remove any adherent cells. The cells were then centrifuged at 2,400g for 5 min, the supernatant removed, and the cells resuspended in 25 ml of the culture media. The cells were counted and subjected to further analysis.

Analysis of Transduction Efficiency

PCR Analysis of CFU-GM Colonies

Clonogenic Assays. Transduced $CD34^+$ cells were plated at a concentration of 1×10^4 cells/ml in methyllcellulose with IL-3 and GM-CSF. Individual CFU-GM colonies generated after 14 days were plucked and analyzed by PCR for the GC gene in the retroviral vector.

PCR Technique. PCR was carried out on genomic DNA samples in a final volume of 50 μl. The reaction mixture contained 200 μM of each dNTP, 0.5 units Amplitaq (Perkin-Elmer), 2 mM $MgCl_2$, and 0.2 μM of each primer in Amplitaq buffer. One primer hybridizes within the GC cDNA region and the second within the viral sequence, yielding a unique 407 bp amplification product (AB1: 5′ ACG GCA TGG CAG CTT GGA TA 3′; AB2: 5′ AGT AGC AAA TTT TGG GCA GG 3′). Thermal cycling was performed on Gene Amp PCR System 9600 as follows: 94°C $\times$ 5 min for an initial denaturing cycle, followed by 30 cycles of 94°C $\times$ 30 sec, 58°C $\times$ 30 sec, and 72°C $\times$ 30 sec. The PCR products were resolved on a 6% acrylamide or 2% agarose gel and the bands visualized by ethidium bromide and UV light.

PCR of Methylcellulose Colonies. Because of the viscosity of the methylcellulose media, extraction of DNA from these GM-CFU colonies required further preparation. Isolated, single colonies were removed by a pipetman and placed into a sterile, nuclease-free microcentrifuge tube. The DNA extraction method was adapted as previously described (Nimgaonkar et al., 1995) (30). The lysis solution consists of 1.5 μl of glycogen, 10 μl 2M sodium acetate (pH 4.5), 20 μl of sterile distilled water, 100 μl phenol, and 20 μl chloroform. To each tube 100 μl of this mixture was added and left to incubate for at least 30 min at 4°C. The tubes were then spun in a microcentrifuge at 12,000g for 10 min. The aqueous phase was transferred to a clean tube and precipitated with 100 μl isopropanol at −20°C for 60 min or overnight. The samples were spun at 12,000g for 10 min, and the pellets washed in cold 70% ethanol, allowed to briefly air dry, and resuspended in 25 μl sterile distilled water. This sample was then ready for PCR analysis, as described.

PCR Analysis of LTC-IC. Long term bone marrow cultures were maintained as previously described (Eaves et al., 1991). Transduced $CD34^+$ cells were placed on preformed irradiated allogeneic bone marrow stroma at a minimum concentration of 5×10^5 cells per T25 flask. Half the media was replaced weekly. The cultures were maintained at a temperature of 33°C. Cells were removed after 4 and 5 weeks and plated in methylcellulose for clonogenic assays. At week 6, nonadherent cells and adherent cells which were removed using trypsin were plated individually in methylcellulose. Individual GM-CFU were plucked at 14 days and analyzed by PCR for the GC gene as described above.

GC Enzyme Assay in Expanded Cells. The cells were expanded for 1 week in cytokines and assayed for enzyme activity as previously described (Bahnson et al., 1994).

Transduction Efficiency. The transduction efficiency was measured in clonogenic cells and LTC-IC. Using the clinically approved viral supernatant LM30.2.7, a mean transduction efficiency of 32% was demonstrated in clonogenic cells. A transduction efficiency of 25% for patient 1 and 50% for patient 3 in LTC-IC was noted at week 6.

Measurement of GC Enzyme Levels. Cells were maintained in long-term bone marrow culture media and assayed for their GC enzyme levels at 6 days. Up to a 50-fold increase in enzyme levels above deficient levels was noted on day 6.

CLINICAL TRIAL OF GENE THERAPY FOR GD

Description

Allogeneic BMT has been used successfully to treat several patients with GD. ERT has been available since 1991. More than 1,500 patients with type 1 GD have been treated, and, with few exceptions, the patients demonstrate a reversal of disease symptoms and arrest of the progressive disease.

The risk of mortality associated with allogeneic BMT (~30%), the need for an HLA-matched donor, and the availability of ERT render this approach to treatment of GD obsolete. ERT is efficacious; however, it involves lifelong infusions and is expensive. *Ex vivo* gene transfer and autologous BMT could result in a permanent treatment for GD without the requirement for a matched donor, risk of mortality, and expense of ERT.

Gene therapy involves the insertion of a normal copy of a gene into the cells of a patient with an inherited defect in the corresponding gene. Hematopoietic stem cells are a pathobiologically important target cell for gene transfer in GD because macrophages are derived from the bone marrow. In this approach, hematopoietic stem cells are collected, genetically corrected by inserting the gene, and then re-colonized in the patient. Gene transfer/autologous BMT avoids the immunological problems of graft rejection and graft-versus-host disease, which occur with high frequency in allogeneic transplantations. To be successful, gene therapy requires high-efficiency gene transfer into cells, followed by persistent expression of the transferred gene at an appropriate level.

In this study, we transduced $CD34^+$ cells obtained from the blood of GD patients using a replication-defective retroviral vector called R-GC. The vector carries the human GC cDNA. Genetically corrected $CD34^+$ cells are returned intravenously to the patient who donated the cells. This process is referred to as *ex vivo gene transfer* and *autologous BMT*. The primary aim of the study is to evaluate the safety of this approach. Other aims include estimating the extent of competitive engraftment of genetically corrected $CD34^+$ cells, measuring the endurance of bone marrow engraftment with $CD34^+$ cells, measuring the ability of these genetically corrected cells to sustain expression of GC, and examining the patients for any clinical response.

The hypothesis of this study is that genetically corrected peripheral blood stem cells (PBSC) will engraft and result in the supply of enzymatically competent progeny sufficient to reverse the phenotype in patients with GD. The specific aims to be achieved are

1. Evaluation of the safety and feasibility of correcting the basic genetic defect of GD by infusing patients with transduced $CD34^+$ cells
2. Transfer of the human GC gene into PBSC obtained from patients with GD
3. Autologous transplantation of transduced PBSC to patients
4. Measurement of the carriage and expression of the transferred gene and its duration in PBL
5. Assessment of the clinical effects of transplanting genetically corrected PBSC in patients with GD

The outcome of the study will influence future plans by providing information about the safety and capability of gene transfer using our proposed method. Further developments could influence our decisions about different vectors, different preparations of hematopoietic stem cells, and different methods of preparation of the patients.

Methodology

This clinical trial is a phase I study to evaluate the safety and limited efficacy of the gene therapy approach as a treatment for type1 GD. A total of five patients with moderate to moderately severe disease will be enrolled.

Patients with type I GD were recruited from clinics at the University of Pittsburgh and from referrals. They were counseled and informed of the risks and inconveniences of the study during three separate interviews. They were required to sign a consent form. Both patients who are undergoing ERT and untreated patients may be selected for the study.

G-CSF mobilization (10 μg/kg) was used in patients admitted to the study. Leukapheresis procedures began on day 6 of the G-CSF mobilization and continued until a total of 7×10^8/kg mononuclear cells were obtained. The cells were transduced with the R-GC vector to deliver the GC gene. Patients were transplanted with autologous genetically corrected cells at a dose of 2×10^6/kg $CD34^+$ cells. The patient's blood was assessed for the carriage and expression of the transduced gene in PBL by PCR. Measurement of GC activity was used to quantify the extent of restoration of enzyme in these cells. After the restoration of enzymatic activity and carriage of the transferred gene has been established, patients are studied for the clinical responses to the therapy.

We will determine the extent to which $CD34^+$ cells can engraft without a myeloablative preparative step. Engraftment of $CD34^+$ cells is rapid and occurs within 1 month. Therefore, assay of PBL for carriage of the GC gene and GC activity was performed after that interval of time. The results in these patients will determine the approach to be used in subsequent studies during the first year. If the amount of GC leukocytes is not increased twofold above the deficient level, it will be concluded that an inadequate number of corrected $CD34^+$ cells have engrafted, and the procedure will be repeated up to a maximum of four times in the first year. After having established the restoration of enzymatic activity and carriage of the transferred gene, patients will be studied for the clinical response to the therapy. This evaluation includes repeated measurements of clinical and laboratory parameters.

If studies of PBL do not indicate sufficient engraftment (i.e., greater than two times the deficient level of GC), we will repeat the transplantation of genetically corrected $CD34^+$ cells up to a maximum of four times during the course of the year. Each time an infusion is performed, the leukapheresis procedure, enrichment of $CD34^+$ cells, and transduction of $CD34^+$ cells will be repeated. The same certifications of the transduced cells will be performed prior to their use. A flow sheet summarizing the study is shown in Figure 1.

The outcome of the study will influence the development plans by providing additional information about the safety and capability of gene transfer using our proposed method. Further developments could influence our decisions about other vectors, different preparations of hematopoietic stem cells, and different methods of preparation for transplantation. The potential risks associated with the study include those associated with the collection and reinfusion of the patient's $CD34^+$ cells and the minor risks associated with the patient evaluation procedures. However, we believe these risks are minimal given the care we have taken in constructing the vector and designing the protocol. We believe that potential risks are outweighed by the critical data we will obtain regarding safety and clinical efficacy. The results of this study will guide the design of future studies in gene therapy for GD.

G-CSF Priming and Collection of a Mononuclear Cell Fraction

G-CSF mobilization was used in patients with GD who are participants in this study to increase the number of $CD34^+$ cells in their blood. Study candidates received G-CSF at a dosage of 10 μg/kg/day by subcutaneous injection on consecutive afternoons up to 10 days. Injections were scheduled to begin 5 days prior to the scheduled first leukapheresis procedure. Daily monitoring of white blood cell count and differential count was performed. Daily monitoring is essential, since white blood cell count can rise rapidly in response to G-CSF. If the count is >75,000/ml, the G-CSF dose is to be reduced.

Leukapheresis procedures started on day 6 of G-CSF administration and continued on a daily morning schedule until a total mononuclear cell yield of 7×10^8/kg is obtained. It is anticipated that two to three collections will be required of the majority of patients to achieve this yield. If a sufficient number of mononuclear cells is not collected, daily G-CSF may be continued at the discretion of the attending physician until an adequate mononuclear cell dose is obtained. If cytopenia (white blood cell count <3,000 μl or platelets <50,000) develops during or as a result of leukapheresis, the procedure will be postponed until recovery.

Approximately 15 L of the patient's blood was processed during each leukapheresis procedure. Samples from each leukapheresis product were obtained for hemoglobin, hematocrit, total white blood cell count, and differential, platelet count, colony assays, flow cytometry, and microbiological assays. Each leukapheresis product was enriched for $CD34^+$ cells on the day of collection and pooled.

Dosage

Each patient received 2–4 $\times$ 10^6 transduced $CD34^+$ cells/kg body weight. This is the dose used to reconstitute the bone marrow in patients who have been

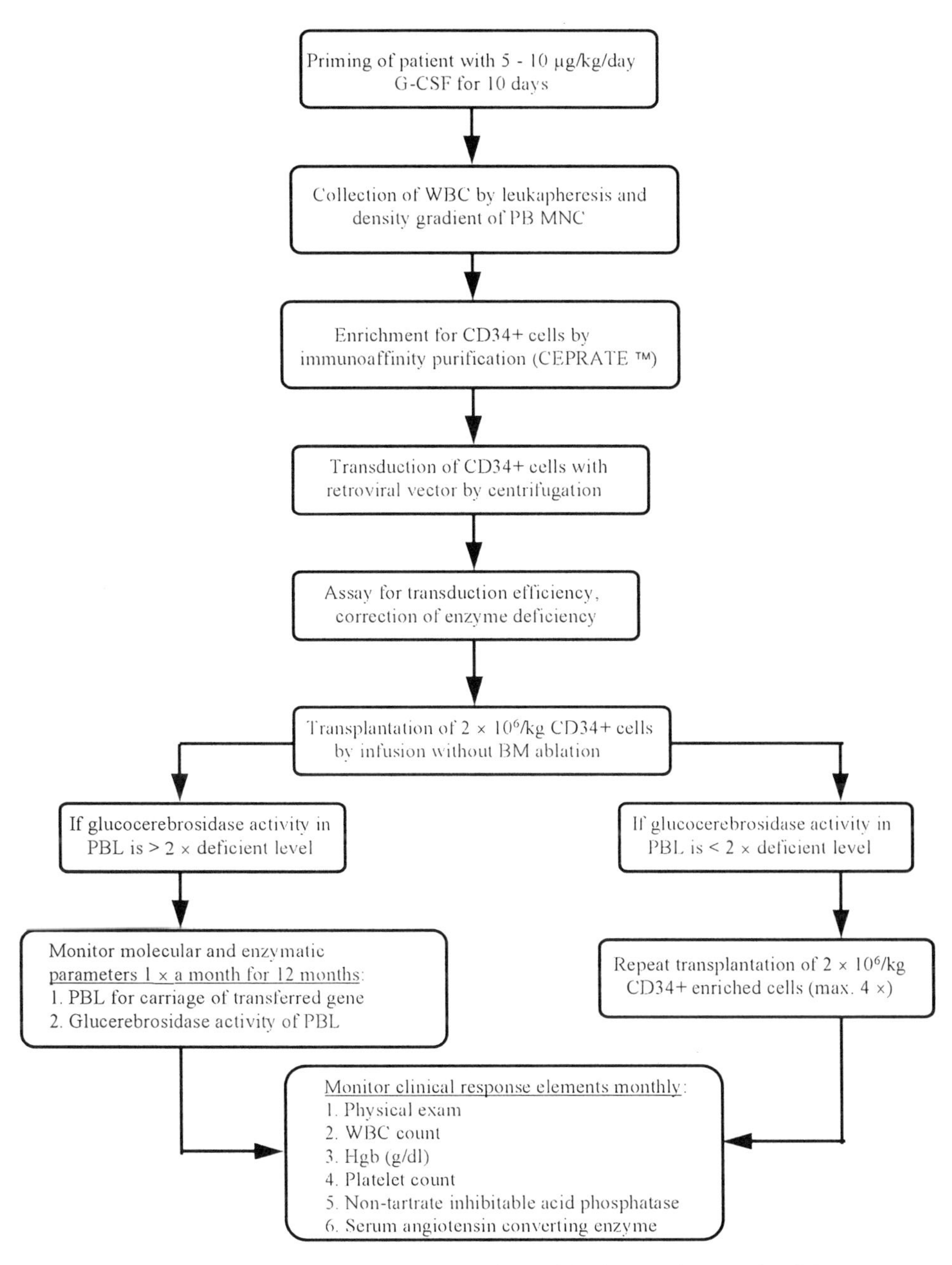

FIGURE 1. Flow diagram for the clinical trial of gene therapy for GD.

myeloablated. The dose needed to result in engraftment in an unprepared (nonmyeloablated) recipient is unknown. Studies of bone marrow transplants in mice and dogs suggest that competitive engraftment in the host animal without preparative myeloablation can be very efficient. We have selected a dose of $CD34^+$ cells for this study that has been used to reconstitute the bone marrow in myeloablated cancer patients.

Infusion of the Transduced Cells

After infection with R-GC, the $CD34^+$ cells were administered to the patient. Aliquots were sampled for biological activity and tested for adventitious agents.

Postinfusion Laboratory Parameters

In this protocol, the assessment of engraftment of the transduced cells into bone marrow depends on the estimation of carriage of the transgene PCR and enzymatic activity in PBL and bone marrow. At 1 month postinfusion, PBL from all patients were assayed for GC activity and carriage of the GC gene. These molecular and enzymatic parameters continue to be monitored according to the schedule. Indicators of clinical response (excluding magnetic resonance imaging, bone marrow biopsy, and x-rays) are conducted monthly for 24 months.

Preliminary Findings

Three study candidates have been entered into the study. One patient experienced a decline in platelet count below 100,000 during G-CSF stimulation. No other side effects of the procedure have been noted in the 9 months duration of the study. All of the candidates have been transplanted at least twice with genetically corrected CD34 cells at a dose of $2–4 \times 10^6$/kg. The GC activity of the corrected cells was 2–26-fold higher than the deficient level. Positive signals for the transgene have been noted in the PBL of each subject, and enzymatic activity in circulating white cells has risen to as high as carrier levels.

HEMATOPOIETIC CELLS—NEW STRATEGIES

Efficient retroviral-mediated transfer of human GC gene into murine hematopoietic stem cells has been accomplished (Ohashi et al., 1992). Despite these successes in mice, it is still unclear how efficiently human hematopoietic stem cells can be transduced, although efficient gene transfer has been demonstrated in human hematopoietic progenitor cells, GM-CFU, and in more primitive cells, LTC-IC (Xu et al., 1994). A number of approaches have been tested to solve this problem.

The titer of retroviral vectors is one of the keys to increase efficiency of gene transfer, and another is the cell cycling of target cells. Generally, retroviral vectors do not integrate into noncycling cells because reverse transcription is blocked. Hematopoietic stem cells, which are mostly quiescent, that is, in the G0 phase of the cell cycle, are difficult targets for gene transfer by retroviral vectors. To stimulate quiescent hematopoietic cells to enter active cell cycle, various cytokines have been used. Among them, a combination of SCF, IL-3, and IL-6 has been shown to increase retroviral-mediated gene transfer into hematopoietic cells effectively (Luskey et al., 1992; Van Beusechem et al., 1995).

Bone marrow stroma, which constitutes the hematopoietic microenvironment, and fibronectin, an extracellular matrix protein, have also been shown to increase retroviral-mediated gene transfer into hematopoietic cells, although the mechanism underlying this effect is not known well (Moore et al., 1992; Nolta et al., 1995; Moritz et al., 1994). However, obtaining bone marrow for stroma cultures, sufficient to support transduction of the large number of cells required for clinical trials, is difficult even from normal individuals; for GD patients, whose cells often grow poorly, this may be a major impediment.

Centrifugation is a simple and safe technique to enhance retroviral-mediated gene transfer. We have already demonstrated that centrifugation increases transduction efficiency for human $CD34^+$-enriched CB (Bahnson et al., 1995). There is a direct correlation between length of centrifugation and transduction efficiency and an inverse relationship between cell number in the infection and efficiency. This technique has advantages and can be applied to clinical trials using blood collection bags as we have described (Nimgaonkar et al., 1995).

The use of adeno-associated virus (AAV) is another approach. AAV is not pathogenic and integrates with some preference into a site on chromosome 19 and is not dependent on active cell cycling. It is a promising vector for gene transfer into hematopoietic cells (Wei et al., 1994). However, production and purification of high titer recombinant AAV vector is complicated, and this problem should be solved.

Enrichment of hematopoietic stem cells increases virus-to-target cell ratio and thereby increases the transduction efficiency. $CD34^+$ enrichment by monoclonal antibody-based methods are commonly used. We have demonstrated that $CD34^+$-enriched cells can be collected from peripheral blood in patients with GD by leukapheresis following G-CSF mobilization and can be efficiently transduced using a centrifugation-enhanced technique (Nimgaonkar et al., 1995). Recently, several groups have reported that hematopoietic stem cells from bone marrow or peripheral blood could be expanded *in vitro* with various combinations of cytokines (Spour et al., 1993). Although the reconstitution ability of these expanded cells needs to be clarified, *in vitro* expansion of hematopoietic cells is an attractive method that may be used in gene therapy approaches.

Results

Cytokines. First, we examined the effect of pre-stimulation with cytokines (IL-3, IL-6, SCF) on transduction of $CD34^+$ cells. CB $CD34^+$-enriched cells were precultured with or without 10 ng/ml each of IL-3, IL-6, and SCF for 0, 16, and 48 hs. After pre-stimulation, the cells were transduced with the GC-containing retroviral vector by centrifugation for 2 hrs. GC activity was measured 6 days post-transduction, and transduction efficiency was analyzed by PCR for the GC transgene in individual GM-CFU. The results showed that $CD34^+$ cells prestimulated with cytokines had high transduction efficiency: 88% and 92% for 16 and 48 h pre-stimulation, respectively. On the other hand, transduction efficiency of nonpre-stimulated $CD34^+$ cells increased with time of preculture: 44%, 63%, and 75% for 0, 16, and 48 h preculture, respectively.

Next, we examined the effect of cytokines during transduction. Transduction of $CD34^+$ cells in the presence of cytokines (IL-3, IL-6, SCF) showed higher GC activity and higher transduction efficiency than that in the absence of cytokines (345 vs. 497 U/mg and 44% vs. 88%, respectively). The use of cytokines before and during transduction is useful in retroviral-mediated gene transfer into hematopoietic cells.

Stroma and Fibronectin. The effect of stroma and fibronectin on transduction of $CD34^+$ cells was assessed. CB $CD34^+$ cells were put in culture plates in which some wells were pre-coated with stroma cell layer or fibronectin. The cells were transduced with the GC-containing retroviral vector for 24 h with or without 2 h centrifugation. GC activity and transduction efficiency were measured by the method described above. Both stroma and fibronectin increased transduction efficiency: 50% with stroma and 63% with fibronectin. Centrifugation alone could increase transduction efficiency up to 69%. A combination of centrifugation and stromal support showed high transduction efficiency (75%). Although stroma and fibronectin enhance retroviral-mediated gene transfer into hematopoietic cells, the safety and logistics of these approaches have not yet been defined.

Expansion. It has already been shown that hematopoietic cells can be expanded *in vitro* in long-term bone marrow culture media containing cytokines. To assess the requirement for addition of cytokines and catalase (antioxidative), CB mononuclear cells were incubated for 21 days in long-term bone marrow culture media, long-term bone marrow culture media containing 10 ng/ml each of the cytokines (IL-3, IL-6, SCF), or long-term bone marrow culture media containing cytokines plus 0.2 mg/ml catalase. The number of total cells increased in the presence of cytokines, and the number of GM-CFU increased in the presence of cytokines and catalase.

$CD34^+$ cells were obtained from peripheral blood of GD patients using G-CSF mobilization and were expanded in long-term bone marrow culture media containing cytokines (IL-3, IL-6, SCF) and catalase. $CD34^+$ cells were expanded 222-fold in cell number on day 23, and 5,310 GM-CFU colonies were noted from 1×10^5 cells on day 14. *In vitro* expansion of $CD34^+$ cells from the patient's peripheral blood was demonstrated, but it remains unclear whether repopulating hematopoietic stem cells can be expanded. Studies in NOD-SCID mice and the SCID-hu bone model should be helpful in assessing this question.

PRENATAL GENE THERAPY

An ever-increasing number of genetic disorders can now be identified *in utero* at an early stage in pregnancy. In current clinical practice, a prenatal diagnosis of a serious genetic disease leaves the parents with very few options. The vast majority of these diseases are not amenable to simple surgical or pharmacological treatments. A variety of experimental approaches aimed at replacement of the deficient gene function or modification of function are being pursued. In general, these involve either cell or organ transplantation or somatic cell gene therapy.

The most common organ or cell type to be targeted for transplantation or gene delivery is the bone marrow or, more specifically, the hematopoietic stem cell. One major hurdle has been the inability to unequivocally identify the pluripotent stem cell. Consequently, it has been difficult to prove that genes can be delivered to the pluripotent stem cell itself with persistent expression of the gene product at therapeutic levels in its differentiated progeny. However, a subpopulation of hematopoietic cells containing the bulk of stem cells and progenitors can be readily identified with the use of specific antibodies, i.e., CD34. In animal studies and in clinical trials, genes responsible for a variety of hematological, immunological, and metabolic disorders have been successfully introduced into this subpopulation using a number of gene delivery vehicles (Ohashi et al., 1992; Adams et al., 1992; Wolfe et al., 1992; Miller, 1992). The success of somatic cell gene therapy in animals led a number of investigators to explore the feasibility of gene transfer *in utero* into fetal hematopoietic cells (Anderson et al., 1986; Soriano et al., 1993; Touraine et al., 1987). These studies suggest that not only is it possible to introduce genes into fetal tissues *in utero* but it may in fact have certain advantages over postnatal gene therapy. Fetal hematopoietic cells are more readily transduced than adult hematopoietic progenitors (al-Lebban et al., 1990). Transplantation *in utero* may obviate the need for bone marrow ablation to create space for engraftment. Early treatment may avert the occurrence of significant functional compromise before clinical diagnosis and treatment. Finally, the neurological manifestations of disease may be more amenable to correction by prenatal hematopoietic stem cell transplantation or somatic gene therapy for several reasons. The blood–brain barrier presents a formidable impediment to the transfer of cells into the brain. In many disorders, the disease process begins in the fetus, leading to potentially irreversible neurological damage before birth (Poduslo et al., 1976; Baier and Harzer, 1983; Adachi et al., 1974; Schneider et al., 1972). The marrow-derived elements in the central nervous system, if present, account for less than 1% of brain tissue, and their turnover is slow. The fetal central nervous system, on the other hand, is in an active state of growth and development, lacks a blood–brain barrier, and may therefore be more accessible to genetically modified cells of hematopoietic origin.

During fetal life, hematopoiesis begins in the yolk sac, transfers to the liver, and initiates in the bone marrow at mid to late gestation (Metcalf and Moore, 1971). Liver is the primary site of hematopoiesis between 7 and 19 weeks of gestation (O'Reilly et al., 1983) and an attractive source of fetal hematopoietic progenitors and stem cells. Fetal liver cells are comparable to adult bone marrow cells in their capacity to self-renew and differentiate into a complete repertoire of erythroid, myeloid, and lymphoid cell lineages (Wu et al., 1968). Fetal tolerance resulting in permanent chimerism has been documented to occur naturally in animal and human twins with shared placental circulation (Turner et al., 1973; Szymanski et al., 1977) and has been experimentally documented in several animal models (Billingham et al., 1953; Binns, 1967; Roncaro and Vandekerckhove, 1992). Touraine et al. (1993) reported on five fetuses transplanted prenatally at 12 to 28 weeks of gestation with fetal liver cells. Three had immunodeficiency disorders, and two had thalessemia major. Three survived to term with evidence of engraftment and clinical improvement. More recently, a fetus with Krabbe's disease was transplanted *in utero* with paternal $CD34^+$ cells. The fetus died *in utero* but demonstrated complete repopulation of hematopoietic tissues with donor cells. Another fetus with SCID has been transplanted with paternal $CD34^+$ cells. This child was born in good condition with significant chimerism in peripheral blood (Flake et al., 1995) and is well at 4 months of age. Use of paternal $CD34^+$ cells for transplantation *in utero* induces tolerance to paternal histocompatibility antigens and thereby provides a ready source of hematopoietic cells for subsequent transplantations.

Genes may be delivered to fetal hematopoietic stem cells and progenitors by two approaches: (1) *ex vivo* transduction of allogeneic or autologous fetal hematopoietic stem cells followed by transplantation *in utero* (2) *in vivo* gene delivery by direct injection of the recombinant vector such as a retrovirus into fetal liver during the period of hepatic hematopoiesis. Direct *in vivo* gene delivery is technically easier and does not require fetal liver harvest or depend on the success of cell culture and subsequent survival of transplanted cells. In the initial studies, a replication-competent moloney murine leukemia virus (MoMuLV) was injected into mice embryos (Jaenisch, 1980). Compere et al. (1989) injected an H-ras–carrying replication-incompetent retrovirus into the amniotic sac of D-8 mouse fetuses. H-ras–expressing cells underwent neoplastic transformation. Fetal rats injected *in utero* with a replication-incompetent retroviral vector carrying the chimeric phosphoenolpyruvate carboxykinase–BGH gene demonstrated the presence of the provirus in the liver for up to 1 year. In two of the rats killed at 3 months of age, BGH gene was identified in the bone marrow (Hatzoglou et al., 1990). Clapp et al. (1991) injected cell-free supernatants containing a replication-defective retroviral vector into rat fetal livers at 11, 14, 16, and 18 days of gestation. In 10 of 12 fetuses injected at days 14–16 of gestation, the provirus was detected in the bone marrow by PCR or Southern analysis. In two animals the provirus persisted for 26 weeks.

As expected, fetal demise following these *in utero* injections was proportional to the volume of supernatant injected. The greater the injection volume, the higher the mortality. Attempts to concentrate the virus by filtration or ultracentrifugation have generally resulted in loss of infectious virus presumably due to the instability of the envelope proteins. Furthermore, MoMuLV-based retroviral vectors activate complement via the alternate pathway. To overcome these difficulties, pseudotyped retroviral vectors were developed. These have the same retroviral core but a different envelope protein. As gibbon ape leukemia virus (GALV) receptor (Glvr-l) expression in human hematopoietic cells is higher than the expression of the amphotropic receptor (Ram-l) (Kavanaugh et al., 1994), MoMuLV was pseudotyped with the GALV envelope (Miller et al., 1991). In other studies, moloney-based retroviruses were pseudotyped with G protein of the vesicular stomatitis virus (VSV). VSV-G pseudotyped virus can be concentrated by ultracentrifugation to titers up to 10^9 pfu/ml (Friedman et al., 1994; Yee et al., 1994). Despite the early promise, studies comparing their transduction efficiency in human hematopoietic cells with that of amphotropic MoMuLV-based vectors have been disappointing. Amphotropic retroviral vectors would appear to be more efficient than VSV-G pseudotyped virus. Tsukamoto et al. (1995) injected SV40-chloramphenicol acetyltransferase (CAT) : cationic lipopolyamine complexes intravenously into pregnant mice. The CAT plasmid thus introduced at day 9 of gestation was detected by slot blot analysis in the organs of the progeny at least 1 month after birth or 40 days after injection. Heart, lung, liver, and skeletal muscle contained larger amounts of the transferred genome. Bone marrow and blood were not examined, but levels in spleen were low. Unintegrated plasmid was detected at low levels in the brain tissue, and none was present in the germ cells (Tsukamoto et al., 1995).

Most studies of *in utero* gene transfer have used an *ex vivo* approach of preincubation with cytokines followed by gene transfer using a replication-incompetent retroviral vector. Although a greater fraction of hematopoietic progenitors in fetal liver as opposed to adult bone marrow is actively proliferating, preincubation with cytokines remains crucial for retrovirus-mediated gene transfer (Gardner et al., 1990; Bodine et al., 1989). Fetal liver cells are thought to resemble adult bone marrow cells in their response to cytokines; the issue, however, is not yet fully resolved (Yu et al., 1983; Zanjani et al., 1992, 1991; Flake and Zanjani, 1993). It is also unclear if the retroviral receptor expression in fetal liver cells is similar to adult hematopoietic cells. It is possible that amphotropic receptor expression in midgestational murine fetal liver cells may be less than adult levels (Richardson et al., 1994). It is, however, present, and murine fetal liver cells are susceptible to ecotropic and amphotropic retroviral vectors (Bansal et al., 1995a). Several *ex vivo* studies of gene transfer into fetal liver cells using retroviral vectors suggest that fetal liver cells may in fact be transduced more readily than adult hematopoietic cells. Even so, the transduction efficiency in humans is variable and low. Efforts are underway to develop a variety of different strategies aimed at increasing the efficiency of gene transfer using retroviral vectors. Centrifugation has been successfully employed to enhance viral infection of cell cultures mainly in detection at low titers of a variety of viruses such as CMV (Espy and Smith, 1987), herpes (Gleaves et al., 1985), and adenovirus (Espy et al., 1987). It has only recently been demonstrated to be effective for retroviruses such as HIV (Ho et al., 1993) and MoMuLV (Kotani et al., 1994; Bahnson et al., 1997). We have examined the effect of centrifugation during viral inoculation on the efficiency of human GC gene transfer with a MoMuLV-based retroviral vector (MFG-GC) into midgestational fetal liver cells. After incubation with cytokines for 24–48 h, cell-free viral supernatants were added to fetal liver cells and centrifuged at 2,200 g for 2 h at 22°C followed by incubation at 37°C. GM-CFUs in methylcellulose were analyzed individually by PCR for the presence of vector-specific sequences. Transduction efficiency in GM-CFUs by the conventional protocol of incubation with the retrovirus containing supernatant was on the order fo 16.7% with ecotropic MFG-GC and 8.3% with the amphotropic form of MFG-GC. Centrifugation led to a 5–10-fold increase in transduction efficiency, being 75% and 87.5%, respectively, with the ecotropic and amphotropic forms of MFG-GC. After transduction by conventional incubation, about 10%–15% of the fetal liver mononuclear cells expressed the immunoreactive GC protein. In contrast, when centrifugation was used for viral inoculation, three to four times as many cells stained positive for the human GC protein. This was reflected in the GC enzyme activity, which was 1,572 U/mg, 8.5 times above controls (166 U/mg) and three times higher than cells transduced by conventional incubation (509 U/mg). There was no significant change in cell number, cell viability, or colony count in methylcellulose when fetal liver cells were transduced by centrifugation compared with cells tranduced by conventional incubation. We then transplanted nine C57Bl/6J fetal mice at day 13 of gestation by intraplacental injections of fetal liver cells transduced with MFG-GC by centrifugation; 5–10 $\times$ 10^5 cells suspended in 5 μl volume were injected into each fetus. Six out of the nine (66%) fetuses were born at term. The neonates survived to adulthood (8 weeks) with no obvious adverse effects from fetal intervention. The animals were sacrificed at 8 weeks of age, and proviral sequences were detected in the bone marrow by PCR in six of six animals. Our preliminary studies suggest that centrifugation during viral inoculation may be one way of increasing the efficiency of gene transfer into fetal liver cells using retroviral vectors. Whether this increase in gene transfer into hematopoietic progenitors is accompanied by a significant improvement in transduction of pluripotent hematopoietic stem cells will be examined by studies of long-term repopulation of fetal mice with transduced fetal liver cells.

ACKNOWLEDGMENTS

The work reported in this chapter was supported by grants from the National Institutes of Health (DK43709-06, DK48436-02, DK44935-04, and RR00056-35), the National Gaucher Foundation, and Genzyme Corporation. The generous gifts of Neupogen from Amgen Corp. and Ceprate columns from Cellpro are greatly appreciated. We thank the UPMC, GCRC Nursing Staff for their help, Joseph E. Kiss and the Apheresis Unit, and Nicole Moore for typing the manuscript. The courage and cooperation of the patients involved in these studies have made these contributions possible.

REFERENCES

Adachi M, Schneck L, Volk BW (1974): Ultrastructural studies of eight cases of fetal Tay-Sachs disease. *Lab Invest* 30:102–112.

Adams RM, Soriano HE, Wang M, et al. (1992): Transduction of primary human hepatocytes with amphotropic and xenotropic retroviral vectors. *Proc Natl Acad Sci USA* 89:8981.

al-Lebban ZS, Henry JM, Jones JB, et al. (1990): Increased efficiency of gene transfer with retroviral vectors in neonatal hematopoietic progenitor cells. *Exp Hematol* 18:180.

Anderson WF (1986): Prospects for human gene therapy in the born and unborn patient. *Clin Obstet Gynecol* 29:586.

Bahnson AB, Dunigan JT, Baysal BE, et al. (1997): Centrifugal enhancement of retroviral-mediated gene transfer. *J Virol Methods* 54:131–143.

Bahnson AB, Dunigan JT, Baysal BE, Mohney T, Atchison RW, Nimgaonkar MT, Ball ED, Barranger JA (1995): Centrifugal enhancement of retroviral-mediated gene transfer. *J Virol Methods* 54:131–143.

Bahnson A, Nimgaonkar M, Fei Y, Boggs S, Robbins P, Ohashi T, Dunigan J, Li J, Ball ED, Barranger JB (1994): Transduction of $CD34^+$ enriched cord blood and Gaucher bone marrow cells by a retroviral vector carrying the glucocerebrosidase gene. *Gene Ther* 1:176–184.

Baier W, Harzer K (1983): Sulfatides in prenatal metachromatic leukodystrophy. *J Neurochem* 41:1766–1768.

Bansal V, Bahnson AB, Barranger JA (1995a): Ram-I is expressed by mid-gestational murine fetal liver cells for successful transduction with an amphotropic retroviral vector. *Stem Cell Gene Therapy: Biology and Technology.*

Bansal V, Bahnson AB, Barranger JA, Hogge WA (1997): Centrifugation enhances retrovirus mediated gene transfer into fetal liver cells. (in preparation).

Barranger JA, Ginns EI (1989): Glucosylceramide lipidoses: Gaucher disease. In Scriver CR, Beaudet AL, Sly WS, Valle D (eds): *The Metabolic Basis of Inherited Disease,* 6th ed. New York: McGraw-Hill, pp 1677–1698.

Barranger JA, Ohashi T, Hong CM, et al. (1989): Molecular pathology and therapy for Gaucher disease. *Jpn J Inherit Metab Dis* 51:45–71.

Barton NW, Brady RO, Dambrosia JM, et al. (1991): Replacement therapy for inherited enzyme deficiency—Macrophage-targeted glucocerebrosidase for Gaucher's disease. *N Engl J Med* 324:1464–1470.

Berenson RJ, Bensinger WI, Hill RS, Andrews RG, Garcia-Lopez J, Kalamasz DF, Still BJ, Spitzer G, Buckner CD, Bernstein ID, Thomas ED (1991): Engraftment after infusion of $CD34^+$ marrow cells in patients with breast cancer or neuroblastoma. *Blood* 77:1717.

Beutler E, Kay A, Saven A, et al. (1991): Enzyme replacement therapy for Gaucher disease. *Blood* 78:1183–1189.

Billingham RE, Brent L, Medawai PB (1953): Actively acquired tolerance of foreign cells. *Nature* 172:603–606.

Binns RM (1967): Bone marrow and lymphoid cell injection of the pig fetus resulting in transplantation tolerance or immunity and immunoglobulin production. *Nature* 214:179–180.

Bodine DM, Karlsson S, Neinhuis AW (1986): Combination of interleukins 3 and 6 preserves stem cell function in culture and enhances retrovirus-mediated gene transfer tino hematopoietic stem cells. *Proc Natl Acad Sci USA* 86:8897–8901.

Bodine DM, Karlsson S, Nienhuis AW (1989): Combination of interleukin-3 and 6 preserves stem cell function in culture and enhances retrovirus-mediated gene transfer into hematopoietic stem cells. *Proc Natl Acad Sci USA* 86:8897–8901.

Choudary PV, Barranger JA, Tsuji S, et al. (1986): Retrovirus mediated transfer of the human glucocerebrosidase gene to Gaucher fibroblasts. *Mol Biol Med* 3:293.

Clapp DW, Luba LD, Hatzoglou M, Gerson SL (1991): Fetal liver hematopoietic stem cells as a target for *in utero* retroviral gene transfer. *Blood* 78:1132–1139.

Compere SJ, Baldacci PA, Sharpe AH, Jaenisch R (1989): Retroviral transduction of the human c-Has-ras-1 oncogene into midgestation mouse embryos promotes rapid epithelial hyperplasia. *Mol Cell Biol* 9:1141.

Dunbar CE, O'Shaughnessy JA, Cottler-Fox M, Carter CS, Doren S, Cowan KH, Leitman SF, Wilson W, Young NS, Nienhuis AW (1993): Transplantation of retrovirally-marked $CD34^+$ bone marrow and peripheral blood cells in patients with multiple myeloma or breast cancer. *Blood* 82(suppl 1):217a.

Eaves CJ, Cashman JD, Eaves AC (1991): Methodology of long-term culture of human hematopoietic cells. *J Tissue Culture Methods* 13:55–62.

Espy M, Hierholzer J, Smith T (1987): The effect of centrifugation on the rapid detection of adenovirus in shell vials. *Am J Clin Pathol* 88:358–360.

Espy MJ, Smith TF (1987): Simultaneous seeding and infecting of shell vials for rapid detection of cytomegalovirus infection. *J Clin Microbiol* 25:940–941.

Fallet S, Sibille A, Mendelson R, Shapiro D, Hermann G, Grabowski GA (1992): Gaucher disease: Enzyme augmentation in moderate to life-threatening disease. *Pediatr Res* 31:496–502.

Flake AW, Puck JM, Almedia-Porada G, Evans MI, Johnson MP, Roncarolo MG, Zanjani ED (1995): Successful treatment of X-linked recessive combined immunodeficiency (X-SCID) by *in utero* transplantation of CD34 enriched paternal bone marrow cells. *Blood* 86:125a.

Flake AW, Zanjani ED (1993): Retention and multilineage expression of human HSC within human-sheep chimera. *Workshop on Hematopoietic Stem Cell Purification and Biology*. Washington, DC, p 31.

Friedman T, Bouic K, LaPorte P, Yee JK, Miyanohara A (1994): Pseudotyped retroviral vectors and efficient gene transfer to the liver. Gene Therapy Meeting, Cold Spring Harbor Laboratory, Cold Spring Harbor, NY.

Furbish FS, Steer CJ, Krett NL, Barranger JA (1981): Uptake and distribution of placental glucocerebrosidase in rat hepatic cells and effects of sequential deglycosylation. *Biochim Biophys Acta* 673:425–434.

Gardner JD, Liechly KW, Christinsen RD (1990): Effects of interleukin-6 on fetal hematopoietic progenitors. *Blood* 75:2150–2155.

Ginns EI, Choudary PV, Martin BM, et al. (1984): Isolation of cDNA clones for human β-glucocerebrosidase using the ht11 expression. *Biochem Biophys Res Commun* 123:574–580.

Gleaves CA, Wilson DJ, Wold AD, Smith TF (1985): Detection and serotyping of herpes simplex virus in MRC-5 cells by use of centrifugation and monoclonal antibodies 16 h postinoculation. *J Clin Microbiol* 21:29–32.

Hatzoglou M, Lamers W, Bosch F, Wynshaw-Boris A, Clapp DW, Hanson RW (1990): Models for hepatic gene therapy. *J Biol Chem* 265:17285.

Ho WZ, Cherukuri R, GE HD, et al. (1993): Centrifugal enhancement of human immunodeficiency virus type 1 infection and human cytomegalovirus gene expression in human primary monocyte/macrophages *in vitro*. *J Leukocyte Biol* 53:208–212.

Jaenisch R (1980): Retroviruses and embryogenesis: Microinjection of moloney murine leukemia virus into midgestation mouse embryos. *Cell* 19:181.

Kavanaugh MP, Miller DG, Zhang W, Law W, Kozak SL, Kabat D, Miller AD (1994): Cell surface receptors for gibbon ape leukemia virus and amphotropic murine retrovirus are inducible sodium-dependent phosphate symporters. *Proc Natl Acad Sci USA* 91:7071.

Kitamura T, Tange T, Terasawa T, Chiba S, Kuwaki T, Miyagawa K, Piao Y, Miyazono K, Urabe A, Takaku F (1989): Establishment and characterization of a unique human cell line that proliferates dependently on GM-CSF, IL-3, or erythropoietin. *J Cel Physiol* 140:323.

Kohn DB, Weinberg KI, Parkman R, Lenarsky C, Crooks GM, Shaw K, Hanley ME, Lawrence K, Annett G, Brooks JS, Wara D, Elder M, Bowen T, Hershfield MS, Berenson RI, Moen RC, Mullen CA, Blaese RM (1993): Gene therapy for neonates with ADA-deficient SCID by retroviral-mediated transfer of the human ADA cDNA into umbilical cord $CD34^+$ cells. *Blood* 82(suppl 1):315a.

Kotani H, Newton PB, Zhang S, et al. (1994): Improved methods of retroviral transduction and production for gene therapy. *Hum Gene Ther* 5:19–28.

Luskey BD, Rosenblatt M, Zsebo K, Williams DA (1992): Stem cell factor, interleukin-3, and interleukin-6 promote retroviral-mediated gene transfer into murine hematopoietic stem cells. *Blood* 80:396–402.

Mannion-Henderson J, Kemp A, Mohney T, Mingaonkar M, Lancia I, Beeler MT, Mierski J, Bahnson AB, Ball ED, Barranger JA (1995): Efficient retroviral mediated transfer of the glucocerebrosidase gene in CD34+ enriched umbilical cord blood human hematopoietic progenitors. *Exp Hematol* 23:1623–1632.

Metcalf D, Moore MAS (1971): Embryonic aspects of hematopoiesis. In Neuberger A, Tatum EL (eds): *Frontiers of Biology—Hematopoietic Cells.* Amsterdam: North Holland Publishing Co, p 172.

Miller AD (1992): Human gene therapy comes of age. *Nature* 357:455–460.

Miller AD, Garcia JV, von Suhr N, Lynch CM, Wilson C, Eiden MV (1991): Construction and properties of retrovirus packaging cells based on gibbon ape leukemia virus. *J Virol* 65:2220.

Moore KA, Deisserroth AB, Reading CL, Williams DE, Belmont JW (1992): Stromal support enhances cell-free retroviral vector transduction of human bone marrow long-term culture-initiating cells. *Blood* 79:1393–1399.

Moritz T, Patel VP, Williams DA (1994): Bone marrow extracellular matrix molecules improve gene transfer into human hematopoietic cells via retroviral vectors. *J Clin Invest* 93:1451–1457.

Murray L, Chen B, Galy A, Chen S, Tushinski R, Uchida N, Negrin R, Tricot G, Jagannath S, Vesole D, Barlogie B, Hoffman R, Tsukamoto A (1995): Enrichment of human hematopoietic stem cell activity in the CD34+Thy-1+Lin− subpopulation from mobilized peripheral blood. *Blood* 85:368.

Nimgaonkar M, Bahnson A, Boggs S, Ball ED, Barranger JA (1994): Transduction of mobilized peripheral blood CD34+ cells with the glucocerebrosidase gene. *Gene Ther* 1:201–207.

Nimgaonkar M, Mierski J, Beeler M, Kemp A, Lancia J, Mannion-Henderson J, Mohney T, Bahnson A, Rice E, Ball ED, Barranger JA (1995): Cytokine mobilization of peripheral blood stem cells in patients with Gaucher disease with a view to gene therapy. *Exp Hematol* 23:1633–1641.

Nolta JA, Sender LS, Barranger JA, Kohn D (1990): Expression of human glucocerebrosidase in murine long-term bone marrow cultures following retroviral vector-mediated transfer. *Blood* 75:787–791.

Nolta JA, Smogorzewska EM, Kohn DB (1995): Analysis of optimal conditions for retroviral-mediated transduction of primitive human hematopoietic cells. *Blood* 86:101–110.

Ohashi T, Boggs S, Robbins P, et al. (1992): Efficient transfer and sustained high expression of the human glucocerebrosidase gene in mice and their functional macrophages following transplantation of bone marrow transduced by a retroviral vector. *Proc Natl Acad Sci USA* 89:11332–113336.

O'Reilly RJ, Pollack MS, Kapoor N, et al. (1983): Fetal liver transplantation in man and animals. In Gale RG (ed): *Recent Advances in Bone Marrow Transplantation.* New York: Alan R Liss, Inc, p 779.

Pear WS, Nolan GP, Scott ML, Baltimore D (1993): Production of high-titer helper-free retroviruses by transient transfection. *Proc Natl Acad Sci USA* 90:8392–8396.

Peters SO, Kittler ELW, Ramshaw HS, Quesenberry PJ (1996): *Ex vivo* expansion of murine marrow cells with interleukin-3, interleukin-6, interleukin-11 and stem cell factor leads to impaired engraftment in irradiated hosts. *Blood* 87:30–37.

Poduslo SE, Tennekoon G, Miller K, McKhann GM (1976): Fetal metachromatic leukodystrophy: Pathology, biochemistry and a study of *in vitro* enzyme replacement in CNS tissue. *J Neuropathol* 35:622–632.

Richardson C, Ward M, Podda S, Bank A (1984): Mouse fetal liver cells lack functional amphotropic retroviral receptors. *Blood* 84:433–439.

Roncaro MG, Vandekerckhove B (1992): SCID-hu mice as a model to study tolerance after fetal stem cell transplantation. *Bone Marrow Transplantation* 9(suppl 1):83–84.

Schneider EL, Ellis WG, Brady RO, McCulloch JR, Epstein CJ (1972): Infantile (type II) Gaucher's disease: *In utero* diagnosis and fetal pathology. *J Pediatr* 81:1134–1139.

Soriano HE, Gest AL, Bain DK, et al. (1993): Feasibility of hepatocellular transplantation via the umbilical vein in prenatal and perinatal lambs. *Fetal Diagn Ther* 8:293–304.

Spour EF, Brandt JE, Bridell RA, Grigsby S, Leemhuis T, Hoffman R (1993): Long-term generation and expansion of human primitive hematopoietic progenitor cells *in vitro*. *Blood* 81:661–669.

Szymanski IO, Tilley CA, Crookston ME, et al. (1977): Case reports. A further example of human blood group chimerism. *J Med Genet* 14:279–281.

Touraine JL, Roncaro MG, Bacchetta R, Raudrant D, Rebaud A, Laplace S, Cesbron P, Gebuhrer L, Zabot MT, Touraine F, Frappaz D, Souillet G, Vullo C (1993): Fetal liver transplantation biology and clinical results. *Bone Marrow Transplantation* II(suppl 1):119–122.

Touraine JL, Roncarolo MG, Royo C, et al. (1987): Fetal tissue transplantation, bone marrow transplantation and prospective gene therapy in severe immunodeficiencies and enzyme deficiencies. *Thymus* 10:75.

Tsukamoto M, Takahiro O, Sho Y, Takashi S, Masaaki T (1995): Gene transfer and expression in progeny after intravenous DNA injection into pregnant mice. *Nat Genet* 9:243–248.

Turner JH, Hutchinson DL, Petricciani JC (1973): Chimerism following fetal transfusion. Report of leukocyte hybridization in an infant with acute lymphocytic leukemia. *Scand J Hematol* 10:358–366.

Van Beusechem VW, Bart-Baumeiser JAK, Hoogerbrugge PM, Valerio D (1995): Influence of interleukin-3, interleukin-6, and stem cell factor on retroviral transduction of rhesus monkey $CD34^+$ hematopoietic progenitor cells measured *in vitro* and *in vivo*. *Gene Ther* 2:245–255.

Wei J-F, Wei F-S, Samulski RJ, Barranger JA (1994): Expression of the human glucocerebrosidase and arylsulfatase A genes in murine and patient primary fibroblasts transduced by an adeno-associated virus vector. *Gene Ther* 261–268.

Wolfe JH, Sands MS, Barker JE, Gwynn B, Rowe LB, Vogler CA, Birkenmeier EH (1992): Reversal of pathology in murine mucopolysaccharidosis type VII by somatic cell gene transfer. *Nature* 360:749–753.

Wu AM, Till JE, Siminovitch, McCulloch EA (1968): Cytological evidence for a relationship between normal hematopoietic colony forming cells and cells of the lymphoid system. *J Exp Med* 127:455–564.

Xu L, Stahl SK, Dave HPG, Karlsson S (1994): Correction of the enzyme deficiency in hematopoietic cells of Gaucher patients using clinically acceptable retroviral supernatant transduction protocol. *Exp Hematol* 22:223–230.

Yee JK, Miyanohara A, Burns JC, Friedman T (1994): Generation and infectivity of retroviral vectors pseudotyped with the G protein of vesicular stomatitis virus. Gene Therapy Meeting, Cold Spring Harbor Laboratory, Cold Spring Harbor, NY.

Yu H, Bauer B, Lipke GK, Phillips RL, Vanzant G (1983): Apoptosis and hematopoiesis in murine fetal liver. *Blood* 81:373–384.

Zanjani ED, Ascensao JL, Harrison MR, Tavassoli M (1992): *Ex vivo* incubation with growth factors enhances the engraftment of fetal hematopoietic cells transplanted in sheep fetuses. *Blood* 79:3045–3049.

Zimran A, Gelbart T, Westwood B, Grabowski GA, Beutler E (1991): High frequency of the Gaucher disease mutation at nucleotide 1226 among Ashkenazic Jews. *Am J Hum Genet* 49:855–859.

14

CLINICAL APPLICATIONS OF GENE THERAPY: CORRECTION OF GENETIC DISEASE AFFECTING HEMATOPOIETIC CELLS

JEFFREY A. MEDIN, JOHAN RICHTER, AND STEFAN KARLSSON

Section of Hematology/Oncology, University of Illinois at Chicago, Chicago, IL 60607-7173 (J.A.M.); Gene Therapy Center, Molecular Medicine Section and The Department of Medicine, University Hospital, Lund SE-22362, Sweden (J.R., S.K.)

INTRODUCTION

Treatment of single-gene defects by corrective gene transfer or gene augmentation has not been the immediate medical panacea that some had hoped for. There have been a number of limitations to the success of the approach. Most of the problems have been directly attributable to deficiencies in the gene delivery technology, while other difficulties have arisen that are due to the limited accessibility or limited numbers of important target cells that can be obtained and modified genetically. Nonetheless, there have been some definitive successes in preclinical studies, and many protocols have been initiated by a number of investigators throughout the world. Corrective gene transfer provides a method of directly addressing the defects that lead to various inherited diseases and not just a treatment of the resulting symptoms. It is this promise or idea of a permanent cure originating at the point of the defect that drives this science.

Stem Cell Biology and Gene Therapy, Edited by Peter J. Quesenberry, Gary S. Stein, Bernard G. Forget, and Sherman M. Weissman
ISBN 0-471-14656-0

In this chapter are examined clinical and preclinical data from the genetic correction of a selection of single-gene disorders that affect the hematopoietic system. Because more than 4,000 diseases are thought to be caused by single-gene defects (Beaudet et al., 1995), this sample represents a minute fraction of available treatment targets. As previously discussed (Karlsson, 1991), and also presented in other chapters of this text, hematopoietic cells have been prime targets for initial studies in gene therapy research. This is because of the high accessibility of these cells and because bone marrow transplantation (BMT), as a cure for some cases of some disorders and as a key component of the gene transfer regimen, is an established clinical procedure. In addition, retroviral vectors have been shown to transduce primary hematopoietic cells at an appreciable level and effect permanent genomic alteration. Hematopoietic stem cells are also long lived and have the capability to differentiate into every cell lineage of the blood system. The main challenge in the field is therefore to develop strategies to transduce these pluripotent cells. As is shown, there are other long-lived cells that are also interesting targets for therapy, for example, T cells. Lastly, hematopoietic stem and progenitor cells and their progeny also interact with a number of other organ systems through circulation and migration, providing access to the majority of the body.

GENE THERAPY RESEARCH AIMED AT GENETIC DISORDERS OF THE HEMATOPOIETIC SYSTEM

Immunodeficiencies

Adenosine Deaminase Deficiency. The lack of functional adenosine deaminase (ADA) results in a rare autosomal recessive disorder that is responsible for about 20% of all cases of severe combined immunodeficiency (SCID) (Kredich and Hershfield, 1989). The enzyme catalyzes the conversion of ribo- and deoxyriboadenine nucleosides into the corresponding inosine moieties. In the absence of this catalytic activity, toxic metabolites, primarily deoxyadenosine triphosphate, accumulate in cells. This results in a systemic deficiency of T and B cells and severely diminished or nonexistent immunological functions. Patients die at an early age from opportunistic infections. Both BMT (Kantoff et al., 1988) and enzyme replacement therapy (Hershfield et al., 1987) have been successful in some cases but are not always completely effective or without risks, in the case of the former treatment, or expense and inconvenience, in the case of the latter.

ADA deficiency has the landmark distinction of being the first characterized single-gene disorder for which a therapeutic gene transfer protocol was initiated. This occurred in 1990 at the National Institutes of Health, and choice of this disease was based on a number of observations (Blaese, 1993). The ADA gene had been cloned and well characterized, and also a wide range in enzyme expression levels had been found to be tolerated (Valentine et al.,

1977). A substantial amount of background experimental work had been performed prior to approval of this trial. Safe and stable retroviral vectors had been constructed for the transfer of the wild-type human ADA cDNA sequence, and effective levels of enzyme had been observed in culture and in mice (Williams et al., 1986; Kaleko et al., 1990). Especially important, however, was the finding from allogeneic BMT and experiments using transduced patient cells injected into immunodeficient mouse models (Ferrari et al., 1991) that ADA^+ cells have a growth advantage over cells lacking the enzyme.

Peripheral blood T lymphocytes were chosen as target cells for the initial trial (Blaese et al., 1995), and two patients were identified who had demonstrated less than optimal outcomes with the enzyme replacement therapy (pegylated ADA) [PEG-ADA]). T cells were collected from these patients after multiple rounds of leukapheresis, stimulated to proliferate with interleukin (IL)-2 and OKT3 T-cell receptor antibody, and then transduced three to five times with a recombinant retroviral vector containing the human ADA cDNA and the neomycin phosphotransferase gene as a marker. After the cells were expanded in culture, numerous infusions of the transduced cells were given to each patient in multiple procedures for a period of up to 2 years after the start of the study. In these breakthrough studies, beneficial clinical outcomes for each patient have been documented that have lasted over time. In patient 1, peripheral blood T cell levels rose to normal early in the treatment course, and the levels have been maintained. Lymphocyte ADA enzyme activity also increased from very low levels initially to half of the observed enzymatic activity in heterozygote carriers of the disease. In patient 2, the peripheral blood T-cell count was increased following transfusions, and this level was sustained over the course of the study although the peripheral blood lymphocytes did not show any increases in ADA enzymatic activity in comparison to levels observed prior to gene transfer. Despite these differences and the fact that the overall level of gene transfer was fairly low (especially in patient 2), both patients have reconstituted cellular and humoral immune responses as measured by several criteria. Also contributing to the reduced effect in patient 2 is that there appears actually to be an immune response generated against the cells containing the neomycin phosphotransferase gene (M. Blaese, unpublished observation).

The dose of PEG-ADA, which was maintained as part of the therapy throughout the gene transfer protocol for ethical reasons, has been decreased in both patients. Most importantly, however, is the fact that the patients are both able to lead normal lives, and no detrimental effects of the gene transfer methodology have been documented. Furthermore, in these studies, it has also has been demonstrated that the retroviral long terminal repeat (LTR) is capable of directing expression over time without transcriptional silencing occurring and also that certain T cells are extremely long lived in the body.

Another gene therapy protocol initiated for ADA deficiency used different retroviral vectors carrying the corrective ADA gene to transduce both peripheral blood lymphocytes and bone marrow stem cells from two patients

(Bordignon et al., 1995). Use of the two vectors, which had different restriction enzyme sites in nonfunctional regions of the retroviral LTRs, allowed analyses to be performed on the longevity and repopulation of the patient by corrected cells from both transduced source compartments. As in the aforementioned study, both patients had failed PEG-ADA enzyme replacement treatment, and, likewise, multiple infusions of *ex vivo*–transduced cells were again the therapeutic regimen. In this study "double-copy" retroviral vectors were used that expressed the human cDNA driven by its native promoter. Cells were collected from the patients and exposed to retroviral-containing supernatants or incubated with irradiated viral producing cells directly. Gene transfer efficiency was gauged by polymerase chain reaction (PCR) analyses for the presence of the proviral DNA in progenitor colonies grown in semisolid media. One patient received five infusions of transduced cells, and the other patient received a total of nine infusions.

Marked immune system reconstitution, as measured by function and by cell count, was demonstrated for both patients. No serious infections have been seen in either patient in this study. The PCR analyses determined that after about 1 year long-lived, transduced peripheral blood cells were replaced by cells originating from the transduced bone marrow population. Physical growth parameters in the patients were normalized following treatment, and no adverse effects of the therapy or recombinant helper virus were seen.

Besides the two studies described above, gene augmentation therapy with the corrective ADA cDNA has been initiated for three children diagnosed *in utero* with ADA deficiency (Kohn et al., 1995). Cord blood stem cells were harvested for each neonate and used as a source to obtain an enriched population of $CD34^+$ cells by immunoaffinity separation technology. These cells were then transduced three times in the presence of recombinant IL-3, IL-6, and stem cell factor (SCF) (to enhance the efficiency of gene transfer) with the same retroviral vector utilized in the first gene therapy clinical protocol (see above). Gene transfer efficiency, as measured by the ability of progenitor colonies to survive selection due to the added neomycin phosphotransferase gene, was estimated to be in the range of 12%–22%. The transduced cells were then reinfused into the recipients on the fourth day of life. Semiquantitative PCR analyses were used to detect and enumerate the presence of the retroviral vector sequence in circulating and bone marrow leukocytes for time periods of longer than 1 year. In the bone marrow–derived cells, approximately 1/10,000 cells were shown to contain the proviral sequence, and similar frequencies were seen for the circulating cells. ADA enzymatic activity was found in cells expanded from selected progenitor cells to be much increased over that seen in control cells.

Another important result was also shown in that study. Again after 1 year, clonal analysis determined that, for each patient, numerous different transduced progenitor cells were present, and not just those derived from the transduction of a single cell. This indicates that multiple infection/integration events occurred for the cells from each patient. As in the above studies, the patients were maintained on PEG-ADA enzyme replacement therapy throughout the course of treatment. This has enabled the patients to develop normal immune function. Encouragingly, though, based on the low but significant levels of positive cells that has persisted for longer than 1 year, the doses of the PEG-ADA have been able to be reduced for each patient, which may actually give the transduced cells a selective advantage *in vivo*. As in the first study, a reduction due to the expression of the neomycin phosphotransferase gene as cells matured may be evident in this study also.

As of December 1996, a number of protocols originating from various medical centers in the United States, Europe, and Japan have been initiated for the genetic correction of ADA deficiency (Marcel and Grausz, 1996). At least 13 patients have been recruited, and studies are underway to determine the long-term effects on immune function of genetic correction in these individuals. In all the studies published thus far, PEG-ADA has been used concurrently with the gene therapy. Therefore, a direct proof of efficacy remains to be conclusively demonstrated.

X-Linked SCID. X-linked SCID is the most common form of the severe primary immunodeficiencies accounting for roughly half of all the cases. It is due to mutations in the gene that encodes for the common γ-chain present in the receptors for IL-2, IL-4, IL-7, IL-9, and IL-15. This results in defective T-lymphocyte and natural killer cell differentiation; T cells in blood and lymphoid organs are extremely infrequent or absent. B cells are usually present but functionally impaired. The afflicted individuals usually succumb early in life to overwhelming opportunistic infections. BMT may be curative, but a lack of suitable donors and the risks associated with this treatment are its major drawbacks.

In 1996, three groups simultaneously reported the generation of recombinant retroviral vectors encoding the human cDNA for the common γ-chain and the successful transduction of Epstein-Barr virus–transformed cell lines derived from X-linked SCID patients (Candotti et al., 1996; Taylor et al., 1996; Hacein-Bey et al., 1996). Intracellular signaling via the IL-2 and IL-4 receptors was restored in the transduced cells as shown by phosphorylation of the JAK1 and JAK3 members of the Janus family of tyrosine kinases upon exposure of the cells to IL-2 or Il-4.

Recently a clinical protocol for treatment of X-linked SCID patients with retroviral-mediated gene transfer of the cDNA for the common γ-chain into $CD34^+$ selected peripheral stem cells or umbilical cord stem cells was approved (Weinberg, 1996). As of December 1996, no patients had yet been entered in this trial.

Purine Nucleoside Phosphorylase Deficiency. Clinical trials for corrective gene transfer using retroviral vectors have also been initiated for a third immune function disorder (Marcel and Grausz, 1996), the purine nucleoside phosphorylase (PNP) deficiency. This malady, with some symptoms similar

TABLE 1 Clinical Gene Therapy Trials Targeting Hematopoietic Cells

Gene Product	Disease	Cells Targeted	No. of Trials[1]	No. of Patients[1]
Adenosine deaminase	Severe combined immunodeficiency due to adenosine deaminase deficiency	T-lymphocytes, bone marrow stem cells, $CD34^+$ umbilical cord blood cells	6	13
Common γ-chain of IL-2, -4, -7, -9, and -15 receptors	X-linked severe combined immunodeficiency	$CD34^+$ cells from cord blood or bone marrow	1	0
Glucocerebrosidase	Gaucher's disease	CD34-selected peripheral or bone marrow cells	3	9
Fanconi anemia complementation group C	Fanconi anemia group C	CD34-selected peripheral stem cells	1	3
Idunorate-2-sulfatase	Hunter (mucopolysaccharidosis type II)	Peripheral blood lymphocytes	1	1
α-Iduronidase	Hurler (mucopolysaccharidosis type I)	Bone marrow stem cells?	1	0
p47-phox	Chronic granulomatous disease	CD34-selected peripheral stem cells	1	5
Purine nucleoside phosphorylase	Purine nucleoside phosphorylase deficiency	Peripheral blood lymphocytes	1	0

[1]Number of patients enrolled according to December 1996 Worldwide Gene Therapy Enrollment Report (Marcel and Grausz, 1996).

to ADA deficiency except for presentations of neurological involvement, is due to a reduction in the enzymatic activity that catalyzes the conversion of inosine and deoxyinosine to hypoxanthine. In this disorder patients have defective T cell–mediated immunity and intact or even hyperactive B-cell function. About 30 patients have been described to date (Hershfield and Mitchell, 1995).

It was recognized early that this disorder was also a prime target for gene therapy. Retroviral vectors were created (McIvor et al., 1987; Osborne and Miller, 1988), and cultured skin fibroblasts were corrected for the enzyme deficiency. Actual metabolic enhancement, as measured by decreased sensitivity to deoxyguanosine, was demonstrated in a mouse cell model of PNP deficiency by transfer of both the human and the mouse cDNA sequences (Foresman et al., 1992). Later studies were directed to T cells and primitive progenitors, as these are the most salient target cell populations (Nelson et al, 1995). In that study, increased immune function, above background levels, was seen in transduced patient T lymphocytes. Newer generation retroviral vectors were also recently created that incorporated intronic sequences into the expression cassette (Jonsson et al., 1995). As of the December 1996 enrollment report, no patients had yet been recruited for the clinical trial (Marcel and Grausz, 1996).

Lysosomal Storage Disorders

Introduction. An important collective target for therapeutic strategies based on gene augmentation or correction are the lysosomal storage disorders. At least 30 have been identified (for review, see Neufeld, 1991) and the disease frequency is fairly high (1/1,500 for all types; Beaudet et al., 1995). These disorders are good candidates for a number of reasons. Unlike many metabolic defects that arise from mutations that are lethal early to developing organisms, many of these single-gene alterations result in viable although severely compromised offspring. The disease conditions from this class of disorders are a result of the accumulation of incompletely processed metabolites. In many cases it takes time to accumulate the offending catabolite to detrimental or toxic levels. For many of these metabolic disorders the gene defect has been well characterized years ago, and also for many there is, at present, no curative therapy that directly addresses the disorder. A number of the relevant genes have been isolated, sequenced, and cloned, and overexpression has been shown to correct the metabolic defect in patient-derived cells *in vitro*. In addition to the previously mentioned reasons, many of these disorders also have components or manifestations that are directly attributable or accessible to hematopoietic cells—thus meeting the target accessibility criterion for gene transfer therapy to be a reality. Enzyme replacement therapy has been shown to improve disease outcomes, and for some of these disorders BMT has proven to be successful. Besides correction in patient cells *in vitro*, many studies on gene transfer into murine models have been performed for this class of defects, demonstrating that long-term expression can be generated and maintained. A few studies have been done in other animal model systems, including dogs, primates, and sheep.

After a short description of the biology of some of these lysosomal storage disorders, results from preclinical experiments in animal models and in human cells in culture are presented. If available, data from clinical trials for this important class of metabolic disorders is also discussed.

Gaucher's Disease. A deficiency in the enzyme glucocerebrosidase (GC) results in Gaucher's disease (GD), the most prevalent lysosomal storage disorder, which can be classified into 3 types based on the degree of neuronal involvement (Beutler, 1993). Types 1 and 3 are the most likely immediate candidates for gene therapy, as type 2 patients have severe neuronal impairment and die before the age of 2. GD patients have enlarged spleens and livers, anemia, and bone frailty. The accumulation of the β-glucoside substrates occurs in various tissues, but the clinical effects are most pronounced due to accumulation in macrophages. BMT has been shown to be successful in the treatment of GD if an HLA-matched donor can be found (Ringden et al., 1988; Erikson et al., 1990). Enzymatic replacement therapy has been shown to be effective once the delivered GC was altered to allow uptake through the mannose receptor-mediated endocytosis system (Barton et al., 1990).

The cDNA sequence encoding GC was isolated in 1985 (Sorge et al., 1985), and shortly thereafter numerous retroviral constructs were made by different groups interested in therapeutic transfer of the gene (Choudary et al., 1986; Sorge et al., 1987; Correll et al., 1989). Transfer of the gene and expression of the protein to corrective levels were seen in experiments employing those recombinant viruses. Gene transfer of the GC cDNA into murine hematopoietic stem/progenitor cells has also been fully documented. Transfer has been shown into murine long-term culture-initiating cells (LTC-IC) and spleen colony-forming units (S-CFU) (Nolta et al, 1990) and into long-term repopulated mice (Correll et al., 1990; Ohashi et al., 1992).

Studies on the transfer of the human GC gene into human hematopoietic cells have been extensive. The first report to document this transfer was published in 1990, (Fink et al., 1990), and that account described fairly efficient transduction of progenitor cells (assayed by PCR detection) obtained from patients with GD. However, selection was needed to see increased levels of GC enzyme activity due to the relative inefficiency of expression driven by the viral promoter used. In another study, long-term cell cultures were established from total marrow samples of a type 3 GD patient or from $CD34^+$ cells derived from immunomagnetic purification of cells from the same source (Nolta et al., 1992). These cells were infected by co-culture with recombinant virus that contained the GC cDNA or a virus that contained the neomycin phosphotransferase gene as a marker. As the LTC-ICs matured over time, it was shown that the nonadherent cells had GC intracellular enzyme activities that were at least half of normal activity after 1 month of culture and that

GC mRNA levels (derived from the added provirus) were maintained for at least 5 months. In another study, LTC-ICs were transduced using a clinically acceptable supernatant transduction protocol (Xu et al., 1994).

Some other studies involving mouse models are especially noteworthy, not only for the genetic treatment of GD, but also for the treatment of other disorders of this group. In the first study (Krall et al., 1994), it was shown that transduced murine bone marrow cells, containing the GC cDNA, infiltrated lung, liver, and especially brain tissues of recipients by 3–4 months. Indeed, a high percentage of brain microglia were found to be vector positive. This result points to a potential therapy for type 2 GD and other disorders affecting the central nervous system. It should be noted, however, that other investigators have not seen similar renewal effects that were as extensive. In the next important study, transfer of the human GC cDNA by multiple infusions of infected cells was demonstrated in mice not given any myeloablative treatment prior to the transplantation (Schiffmann et al., 1995). This showed, at a low level, that transduced cells containing the corrective gene could engraft and repopulate nonablated hosts. This validated the clinical approach proposed for GD patients, since total marrow ablation will not be part of the therapy regimen.

Clinical trials utilizing gene transfer for the correction of GD have now been initiated. Three separate protocols that target peripheral blood or bone marrow stem/progenitor cells have been started, and, as of December 1996 (Marcel and Grausz, 1996), nine patients have been enrolled. To date, two patients have been enrolled in the study that is being conducted at the NIH (Dunbar et al., 1996). Two different protocols that both involve transduction in the presence of autologous stromal cells were used for the therapeutic intervention strategy. For one patient, cells were transduced without cytokines, while the second transduction was performed with recombinant IL-3, IL-6, and SCF present. The gene transfer efficiency, as measured by PCR analyses of progenitor colonies grown in semisolid media, was found to be 2% and 10%, respectively. In both cases the transduced cells were returned to the patients in a single infusion that was performed without administration of myeloablative therapy. The initial results of these two interventions have been disappointing, as in neither patient has it been possible to detect the vector containing the corrective cDNA in either the peripheral blood or in the bone marrow by PCR analysis (S. Karlsson, personal communication). Clinical benefit, if any, of the intervention has not been fully assessed.

The absence of detectable vector in the patients of this trial is due to at least two factors: (1) low transduction efficiency and (2) reduced engraftment efficiency of the transduced cells due to transplantation into nonablated recipients. Most importantly, however, is that transduced GD patient cells are not expected to have the same growth advantage *in vivo* that deficient cells receiving the ADA gene have. Future studies will determine if the transduction efficiency can be increased significantly or if multiple infusions of transduced and corrected cells will lead to an increase in the systemic population of cells that carry the vector. It may also be possible to select and sort transduced GD patient cells prior to re-infusion using recombinant retroviruses that contain the GC cDNA and cell surface selectable marker genes (Medin et al., 1996a).

Mucopolysaccharidoses. Ten different enzymes are needed in humans for the catabolism of glycosaminoglycans such as heparin sulfate, keratan sulfate, and others that are derived from the break down of proteoglycans. Decreases in these enzymatic functions cause accumulation of problematic substrates in the lysosomes (for review, see Neufeld 1995). The disorders have been collectively called the *mucopolysacchararidoses* (MPS). A wide variety of clinical effects are manifested that differ greatly even among identical enzymatic deficiencies. Some of these disorders can be cured by BMT, or the symptoms can be reduced because the normal cells can reduce the substrate load or they can provide a metabolic cooperativity function through low levels of secretion of the protein that affects bystander cells. Thus, since BMT has been shown to be effective, interest is high to apply the techniques of corrective gene transfer to metabolic disorders of this class. Indeed, enrollment for approved clinical trials for gene therapy for some of these disorders has begun.

A deficiency in the enzyme α-iduronidase results in MPS I, commonly called Hurler's or Scheie's syndrome (or a hyphenated amalgam of the two) depending on the severity of the defect. Symptoms include enlarged liver and spleen, skeletal abnormalities, and cardiovascular involvement. Patients with the more severe Hurler syndrome have a deterioration in central nervous system function and a short life span. Some studies have been initiated to develop gene therapy for these conditions. In 1992, recombinant retroviral vectors were created for the expression of the α-iduronidase cDNA (Anson et al., 1992). These vectors also contained the gene for neomycin phosphotransferase as a marker. Skin fibroblasts from MPS I patients were corrected for the disorder in these transfer experiments.

In a later study neo-organs were created using retroviral vectors to transduce fibroblast cells with the α-iduronidase cDNA, which were then implanted into recipient mice (Salvetti et al., 1995). After some months, enzyme derived from the human cDNA and secreted from the neo-organs was found in both the liver and spleen of recipients. It is unlikely, though, that this approach will address the severe neurological manifestations of this deficiency. Since BMT has been shown to be beneficial in this regard if administered early in life, correction has been directed to cells of the hematopoietic system. Long-term expression of the α-iduronidase cDNA in patient hematopoietic cells was demonstrated in a recently published report. Fairbairn et al. (1996) transduced bone marrow cells and maintained them in culture for greater than 15 weeks. High levels of enzyme expression were seen in nonadherent, differentiated progeny cells as was increased levels of secreted α-iduronidase that were maintained over the course of the study. Correction also lessened the lysosomal distentions in morphological analyses of macrophage cells removed from the long-term culture. In England, a clinical trial has been initiated for this disorder (Marcel and Grausz, 1996).

MPS II results from a deficiency in the enzyme iduronate sulfatase. It is the only MPS defect that is X-linked. Clinical symptoms range from mild to severe and often mirror those seen in MPS I. A retroviral vector for the correction of this disorder was made in 1993 (Braun et al., 1993), and correction of lymphoblastoid cell lines from Hunter syndrome was demonstrated. In that study a cross-corrective effect was also shown in that co-culturing the corrected lymphoid cells with uncorrected fibroblasts could affect substrate accumulation in the naive cells. That this cross-correction result was most likely due to intercellular enzyme transfer was shown by the same group in follow-up preclinical studies that described the construction of another vector containing the iduronate sulfatase cDNA under control of a different promoter (Braun et al., 1996). A clinical trial based on the transduction of lymphocytes has been initiated (Whitley et al., 1996), and at the time of the worldwide gene therapy enrollment report in December 1996 (Marcel and Grausz, 1996) one patient had entered this trial.

A few studies toward gene therapy have been done for MPS VI (Maroteaux-Lamy syndrome). This extremely rare disorder is due to deficiencies in the enzyme arylsulfatase B. Affected humans have similar symptoms as those with Hurler syndrome except for the extensive neurological involvement. In 1991, a recombinant retroviral vector was created for the transfer of the arylsulfatase B gene (Peters et al., 1991). Patient fibroblasts showed expression in culture up to a level of 36 times the enzyme activity that is seen even in normal fibroblasts. Recent studies have shown transfer of the human cDNA not only into patient fibroblasts, chondrocytes, and bone marrow cells but also into rat and cat MPS VI cells since these animal models of the disorder have been characterized (Fillat et al., 1996). Interestingly, in that study metabolic correction of animal model fibroblasts occurred, but correction of the human patient cells did not. Uptake into important cell types as a result of overexpression and secretion was also seen.

Even though only about 20 cases of MPS VII (Sly's syndrome) have been described in the literature (Neufeld and Muenzer, 1995), some of the most intense study toward gene therapy in the MPSs has occurred in this context. MPS VII is due to a deficiency in the enzyme β-glucuronidase, and the disorder has manifestations that directly imply hematopoietic cell involvement, much like GD. In this case, granulocytes and not macrophages show coarse granule accumulations. Neurological involvement is also seen, and some physical symptoms of the disorder resemble those seen in MPS I. Deficiencies in β-glucuronidase have been described in mice and dogs, providing excellent models for study. Early retroviral vectors were constructed that contained the rat cDNA sequence, and full enzymatic correction of human patient and dog fibroblasts was seen (Wolfe et al., 1990). Expression was also seen in transduced dog bone marrow cells. These studies were succeeded by a demonstration, in the mouse model of the disease, that gene transfer into deficient hematopoietic stem cells could correct the defect in some organs by reducing metabolite storage in the lysosomes of transplanted hosts (Wolfe et al., 1992). Studies with the human form of the β-glucuronidase gene were done by retroviral-mediated transfer into deficient hematopoietic cells followed by reimplantation into sublethally irradiated mice (Marechal et al., 1993). Even with low levels of engraftment of corrected cells, expression of the enzyme could be detected and reductions in storage could be seen in the liver and spleen. Neo-organs were created from transduced fibroblasts in another study and transplanted into hosts with the disorder (Moullier et al., 1993). Cross-corrective effects were followed using transduced fibroblasts as enzyme secretion apparatuses in studies involving rat and human cDNA sequences and human, dog, and mouse cells (Taylor and Wolfe, 1994). Lastly, a "double copy" retroviral vector was made that expressed human β-glucuronidase at high levels driven by both the natural β-glucuronidase promoter and by a thymidine kinase transcriptional unit (Wolfe et al., 1995). No mention of clinical trials for therapy of this disorder was made in the December 1996 enrollment report (Marcel and Grausz, 1996).

Fabry's Disease

Fabry's disease results from a deficiency of the lysosomal enzyme α-galactosidase A. Patients have low or nonexistent levels of this protein, which leads to progressive accumulation throughout the body of sphingolipids with α-terminal galactose residues. The most abundant lipid, ceremide trihexoside, is derived from the break down of senescent blood cells and from secretion by cells lining the vasculature. The build up of lipid leads to a number of clinical effects, including severe pain crises, hypohidrosis, corneal opacity, and anemia (Desnick et al., 1995). These, however, are not the most damaging consequences. Patients build up high levels of the detrimental lipids in cells that line the vasculature. This leads to occlusion of tubules and vessels. Patients with this X-linked disorder succumb in mid-life to cerebrovascular, cardiovascular, or renal disease. In this disorder, enzyme replacement therapy has not become part of the treatment regimen, although phase I trials toward this goal have begun (Brady, R.O., personal communication). Also, curative BMT has not been described for this disease. Thus treatments of this disorder have been directed to management of symptoms, and the overall clinical prognosis of these patients has been quite poor.

Studies have been initiated to bring about gene transfer therapy for Fabry's disease. Retroviral vectors have been constructed to direct expression of the human α-galactosidase A cDNA. In 1995, the first recombinant vector for the correction of this defect was described (Sugimoto et al., 1995). This was a bicistronic vector that also contained the multidrug resistance gene to allow selection of transduced cells by the addition of a cytotoxic drug. Increased levels of enzyme activity were seen in test NIH-3T3 cells that had been transduced and selected with vincristine. In 1996 a vector designed toward therapeutic utility in patients was reported (Medin et al., 1996b). This high titer vector generated increased levels of human α-galactosidase A in Fabry's disease patient skin fibroblasts and in B-cell hematological models derived from patients. Furthermore, cross-correction effects were evident as secreted enzyme activity from overexpressing cells could be demonstrated to be taken up into uncorrected bystander cells by a mannose receptor-mediated mechanism. Recently, in preclinical studies, Fabry's disease patient hematopoietic stem/progenitor cells were transduced at a fairly high frequency, and enzymatic correction was demonstrated that was maintained for extended periods of time in tissue culture (Medin et al., submitted). Clinical trials using gene transfer to ameliorate this disorder have been planned, but no actual protocols have been reported yet.

Other Lysosomal Defects. Other lysosomal disorders besides those mentioned above are considered to be targets for gene augmentation therapy. Recently, recombinant retroviral vectors have been made to enable these preliminary studies. Vectors have been made toward the treatment of metachromatic leukodystrophy by transfer of the arylsulfatase A gene (Rommerskirch et al., 1991; Ohashi et al., 1993), Tay-Sachs disease by transfer of the human β-hexosaminidase α-subunit (Lacorazza et al., 1996), Krabbe's disease by transfer of the GC gene (Gama Sosa et al., 1996; Rafi et al., 1996), aspartylglucosaminuria by the transfer of the aspartylglucosaminidase gene (Enomaa et al., 1995), and fucosidosis by the transfer of the α-L-fucosidase gene (Occhiodoro et al., 1992), among others. The metabolic cooperativity or cross-correction that is seen for some of these enzymes on overexpression, combined with the fact that many of these disorders have hematopoietic or hematopoietic-derived components, makes this class of metabolic disorders a suitable goal for studies of gene transfer into hematopoietic stem cells.

Chronic Granulomatous Disease

NADPH oxidase is a multisubunit enzyme complex that catalyzes the conversion, in activated phagocytic cells, of oxygen into a highly microbicidal form (for review, see Curnutte, 1993). Four components of the oxidase that generate the respiratory burst have been identified, and defects in each component have been characterized. Collectively, these disorders are called the *chronic granulomatous diseases* (CGD; total frequency, 1/750,000). CGD patients have plentiful and recurring infections that can be fatal that are caused by opportunistic catalase-positive bacteria or fungi. The infections result in a number of clinical manifestations, including granuloma formation, pneumonia, lymphadenopathy, dermatitis, and abscesses in various organs. Patients also have other severe systemic effects due to being in a continual inflammatory state. The most common form of CGD is X-linked and due to a deficiency in the gp91-phox cytochrome b_{558} component of the oxidase. It is responsible for about 65% of the cases worldwide. The next most prevalent defect is a deficiency in the p47-phox cytosolic component representing about 25% of the reported cases. Mutations in the other two identified genes comprising the NADPH oxidase, the p22-phox subunit of cytochrome b_{558} and the p67-phox cytosolic component, are extremely rare.

CGD is a prime candidate for the clinical application of gene therapy for a number of reasons. These include the fact that effects are directly manifested due to a deficiency in cells of the hematopoietic system. Indeed, BMT has been successfully performed (Hobbs, 1990). Low levels of factor correction may also be sufficient for curative effects, as healthy female CGD carriers have been identified who have <10% of the normal levels of neutrophils (Woodman et al., 1995). Also, extreme overexpression of a single component of the oxidase is not likely to be detrimental to the cell, since all components are necessary for enzyme activity. Concerns of overexpression of a deficient activity is a consideration for gene therapy that has not been fully explored except in the case of ADA SCID. Most importantly, however, is that corrected cells of CGD patients may have a selective growth advantage *in vivo* as phagocytes will be able to break down and not harbor engulfed pathogens. Problematic in considerations of therapy for this disorder, though, is that the primary target cells are fairly short lived and thus transduction of stem cells or repeated transductions will probably be required. Toward genetic therapy recombinant retroviral vectors have been engineered that include the DNA-coding sequences for all four of the NADPH oxidase components. Studies utilizing these constructs are discussed separately in the following paragraphs.

Three independent groups initially reported construction of retroviral vectors containing the gp91-phox cDNA sequence (Porter et al., 1993; Li et al., 1994; Kume and Dinauer, 1994). To reiterate, deficiencies in this sequence comprise the most common inherited forms of CGD, and the clinical presentation has been thought to be the most severe. In the first study (Porter et al., 1993), a vector that also contained the cDNA for the puromycin acetyltransferase gene as a selectable marker was tested in different model B-cell lines made by Epstein-Barr virus transformation of patient lymphocytes. The B-cell model is convenient and directly applicable, since normal B cells have been found to demonstrate NADPH oxidase activity. The recombinant virus governed reconstitution of the oxidase activity at some level that was stable over time. In the second study (Li et al., 1994), a retroviral vector that did not contain a selectable marker gene was made. Both a CGD patient B-cell oxidase-deficient model and also patient neutrophils derived from peripheral blood $CD34^+$ cells were corrected for the defect, without selection of transduced cells, to oxidase activity levels that may be clinically relevant. Another study described the construction of a retroviral vector containing a phosphoglycerokinase (PGK) internal promoter that directed expression of gp91-phox in a human myeloid cell line that was made deficient for the oxidase activity by the targeted disruption of the gp91-phox gene (Kume and Dinauer, 1994). Superoxide generation prompted by the oxidase activity was increased to levels of 50% of normal in corrected cells even though gp91-

phox protein expression was estimated to be only 2%–12% of normal levels. A later study on the X-linked form of CGD involved transfer of the gp91-phox gene into purified human CD34+ cells and into LTCIC (Porter et al., 1996). The gene transfer efficiency was found to be quite high (50% and above) as measured by PCR analyses of progenitor colonies, but very low levels of oxidase reconstitution were seen that did not seem to correlate well with the observed transduction efficiencies. Recently, experiments examining transfer into hematopoietic cells of X-linked CGD mice (Pollock et al., 1995) have been performed, and it was established that corrected cells generate resistance in the host to fungal challenge, whereas all control mice are susceptible to the pathogen (Bjorgvinsdottir et al., 1997). The December 1996 enrollment report (Marcel and Grausz, 1996) made no mention of clinical trials for this disorder.

The other membrane-bound component of the oxidase, p22-phox, has been expressed in cells as a result of recombinant retroviral transduction as demonstrated by two groups (Li et al., 1994; Porter et al., 1994). Both studies involved the use of Epstein-Barr virus–transformed B-cell lines derived from patients with a deficiency of that subunit, and one of the studies demonstrated oxidase correction in patient $CD34^+$ cells (Li et al., 1994). To date, no further studies or clinical protocols have been described for this rare defect.

The remaining autosomal CGD disorder, due to a defect in the p47-phox cytosolic component, has also received some study and is a target disease for an ongoing gene transfer clinical trial. Retroviral vectors were engineered in 1992 for the transfer of the p47-phox gene (Cobbs et al., 1992; Thrasher et al., 1992). B-cell lines were established from patients, and protein expression of the added factor was demonstrated by Western blot analyses; superoxide production was also facilitated by the gene transfer. A clinical trial was initiated at NIH for the correction of this disorder using $CD34^+$ cells infected with a retrovirus. Five patients have been recruited (Marcel and Grausz, 1996) as of December 1996. In the trial, peripheral blood cells were selected for $CD34^+$ expression, transduced, and then reinfused into nonablated hosts. Circulating blood cells that were found to be positive by PCR analyses for the vector sequence and also exhibited oxidase activity could be found within a short period after infusion, but these cells were undetectable within 6 months (Malech et al., 1996).

Very recently, preliminary work toward genetic correction of the last of the four described CGD deficiencies, that of a defective p67-phox subunit, by use of a recombinant retroviral vector was described (Weil et al., 1997). $CD34^+$ peripheral blood progenitor cells from a p67-phox–deficient CGD patient were transduced with a highly efficiency MFG-based vector encoding the corrective cDNA sequence. After differentiation of the cells in culture, it was shown that they produced 25% of the total superoxide that is produced by normal $CD34^+$ cells. However, in the fraction of cells that were corrected (up to 32%), oxidase function was found to be 85% of normal.

FUTURE PERSPECTIVES

It is clear that considerable advances have been made in the development of gene therapy for hematopoietic stem cells during the last 10 years. However, fundamental problems remain. The greatest advance has been made in the transduction of human hematopoietic progenitor cells, with retroviral-mediated transduction efficiencies now approaching 50% on average, even for multipotential progenitors as assayed in *in vitro* cultures. In contrast, clinical protocols that have been designed to ask whether human hematopoietic stem cells can be transduced have invariably led to poor gene marking efficiency by the vector. These clinical experiments have involved autologous transplantation, and it has been impossible to determine whether the lack of efficient marking is due to very low transduction of repopulating stem cells or whether poor engraftment of transduced cells can explain these findings. It is almost certain that poor gene marking of stem cells is an important problem here because hematopoietic stem cells are quiescent and standard retroviral vectors will not transduce them efficiently. To support this hypothesis, it has recently been shown that the average time for primitive hematopoietic cells to reach S-phase *in vitro* is 9 days (Petzger et al., 1996). Therefore, it is likely that the culture time during transduction in standard transduction protocols is too short.

To overcome these difficulties, three main approaches are now possible or already under investigation. First, selectable vectors that allow selection and sorting of retrovirally transduced cells have been generated and have been used to sort human hematopoietic progenitors (Migita et al., 1995; Medin et al., 1996a). Second, it is possible that methods can be developed to stimulate proliferation ("self-renewal" or true stem cell expansion) of primitive hematopoietic cells *in vitro,* prior to transduction with retroviral vectors. Third, hybrid viral vectors that can transduce quiescent cells have already been developed, for example pseudotyped human immunodeficiency virus-1 particles (Naldini et al., 1996; Reiser et al., 1996), although it remains to be seen whether these hybrid vectors can transduce quiescent hematopoietic stem cells and integrate into their chromosomal DNA. All these approaches are currently being explored. If successful, these efforts will lead to marking of hematopoietic stem cells, which will be much higher than presently possible, and this in turn may lead to therapeutic effects from stem cell–directed gene transfer.

REFERENCES

Anson DS, Bielicki J, Hopwood JJ (1992): Correction of mucopolysaccharidosis type I fibroblasts by retroviral-mediated transfer of the human α-L-iduronidase gene. *Hum Gene Ther* 3:371–379.

Barton NW, Furbish FS, Murray GJ, et al. (1990): Therapeutic response to intravenous infusions of glucocerebrosidase in a patient with Gaucher disease. *Proc Natl Acad Sci USA* 87:1913–1916.

Beaudet AL, Scriver CR, Sly WS, Valle D (1995): Genetics, biochemistry, and molecular basis of variant human phenotypes. In *The Metabolic Basis of Inherited Disease,* 7th ed. New York: McGraw-Hill, pp 53–118.

Beutler E (1993): Gaucher disease as a paradigm of current issues regarding single gene mutations of humans. *Proc Natl Acad Sci USA* 90:5384–5390.

Bjorgvinsdottir H, Ding C, Pech N, et al. (1997): Retroviral-mediated gene transfer of gp91-phox into bone marrow cells rescues defect in host defense against *Aspergillus fumigatus* in murine X-linked chronic granulomatous disease. *Blood* 89:41–48.

Blaese RM (1993): Development of gene therapy for immunodeficiency: Adenosine deaminase deficiency. *Pediatr Res* 33(Suppl): S49–S55.

Blaese RM, Culver KW, Miller AD, et al. (1995): T lymphocyte–directed gene therapy for ADA-SCID: Initial trial results after 4 years. *Science* 270:475–480.

Bordignon C, Notarangelo LD, Nobili N, et al. (1995): Gene therapy in peripheral blood lymphocytes and bone marrow for ADA-immunodeficient patients. *Science* 270:470–474.

Braun SE, Aronovich EL, Anderson RA, et al. (1993): Metabolic correction and cross-correction of mucopolysaccharidosis type II (Hunter syndrome) by retroviral-mediated gene transfer and expression of human iduronate-2-sulfatase. *Proc Natl Acad Sci USA* 90:11830–11834.

Braun SE, Pan D, Aronovich EL, et al. (1996): Preclinical studies of lymphocyte gene therapy for mild Hunter syndrome (mucopolysaccharidosis type II). *Hum Gene Ther* 7:283–290.

Candotti F, Johnston JA, Puck JM, et al. (1996): Retroviral-mediated gene correction for X-linked severe combined immunodeficiency. *Blood* 87:3097–3102.

Choudary PV, Barranger JA, Tsuji S, et al. (1986): Retrovirus-mediated transfer of the human glucocerebrosidase gene to Gaucher fibroblasts. *Mol Biol Med* 3:293–299.

Cobbs CS, Malech HL, Leto TL, et al. (1992): Retroviral expression of recombinant p47-phox protein by Epstein-Barr virus–transformed B lymphocytes from a patient with autosomal chronic granulomatous disease. *Blood* 79:1829–1835.

Correll PH, Fink JK, Brady RO, et al. (1989): Production of human glucocerebrosidase in mice after retroviral gene transfer into multipotential hematopoietic progenitor cells. *Proc Natl Acad Sci USA* 86:8912–8916.

Correll PH, Kew Y, Perry LK, et al. (1990): Expression of human glucocerebrosidase in long-term reconstituted mice following retroviral-mediated gene transfer into hematopoietic stem cells. *Hum Gene Ther* 1:277–287.

Curnutte JT (1993): Chronic granulomatous disease: The solving of a clinical riddle at the molecular level. *Clin Immunol Immunopathol* 67:S2–S15.

Desnick RJ, Ioannou YA, Eng CM (1995): α-Galactosidase A deficiency. Fabry disease. In Scriver CR, Beaudet AL, Sly WS, Valle D (eds.), *The Metabolic Basis of Inherited Disease,* 7th ed. New York: McGraw-Hill, pp 2741–2784.

Dunbar C, Kohn D, Karlsson S, et al. (1996): Clinical protocol. Retroviral mediated transfer of the cDNA for human glucocerebrosidase gene into hematopoietic stem cells of patients with Gaucher disease. A phase I study. *Hum Gene Ther* 7:231–253.

Enomaa N, Danos O, Peltonen L, Jalanko A (1995): Correction of deficient enzyme activity in a lysosomal storage disease, aspartylglucosaminuria, by enzyme replacement and retroviral gene transfer. *Hum Gene Ther* 6:723–731.

Erikson A, Groth C-G, Mansson J-E, et al, (1990): Clinical and biochemical outcome of marrow transplantation for Gaucher disease of the Norrbottnian type. *Acta Paediatr Scand* 79:680–685.

Fairbairn LJ, Lashford LS, Spooncer E, et al. (1996): Long term *in vitro* correction of α-L-iduronidase deficiency (Hurler syndrome) in human bone marrow. *Proc Natl Acad Sci USA* 93:2025–2030.

Ferrari G, Rossini S, Giavazzi R, et al. (1991): An *in vivo* model of somatic cell gene therapy for human severe combined immunodeficiency. *Science* 251:1363–1366.

Fillat C, Simonaro CM, Yeyati PL, et al. (1996): Arylsulfatase B activities and glycosaminoglycan levels in retrovirally transduced mucopolysaccharidosis type VI cells. Prospects for gene therapy. *J Clin Invest* 98:497–502.

Fink JK, Correll PH, Perry LK, et al. (1990): Correction of glucocerebrosidase deficiency after retroviral-mediated gene transfer into hematopoietic progenitor cells from patients with Gaucher disease. *Proc Natl Acad Sci USA* 87:2334–2338.

Foresman MD, Nelson DM, McIvor RS (1992): Correction of purine nucleoside phosphorylase deficiency by retroviral-mediated gene transfer in mouse S49 T cell lymphoma: A model for gene therapy of T cell immunodeficiency. *Hum Gene Ther* 3:625–631.

Gama Sosa MA, de Gasperi R, Undevia S, et al. (1996): Correction of the galactocerebrosidase deficiency in globoid cell leukodystrophy-cultured cells by SL3-3 retroviral-mediated gene transfer. *Biochem Biophys Res Commun* 218:766–771.

Hacein-Bey S, Cavazzana-Calvo F, Le Deist F, et al. (1996): Common gamma chain gene transfer into SCID X1 patients' B-cell lines restores normal high-affinity interleukin-2 receptor expression and function. *Blood* 87:3108–3116.

Hershfield MS, Buckley RH, Greenburg ML, et al. (1987): Treatment of adenosine deaminase deficiency with polyethylene glycol-modified adenosine deaminase. *N Engl J Med* 16:589–596.

Hershfield MS, Mitchell BS (1995): Immunodeficiency diseases caused by adenosine deaminase deficiency and purine nucleoside phosphorylase deficiency. In Scriver CR, Beaudet AL, Sly WS, Valle D (eds.), *The Metabolic Basis of Inherited Disease,* 7th ed. New York: McGraw-Hill, pp 1725–1768.

Hobbs JR (1990): Displacement bone marrow transplantation for some inborn errors. *J Inherit Metab Dis* 13:572–576.

Jonsson JJ, Habel DE, McIvor RS (1995): Retrovirus-mediated transduction of an engineered intron-containing purine nucleoside phosphorylase gene. *Hum Gene Ther* 6:611–623.

Kaleko M, Garcia JV, Osborne WRA, Miller AD (1990): Expression of human adenosine deaminase in mice after transplantation of genetically-modified bone marrow. *Blood* 75:1733–1741.

Kantoff PW, Freeman SM, Anderson WF (1988): Prospects for gene therapy for immunodeficiency diseases. *Annu Rev Immunol* 6:581–594.

Karlsson S (1991): Treatment of genetic defects in hematopoietic cell function by gene transfer. *Blood* 78:2481–2492.

Kohn DB, Weinberg KI, Nolta JA, et al. (1995): Engraftment of gene-modified umbilical cord blood cells in neonates with adenosine deaminase deficiency. *Nat Med* 1:1017–1023.

Krall WJ, Challita PM, Perlmutter LS, et al. (1994): Cells expressing human glucocerebrosidase from a retroviral vector repopulate macrophages and central nervous system microglia after murine bone marrow transplantation. *Blood* 83:2737–2748.

Kredich NM, Hershfield MS (1989): Immunodeficiency diseases caused by adenosine deaminase deficiency and purine nucleoside phosphorylase deficiency. In Scriver CR, Beaudet AL, Sly WS, Valle D (eds.), *The Metabolic Basis of Inherited Disease,* 6th ed. New York: McGraw-Hill, pp 1045–1075.

Kume A, Dinauer MC (1994): Retrovirus-mediated reconstitution of respiratory burst activity in X-linked chronic granulomatous disease cells. *Blood* 84:3311–3316.

Lacorazza HD, Flax JD, Snyder EY, Jendoubi M (1996): Expression of human β-hexosaminidase α-subunit gene (the gene defect of Tay-Sachs disease) in mouse brains upon engraftment of transduced progenitor cells. *Nat Med* 2:424–429.

Li F, Linton GF, Sekhsaria S, et al. (1994): $CD34^+$ peripheral blood progenitors as a target for genetic correction of the two flavocytochrome b_{558} defective forms of chronic granulomatous disease. *Blood* 84:53–58.

Malech HL, Sekhsaria S, Whiting-Theobald N, et al. (1996): Prolonged detection of oxidase-positive neutrophils in the peripheral blood of five patients following a single cycle of gene therapy for chronic granulomatous disease. *Blood* 88(suppl 10):486a.

Marcel T, Grausz JD (1996): *The RAC report and the TMC worldwide gene therapy enrollment report.* Paris, France: TMC Development.

Marechal V, Naffakh N, Danos O, Heard JM (1993): Disappearance of lysosomal storage in spleen and liver of mucopolysaccharidosis VII mice after transplantation of genetically modified bone marrow cells. *Blood* 82:1358–1365.

McIvor RS, Johnson MJ, Miller AD, et al. (1987): Human purine nucleoside phosphorylase and adenosine deaminase: Gene transfer into cultured cells and murine hematopoietic stem cells by using recombinant amphotropic retroviruses. *Mol Cell Biol* 7:838–846.

Medin JA, Migita M, Pawliuk R, et al. (1996a): A bicistronic therapeutic retroviral vector enables sorting of transduced $CD34^+$ cells and corrects the enzyme deficiency in cells from Gaucher patients. *Blood* 87:1754–1762.

Medin JA, Tudor M, Simovitch R, et al. (1996b) Correction in trans for Fabry disease: Expression, secretion, and uptake of α-galactosidase A in patient-derived cells driven by a high-titer recombinant retroviral vector. *Proc Natl Acad Sci USA* 93:7917–7922.

Migita M, Medin JA, Pawliuk R, et al. (1995): Selection of transduced $CD34^+$ progenitors and enzymatic correction of cells from Gaucher patients, with bicistronic vectors. *Proc Natl Acad Sci USA* 92:12075–12079.

Moullier P, Bohl D, Heard JM, Danos O (1993): Correction of lysosomal storage in the liver and spleen of MPS VII mice by implantation of genetically modified skin fibroblasts. *Nat Genet* 4:154–159.

Naldini L, Blömer U, Gallay P, et al. (1996): In vivo gene delivery and stable transduction of nondividing cells by a lentiviral vector. *Science* 272:263–267.

Nelson DM, Butters KA, Markert ML, et al. (1995): Correction of proliferative responses in purine nucleoside phosphorylase (PNP)-deficient T lymphocytes by retroviral-mediated PNP gene transfer and expression. *J Immunol* 154:3006–3014.

Neufeld EF (1991): Lysosomal storage diseases. *Annu Rev Biochem* 60:257–280.

Neufeld EF, Muenzer J (1995): The mucopolysaccharidoses. In Scriver CR, Beaudet AL, Sly WS, Valle D (eds.), *The Metabolic Basis of Inherited Disease,* 7th ed. New York: McGraw-Hill, pp. 2465–2494.

Nolta JA, Sender LS, Barranger JA, Kohn DB (1990): Expression of human glucocerebrosidase in murine long-term bone marrow cultures after retroviral vector-mediated transfer. *Blood* 75:787–797.

Nolta JA,Yu XJ, Bahner I, Kohn DB (1992): Retroviral-mediated transfer of the human glucocerebrosidase gene into cultured Gaucher bone marrow. *J Clin Invest* 90:342–348.

Occhiodoro T, Hopuood JJ, Morvis CP, Anson DS (1992): Correction of α-L-fucosidase deficiency in fucosidosis fibroblasts by retroviral vector-mediated gene transfer. *Hum Gene Ther* 3:365–369.

Ohashi T, Boggs S, Robbins P, et al. (1992): Efficient transfer and sustained high expression of the human glucocerebrosidase gene in mice and their functional macrophages following transplantation of bone marrow transduced by a retroviral vector. *Proc Natl Acad Sci USA* 89:11332–11336.

Ohashi T, Eto Y, Learish R, Barranger JA (1993): Correction of enzyme deficiency in metachromatic leukodystrophy fibroblasts by retroviral mediated transfer of the human arylsulphatase A gene. *J Inherit Metab Dis* 16:881–885.

Osborne WRA, Miller AD (1988): Design of vectors for efficient expression of human purine nucleoside phosphorylase in skin fibroblasts from enzyme-deficient humans. *Proc Natl Acad Sci USA* 85:6851–6855.

Peters C, Rommerskirch W, Modaressi S, von Figura K (1991): Restoration of arysulphatase B activity in human mucopolysaccharidosis-type-VI fibroblasts by retroviral-vector–mediated gene transfer. *Biochem J* 276:499–504.

Petzger AL, Hogge DE, Landsdorp PM, et al. (1996): Self-renewal of primitive human hematopoietic cells (long-term-culture-intiating cells) *in vitro* and their expansion in defined medium. *Proc Natl Acad Sci USA* 93:1470–1474.

Pollock JD, Williams DA, Gifford MA, et al. (1995): Mouse model of X-linked chronic granulomatous disease, an inherited defect in phagocyte superoxide production. *Nat Genet* 9:202–209.

Porter CD, Parkar MH, Levinsky RJ, et al. (1993): X-linked chronic granulomatous disease: Correction of NADPH oxidase defect by retrovirus-mediated expression of gp91-phox. *Blood* 82:2196–2202.

Porter CD, Parkar MH, Verhoeven AJ, et al. (1994): p22-phox–Deficient chronic granulomatous disease: Reconstitution by retrovirus-mediated expression and identification of a biosynthetic intermediate of gp91-phox. *Blood* 84:2767–2775.

Poutev CD, Parkar MH, Collin MKL, et al. (1996): Efficient retroviral transduction of human bone marrow progenitor and long-term culture-initiating cells: Partial reconstitution of cells from patients with X-linked chronic granulomatous disease by gp91-phox expression. *Blood* 87:3722–3730.

Rafi MA, Fugaro J, Amini S, et al. (1996): Retroviral vector-mediated transfer of the galactocerebrosidase (GALC) cDNA leads to overexpression and transfer of GALC activity to neighboring cells. *Biochem Mol Med* 58:142–150.

Reiser J, Harmison G, Kluepfel-Stahl S, et al. (1996): Transduction of nondividing cells using pseudotyped defective high titer HIV type 1 particles. *Proc Natl Acad Sci USA* 93:15266–15271.

Ringden O, Groth C-G, Erikson A, et al. (1988): Long-term follow-up of the first successful bone marrow transplantation in Gaucher disease. *Transplant* 46:66–69.

Rommerskirch W, Fluharty AL, Peters C, et al. (1991): Restoration of arylsulphatase A activity in human metachromatic leucodystrophy fibroblasts via retroviral vector mediated gene transfer. *Biochem J* 280:459–461.

Salvetti A, Moullier P, Cornet V, et al. (1995): *In vivo* delivery of human α-L-iduronidase in mice implanted with neo-organs. *Hum Gene Ther* 6:1153–1159.

Schiffmann R, Medin JA, Ward JM, et al. (1995): Transfer of the human glucocerebrosidase gene into hematopoietic stem cells of nonablated recipients: Successful engraftment and long-term expression of the transgene. *Blood* 86:1218–1227.

Sorge J, Kuhl W, West C, Beutler E. (1987): Complete correction of the enzymatic defect of type I Gaucher disease fibroblasts by retroviral-mediated gene transfer. *Proc Natl Acad Sci USA* 84:906–909.

Sorge J, West C, Westwood B, Beutler E (1985): Molecular cloning and nucleotide sequence of human glucocerebrosidase cDNA. *Proc Natl Acad Sci USA* 82:7289–7293.

Sugimoto Y, Aksentijevich I, Murray GJ, et al. (1995): Retroviral coexpression of a multidrug resistance gene (MDR1) and human α-galactosidase A for gene therapy of Fabry disease. *Hum Gene Ther* 6:905–915.

Taylor N, Uribe L, Smith S, et al. (1996): Correction of interleukin-2 receptor function in X-SCID lymphoblastoid cells by retrovirally mediated transfer of the common gamma chain gene. *Blood* 87:3103–3107.

Taylor RM, Wolfe JH (1994): Cross-correction of β-glucuronidase by retroviral vector-mediated gene transfer. *Exp Cell Res* 214:606–613.

Thrasher A, Chetty M, Casimir C, Segal AW (1992): Restoration of superoxide generation to a chronic granulomatous disease-derived B-cell line by retrovirus mediated gene transfer. *Blood* 80:1125–1129.

Valentine WN, Paglia DE, Tartaglia AP, Gisanz F (1977): Hereditary hemolytic anemia with increased adenosine deaminase (45-70 fold) and decreased adenosine triphosphates. *Science* 195:783–785.

Weil WM, Linton GF, Whiting-Theobald N, et al. (1997): Genetic correction of p67phox deficient chronic granulomatous disease using peripheral blood progenitor cells as target for retrovirus mediated gene transfer. *Blood* 89:1754–1761.

Weinberg KI (1996). Gene therapy for X-linked severe combined immunodeficiency using retroviral mediated transduction of the common gamma chain cDNA into CD34^{+} cells. *Human Gene Therapy Protocols 9604-152.* ORDA. NIH Bethesda, MD.

Whitley CB, McIvor RS, Aronovich EL, et al. (1996): Clinical protocol. Retroviral-mediated transfer of the iduronate-2-sulfatase gene into lymphocytes for treatment of mild Hunter syndrome (mucopolysaccharidosis type II). *Hum Gene Ther* 7:537–549.

Williams DA, Orkin SH, Mulligan RC (1986): Retrovirus-mediated transfer of human adenosine deaminase gene sequences into cells in culture and into murine hematopoietic cells in vivo. *Proc Natl Acad Sci USA* 83:2566–2570.

Wolfe JH, Kyle JW, Sands MS, et al (1995): High level expression and export of β-glucuronidase from murine mucopolysaccharidosis VII cells corrected by a double-copy retrovirus vector. *Gene Ther* 2:70–78.

Wolfe JH, Sands MS, Barker JE, et al. (1992): Reversal of pathology in murine mucopolysaccharidosis type VII by somatic cell gene transfer. *Nature* 360:749–753.

Wolfe JH, Schuchman EH, Stramm LE et al. (1990): Restoration of normal lysosomal function in mucopolysaccharidosis type VII cells, by retroviral vector-mediated gene transfer. *Proc Natl Acad Sci USA* 87:2877–2881.

Woodman RC, Teoh D, Johnston F, et al. (1995): X-linked carriers of gp91-phox chronic granulomatous disease (CGD): Clinical and potential therapeutic implications of extreme unbalanced X chromosome inactivation. *Blood* 86(suppl 1):27a

Xu L, Stahl SK, Dave HPG, Schiffmann R, et al. (1994): Correction of the enzyme deficiency in hematopoietic cells of Gaucher patients using a clinically acceptable retroviral supernatant transduction protocol. *Exp Hematol* 22:223–230.

15

GENE THERAPY FOR HEMOPHILIA

KATHERINE A. HIGH
The Children's Hospital of Philadelphia, Abramson Research Center, Philadelphia PA 19104

INTRODUCTION

Hemophilia is the X-linked bleeding diathesis that results from a deficiency of functional Factor VIII (hemophilia A) or Factor IX (hemophilia B). The disease has been known since antiquity, and passages in the Talmud warn that if a woman has two sons who die from bleeding following circumcision, subsequent sons should not be circumcised. The disease occurs worldwide and affects approximately 1 in 10,000 male births. Hemophilia A and hemophilia B are indistinguishable clinically; the diseases were first established as separate clinical entities in 1952, when Aggeler and coworkers and Biggs and colleagues demonstrated that the plasma of one patient with hemophilia was able to correct the functional clotting defect of a second patient's plasma. Hemophilia A is approximately five times as common as hemophilia B; it has been shown for X-linked diseases that the frequency of disease bears a direct relationship to the size of the affected gene. This would appear to hold true for the hemophilias, where the Factor VIII (F.VIII) gene is 186 kb in length and the Factor IX (F.IX) gene is 35 kb.

F.IX and F.VIII are both critical components of the enzyme complex that catalyzes the conversion of F.X to activated F.X (F.Xa). (The other components of the complex are calcium and phospholipids.) F.IX is a vitamin K–dependent protein that circulates in zymogen form. When cleaved by F.XIa,

Stem Cell Biology and Gene Therapy, Edited by Peter J. Quesenberry, Gary S. Stein, Bernard G. Forget, and Sherman M. Weissman
ISBN 0-471-14656-0

the zymogen is converted to an active enzyme. Like the other enzymes in the clotting cascade, F.IXa is a trypsin-like serine protease, with remarkable specificity for its substrate F.X (Fig. 1). F.VIIIa is a cofactor in this reaction; it probably plays a critical role in aligning the enzyme IXa with the substrate F.X (Brandstetter et al., 1995). Although the activation of F.X can proceed without F.VIIIa, the presence of the cofactor accelerates the reaction by 100,000-fold (Van Dieijen et al., 1981). Thus, under physiological conditions, the reaction does not proceed in the absence of functional F.VIIIa. The cDNAs encoding F.IX and F.VIII were cloned in 1982 (Kurachi, 1982) and 1984 (Gitschier et al., 1984; Toole et al., 1984), respectively, with sequences of genomic clones available a short time later (Gitschier et al., 1984; Yoshitake et al., 1985). The availability of these data has allowed rapid advances in this field, including the elucidation of molecular defects responsible for the hemophilias (Giannelli et al., 1996; Tuddenham et al., 1994), the preparation of recombinant clotting factors for therapeutic use (White et al., 1989, 1997), the generation of "knockout" animals as disease models for hemophilia (Bi et al., 1995; Lin et al., 1996), and the production of wild-type and mutant recombinant proteins for structure–function analysis (Hamaguchi et al., 1993; Pittman et al., 1992; Toole et al., 1986). In addition, the F.IX and F.VIII cDNAs are critical reagents in the development of gene therapy strategies for the hemophilias. The structures of the F.IX and F.VIII genes and the proteins they encode are outlined in Figures 2 and 3.

Clinically, hemophilia is characterized by frequent spontaneous hemorrhages; the most common sites of bleeding are the joint spaces and soft tissues, but other sites, such as intracranial, retroperitoneal, and retropharangeal

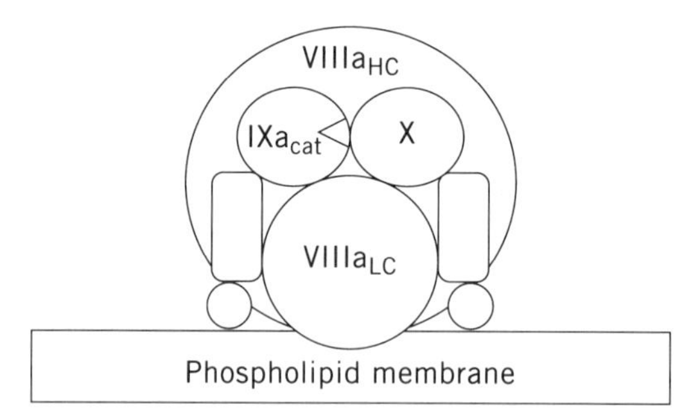

FIGURE 1. Cleavage of Factor X by the IXa-VIIIa complex. Factor IXa is the enzyme, and Factor VIIIa the cofactor, that catalyzes the conversion of Factor X to Factor Xa. The reaction also requires phospholipid (it occurs on a phospholipid surface) and calcium. Activation of Factor X involves the cleavage of an Arg^{194}-Ile^{195} bond within the heavy chain. The cofactor probably serves to orient the enzyme and substrate (Brandstetter et al., 1995); its presence increases k_{cat} by a factor of $>10^4$ (van Dieijen et al., 1981). $VIIa_{HC}$ = heavy chain of factor VIIIa; $VIIIa_{LC}$ = light chain; IXa_{cat} = catalytic domain of F.IXa.

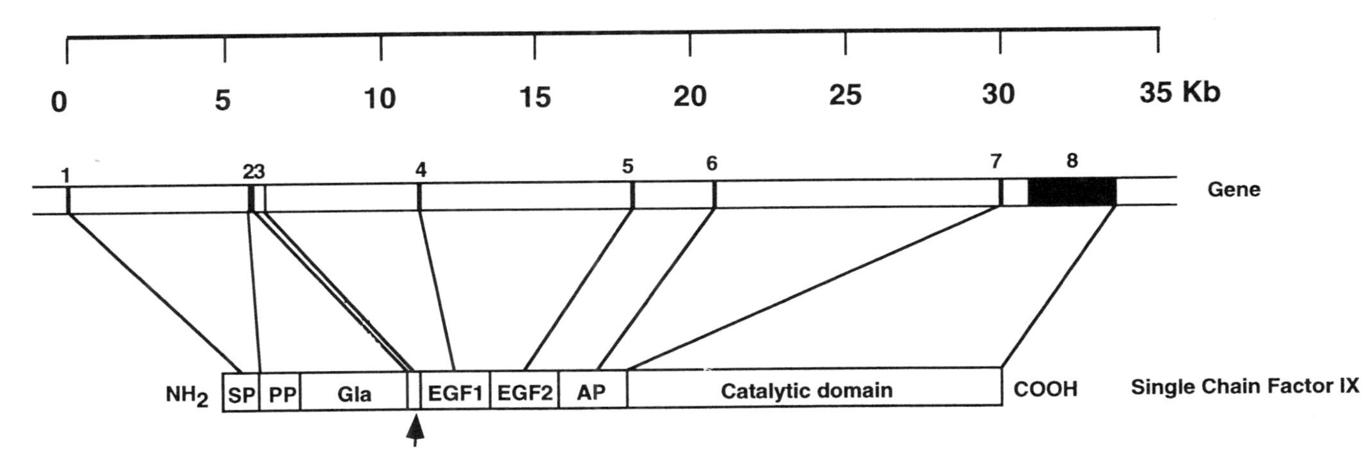

FIGURE 2. Structure of F.IX gene and protein. The top line indicates the scale in kilobases. The second shows the genomic organization of the human F.IX gene, which consists of eight exons and seven introns. Exons are indicated by black bars with Arabic numerals and introns by the white (intervening) spaces. The exons correlate in an approximate fashion with the domains of the protein, as shown in the third line; thus, exon 1 encodes the signal peptide (SP), exon 2 the propeptide (PP) and Gla domain, exon 3 the short aromatic acid stack, exons 4 and 5 the EGF-like domains, exon 6 the activation peptide (AP), and exons 7 and 8 the catalytic domain and the long 3′-untranslated region.

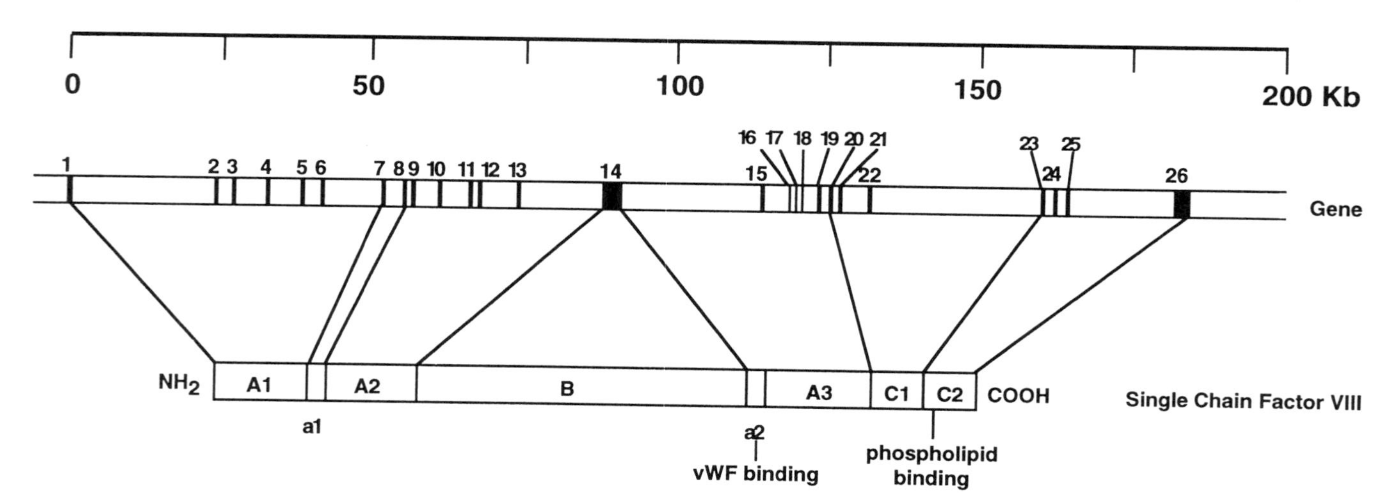

FIGURE 3. Structure of Factor VIII gene and protein. The top line indicates the scale in kilobases. The Factor VIII gene consists of 26 exons (second line) encoding a protein of 2,351 amino acids (third line). The A domains, consisting of ~330 amino acids each, share ~30% homology with each other and with the copper-binding protein ceruloplasmin. The heavily glycosylated B domain is homologous to the B domain of Factor V but not to other proteins in the database. The duplicated C domains share about 40% homology with each other. Two short acidic regions, designated a1 and a2, are rich in Asp and Glu residues. The residues in a2 are involved in binding to von Willebrand factor.

spaces, can be affected as well and can lead to life-threatening or fatal outcomes. Current treatment for hemophilia consists of intravenous infusion of clotting factor concentrates in response to a bleeding episode. The development of plasma-derived clotting factor concentrates in the early 1970s resulted in a dramatic improvement in life expectancy and quality of life for patients with hemophilia (Rosendaal et al., 1991), but the subsequent contamination of the blood supply with human immunodeficiency virus (HIV) pointed up one of the major disadvantages of the plasma-derived concentrates, the risk of transmitting blood-borne diseases. Current methods of viral inactivation have eliminated the risk of HIV transmission by clotting factor concentrates, but have not been 100% effective at eradicating all blood-borne viruses, especially nonenveloped viruses (Mannucci, 1993). Recombinant clotting factor concentrates, developed within the past decade, eliminate the risks associated with plasma, but are even more expensive than the plasma-derived concentrates. A typical patient with severe disease may spend $50,000–$100,000 on concentrate in a single year. The expense of both plasma-derived and recombinant concentrates has made it difficult to justify using them prophylactically, and most patients are treated only in response to bleeding. During the inevitable delay between the onset of a bleed and the infusion of concentrate, tissue damage may occur and over a period of years can result in considerable morbidity. The most frequent cause of morbidity in patients with hemophilia is hemophilic arthropathy, which can eventually result in severe limitation of range of motion (knees, ankles, and elbows are the most commonly affected joints) that may require surgical joint replacement. The necessity for intravenous infusion of concentrates, and for refrigeration of concentrates, imposes additional constraints on the lives of patients with hemophilia.

These disadvantages of protein-based therapy have fueled interest in a gene therapy approach to hemophilia. The most important advantage of this type of approach in hemophilia is that a constant level of clotting factor is maintained so that bleeds are prevented rather than merely treated *post facto*. In addition, the risks and inconvenience of factor infusion are avoided.

Gene therapy has been proposed as a treatment modality for a number of inherited and acquired disorders, but several factors make hemophilia a particularly attractive model in which to undertake gene therapy. First, tissue-specific expression of the transgene is not required. Clotting factors are normally synthesized in the liver, but work by a number of investigators has shown that biologically active F.IX can be produced in other cell types, including fibroblasts (Palmer et al., 1989), myoblasts (Yao and Kurachi, 1992), and endothelial cells (Yao et al., 1991). Thus the choice of target cell is not restricted, and the transgene can be inserted into any one of a number of tissues as long as the gene product can gain access to the circulation. This is in sharp contrast to many other inherited diseases, e.g., sickle cell disease or the thalassemias, where expression of the transgene must occur in a single cell type, the red cell precursor, to have therapeutic efficacy.

A second advantage of hemophilia as a model for gene therapy is that precise regulation of expression of the transgene is not required. Most patients with hemophilia have severe disease, with <1% normal factor levels. Patients with levels of 5% have much milder disease, and only rarely experience spontaneous bleeding episodes (although they exhibit abnormal bleeding in response to hemostatic challenges such as surgery or trauma). Thus, raising a patient's level from <1% to 5% results in a dramatic improvement in phenotype. Similarly, it is clear based on data from infusion of concentrates into patients with hemophilia that levels as high as 150% are not associated with ill effects, since the proteins circulate as zymogens. Thus a remarkably wide range of levels of expression (from 3% to 150% of normal) fall within a therapeutic range. Again this is in contrast to the requirements of certain other enzyme deficiencies (e.g., diabetes) where the level of the enzyme is very tightly regulated and an excess is as harmful as a deficiency.

A third advantage of hemophilia compared with other diseases is the availability of large and small animal models that faithfully mirror the human disease. Clearly animal models are a major asset in efforts to establish an experimental basis for gene therapy. In the case of hemophilia, there are well-characterized, naturally occurring canine models of the disease and genetically engineered hemophilic mice. Dogs with hemophilia were identified in the 1950s, and colonies have been maintained at several centers in North America since that time. Both hemophilia A and hemophilia B occur in dogs; the animals maintained in colonies have severe disease with <1% activity. The canine disease is X linked and is characterized by frequent spontaneous bleeds, including hemarthroses. The canine F.IX cDNA has been isolated and characterized (Evans et al., 1989b) so that species-specific vectors can be prepared for these animals. This is an important consideration, since work by a number of investigators has demonstrated that hemophilic dogs develop antibodies to human clotting factors within a few weeks of initial exposure (Connelly et al., 1996b). This factor has hampered use of the hemophilia A dogs for gene transfer experiments, but a full-length canine F.VIII cDNA clone has recently become available (Cameron et al., 1998). The point mutation responsible for hemophilia B in the Chapel Hill colony has been determined (Evans et al., 1989a) and is a missense mutation in a highly conserved residue in the catalytic domain of the molecule. A G → A transition at the middle nucleotide of the codon for amino acid 379 changes a glycine to a glutamic acid. Molecular modeling studies based on the known high-resolution structure of bovine pancreatic trypsin suggest that the folding of the mutant protein is likely to be severely disrupted. No F.IX antigen is detectable in the plasma of the hemophilic dogs; this has rendered them useful for studies of gene transfer, since expression of the transgene can be followed either by activity assays or by immunological assays (or both).

A mouse model of hemophilia A has been produced by targeted disruption of the murine F.VIII gene. Bi and colleagues (1995) created two lines with two distinct mutations by inserting a neo cassette into exon 16 or exon 17 of the murine gene. Both lines exhibited a severe phenotype with F.VIII activity levels of <1% by a chromogenic assay in which normal littermates exhibited 100% activity. A similar strategy has been used by at least two groups to create mice with hemophilia B (Lin et al., 1997; Wang et al., 1997). The availability of these small animal models is also an important asset, since they are considerably cheaper to maintain than the hemophilic dogs.

Although hemophilia A and hemophilia B are indistinguishable clinically, there are several differences between these two entities that bear on their suitability as models for gene therapy, and it is worth pointing these out before proceeding to a more detailed summary of experimental results. First, the F.IX cDNA at 3 kb is considerably smaller than the F.VIII cDNA (~7 kb). Since most viral vectors have limitations on the size of the foreign insert that can be accommodated, this factor favors hemophilia B as a model. For F.VIII, this problem has been circumvented by constructing vectors in which the B domain has been deleted (see Fig. 3), yielding a F.VIII cDNA of ~4.4 kb, but this size still places limits on the construction of the minigene cassette, e.g., in choice of promoter to include in the cassette. On the other hand, the F.IX cDNA, though it has a total length of 3 kb, actually contains only 1.5 kb of coding sequence. The remainder of the cDNA consists of a long 3′-untranslated region. The function of this region is unknown, but a number of viral vectors have been produced that do not include this region as part of the transgene, so it appears to be dispensable. This has allowed for considerable flexibility in the design of F.IX cassettes to insert into viral vectors, and the testing of a wide range of promoters within the cassette has yielded important insights (*vide infra*). A second difference between these two disease entities that has relevance to gene therapy is the frequency with which antibodies to the clotting factor arise on exposure to what may be a "foreign" protein in the hemophilic patient. In large series of patients with hemophilia A, inhibitors (antibodies) occur in ~20% of individuals, whereas in hemophilia B the frequency is much lower, in the range of 1%–5% (High, 1995). The reasons for this discrepancy are not clear, although it has been noted that individuals with molecular defects in which no protein is produced (e.g., large gene deletions, stop codons in the 5′ portion of the transcript, gene inversions) are more likely to develop inhibitors than are those with mutations (e.g., missense mutations) in which small amounts of protein are produced (Green et al., 1991). Thus, the fact that ~40% of individuals with severe hemophilia A have a gene inversion defect associated with a complete absence of protein (Lakich et al., 1993) may account in part for the higher frequency of inhibitor formation in this group. Whatever the cause, it is clear that the formation of inhibitors is a major drawback in a clinical trial, since levels of expression of the transgene product cannot be reliably determined in this setting. This feature thus favors hemophilia B over hemophilia A as a model for studying gene therapy. Finally, it should be noted that the plasma levels of the circulating protein are much higher in the case of F.IX than in the case of F.VIII. Normal plasma levels of F.IX are ~5 μg/ml and of F.VIII are ~100 ng/ml. Thus levels of protein expression required for therapeutic efficacy in the case of F.IX are ~50-fold higher than those required for F.VIII deficiency. This feature would appear to constitute an advantage for hemophilia A as a model for gene therapy, although this potential advantage is largely offset by sequences within the F.VIII cDNA that inhibit RNA accumulation (*vide infra*).

EXPRESSION OF TRANSFERRED GENES *IN VITRO* AND IN ANIMAL MODELS

From the foregoing discussion it should be apparent that investigators have had wide latitude in the selection of target cells and of vectors in attempts to establish an experimental basis for gene therapy of hemophilia. It should be noted at the outset that most of the strategies attempted to date and reviewed herein have not resulted in a level of success that would warrant human trials. Nonetheless, there has been steady progress in the effort to produce sustained expression of clotting factor protein at a level that would be therapeutic, and recent data using adeno-associated viral vectors in muscle and liver have resulted in sustained expression of F.IX at therapeutic levels.

For purposes of summarizing the data in an organized fashion, the work in the field will be arbitrarily divided based on the vector used. In general, emphasis is on *in vivo* data rather than *in vitro* data, since a wealth of experience documents that, in the case of gene transfer experiments, the latter is a poor predictor of the former. Since F.IX has been used much more extensively than F.VIII as a transgene, most of the data presented here involves F.IX and hemophilia B. F.VIII expression is subject to a unique set of problems, and these data are reviewed in a separate section.

DATA WITH RETROVIRUSES

As is the case for most transgenes, the preponderance of the data has been generated with retroviruses. Early attempts (Palmer et al., 1989, 1991; Scharfmann et al., 1991; Yao et al., 1991) to produce F.IX in animal models by retrovirally mediated gene transfer techniques were plagued by low levels and short duration of expression. More recently, several groups have achieved long-term expression of F.IX in experimental animals using retroviral vectors, but in general the levels have been too low to achieve therapeutic efficacy (Dai et al., 1992; Kay et al., 1993). One group using F.X as the transgene has very recently produced levels in the therapeutic range through *in vivo* transduction of liver with a retroviral vector (Le et al., 1997). In addition to liver, retroviruses have been used to transduce myoblasts, hematopoietic cells, and fibroblasts.

Retroviral Transduction of Myoblasts

Myoblasts are an appealing target cell type for a number of reasons. The tissue is well vascularized and thus has the potential for efficient transfer of gene products to the circulation. Muscle tissue is abundant and can be readily accessed (to serve as a source for autologous transplant) and cultured. Re-injection of myoblasts results in an efficient fusion process with existing muscle fibers. Unfortunately, in experiments where species boundaries are crossed, introduction of F.IX into muscle cells appears to generate antibodies in a highly efficient fashion (in contrast to, e.g., intravenous injection of adenoviral vectors expressing human F.IX; *vide infra*), and this has hampered experimental work with myoblasts in hemophilia.

Yao and Kurachi (1992) were the first to demonstrate the utility of this approach by using a retroviral vector to transduce a mouse muscle cell line (C2C12 cells) and then re-implanting the transduced cells into the hindlimb muscles of cyclosporine-treated adult C3H mice. Mice injected with $\sim 3 \times 10^7$ transduced cells expressed F.IX levels as high as 1,000 ng/ml, but by 30 days postinjection levels of human F.IX were undetectable and antibodies to human F.IX were present in mouse serum. However, marking experiments with a β-galactosidase gene established that the transplanted transduced C2C12 cells fused with existing muscle fibers and persisted without diminution in numbers for periods of up to 4 months. This suggests that the strategy is sound if cross-species transgenes are avoided.

In a study published a few months later, Dai et al. (1992) achieved long-term expression of canine F.IX by transplanting myoblasts (primary culture cells from Swiss-Webster mice) transduced with a F.IX-expressing retroviral vector into nude mice. In this case, however, levels of expression were low, on the order of 10–20 ng/ml. An additional finding of interest reported by these investigators was that, of three different retroviral vectors tested, only one, containing the cytomegalovirus (CMV) promoter and the muscle creatine kinase (MCK) enhancer, yielded long-term expression (180 days). With the two other vectors, one containing the CMV promoter alone and the other containing the MCK enhancer coupled with the human α-globin promoter and the *Xenopus* β-globin 5′-untranslated region, expression *in vivo* was lost after ~5 days. This result is intriguing but defies simple interpretation, since subsequent data from Yao et al. (1994) showed long-term expression (180 days) of human F.IX using a similar strategy (re-implantation of transduced primary myoblasts into syngeneic severe combined immunodeficiency [SCID] mice) employing a vector in which the viral long terminal repeat (LTR) drives F.IX expression. Kurachi and colleagues have isolated and characterized human F.IX produced in mouse myoblasts and have shown, using ELISA and one-stage clotting assays, that the material has a specific activity of 95%–125%, despite differences in glycosylation compared to plasma-derived human F.IX (Yao and Kurachi, 1992).

The data suggest, then, that a strategy based on *ex vivo* transduction of primary culture (autologous) myoblasts followed by re-implantation of these cells could result in efficient transfer of F.IX into the circulation (Yao et al., 1994) and persistence of the transduced myoblasts if cross-species boundaries are not transgressed. Note that none of these data bear directly on *in vivo* transduction of existing muscle fibers.

Retroviral Transduction of Liver

Not surprisingly, liver is the favored target organ for production of clotting factors, since it is the normal site of synthesis. Proper execution of the complex series of steps involved in post-translational processing and access of the gene product to the circulation are ensured in this setting. Retroviral transduction of hepatocytes can take place either as an *ex vivo* or an *in vivo* procedure. In the *ex vivo* procedure, a portion of liver is excised, plated out in culture, and transduced with the vector of interest. Cells are subsequently re-infused (Grossman et al., 1995). Kay and colleagues (1993) demonstrated several years ago that retroviral vectors can be used to transduce hepatocytes *in vivo*. This experiment is instructive, as it demonstrates both the promises and the pitfalls of this approach. Dogs with hemophilia B were subjected to partial hepatectomy, and retroviral vector expressing canine F.IX was infused into the portal circulation 24, 48, and 72 h after partial hepatectomy. F.IX levels in the plasma rose to ~6 ng/ml and remained in that range for a period of ~2 years, the duration of the experiment (personal communication, Mark Kay). Parallel experiments with a vector expressing β-galactosidase demonstrated that only 0.3%–1.0% of hepatocytes were transduced in this procedure, despite a 60% hepatectomy to induce hepatocyte cell division in the remaining liver.

The F.IX levels achieved in this experiment are considerably less than 1% (normal levels are 5,000 ng/ml) and thus are not within a therapeutic range. Nevertheless, long-term expression was achieved. The two major obstacles that must be overcome to make this strategy a therapeutic reality are the low level of gene expression per integrated provirus and the low efficiency of transduction. Recent studies have demonstrated progress in both areas. Okuyama et al. (1996) have generated a number of retroviral vectors with liver-specific promoters and enhancers and have tested these in a model system in which rats undergo partial hepatectomy and subsequent infusion of vector. These experiments have shown that the human α_1-antitrypsin (hAAT) promoter coupled with the apolipoprotein E enhancer yields levels of transgene expression at least one log higher than other viral or eukaryotic promoters tested. Thus it is possible to increase levels of expression per integrated provirus through judicious choice of promoter. (It should be noted that this vector also contained a retroviral backbone with an intact 3′ LTR that yielded higher titer vector than constructions with deletions in the viral LTR. This factor may have contributed to the higher levels of expression seen here.) In subsequent experiments (Le et al., 1997), the hAAT promoter was coupled with a human F.X cDNA and tested in the rat model. Sustained expression of F.X levels of >1,000 ng/ml was achieved. Since plasma levels of F.X are comparable with those for F.IX, this would seem to be a promising approach for hemophilia B, although it will be important to determine whether similar results can be achieved in species other than the rat. (The transduction efficiency for infusion of retroviral vector through the portal circulation preceded by partial hepatectomy is typically 1% in mice and dogs, but 5%–15% in rats [Ferry et al., 1991; Kolodka et al., 1993; Moscioni et al., 1993].)

A number of approaches have been explored in an effort to increase transduction efficiency and to obviate the need for partial hepatectomy, a requirement that makes this strategy unappealing for human trials. Lieber et al. (1995) have used an adenoviral vector expressing urokinase-type plasminogen activator (uPA) to induce hepatocyte necrosis and regeneration and have shown that retroviral transduction efficiency is increased 5–10-fold in mice following prior infusion of the uPA-expressing adenoviral construct. This effect is seen *without* any prior hepatectomy. In another approach, Bowling et al. (1996) have used portal branch occlusion to achieve retroviral transduction efficiencies comparable with those seen following partial hepatectomy. In this procedure, branches of the portal vein are occluded, resulting in apoptosis of hepatocytes in the occluded lobes and compensatory replication in the remaining lobes. Another strategy has been the use of keratinocyte growth factor to induce hepatocyte proliferation; this results in transduction efficiencies on the order of 2% (Bosch et al., 1996). Finally, the generation of higher titer vector preparations (Bowles et al., 1996) allows increased transduction efficiency by increasing the effective multiplicity of infection.

Retroviral Transduction of Hematopoietic Cells

The use of hematopoietic cells as targets has considerable appeal as a treatment strategy for gene therapy of hemophilia B. Experience with autologous bone marrow transplantation is extensive, and one can readily envision a strategy where a patient's bone marrow is removed, transduced with a retroviral vector expressing F. IX, and then re-implanted. It has been shown that biologically active F.IX can be synthesized in these cells (Hao et al., 1995), and the gene product has ready access to the circulation. Expression of F.IX in platelet precursors is particularly attractive, since platelets accumulate at the site of a bleed. Thus, even if circulating levels of F.IX are below the therapeutic range, local concentrations at the bleeding site may be adequate to ensure hemostasis. Little is known about certain critical aspects of the biosynthesis of F.IX in megakaryocytes and other hematopoietic cells, e.g., whether F.IX is packaged into platelet granules or constitutively secreted; these areas are being actively investigated. A theoretical objection to this strategy (expression in hematopoietic cells) centers on the potential for thrombosis that may be generated by secretion of F.IX from monocytes. Activated monocytes express tissue factor on the surface; when combined with VII/VIIa in the circulation, the VIIa/tissue factor complex catalyzes the conversion of F.IX to the active species IXa. The thrombogenicity of the older plasma-derived F.IX preparations has been well-documented and was attributed to contamination of the products with activated factors (Lusher, 1991). Whether an approach targeting hematopoietic cells would be thrombogenic is unclear and needs to be addressed. The use of platelet-specific promoters in the viral vectors may allow one to circumvent this problem (Ravid et al., 1991).

The potential for treatment of hemophilia by transduction of hematopoietic cells has been thwarted by the same obstacles that have plagued other attempts to treat disease by genetic modification of these cells, including that transduction efficiency is quite low, on the order of 1%. To achieve long-term expression, a critical requirement for genetic diseases like hemophilia, it is necessary to transduce stem cells, but since most of these are quiescent, they are not efficiently transduced by retroviral vectors, which require dividing target cells. In the murine system it has been possible to achieve very high efficiency transduction (10%–80%) of hematopoietic stem cells (HSCs) through the use of recombinant hematopoietic growth factors to stimulate proliferation (Williams, 1990), but similar approaches in canine and simian recipients have failed to achieve HSC transduction efficiencies of >1%–2% (Kohn, 1995), and results from gene marking studies in humans have yielded similarly disappointing results (Brenner et al., 1993; Deisseroth et al., 1994; Dunbar et al., 1995). The factors that account for the differences between the murine system and larger mammals have not been clearly established, but may relate to differences in the density of receptors for retroviral receptors on the cell surface or to differences in the cell cycle kinetics of HSCs among species. Analysis of the potential utility of this strategy in humans will require progress in the ability to manipulate stem cells or improvements in vector technology.

ADENOVIRAL VECTORS IN F.IX GENE TRANSFER

Adenoviral vectors have been the subject of considerable interest to investigators working in the area of hemophilia. Advantages of these vectors include the ease with which they can be prepared in high titer, their wide host range, and the ability to transduce nondividing target cells such as liver. However, additional experience with these vectors has shown that expression levels, though initially high, undergo an inexorable decline and that repeat administration of the vector does not, under most circumstances, result in expression of the transgene. The decline in expression following the first administration of vector appears to be due to elimination of vector-transduced cells by the host immune response (Dai et al., 1995; Yang et al., 1994a,b; Yang and Wilson, 1995); the immune response also mediates the failure of expression following repeat administration of vector. A number of strategies, including immunomodulation of the host and engineering of less immunogenic adenoviral vectors, have been developed to circumvent this obstacle, but none have yet met with unqualified success.

The first report of successful F.IX gene transfer using adenoviral vectors was by Smith and coworkers (1993). Using a so-called first-generation adenoviral vector expressing human F.IX, these investigators injected (immunocompetent) C57Bl/6 mice via the tail vein and demonstrated therapeutic plasma levels of F.IX (350 ng/ml) within 1 week following injection. Plasma levels of F.IX declined slowly and were undetectable by 7–9 weeks after injection, and readministration of vector did not result in any rise in levels. Quantitation of vector DNA by Southern blots on liver DNA demonstrated a marked loss of vector DNA between week 4 and week 7 (from 55 to 3 copies/cell), consistent with destruction of adenoviral-transduced cells, and neutralizing antibodies to adenoviral proteins were also present, with a boost in titers by at least 32-fold following re-administration of vector. These results were confirmed and extended in the hemophilia B dog model by Kay et al. (1994). Using portal vein infusion of a first-generation adenoviral vector expressing canine F.IX, these workers achieved F.IX levels in the hemophilic dogs of 250%–300% normal, but expression was transient, falling to 1% of normal 3 weeks following injection and undetectable levels at 100 days after injection. (Of note is the fact that the number of transduced cells is very high, on the order of 20%–50% [using an ad-lacZ vector]. This is in marked contrast to the situation with retroviral vectors, *vide supra,* where the number of transduced cells is consistently around 1%.)

The role of immunomodulation in extending expression of F.IX from adenoviral vectors has been explored by several groups. Dai et al. (1995) showed that an adenoviral vector could direct long-term expression (>1 year) of F.IX in nude mice (mice that lack both T- and B-cell function), whereas expression was virtually undetectable in (immunocompetent) Swiss Webster mice 12 days after injection. To further dissect the contribution of cellular and humoral immunity, these workers injected the same vector into IgM heavy chain knockout mice (these mice have no antibody response) and into β_2-microglobulin knockout mice (no cytotoxic T-lymphocyte response). In both groups, expression fell to a subtherapeutic range (<10 ng/ml) by 3 weeks following injection, indicating that both cellular and humoral responses play a role in extinguishing expression from this adenoviral vector expressing canine F.IX in mice. Attempts to prolong expression by administration of immunosuppressive agents resulted in only modest success in this study. Administration of cyclosporin A, which should suppress cell-mediated immunity, at a dose of 10 mg/kg, had no effect, but administration of both cyclosporin A and cyclophosphamide (50 mg/kg), which kills rapidly dividing cells, resulted in a more gradual decline in plasma F.IX levels so that F.IX was still detectable (albeit at subtherapeutic levels) 150 days after injection.

The results obtained by Fang et al. (1995a,b) in the hemophilic dog model were somewhat different. These experiments are in a sense more straightforward to interpret, because, in contrast to those of Dai et al. (1995), species boundaries are not transgressed, i.e., the investigators used a vector expressing canine F.IX in a hemophilia B dog. Continuous administration of cyclosporin

A at a daily dose of 19.5 mg/kg resulted in expression of canine F.IX at therapeutic levels (>100 ng/ml) for approximately 180 days following a single administration of vector, but subsequently levels fell to a nontherapeutic range (~10 ng/ml). In contrast, untreated animals showed a rapid decline in F.IX levels, dropping to 10 ng/ml within ~40 days following injection of vector. These results are encouraging, but the prospect of prolonged treatment with cyclosporin A has dampened enthusiasm for this approach.

More appealing is a strategy reported by Kay et al. (1995) that relies on co-administration of a single dose of an antibody that blocks a necessary co-stimulatory signal. T cell–mediated immune response requires that cells present peptide fragments of the foreign protein in association with major histocompatibility molecules to the T-cell receptor. Additional co-stimulatory signals, in which the B7-1 and B7-2 ligands on antigen-presenting cells bind to CTLA4 receptors on T cells, are required for T-cell activation (Fig. 4). Administration of CTLA4Ig, a chimeric molecule that blocks the murine CTLA4 receptor, to mice injected with an adenoviral vector expressing hAAT, results in prolongation of expression to ~5 months vs. 1–2 months in controls. Since only a single dose of CTLA4Ig was required, there was no long-term immunosuppression, and treated mice demonstrated the ability to mount an effective T cell–mediated response to a bacteriophage infused intravenously 10 weeks after receiving CTLA4Ig. It will be of considerable interest to determine whether this result can be replicated in hemophilic dogs, but this will require the availability of canine CTLA4Ig.

Another approach based on manipulation of the host immune response is the induction of tolerance to adenoviral proteins. The engineering of transgenic animals demonstrates that animals can acquire tolerance to "foreign" proteins provided they are exposed to them early in development, but attempts

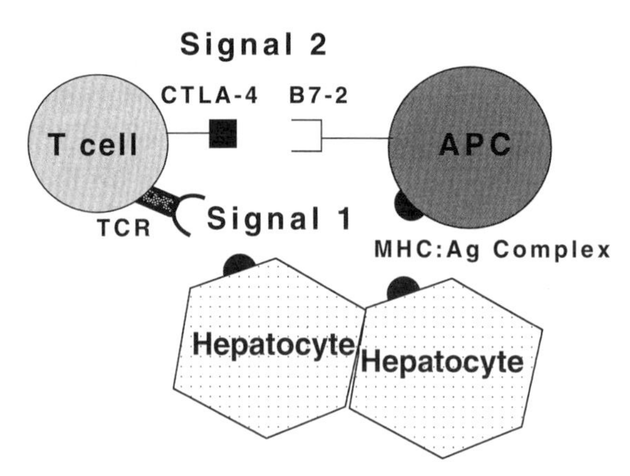

FIGURE 4. T-cell activation. T-cell activation requires two signals. Signal 1 is the presentation of a peptide fragment from the foreign protein, displayed in the context of an MHC molecule, to the T-cell receptor (TCR). Signal 2 is a co-stimulatory signal from an antigen-presenting cell (APC). See text for further details.

to induce tolerance by injecting adenoviral vector *in utero* have been mostly disappointing, probably because exposure to vector did not occur early enough in development (McCray et al., 1995). Of interest is the fact that injection of adenoviral vector in the neonatal period results in a pattern different from that seen in adult animals, a result confirmed by two groups. Stratford-Perricaudet and colleagues (1992) demonstrated long-term expression (up to 12 months) of an adenoviral vector expressing β-galactosidase in a small percentage of mice that were injected on day 2 of life. Walter et al. (1996), on the other hand, detected no difference in duration of expression of an adenoviral vector expressing human F.IX between adult CD-1 mice and newborn CD-1 mice injected on day 1 of life, but the mice injected as newborns could be successfully re-injected at least once, whereas mice initially injected as adults could not.

The effort to engineer less immunogenic adenoviral vectors began with the use of a temperature-sensitive adenoviral mutant that had initially been described in the 1970s (Ginsberg et al., 1974). The H5ts125 mutant contains a missense mutation in the E2a gene encoding the DNA-binding protein that renders the function of the protein temperature dependent. The E2a gene product is required for expression of fiber protein, an immunogenic structural protein of the vector. Recombinant vector can be prepared at the permissive temperature, 32°C, a temperature at which E2a function is normal in the mutant, but at the restrictive temperature of 39°C E2a is nonfunctional, and fiber protein is not produced (Engelhardt et al., 1994). Unfortunately, at 37°C, E2a function is reduced but not absent, and efforts to achieve longer duration of expression of F.IX in experimental animals with this vector have been unsuccessful (Fang et al., 1995b).

The logical culmination of this strategy has been realized in the attempts to produce so-called gutless adenoviral vectors, i.e., adenoviral vectors that contain only the minimal *cis* elements required for replication and packaging but are devoid of any elements encoding viral proteins (Fisher et al., 1996; Hardy et al., 1997; Kochanek et al., 1996). The bacterial cre-lox recombination system has been successfully used to eliminate replication and packaging of the helper virus necessary in the production of these gutless vectors (Hardy et al., 1997). An advantage of these vectors is that they allow incorporation of large expression cassettes. In fact, the vector genome is destabilized if it is engineered to be much smaller than wild-type adenovirus (Fisher et al., 1996; Hardy et al., 1997). Extensive *in vivo* testing of gutless vectors has not been documented yet, but studies involving large deletions of the adenoviral genome showed that such vectors might be very unstable in transduced cells (Lieber et al., 1996).

ADENO-ASSOCIATED VIRAL VECTORS

AAV vectors are engineered from human AAV, a nonpathogenic, replication-defective parvovirus. Wild-type AAV has a single-stranded DNA genome of

4,680 nucleotides; the virus has both lytic and latent phases in the life cycle. The lytic phase requires a helper virus, e.g., adenovirus or herpes virus; in the absence of helper virus, wild-type AAV integrates into the host cell genome. AAV integration in humans manifests site-specificity, with a high proportion of integration events occurring at a specific site on chromosome 19. To generate recombinant vector from the virus, the viral coding sequences are replaced by the gene of interest in a plasmid containing only the viral inverted terminal repeats (ITRs), and the necessary viral genes are supplied in *trans* on a second plasmid. Following introduction of these plasmids into 293 cells, the cells are infected with helper adenovirus and lysed 48 h later, and recombinant AAV is purified from cell lysate (Samulski, 1993; Samulski et al., 1989).

The difficulties involved in generating high titer preparations of purified recombinant AAV have limited the accumulation of *in vivo* data with this vector. Nonetheless, the vector has a number of advantages that have motivated continued interest despite this obstacle. First, it has the ability to transduce nondividing target cells, including brain, muscle, and liver (Blau and Khavari, 1997; Fisher et al., 1997; Kaplitt et al., 1994; Kessler et al., 1996; Snyder et al., 1997; Xiao et al., 1996). Second, the structure of the recombinant vector is such that no viral coding sequences are transferred to the recipient. In addition, the parent virus is nonpathogenic in humans, and the vector does not appear to be highly immunogenic, as is the case for adenoviral vectors. The major disadvantages are that there is a size constraint of ~4.7 kb on the cassette that can be packaged into the vector, and the vector is, as noted above, difficult to prepare.

Published *in vivo* data about using an AAV vector to express F.IX in experimental animals are limited, but appear to be very promising. Koeberl et al. (1997) have shown in mice that intravenous injection of AAV F.IX results in transduction of hepatocytes and that transduction efficiency can be improved by prior γ-irradiation. A subsequent report demonstrates therapeutic plasma levels of F.IX in C57BL/6 mice using a similar approach (portal vein injection into mice *without* γ-irradiation). Differences in the design of the F.IX expression cassette, and possibly differences in the route of administration, may have accounted for the much higher plasma levels obtained (Snyder et al., 1997). Herzog et al. (1997), using muscle as the target cell, have demonstrated stable expression of therapeutic levels of F.IX in the plasma of mice injected intramuscularly with an AAV F.IX vector (Fig. 5). The data in these latter two reports are perhaps the most promising yet generated; if the results can be replicated in larger animal models (i.e., hemophilic dogs), then this treatment strategy would seem to be feasible for human trials.

FACTOR VIII GENE TRANSFER

F.VIII deficiency affects about five to six times as many individuals as F.IX deficiency. Despite its higher prevalence, progress in gene transfer has been

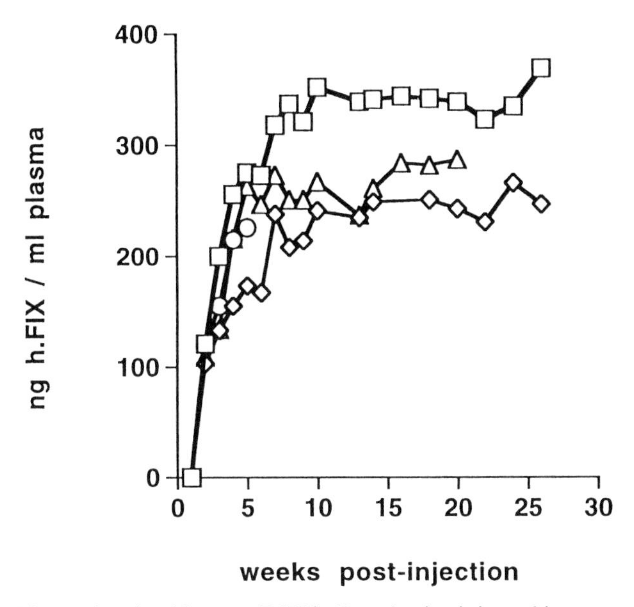

FIGURE 5. Plasma levels of human F.IX in Rag-1 mice injected intramuscularly with 2×10^{11} vector genomes of an AAV vector expressing human F.IX. Each line represents an individual animal.

slower because of problems peculiar to F.VIII expression and secretion. The F.VIII cDNA is too large to be accommodated by most viral vectors, but this problem has been circumvented by using a truncated form (B-domain deleted) that has full biological activity. Other problems have been more difficult to overcome. First, the stability of F.VIII in plasma is dependent on interaction with von Willebrand factor (vWF); in addition, the F.VIII protein is large (MW ~250,000) and does not easily traverse the extracellular spaces to reach the circulation. These factors limit the choice of sites of synthesis. Moreover, sequences present in the coding region of F.VIII appear to act as dominant inhibitors of RNA accumulation; this results not only in lower levels of expression but in lower titers of retroviral vector as well. Lynch et al. (1993) compared titers of retrovirus generated using a series of different inserts and demonstrated that the titer of F.VIII retroviral vector was consistently two logs lower than titers generated using other inserts. This did not appear to result solely from the size of the F.VIII insert, since a vector expressing a transgene of similar size (the insulin-like growth factor I receptor) could be prepared in high titer (2×10^6 G418^r cfu/ml vs. 1×10^4 G418^r cfu/ml for F.VIII). These investigators identified a 1.2 kb fragment of the F.VIII cDNA that encodes portions of both the A2 and A3 domains (Fig. 3) that appears to be responsible for the reduction in RNA accumulation. Deletion of the fragment results in a 10-fold rise in titer (and of course in a functionally inactive F.VIII species), and inclusion of the fragment in other unrelated inserts results in a drop in titer. Attempts to define the inhibitory sequence more precisely using 5′, 3′, and internal deletions were unsuccessful (the inhibitory effect was lost with *any* truncation of the fragment). On the assumption that this effect was sequence-specific, Chuah et al. (1995) attempted to overcome this inhibitory effect by changing third codon bases (silent at the amino acid level) within the 1.2 kb fragment, but this appeared to have no effect on retroviral titers or levels of expression.

Hoeben et al. (1995) also reported the presence of sequences within the F.VIII gene that inhibit RNA accumulation; these sequences overlap but do not precisely coincide with the sequences identified by Lynch et al. (1993). There is no clear consensus on the mechanism of inhibition; based on nuclear run-on data and steady-state levels of RNA in transfected CHO cells, Kaufman et al. (1989) have suggested that the effect is post-transcriptional, whereas Hoeben et al. (1995) using similar approaches in retroviral-transduced RAT2 cells, showed repression at the level of transcription.

There have been a few reports in the literature of successful *in vitro* and *in vivo* expression of human F.VIII using a retroviral vector. Dwarki and colleagues (1995) at Somatix Corporation used the MFG vector, originally described by Mulligan and coworkers (see Dranoff et al., 1993; Ohashi et al., 1992; Sekhsaria et al., 1993), to achieve levels of 500–2,000 ng/ml in virally transduced cells in culture. (Titers of this vector were not reported; the absence of a selectable marker in the vector makes determination of titers difficult.) The authors analyzed expression *in vivo* by transducing cell lines (C2C12 and NIH 3T3) with the vector and implanting transduced cells into SCID mice either intramuscularly or via intraperitoneal injection. These experiments showed that F.VIII could reach the circulation from cells implanted in the intraperitoneal site but not from cells in the intramuscular site. (Parallel experiments with a vector expressing F.IX showed that F.IX can reach the circulation from either site.) The failure of F.VIII to reach the circulation from intramuscular sites may reflect susceptibility of the protein to proteases in the extracellular space or the requirement for rapid association with vWF. Intraperitoneal implantation of transduced primary fibroblasts into SCID mice resulted in expression of human F.VIII at therapeutic levels for ~7 days, after which expression declined to undetectable levels. Explants of the transplanted cells showed an overall loss of cell number, the presence of vector DNA in the remaining cells, and the absence (by reverse transcriptase PCR) of F.VIII mRNA. These results may reflect either extinction of viral LTR-driven transcription, as has been previously reported (Palmer et al., 1991; Scharfmann et al., 1991), or limited survival of the transduced cells, or both.

Problems with the development of high titer F.VIII-expressing retroviral vectors have fueled efforts to explore other gene delivery vehicles. Connelly et al. (1995) have produced adenoviral vectors that direct high level expression of human F.VIII. Intravenous administration of adenovirus F.VIII vector in C57/Bl6 mice resulted initially in expression in the therapeutic range (50 ng/

ml), but F.VIII levels declined steadily and were no longer detectable after 7 weeks. Consistent with results reported by other groups using adenoviral vectors, the gradual loss of expression was accompanied by loss of vector DNA from the liver, presumably due to host immune response to adenoviral antigens (*vide supra*). In the case of F.VIII, several groups have pursued a variety of strategies to generate a more potent F.VIII expression cassette, which would allow administration of lower doses of viral vector (which would presumably result in a less vigorous immune response). Connelly et al. (1996a) compared an adenoviral vector containing a mouse albumin promoter to one that included both the albumin promoter and the 5′-untranslated region from the human apolipoprotein A1 gene (first untranslated exon, first intron, and second exon up to translation initiation site) and showed that inclusion of the genomic elements enhanced expression in a mouse model. At a dose of 4×10^9 pfu, the second vector resulted in two- to fivefold enhanced expression (levels of 600–1,000 ng/ml vs. 200–300 ng/ml); at a lower dose of 5×10^8 pfu, the effect was even more marked, with the second vector achieving levels of expression of 200–400 ng/ml, while F.VIII expression was undetectable with the vector containing the albumin promoter alone. In subsequent experiments these investigators compared duration of expression of the second vector as a function of dose and showed that expression could be achieved for a longer period (F.VIII levels of 40 ng/ml still detectable 22 weeks following injection) at the lower dose of $\sim 10^8$ pfu than at a higher dose of $\sim 10^9$ pfu (F.VIII no longer detectable 15 weeks after injection). Pipe and Kaufman (1996) have engineered a F.VIII variant with six- to eightfold higher specific activity than wild-type F.VIII; this was achieved by altering sites at which F.VIIIa is normally cleaved by activated protein C and by altering a thrombin cleavage site so that the A2 domain is no longer dissociable from A3 (a mechanism by which F.VIIIa is inactivated). The availability of a higher specific activity variant should allow administration of even lower doses of adenoviral vector; such a maneuver may prolong the duration of expression but is unlikely to result in stable long-term expression unless other strategies (e.g., the use of a re-engineered vector with more viral sequences removed, transient immunosuppression when vector is administered) are employed as well.

NONVIRAL APPROACHES

A variety of nonviral approaches are currently under investigation to develop a gene therapy for hemophilia. Therapeutic levels of human F.IX have been achieved in mouse plasma after implantation of stably transfected myoblasts in skeletal mouse muscle. The myoblasts were expanded clones of cells stably transfected in *ex vivo* experiments with a human F.IX-expressing plasmid (Wang et al., 1997). In another study, naked plasmid DNA incorporated into liposomes was intravenously injected into mice, but expression levels were very low (<1 ng/ml plasma), and it is unclear whether persistent expression can be achieved with this approach (Baru et al., 1995). Liposomes are considered attractive gene delivery vehicles because they are neither immunogenic nor toxic. Synthetic DNA–ligand complexes for receptor-mediated gene transfer have been described as a promising basis for gene therapy. Complexes of galactosidated poly(L-lysine) and plasmid DNA including a human F.IX expression cassette were manufactured by titration with NaCl and subsequently injected into the caudal vena cava of adult rats (Perales et al., 1994). Receptor-mediated uptake of these complexes by the asialoglycoprotein receptor in hepatocytes ensured liver-specific targeting of gene delivery. Expression of human F.IX was stable for at least 140 days, although with significant variation of the expression levels among different animals (15 ng to 1 μg/ml plasma). The stability of the introduced plasmid DNA (which is present in episomal form after cellular uptake) remains an issue for further investigation. In addition, this vehicle is very difficult to prepare, a factor that has limited attempts to extend and replicate these results (Hanson, personal communication).

ACKNOWLEDGMENT

This work was supported by grants R01 HL53668 and P50 HL54500.

REFERENCES

Aggeler PM, White SG, Glendenning MB, Page EW, Leake TB, Bates G (1952): Plasma thromboplastin component (PTC) deficiency: A new disease resembling hemophilia. *Proc Soc Exp Biol Med* 79:692–694.

Baru M, Axelrod JH, Nur I (1995): Liposome-encapsulated DNA-mediated gene transfer and synthesis of human factor IX in mice. *Gene* 161:143–150.

Bi L, Lawler AM, Antonarakis SE, High KA, Gearhart JD, Kazazian HH (1995): Targeted disruption of the mouse factor VIII gene: A model for hemophilia A. *Nat Genet* 10:119–121.

Biggs R, Douglas AS, Macfarlane RG, et al. (1952): Christmas disease: A condition previously mistaken for hemophilia. *BMJ* 2:1378–1382.

Blau H, Khavari P (1997): Gene therapy: Progress, problems, prospects. *Nat Med* 3:612–613.

Bosch A, McCray PB, Chang SMw, Ulich TR, Simnet WS, Jolly DJ, Davidson BL (1996): Proliferation induced by keratinocyte growth factor enhances *in vivo* retroviral-mediated gene transfer to mouse hepatocytes. *J Clin Invest* 98:2683–2687.

Bowles NE, Eisensmith RC, Mohuiddin R, Pyron M, Woo SL (1996): A simple and efficient method for the concentration and purification of recombinant retrovirus for increased hepatocyte transduction in vivo. *Hum Gene Ther* 7:1735–1742.

Bowling WM, Kennedy SC, Cai SR, Duncan JR, Gao C, Flye MW, Ponder KP (1996): Portal branch occlusion safely facilitates *in vivo* retroviral vector transduction of rat liver. *Hum Gene Ther* 7:2113–2121.

Brandstetter H, Bauer M, Huber R, Lollar P, Bode W (1995): X-ray structure of clotting factor IXa: Active site and module structure related to Xase activity and hemophilia B. *Proc Natl Acad Sci USA* 92:9796–9800.

Brenner MK, Rill DR, Moen RC, et al. (1993): Gene-marking to trace origin of relapse after autologous bone-marrow transplantation. *Lancet* 341:85–86.

Cameron C, Notley C, Hoyle S, McGlynn L, Hough C, Kamisue S, Giles A, Lillicrap D (1998): The canine factor VIII cDNA and 5′ flanking sequence. *Thromb Haemost* 79:317–322.

Chuah MK, Vandendriessche T, Morgan RA (1995): Development and analysis of retroviral vectors expressing human factor VIII as a potential gene therapy for hemophilia A. *Hum Gene Ther* 6:1363–1377.

Connelly S, Gardner JM, McClelland, Kaleko M (1996a): High level tissue-specific expression of functional human FVIII in mice. *Hum Gene Ther* 7:183.

Connelly S, Mount J, Mauser A, Gardner JM, Kaleko M, McClelland A, Lothrop CD (1996b): Complete short-term correction of canine hemophilia A by *in vivo* gene therapy. *Blood* 88:3846–3853.

Connelly S, Smith TAG, Dhir G, Gardner JM, Mehaffey MG, Zaret KS, McClelland A, Kaleko M (1995): *In vivo* gene delivery and expression of physiological levels of functional human factor VIII in mice. *Hum Gene Ther* 6:185.

Dai Y, Roman M, Naviaux RK, Verma IM (1992): Gene therapy via primary myoblasts: Long-term expression of Factor IX protein following transplantation *in vivo. Proc Natl Acad Sci USA* 89:10892–10895.

Dai Y, Schwarz EM, Gu D, Zhang WW, Sarvetnick N, Verma IM (1995): Cellular and humoral immune responses to adenoviral vectors containing Factor IX gene: Tolerization of Factor IX and vector antigens allows for long-term expression. *Proc Natl Acad Sci USA* 92:1401–1405.

Deisseroth AB, Zu Z, Claxton D, et al. (1994): Genetic marking shows that Ph^+ cells present in autologous transplants of chronic myelogenous leukemia (CML) contribute to relapse after autologous bone marrow in CML. *Blood* 83:3068–3076.

Dranoff G, Jaffee E, Lazenby A, Golumbek P, Levitsky H, Brose K, Jackson V, Hamada H, Pardoll D, Mulligan RC (1993): Vaccination with irradiated tumor cells engineered to secrete murine granulocyte-macrophase-colony-stimulating factor stimulates potent, specific and long-lasting anti-tumor immunity. *Proc Natl Acad Sci USA* 90:3539–3543.

Dunbar CE, Cottler-Fox M, O'Shaughnessy JA, et al. (1995): Retrovirally marked CD34-enriched peripheral blood and bone marow cells contribute to long-term engraftment after autologous transplantation. *Blood* 85:3048–3057.

Dwarki VJ, Belloni P, Niijar T, Smith J, Couto L, Rabier M, Clift S, Berns A, Cohen LK (1995): Gene therapy for hemophilia A: Production of therapeutic levels of human Factor VIII *in vivo* in mice. *Proc Natl Acad Sci USA* 92:1023.

Engelhardt JF, Ye X, Doranz B, Wilson JM (1994): Ablation of E2A in recombinant adenoviruses improves transgene persistence and decreases inflammatory response in mouse liver. *Proc Natl Acad Sci USA* 91:6196–6200.

Evans JP, Brinkhous KM, Brayer GD, Reisner HW, High KA (1989a): Canine hemophilia B resulting from a point mutation with unusual consequences. *Proc Natl Acad Sci USA* 86:10095–10099.

Evans JP, Watzke HH, Ware JL, Stafford DW, High KA (1989b): Molecular cloning of cDNA encoding canine Factor IX. *Blood* 74:207–212.

Fang B, Eisensmith RC, Wang H, Kay Ma, Cross RE, Landen CN, Gordon G, Bellinger DA, Read MS, Hu PC, et al. (1995a): Gene therapy for hemophilia B: Host immunosuppression prolongs the therapeutic effect of adenovirus-mediated factor IX expression. *Hum Gen Ther* 6:1039–1044.

Fang B, Wang H, Gordon G, Bellinger DA, Read MS, Brinkhous KM, Woo SLC, Eisensmith RC (1995b): Lack of persistence of E1-recombinant adenoviral vectors containing a temperature-sensitive E2A mutation in immunocompetent mice and dogs. *Gene Ther* 3:217–222.

Ferry N, Dupleissis O, Houssin D, Danos O, Heard J-M (1991): Retroviral-mediated gene transfer into hepatocytes *in vivo. Proc Natl Acad Sci USA* 88:8377–8381.

Fisher KJ, Choi H, Burda J, Chen S, Wilson JM (1996): Recombinant adenoviruses deleted of all viral genes for gene therapy of cystic fibrosis. *Virology* 217:11–22.

Fisher KJ, Jooss K, Alston J, Yang Y, Haecker SE, High K, Pathak R, Raper SE, Wilson JM (1997): Recombinant adeno-associated virus for muscle directed gene therapy. *Nat Med* 3:306–312.

Giannelli F, Green PM, Sommer SS, Poon MC, Ludwig M, Schwaab R, Reitsma PH, Goossens M, Yoshioka A, Brownlee GG (1996): Haemophilia B (sixth edition): A database of point mutations and short additions and deletions. *Nucleic Acids Res* 24:1:103–118.

Ginsberg HS, Ensinger MJ, Kauffman RS, Mayer AJ, Lundholm U (1974): Cell transformation: A study of regulation with types 5 and 12 adenovirus temperature-sensitive mutants. Cold Spring Harbor Symp Quant Biol, Vol. 39, Pt. 1:419–426.

Gitschier J, Wood WI, Goralka TM, Wion KL, Chen EV, Eaton DH, Vehar GA, Capon DJ, Lawn RM (1984): Characterization of the human Factor VIII gene. *Nature* 312:326–330.

Green PM, Montandon AJ, Bentley DR, Giannelli F (1991): Genetics and molecular biology of haemophilias A and B. *Blood Coag Fibrinol* 2:539.

Grossman M, Rader DJ, Muller DW, Kolansky DM, Kozarsky K, Clark BJ, Stein EA, Lupien PJ, Brewer HB, Raper SE, et al. (1995): A pilot study of *ex vivo* gene therapy for homozygous familial hypercholesterolaemia. *Nat Med* 1:1148–1154.

Hamaguchi N, Roberts HR, Stafford DW (1993): Mutations in the catalytic domain of Factor IX which are related to the subclass hemophilia B. *Biochemistry* 32:6324–6329.

Hao Q-L, Malik P, Salazar R, Tang H, Gordon EM, Kohn DB (1995): Expression of biologically active human Factor IX in human hematopoietic cells after retroviral vector-mediated gene transduction. *Hum Gen Ther* 6:873–880.

Hardy S, Kitamura M, Harris-Stansil T, Dai Y, Phipps L (1997): Construction of adenovirus vectors through cre-lox recombination. *J Virol* 71:1842–1849.

Herzog RW, Hagstrom JN, Kung Z-H, Tai SJ, Wilson JM, Fisher KJ, High KA (1997): Stable gene transfer and expression of human blood coagulation Factor IX after intramuscular injection of recombinant adeno-associated virus. *Proc Natl Acad Sci USA* 94:5804–5809.

High KA (1995): Factor IX: Molecular structure, epitopes, and mutations associated with inhibitor formation. *Adv Exp Med Biol* 386:79–86.

Hoeben RC, Fallaux FJ, Cramer SJ, van den Wollenberg DJM, van Ormondt H, Briöt E, van der Eb AJ (1995): Expression of the blood-clotting Factor-VIII cDNA is repressed by a transcriptional silencer located in its coding region. *Blood* 85:2447–2454.

Kaplitt MG, Leone R, Samulski RJ, Xiao X, Pfaff DW, O'Malley KL, During MJ (1994): Long-term expression and phenotypic correction using adeno-associated virus vectors in the mammalian brain. *Nat Genet* 8:148–154.

Kaufman RJ, Wasley LC, Davies MV, Wise RJ, Israel DI, Dorner AJ (1989): Effect of von Willebrand factor coexpression on the synthesis and secretion of Factor VIII in Chinese hamster ovary cells. *Mol Cell Biol* 9:1233.

Kay MA, Holterman AX, Meuse L, Gown A, Ochs HD, Linsley PS, Wilson CB (1995): Long-term hepatic adenovirus-mediated gene expression in mice following CTLA4Ig administration. *Nat Genet* 11:191–197.

Kay MA, Landen CN, Rothenberg SR, Taylor LA, Leland F, Wiehle S, Fang B, Bellinger D, Finegold M, Thompson AR, et al. (1994): *In vivo* hepatic gene therapy: Complete albeit transient correction of Factor IX deficiency in hemophilia B dogs. *Proc Natl Acad Sci USA* 91:2353–2357.

Kay MA, Rothenberg S, Landen CN, Bellinger DA, Leland F, Toman C, Finegold M, Thompson AR, Read MS, Brinkhous KM, et al. (1993): *In vivo* gene therapy of hemophilia B: Sustained partial correction in Factor IX–deficient dogs. *Science* 262:117–119.

Kessler PD, Podsakoff GM, Chen X, McQuiston SA, Colosi PC, Matelis LA, Kurtzman GJ, Bryne BJ (1996): Gene delivery to skeletal muscle results in sustained expression and systemic delivery of a therapeutic protein. *Proc Natl Acad Sci USA* 93:14082–14087.

Kochanek S, Clemens PR, Mitani K, Chen H-H, Chan S, Caskey CT (1996): A new adenoviral vector: Replacement of all viral coding sequences with 28 kb of DNA independently expressing both full-length dystrophin and beta-galactosidase. *Proc Natl Acad Sci USA* 93:5331–5736.

Koeberl DD, Alexander IE, Halbert CL, Russell DW, Miller AD (1997): Persistent expression of human clotting factor IX from mouse liver after intravenous injection of adeno-associated virus vectors. *Proc Natl Acad Sci USA* 94:1426–1431.

Kohn DB (1995): The current status of gene therapy using hematopoietic stem cells. *Curr Opin Pediatr* 7:56–63.

Kolodka TM, Finegold M, Kay Ma, Woo SLC (1993): Hepatic gene therapy: Efficient retroviral-mediated gene transfer into rat hepatocytes *in vivo. Somat Cell Mol Genet* 19:491–497.

Kurachi K, Davie EW (1982): Isolation and characterization of a cDNA coding for human factor IX. *Proc Natl Acad Sci USA* 79:6461–6464.

Lakich D, Kazazian HH Jr, Antonarkis SE, Gitschier J (1993): Inversions of disrupting the Factor VIII gene are a common cause of severe haemophilia A. *Nat Genet* 5:236–241.

Le M, Okuyama T, Cai SR, Kennedy SC, Bowling WM, Flye MW, Ponder KP (1997): Therapeutic levels of functional human Factor X in rats after retroviral-mediated hepatic gene therapy. *Blood* 89:1254–1259.

Lieber A, He C-Y, Kirivolla I, Kay MA (1996): Recombinant adenoviruses with large deletions generated by cre-mediated excision exhibit different biological properties compared with first-generation vectors *in vitro* and *in vivo*. *J Virol* 70:8944–8960.

Lieber A, Vrancken Peeters M-JTFD, Fausto N, Perkins J, Kay MA (1995): Adenovirus-mediated urokinase gene transfer induces liver regeneration and allows for efficient retrovirus transduction of hepatocytes *in vivo. Proc Natl Acad Sci USA* 92:6210–6214.

Lin HF, Maeda N, Smithies O, Straight DL, Stafford DW (1997): Coagulation Factor IX–deficient mice developed by gene targeting. *Blood* 90:3962–3966.

Lusher JM (1991): Thrombogenicity associated with Factor IX complex concentrates. *Semin Hematol* 28:3–5.

Lynch CM, Israel DI, Kaufman RJ, Miller AD (1993): Sequences in the coding region of clotting Factor VIII act as dominant inhibitors of RNA accumulation and protein production. *Hum Gene Ther* 4:259.

Mannucci PM (1993): Modern treatment of hemophilia: From the shadows towards the light. *Thromb Haemost* 70:17–23.

McCray PB Jr., Armstrong K, Zabner J, Miller DW, Koretzky GA, Couture L, Robillard JE, Smith AE, Welsh MJ (1995): Adenoviral-mediated gene transfer to fetal pulmonary epithelia *in vitro* and *in vivo. J Clin Invest* 95:2620–2632.

Moscioni AD, Rozga J, Neuzil DF, Overell RW, Holt JT, Demetriou Aa (1993): *In vivo* regional delivery of retrovirally mediated foreign genes to rat liver cells: Need for partial hepatectomy for successful foreign gene expression. *Surgery* 113:304–311.

Ohashi T, Boggs S, Robbins P, Bahnson A, Patrene K, Wei FS, Wei JF, Li J, Lucht L, Fei Y, et al. (1992): Efficient transfer and sustained high expression of the human glucocerebrosidase gene in mice and their functional macrophages following transplantation of bone marrow transduced by a retroviral vector. *Proc Natl Acad Sci USA* 89:11332–11336.

Okuyama T, Huber RM, Bowling W, Pearline R, Kennedy SC, Flye MW, Ponder KP (1996): Liver-directed gene therapy: A retroviral vector with a complete LTR and the ApoE enhancer-alpha-1-antitrypsin promoter dramatically increases expression of human alpha 1-antitrypsin *in vivo. Hum Gene Ther* 7:637–645.

Palmer TD, Rosman GJ, Osborne WRA, Miller AD (1991): Genetically modified skin fibroblasts persist long after transplantation but gradually inactivate introduced genes. *Proc Natl Acad Sci USA* 88:1330–1334.

Palmer TD, Thompson AR, Miller AD (1989): Production of human Factor IX in animals by genetically modified skin fibroblasts: Potential therapy for hemophilia B. *Blood* 73:438–445.

Perales JC, Ferkol T, Molas M, Hanson RW (1994): An evaluation of receptor-mediated gene transfer using synthetic DNA–ligand complexes. *Eur J Biochem* 226:255–266.

Pipe SW, Kaufman RJ (1996): Construction and characterization of inactivation resistant Factor VIII (R3361/R562K/R740A/DEL795-1688). *Blood* 88(suppl 1):441a.

Pittman D, Millenson M, Marquette K, Bauer K, Kaufman R (1992): A2 domain of human recombinant-derived FVIII is required for procoagulant activity but not for thrombin cleavage. *Blood* 79:389–397.

Ravid K, Beeler D, Rabin M, Ruley HE, Rosenberg RD (1991): Selective targeting of gene products with the megakaryocyte platelet factor 4 promoter. *Proc Natl Acad Sci USA* 88:1521–1525.

Rosendaal FR, Smit C, Briët E (1991): Hemophilia treatment in historical perspective: A review of medical and social developments. *Ann Hematol* 62:5.

Samulski RJ (1993): Adeno-associated virus: Integration at a specific chromosomal locus. *Curr Opin Genet Dev* 3:74–80.

Samulski RJ, Chang L-S, Shenk T (1989): Helper-free stocks of recombinant adeno-associated viruses: Normal integration does not require viral gene expression. *J Virol* 63:3822–3828.

Scharfmann R, Axelrod JH, Verma IM (1991): Long-term *in vivo* expression of retrovirus-mediated gene transfer in mouse fibroblast implants. *Proc Natl Acad Sci USA* 88:4626–4630.

Sekhsaria S, Gallin JJ, Linton GF, Mallory RM, Mulligan RC, Malech HL (1993): Peripheral blood progenitors as a target for genetic correction of p47phox-deficient chronic granulomatous disease. *Proc Natl Acad Sci USA* 90:7446–7450.

Smith TAG, Mehaffey MG, Kayda DB, Saunders JM, Yei S, Trapnell BC, McClelland A, Kaleko M (1993): Adenovirus mediated expression of therapeutic plasma levels of human Factor IX in mice. *Nat Genet* 5:397–402.

Snyder RO, Miao CH, Patijn GA, Spratt SK, Danos O, Nagy D, Gown AM, Winther B, Meuse L, Cohen LK, et al. (1997): Persistent and therapeutic concentrations of human Factor IX in mice after hepatic gene transfer of recombinant AAV vectors. *Nat Genet* 16:270–276.

Stratford-Perricaudet LD, Makeh I, Perricaudet M, Briand P (1992): Widespread long-term gene transfer to mouse skeletal muscles and heart. *J Clin Invest* 90:626–630.

Toole J, Pittman D, Orr E, Murtha P, Walsey L, Kaufman R (1986): A large region (=95 kDa) of human Factor VIII is dispensable for *in vitro* procoagulant activity. *Proc Natl Acad Sci USA* 83:5939–5942.

Toole JJ, Knopf JL, Wozney JM, et al. (1984): Molecular cloning of a cDNA encoding human antihaemophiliac factor. *Nature* 312:342–347.

Tuddenham EG, Schwaab R, Seehafer J, Millar DS, Gitschier J, Higuchi M, Bidichandani S, Connor JM, Hoyer LW, Yoshioka A, et al. (1994): Haemophilia A: Database of nucleotide substitutions, deletions, insertions and rearrangements of the Factor VIII gene, second edition [corrected and republished article originally printed in *Nucleic Acids Res* 1994 22(17):3511–3533]. *Nucleic Acids Res* 22:4851–4868.

Van Dieijen G, Trans G, Rosing J, Hemker HC (1981): The role of phospholipid and Factor VIIIa in the activation of bovine Factor X. *J Biol Chem* 256:3433–3442.

Walter J, You Q, Hagstrom N, Sands M, High KA (1996): Successful expression of human Factor IX following repeat administration of an adenoviral vector in mice. *Proc Natl Acad Sci USA* 93:3056–3061.

Wang J-M, Zheng H, Blaivas M, Kurachi K (1997): Persistent systemic production of human Factor IX in mice by skeletal myoblast-mediated gene transfer: Feasibility of repeat application to obtain therapeutic levels. *Blood* 90:1075–1082.

Wang L, Zoppe M, Hackeng TM, Griffin JH, Lee K-F, Verma IM (1997): A factor IX-deficient mouse model for haemophilia B gene therapy. *PNAS* 94:11563–11566.

White G, Pasi KJ, Lusher J, Brackman HH, Magill M, Courter S (1997): Recombinant factor IX in the treatment of previously-treated patients with hemophilia B. *Thromb Haemost* 77:52.

White GC, McMillan CW, Kingdon HS, Shoemaker CB (1989): Use of recombinant antihemophilia factor in the treatment of two patients with classic hemophilia. *N Engl J Med* 321:329–330.

Williams DA (1990): Expression of introduced genetic sequences in hematopoietic cells following retroviral-mediated gene transfer. *Hum Gene Ther* 1:229–239.

Xiao X, Li J, Samulski RJ (1996): Efficient long-term gene transfer into muscle tissue of immunocompetent mice by adeno-associated virus vector. *J Virol* 70:8098–8108.

Yang Y, Ertl HC, Wilson JM (1994a): MCH class I–restricted cytotoxic T lymphocytes to viral antigens create barriers to lung-directed gene therapy with recombinant adenoviruses. *Immunity* 1:433–442.

Yang Y, Nunes FA, Berencsi K, Furth EE, Gonczol E, Wilson JM (1994b): Cellular immunity to viral antigens limits E1-deleted adenoviruses for gene therapy. *Proc Natl Acad Sci USA* 91:4407–4411.

Yang Y, Wilson JM (1995): Clearance of adenovirus-infected hepatocytes by MHC class I–restricted CD4 CTLs *in vivo. J Immunol* 155:2564–2570.

Yao S-N, Wilson JM, Nabel EG, Kurachi S, Hachiya HL, Kurachi K (1991): Expression of human Factor IX in rat capillary endothelial cells: Toward somatic gene therapy for hemophilia B. *Proc Natl Acad Sci USA* 88:8101–8105.

Yao SN, Kurachi K (1992): Expression of human Factor IX in mice after injection of genetically modified myoblasts. *Proc Natl Acad Sci USA* 89:3357–3361.

Yao SN, Smith KJ, Kurachi K (1994): Primary myoblast-mediated gene transfer; persistent expression of human Factor IX in mice. *Gene Ther* 1:99–107.

Yoshitake S, Schach BG, Foster DC, Davie EW, Kurachi K (1985): Nucleotide sequence of the gene for human Factor IX (antihemophilic factor B). *Biochemistry* 24:3736–3750.

16

CLINICAL APPLICATIONS OF GENE THERAPY: ANEMIAS

GEORGE F. ATWEH AND BERNARD G. FORGET

Division of Hematology, Mount Sinai School of Medicine, New York, NY 10029 (G.F.A.); Hematology Section, Yale University School of Medicine, New Haven, CT 06510 (B.G.F.)

GENE THERAPY OF GLOBIN DISORDERS

Disorders of the β-globin gene have long been considered excellent models for human gene therapy. The cloning of all human globin genes almost two decades ago and the intensive studies of their regulation that followed provided the foundations on which genetic approaches to therapy could be based. Thus, not surprisingly, the β-globin gene was one of the first human genes to be used in pioneering gene transfer experiments (Cone et al., 1987). However, it soon became apparent that the application of this evolving gene transfer technology to globin genes posed some unique and formidable challenges. In this chapter, we discuss those challenges, the approaches that are being developed to overcome them, and the remaining hurdles to the successful application of corrective genetic therapy in the treatment of patients afflicted by these disorders.

Conceptually, current approaches to globin gene therapy consist of the introduction of a functional globin gene into hematopoietic stem cells of a homozygous affected individual as a substitute for the defective globin gene(s). One such approach, as illustrated in Figure 1, would involve the transfer of

Stem Cell Biology and Gene Therapy, Edited by Peter J. Quesenberry, Gary S. Stein, Bernard G. Forget, and Sherman M. Weissman
ISBN 0-471-14656-0

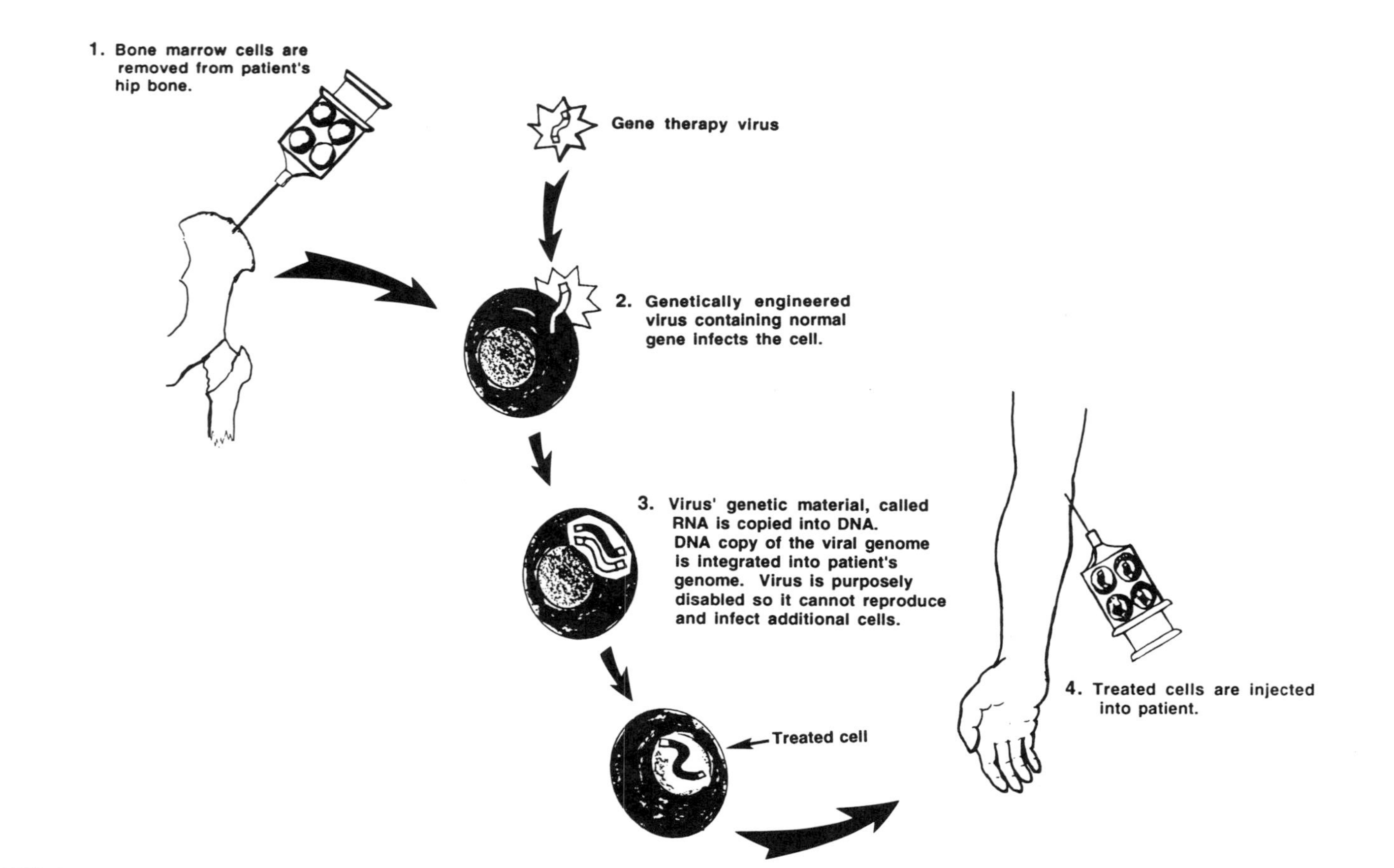

FIGURE 1. Schematic representation of the process of gene therapy using transduction of bone marrow stem cells by recombinant retroviruses. (Reproduced from Forget, 1994, with permission of the publisher.)

a normal β^A-globin gene into hematopoietic stem cells of an affected patient's bone marrow *ex vivo,* using a recombinant retrovirus as the vehicle, followed by the reinfusion of the transduced marrow into the patient in the same manner that one performs bone marrow transplantation. However, a number of obstacles must be overcome before gene therapy becomes an option in the treatment of hemoglobinopathies. First, one needs to have regulated expression of the transferred gene. When one transfers a globin gene into hematopoietic stem cells, one wants it to be expressed only in the erythroid cells and not in the nonerythroid progeny of the stem cells. Although this potential problem was an early concern, it has not turned out to be a serious issue in practice. However, one also needs to have consistently high levels of expression of the transferred gene. The amount of hemoglobin made in the red cell is considerable, and one has to have very high levels of expression of the transferred β^A-globin gene to compensate for the deficit in gene expression in the case of β-thalassemia or to dilute the endogenous β^S-globin chains in the case of sickle cell anemia. This problem also raises the issue of whether one should attempt, in gene therapy of the sickle syndromes, to concomitantly inactivate one (or both) of the endogenous β^S-globin genes by targeted integration of the transferred β^A-globin gene at the site of a β^S gene, thereby disrupting it. Another important consideration is that the gene must be transferred into the self-renewing or reconstituting hematopoietic stem cell to obtain long-lasting therapeutic results, avoiding the need to periodically repeat the procedure. Finally, a number of issues need to be addressed regarding the safety of the procedure. Nevertheless, considerable progress has been made in recent years in the technical ability to transfer globin (and nonglobin) genes into primary hematopoietic cells, including the reconstituting stem cells of a number of animals, including nonhuman primates and more recently humans (reviewed by Friedmann, 1989; Verma, 1990; Apperley and Williams, 1990; Nienhuis et al., 1991; Karlsson, 1991; Hesdorffer et al., 1991; Anderson, 1992; Miller, 1990, 1992).

Retroviral Gene Transfer Vectors

Retroviral vectors remain the most popular vehicles for the transfer of globin genes into hematopoietic stem cells (reviewed by Cornetta, 1992; Berenson et al., 1991; Forrester et al., 1986). The properties of these retroviral gene transfer vectors are discussed in detail in another chapter. Briefly, the gene of interest (a globin gene in this case) is inserted into a DNA copy of the virus from which most of the viral genes have been removed to render it incapable of replication (i.e., replication defective). The DNA is then transfected into a viral packaging cell line that produces all of the necessary components to package RNA copies of the modified retroviral genome into infectious viral particles. After entry of these particles into a target cell, a DNA copy of the viral RNA genome is produced by the reverse transcriptase enzyme of the virus. The DNA copy is then integrated into a random site of the chromosomal DNA of the host cell. Because the modified retroviral genome lacks essential structural genes such as those for viral envelope proteins, no further cycles of viral production can occur, but the integrated viral genome can be transcribed to produce mRNA copies of the gene inserted into it for the purpose of gene therapy. In the case of globin gene transfer, the retrovirus is constructed in such a way that the globin gene is transcribed from its own promoter in a direction that is opposite from that of viral gene transcription, which occurs from the promoter of the 5′ viral long terminal repeat (LTR) sequence.

The general approach to gene therapy, as illustrated in Figure 1, would consist of the aspiration of bone marrow from the affected patient followed by the *in vitro* exposure of the total marrow, or a subpopulation of marrow cells enriched for stem cells by antibody selection (the $CD34^+$ cell population [Berenson et al., 1991]), to defective or disabled retroviral particles containing a copy of a normal β^A-globin gene. After an appropriate period of time that allows entry and integration of this virus into the cells, the transduced cells would then be reinfused intravenously into the patient where they could hone to the bone marrow space and continue to proliferate normally. The transferred β^A-globin gene should be expressed exclusively, and hopefully at a high level, in the erythroid cell progeny of the transduced stem cells.

Gene Transfer Into Erythroid Cells in Culture

In the first retroviral globin gene transfer experiments performed by Cone et al. (1987), a genomic β-globin gene and a neomycin resistance gene were inserted into the genome of a retroviral vector. A recombinant retrovirus carrying the human β-globin gene in the reversed transcriptional orientation relative to the LTR produced retroviral stocks of a reasonable titer, while the recombinant retrovirus that carried a β-globin gene in the same orientation did not produce any retroviruses. When the globin retroviral vectors were used to transduce mouse erythroleukemia (MEL) cells in culture, retroviral DNA integrated in the genome at one copy per cell and generated human β-globin mRNA in an inducible manner. The level of expression of the human β-globin mRNA, however, was only 0.01% of the endogenous mouse β^{maj}-globin mRNA. The pattern of expression appeared to be erythroid specific, as no β-globin mRNA was detectable when these retroviruses were used to transduce NIH 3T3 fibroblasts. The highest titers of retroviral stocks produced from the ecotropic packaging cell line, Ψ2, was 2×10^5 cfu/ml. The highest titers from the amphotropic packaging cell line, ΨAM, however, was one to two orders of magnitude lower. It is also important to note that both Ψ2 and ΨAM were shown in later studies to produce retroviral stocks that contained replication-competent helper virus. This precludes the use of such retroviral stocks in human gene therapy applications (Cone et al., 1987).

A similar strategy was used by Karlsson et al. (1987) to transfer the human β-globin gene to MEL cells. The level of expression of human β-globin mRNA and protein in the transduced cells was significantly higher than that reported by Cone et al. (1987) (10% of the endogenous mouse globin expression). However, only 3 of 12 clones of amphotropic producer cell lines contained an unrearranged copy of the β-globin gene, and the range of titers of their viral stocks was between 2 × 10^3 and 2 × 10^4 cfu/ml. These retroviral titers are probably too low to achieve efficient gene transfer into hematopoietic stem cells. Bender et al. (1988) prepared similar amphotropic retroviral stocks that included a human β-globin gene that was marked by a 6 bp insertion in its 5′-untranslated sequences. Only 3 of 20 clones of producer cell lines integrated an unrearranged copy of the β-globin gene, and the highest retroviral titer generated by these producers was 5 × 10^4 cfu/ml. When one of these producer cell lines was used to infect human erythroid blast-forming units (E-BFU) colonies in culture by co-cultivation, the level of expression of the marked β-globin mRNA was about 5% of the endogenous β-globin mRNA. The infection frequency, however, was only 0.04%, which is at least in part a reflection of the low retroviral titers.

Globin Gene Transfer Into Murine Hematopoietic Stem Cells

The first successful globin gene transfer experiments in a long-term animal model system were performed by Dzierzak et al. (1988). These experiments illustrate many of the practical difficulties of this approach that must be overcome before it can be applied in human gene therapy. The human β-globin gene was successfully transferred by means of a retrovirus into self-renewing hematopoietic stem cells of mice, as evidenced by the fact that DNA of the transferred globin retrovirus was still detectable in blood cells of the mice 4–9 months after transplantation of the virally transduced bone marrow cells. The extent of donor marrow cell engraftment varied between 40% and 98%, whereas the efficiency of infection of the engrafted donor marrow cells ranged between 2% and 100%. The transferred human globin gene was expressed in virtually a tissue-specific manner, globin mRNA being detected in significant amounts essentially only in erythroid cells, although trace amounts were observed in lymphoid cells and macrophages of some of the animals. However, a disappointing result was the very low level of human β-globin mRNA expression that was estimated to correspond to only 0.4%–4.0% of the level of the endogenous murine β-globin mRNA, when considered on a single copy per genome basis. Another disappointing observation was that the human β-globin genes were expressed in only 17% of the transplanted animals. This low efficiency of transduction was attributed to low viral titers and suboptimal marrow infection conditions. Nonetheless, these experiments demonstrated the feasibility of long-term expression of a transduced β-globin gene in erythroid cells *in vivo,* albeit at a level that is not sufficient to ameliorate the clinical manifestations of human hemoglobin disorders (Dzierzak et al., 1988). These findings were also confirmed independently by other groups (Karlsson et al., 1988; Bender et al., 1989). The transduction of the pluripotent hematopoietic stem cell in those studies was demonstrated by serial marrow reconstitution in secondary recipients (Bender et al., 1989; Bodine et al., 1989).

Approaches To Increase Globin Gene Expression in Transduced Cells

One important feature of the normal chromosomal environment that confers a high level of expression to globin genes is the locus control region (LCR) (reviewed by Townes and Behringer, 1990). The LCR consists of an interesting region of DNA, located far upstream from the β-globin gene, that was initially identified because of the presence of sites hypersensitive to digestion by DNase I (Tuan et al., 1985; Forrester et al., 1986). The importance of these DNA sequences was first suggested by the finding of a number of mutations in which the β-globin gene itself was structurally normal but not expressed when located on a chromosome that carried a deletion of these sequences. When DNA sequences containing these DNase I hypersensitive sites were linked to a β-globin gene in transgenic mice, the level of expression of the transferred β-globin gene was virtually equal to that of the endogenous gene (Grosveld et al., 1987). These sequences have therefore been called the *locus control region.*

The LCR confers high levels of erythroid-specific expression to transferred β-globin genes in a relatively copy-dependent and position-independent manner, i.e., the level of expression is proportional to the number of copies of the gene that are transferred and is uniformly high irrespective of the chromosomal site of integration of the transferred gene(s). The DNA sequences that are critical for the functional activity of the LCR have been localized to a small region of DNA surrounding the DNase I hypersensitive sites. Therefore, it was initially thought that incorporation of these LCR core sequences into the retroviral vectors would overcome the problem of low expression levels. However, initial attempts to include LCR sequences in gene transfer vectors have had unexpected deleterious effects on the ability to obtain retroviral stocks of sufficiently high titer to be efficient vehicles for gene transfer into bone marrow hematopoietic stem cells.

Novak et al. (1990) constructed several retroviral vectors that consisted of a human β-globin gene linked to one of the four hypersensitive sites of the β-LCR. These retroviral vectors were used to infect MEL cells in culture. One retrovirus that contained HS2 resulted in high level expression of β-globin mRNA comparable with the level of the endogenous mouse β^{maj} mRNA, but only in a small number of MEL cell clones. The level of human β-globin mRNA in the different clonal isolates of MEL cells transduced with this retrovirus was quite variable (10%–310% of the mouse levels). In addition, rearrangements of the viral sequences were noted in most of the packaging cell lines and the majority of these producer cell lines, including those that initially integrated unrearranged copies of the viral genome, transmitted rearranged retroviral sequences. The titers of the viral stocks from these producer cell lines ranged from 2×10^4 to 1×10^5 cfu/ml (Novak et al., 1990).

Chang et al. (1992) tried to overcome the problems resulting from the genetic instability of recombinant retroviruses that contain HS2 by using progressively smaller segments of this hypersensitive site to enhance β-globin expression. Unfortunately, the resulting retroviruses remained unstable, and the human β-globin gene was rearranged or deleted in all the packaging cell lines that were examined. When HS2 sequences were reduced to a 36 bp fragment containing the NF-E2/AP-1 core enhancer sequence, 6 of 12 producer cell lines integrated unrearranged vector sequences. The stability of these retroviral producing cell lines appeared to be similar to that of retroviral producing cell lines that contained the β-globin gene without HS2. The titers of retroviruses generated from unrearranged producer cell lines were in the 10^4 cfu/ml range. The expression of the human β-globin gene in transduced MEL cells increased from an average of 6% to an average of 12% of the mouse β^{maj}-globin expression as a result of the addition of the 36 bp fragment.

Plavec et al. (1993) constructed retroviral vectors that carried a cassette consisting of the four major HS sites of the β-LCR fused to the human β-globin gene. The highest titer achieved in a helper-free amphotropic retroviral stock was approximately 1×10^4 cfu/ml. When these retroviral vectors were used to transduce MEL cells, the level of human β-globin mRNA ranged from 61% to 79% of the mRNA of a single endogenous mouse β-globin gene. To produce a viral stock of a higher titer that would be capable of infecting mouse hematopoietic stem cells, the supernatant from the highest titer amphotropic retroviral stock was used to infect the Ψ ecotropic packaging cell line. Using this strategy, it was possible to obtain a stock with a viral titer of 10^5 cfu/ml. However, these retroviral stocks consisted of ecotropic retroviruses that cannot transduce human cells. In addition, the stocks were contaminated with helper virus. These retroviral stocks were used to infect mouse bone marrow cells in culture, and the infected cells were injected into lethally irradiated mice. Essentially all of the transplanted mice showed expression of the transferred human gene in their red cells. However, only 0.4%–12% of the circulating red blood cells of these mice stained positive for human globin chains, and the overall level of human β-globin mRNA in their red blood cells ranged from 0.01% to 2% of the level of the mouse globin mRNA (Plavec et al., 1993). These low efficiencies of transduction of hematopoietic stem cells may be a consequence of the low titers of the viral stocks.

Newer Generation of Globin Retroviral Vectors

Over the last 2 years, three different groups described the production of genetically stable retroviral vectors that can give rise to high level expression of transferred globin genes in transduced erythroid cells. Previous studies by Miller et al. (1988) had suggested that sequences within IVS-2 may be responsible for the genetic instability of retroviruses that carry the β-globin gene. When Leboulch et al. (1994) deleted a 372 bp fragment from IVS-2 of the β-globin gene, the titer of the corresponding retroviruses increased about 10-

fold with some improvement in genetic stability in the presence of simple β-LCR cassettes. Leboulch et al. (1994) also performed extensive site-directed mutagenesis of the human β-globin gene to remove all potential splice sites and polyadenylation sites that could interfere with the stable propagation of these retroviruses. Although these additional modifications did not increase the titers further, they appeared to improve stability in the presence of complex β-LCR cassettes. The viral titers ranged from 10^4 to 10^5 cfu/ml, and the expression of the β-globin gene in the transduced cells ranged from 54% to 108% of the mouse β^{maj}-globin gene (Leboulch et al., 1994).

Sadelain et al. (1995) used a similar strategy in which a portion of IVS-2 was deleted from the β-globin gene in retroviral vectors that carried multiple core hypersensitive sites from the β-LCR in different configurations. They examined a large number of such recombinant retroviruses and were able to identify only one genetically stable retrovirus that contained the modified β-globin gene linked to the core elements of HS2, HS3, and HS4. Although stocks of this retroviral vector had titers greater than 10^6 cfu/ml, the expression of the integrated β-globin gene in transduced erythroid cells was clearly not position independent. The level of expression of the transduced human β-globin gene in MEL cells ranged from 4% to 146% of the murine β-globin gene. They concluded that even though their β-LCR cassette demonstrated significant enhancer activity, it clearly did not satisfy the criteria of position independence that defines LCR function (Sadelain et al., 1995).

Ren et al. (1996) used a totally different approach to solve the problems of genetic instability and low level expression of globin genes in transduced erythroid cells. They incorporated the human γ-globin gene instead of the β-globin gene in their retroviral vectors. They also replaced the β-LCR element with the much simpler α-LCR (i.e., HS − 40) to stimulate the expression of the target globin gene in an erythroid-specific manner. These two modifications allowed the generation of retroviral vectors whose titers exceeded 7×10^6. These retroviruses were capable of expressing the transferred γ-globin gene at a high level in a position-independent manner in MEL cells. Since these retroviruses include the human γ-globin gene instead of the β-globin gene, they may be better suited for the gene therapy of sickle cell anemia due to the better antisickling activity of fetal hemoglobin (Ren et al., 1996). Although the performance of all three retroviral vectors described above (Leboulch et al., 1994; Sadelain et al., 1995; Ren et al., 1996) is clearly superior to that of the earlier generation of vectors in MEL cells, their transduction efficiency and the level of expression of their globin genes have not been evaluated in an *in vivo* system.

Adeno-Associated Viral Globin Vectors

Another newly developed viral gene transfer system that may be well suited for gene transfer into hematopoietic stem cells is the adeno-associated virus (AAV), a defective nonpathogenic member of the parvovirus family (Berns and Hauswirth, 1979). One attractive feature of this virus is the site-specific integration of the wild-type virus on chromosome 19 in infected cells (Kotin et al., 1990). Unfortunately, this feature is lost in recombinant viruses that appear to integrate at random in the genome (Walsh et al., 1992). AAV has been successfully used to transfer a marked $^{A}\gamma$-globin gene linked to HS2 into K562 cells (Walsh et al., 1992). This resulted in regulated high level expression of the $^{A}\gamma$-globin gene with very little evidence of sequence rearrangements. More recently, an AAV vector that carries a human $^{A}\gamma$-globin gene with β-LCR elements in the absence of any selectable marker was used to infect $CD34^{+}$ human hematopoietic cells (Miller et al., 1994). From 20% to 40% of the BFU-e derived colonies expressed the γ-globin gene at levels of 4%–71% of the endogenous β-globin gene. This AAV gene transfer system, however, still suffers from low titers of viral stocks and contamination with helper virus. It is also still not clear whether recombinant AAV can integrate efficiently in the genome of transduced cells. The development of suitable packaging cell lines may be necessary to simplify the use of AAV vectors and solve some of the problems associated with their use. It also remains to be seen whether this gene transfer system will work efficiently *in vivo* in an animal model.

Gene Targeting Approaches

An alternative approach to the problem of low expression levels of transferred globin genes is to attempt targeted insertion by homologous recombination of the transferred gene into the locus of the endogenous defective globin gene, with the anticipation that the transferred β^{A}-globin gene, once located in the proper chromosomal environment, would be expressed at the same level as a normal endogenous β-globin gene. In the case of gene therapy strategies for sickle cell anemia, in particular, gene targeting also provides the advantage that the recombination event would inactivate the mutant β^{S}-globin gene. Thus, one could theoretically effectively replace the deleterious gene product (β^{S}-chains) with the beneficial gene product (β^{A}-chains) instead of simply diluting the endogenous β^{S}-chains with added β^{A}-chains derived from a randomly integrated transferred β^{A}-globin gene. In fact, one might speculate that expression of additional hemoglobin into a sickle red cell without a concomitant reduction in the amount of the HbS could be deleterious because of the effect of hemoglobin concentration on HbS polymerization. Thus, there is an added theoretical advantage to gene targeting approaches in gene therapy for sickle cell anemia.

Although targeted transfer of β-globin (Smithies et al., 1985; Nandi et al., 1988; Popovich et al., 1991) and other genes has been successfully accomplished in tissue culture cells as well as in mouse embryonic stem cells, such homologous recombination events occur at a very low frequency. To be useful in practical terms for gene therapy, this approach requires the ability to carry out selection and subsequent *in vitro* expansion of the rare cell in which the event has taken place. Unfortunately, *in vitro* techniques for selection and expansion of reconstituting and self-renewing hematopoietic stem cells are not yet readily available, although progress along these lines can be anticipated in the near future.

Issues Related to Stem Cell Biology

For gene therapy to be effective, the gene must be transferred into the self-renewing hematopoietic stem cell. If the gene is transferred only into the more differentiated, nonself-renewing progenitors or precursor cells, then the cell lineage containing the transferred gene will persist only transiently. Unfortunately, the efficiency of gene transfer into hematopoietic stem cells is generally much lower than the efficiency of transfer into committed progenitors. This may be in part a result of the presence of very low levels of amphotropic retrovirus receptors in the murine and human hematopoietic stem cells (Orlic et al., 1996). In addition, gene transfer with chromosomal integration using retroviral vectors requires cell division. The majority of the reconstituting stem cells in the bone marrow are usually in the nondividing (G0) phase of the cell cycle and therefore relatively refractory to successful gene transfer. Although transfer of globin (and other) genes into the reconstituting hematopoietic stem cell of the mouse has been successfully accomplished, experiments in larger animal species have been generally disappointing and suggest that gene transfer into reconstituting hematopoietic stem cells of these species occurs at too low a level to achieve any therapeutic impact. One approach to this problem is the exposure of the bone marrow to various combinations of hematopoietic growth factors, prior to infection with the retroviral vector, to stimulate the resting reconstituting stem cells to enter into the cell cycle and divide. Experiments in the murine model system indicate that a higher proportion of reconstituting stem cells can be infected following pre-stimulation of the marrow *in vitro* with hematopoietic growth factors such as interleukin (IL)-3 and IL-6 (Bodine et al., 1989, 1991a). The addition of stem cell factor (kit ligand) to IL-3 and IL-6 may also prove to be beneficial (Luskey et al., 1992). Such pre-stimulation approaches, combined with the development of very high titer producer cell lines (Bodine et al., 1990) have resulted in the successful transfer of a retroviral vector into reconstituting stem cells of the rhesus monkey, but at a very low frequency of approximately 1% of the stem cells (Bodine et al., 1991b; vanBeusechem et al., 1992). More recently, the efficiency of transfer into rhesus monkey hematopoietic stem cells was shown to be significantly enhanced by priming with stem cell factor and granulocyte colony-stimulating factor *in vivo*. More than 5% of the circulating cells in these monkeys contained the retroviral genome up to 1 year after transplantation (Dunbar et al., 1996). A most exciting breakthrough in this field is the development of a new lentiviral gene transfer system that offers considerable promise as a tool for gene transfer into hematopoietic stem cells (Naldini et al., 1996). In contrast to other types of retroviruses, this human immunodeficiency virus-like retrovirus does not require cell division for stable integration in the genome of an infected cell. However, before such a system could be used in human gene therapy, appropriate packaging cell lines that address all potential safety concerns will have to be developed. Such an important advance has the potential of overcoming the last major hurdle to the successful application of gene therapy to hemoglobin disorders.

Safety Issues

Potential safety problems associated with the use of retroviral vectors as gene transfer vehicles include the following (reviewed by Cornetta, 1992): (1) insertional mutagenesis due to the random nature of the DNA integration event, resulting in either inactivation of an essential gene or inappropriate activation of a nearby latent potentially harmful gene such as a protooncogene; and (2) inadvertent infection of the marrow, and therefore the patient, with oncogenic wild-type retrovirus that may contaminate the gene transfer virus stock due to recombination with the viral genome in the packaging cell line. In fact, such a contamination resulted in the development of lymphomas in a number of nonhuman primates in one series of gene transfer experiments (Donahue et al., 1992). Although these are definite risks, they occur at a very low frequency. In particular, the development of a newer generation of packaging cell lines has greatly reduced the risk of contamination of gene therapy vectors by wild-type retroviruses (reviewed by Hesdorffer et al., 1991; Cornetta, 1992).

GENE THERAPY OF OTHER ANEMIAS

Although disorders of the globin genes have received a great deal of attention as models for human gene therapy, a number of other inherited anemias may be equally amenable to such therapeutic strategies. Although anemia may be one of the important clinical manifestations of a large number of genetic disorders, we limit the discussion in this chapter to disorders where the primary genetic defect is manifested in the erythroid lineage of the hematopoietic stem cells. This excludes several important disorders such as Fanconi's anemia where the genetic defect affects multiple organ systems, including the bone marrow. Although efforts to develop gene therapy for such disorders are already underway in humans (Walsh et al., 1994a,b), somatic gene therapy directed at the hematopoietic stem cells will clearly not have an effect on the other developmental abnormalities that complicate these disorders. Another important stem cell disorder where anemia is a salient feature is paroxysmal nocturnal hemoglobinemia. In this case, transfer of the PIG-A gene into hematopoietic stem cells may be effective in ameliorating the anemia. However, the effect of such therapy on the propensity to develop aplastic anemia and/or leukemia is not clear at this stage. The disorders that are discussed in the remainder of this chapter can be classified as disorders of the red cell membrane skeleton, disorders of heme biosynthesis and red cell disorders resulting from enzyme deficiencies.

Disorders of the Erythrocyte Membrane Skeleton

Disorders affecting proteins of the erythrocyte membrane skeleton have been found to be associated with hereditary hemolytic anemias such as hereditary elliptocytosis and hereditary spherocytosis (reviewed by Gallagher et al., 1997; Becker and Lux, 1995; Lux and Palek, 1995; Benz, 1994). These autosomal dominant disorders are relatively common but are only rarely associated with severe anemia. Nevertheless, in cases of homozygosity or double heterozygosity for certain mutations, severe fatal or near-fatal hemolytic anemia can result, leading to hydrops fetalis or transfusion dependence at an early age (Arcasoy and Gallagher, 1997). In families at risk for such severe disease, gene therapy may be indicated in the future if effective vectors can be developed. The genes most frequently mutated in this family of disorders include those encoding the following membrane proteins: α-spectrin, β-spectrin, ankyrin, band 3 (the anion transporter), and protein 4.1 (Gallagher et al., 1997; Becker and Lux, 1995; Lux and Palek, 1995; Benz, 1994).

The difficulties facing gene therapy for these disorders include some of the same challenges involved in globin gene therapy, such as the level of expression that is required, as well as new challenges, including the size of the cDNA that must be incorporated into the gene therapy vector. Proteins of the membrane skeleton, in particular α- and β-spectrin and band 3, are expressed at a relatively high level in erythroid cells, and therefore successful future gene therapy will require high levels of expression of the transduced or transferred gene. The lessons learned from the globin gene therapy experiments and the use of LCR-containing vectors will be particularly useful in this regard. The promoters of many of the membrane protein genes have been characterized and could be linked to the cDNAs to provide minigenes that will hopefully be regulated in the erythroid progeny of transduced stem cells in a similar fashion to that of a normal endogenous gene. The size of the full-length cDNA encoding certain membrane skeleton proteins, in particular α-spectrin and β-spectrin, limits the use of conventional retroviral vectors because these vectors cannot accommodate inserts over a certain size. In these cases, one will need to await the development of newer retroviral vectors with larger capacity or explore the use of alternative vectors such as adenoviruses.

There exists a number of genetically distinct mouse models for these disorders that are characterized by defective expression of the spectrin or ankyrin genes (Bodine et al., 1984). These mouse model systems provide very useful animal systems for studying the feasibility and efficiency of gene transfer as an approach to future attempts to correct or ameliorate human disorders of the erythrocyte membrane. Preliminary results have already been reported on the successful transfer and expression of a hybrid ankyrin minigene into

MEL cells and murine bone marrow progenitor cells using a retroviral vector (Becker et al., 1995) as an approach for gene therapy of the murine *nb/nb* mutation that is characterized by marked ankyrin deficiency (White et al., 1990).

Disorders of Heme Biosynthesis

This is a heterogeneous group of disorders that result from inherited defects in each of the eight enzymes involved in heme biosynthesis. Although heme is a very important component of the hemoglobin molecule, deficiencies of the majority of the enzymes required for its biosynthesis do not result in anemia. The clinical presentation in the majority of patients with these conditions may be dominated by cutaneous photosensitivity, abdominal pains, and/or peripheral neuropathy. These nonanemic disorders are sometimes referred to as *hepatic porphyrias.* Gene therapy directed at the hematopoietic stem cells would not be expected to have a large impact on the clinical course of patients with these conditions. In contrast, the clinical picture in most patients with congenital erythropoietic porphyria (CEP) and in a few patients with erythropoietic protoporphyria (EPP) may be dominated by anemia, while the only clinical manifestations in patients with X-linked sideroblastic anemia resulting from deficiency of δ-aminolevulinic acid synthase (δ-ALAS) is anemia.

CEP is an autosomal recessive disorder resulting from a deficiency of uroporphyrinogen III cosynthase. This is a severe but uncommon disorder that should be amenable to stem cell–directed gene therapy. Two different groups have published preliminary studies that describe the generation of retroviral vectors for use in this disorder (Moreau-Gaudry et al., 1995; Glass et al., 1996). Photosensitivity, the other major manifestation of this disease, may improve as a result of the decrease in the production of porphyrin in the bone marrow of these patients. EPP, on the other hand, is the most common form of erythropoietic porphyria and the second most common form of human porphyria. It results from an autosomal dominant deficiency of ferrocheletase, the last enzyme in the heme biosynthetic pathway. This disorder is sometimes referred to as *erythrohepatic protoporphyria* to reflect the involvement of the bone marrow and liver in these patients. Although mild anemia is occasionally seen in some of these patients, hemolysis is very uncommon. Liver function is usually preserved, and the clinical picture is predominantly that of a unique type of cutaneous photosensitivity. The source of excess hepatic protoporphyrin in EEP with hepatic complications is not clear yet. At least in one patient, the hepatic complication and the hemolysis improved after splenectomy. This argues that the bone marrow may be the major source of protoporphyrin accumulation, while the liver injury may be secondary to this accumulation. In principle, with further advances in gene transfer technology, hematopoietic stem cell–directed gene therapy may become a reasonable therapeutic option for the rare patient with severe EPP.

Deficiency of δ-ALAS, the first enzyme in the heme biosynthesis pathway, results in sideroblastic anemia rather than porphyria. A number of mutations in the erythroid-specific ALAS2 gene have been identified in patients with X-linked sideroblastic anemia, including the mutation in the original family described by Cooley (Cotter et al., 1994). The degree of anemia can be quite variable in patients with this disorder, and the response to pyridoxine may be partial. Although the anemia is generally not life-threatening, gene therapy approaches may be appropriate with the aim of improving the quality of life of patients with the more severe anemia. In this case also, this should only be considered after major improvements in the efficacy and safety of gene transfer into hematopoietic stem cells.

Anemias of Enzymatic Deficiencies

Deficiency of the X-chromosome encoded enzyme glucose-6-phosphate dehydrogenase (G-6-PD) is by far the most common red cell enzyme deficiency that can result in anemia. Hundreds of G-6-PD variants have been identified, the majority of which occur in patients who never manifest any hemolytic anemia. The A^- variant that is commonly seen in people of African descent can result in episodic hemolysis upon exposure to oxidant drugs/compounds or during various infectious episodes. In contrast, the Mediterranean G-6-PD deficiencies that are associated with a variety of mutations of the G-6-PD gene can result in severe life-threatening hemolysis upon exposure to fava beans (i.e., favism) and other oxidants. Some of these variants can also be associated with chronic congenital nonspherocytic anemia. In the case of the episodic hemolysis of drug exposure in the majority of patients with G-6-PD, early screening and education may be sufficient to avoid hemolytic complications. In the more severe enzymatic deficiencies, particularly those associated with chronic hemolytic anemia, hematopoietic stem cell–targeted gene therapy may become a viable therapeutic option in the future.

There is a very long list of other deficiency states affecting red cell enzymes. Fortunately, the vast majority of these are not associated with anemia. At least 10 different enzyme deficiency states have been associated with mild to moderate anemia. The incidences of these deficiencies range from rare to very rare. Some of these disorders may at times be associated with neurological, neuromuscular, or hepatic complications. The most common of these rare disorders is pyruvate kinase deficiency. In some patients with pyruvate kinase and severe anemia, splenectomy has been associated with partial amelioration of the anemia. In view of the very rare occurrence of these disorders and their generally benign course, development of gene therapy is difficult to justify at this time given the current state-of-the-art of gene transfer into hematopoietic stem cells. This, however, may change in the future once the efficiency and safety of gene transfer into hematopoietic stems cells are better established.

REFERENCES

Anderson WF (1992): Human gene therapy. *Science* 256:808–813.

Apperley JF, Williams DA (1990): Gene therapy: Current status and future directions. *Br J Haematol* 75:148–155.

Arcasoy MO, Gallagher PG (1997): Hematologic disorders and nonimmune hydrops fetalis. *Semin Perinatol* 19:502–515.

Becker PS, Debatis ME, Riel GJ, Barker JE, Forget BG, Quesenberry PJ (1995): Retroviral mediated transfer of a hybrid mouse/human erythroid ankyrin minigene to murine erythroleukemia (MEL) and normal murine bone marrow cells. *Blood* 88:6a.

Becker PS, Lux SE (1995): Disorders of the red cell membrane skeleton: Hereditary spherocytosis and hereditary elliptocytosis. In Scriver CR, Beaudet AL, Sly WS, Valle D (eds): *Metabolic and Molecular Basis of Inherited Disease,* 7th ed. New York: McGraw-Hill, pp 3513–3560.

Bender MA, Gelinas RE, Miller AD (1989): A majority of mice show long-term expression of a human β-globin gene after retrovirus transfer into hematopoietic stem cells. *Mol Cell Biol* 9:1426–1434.

Bender MA, Miller AD, Gelinas RE (1988): Expression of the human β-globin gene after retroviral transfer into murine erythroleukemia cells and human BFU-E cells. *Mol Cell Biol* 8:1725–1735.

Benz EJ Jr (1994): The erythrocyte membrane and cytoskeleton: Structure, function and disorders. In Stamatoyannopoulos G, Nienhuis AW, Majerus PW, Varmus H (eds): *Molecular Basis of Blood Diseases.* Philadelphia: WB Saunders, pp 257–292.

Berenson RJ, Bensinger WI, Hill RS (1991): Engraftment after infusion of $CD34^+$ marrow cells in patients with breast cancer or neuroblastoma. *Blood* 77:1717–1722.

Berns KI, Hauswirth WW (1979): Adeno-associated viruses. *Adv Virus Res* 25:407–449.

Bodine D, Birkenmeier C, Barker JE (1984): Spectrin deficient inherited hemolytic anemias in the mouse: Characterization by spectrin synthesis and mRNA activity in reticulocytes. *Cell* 37:721–729.

Bodine DM, Crosicr PS, Clark SC (1991a): Effects of hematopoietic growth factors on the survival of primitive stem cells in liquid suspension culture. *Blood* 78:914–920.

Bodine DM, Karlsson S, Nienhuis AW (1989): Combination of interleukins 3 and 6 preserves stem cell function in culture and enhances retrovirus-mediated gene transfer into hematopoietic stem cells. *Proc Natl Acad Sci USA* 86:8897–8901.

Bodine DM, McDonagh KT, Brandt SJ, Ney PA, Agricola B, Byrne E, Nienhuis AW (1990): Development of a high-titer retrovirus producer cell line capable of gene transfer into rhesus monkey hematopoietic stem cells. *Proc Natl Acad Sci USA* 87:3738–3742.

Bodine DM, McDonagh KT, Seidel NE, Nienhuis AW (1991b): Retrovirus mediated gene transfer to pluripotent hematopoietic stem cells. *Regul Hemoglobin Switching* 505–525.

Chang JC, Liu D, Kan YW (1992): A 36-base-pair core sequence of locus control region enhances retrovirally transferred human β-globin gene expression. *Proc Natl Acad Sci USA* 89:3107–3110.

Cone RD, Weber-Benarous A, Baorto D, Mulligan RC (1987): Regulated expression of a complete human β-globin gene encoded by a transmissible retrovirus vector. *Mol Cell Biol* 7:887–897.

Cornetta K (1992): Safety aspects of gene therapy. *Br J Haematol* 80:421–426.

Cotter PD, Rucknagel DL, Bishop DF (1994): X-linked sideroblastic anemia: Identification of the mutation in the erythroid-specific delta-aminolevulinate synthase gene (ALAS2) in the original family described by Cooley. *Blood* 84:3915–3924.

Donahue RE, Kessler SW, Bodine D, McDonagh K, Dunbar C, Goodman S, Agricola B, Byrne E, Raffeld M, Moen R, Bacher J, Zsebo KM, Nienhuis AW (1992): Helper virus induced T cell lymphoma in non-human primates after retroviral mediated gene transfer. *J Exp Med* 176:1125–1135.

Dunbar CE, Seidel NE, Doren S, Sellers S, Cline AP, Metzger ME, Agricola BA, Donahue RE, Bodine DM (1996): Improved retroviral gene transfer into murine and Rhesus peripheral blood or bone marrow repopulating cells primed *in vivo* with stem cell factor and granulocyte colony-stimulating factor. *Proc Natl Acad Sci USA* 93:11871–11876.

Dzierzak EA, Papayannopoulou T, Mulligan RC (1988): Lineage-specific expression of a human β-globin gene in murine bone marrow transplant recipients reconstituted with retrovirus-transduced stem cells. *Nature* 331:35–41.

Forget BG (1994): In Embury SH, et al. (eds): "Gene Therapy" *Sickle Cell Disease: Basic Principles and Clinical Practice.* New York: Raven Press, pp 853–860.

Forrester WC, Thompson C, Elder JT (1986): A developmentally stable chromatin structure in the human β-globin gene cluster. *Proc Natl Acad Sci USA* 83:1359–1363.

Friedmann T (1989): Progress toward human gene therapy. *Science* 244:1275–1281.

Gallagher PG, Forget BG, Lux SE (1997): Disorders of the cell membrane. In Nathan DG, Orkin SH (eds): *Hematology of Infancy and Childhood,* 5th ed. Philadelphia: WB Saunders.

Glass IA, Kauppinen R, Atweh G, Desnick RJ (1996): Toward gene therapy for congenital erythropoietic porphyria. *Am J Hum Genet* 59:A198.

Grosveld F, van Assendelft GB, Greaves DR, Kollias G (1987): Position-independent, high level expression of the human β-globin gene in transgenic mice. *Cell* 51:975.

Hesdorffer C, Markowitz D, Ward M, Bank A (1991): Somatic gene therapy. *Hematol Oncol Clin North Am* 5:423–432.

Karlsson S (1991): Treatment of genetic defects in hematopoietic cell function by gene transfer. *Blood* 78:2481–2492.

Karlsson S, Bodine DM, Perry L, Papayannopoulou T, Nienhuis AW (1988): Expression of the human β-globin gene following retroviral-mediated transfer into multipotential hematopoietic progenitors of mice. *Proc Natl Acad Sci USA* 85:6062-6066.

Karlsson S, Papayannopoulou T, Schweiger SG, Stamatoyannopoulos G, Nienhuis AW (1987). Retroviral-mediated transfer of genomic globin genes leads to regulated production of RNA and protein. *Proc Natl Acad Sci USA* 84:2411–2415.

Kotin RM, Siniscalco M, Samulski RJ, Zhu X, Hunter L, Laughlin CA, McLaughlin S, Muzyczka N, Rocchi M, Berns KI (1990): Site-specific integration by adeno-associated virus. *Proc Natl Acad Sci USA* 87:2211–2215.

Leboulch P, Huang GMS, Humphries RK, Oh YH, Eaves CJ, Tuan DYH, London IM (1994): Mutagenesis of retroviral vectors transducing human β-globin gene and

β-globin locus control region derivatives results in stable transmission of an active transcriptional structure. *EMBO J* 13:3065–3076.

Luskey BD, Rosenblatt M, Zsebo K, Williams DA (1992): Stem cell factor, interleukin-3, and interleukin-6 promote retroviral-mediated gene transfer into murine hematopoietic stem cells. *Blood* 80:396–402.

Lux SE, Palek J (1995): Disorders of the red cell membrane. In Handin RI, Lux SE, Stossel TP (eds): *Blood: Principles and Practice of Hematology.* Philadelphia: JB Lippincott, pp 1701–1818.

Miller AD (1990): Progress toward human gene therapy. *Blood* 76:271–278.

Miller AD (1992): Human gene therapy comes of age. *Nature* 357:455–460.

Miller JL, Donahue RE, Sellers SE, Samulski RJ, Young NS, Nienhuis AW (1994): Recombinant adeno-associated virus (rAAV)–mediated expression of a human γ-globin gene in human progenitor-derived erythroid cells. *Proc Natl Acad Sci USA* 91:10183–10187.

Moreau-Gaudry F, Ged C, Barbot C, Mazurier F, Boiron JM, Bensidhoum M, Reiffers J, deVerneuil H (1995): Correction of the enzyme defect in cultured congenital erythropoietic porphyria disease cells by retrovirus-mediated gene transfer. *Hum Gene Ther* 6:13–20.

Naldini L, Blömer U, Gallay P, Ory D, Mulligan R, Gage FH, Verma IM, Trono D (1996): *In vivo* gene delivery and stable transduction of non-dividing cells by a lentiviral vector. *Science* 272:263–267.

Nandi AK, Roginski RS, Gregg RG, Smithies O, Skoultchi AI (1988): Regulated expression of genes inserted at the human chromosomal beta-globin locus by homologous recombination. *Proc Natl Acad Sci USA* 85:3845–3849.

Nienhuis AW, McDonagh KT, Bodine DM (1991): Gene transfer into hematopoietic stem cells. *Cancer* 67:2700–2704.

Novak U, Harris EAS, Forrester W, Groudine M, Gelinas R (1990): High-level β-globin expression after retroviral transfer of locus activation region-containing human β-globin gene derivatives into murine erythroleukemia cells. *Proc Natl Acad Sci USA* 87:3386–3390.

Orlic D, Girard LJ, Jordan CT, Anderson SM, Cline AP, Bodine DM (1996): The level of mRNA encoding the amphotropic retrovirus receptor in mouse and human hematopoietic stem cells is low and correlates with the efficiency of retrovirus transduction. *Proc Natl Acad Sci USA* 93:11097–11102.

Plavec I, Papayannopoulou T, Maury C, Meyer F (1993): A human β-globin gene fused to the human β-globin locus control region is expressed at high levels in erythroid cells of mice engrafted with retrovirus-transduced hematopoietic stem cells. *Blood* 81:1384–1392.

Popovich BW, Kim HS, Shesely EG, Shehee WR, Papayannopoulou T, Smithies O (1991): Correction of the human sickle cell gene using targeted gene replacement. *Regul Hemoglobin Switching* 526–540.

Ren S, Wong BY, Li J, Luo XN, Wong PMC, Atweh GF (1996): Production of genetically stable high-titer retroviral vectors that carry a human γ-globin gene under the control of the α-globin locus control region. *Blood* 87:2518–2524.

Sadelain M, Wang CHJ, Antoniou M, Grosveld F, Mulligan RC (1995): Generation of a high-titer retroviral vector capable of expressing high levels of the human β-globin gene. *Proc Natl Acad Sci USA* 92:6728–6732.

Smithies O, Gregg RG, Boggs SS, Koralewski MA, Kucherlapati RS (1985): Insertion of DNA sequences into the human chromosomal beta-globin locus by homologous recombination. *Nature* 317:230–234.

Townes TM, Behringer RR (1990) Human globin locus activation region (LAR) role in temporal control. *Trends Genet* 6:219–223.

Tuan D, Solomon W, Li Q, London IM (1985): The "β-like" globin gene domain in human erythroid cells. *Proc Natl Acad Sci USA* 82:6384–6388.

vanBeusechem VW, Kukler A, Heidt PJ, Valerio D (1992): Long term expression of human adenosine deaminase in rhesus monkeys transplanted with retrovirus-infected bone marrow cells. *Proc Natl Acad Sci USA* 89:7640–7644.

Verma IM (1990): Gene therapy. *Sci Am* 263:68–84.

Walsh CE, Grompe M, Vanin E, Buchwald M, Young NS, Nienhuis AW, Liu JM (1994a): A functionally active retrovirus vector for gene therapy in Fanconi anemia group C. *Blood* 84:453–459.

Walsh CE, Liu JM, Xiao X, Young NS, Nienhuis AW, Samulski RJ (1992): Regulated high level expression of a human γ-globin gene introduced into erythroid cells by an adeno-associated virus vector. *Proc Natl Acad Sci USA* 89:7257–7261.

Walsh CE, Nienhuis AW, Samulski RJ, Brown MG, Miller JL, Young NS, Liu JM (1994b): Phenotypic correction of Fanconi anemia in human hematopoietic cells with a recombinant adeno-associated virus vector. *J Clin Invest* 94:1440–1448.

White RA, Birkenmeier CS, Lux SE, Barker JE (1990): Ankyrin and the hemolytic anemia mutation, *nb,* map to mouse chromosome 8: Presence of the *nb* allele is associated with a truncated erythrocyte ankyrin. *Proc Natl Acad Sci USA* 87:3117–3121.

17

CLINICAL APPLICATIONS OF GENE THERAPY IN CANCER: MODIFICATION OF SENSITIVITY TO THERAPEUTIC AGENTS

THOMAS LICHT, MICHAEL M. GOTTESMAN, AND IRA PASTAN

Laboratory of Molecular Biology (T.L., I.P.) and Laboratory of Cell Biology (M.M.G.), National Cancer Institute, NIH, Bethesda, MD 20892-4255

INTRODUCTION

Chemotherapy bears the potential of curing certain types of cancer but is widely limited in its efficacy by the occurrence of resistance to anticancer drugs. Administration of chemotherapeutic agents can result in lasting remissions in patients suffering from neoplasms of the hematopoietic system, e.g., malignant lymphomas and acute leukemias, certain solid tumors such as some cancers of childhood, and cancers originating from germ cells in adults. Conversely, even high-dose chemotherapy combined with transplantation of bone marrow or reinfusion of stem cells mobilized into peripheral blood fails to cure cancers of brain, kidney, prostate, or the gastrointestinal tract. Moreover, even after initial response to chemotherapy, many tumors will regrow and metastasize. These relapsed cancers are frequently unresponsive to any further chemotherapy. As a consequence, the majority of patients with cancers that extend beyond local control will eventually die from their underlying disease.

Several genes have been characterized that attenuate the susceptibility of malignant cells to the toxicity of various anticancer drugs. Some of the encoded

Stem Cell Biology and Gene Therapy, Edited by Peter J. Quesenberry, Gary S. Stein, Bernard G. Forget, and Sherman M. Weissman
ISBN 0-471-14656-0

proteins interact with a single class of substrates, e.g. dihydrofolate reductase, a gene that upon mutation causes resistance to methotrexate and trimetrexate (Haber et al., 1981; Simonsen et al., 1983). Others render cells resistant to multiple drugs. For instance, P-glycoprotein, also known as the *multidrug transporter,* inactivates natural toxins and antibiotics, e.g. anthracyclines, Vinca alkaloids, epipodophyllotoxins, and taxanes. The MDR1 (multidrug resistance) gene is overexpressed in up to 50% of clinical tumor specimens (Goldstein et al., 1989; Gottesman et al., 1995). Overexpression of P-glycoprotein or of MDR1 transcripts, respectively, has been associated with treatment failure in various cancers such as sarcomas of soft tissue and osteosarcomas, and neuroblastomas (Chan et al., 1990; Goldstein et al., 1989; Baldini et al., 1996). P-glycoprotein expression is not infrequent in acute leukemias and has been identified to be an independent risk factor for treatment failure (Musto et al., 1991; Campos et al., 1992; Pirker et al., 1992). Similarly, expression of P-glycoprotein in multiple myelomas is predictive for poor outcome (Musto et al., 1991). In multiple myeloma P-glycoprotein expression has been shown to occur more frequently after chemotherapy than in untreated patients (Salmon et al., 1989b). Expression levels appear to be associated with the intensity of prior chemotherapy (Grogan et., 1993), suggesting an induction of MDR1 gene expression during drug treatment.

A genetically and functionally related type of chemoresistance is mediated by the multidrug-resistance-associated protein (MRP) gene. Like the MDR1 gene, MRP encodes a transporter that actively extrudes anticancer drugs out of cells. However, their substrate specificities for anticancer drugs as well as for certain inhibiting compounds are different. Other mechanisms for resistance to multiple drugs are associated with overexpression of glutathione-S-transferases and other enzymes involved in glutathione metabolism. Decreased expression of DNA topoisomerases is also a well-documented cause of multidrug resistance. Resistance to antifolates like methotrexate is conferred by different mechanisms. In clinical treatment the folate transporter seems to play a major role; in gene therapy mutated dihydrofolate reductase (DHFR) molecules have been utilized. These molecules display altered affinities to the toxic substrate methotrexate, thereby conferring resistance against this drug.

An increasing number of mechanisms that render cells chemoresistant are being uncovered. We focus on mechanisms of drug resistance that can potentially be overcome. Strategies to modify chemoresistance are aimed at either interfering with transcription or translation of the respective genes or at modulating the function of their products. More recently, the potential use of drug resistance genes in gene therapy has been analyzed. Transfer of drug resistance genes to cells can protect them from the toxicity of anticancer drugs. Thereby the adverse effects of chemotherapy are expected to become ameliorated. Clinical trials on the transfer of the MDR1 gene to hematopoietic progenitor cells of cancer patients are currently being performed.

By transfer of drug resistance genes, target cells obtain a selective advantage while nontransduced cells remain susceptible to drug toxicity. If drug resistance genes are coexpressed with other genes, both genes may be overexpressed in target cells. Thus, it appears possible to increase expression of transgenes that correct genetically determined disorders by administration of anticancer drugs. This may improve the efficiency of somatic gene therapy, which is commonly hampered by low or unstable expression of proteins encoded by transgenes.

GENES CONFERRING RESISTANCE TO ANTICANCER DRUGS

The Multidrug Transporter P-Glycoprotein Encoded by the MDR1 Gene

The chemoresistance gene investigated most extensively is MDR1. Overexpression of its product, the multidrug transporter, also referred to as *P-glycoprotein,* confers resistance to a broad variety of structurally unrelated natural toxins and synthetic drugs derived from them. Most of its substrates are hydrophobic, and many of them contain aromatic ring structures (reviewed by Endicott and Ling, 1989; Gottesman and Pastan, 1993). Following passive diffusion of drugs through the plasma membrane into the cell, P-glycoprotein binds them as shown by photoaffinity labeling experiments (Cornwell et al., 1986; Bruggeman et al., 1992). Consequently, drugs are transported from the cytoplasm or the membrane into the extracellular space in an energy-dependent manner. Hydrolysis of ATP or, under certain circumstances, GTP is stimulated by anticancer drugs that bind to the multidrug transporter (Horio et al., 1988; Ambudkar et al., 1992). In addition, P-glycoprotein reduces drug influx into cells (Stein et al., 1994). As a result, the intracellular accumulation of drugs is diminished below levels sufficient to kill the cell. Due to its very wide substrate specificity, P-glycoprotein also interacts with compounds that display no cytotoxicity, e.g., calcium channel blockers, immunosuppressants, steroid hormones, and neuroleptic drugs. As discussed below, such agents can be useful as inhibitors of drug transport.

The multidrug transporter consists of two highly homologous halves, each consisting of six transmembrane domains. It has been suggested that transmembrane loops form a core structure within the molecule through which substrates are transported (Germann, 1993). A highly homologous molecule, the product of the MDR2 gene (also designated MDR3), which is located in close proximity to MDR1 on chromosome 7, does not confer drug resistance (Lincke et al., 1991; Schinkel et al., 1994) but functions as a transporter for phosphatidyl-choline in liver cells (Smit et al., 1993). Both genes are members of a gene superfamily known as the ATP-binding cassette (ABC) family. ABC proteins function as transporter molecules and channel proteins. The multidrug-resistance-associated protein (MRP), the cystic fibrosis transmembrane conductance regulator (CFTR) protein, and the Tap-1 and Tap-2 transporter proteins, which are involved in antigen presentation to immunocompetent cells, belong to this family (Hyde et al., 1990; Ames et al., 1992; Higgins, 1993).

The multidrug transporter appears to be a multifunctional protein. Because of conservation of resistance-mediating proteins during evolution (Gros et al., 1986; Wilson et al., 1986; Henderson et al., 1992) and physiological expression in tissues like kidney, gut, liver, and bile duct (Thiebaut et al., 1987; Sugawara et al., 1988), it has been suggested that P-glycoprotein may serve as a natural detoxifying system for ingested xenobiotics (Gottesman et al., 1995). The observations that P-glycoprotein is involved in the blood–brain barrier further support this hypothesis (Schinkel et al., 1994). Noteworthy is that tumors originating from organs in which P-glycoprotein is physiologically expressed are usually insensitive to chemotherapy. Recently published investigations revealed that there may be additional functions of P-glycoprotein. The multidrug transporter can translocate lipids across the plasma membrane (van Helvoort et al., 1996).

In the hematopoietic system, the multidrug transporter is expressed on hematopoietic stem cells (Chaudhary and Roninson, 1991), on the majority of T lymphocytes (Neyfakh et al., 1989), and natural killer cells as well as B lymphocytes (Chaudhary et al., 1992). In contrast, there is very little if any P-glycoprotein present in granulocytes and their myeloid progenitor cells and in erythroid cells (Drach et al., 1992; Klimecki et al., 1994). These low expression levels are not capable of rendering myeloid cells resistant to the cytotoxicity of anticancer drugs. Thus, granulocytopenia caused by myelosuppression is a common side effect of anticancer chemotherapy. While expression on hematopoietic stem cells may reflect a need for protection of this valuable cell population, the physiological role of P-glycoprotein in lymphoid cells is still to be elucidated.

The Multidrug-Resistance-Associated Protein

A protein genetically and biochemically related to P-glycoprotein was initially isolated from chemoresistant lung cancer cells in which MDR1 expression was undetectable (Cole et al., 1992). This 190 kD transporter protein, designated MRP (multidrug-resistance-associated) transporter, is also a drug-efflux pump located in the plasma membrane (Zaman et al., 1994). MRP is expressed not only in cancer cells but also in a variety of normal tissues (Zaman et al., 1993; Schneider et al., 1994; Flens et al., 1996). Like P-glycoprotein, the MRP transporter inactivates a broad variety of drugs, including natural toxins such as Vinca alkaloids, anthracyclines, and epipodophyllotoxins by ATP-dependent extrusion out of tumor cells (Cole et al., 1994; Grant et al., 1994; Kruh et al., 1994; Paul et al., 1996). However, in contrast to P-glycoprotein, the MRP transporter fails to extrude taxol (Breuninger et al., 1995). In addition to anticancer drugs, leukotrienes are substrates of this transporter (Leier et al., 1994). Cells expressing MRP are also resistant to some heavy metal anions, including arsenite, arsenate, and antimonials (Cole et al., 1994). MRP extrudes compounds conjugated with glutathione, glucuronate, or sulfate out of cells (Muller et al., 1994; Zaman et al., 1995; Jedlitschky et al., 1996). Drug transport in MRP-expressing cells may be regulated by intracellular glutathione levels (Versantvoort et al., 1995).

MRP is frequently expressed in non-small cell lung cancers (Giaccone et al., 1996; Sugawara et al., 1995). The clinical relevance of this mechanism of drug resistance in certain tumor entities is underlined by recent investigations showing a correlation between MRP expression and treatment failure in childhood neuroblastomas (Norris et al., 1996). Conversely, deletion of this gene in acute myelomonocytic leukemia with inversion of chromosome 16 is associated with a favorable prognosis (Kuss et al., 1994).

The Lung-Resistance-Related Protein

Another gene associated with drug resistance, LRP (lung-resistance-related protein) is located in close proximity to MRP on chromosome 16 (Slovak et al., 1995). This 110 kD protein is detectable in normal epithelial cells and tissues chronically exposed to xenobiotics and toxic agents, e.g., bronchial epithelium, kidney tubules, and cells lining the intestines. The molecule was found to be closely associated with vesicular or lysosomal structures (Scheper et al., 1993). From its deduced amino acid sequence, it is hypothesized to be involved in nucleocytoplasmic transport, presumably acting as a vault protein (Scheffer et al., 1995). LRP may play a role in multidrug-resistant malignant melanoma. Whereas in several investigations P-glycoprotein was not commonly detectable in most melanoma samples (Fuchs et al., 1991; Schadendorf et al., 1995), LRP expression was detected in melanoma cell lines and in the majority of patient samples investigated (Schadendorf et al., 1995). Expression of LRP has been found in cancers of various histological origins. For instance, patients with LRP-positive ovarian cancers had poorer response to chemotherapy and shorter survival than LRP-negative patients (Izquierdo et al., 1995). Expression of LRP has been associated with poor treatment outcome in acute myeloid leukemia (List et al., 1996).

Chemoresistance Related to Altered Cellular Content of Glutathione

As mentioned, the MRP transporter is involved in excretion of glutathione-conjugated anticancer drugs. Other transporter systems may also be involved in removal of glutathione-bound drugs (Chuman et al., 1996). Another major reason for resistance to anticancer drugs is related to the intracellular content of glutathione. High levels of glutathione cause reduced cytotoxicity of alkylating agents, cisplatin, and anthracyclines (Ozols et al., 1987), and high resistance to cisplatin in human ovarian cancer cell lines is associated with marked increase of glutathione synthesis (Godwin et al., 1992). Alkylating agents become more water soluble when conjugated to glutathione, resulting in enhanced efflux out of cells. In addition, glutathione can interact with oxydizing agents such as free radicals, a reaction that interferes with the action of anthracyclines (Arrick et al., 1984).

Several groups of enzymes are involved in regulation of intracellular glutathione levels. Besides glutathione synthetase and glutathione peroxidase, glutathione-S-transferase isoenzymes π, μ, and α seem to play a major role (for reviews, see Waxman, 1990; Black et al., 1991; van Bladeren et al., 1991). Direct evidence is provided by investigations on transfer of glutathione-S-transferases to cancer cells by which they are rendered drug resistant (Puchalski et al., 1990; Schecter et al., 1993).

Clinically, expression of glutathione-S-transferase isoenzymes appears to be correlated with resistance to drugs in cancers of various origins, e.g., ovary, stomach, and lung (Hamada et al., 1994; Kodera et al., 1994; Bai et al., 1996). However, several mechanism of drug resistance may simultaneously be expressed in cancers (Volm et al., 1993). In lymphomas, multiple myelomas, and acute leukemias, glutathione-S-transferases were frequently found to be coexpressed with the MDR1 gene (Cheng et al., 1993, Russo et al., 1994; Zhou et al., 1994; Petrini et al., 1995). In conclusion, elevated levels of glutathione or of enzymes involved in glutathione metabolism, respectively, are associated with increased resistance to anticancer drugs. Because glutathione-S-transferases are frequently coexpressed with other mechanisms of drug resistance, common mechanisms of activation are likely. The MRP transporter protein has been found to play a crucial role in elimination of glutathione-conjugated anticancer drugs.

Mutated Dihydrofolate Reductase

The dihydrofolate reductase (DHFR) gene displays reduced affinity to methotrexate when mutated (Haber et al., 1981; Simonsen et al., 1983; Srimatkandada et al., 1989; Dicker et al., 1990). Different from the previously discussed mechanisms of drug resistance, mutations of the DHFR gene affect only antifolates. Different point mutations result in different degrees of chemoresistance. With mutated DHFR genes vectors for gene transfer have been engineered (Miller et al., 1985).

Multiple Causes of Chemoresistance

While additional causes for drug resistance are currently being identified, it is not alwalys possible to clearly distinguish between their respective roles in a given cancer. Simultaneous expression of several mechanisms of drug resistance appears to be a common phenomenon. For instance, MDR1, MRP, and LRP are frequently coexpressed within chemoresistant cancer cells (Izquierdo et al., 1996). In addition, overexpression of genes encoding other transporter molecules may result in chemoresistance under certain circumstances. In a recent investigation, the cystic fibrosis transmembrane conductance regulator (CFTR), a member of the ABC gene superfamily, was found to alter biophysical membrane properties and intracellular pH. Fibroblasts transfected with CFTR appeared to display resistance to doxorubicin, vincristine, and colchicine (Wei et al., 1995).

MODULATION OF CHEMORESISTANCE BY INTERACTION WITH THE FUNCTION OF PROTEINS ENCODED BY DRUG RESISTANCE GENES

Reversal of P-Glycoprotein–Mediated Multidrug Resistance

Interest in chemoresistance genes has mainly been focused on strategies to overcome the underlying mechanisms of treatment failure. Verapamil, a calcium channel blocker, was one of the first compounds that were identified to circumvent resistance to vincristine (Tsuruo et al., 1981). Soon thereafter cyclosporine was found to be a modulator of resistance (Slater et al., 1986). Both drugs are inhibitors of P-glycoprotein. Early clinical trials with verapamil revealed the feasibility of chemosensitization in certain patients. However, cardiac toxicity was unacceptable for routine anticancer treatment (Salmon et al., 1991). Cyclosporine A caused side effects such as hypertension, nephrotoxicity, and hyperbilirubinemia (Yahanda et al., 1992). Efficient chemosensitization of refractory plasmocytoma by cyclosporine A was demonstrated by Sonneveld et al. (1992). In that study, cells expressing P-glycoprotein were eliminated from bone marrow. Similarly, verapamil suppresses the emergence of P-glycoprotein–mediated multidrug resistance in multiple myeloma cells (Futscher et al., 1996).

Because of the toxicity of chemosensitizing agents, new compounds were developed. For instance, D-verapamil, a compound that displays 10-fold less cardiotoxicity than the racemate verapamil at equal chemosensitizing potential, was shown to be effective against refractory lymphomas or sarcomas (Wilson et al., 1995). Other structural analogs of verapamil display even more potent chemosensitizing potential (Pirker et al., 1990). Similarly, derivatives of cyclosporines have been identified that combine improved chemomodulation with reduced immunosuppressive adverse effects. SDZ PSC 833, a derivative of cyclosporine D, has high reversing potency (Boesch et al., 1991; Gaveriaux et al., 1991). This compound is currently being investigated in clinical trials.

Due to its wide substrate specificity, P-glycoprotein interacts with numerous nontoxic compounds. Inhibitors of P-glycoprotein are structurally unrelated molecules, among them being steroid hormones and their antagonists (Ramu et al., 1984; Fleming et al., 1992), neuroleptic agents such as thioxanthenes and phenothiazines (Ford et al., 1989, 1990), antibiotics, and antifungal drugs (Gosland et al., 1989; Hofsli et al., 1989; Siegsmund et al., 1994; Kurosawa et al., 1996). In addition to single agents, combinations of inhibitors at suboptimal doses may display additive effects (Hwang et al., 1996). Thereby concentrations of inhibitors might be reduced to a nontoxic level.

Glutathione Metabolism and Inhibition of the MRP Transporter

Elevated levels of glutathione reduce the toxicity of alkylating agents, cisplatin, and anthracyclines. Conversely, depletion of glutathione can restore the chemosensitivity of cancer cells (Ozols et al., 1987). Compounds like buthionine sulfoximine, a synthetic amino acid, have been used to overcome drug resistance due to glutathione (Ozols et al., 1987; Sielmann et al., 1993; Saikawa et al., 1993). In recent phase I trials, buthionine sulfoximine was tolerated well, but the decrease of intracellular glutathione levels was only modest (Bailey et al., 1994; O'Dwyer et al., 1996). However, some drugs like pentoxifylline and ethacrynic acid act synergistically with buthionine sulfoximine (Lai et al., 1995; Chen et al., 1994). Thus, it appears to be possible to further increase the efficiency of chemosensitization by combining several agents for depletion of glutathione. On the other hand, it is possible that resistance to treatment with buthionine sulfoximine may occur. In a recent investigation, development of resistance to buthionine sulfoximine treatment was related to decreased expression of GST π (Yokomizo et al., 1995).

The recent understanding of the role of the MRP transporter in glutathione-related chemoresistance may allow new strategies to circumvent drug resistance. A dihydropyridine analog, NIK250, was found to reverse both MDR1- and MRP-mediated multidrug resistance (Abe et al., 1995; Tasaki et al., 1995). Many inhibitors of P-glycoprotein fail to reverse chemoresistance due to overexpression of MRP, but recently several drugs were identified that inhibit this transport molecule. For instance, genistein is capable of reversing MRP-related drug transport (Versantvoort et al., 1994). A quinolone antimicrobial, difloxacin, reverses drug resistance in resistant HL-60 leukemia cells, which express the MRP gene (Gollapudi et al., 1995), and a leukotriene LTD4 receptor antagonist, MK571, modulates MRP-associated, but not P-glycoprotein–related, multidrug resistance (Gekeler et al., 1995). Preclinical and clinical investigations are needed to evaluate the possibility to reverse drug resistance due to this transport system *in vivo.*

INTERFERENCE WITH EXPRESSION OF CHEMORESISTANCE GENES IN CANCER CELLS

While the latter approaches are aimed at inhibiting or blocking the function of drug resistance proteins, molecular strategies have been developed to interfere with synthesis of these proteins. This has been attempted with antisense oligonucleotides, ribozymes, and proteins regulating differentiation of cancer cells.

Transcription of the human MDR1 gene has been blocked with antisense oligonucleotides (Corrias et al., 1992; Efferth et al., 1993). Similarly, reduction of MRP expression can be accomplished with the use of antisense oligonucleotides (Stewart et al., 1996).

Like antisense oligonucleotides, ribozymes designed to cleave the MDR1 mRNA have the ability to restore chemosensitivity in multidrug-resistant cells (Kobayashi et al., 1993; Scanlon et al., 1994). Interestingly, not only MDR1-directed ribozymes but also ribozymes that interact with c-fos decrease expression of P-glycoprotein (Scanlon et al., 1994). The observation that ribozymes against transcription factors can modulate drug resistance relates to the complex regulation of MDR1 expression.

The promoters of the MDR1 gene contain several binding sites for transcription factors (Ueda et al., 1987; Gottesman et al., 1996). Because transcription factors are involved in regulation of the expression of drug resistance genes, they appear to be suitable targets for strategies to circumvent drug resistance. For instance, recent investigations have provided evidence that SP1 modulates transcriptional activity of the MRP gene (Zhu et al., 1996). Activity of the MDR1 promoter is modulated by ras oncogenes and p53 (Chin et al., 1992). Factors regulating differentiation of malignant cells may contribute to the complex multidrug-resistant phenotype. In certain cell lines, e.g., colon cancer lines, cytokines such as tumor necrosis factor-α can downregulate the expression of P-glycoprotein (Walther et al., 1994). In addition, exogenous tumor necrosis factor-α decreases proliferation of multidrug-resistant but not of chemosensitive myeloma cells (Salmon et al., 1989a).

A different approach is to eliminate selectively multidrug-resistant cells from heterogeneous populations by monoclonal antibodies or immunotoxins. This has been shown for P-glycoprotein. Multidrug-resistant cells can be removed immunomagnetically with monoclonal antibodies (Padmanabhan et al., 1993) or with antibodies and complement (Kulkarni et al., 1989). This may be useful in autologous transplantation if drug-resistant cancer cells contaminate bone marrow or apheresates. Treatment of multidrug-resistant cancers with monoclonal antibodies is also feasible *in vivo* as shown in xenograft models of human tumors (Tsuruo et al., 1989; Pearson et al., 1991). An immunotoxin consisting of a monoclonal P-glycoprotein-antibody attached to *Pseudomonas* exotoxin was used to target renal carcinoma cells (Mickisch et al., 1993). This approach was also used *in vivo* in a transgenic mouse model (Mickisch et al., 1993).

TRANSFER OF DRUG RESISTANCE GENES TO HEMATOPOIETIC CELLS FOR GENE THERAPY OF CANCER

Although drug resistance genes are mainly studied because of their association with failure of anticancer treatment, such genes may be useful tools for gene therapy. Transfer of drug resistance genes to chemosensitive cells may render them resistant and protect them from the adverse effects of chemotherapy. This may be particularly helpful for the hematopoietic system because most hematopoietic cells are highly susceptible to antineoplastic agents. In addition, it has been suggested that drug resistance genes may act as selectable markers *in vivo,* facilitating increased gene expression in transduced cells after exposure to drugs. Both applications have been investigated, and *in vivo* models have been established.

Transgenic Animal Models

Overexpression of chemoresistance genes in hematopoietic organs has initially been studied in transgenic animal models. To investigate the effects of overexpression of P-glycoprotein in bone marrow, transgenic mice were generated that expressed a human MDR1 cDNA under the control of a chicken β-actin promoter in virtually all bone marrow cells. While the normal function of bone marrow was not altered, severalfold higher doses of taxol and daunomycin could be administered safely as compared with normal animals of the respective background strains (Galski et al., 1989; Mickisch et al., 1991a). Protection of hematopoietic cells in these animals was specifically due to overexpression of the MDR1 transgene because it was circumvented by simultaneous administration of chemosensitizing agents like verapamil (Mickisch et al., 1991b). Subsequently, bone marrow cells of MDR1-transgenic mice were transplanted into lethally irradiated recipient mice. It was demonstrated that the recipients were also protected from myelosuppression following chemotherapy (Mickisch et al., 1992). Similarly, mice transgenic for DHFR were found to be protected from methotrexate administered at doses that were lethal for animals of the respective background strain (Isola et al., 1986). Transplantation of transgenic bone marrow protected recipient mice from lethal doses of methotrexate (May et al., 1995).

The availability of transgenic mouse models allowed the identification of additional potential functions of drug resistance genes. They may be involved in carcinogenesis by protecting cells from mutagenic compounds (Gottesman, 1988; Ferguson et al., 1993). For instance, rapid repair of O^6-methylguanine–DNA adducts in transgenic mice protects them from *N*-methyl-nitrosourea–induced thymic lymphomas (Liu et al., 1994). This protection can be targeted to other organs like liver by suitable promoter systems (Nakatsuru et al., 1993). It is important to note that growth and differentiation of hematopoietic cells revealed no major disturbances in transgenic animals.

Transfer of Chemoresistance Genes to Hematopoietic Cells

Based on investigations of transgenic animals, methods for transfer of chemoresistance genes to hematopoietic progenitor cells of normal recipients were developed. It has been shown that transfer of the multidrug resistance gene to normal hematopoietic precursor cells protects them from the toxicity of anticancer agents. K562 erythroleukemia cells (DelaFlor-Weiss et al., 1992) and primary bone marrow cells (McLachlin et al., 1990) were transduced with a vector containing a full-length MDR1 cDNA under control of Harvey sarcoma virus long terminal repeats (Pastan et al., 1988). Transduced cells were found to be resistant to multiple drugs, including taxol, colchicine, and daunomycin. Similar studies were performed with the use of a retroviral vector that contained the MRP transporter sequence (D'Hondt et al., 1995). Vectors containing an MRP cDNA may be useful for protection of hematopoietic cells if MDR1-mediated chemoresistance of tumor cells should be reversed simultaneously. The MRP transporter is less susceptible to most agents that reverse the function of P-glycoprotein. While transfer of these drug transporters confers resistance primarily to natural toxins, transfer of glutathione-S-transferase or O^6-alkylguanine DNA alkyltransferase genes can render cells resistant to alkylating agents such as chlorambucil, nitrosoureas, and mechlorethamine (Greenbaum et al., 1994; Liu et al., 1994). Moreover, a broad range of chemoresistance can be transferred by a retroviral vector in which glutathione-S-transferase π and MDR1 are coexpressed from separate promoters (Doroshow et al., 1995).

Animal Models for the Use of Drug Resistance Genes in Gene Therapy

Animal models aimed at studying the potential use of drug resistance genes in gene therapy were subsequently developed. MDR1-transduced murine bone marrow cells were transplanted into anemic W/W^v mice (Sorrentino et al., 1992) or lethally irradiated normal syngeneic mice (Podda et al., 1992). Both models detected elevated levels of MDR1 expression after treatment of recipient mice with taxol. These experiments favored the idea of a selective advantage *in vivo* of hematopoietic cells overexpressing the MDR1 transgene.

Further support was provided by experiments in which MDR1-transduced bone marrow was first transplanted into recipient mice. After taxol treatment of recipient mice their bone marrow was then transplanted into a second generation of recipient mice. In several cycles of retransplantation and taxol treatment of recipient mice, increasingly high levels of drug resistance were generated *in vivo*. Mice of the fifth and sixth generations survived doses of taxol that were lethal for normal, untreated mice (Hanania et al., 1994).

Likewise, vectors containing mutated DHFR cDNAs have been constructed for transduction of hematopoietic progenitor cells (Miller et al., 1985; Li et al., 1994; Zhao et al., 1994). Williams et al. (1987) and Cline et al. (1980) demonstrated protection of recipient animals from lethal doses of methotrexate. Retransplantation experiments performed with DHFR (Corey et al., 1990) gave results comparable to those obtained with MDR1: Both MDR1 and DHFR may act as selectable marker genes *in vivo*.

More recently, chemoresistance genes were transferred to isolated subpopulations of progenitor cells, namely, to hematopoietic stem cells from various origins. Hematopoietic stem cells are thought to be ideal target cells for gene therapy. Their life span is thought to be unlimited because they have the capacity of self-renewal, whereas more mature progenitor cells become more differentiated with each step of cell duplication and eventually undergo terminal differentiation followed by apoptosis. Targeting of more differentiated, lineage-committed progenitor cells may therefore result in merely transient gene expression (Lemischka, 1991). Incidental transduction of stem cells has been confirmed in numerous investigations by gene transfer to unfractionated bone marrow. Transduction of true pluripotent stem cells, though feasible, has usually been found to be elusive (Nolta et al., 1996). A major impediment to efficient transduction is the quiescence of the majority of stem cells (Ogawa, 1993), while cell proliferation is required for integration of retroviral sequences into the genome (Miller et al., 1992). The efficiency of gene transfer has been optimized with the use of growth factors and cytokines and by chemotherapeutic pretreatment of bone marrow donors.

We have transferred an MDR1 cDNA to a small population of mouse bone marrow cells that expressed neither lineage-specific antigens nor the MHC class II–associated I-a antigen, but had high levels of Sca-1, also referred to as Ly6A/E (Licht et al., 1995). Stem cell properties of this cell fraction were confirmed by transplantation into sublethally irradiated SCID mice, revealing the sustained presence of an MDR1 marker cDNA in recipient mice, multilineage engraftment, and the presence of the marker gene after retransplantation into a second generation of recipient mice. The isolated cells were expanded *ex vivo* with the use of growth factors while gene transfer was performed by coculturing with retrovirus producing GP + E86 fibroblasts. Functional human P-glycoprotein was detected in approximately 60% of expanded cells. Following transplantation of the *ex vivo*–transduced cell population into SCID mice, P-glycoprotein was expressed in a high proportion of bone marrow cells of recipient animals at levels comparable to those observed in multidrug-resistant cancers in the clinic.

Stem cells mobilized into peripheral blood of splenectomized mice by administration of granulocyte colony-stimulating factor and stem cell factor have been found to be useful targets for MDR1 gene transfer (Bodine et al., 1994). Additional sources for activated stem cells of which many are in cell cycle are cord blood and fetal liver. These investigations revealed biological differences between different stem cell populations. Transfer of the MDR1 gene to fetal liver stem cells was feasible only with ecotropic retroviruses since these cells lack receptors for amphotropic viruses (Richardson et al., 1994).

In a recent article, retroviral MDR1 transfer to MO-7e cells was reported. As compared with nontransduced cells of this factor-dependent human leukemia cell line, they became 20-fold more resistant to taxol following MDR1-gene transfer. After transplantation of MO-7e cells into immunodeficient NOD/SCID mice, cells expressing the MDR1 gene were able to survive taxol treatment, which was not observed with nontransduced cells (Schwarzenberger et al., 1996).

In conclusion, methods have been developed to overexpress P-glycoprotein and other drug resistance genes in hematopoietic progenitor cells *in vivo.* These investigations demonstrate protection of hematopoietic cells and the function of MDR1 as a drug-selectable marker gene *in vivo.*

Clinical Investigations on MDR1 Gene Transfer to Hematopoietic Progenitor Cells

Animal experiments demonstrate the feasibility of MDR1 gene transfer to hematopoietic progenitor cells. These investigations also suggest that P-glycoprotein expression can be maintained for several months *in vivo*. Transfer of the MDR1 gene to human $CD34^+$ progenitor cells was performed by Ward et al. (1994). Based on these results, clinical trials on transfer of the MDR1 gene to bone marrow cells have been initiated (O'Shaughnessy et al., 1994; Hesdorffer et al., 1994; Deisseroth et al., 1994). Bone marrow or peripheral blood progenitor cells from patients suffering from advanced breast or ovarian cancers or non-Hodgkin's lymphomas were retrovirally transduced and reinfused after high-dose chemotherapy. These phase I studies were aimed at proving efficiency, feasibility, and safety of MDR1 gene transfer. In some trials patients were treated with taxol following reinfusion of transduced cells. However, technical problems remain to be solved. In the first published clinical study, MDR1 expression levels were found to be disappointingly low (Hanania et al., 1996).

Coexpression of the Multidrug Resistance Gene and Nonselectable Genes

Low or unstable expression of transferred genes is still a major cause for failure of clinical gene therapy. To overcome the limitations of gene therapy the use of drug resistance genes for enrichment of cells that express encoded proteins has been suggested (Gottesman et al., 1991; Banerjee et al., 1994; Licht et al., 1997). If a selectable marker gene is coexpressed with a nonselectable gene, drug selection should elevate the expression levels of two genes.

While several approaches can be used to simultaneously express two genes, coexpression in translational or transcriptional fusion vectors has been found to be superior to cotransfection or the use of separate promoters (Gottesman et al., 1994). A translational fusion gene encoding a chimeric protein has been constructed by Germann et al. (1990). This protein consists of P-glycoprotein and adenosine deaminase, an enzyme that upon mutation causes a SCID syndrome. The bifunctional protein displayed both drug transport and adenosine deaminase activity.

Alternatively, polycistronic vectors have been engineered that facilitate transcription of multiple genes by internal ribosomal entry sites. In vectors containing the MDR1 cDNA and cDNAs that correct inherited metabolic disorders, a single mRNA containing both genes is transcribed from one retroviral promoter. This transcript is translated from two open reading frames into two separate proteins (Sugimoto et al., 1994). Transcriptional fusions between MDR1 and the glucocerebrosidase gene or the α-galactosidase gene, which are defective in Gaucher's and Fabry's diseases, respectively, allowed enrichment of transduced cells by drug selection (Aran et al., 1994; Sugimoto et al., 1995a). With a suitable selection strategy, complete restoration of the underlying enzyme deficiency can be accomplished in patient cells (Aran et al., 1996). Application of these vectors in clinical treatment and suitable selection strategies *in vivo* have still to be optimized. For *in vivo* characterization, a gene-marking vector has been constructed (Aran et al., 1995). This vector contains the lacZ gene; transduced cells are easily detectable by their β-galactosidase activity.

Another application of bicistronic vectors is to combine the MDR1 gene with a negative-dominant selectable marker gene such as thymidine kinase from herpes simplex virus (HSV-TK). If preparations of hematopoietic progenitor cells are contaminated with cancer cells, there is a risk of rendering the malignant cells chemoresistant. If this should happen, it would be advantageous to have a system that allows selective elimination of transduced cells. To this end, a vector containing HSV-TK and MDR1 has been constructed (Sugimoto et al., 1995b). HSK-TK is a suicide gene that confers hypersensitivity to the antiviral agent ganciclovir. Treatment with ganciclovir would thus kill transduced cancer cells. Such constructs may help to increase the safety of clinical gene therapy using the MDR1 gene.

CONCLUSION

Resistance to anticancer drugs is multicausative. An increasing number of genes have been identified that are associated with chemoresistance. While strategies developed to circumvent P-glycoprotein–mediated resistance have already been introduced into clinical treatment, approaches using other genes are still at the stage of preclinical models.

In contrast to attempts to overcome drug resistance, more recently research has focused on potential applications for chemoresistance genes in gene therapy. Transfer of drug resistance genes may protect hematopoietic cells from the toxicity of anticancer drugs. In addition, they may be useful tools to introduce and overexpress otherwise nonselectable genes into bone marrow. Thereby unstable expression of therapeutic genes that hampers efficient gene therapy of hematopoietic disorders may be improved.

REFERENCES

Abe T, Koike K, Ohga T, et al. (1995): Chemosensitisation of spontaneous multidrug resistance by a 1,4-dihydropyridine analogue and verapamil in human glioma cell lines overexpressing MRP or MDR1. *Br J Cancer* 72:418–423.

Ambudkar SV, Lelong IH, Zhang JP, et al. (1992): Partial purification and reconstitution of the human multidrug-resistance pump—Characterization of drug-stimulatable ATP hydrolysis. *Proc Natl Acad Sci USA* 89:8472–8476.

Ames GF-L, Lecar H (1992): ATP-dependent bacterial transporters and cystic fibrosis: Analogy between channels and transporters. *FASEB J* 6:2660–2666.

Aran JM, Gottesman MM, Pastan I (1994): Drug-selected coexpression of human glucocerebrosidase and P-glycoprotein using a bicistronic vector. *Proc Natl Acad Sci USA* 91:3176–3180.

Aran JM, Licht T, Gottesman MM, Pastan M (1995): Design of a multidrug-resistance–marking vector. *Proc Am Assoc Cancer Res* 36:336.

Aran JM, Licht T, Gottesman MM, Pastan I (1996): Complete restoration of glucocerebrosidase deficiency in Gaucher fibroblasts using a bicistronic MDR retrovirus and a new selection strategy. *Hum Gene Ther* 7:2165–2175.

Arrick BA, Nathan CF (1984): Glutathione metabolism as a determinant of therapeutic efficacy. *Cancer Res* 44:4224–4234.

Bai F, Nakanishi Y, Kawasaki M, et al. (1996): Immunohistochemical expression of glutathione S-transferase-pi can predict chemotherapy response in patients with nonsmall cell lung carcinoma. *Cancer* 78:416–421.

Bailey HH, Mulcahy RT, Tutsch KD, et al. (1994): Phase I clinical trial of intravenous L-buthionine sulfoximine and melphalan: An attempt at modulation of glutathione. *J Clin Oncol* 12:194–205.

Banerjee D, Zhao SC, Li MX, et al. (1994): Gene therapy utilizing drug resistance genes: A review. *Stem Cells* 12:378–385.

Baldini N, Scotlandi K, Barbanti-Bròdano G, et al. (1995): Expression of P-glycoprotein in high-grade osteosarcomas in relation to clinical outcome. *N Engl J Med* 333:1380–1385.

Black SM, Wolf CR (1991): The role of glutathione-dependent enzymes in drug resistance. *Pharmacol Ther* 51:139–154.

Bodine DM, Seidel NE, Gale MS, et al. (1994): Efficient retrovirus transduction of mouse pluripotent hematopoietic stem cells mobilized into the peripheral blood by treatment with granulocyte colony-stimulating factor and stem cell factor. *Blood* 84:1482–1491.

Boesch D, Müller K, Pourtier-Manzanedo A, et al. (1991): Restoration of daunomycin retention in multidrug-resistant P388 cells by submicromolar concentrations of SDZ PSC 833, a non-immunosuppressive cyclosporine derivative. *Exp Cell Res* 196:26–32.

Breuninger LM, Paul S, Gaughan K, et al. (1995): Expression of multidrug resistance–associated protein in NIH/3T3 cells confers multidrug resistance associated with increased drug efflux and altered intracellular drug distribution. *Cancer Res* 55:5342–5347.

Bruggemann EP, Currier SJ, Gottesman MM, Pastan I (1992): Characterization of the azidopine and vinblastine binding site of P-glycoprotein. *J Biol Chem* 264:15483–15488.

Campos L, Guyotat D, Archimbaud E, et al. (1992): Clinical significance of multidrug resistance P-glycoprotein expression on acute nonlymphoblastic leukemia cells at diagnosis. *Blood* 79:473–476.

Chan HSL, Thorner PS, Haddad G, et al. (1990): Immunohistochemical detection of P-glycoprotein: Prognostic correlation in soft tissue sarcoma of childhood. *J Clin Oncol* 8:689–704.

Chaudhary PM, Roninson IB (1991): Expression and activity of P-glycoprotein, a multidrug efflux pump, in human hematopoietic stem cells. *Cell* 66:85–94.

Chaudhary PM, Mechetner EB, Roninson IB (1992): Expression and activity of the multidrug resistance P-glycoprotein in human peripheral blood lymphocytes. *Blood* 80:2729–2734.

Chen G, Waxman DJ (1994): Role of cellular glutathione and glutathione S-transferase in the expression of alkylating agent cytotoxicity in human breast cancer cells. *Biochem Pharmacol* 47:1079–1087.

Cheng AL, Su IJ, Chen YC, et al. (1993): Expression of P-glycoprotein and glutathione-S-transferase in recurrent lymphomas: The possible role of Epstein-Barr virus, immunophenotypes, and other predisposing factors. *J Clin Oncol* 11:109–115.

Chin K-V, Ueda K, Pastan I, Gottesman MM (1992): Modulation of the activity of the promoter of the human MDR1 gene by ras and p53. *Science* 255:459–462.

Chuman Y, Chen ZS, Sumizawa T, et al. (1996): Characterization of the ATP-dependent LTC4 transporter in cisplatin-resistant human KB cells. *Biochem Biophys Res Commun* 226:158–165.

Cline MJ, Stang H, Mercola K, et al. (1980): Gene transfer in intact animals. *Nature* 284:422–425.

Cole SPC, Bhardwaj G, Gerlach JH, et al. (1992): Overexpression of a transporter gene in a multidrug-resistant human lung cancer cell line. *Science* 258:1650–1654.

Cole SP, Sparks KE, Fraser K, et al. (1994): Pharmacological characterization of multidrug resistant MRP-transfected human tumor cells. *Cancer Res* 54:5902–5910.

Corey CA, DeSilva AD, Holland CA, Williams DA (1990): Serial transplantation of methotrexate resistant bone marrow: Protection of murine recipients from drug toxicity by progeny of transduced stem cells. *Blood* 76:337–343.

Cornwell MM, Safa AR, Felsted RL, et al. (1986): Membrane vesicles from multidrug-resistant human cancer cells contain a specific 150–170 kDa protein detected by photoaffinity labeling. *Proc Natl Acad Sci USA* 83:3847–3890.

Corrias MV, Tonini GP (1992): An oligomer complementary to the 5′ end region of the MDR1 gene decreases resistance to doxorubicin of human adenocarcinoma multidrug-resistant cells. *Anticancer Res* 12:1431–1438.

Deisseroth AB, Kavanagh J, Champlin R (1994): Use of safety-modified retroviruses to introduce chemotherapy resistance sequences into normal hematopoietic cells for chemoprotection during the therapy of ovarian cancer: A pilot trial. *Hum Gene Ther* 5:1507–1522.

DelaFlor-Weiss E, Richardson C, Ward M, et al. (1992): Transfer and expression of the human multidrug resistance gene in mouse erythroleukemia cells. *Blood* 80:3106–3111.

D'Hondt V, Caruso M, Ward M, et al. (1995): Expression of the multidrug resistance gene using retroviral vectors. *Blood* 86:466A.

Dicker AP, Volkenandt M, Schweitzer BI, et al. (1990): Identification and characterization of a mutation in the dihydrofolate reductase gene from the methotrexate resistant CHO cell line Pro^{-3} MTX^{RIII}. *J Biol Chem* 265:8317–8321.

Doroshow JH, Metz MZ, Matsumoto L, et al. (1995): Transduction of NIH 3T3 cells with a retrovirus carrying both human MDR1 and glutathione S-transferase pi produces broad-range multidrug resistance. *Cancer Res* 55:4073–4078.

Drach D, Zhao S, Drach J, et al. (1992): Subpopulations of normal and peripheral blood and bone marrow cells express a functional multidrug resistant phenotype. *Blood* 80:2735–2739.

Efferth T, Volm M (1993): Modulation of P-glycoprotein mediated multidrug resistance by monoclonal antibodies, immunotoxins or antisense oligodeoxynucleotides in kidney carcinoma and normal kidney cells. *Oncology* 50:303–308.

Endicott JA, Ling V (1989): The biochemistry of P-glycoprotein–mediated multidrug resistance. *Annu Rev Biochem* 58:137–171.

Ferguson L, Baguley BC (1993): Multidrug resistance and mutagenesis. *Mutat Res* 285:79–90.

Fleming GF, Amato JM, Agresti M, et al. (1992): Megestrol acetate reverses multidrug resistance and interacts with P-glycoprotein. *Cancer Chemother Pharmacol* 29:445–449.

Flens MJ, Zaman GJ, van der Valk P, et al. (1996): Tissue distribution of the multidrug resistance protein. *Am J Pathol* 148:1237–1247.

Ford JM, Bruggemann EP, Pastan I, et al. (1990): Cellular and biochemical characterization of thioxanthenes for reversal of multidrug resistance in human and murine cell lines. *Cancer Res* 50:1748–1756.

Ford JM, Prozialeck WC, Hait WN (1989): Structural features determining activity of phenothiazines and related drugs for inhibition of cell growth and reversal of multidrug resistance in human and murine cell lines. *Mol Pharmacol* 35:105–115.

Fuchs B, Ostmeier H, Suter L (1991): P-glycoprotein expression in malignant melanoma. *J Cancer Res Clin Oncol* 117:168–171.

Futscher BW, Foley NE, Gleason-Guzman MC, et al. (1996): Verapamil suppresses the emergence of P-glycoprotein–mediated multi-drug resistance. *Int J Cancer* 66:520–525.

Galski H, Sullivan M, Willingham MC, et al. (1989): Expression of a human multidrug resistance cDNA (MDR1) in the bone marrow of transgenic mice: Resistance to daunomycin induced leukopenia. *Mol Cell Biol* 9:4357–4363.

Gaveriaux C, Boesch D, Jachez B (1991): SDZ PSC 833, a non-immunosuppressive cyclosporin analog is a very potent multidrug-resistant modifier. *J Cell Pharmacol* 2:225–234.

Germann UA (1993): Molecular analysis of the multidrug transporter. *Cytotechnology* 12:33–62.

Germann UA, Chin K-V, Pastan I, Gottesman MM (1990): Retroviral transfer of a chimeric multidrug resistance-adenosine deaminase gene. *FASEB J* 4:1501–1507.

Gekeler V, Ise W, Sanders KH, et al. (1995): The leukotriene LTD4 receptor antagonist MK571 specifically modulates MRP associated multidrug resistance. *Biochem Biophys Res Commun* 208:345–352.

Giaccone G, van Ark-Otte J, Rubio GJ, et al. (1996): MRP is frequently expressed in human lung cancer cell lines, in non-small-cell lung cancer and in normal lungs. *Int J Cancer* 66:760–767.

Godwin AK, Meister A, O'Dwyer PJ, et al. (1992): High resistance to cisplatin in human ovarian cancer cell lines is associated with marked increase of glutathione synthesis. *Proc Natl Acad Sci USA* 89:3070–3074.

Gollapudi S, Thadepalli F, Kim CH, Gupta S (1995): Difloxacin reverses multidrug resistance in HL-60/AR cells that overexpress the multidrug resistance-related protein (MRP) gene. *Oncol Res* 7:213–225.

Goldstein L, Galski H, Fojo A, et al. (1989): Expression of a multidrug resistance gene in human cancers. *J Natl Cancer Inst* 81:116–124.

Gosland MP, Lum BL, Sikic BI (1989): Reversal by cefoperazone of resistance to etoposide, doxorubicin and vinblastine in multidrug resistant human sarcoma cells. *Cancer Res* 49:6901–6905.

Gottesman MM (1988): Multidrug-resistance during chemical carcinogenesis: A mechanism revealed? *J Natl Cancer Inst* 80:1352–1353.

Gottesman MM, Ambudkar SV, Cornwell MM, Pastan I, Germann UA (1996): Multidrug resistance transporter. In Schultz SG, et al. (eds): *Molecular Biology of Membrane Transport Disorders.* New York; Plenum Press, pp 243–257.

Gottesman MM, Germann UA, Aksentijevich I, Sugimoto Y, Cardarelli CO, Pastan I (1994): Gene transfer of drug resistance genes. *NY Acad Sci* 716:126–139.

Gottesman MM, Hrycyna CA, Schoenlein PV, et al. (1995): Genetic analysis of the multidrug transporter. *Annu Rev Genet* 29:607–649.

Gottesman MM, Pastan I (1991): The multidrug resistance (MDR1) gene as a selectable marker in gene therapy. In: Cohen-Haguenauer O, Boiron M (eds): *Human Gene Transfer,* vol. 219. London: Coloque INSERM/John Libbey Eurotext, pp 185–191.

Gottesman MM, Pastan I (1993): Biochemistry of multidrug resistance mediated by the multidrug transporter. *Annu Rev Biochem* 62:385–427.

Grant CE, Valdimarsson G, Hipfner DR, et al. (1994): Overexpression of multidrug resistance–associated protein (MRP) increases resistance to natural product drugs. *Cancer Res* 54:357–361.

Greenbaum M, Letourneau S, Assar H, et al. (1994): Retrovirus-mediated gene transfer of rat glutathione S-transferase Yc confers alkylating drug resistance in NIH 3T3 mouse fibroblasts. *Cancer Res* 54:4442–4447.

Grogan TM, Spier CM, Salmon SE, et al. (1993): P-glycoprotein expression in human plasma cell myeloma: Correlation with prior chemotherapy. *Blood* 81:490–495.

Gros P, Croop J, Housman DE (1986): Mammalian multidrug resistance gene: Complete cDNA sequence indicates strong homology with bacterial transport proteins. *Cell* 47:371–380.

Haber DA, Beverly SM, Kiely ML, et al. (1981): Properties of an altered dihydrofolate reductase encoded by amplified genes in cultured mouse fibroblasts. *J Biol Chem* 256:9501–9510.

Hamada S, Kamada M, Furumoto H, et al. (1994): Expression of glutathione S-transferase-pi in human ovarian cancer as an indicator of resistance to chemotherapy. *Gynecol Oncol* 52:313–319.

Hanania EG, Deisseroth AB (1994): Serial transplantation shows that early hematopoietic precursor cells are transduced by MDR-1 retroviral vector in a mouse gene therapy model. *Cancer Gene Ther* 1:21–25.

Hanania EG, Giles RE, Kavanagh J, et al. (1996): Results of MDR-1 vector modification trial indicate that granulocyte-macrophage colony-forming unit cells do not contribute to posttransplant hematopoietic recovery following intensive chemotherapy. *Proc Natl Acad Sci USA* 93:15346–15351.

Henderson DM, Sifri CD, Rodgers M (1992): Multidrug resistance in *Leishmania donovani* is conferred by amplification of a gene homologous to the mammalian *mdr*-1 gene. *Mol Cell Biol* 12:2855–2865.

Hesdorffer C, Antman K, Bank A, et al. (1994): Human MDR1 gene transfer in patients with advanced cancer. *Hum Gene Ther* 5:1151–1160.

Higgins CF (1993): The ABC transporter channel superfamily—An overview. *Semin Cell Biol* 4:1–5.

Hofsli E, Nissen-Meyer J (1989): Effect of erythromycin and tumor necrosis factor on the drug resistance of multidrug-resistant cells: Reversal of drug resistance by erythromycin. *Int J Cancer* 43:520–525.

Horio M, Gottesman MM, Pastan I (1988): ATP-dependent transport of vinblastine in vesicles from human multidrug-resistant cells. *Proc Natl Acad Sci USA* 85:3580–3584.

Hwang M, Ahn C-H, Pine PS, et al. (1996): Effect of combination of suboptimal concentrations of P-glycoprotein blockers on the proliferation of MDR1 gene expressing cells. *Int J Cancer* 65:389–397.

Hyde SC, Emsley P, Hartshorn MJ, et al. (1990): Structural and functional relationships of ATP-binding proteins associated with cystic fibrosis, multidrug resistance and bacterial transport. *Nature* 346:362–365.

Isola LM, Gordon JW (1986): Systemic resistance to methotrexate in transgenic mice carrying a mutant dihydrofolate reductase gene. *Proc Natl Acad Sci USA* 83:9621–9625.

Izquierdo MA, van der Zee AG, Vermorken JB, et al. (1995): Drug resistance-associated marker Lrp for prediction of response to chemotherapy and prognoses in advanced ovarian carcinoma. *J Natl Cancer Inst* 87:1230–1237.

Izquierdo MA, Shoemaker RH, Flens-MJ, et al. (1996): Overlapping phenotypes of multidrug resistance among panels of human cancer-cell lines. *Int J Cancer* 65:230–237.

Jedlitschky G, Leier I, Buchholz U, et al. (1996): Transport of glutathione, glucuronate, and sulfate conjugates by the MRP gene-encoded conjugate export pump. *Cancer Res* 56:988–994.

Klimecki WT, Futscher BW, Grogan TM, et al. (1994): P-glycoprotein expression and function in circulating blood cells from normal volunteer. *Blood* 83:2451–2458.

Kobayashi H, Dorai T, Holland JF, Ohnuma T (1993): Cleavage of human MDR1 by a hammerhead ribozyme. *FEBS Lett* 319:71–74.

Kodera Y, Isobe K, Yamauchi M, et al. (1994): Expression of glutathione-S-transferases alpha and pi in gastric cancer: A correlation with cisplatin resistance. *Cancer Chemother Pharmacol* 34:203–208.

Kruh GD, Chan A, Myers K, et al. (1994): Expression complementary DNA library transfer establishes mrp is a multidrug resistance gene. *Cancer Res* 54:1649–1652.

Kulkarni SS, Wang Z, Spitzer G, et al. (1989): Elimination of drug-resistant myeloma tumor cell lines by monoclonal anti-P-glycoprotein antibody and rabbit complement. *Blood* 74:2244–2251.

Kurosawa M, Okabe M, Hara N, et al. (1996): Reversal effect of itraconazole on adriamycin and etoposide resistance in human leukemia cells. *Ann Hematol* 72:17–21.

Kuss BJ, Deeley RG, Cole SP, et al. (1994): Deletion of gene for multidrug resistance in acute myeloid leukaemia with inversion in chromosome 16: Prognostic implications. *Lancet* 343:1531–1534.

Lai SL, Hwang J, Perng RP, et al. (1995): Modulation of cisplatin resistance in acquired-resistant nonsmall cell lung cancer cells. *Oncol Res* 7:31–38.

Leier I, Jedlitschky G, Buchholz U, et al. (1994): The MRP gene encodes an ATP-dependent export pump for leukotriene C4 and structurally related conjugates. *J Biol Chem* 69:27807–27810.

Lemischka IR (1991): Clonal, *in vivo* behavior of the totipotent hematopoietic stem cell. *Semin Immunol* 3:349–355.

Li M-X, Banarjee D, Zhao S-C, et al. (1994): Development of a retroviral construct containing a human mutated dihydrofolate reductase cDNA for hematopoietic stem cell transduction. *Blood* 83:3403–3408.

Licht T, Aksentijevich I, Gottesman MM, Pastan I (1995): Efficient expression of functional human MDR1 gene in murine bone marrow after retroviral transduction of purified hematopoietic stem cells. *Blood* 86:111–121.

Licht T, Herrmann F, Gottesman MM, Pastan I (1997): *In vivo* drug-selectable genes: A new concept in gene therapy. *Stem Cells* 15:104–111.

Lincke CR, Smit JJM, van der Velde-Koerts T, Borst P (1991): Structure of the human MDR3 gene and physical mapping of the MDR locus. *J Biol Chem* 266:5303–5310.

List AF, Spier CS, Grogan TM, et al. (1996): Overexpression of the major vault transporter protein lung-resistance protein predicts treatment outcome in acute myeloid leukemia. *Blood* 87:2464–2469.

Liu L, Allay E, Dumenco LL, et al. (1994): Rapid repair of O^6-methylguanine–DNA adducts protects transgenic mice from *N*-methylnitrosourea–induced thymic lymphomas. *Cancer Res* 54:4648–4652.

May C, Gunther R, McIvor RS (1995): Protection of mice from lethal doses of methotrexate by transplantation with transgenic marrow expressing drug-resistant dihydrofolate reductase activity. *Blood* 86:2439–2448.

McLachlin JR, Eglitis MA, Ueda K, et al. (1990): Expression of a human complementary DNA for the human multidrug resistance gene in murine hematopoietic precursor cells with the use of retroviral gene transfer. *J Natl Cancer Inst* 82:1260–1263.

Mickisch GH, Aksentijevich I, Schoenlein PV, et al. (1992): Transplantation of bone marrow cells from transgenic mice expressing the human MDR1 gene results in long-term protection against the myelosuppressive effect of chemotherapy in mice. *Blood* 79:1087–1093.

Mickisch GH, Licht T, Merlino GT, et al. (1991a): Chemotherapy and chemosensitization of transgenic mice which express the human multidrug resistance gene in bone marrow: Efficacy, potency and toxicity. *Cancer Res* 51:5417–5424.

Mickisch GH, Merlino GT, Galski H, et al. (1991b): Transgenic mice which express the human multidrug resistance gene in bone marrow enable a rapid identification of agents which reverse drug resistance. *Proc Natl Acad Sci USA* 88:547–551.

Mickisch GH, Pai LH, Siegsmund M, et al. (1993): *Pseudomonas* exotoxin conjugated to monoclonal antibody MRK16 specifically kills multidrug resistant cells in cultured renal cell carcinomas and in MDR-transgenic mice. *J Urol* 149:174–178.

Miller AD, Adam MA, Miller AD (1992): Gene transfer by retrovirus vectors occurs only in cells that are actively replicating at the time of infection. *Mol Cell Biol* 10:4239–4242.

Miller AD, Law M-F, Verma IM (1985): Generation of a helper free amphotropic retrovirus that transduces a dominant acting methotrexate resistant dihydrofolate reductase gene. *Mol Cell Biol* 5:431–437.

Muller M, Meijer C, Zaman GJ, et al. (1994): Overexpression of the gene encoding the multidrug resistance-associated protein results in increased ATP-dependent glutathione S-conjugate transport. *Proc Natl Acad Sci USA* 91:13033–13037.

Musto P, Lombardi G, Matera R, et al. (1991): The expression of the multidrug transporter P-170 glycoprotein in remission phase is associated with early and resistant relape in multiple myeloma. *Haematologica* 76:513–516.

Nakatsuru Y, Matsukuma S, Nemoto N, et al. (1993): O^6-methylguanine DNA methyltransferase protects against nitrosamine-induced hepatocarcinogenesis. *Proc Natl Acad Sci USA* 90:6468–6472.

Neyfakh AA, Serpinskaya AS, Chervonsky AV, et al. (1989): Multidrug-resistance phenotype of a subpopulation of T-lymphocytes without drug selection. *Exp Cell Res* 185:496–505.

Nolta JA, Dao MA, Wells S, et al. (1996): Transduction of pluripotent human hematopoietic stem cells demonstrated by clonal analysis after engraftment in immune-deficient mice. *Proc Natl Acad Sci USA* 93:2414–2419.

Norris MD, Bordow SB, Marshall GM, et al. (1996): Expression of the gene for multidrug-resistance associated-protein and outcome in patients with neuroblastoma. *N Engl J Med* 334:231.

O'Dwyer PJ, Hamilton TC, LaCreta FP, et al. (1996): Phase I trial of buthionine sulfoximine in combination with melphalan in patients with cancer. *J Clin Oncol* 14:249–256.

Ogawa M (1993): Differentiation and proliferation of stem cells. *Blood* 81:2844–2853.

O'Shaughnessy JA, Cowan KH, Nienhuis AW, et al. (1994): Retroviral mediated transfer of the human multidrug resistance gene (MDR-1) into hematopoietic stem cells during autologous transplantation after intensive chemotherapy for metastatic breast cancer. *Hum Gene Ther* 5:891–911.

Ozols RF, Louie KG, Plowman BC, et al. (1987): Enhanced melphalan cytotoxicity in human ovarian cancer *in vitro* and in tumor-bearing nude mice by buthionine sulfoximine depletion of glutathione. *Biochem Pharmacol* 36:147–153.

Padmanabhan R, Padmanabhan R, Howard T, et al. (1993): Magnetic affinity cell sorting to isolate transiently transfected cells, multidrug-resistant cells, somatic cell hybrids, and virally infected cells. *Methods Enzymol* 218:637–651.

Pastan I, Gottesman MM, Ueda K, et al. (1988): A retrovirus carrying an MDR1 cDNA confers multidrug resistance and polarized expression of P-glycoprotein in MDCK cells. *Proc Natl Acad Sci USA* 85:4486–4490.

Paul S, Breuninger LM, Tew KD, et al. (1996): ATP-dependent uptake of natural product cytotoxic drugs by membrane vesicles establishes MRP as a broad specificity transporter. *Proc Natl Acad Sci USA* 93:6929–6934.

Pearson JW, Fogler WE, Volker K, et al. (1991): Reversal of drug resistance in a human colon cancer xenograft expressing MDR1 complementary DNA by *in vivo* administration of MRK-16 monoclonal antibody. *J Natl Cancer Inst* 83:1386–1391.

Petrini M, Di Simone D, Favati A, et al. (1995): GST-pi and P-170 co-expression in multiple myeloma. *Br J Haematol* 90:393–397.

Pirker R, Keilhauer G, Raschak M, et al. (1990): Reversal of multi-drug resistance in human KB cell lines by structural analogs of verapamil. *Int J Cancer* 45:916–919.

Pirker R, Wallner J, Gotzl M, et al. (1992): MDR1 RNA expression is an independent prognostic factor in acute myeloid leukemia. *Blood* 80:557–559.

Podda S, Ward M, Himelstein A, et al. (1992): Transfer and expression of the human multiple drug resistance gene into live mice. *Proc Natl Acad Sci USA* 89:9676–9680.

Puchalski RB, Fahl WE (1990): Expression of recombinant glutathione S-transferase pi, Ya, or Yb1 confers resistance to alkylating agents. *Proc Natl Acad Sci USA* 87:2443–2447.

Ramu A, Glaubiger D, Fuks Z (1984): Reversal of acquired resistance to doxorubicin in P388 murine leukemia cells by tamoxifen and other triparanol analogues. *Cancer Res* 44:4392–4395.

Richardson C, Ward M, Podda S, Bank A (1994): Mouse fetal liver cells lack amphotropic retroviral receptors. *Blood* 84:433–439.

Russo D, Marie JP, Zhou DC, et al. (1994): Evaluation of the clinical relevance of the anionic glutathione-S-transferase (GST pi) and multidrug resistance (mdr-1) gene coexpression in leukemias and lymphomas. *Leuk Lymphoma* 15:453–468.

Saikawa Y, Kubota T, Kuo TH, et al. (1993): Enhancement of antitumor activity of cisplatin on human gastric cancer cells *in vitro* and *in vivo* by buthionine sulfoximine. *Jpn J Cancer Res* 84:787–793.

Salmon SE, Dalton WS, Grogan TM, Plezia P, Lehnert M, Roe DJ, Miller TP (1991): Multidrug-resistant myeloma: Laboratory and clinical effects of verapamil as a chemosensitizer. *Blood* 78:44–50.

Salmon SE, Grogan TM, Miller T, et al. (1989b): Prediction of doxorubicin resistance *in vitro* in melanoma, lymphoma, and breast cancer by P-glycoprotein staining. *J Natl Cancer* 81:696–701.

Salmon SE, Soehnlen B, Dalton WS, et al. (1989a): Effects of tumor necrosis factor on sensitive and multidrug resistant human leukemia and myeloma cell lines. *Blood* 74:1723–1727.

Scanlon KJ, Ishida H, Kashani-Sabet M (1994): Ribozyme-mediated reversal of the multidrug-resistant phenotype. *Proc Natl Acad Sci USA* 91:11123–11127.

Schadendorf D, Makki A, Stahr C, et al. (1995): Membrane transport proteins associated with drug resistance expressed in human melanoma. *Am J Pathol* 47:1545–1552.

Schecter RL, Alaoui-Jamali MA, Woo A, et al. (1993): Expression of a rat glutathione-S-transferase complementary DNA in rat mammary carcinoma cells: Impact upon alkylator-induced toxicity. *Cancer Res* 53:4900–4906.

Scheffer GL, Wijngaard PL, Flens MJ, et al. (1995): The drug resistance-related protein LRP is the human major vault protein. *Nat Med* 1:578–582.

Scheper RJ, Broxterman HJ, Scheffer GL, et al. (1993): Overexpression of a M(r) 110,000 vesicular protein in non-P-glycoprotein–mediated multidrug resistance. *Cancer Res* 53:1475–1479.

Schinkel AH, Smit JJM, van Tellingen O, et al. (1994): Disruption of the mouse mdr1a leads to a deficiency in the blood–brain barrier and to increased sensitivity to drugs. *Cell* 77:491–502.

Schneider E, Horton JK, Yang CH, et al. (1994): Multidrug resistance associated protein gene overexpression and reduced drug sensitivity of topoisomerase II in a human breast carcinoma MCF7 cell line selected for etoposide resistance. *Cancer Res* 54:152–158.

Schwarzenberger P, Spence S, Lohrey N, et al. (1996): Gene transfer of multidrug resistance into a factor-dependent human progenitor cell line: *In vivo* model for genetically transferred chemoprotection. *Blood* 87:2723–2731.

Siegsmund MJ, Cardarelli C, Aksentijevich I, et al. (1994): Ketoconazole effectively reverses multidrug resistance in highly resistant KB cells. *J Urol* 151:485–491.

Siemann DW, Beyers KL (1993): *In vivo* therapeutic potential of combination thiol depletion and alkylating chemotherapy. *Br J Cancer* 68:1071–1079.

Simonsen CC, Levinson AD (1983): Isolation and expression of an altered mouse dihydrofolate reductase cDNA. *Proc Natl Acad Sci USA* 80:2495–2499.

Slater LM, Sweet P, Stupeky M, Gaptz S (1986): Cyclosporin A reverses vincristine and daunorubicin resistance in acute lymphatic leukemia *in vitro*. *J Clin Invest* 77:1405–1408.

Slovak ML, Ho P, Cole SP, et al. (1995): The LRP gene encoding a major vault protein associated with drug resistance maps proximal to MRP on chromosome 16: Evidence that chromosome breakage plays a key role in MRP or LRP gene amplification. *Cancer Res* 55:4214–4219.

Smit, JJM, Schinkel AH, Oude Elferink RPJ, et al. (1993): Homozygous disruption of the murine mdr2 P-glycoprotein gene leads to a complete absence of phospholipid from bile and to liver disease. *Cell* 75:451–462.

Sonneveld P, Durie BDM, Lokhorst HM, et al. (1992): Modulation of multidrug-resistant multiple myeloma by cyclosporine. *Lancet* 340:255–259.

Sorrentino BP, Brandt SJ, Bodine D, et al. (1992): Selection of drug-resistant bone marrow cells *in vivo* after retroviral transfer of human MDR1. *Science* 257:99–103.

Srimatkandada S, Schweitzer BI, Moroson BA, et al. (1989): Amplification of a polymorphic dihydrofolate reductase gene expressing an enzyme with a decreased binding to MTX in a human colon carcinoma cell line, HCT-8R4 resistant to this drug. *J Biol Chem* 264:3524.

Stein W, Cardarelli C, Pastan I, Gottesman MM (1994): Kinetic evidence suggesting that the multidrug transporter differentially handles influx and efflux of its substrates. *Mol Pharmacol* 45:763–772.

Stewart AJ, Canitrot Y, Baracchini E, et al. (1996): Reduction of expression of the multidrug resistance protein (MRP) in human tumor cells by antisense phosphorothioate oligonucleotides. *Biochem Pharmacol* 51:461–469.

Sugawara I, Kataoka I, Morishita Y, et al. (1988): Tissue distribution of P-glycoprotein encoded by a multidrug-resistant gene as revealed by a monoclonal antibody, MRK16. *Cancer Res* 48:1926–1929.

Sugawara I, Yamada H, Nakamura H, et al. (1995): Preferential expression of the multidrug-resistance-associated protein (MRP) in adenocarcinoma of the lung. *Int J Cancer* 64:322–325.

Sugimoto Y, Aksentijevich I, Gottesman MM, Pastan I (1994): Efficient expression of drug-selectable genes under control of an internal ribosome entry site. *Biotechnology* 12:694–698.

Sugimoto Y, Aksentijevich I, Murray G, Brady RO, Pastan I, Gottesman MM (1995a): Retroviral co-expression of a multidrug-resistance gene (MDR1) and human alpha-galactosidase A for gene therapy of Fabry disease. *Hum Gene Ther* 6:905–915.

Sugimoto Y, Hrycyna CA, Aksentijevich I, Pastan I, Gottesman MM (1995b): Co-expression of a multidrug resistance gene (MDR1) and herpes simplex virus thymidine kinase gene as a part of a bicistronic mRNA in a retrovirus vector allows selective killing of MDR1-transduced cells. *Clin Cancer Res* 2:447–457.

Tasaki Y, Nakagawa M, Ogata J, et al. (1995): Reversal by a dihydropyridine derivative of non-P-glycoprotein–mediated multidrug resistance in etoposide-resistant human prostatic cancer cell line. *J Urol* 154:1210–1216.

Thiebaut F, Tsuruo T, Hamada H, et al. (1987): Cellular localization of the multidrug-resistance gene product in normal human tissues. *Proc Natl Acad Sci USA* 84:7735–7738.

Tsuruo T, Hamada H, Sato S, Heike Y (1989): Inhibition of multidrug-resistant human tumor cell growth in athymic mice by anti-P-glycoprotein monoclonal antibodies. *Jpn J Cancer Res* 80:627–631.

Tsuruo T, Iida K, Tsukagoshi S, et al. (1981): Overcoming of vincristine resistance in P388 leukemia *in vivo* and *in vitro* through enhanced cytotoxicity of vincristine and vinblastine by verapamil. *Cancer Res* 43:808–813.

Ueda K, Pastan I, Gottesman MM (1987): Isolation and sequence of the promoter region of the human multidrug-resistance (P-glycoprotein) gene. *J Biol Chem* 262:17432–17436.

van Bladeren PJ, van Ommen B (1991): The inhibition of glutathione S-transferases: Mechanisms, toxic consequences and therapeutic benefits. *Pharmacol Ther* 51:35–46.

van Helvoort A, Smith AJ, Sprong H, et al. (1996): MDR1 P-glycoprotein is a lipid translocase of broad specificity, while MDR3 P-glycoprotein specifically translocates phosphatidylcholine. *Cell* 87:507–515.

Versantvoort CH, Broxterman HJ, Lankelma J, et al. (1994): Competitive inhibition by genistein and ATP dependence of daunorubicin transport in intact MRP overexpressing human small cell lung cancer cells. *Biochem Pharmacol* 48:1129–1136.

Versantvoort CH, Broxterman HJ, Bagrij T, et al. (1995): Regulation by glutathione of drug transport in multidrug-resistant human lung tumour cell lines overexpressing multidrug resistance-associated protein. *Br J Cancer* 72:82–89.

Volm M, Kastel M, Mattern J, et al. (1993): Expression of resistance factors (P-glycoprotein, glutathione S-transferase-pi, and topoisomerase II) and their interrelationship to protooncogene products in renal cell carcinomas. *Cancer* 71:3981–3987.

Walther W, Stein U (1994): Influence of cytokines on mdr1 expression in human colon carcinoma cell lines: Increased cytotoxicity of MDR relevant drugs. *J Cancer Res Clin Oncol* 120:471.

Ward M, Richardson C, Pioli P, et al. (1994): Transfer and expression of the human multiple drug resistance gene in human CD34$^+$ cells. *Blood* 84:1408–1414.

Waxman DJ (1990): Glutathione S-transferases: Role in alkylating agent resistance and possible target for modulation chemotherapy—A review. *Cancer Res* 50:6449–6454.

Wei LY, Stutts MJ, Hoffman MM, Roepe PD (1995): Overexpression of the cystic fibrosis transmembrane conductance regulator in NIH 3T3 cells lowers membrane potential and intracellular pH and confers a multidrug resistance phenotype. *Biophys J* 69:883–895.

Williams DA, Hsieh K, DeSilva A, Mulligan RC (1987): Protection of bone marrow transplant recipients from lethal doses of methotrexate by the generation of methotrexate resistant bone marrow. *J Exp Med* 166:210–218.

Wilson CM, Serrano A, Wasley A, et al. (1986): Amplification of a gene related to mammalian mdr genes in drug-resistant *Plasmodium falciparum. Science* 244:1184–1186.

Wilson WH, Jamis-Dow C, Bryant G, et al. (1995): Phase I and pharmacokinetic study of the multidrug resistance modulator dexverapamil with EPOCH chemotherapy. *J Clin Oncol* 13:1985–1994.

Yahanda AM, Adler KM, Fisher GA, et al. (1992): Phase I trial of etoposide with cyclosporine as a modulator of multidrug resistance. *J Clin Oncol* 10:1624–1634.

Yokomizo A, Kohno K, Wada M, et al. (1995): Markedly decreased expression of glutathione S-transferase pi gene in human cancer cell lines resistant to buthionine sulfoximine, an inhibitor of cellular glutathione synthesis. *J Biol Chem* 270:19451–19457.

Zaman GJ, Flens MJ, van Leusden MR, et al. (1994): The human multidrug resistance-associated protein MRP is a plasma membrane drug-efflux pump. *Proc Natl Acad Sci USA* 91:8822–8826.

Zaman GJ, Lankelma J, van Tellingen O, et al. (1995): Role of glutathione in the export of compounds from cells by the multidrug-resistance-associated protein. *Proc Natl Acad Sci USA* 92:7690–7694.

Zaman GJR, Versantvoort CHM, Smit JJM, et al. (1993): Analysis of the expression of MRP, the gene for a new putative transmembrane drug transporter, in human multidrug resistant lung cancer cell lines. *Cancer Res* 53:1747–1750.

Zhao SC, Li M-X, Banerjee D, et al. (1994): Long term protection of recipient mice from lethal doses of methotrexate by marrow infected with a double copy vector retrovirus containing a mutant dihydrofolate reductase. *Cancer Gene Ther* 1:27–33.

Zhou DC, Hoang-Ngoc L, Delmer A, et al. (1994): Expression of resistance genes in acute leukemia. *Leuk Lymphoma* 13(suppl 1):27–30.

Zhu Q, Center MS (1996): Evidence that SP1 modulates transcriptional activity of the multidrug resistance-associated protein gene. *DNA Cell Biol* 15:105–111.

18

CLINICAL APPLICATIONS OF GENE THERAPY: BRAIN TUMORS

KENNETH W. CULVER AND JOHN C. VAN GILDER

Gene Therapy Research and Clinical Affairs, Codon Pharmaceuticals, Inc., Gaithersburg, MD 20877 (K.W.C.), Division of Neurosurgery, University of Iowa School of Medicine, (J.C.V.G.) Iowa City 52242

INTRODUCTION

Both primary and metastatic brain tumors are a major cause of morbidity and mortality in the general population. Astroglial brain tumors, including the highly malignant glioblastoma multiforme (GBM), are the most common primary brain tumors. Despite aggressive therapy, which includes surgical removal of the tumor, postoperative high-dose radiation, and chemotherapy, the prognosis of patients with GBM is very grim. Therefore, researchers are working hard on the development of a variety of novel experimental therapies; however, none of them has substantially changed the dismal prognosis for patients with GBM (Culver et al., 1996b).

In addition to primary brain tumors, the central nervous system (CNS) is a frequent site of metastasis for systemic malignancies. In fact, cerebral metastases occur in 25%–35% of the 1.1 million new cases of cancer per year in the United States. Tumors that most frequently metastasize to the brain are melanoma, lung, breast, colorectal, and renal cell cancers. Surgery combined with radiation therapy is the treatment of choice for a surgically accessible, single brain metastasis. Median survival using these two modalities approaches

Stem Cell Biology and Gene Therapy, Edited by Peter J. Quesenberry, Gary S. Stein, Bernard G. Forget, and Sherman M. Weissman
ISBN 0-471-14656-0

40 weeks. Unfortunately, most patients have multiple lesions and/or their location in the CNS prohibits surgical intervention and limits therapy to radiation alone with a median survival of about 15 weeks.

Despite the many advances in medicine over the past decade, radiotherapy and adjuvant chemotherapy have minimally improved the prognosis for most patients with GBM and symptomatic CNS metastases. Gene transfer has a number of novel new approaches to the selective elimination of tumor cells that may be beneficial for the treatment of brain tumors. Unlike many standard cancer therapies, gene transfer approaches are not immunosuppressive, allowing their potential concomitant use with each other or with traditional therapies. Hopefully the further development of new gene therapy strategies for the destruction of malignant cells in the CNS will provide significant improvements in the prognosis for patients with brain tumors.

RATIONALE FOR BRAIN TUMOR GENE THERAPY EXPERIMENTATION

Four general approaches are currently approved for clinical experimentation (Table 1). These are based on genes that selectively confer a sensitivity to an otherwise nontoxic drug (e.g., the herpes simplex virus thymidine kinase [HSV-TK] gene), genes that alter the immunogenicity of the tumor by secretion of cytokines from transduced autologous fibroblasts admixed with autologlous tumor cells or transduced tumor cells alone, inhibition of tumor elaborated factors that functionally immunosuppress the host immune response (i.e., antisense insulin growth factor type 1 [IGF-1], antisense transforming growth factor-1 [TGF-1]), and the insertion of the multiple drug resistance type I (MDR1) gene into hematopoietic stem cells (HSC) in an attempt to protect them from the toxicities of systemic chemotherapy. These four categories used alone or together may provide unique opportunities to destroy residual tumor cells that escape standard modalities of treatment.

TABLE 1 Antitumor Methodologies Approved for Use in Brain Tumor Gene Therapy Trials

Insertion of the HSV-TK "sensitivity" gene into tumor cells *in vivo* followed by intravenous treatment with ganciclovir
Enhancement of tumor immunogenicity by the *ex vivo* gene transfer of cytokine genes that are reinjected subcutaneously
Inhibition of tumor-derived immunosuppressive substances resulting from the *ex vivo* transfer of antisense genes into autologous tumor cells followed by reimplantation subcutaneously
Protection of hematopoietic stem cells from the adverse effects of systemic chemotherapy following retroviral-mediated gene transfer of the MDR1 gene

APPROVED GENE THERAPY TRIALS FOR BRAIN TUMORS USING THE HSV-TK GENE

The HSV-TK Sensitivity Gene

The HSV-TK gene was first cloned in 1979 from the type 1 virus (Colbere-Garapin et al., 1979). This gene is now the most used sensitivity gene in preclinical and clinical studies because it produces target cell destruction with limited nonspecific tissue toxicity. HSV-TK catalyzes the monophosphorylation of several drugs, including the FDA-approved antiherpes drugs acyclovir (ACV) and ganciclovir (GCV) (Balzarini et al., 1995; Cheng et al., 1983; Elion, 1980; Faulds and Heel, 1990). Cellular kinases then convert the monophosphate (MP) forms of the drugs to the diphosphate (DP) and triphosphate (TP) forms that incorporate into DNA as a nucleoside analog. The ACV-TP and GCV-TP inhibit DNA polymerase from continuing beyond the analog leading to DNA fragmentation. As a result, when a HSV-infected cell is exposed to ACV or GCV, the phosphorylated, incorporated drug product leads to fragmentation of the DNA and apoptosis, thereby killing the HSV-infected cell (Samejima and Meruelo, 1995; Smee et al., 1983; Terry et al., 1991). Host cellular TKs do not phosphorylate ACV or GCV, only the MP and DP forms, limiting cellular toxicity (Field et al., 1983).

The same method can be used to destroy malignant cells (Moolten, 1986). Transfer of the HSV-TK gene into tumor cells *in vitro* results in their destruction with treatment by ACV or GCV either *in vitro* or after reimplantation into mice (Moolten et al., 1990). GCV has a more potent antitumor effect than ACV and other analogs in animal tumor model systems (Smee et al., 1985). This is probably due to a greater uptake of GCV into HSV-infected cells and the fact that GCV is a better substrate for both viral and host kinase enzymes, while being only minimally phosphorylated in HSV-TK–negative cells (Cheng et al., 1983).

This method of cell destruction requires that the cell be actively dividing in order to incorporate the triphosphate derivatives in the cellular DNA (Chen et al., 1994). Since tumor cells are usually the most actively dividing cell type in most tissues, this feature provides for some level of selectivity for tumor cell destruction *in vivo.* Based on these types of preclinical studies and the fact that GCV is FDA approved for the treatment of HSV infections, GCV has been used exclusively with HSV-TK gene transfer in approved human gene therapy clinical trials.

Gene Transfer Methods Under Investigation for CNS Tumors

Gene transfer is divided broadly into *ex vivo* and *in vivo* gene transfer approaches. Both *ex vivo* (the transfer of genes outside of the body) and in vivo (transfer of the genes inside the body) methods have been approved for clinical trial application for brain tumors (Table 2). The *ex vivo* approaches utilize

TABLE 2 General Features of RAC-Approved Brain Tumor Gene Therapy Trials

Gene Transferred	Method of Gene Transfer	Type of Transfer (Tissue)	Primary Type of Antitumor Effect Expected
HSV-TK	Retroviral mediated	*In vivo* (tumor)	Induction of apoptosis
HSV-TK	Adenoviral mediated	*In vivo* (tumor)	Induction of apoptosis
IL-2	Plasmid based	*Ex vivo* (fibroblasts)	Immunological
IL-4	Retroviral mediated	*Ex vivo* (tumor)	Immunological
Antisense IGF-1	Plasmid transfection with liposomes	*Ex vivo* (tumor)	Immunological
Antisense TGF-β	Plasmid transfection by electroporation	*Ex vivo* (tumor)	Immunological
MDR-1	Retroviral mediated	*Ex vivo* (HSC)	Chemotherapeutic

electroporation, liposomes, and murine retroviral vectors. The *in vivo* trials use adenoviral and retroviral vectors. Generally speaking, the *ex vivo* approaches focus on attempts at increasing tumor immunogenicity, while *in vivo* approaches attempt to deliver sensitivity genes directly into the tumor mass. The next sections discuss each of these approaches in the context of the gene transfer method, genes to be transferred, and the clinical trial design.

Retroviral-Mediated *In Vivo* Transfer of HSV-TK Sensitivity Gene

Murine retroviral vectors have been the work horse for gene delivery in the first years of human gene therapy experimentation (Culver, 1996a). The primary reason for their favored status stems from their ability to stably integrate their vector genes (transduction), potentially allowing vector gene expression for the life of the cell and in all progeny (Miller, 1990). However, in the case of brain tumor treatment, they have been considered particularly attractive because they require a proliferating cell to achieve transduction (Miller et al., 1990). Since the tumor is the most actively proliferating cell in the brain, this provides for relatively selective delivery of the vector into tumor cells, especially in the brain, where the resident cell population has a very low proliferation index (Culver et al., 1992).

The delivery of the HSV-TK gene by retroviral vectors is the most advanced of the different brain tumor gene therapy clinical trials with nine phase I and II trials in the United States and another three trials in Europe (Table 3). In these protocols, murine fibroblasts (NIH 3T3 cells) that have been genetically engineered to produce murine retroviral vectors (vector producer cells [VPC]) are directly implanted into growing brain tumors in human patients. The gene being transferred into the surrounding brain tumor cells is HSV-TK. This

TABLE 3 Characteristics of Approved Clinical Trials Using Retroviral-Mediated *In Vivo* Transfer of the HSV-TK Sensitivity Gene

Principal Investigators	Centers Enrolling Patients	Trial Design	Tumor Type(s)
Oldfield	NIH	Stereotactic	Recurrent primary and metastatic
Van Gilder	University of Iowa and four other U.S. centers	Combined with surgical resection	Recurrent GBM
Raffel	Mayo Clinic and three other U.S. centers	Combined with surgical resection	Recurrent or progressive tumors
Kun	St. Jude Children's Research Hospital	Stereotactic	Recurrent or progressive tumors
Oldfield	NIH	Direct injection	Leptomeningeal carcinoma
Fetell	Columbia University and six other centers in the U.S. and Israel	Stereotactic	Recurrent GBM and anaplastic astrocytoma
Harsh	Harvard University	Stereotactic	Recurrent glioma
Maria	University of Florida and more than 40 other centers in the U.S., Canada, and Europe	Combined with surgical resection (prospective, randomized phase II trial)	Newly diagnosed, previously untreated GBM
Maria	Multiple centers as in the trial above	Combined with surgical resection (prospective, phase II trial)	Recurrent GBM
Izquierdo	Madrid, Spain	Stereotactic	Recurrent GBM
Klatzmann	Paris, France	Combined with surgical resection	Recurrent GBM
Yla-Herttuala	Kuopio, Finland	Direct injection	Recurrent GBM

method of gene delivery was adopted because the direct injection of the vector particles alone had a low gene transfer efficiency (1%–3%), while the injection of the VPC resulted in a 10%–55% efficiency in animal models (Ram et al., 1993a).

This improvement in gene transfer efficiency had significant ramifications because subsequent treatment with GCV produced complete tumor destruction in those animals that received an intratumoral injection of VPC without evidence of associated toxicity or evidence of systemic spread of the retroviral vectors (Ram et al., 1993a). Tumors continued to grow in the animals that received an intratumoral injection of vector particles alone into their tumor mass. Significantly, complete tumor ablation occurred in experimental tumors despite the fact that <100% of the tumor cells contained the HSV-TK gene. Studies in mice suggested that if at least 10% of the tumor cells contain the HSV-TK gene, then more than 50% of the tumors can be completely eliminated (Culver et al., 1992). This phenomenon, involving the destruction of adjacent, non-HSV-TK–containing tumor cells, is called the *bystander* tumor killing effect.

The etiology of the bystander effect is not completely understood. A number of possibilities have been proposed that result in tumor cell apoptosis, including the transfer of GCV metabolites to adjacent cells through gap junctions (Bi et al., 1993; Colombo et al., 1995; Fick et al., 1995), the uptake of cell fragments from the HSV-TK/GCV–destroyed cells (Freeman et al., 1993) and induction of a host immune response (Gagandeep et al., 1996). It is very possible that these three hypotheses are operative simultaneously *in vivo.* However, *in vitro* studies suggest that the major component of the bystander effect is the transfer of phosphorylated forms of GCV via gap junctions into neighboring tumor cells (Culver, 1996b).

Following the completion of the preclinical safety and efficacy studies and approval by local and national regulatory committees, the first brain tumor gene therapy protocol using this gene transfer system was initiated at the National Institutes of Health in Bethesda, Maryland, in December 1992 (Oldfield et al., 1993). Fifteen patients with recurrent GBM or metastatic tumors were treated with stereotactic injections of HSV-VPC into multiple areas of their tumor (Ram et al., 1997). They had all failed external beam radiation, and most had failed surgery and chemotherapy as well. Each of the 15 patients tolerated multiple stereotactic injections of $0.5–1 \times 10^9$ VPC without evidence of toxicity related to the xenogeneic cells or treatment with GCV. One week after the injection of the VPC, the patients were treated with a 14 day course of intravenous GCV. In this initial phase I trial, the emphasis was on determining the safety of the therapy, so VPC were only injected into the gadolinium-enhancing portion of the tumor so that changes in the tumor and surrounding brain could be visualized with magnetic resonance scanning. In other words, no attempt was made to treat all of the tumor (infiltrating areas) in any of the patients. Gene transfer was determined to occur, albeit at low levels (<1%), by *in situ* hydridization when counterstained with a tumor-specific GFP stain. No clinical adverse reactions were observed related to the direct injection of the xenogeneic VPC, and no significant inflammatory reaction was noticed on biopsy or autopsy materials.

Four of the 13 evaluable patients demonstrated evidence of an antitumor effect based on a >50% in the size of the gadolinium-enhancing portion of the tumor. The change in size was accompanied by cystic changes within the tumor. These findings suggest that, with a low level of gene transfer efficiency and the lack of an inflammatory response, the bystander effect is likely to be operational in some of the patients. Despite this limited treatment area of the tumor, three patients lived more than 12 months after gene therapy, with one of the patients with a recurrent glioblastoma being tumor-free more than 5 years status-post initiation of gene therapy treatment. The reasons for the prolonged survival of this particular patient is currently unknown. Since patients with recurrent glioblastoma in general have a life expectancy of 5.8 months (range 3.4–8.8), we are encouraged that the further optimization of this approach to the treatment of GBM may be an important advance over current therapies (Florell et al., 1992). Due to the lack of toxicity and the suggestion of an antitumor response, additional clinical trials have been approved in the United States using modifications of the NIH trial design with implantation of HSV-TK VPC and GCV (Table 3).

The first modification involves the combination of surgical resection of tumor and gene transfer rather than stereotactic injection (Culver et al., 1993). Unlike the first trial at NIH, this design attempts to deliver the VPC into the nonresectable areas of infiltrating tumor. Once the tumor was maximally resected (complete resection is not possible in these recurrent tumors), 1×10^9 VPC are injected 1 cm deep around the margins of the tumor bed since the majority of infiltrating tumor is in this area. An Ommaya reservoir is then placed in the tumor bed. Two weeks after surgery, the patients receive additional VPC through the Ommaya reservoir in an attempt to transfer the gene into residual tumor cells.

Early results from this phase II trial demonstrated risks associated with the injection of VPC into the Ommaya reservoir. It appears that if the VPC leak around the Ommaya catheter into the subarachnoid space, the patients can sustain an acute meningeal reaction, including high fever, meningismus, severe headache, and severe hypertension. The reaction is self-limited and responds to treatment with analgesics, glucocorticoids, and antihypertensives. It is important to note that these adverse reactions are not related to gene transfer, but appear to be a direct physiological reaction to the infused xenogeneic cells. These side effects were not predicted based on studies in monkeys that evaluated the direct injection of VPC into the subarachnoid space (Ram et al., 1993b). Modifications in the trial design were made to include the injection of technetium into the Ommaya before the VPC to make certain that the resection cavity had sealed. This screening procedure substantially reduced the frequency and severity of reactions. Patients were accruing into the study at the University of Iowa (Iowa City), the University of California, San Francisco, the University of Washington (Seattle), the University of Texas-Southwestern (Dallas) and the University of Cincinnati. Publicly available information and conference presentations have suggested that the direct injection of VPC immediately following resection was safe. A parallel multicenter study in adults was also undertaken using stereotactic delivery as was utilized in the initial NIH study. No published data are available from these two trials.

Two trials were approved for application in children and young adults aged 3–21 years of age. The VPC are delivered using stereotactic injection or the combination of surgical resection (Kun et al., 1995) and direct injection of VPC into nonresectable areas of recurrent supratentorial tumors (Raffel et al., 1994). Between 1×10^8 and 2×10^9 VPC will be injected into multiple sites within the tumor. The centers involved in the pediatric trials are the Mayo Clinic (Rochester, MN), St. Jude Children's Research Hospital (Memphis, TN), the University of Washington, Children's National Medical Center (Washington D.C.), and Children's Hospital of Los Angeles along with centers in Wurzburg and Dusseldorf, Germany.

While no published data are available from these studies, the sponsoring company, Genetic Therapy Inc. (GTI), has advanced two of the adult clinical trials to phase III open label, randomized trials for patients with newly diagnosed, previously untreated GBM or recurrent GBM. In the newly diagnosed group, patients will be randomized (123 patients per group) to receive either standard therapy consisting of surgical resection and external beam radiation or surgical resection with direct injection of VPC into areas of residual tumor, followed by external beam radiation and GCV therapy beginning 14 days after VPC injection. GCV will be administered for 14 days. The Ommaya reservoir retreatment portion of the study design has been eliminated due to the occurrence of multiple adverse reactions to the VPC at most of the trial sites. The approval of this randomized trial in patients with newly diagnosed, untreated tumors suggests that the safety parameters and possibilities for a beneficial antitumor efficacy observed in the previous trials were sufficient to convince the FDA to permit these new studies. These trials are expected to encompass at least 40 centers in North America, Europe, and Israel (Table 3).

In addition to the proliferation of GTI-sponsored trials, there are a number of other trials in the United States and in Europe (Klatzmann et al., 1996) using the implantation of HSV-TK VPC and GCV (Table 3). For instance, Harsh and colleagues in Boston are injecting VPC in a dose escalating fashion beginning at 5×10^5 cells per site into three distinct locations within the tumor at the time of biopsy confirmation of recurrence. Five days later, the patient will receive a single intraoperative injection of GCV, and the tumor will be maximally resected. The primary purpose of this phase I study is to assess the density, extent, and cell types transduced with this gene delivery method.

Delivery methods include the combination of stereotactic injection into the tumor mass or a combination of surgical resection and direct injection as described above. Results from one of these trials have been published (Izquierdo et al., 1996). Five patients received injections of HSV-TK VPC without resection. Seven days later, they each received a 14 day course of intravenous GCV. No toxicities were noted secondary to implantation of the VPC or gene transfer. One of the patients is reported to have had a significant reduction in tumor volume in the lobe that was injected with the VPC. These results appear to be generally comparable with the results from the initial phase I trial conducted at the NIH.

The final approved protocol using *in vivo* delivery of HSV-TK gene by injection of VPC was designed for the treatment of leptomeningeal carcinomatosis (Oldfield et al., 1995). The trial design involved the direct injection of HSV-TK VPC into the CNS ventricular system through an Ommaya catheter in a dose escalation regimen. One patient was treated at the NIH using the starting dose of 1×10^9 VPC. Unexpectedly, the patient developed severe meningismus, neck and back pain, nausea, rigors, and fever that forced discontinuation of the study. These adverse reactions were not predicted based on previous animal studies (Oshiro et al., 1995). The reaction is thought to be related to be the consequence of irritation of the cells lining the subarachnoid space by the murine VPC, as was seen in the brain tumor trials that used the Ommaya reservoir. This trial is now closed.

Taken together, these clinical studies are demonstrating that there is a substantial margin of safety with this delivery method if no cells are injected into the subarachnoid space, that the bystander effect appears to be operative in humans, and that an antitumor effect can be generated. Whether the antitumor effects are significant enough to allow product approval for brain tumor therapy will not be known for several years as these trials progress through phase II and phase III studies around the world.

Adenoviral-Mediated *In Vivo* Transfer of HSV-TK Sensitivity Gene

Adenoviral vectors have also been approved for the *in vivo* delivery of the HSV-TK gene into CNS neoplasms (Table 2). Adenoviruses have a number of advantages as gene transfer vectors, including production at high titer, an ability to infect nearly all human cell types at high efficiency including cells of the CNS (Le Gal La Salle et al., 1993), and no requirement for a proliferating cell target. This is particularly important because many tumor cells within a progressing GBM are not actively proliferating at any one time and therefore may be missed by retroviral vectors (Yoshii et al., 1986). Based on these types of insights, recombinant adenoviral vectors containing the HSV-TK gene were injected into comparable animal models bearing brain tumors as in the retroviral vector experiments discussed above (Chen et al., 1994; Colak et al., 1995). The results are similar to *in vivo* VPC delivery, achieving complete destruction of the tumor in a portion of the animals while sparing normal surrounding tissues that were infected by the vector (Chen et al., 1994). The protection of the normal cells was greater than anticipated since it was known that the vector would transduce both malignant and nonmalignant tissues. This appears to be related to the fact that phosphorylated derivatives of GCV require a proliferating cell to induce cell death. Consequently, tissue-specific promoters may not be required as originally hypothesized.

However, the efficiency of *in vivo* gene delivery into brain tumors remains a significant problem for adenoviral vectors since they appear to only infect cells in the local area of injection like retroviral vectors. However, adenovirus vectors do express HSV-TK at much higher levels than retroviral vectors in the same cell type, which may markedly increase their direct therapeutic effect and the bystander tumor cell killing (Chen et al., 1995; Shewach et al., 1994). Therefore, the use of adenoviral vectors may achieve a greater antitumor effect with the same dose of GCV or allow the use of a decreased dose of GCV if GCV toxicities limit clinical application. Despite this advantage, the major unresolved issue for use of these vectors relates to toxicity. Studies in rodents and nonhuman primates have suggested that the injection of a large number of recombinant adenoviral vector particles into normal brain tissues followed by GCV treatment can have significant adverse effects and induce the production of neutralizing antibodies (Byrnes et al., 1995; Goodman et al., 1996). The clinical design of the dose escalation protocols described below will start at very low doses of vector in an attempt to find an optimum dosage for therapy that is well below the level that induced significant undesirable toxicity in animals.

Three trials have been approved by the Recombinant DNA Advisory Committee, one at the University of Pennsylvania (Philadelphia), a second at Baylor University (Houston, TX), and a third at Mt. Sinai Medical Center in New York. The Pennsylvania trial will enroll patients with recurrent gliomas into two groups (Eck et al., 1996): nine patients with surgically accessible lesions and nine with surgically nonaccessible lesions. Three patients from each group will be treated with the same dose in a dose-escalation protocol using stereotactic injection into multiple sites within the tumor, with the first dose being 10^9 virus particles. GCV is administered intravenously beginning 2 days after injection of the vector. The patients with inoperable lesions will then receive GCV for 14 days. Patients in the surgically resectable group will have a debulking procedure 7 days after the injection of adenovirus so that tumor tissue can be characterized for safety, efficacy, and gene transfer parameters. An additional dose of vector will then be injected into nonresectable areas. GCV will be continued for an additional 2 weeks. No published data are available.

The trial at Baylor University will enroll patients with recurrent high grade astrocytomas, GBM, and metastatic tumors. At the time of stereotactic biopsy, the patient will receive a single injection of adenoviral vector particles into the tumor mass in a dose-escalation style beginning at 1×10^8 vector particles. Five patients per group will be observed for toxicities and an antitumor response. If the dose is well tolerated, the dose will be gradually increased to a maximum of 1.5×10^9 particles. No published data are available.

The study at Mt. Sinai will enroll patients with recurrent malignant glioblastoma. Using a frameless stereotactic surgical procedure, a maximum amount of tumor will be resected and the adenoviral vector will be injected in the resection margins beginning at 1×10^7 pfu. Twenty-four hours later, a seven day course of IV GCV will be initiated. If there are no severe toxicities, the dose of vector will be escalated at 0.5 log increments. No published data are available.

As with the retroviral studies, the determining factor of these direct *in vivo* injection methods will be the ability of adenoviral vectors to distribute the HSV-TK gene sufficiently throughout the tumor to allow the bystander effect to destroy any remaining, nontransduced cells.

Ex Vivo Transduction of the Human IL-2 or IL-4 Gene Into Fibroblasts or Tumor Cells, Respectively, With Retroviral Vectors

The first of two *ex vivo* gene therapy trials is designed to transfer the interleukin (IL)-2 gene into autologous fibroblasts. Fibroblasts were chosen for the retroviral vector target because they are much easier to grow reliably in tissue culture than tumor cells from patients. The transduced fibroblasts are then admixed with 1×10^7 irradiated autologous (nontransduced) tumor cells and injected subcutaneously three times at least 2 weeks apart. The number of transduced fibroblasts will be dose escalated starting from 1.25×10^6 cells in patients with recurrent glioblastoma. Animal experiments have shown that the injection of IL-2 vector–expressing syngeneic tumors in mice results in tumor regression and the development of tumor-specific immunity (Fearon et al., 1990). Therefore, this protocol is designed as an adjunctive therapy for the current surgical treatments for GBM. This trial has been approved for experimentation at the San Diego Regional Cancer Center for patients with recurrent GBM.

The second approved trial using cytokine gene transfer will be conducted at the University of Pittsburgh. Patients with a recurrent glioblastoma or supratentorial anaplastic astrocytoma suitable for subtotal resection will be enrolled. The tumor cells will be grown in culture and transduced with a murine retroviral encoding the human IL-4 gene. After transduction, the gene-modified cells will be returned to the patient in two series of five injections 2 weeks apart. No published data are available on the status of these two trials.

Ex Vivo Transfection of an Antisense IGF-1 Gene With a Plasmid-Based Vector Into Autologous Tumor Cells

This approach to the treatment of brain tumors appears to target one of the methods that tumors employ to hide from the immune system (Table 1). The investigators use an Epstein-Barr virus (EBV) replicating plasmid containing an antisense copy of the IGF-1 gene to block tumor cell production of IGF-1. Their *in vitro* studies demonstrated that the insertion of an antisense IGF-1 gene into IGF-1–producing tumor cells substantially inhibits IGF-1 production (Trojan et al., 1992). The injection of these genetically altered cells into syngeneic animals results in the immunological rejection of both genetically altered and wild-type tumor cells injected simultaneously into the leg and into the brain respectively. Tumor destruction is mediated by $CD8^+$ cytotoxic T cells (Trojan et al., 1992). One human clinical trial has been approved for the treatment of patients with GBM tumors following a course of radiotherapy at the Case Western Reserve University (Cleveland, OH). The study design uses tumor cells obtained at biopsy or during surgical resection. Once a culture has been established, the tumor cells are tested for IGF-1 expression. If the tumor produces IGF-1, the replicating plasmid vector is transferred using cationic liposomes into the tumor cells. The cells are then lethally irradiated and injected subcutaneously in a dose-escalation protocol beginning with 1×10^7 cells. Booster immunizations are to be administered at 4 and 12 weeks. No published data are available from this trial.

Ex Vivo Transfection of an Antisense TGF-β Gene Into Autologous Tumor Cells

It is well known that glioblastoma tumors produce immunosuppressive compounds, transforming growth factor-β (TGF-b) being one of them. Animal brain tumor experiments have demonstrated that the generation of substantial immunologically mediated antitumor efficacy can be induced following the subcutaneous injection of tumors cells expressing an antisense TGF-β gene that eliminated tumor cell-derived TGF-β secretion (Fakhrai et al., 1996). In the one approved dose-escalation clinical protocol (Table 2), scientists will remove tumor from patients with histologically confirmed glioblastoma and assay TGF-β secretion by the cell line. If the cells produce TGF-β, they will be genetically altered with a plasmid containing an antisense TGF-β gene using electroporation. Once sufficient downregulation of TGF-β has been achieved, the cells will be irradiated to prevent regrowth in the patient and reinjected subcutaneously every 3 weeks for four doses. No published data are available on the status of the trial.

Ex Vivo Transduction of the MDR-1 Gene Into Autologous Hematopoietic Stem Cells With Murine Retroviral Vectors

The genetic manipulation of HSC may also be theoretically used to protect HSC from the toxic effects of chemotherapy (Table 1). This is being attempted by inserting the MDR-1 gene into HSC prior to administration of high dose, myelosuppressive chemotherapy. MDR-1 is one method of tumor cell resistance to chemotherapeutic agents because it pumps drugs from the cell (Galski et al., 1989). The use of *ex vivo* retroviral vector-mediated insertion of the MDR-1 gene into murine marrow cells has demonstrated significant HSC protective effects *in vivo* when the animals were treated with high doses of taxol (Sorrentino et al., 1992). Human clinical trials have been approved for the treatment of newly diagnosed or recurrent GBM, anaplastic oligodendroglioma, primary CNS lymphoma, primitive neuroectodermal tumors (PNET), and ependymomas at Columbia University in New York City (Hesdorffer et al., 1994). The investigators plan to harvest bone marrow and then administer high doses of ThioTEPA, VP-16, and carboplatinum. One-third of the marrow harvest will be processed for gene therapy studies. First, $CD34^+$ cells will be selectively removed and transduced with a murine retroviral vector encoding the MDR-1 gene. Second, the transduced cells will be mixed with the nontransduced marrow and reinfused into the patient. The gene transfer efficiency and toxicity of the procedure will be monitored. If the patient goes on to receive taxol, the investigators hope to determine if there is an enrichment of MDR-1 transduced marrow stem cells in humans as was seen in mice.

SUMMARY

We are in the infancy of genetic healing. The early results from the HSV-TK studies and the many different approaches under study in humans are encouraging. However, significant hurdles remain in vector development and gene delivery. Advances in these areas are required to allow the widespread application of gene therapy in general. As discoveries in these areas occur, and researchers learn how to combine various modalities to create efficient, potent gene transfer systems, the first large group of patients to benefit will likely be those with cancer. I expect that over the next 5 years, gene therapy will become a standard form of therapy for certain forms of cancer. Results to date suggest that brain tumors may be one of the first.

REFERENCES

Balzarini J, Morin KW, Knaus EE, Wiebe LI, De Clercq E (1995): Novel (E)-5-(2-iodovinyl)-2′-deoxyuridine derivatives as potential cytostatic agents against herpes simplex virus thymidine kinase gene transfected tumors. *Gene Ther* 2:317–322.

Bi WL, Parysek LM, Warnick R, Stambrook PJ (1993): *In vitro* evidence that metabolic cooperation is responsible for the bystander effect observed with HSV tk retroviral gene therapy. *Hum Gene Ther* 4:725–732.

Byrnes AP, Rusby JE, Wood MJA, Charlton HM (1995): Adenovirus gene transfer causes inflammation in the brain. *Neuroscience* 66:1015–1024.

Chen C-Y, Chang Y-N, Ryan P, Linscott M, McGarrity GJ, Chiang YL (1995): Effect of herpes simplex virus thymidine kinase expression levels on ganciclovir-mediated cytotoxicity and the "bystander effect." *Hum Gene Ther* 6:1467–1476.

Chen S-H, Shine HD, Goodman JC, Grossman RG, Woo SLC (1994): Gene therapy for brain tumors: Regression of experimental gliomas by adenovirus-mediated gene transfer *in vivo. Proc Natl Acad Sci USA* 91:3054–3057.

Cheng Y-C, Grill SP, Dutschman GE, Nakayama K, Bastow KF (1983): Metabolism of 9-(1,3-dihydroxy-2-propoxymethyl)guanine, a new antiherpes virus compound, in herpes-simplex virus–infected cells. *J Biol Chem* 258:12460–12464.

Colak A, Goodman JC, Chen SH, Woo SLC, Grossman RG, Shine HD (1995): Adenovirus-mediated gene therapy in an experimental model of breast cancer metastatic to the brain. *Hum Gene Ther* 6:1317–1322.

Colbere-Garapin F, Chousterman S, Horodniceanu F, Kourilsky P, Garapin A-C (1979): Cloning of the active thymidine kinase gene of herpes simplex virus type 1 in *Escherichia coli* K-12. *Proc Natl Acad Sci USA* 76:3755–3759.

Colombo BM, Benedetti S, Ottolenghi S, Mora M, Pollo B, Poli G, Finocchiaro G (1995): The "bystander" effect: Association of U-87 cell death with ganciclovir-mediated apoptosis of nearby cells and the lack of an effect in athymic mice. *Hum Gene Ther* 6:763–772.

Culver KW (1996a): *Gene Therapy: A Primer for Physicians,* 2nd ed. New York: Mary Ann Liebert, Inc.

Culver KW (1996b): The role of herpes simplex thymidine kinase gene transfer in the treatment of brain tumors. *CNS Drugs* 6:1–11.

Culver KW, Ram Z, Walbridge S, Oldfield EH, Blaese RM (1992): *In vivo* gene transfer with retroviral vector producer cells for treatment of experimental brain tumors. *Science* 256:1550–1552.

Culver KW, Rigolet M, Raffel C (1996): Gene therapy strategies for brain cancer. In Bertino J (ed): *Molecular Biology of Cancer.* San Diego: Academic Press. pp 704–717.

Culver KW, Van Gilder J, Link CJ Jr, Carlstrom T, Buroker T, Yuh W, Koch K, Schabold K, Doornbas S, Wetjen B (1993): Gene therapy for the treatment of malignant brain tumors with *in vivo* tumor transduction with the herpes simplex thymidine kinase gene/ganciclovir system. *Hum Gene Ther* 5: 343–377.

Eck SL, Alavi JB, Davis A, Hackney D, Judy K, Mollman J, Phillips PC, Wheeldon EB, Wilson JM (1996): Treatment of advanced CNS malignancies with the recombinant adenovirus H5.010RSVTK. *Hum Gene Ther* 7:1465–1482.

Elion GB (1980): The chemotherapeutic exploitation of virus-specified enzymes. *Adv Enz Regula* 18:53–66.

Faulds D, Heel RC (1990): Ganciclovir. *Drugs* 39:597–638.

Fakhrai H, Dorigo O, Shawler DL, Lin H, Mercola D, Black KL, Royston I, Sobol RE (1996): Eradication of established intracranial rat gliomas by transforming growth factor b antisense gene therapy. *Proc Natl Acad Sci USA* 93:2909–2914.

Fearon ER, Pardoll DM, Itaya T, Golumbek P, Levitsky HI, Simons JW, Karasuyama H, Vogelstein B, Frost P (1990): Interleukin-2 production by tumor cells bypasses T helper function in the generation of an antitumor response. *Cell* 60:397–403.

Fick J, Barker GII, Dazin P, Westphale EM, Beyer EC, Israel MA (1995). The extent of heterocellular communication mediated by gap junctions is predictive of bystander tumor cytotoxicity *in vitro. Proc Natl Acad Sci USA* 92:11071–11075.

Field AK, Davies ME, DeWitt C, Perry HC, Liou R, Germershausen J, Karkas JD, Ashton WT, Johnston DBR, Tolman RL (1983): 9-{[2-Hydroxy-1-(hyroxymethyl) ethoxy]methyl}guanine: A selective inhibitor of herpes group virus replication. *Proc Natl Acad Sci USA* 80:4139–4143.

Florell RC, MacDonald DR, Irish WD, Bernstein M, Leibel SA, Gutin PH, Cairncross JG (1992): Selection bias, survival, and brachytherapy for glioma. *J Neurosurg* 76:179–183.

Freeman SM, Abboud CN, Whartenby KA, Packman CH, Koeplin DS, Moolten FL, Abraham GM (1993): The "bystander effect": Tumor regression when a fraction of the tumor mass is genetically modified. *Cancer Res* 53:5274–5283.

Gagandeep S, Brew R, Green B, Christmas SE, Klatzmann D, Poston GJ, Kinsella AR (1996): Prodrug-activated gene therapy: Involvement of an immunological component in the "bystander effect." *Cancer Gene Ther* 3:83–88.

Galski H, Sullivan M, Willingham MC, Chin K-V, Gottesman MM, Pastan I, Merlino GT (1989): Expression of a human multidrug resistance cDNA (MDR1) in the

bone marrow of transgenic mice: Resistance to daunomycin-induced leukopenia. *Mol Cell Biol* 9:4357–4363.
Goodman JC, Trask TW, Chen SH, Woo SLC, Grossman RG, Carey KD, Hubbard GB, Carrier DA, Rajagopalan S, Aguilar-Cordova E, Shine HD (1996): Adenoviral-mediated thymidine-kinase gene transfer into the primate brain followed by systemic ganciclovir. *Hum Gene Ther* 7:1241–1250.
Hesdorffer C, Antman K, Bank A, Fetell M, Mears G, Begg M (1994): Human MDR gene transfer in patients with advanced cancer. *Hum Gene Ther* 5:11519–1160.
Izquierdo, M, Martín V, de Felipe P, Izquierdo JM, Pe;a;rez-Higueras A, Cortés ML, Paz JF, Isla A, Blázquez MG (1996): Human malignant brain tumor response to herpes simplex thymidine kinase (HSVtk)/ganciclovir gene therapy. *Gene Ther* 3:491–495.
Klatzmann D, Philippon J, Valery CA, Bensimon G, Salzmann JL (1996): Gene therapy for glioblastoma in adult patients: Safety and efficacy evaluation of an *in situ* injection of recombinant retrovirus producing cells carrying the thymidine kinase gene. *Hum Gene Ther* 7:109–126.
Kun LE, Gajjar A, Muhlbauer M, Heideman RL, Sanford R, Brenner M, Walter A, Langston J, Jenkins J, Facchini S (1995): Stereotactic injection of herpes simplex thymidine kinase vector producer cells (PA317-G1Tk1SvNa.7) and intravenous ganciclovir for the treatment of progressive or recurrent primary supratentorial pediatric malignant brain tumors. *Hum Gene Ther* 6:1231–1255.
Le Gal La Salle G, Robert JJ, Berrard S, Ridoux V, Stratford-Perricaudet LD, Perricaudet M, Mallet J (1993): An adenovirus vector for gene transfer into neurons and glia in the brain. *Science* 259:988–990.
Miller AD (1990): Retrovirus packaging cells. *Hum Gene Ther* 1:5–14.
Miller DG, Adam MA, Miller AD (1990): Gene transfer by retrovirus vectors occurs only in cells that are actively replicating at time of infection. *Mol Cell Biol* 10:4239–4242.
Moolten FL (1986): Tumor chemosensitivity conferred by inserted herpes thymidine kinase genes: Paradigm for a prospective cancer control strategy. *Cancer Res* 46:5276–5281.
Moolten FL, Wells JM, Heyman RA, Evans RM (1990): Lymphoma regression induced by ganciclovir in mice bearing a herpes thymidine kinase transgene. *Hum Gene Ther* 1:125–134.
Oldfield EH, Ram Z, Culver KW, Blaese RM (1993): A clinical protocol: Gene therapy for the treatment of brain tumors using intra-tumoral transduction with the thymidine kinase gene and intravenous ganciclovir. *Hum Gene Ther* 4:39–69.
Oldfield EH, Ram Z, Chiang Y, Blaese RM (1995): Intrathecal gene therapy for the treatment of leptomeningeal carcinomatosis. *Hum Gene Ther* 6:55–85.
Oshiro EM, Viola JJ, Oldfield EH, Walbridge S, Bacher J, Frank JA, Blaese RM, Ram Z (1995): Toxicity studies and distribution dynamics of retroviral vectors following intrathecal administration of retroviral vector-producer cells. *Cancer Gene Ther* 2:87–95.
Raffel C, Culver KW, Kohn D, Nelson M, Siegel S, Gillis F, Link CJ Jr, Villablanca JG (1994): Gene therapy for the treatment of recurrent pediatric malignant astrocytomas using *in vivo* tumor transduction with the herpes simplex thymidine kinase gene/ganciclovir system. *Hum Gene Ther* 5:863–890.
Ram Z, Culver KW, Oshiro EM, Viola JJ, DeVroom HL, Otto E, Long Z, Chiang Y, McGarrity GJ, Muul LM, Katz D, Blaese RM, Oldfield EH (1997): Therapy of

malignant brain tumors by intratumoral implantation of retroviral vector-producing cells. *Nat Med* 3:1354–1361.

Ram Z, Culver KW, Walbridge S, Blaese RM, Oldfield EH (1993a): *In situ* retroviral-mediated gene transfer for the treatment of brain tumors in rats. *Cancer Res* 53:83–88.

Ram Z, Culver KW, Walbridge S, Frank JA, Blaese RM, Oldfield EH (1993b): Toxicity studies of retroviral-mediated gene transfer for the treatment of brain tumors. *J Neurosurg* 79:400–407.

Samejima Y, Meruelo D (1995): Bystander killing induces apoptosis and is inhibited by forskolin: Plasmid pXtk construction for thymidine-kinase gene transfer to investigate bystander killing. *Gene Ther* 2:50–58.

Shewach DS, Zerbe LK, Hughes TL, Roessler BJ, Breakefield XO, Davidson BL (1994): Enhanced cytotoxicity of antiviral drugs mediated by adenovirus directed transfer of the herpes simplex virus thymidine kinase gene in rat glioma cells. *Cancer Gene Ther* 2:107–112.

Smee DF, Boehme R, Chernow M, Binko BP, Matthews TR (1985): Intracellular metabolism and enzymatic phosphorylation of 9-(1,3-dihydroxy-2-propoxymethyl)-guanine and acyclovir in herpes simplex virus-infected and uninfected cells. *Biochem Pharmacol* 34:1049–1056.

Smee DF, Martin JC, Verheyden JPH, Matthews TR (1983): Antiherpesvirus activity of the acyclic nucleoside 9-(1,3-dihydroxy-2-propoxymethyl)guanine. *Antimicrob Agents Chemother* 23:676–682.

Sorrentino BP, Brandt SJ, Bodine D, Gottesman M, Pastan I, Cline A, Nienhuis AW (1992): Selection of drug-resistant bone marrow cells *in vivo* after retroviral transfer of human MDR1. *Science* 257:99–103.

Terry BJ, Cianci CW, Hagen ME (1991): Inhibition of herpes simplex virus type 1 DNA polymerase by [1R(1a,2b,3a)]-9-[2,3-bis(hydroxymethyl)cyclobutyl]guanine. *Mol Pharmacol* 40:591–596.

Trojan J, Blossey BK, Johnson TR, Rudin SD, Tykocinski M, Ilan J, Ilan J (1992): Loss of tumorigenicity of rat glioblastoma directed by episome-based antisense cDNA transcription of the insulin-like growth factor-1. *Proc Natl Acad Sci USA* 89:4874–4878.

Yoshii Y, Maki Y, Tsuboi K, Tomono Y, Nakagawa K, Hoshino T (1986): Estimation of the growth fraction with bromodeoxyuridine in human central nervous system tumors. *J Neurosurg* 65: 659–663.

19

CLINICAL APPLICATIONS OF GENE THERAPY: CARDIOVASCULAR DISEASE

JONATHAN C. FOX
University of Pennsylvania, Department of Medicine, Philadelphia, PA 19104-6100

INTRODUCTION

Gene therapy for cardiovascular diseases presents both great opportunities and significant challenges. Gene therapy provides the opportunity to achieve unprecedented efficacy and specificity in how diseases are treated based on their molecular mechanisms. The rationale for gene therapy relies on manipulating the functions of specific genes to alter pathophysiology by augmenting normal functions, correcting deficiencies, or inhibiting deleterious activities. The challenges to this field fall into three broad categories (Fig. 1). The first relates to the development of gene transfer vector and delivery systems that are suitable for genetic modification of the cardiovascular system, combining efficiency, ease of preparation, and an acceptable safety profile. The second relates to our knowledge of the molecular mechanisms of disease. This requires the identification of the genes and their functions responsible for pathophysiology and the definition of specific gene activities as appropriate targets for manipulation. The third relates to developing suitable animal models that can mimic human disease. As suggested by Figure 1, at the logical intersection of gene delivery vectors and molecular pathophysiology are vectors targeting specific genes. Similarly, reliable animal models are required to develop princi-

Stem Cell Biology and Gene Therapy, Edited by Peter J. Quesenberry, Gary S. Stein, Bernard G. Forget, and Sherman M. Weissman
ISBN 0-471-14656-0

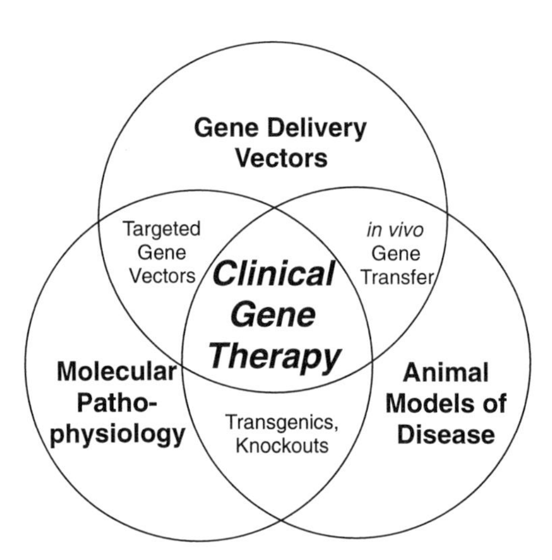

FIGURE 1. Elements of clinical gene therapy. This diagram illustrates the relationships between the three main components considered essential to successful strategies for clinical gene therapy. These include the development of gene delivery vectors and the means for their production and testing, knowledge of molecular pathophysiology and the identification of specific genes implicated in mechanisms of disease, and the adaptation of laboratory animals to models of human disease. The challenges presented in each of these areas, and the intersections between them, are discussed in the text.

ples and protocols for *in vivo* gene transfer. Animal models are used for preclinical testing of specific gene vectors, testing gene targeting strategies, and developing new vector delivery technologies. At the intersection of molecular pathophysiology and animal models of disease are transgenic and knockout animals, wherein specific genes are modified in the germ line to mimic specific molecular defects in human disease, and these can serve as informative models for therapeutic gene transfer. The goals of this chapter are to outline the rationale for gene therapy strategies targeting cardiovascular diseases, give an overview of the major gene transfer technologies currently applied to the cardiovascular system, and review some current approaches to *in vivo* genetic manipulation in the development of therapeutic strategies for cardiovascular disease. Although these examples will soon be outdated, the principles upon which they are based should provide a foundation for understanding newer approaches now under development and in the more distant future.

RATIONALE FOR GENE THERAPY: MOLECULAR MECHANISMS OF DISEASE

Molecular mechanisms of disease often involve defective, inadequate, excessive, or inappropriate gene expression. Several examples of cardiovascular syndromes can be cited to illustrate these genetic influences. Familial hypercholesterolemia is defined by a series of mutations in the low density lipoprotein receptor (LDL-R) gene resulting in either absent or defective LDL-R synthesis. The defective LDL-R function causes accumulation of LDL in the blood, its subsequent deposition in vessel walls and greatly accelerated lesion development. Similarly, many forms of hereditary hypertrophic cardiomyopathy are caused by missense mutations in genes encoding several of the cardiac sarcomeric proteins. These proteins through their abnormal interaction with the rest of the sarcomeric apparatus result in sarcomeric dysfunction and trigger an inappropriate hypertrophic response. The inability to form sufficient collateral vessels in the setting of vascular insufficiency due to atherosclerosis may represent an inadequate angiogenic response of local tissues to ischemia. Some individuals form extensive collaterals, preserving perfusion despite extensive vascular disease, whereas others form collaterals only poorly. One reason for this may be intrinsic differences in the natural production of angiogenic growth factors. Finally, it is generally accepted that restenosis following angioplasty is characterized in part by an inappropriate degree of smooth muscle cell proliferation and extracellular matrix protein synthesis and secretion. A number of growth factors, cytokines, enzymes, and adhesion proteins appear to play key roles in this process.

These examples, which are discussed in greater detail in the following sections, can illustrate how developing rational strategies for cardiovascular gene therapy requires the identification of specific molecular targets for intervention and a detailed understanding of how altering a particular gene function is likely to influence the course of disease. These might include replacing a missing or defective gene product (LDL-R), complementing an abnormal function in *trans* (hypertrophic cardiomyopathy), enhancing gene function through increased expression (angiogenesis), and blocking deleterious gene expression by antisense or dominant-negative approaches (restenosis). All of these strategies have potential application to a variety of clinical problems. Before considering specific strategies for these particular disorders, an overview of some current gene transfer technologies applied to the cardiovascular system is provided.

VECTOR SYSTEMS

Once appropriate molecular targets have been identified, the cellular target of a proposed genetic manipulation must be chosen. Although this usually means the cell type most intimately involved in the disease process under consideration, it can also include a neighboring cell type or an organ distantly removed from the site of disease. Such distant sites might be related either pathophysiologically or simply through the sharing of circulating blood. Once a cellular target has been chosen, the genetic information must be delivered in an efficient, selective fashion and must function in a physiologically relevant manner. Devising efficient and effective means of packaging and delivering genetic information has posed significant technical challenges for developing gene therapy for cardiovascular and other diseases. A variety of approaches have been devised or are under development and generally rely on delivering genetic information (as DNA or RNA) packaged as synthetic (chemical complexes), natural (viruses), or semisynthetic (biochemical conjugates) gene transfer vehicles, usually delivered by some kind of mechanical approach (injection, surgical implantation, or catheter-based delivery). The latter consideration relates to the field of local drug delivery, a detailed discussion of which is beyond the scope of this chapter.

Gene transfer was originally developed to study the function of isolated genes in the context of the living cell. First applied to isolated cells in culture, the later application of gene transfer technology to the development of transgenic mice demonstrated that foreign genes introduced into single cells could be transmitted in the germ line and stably expressed in progeny animals. The later observation that high level expression of cloned genes could be achieved in a regulated fashion both in cultured cells and in transgenic animals suggested that regulated expression of ectopic genes could be accomplished in the somatic cells of otherwise wild-type adult organisms. It was with this aim that the techniques of *in vivo* gene transfer have been developed and refined. These include techniques based on naked DNA, RNA or DNA viruses, and related technologies.

Plasmid DNA

The simplest approach to gene transfer relies on naked plasmid DNA transduction. Plasmids are circular molecules between 4 and 10 kb in size containing sufficient information for their propagation in bacteria and a segment of DNA (usually eukaryotic in origin) encoding the genetic information of interest (Fig. 2). This often includes both structural information (the sequence encoding a protein, e.g., the neomycin resistance gene, neo) as well as regulatory information such as an element or elements specifying expression in eukaryotic cells (e.g., SV40 ori), which may be restricted to particular tissues or regulated by drugs or hormones. The technique of plasmid DNA-mediated gene transfer relies on exposure of a cell to relatively high concentrations of purified ("naked") DNA, with subsequent uptake mediated by nonspecific mechanisms of endocytosis or pinocytosis. *In vivo* gene transfer with plasmid DNA is well tolerated, with little direct or immune-mediated toxicity (Nabel et al., 1992). The advantages of this technique include its simplicity and the relative ease of preparation of the reagents required. The greatest disadvantage is low efficiency of gene transfer and expression. In addition, any degree of persistence of expression depends on integration of the plasmid DNA into the host genome, which requires cell replication, thus limiting this approach to dividing cells. Random chromosomal integration also poses the theoretical risk of insertional mutagenesis. Finally, even following chromosomal integration,

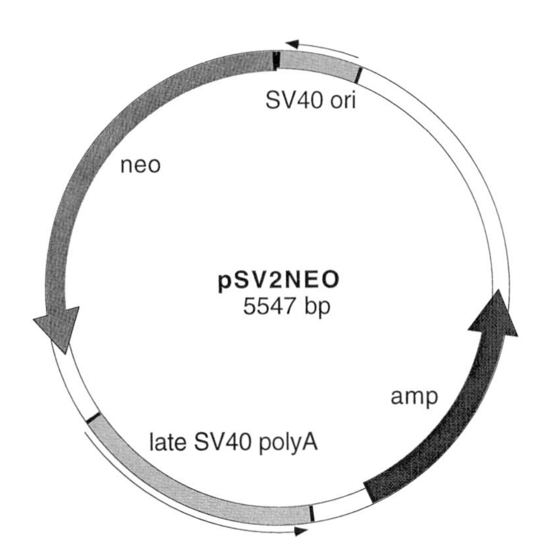

FIGURE 2. Plasmid DNA vector. This diagram illustrates a typical plasmid DNA vector (pSV2Neo), which contains information required for its selective propagation in bacteria (including an ampicillin resistance gene, amp), as well as eukaryotic regulatory (SV40ori) and structural (neo) elements capable of directing expression in mammalian cells.

plasmid DNA-directed gene transfer *in vivo* results in only short-lived expression because of inactivation or elimination by what are as yet poorly understood mechanisms.

Retrovirus

Viruses have evolved efficient mechanisms for delivering genes to target cells. Retroviruses are the first eukaryotic viruses exploited for the efficient delivery of foreign genes to eukaryotic cells (Boris and Temin, 1993; Miller et al., 1993). Originally described for their ability to transform host cells into malignant tumor cells, these viruses are relatively simple in their structure and biology. Retroviruses contain a 5–6 kb linear, double-stranded RNA genome flanked by two copies of a specialized DNA sequence known as the long terminal repeat (LTR), which direct high level expression of the three intervening structural genes, gag, pol, and env (Fig. 3). The viral particle consists of the viral genomic RNA and several associated proteins surrounded by an envelope consisting of host cell membrane and the viral coat glycoprotein (env gene product). The viral coat glycoprotein is recognized by specific cell surface receptors and promotes fusion of the virus with the cell. The RNA genome is copied by viral reverse transcriptase (pol) into DNA that can stably integrate into the host genome as a provirus, a process mediated by the LTR and requiring cell division. The three viral structural genes, which are required

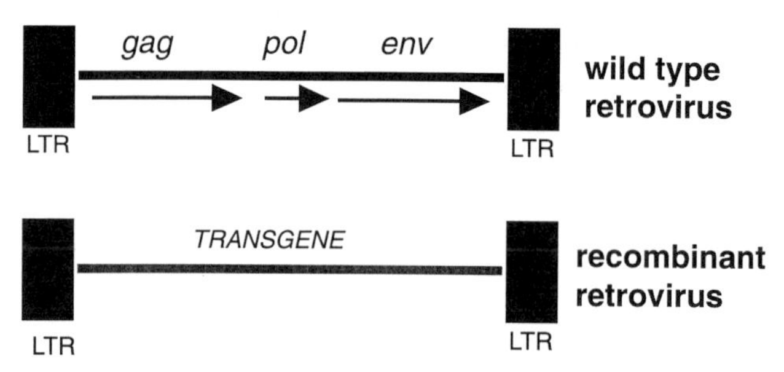

FIGURE 3. Recombinant retrovirus. This diagram illustrates the basic structural organization of the mammalian retroviruses used as gene transfer vectors, including the long terminal repeats (LTRs), and the three structural genes gag, pol, and env. These can be deleted and replaced with a transgene, as shown.

for viral replication but not proviral integration, can be deleted and replaced with foreign DNA. Function of the recombinant retrovirus as a gene transfer vector requires only the preservation of the flanking LTRs to drive constitutive expression of the transgene and integration of the provirus and the packaging signal for proper assembly of the recombinant virion in an environment permissive for replication. Because deletion of the viral structural genes renders the recombinant virus replication defective, propagation of a recombinant requires the presence of a helper virus. Alternatively, propagation can be achieved in cultures of so-called packaging cells, which are mammalian cells engineered to provide the viral structural gene products in *trans.*

Despite these attractive features, recombinant retroviruses present several disadvantages (Ali et al., 1994). As with naked plasmid DNA, persistent transgene expression requires integration and thus cell division (Miller et al., 1990). The selective transfer of recombinant genetic material to dividing cells only may well confer some advantages for targeting gene transfer to rapidly dividing cells, as in the treatment of cancer, but many other potential targets, e.g., cardiomyocytes, divide slowly or not at all, at least in the adult. As with plasmid DNA, integration at random chromosomal locations can theoretically lead to insertional mutagenesis. The retroviral virion is itself inherently unstable so that production of recombinant retrovirus in clinically useful concentrations is technically challenging. Despite these limitations, research on developing improved retroviral gene transfer vectors continues, and their clinical utility has not been excluded. Recently, hybrid or so-called pseudotyped retroviral vectors have been designed that overcome some of these stability problems, allowing concentration of the viral preparation by ultracentrifugation, and extend the host range by packaging the recombinant virus with a pantropic envelope protein (e.g., vesicular stomatitis virus G protein) (Burns et al., 1993; Yee et al., 1994). In addition to their successful use in mammalian cells, pantropic pseudotyped retroviruses have been used to achieve gene transfer to amphibians *in vivo* (Burns et al., 1994) and to generate transgenic zebrafish, an experimental model at the forefront of cardiovascular developmental biology (Lin et al., 1994).

Malignant transformation resulting from retroviral infection in the host remains a concern since productive retroviral infection could theoretically transform host cells into tumor cells (Cornetta et al., 1991; Boris and Temin, 1994). In addition, viral preparations can theoretically be contaminated with replication-competent virus, generated through recombination between recombinant viral sequences and helper or pre-existing proviral sequences to regenerate replication competent virus either during preparation or later, after successful gene transfer to a host cell. The development of improved packaging cell lines that avoid containing the necessary viral genes in a contiguous segment of DNA has lowered the probability of a productive recombination event.

Adenovirus

The development of recombinant adenovirus vectors for *in vitro* and *in vivo* gene transfer represented a significant advance in somatic cell transduction technology (Graham and Prevec, 1992). The main advantages of recombinant adenoviruses are their broad host range, their capacity to infect either replicating or quiescent cells, the relative ease of preparing the virus in high titers (10^{12}–10^{13}/ml), and the lack of an association with human diseases other than mild, self-limited respiratory and gastrointestinal syndromes. Human adenoviruses are a family of nonenveloped, icosahedral viruses containing a linear, 36 kb double-stranded DNA genome. The genome is divided into early (E) and late regions in reference to when they are expressed in relation to viral DNA replication. Mutant adenoviruses deleted of large portions of viral sequence can carry 8 kb or more of foreign DNA. These mutants have typically been engineered by deleting segments of the viral genome (e.g., portions of E3 or E4) that are not required for viral replication, packaging, or infection (Fig. 4). Viral replication requires expression of the E1A and E1B genes, and these genes are also deleted from most mutant strains used for generating recombinant vectors, rendering the virus replication defective. The deleted mutant viruses are propagated in the human embryonal kidney cell line 293,

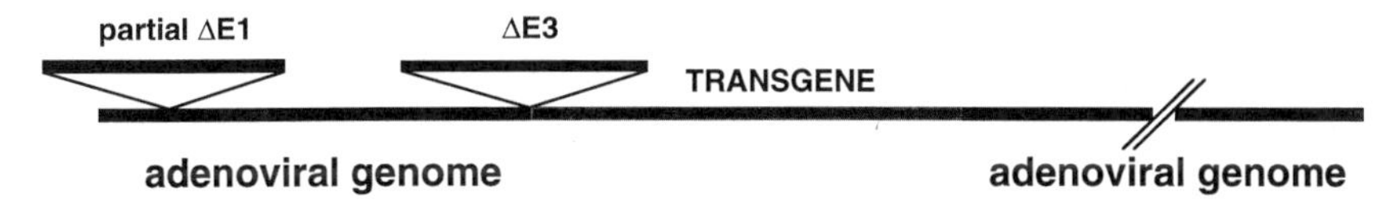

FIGURE 4. Recombinant adenovirus. The most common mutants used to construct recombinant adenovirus vectors contain partial deletions of the E1 region, rendering the vectors replication defective, and additional deletions of either the E3 or E4 regions to accommodate transgenes of up to 8 kb in length.

which is stably transfected with the deleted E1 genes, providing the required gene products in *trans* (Haj and Graham, 1986).

Several different strategies have been developed for inserting a foreign transgene into the adenoviral genome. Different strategies take advantage of the ability of transfected, naked viral DNA to initiate productive infection, resulting in accumulation of infectious virus particles in 293 cells. One popular approach uses a shuttle plasmid, derived from a typical bacterial plasmid, containing the foreign transgene driven by an appropriate expression cassette and flanked by segments of the adenoviral genome. Truncating the E1/E3 deleted mutant virus at the 5′ end deletes the viral packaging signal, generating an adenoviral genome that may replicate but cannot be packaged into virions in 293 cells. The shuttle plasmid cannot itself replicate in 293 cells, but contains the packaging signal missing from the truncated viral DNA. These two purified DNAs are cotransfected into 293 cells, and the adenoviral sequences flanking the transgene in the shuttle vector are able to mediate a homologous recombination event with the truncated adenoviral DNA. The recombination product has the capability of both replicating and being packaged into infectious virions in the 293 cells. Other approaches rely on preparing the entire recombinant viral genome as a single large plasmid DNA molecule, propagating the plasmid for *in vitro* manipulation either in *Escherichia coli* (Ghosh et al., 1986) or in yeast (Ketner et al., 1994). The final recombinant vector is still prepared as infectious virions by transfecting the purified DNA into 293 cells.

There are several distinct advantages that recombinant adenoviral vectors provide compared with retroviral vectors. Perhaps most importantly, adenovirus infects a very broad range of cell types and species, and infection with subsequent expression of the transgene does not depend on replication of the target cell. The recombinant adenovirus remains episomal in most cases, largely avoiding the concerns regarding chromosomal integration, although integration can occur if the target cell is infected with a very large number of virus particles (high multiplicity of infection). Recombinant adenovirus can be prepared in very high titer stocks, permitting its use in a variety of *in vivo* testing situations not amenable to other methods of gene transfer. Although tumorigenicity of this class of viruses cannot be rigorously excluded, it has not been observed to date.

The disadvantages of recombinant adenovirus relate primarily to its immunogenicity, which limits persistence of expression secondary to elimination of the virus via immune clearance of infected cells (Engelhardt et al., 1994a; Kass-Eisler et al., 1994; Yang et al., 1994). In fact, evidence of humoral immunity against these viruses, indicating prior infection, is evident in the great majority of people, which may ultimately limit its clinical utility (Schulick et al., 1997). The mechanisms responsible for the immune clearance of recombinant adenovirus have been defined by experiments using immunocompetent versus immunodeficient or immunosuppressed hosts and identification of activation of specific components of the immune response directed at viral antigens. Virally infected cells may in fact be capable of producing viral proteins in addition to the transgene as a result of "leaky" expression of delayed early genes despite deletion of the immediate early genes. It has been postulated that such expression is mediated by the activity of host genes that are able to complement the missing viral genes. A variety of approaches are currently being used in an attempt to circumvent this problem, including generating new adenoviral vectors containing further deletions or conditional (temperature-sensitive) mutations (Engelhardt et al., 1994b). In fact, vectors devoid of all adenoviral genes may be feasible and confer long-term expression *in vivo* (Chen et al., 1997). Conversely, the presence of the E3 region, once thought dispensable for recombinant adenoviral function, may in fact confer an immunological inhibitory function permitting persistent expression (Ilan et al., 1997a). A unique approach to the problem borrows from the oral tolerization strategy that immunologists are using to treat autoimmune disorders such as rheumatoid arthritis. Enteral administration of adenoviral antigens can permit subsequent parenteral administration of recombinant adenovirus without evidence of immune clearance, and this tolerance can be adoptively transferred (Ilan et al., 1997b). Aside from these immunological issues, nonimmunological properties of the virus can negatively influence expression from heterologous promoter elements inserted as part of the transgene cassette (Armentano et al., 1997). In addition, adenovirus can infect many but not all target cells of interest with high efficiency. Adenoviral infection relies on the expression of specific membrane surface receptor(s) for the virus. It is believed that certain cells express receptors at very low levels or not at all, rendering them relatively or absolutely resistant to infection. One such receptor has only recently been identified (Bergelson et al., 1997), and its identification and characterization should provide additional insight into how adenoviral gene transfer vectors and strategies can be improved.

Adeno-Associated Virus

Adeno-associated virus (AAV) is a small, defective DNA virus (parvovirus) that requires helper adenovirus to propagate *in vitro* and *in vivo* (Berns and Giraud, 1996). The genome consists of a 4.7 kb linear, single-stranded DNA and contains the viral genes *rep* and *cap,* which are required for completion of the wild-type viral life cycle (Fig. 5). The viral genes are flanked by inverted terminal repeat (ITR) sequences, much like the retroviral genomic LTR sequences. By providing the rep and cap functions in *trans,* recombinant AAV can be constructed that contain foreign transgenes flanked only by the ITR sequences. Following infection, the wild-type AAV genome can establish itself as a provirus by predominantly site-specific integration at a site on human chromosome 19 (Samulski, 1993; Linden et al., 1996). However, recent work suggests that recombinant AAV vectors do not integrate at specific site(s) following infection *in vitro* (Kearns et al., 1996; Malik et al., 1997; Ponnazhagan et al., 1997). Site-specific integration relies on rep function, and in its absence integration occurs only with very high multiplicity of infection and at random

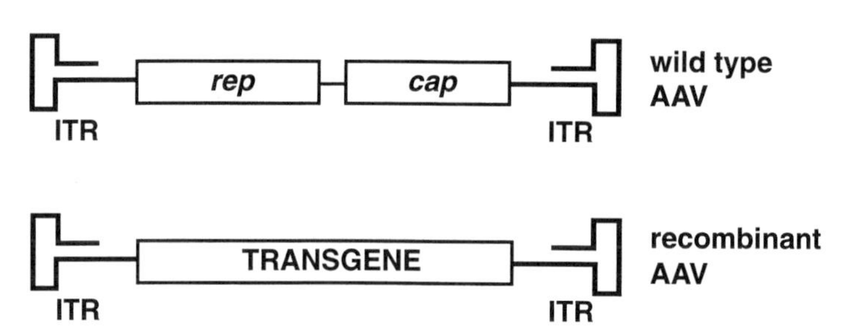

FIGURE 5 Recombinant AAV. This diagrams the structure of the AAV genome, consisting of a 4.7 kb linear, single-stranded DNA. In recombinant AAV, the intervening rep and cap genes are deleted, and the transgene is inserted between the inverted terminal repeat (ITR) sequences.

sites in the host genome. Whether recombinant AAV vectors integrate or remain episomal following infection *in vivo* is currently the focus of intense investigation.

Recombinant AAV can achieve persistent expression of inserted transgenes, especially when delivered to skeletal (Kessler et al., 1996; Xiao et al., 1996; Fisher et al., 1997), cardiac (Kaplitt et al., 1996), and even vascular smooth (Arnold et al., 1997) muscle. In these settings the potential of AAV to mediate gene transfer in the treatment of cardiovascular diseases is readily apparent. Currently the successful application to a wide variety of disorders is limited by difficulties in preparation of the vector in high titer and free of contaminating adenovirus. However, it may soon be useful for disorders that require systemic delivery of even minute quantities of gene product, such as hemophilia (L. Chen et al., 1997; Koeberl et al., 1997). Alternatively, exquisitely targeted therapy may be successful, as recently illustrated by the use of AAV transfer to the hypothalamus for the treatment of hypertension (Phillips, 1997; Phillips et al., 1997). The development of recombinant AAV for gene therapeutics is still in its infancy but should develop rapidly and holds great promise.

Other Viral Vectors

In addition to retrovirus, adenovirus, and AAV, investigators are adapting several other viruses for use in cardiovascular gene therapy. Herpes simplex virus type 1, though typically regarded as a neurotrophic virus most suitable for neuronal gene therapy, has shown some potential for targeting cardiovascular diseases (Mesri et al., 1995; Coffin et al., 1996). The finding that recombinant herpes simplex gene transfer vectors can efficiently transfer genes to the myocardium (Coffin et al., 1996) is perhaps not surprising when viewed in the context of older reports of clinical (McFarlane et al., 1985; Goodman, 1989) and experimentally modeled (Grodums et al., 1981) herpes myocarditis.

Whether herpes will prove to be superior to other viral vectors cannot be judged on the basis of current experience. Other viral vector systems that are in very early phases of development as *in vivo* gene transfer vectors include Sindbis (Herweijer et al., 1995; Dubensky et al., 1996), vaccinia (Jolly, 1994), simian virus 40 (Strayer and Milano, 1996), baculovirus (Hofmann et al., 1995; Boyce and Bucher, 1996) and even human immunodeficiency virus type 1 (Garcia and Miller, 1994; Emerman, 1996). Undoubtedly, other viruses including the both the positive-strand (alphavirus) (Frolov et al., 1996) and negative-strand (Palese et al., 1996) RNA viruses will be developed as vector systems in the years to come. The exquisite efficiency with which they can deliver genetic material to susceptible cells makes them very attractive gene transfer agents, and their practical use is limited by our knowledge of their basic virology, the ease with which their genome can be manipulated, and the safety with which recombinant derivatives can be administered (Smith, 1995).

Semisynthetic Conjugates

Recombinant viral vectors offer unparalleled efficiency of gene transfer to a broad range of cellular targets but there are considerable challenges in manipulating viral genomes and ensuring safety. The general theme of using a specific molecule (in this case the viral coat protein) as a means of receptor-mediated entry into the cell has been approached in some alternative ways (Michael and Curiel, 1994). Semisynthetic conjugates based on adenovirus-mediated cellular entry and escape from lysosomal degradation have been prepared from both human (Cotten et al., 1992) and avian (Cotten et al., 1993) adenoviruses. These gene transfer vehicles use an inactivated or replication defective adenovirus covalently modified with poly-L-lysine to incorporate naked DNA in a complex that is recognized by the viral receptor, thus permitting uptake and subsequent expression of the DNA. The high efficiency of adenoviral infection is due in large part to the ability of the viral capsid protein to disrupt the endosome prior to formation of the endolysosome, thus releasing the viral particle into the cytoplasm and avoiding degradation. The semisynthetic complexes take advantage of this endosomolytic activity. The polylysine moiety serves as a DNA-binding polycation, forming an electrostatic condensate with the DNA (plasmid or other form). A similar approach takes advantage of the ability of the Sendai virus (hemagglutinating virus of Japan) to mediate membrane fusion, and complexes of inactivated Sendai virus, DNA, and liposomes have been used successfully to transfer genes *in vivo* (Dzau et al., 1996). Other approaches have used semisynthetic virus–polylysine : DNA complexes that also incorporate natural ligands to target gene transfer via a specific receptor. For example, complexes containing transferrin (Wagner et al., 1992b) or asialo-orosomucoid (Fisher and Wilson, 1994) have demonstrated such receptor-specific targeting of adenoviral-assisted gene transfer. Newer approaches are being developed that use purified viral peptides that mediate endocytosis and endosomolysis incorporated in a completely synthetic complex with naked DNA (Plank et al., 1992; Wagner et al., 1992a). Such synthetic complexes may someday replace recombinant virions as gene transfer vectors (Remy et al., 1995).

Oligonucleotides

A variant on the naked DNA approach to altering gene expression is the use of short synthetic DNA or RNA molecules to inhibit rather than enhance gene function. These molecules rely principally on their ability to recognize their targets in a sequence-specific manner. Oligodeoxynucleotides, single-stranded DNA molecules typically less than 30 bases in length, have enjoyed the most attention because of their relative ease of preparation and dose-dependent activity in a variety of applications. The most common use of oligonucleotides in the setting of gene therapy is to use the antisense, or minus, strand to interfere with the synthesis, processing, or function of messenger RNA (mRNA), representing the sense, or plus, strand. Antisense oligonucleotides may act to inhibit the expression of the gene by a variety of mechanisms, including interfering with transcription of the chromosomal DNA or the processing of pre-mRNA into mature mRNA, directing premature degradation of the partially double-stranded, mRNA–oligoDNA duplex, or inhibiting translation by preventing the proper interaction of the mRNA with the protein synthetic apparatus of the cell (Stein and Cheng, 1993). Another application of oligonucleotides is to form a triple-stranded DNA complex between the chromosomal DNA of the gene and the single-stranded oligonucleotide, resulting in a block to RNA transcription (Kinniburgh et al., 1994; Reynolds et al., 1994) or recombinatorial mutational gene conversion (Wang et al., 1995; Faruqi et al., 1996). Although an attractive concept, development of the triplex approach is in the early stages. Oligonucleotides can be delivered in solution without modification, applied as a biocompatible gel that slowly dissolves (Simons et al., 1992a), complexed with liposomes or as conjugated with inactivated viruses (Morishita et al., 1993a), or carried with other enhancing chemical modifications such as covalent linkage to cholesterol (Krieg et al., 1993).

Although an attractive concept, the successful application of antisense oligonucleotides to inhibiting gene expression can be problematic. The natural structure of the phosphodiester backbone makes the synthetic oligonucleotides quite sensitive to destruction by cellular nucleases, a problem partially overcome by the use of chemical modifications (phosphorothioate, methylphosphonate, or phosphorodithioate derivatization). One novel approach substitutes the phosphodiester linkages of the backbone with amide bonds, creating so-called peptide nucleic acids that are nuclease resistant (Hanvey et al., 1992). Beyond these issues of stability, efficacy and especially specificity can be difficult to demonstrate. Many cell systems respond to seemingly unrelated sequences in a similar fashion as to the antisense sequence, suggesting interactions with unintended targets (Stein and Cheng, 1993; Burgess et al., 1995;

Villa et al., 1995; Wang et al., 1996). In addition to potentially interacting with unintended DNA or RNA targets, oligonucleotides may also interact with a variety of cytoplasmic and nuclear proteins (Brown et al., 1994; Bergan et al., 1995). If intended, this may represent an effective gene targeting strategy; however, it is not the goal of most oligonucleotide experiments. Oligonucleotides can also interact with proteins that normally do not bind nucleic acid, simply by virtue of chance complementarity in their tertiary structure (so-called aptamers) (Bock et al., 1992). The specificity of this interaction can be quite precise and can be exploited as an alternative approach to drug development. For example, RNA aptamers have been identified through functional screening that inhibit a specific isoform of the signal transduction molecule protein kinase C (Conrad et al., 1994). In fact, phosphorothioate modification of oligonucleotides, intended to simply render these molecules resistant to degradation, may limit their specificity. Compared with minimally modified or unmodified oligonucleotides, phosphorothioate-modified oligonucleotides bind more avidly to a variety of cytoplasmic and nuclear proteins and may bind nuclear transcription factors in a nonspecific fashion (Brown et al., 1994). Despite these caveats that may compromise the usual intended use of oligonucleotides, many studies have successfully applied antisense oligonucleotides to models of gene therapy, including cardiovascular problems such as inhibiting restenosis following angioplasty (see below).

Antisense RNA and Ribozymes

Vectors expressing RNA molecules that do not encode proteins are also potential therapeutic agents. Antisense RNAs are used like antisense oligos to inhibit gene expression but differ substantially in vector design and probable mechanism of action. Delivery of antisense RNA relies on the ability to transfer a comparatively long stretch of DNA encoding the gene of interest, but with the gene inserted in the gene transfer vector in the opposite, or antisense, orientation. Plasmid DNA, retroviruses, and adenoviruses have all been used as antisense RNA gene transfer vectors, and their use is subject to the same caveats and limitations inherent in the general use of these vectors. However, several reports have documented the potential utility of the antisense RNA approach in the cardiovascular system (Du and Delafontaine, 1995; Delafontaine et al., 1996; Fox and Shanley, 1996; Lu et al., 1996; Zhang et al., 1996; Hanna et al., 1997).

Single-stranded RNA molecules may also possess catalytic activity (ribozymes). The structure and activity of therapeutic ribozymes are based on naturally occurring, autocatalytic RNAs that possess both recognition and catalytic domains targeted to another portion of the same RNA molecule (substrate) that requires hydrolytic cleavage for maturation of the entire molecule. Engineered ribozymes are designed to possess recognition domains that target a heterologous transcript (target gene activity to be altered) rather than recognizing another portion of the same molecule (Altman, 1993; Sullivan,

1994). Ribozymes have been developed mainly with the intention of destroying their intended target, and this strategy has been applied to vascular smooth muscle cells (Gu et al., 1995; Jarvis et al., 1996). However, these catalytic RNA molecules can also be engineered to replace stretches of RNA by a *trans*-splicing mechanism, suggesting a unique approach to gene therapy of defective messenger RNA molecules (Sullenger and Cech, 1994). Although this type of defect is not recognized as contributing to a large number of diseases, this approach may well be applied to special problems for which it is particularly well suited.

TARGETING SPECIFIC ORGANS FOR THERAPEUTIC GENE TRANSFER

Skeletal Myoblasts and Muscle Fibers as Gene Therapy Vehicles

Skeletal myoblasts can be removed from adult animals, cultured, and genetically modified *in vitro.* Gene transfer has been accomplished in cultured myoblasts by many of the naked DNA and viral vector methods discussed above and then reimplanted where they fuse with the skeletal myofibers and continue to express the transgene (Blau et al., 1993; Rando and Blau, 1994). The transgene can be therapeutic for the target muscle, as in the heritable muscular dystrophies (Gussoni et al., 1992; Morgan et al., 1992), or can be a gene whose product (for example, growth hormone) is targeted to other tissues (Barr and Leiden, 1991; Dhawan et al., 1991). Plasmid DNA (Jiao et al., 1992; Wolff et al., 1992), recombinant adenoviral vectors, and AAV vectors have all been used to achieve direct *in vivo* gene transfer to skeletal muscle. Some investigators have found that naked DNA may be superior to viral methods because of either low infectivity of viral vectors for terminally differentiated muscle (Davis et al., 1993; Acsadi et al., 1994) or the now well-recognized immune response that often limits the persistence of adenoviral-mediated gene transfer with most of the generally available recombinant adenoviral vectors (Davis et al., 1993). This situation is evolving, however, with the development of AAV and improved adenoviral vectors (see above). Despite these controversies, genetic modification of skeletal muscle *in vivo* has been studied for the treatment of muscular dystrophy (Quantin et al., 1992; Ragot et al., 1993) and in the systemic delivery of erythropoietin (Tripathy et al., 1994) and Factor IX (Yao et al., 1994). More recently, plasmid DNA gene transfer has been used in attempts to express physiologically meaningful levels of vascular endothelial growth factor in skeletal muscle (discussed in more detail below).

Vascular Gene Transfer *In Vivo:* Plasmid DNA, Retrovirus

Early attempts at *in vivo* vascular gene transfer utilized plasmid DNA–liposome transfer vehicles. These experiments demonstrated that marker genes (lacZ or the firefly luciferase genes) could be transferred to coronary or peripheral arteries of the dog first using surgical exposure of the vessels (Lim et al., 1991) and later a percutaneous catheter-based approach (Chapman et al., 1992) to achieve delivery of the genetic material. Subsequent work demonstrated that plasmid DNA–liposomal gene transfer by a percutaneous approach was applicable to atherosclerotic as well as normal arteries in a rabbit model (Leclerc et al., 1992). More recent work has shown that the efficiency of transfection with plasmid DNA can be improved through its delivery as a gel coating an intravascular balloon catheter (Riessen et al., 1993). Although considered safe from the perspective of toxicity and side effects, gene transfer using DNA–liposomes is of generally low efficiency.

In an effort to improve on these results and extend the experience gained with vascular cells *ex vivo,* protocols using recombinant retrovirus have been developed. As discussed above, retroviral-mediated gene transfer requires host cell division at the time of infection, limiting this approach to actively dividing cells. In addition, *in vivo* gene transfer requires very high titer solutions of recombinant retrovirus, which are rapidly inactivated by serum. Despite these technical hurdles, recombinant retrovirus expressing lacZ has been used to demonstrate direct *in vivo* gene transfer to the iliofemoral artery of pigs (Nabel et al., 1990). These investigators used a double-balloon catheter to create an isolated luminal segment of a surgically exposed vessel, which could then be emptied of blood and serum prior to the instillation of the viral solution. The same objective of exposing the arterial wall to recombinant retrovirus while avoiding exposure to serum was achieved in the rabbit aorta using an angioplasty balloon containing microscopic perforations (Wolinsky balloon) that allowed the delivery of the viral solution as high pressure jets directed orthogonal to the vessel wall (Flugelman et al., 1992). Despite variable and generally low efficiency of gene transfer, these results demonstrated the feasibility of retroviral gene transfer *in vivo* and were followed by a series of experiments designed to show that specific genes transferred using retrovirus could play a pathogenic role in the response to vascular injury. These are discussed in detail in the section below describing gene transfer specifically targeted to the problem of vascular injury.

Vascular Gene Transfer *In Vivo:* Recombinant Adenovirus, Other Viruses

The most efficient means of achieving *in vivo* vascular gene transfer developed to date is recombinant adenovirus. As reviewed above, these vectors can be prepared in very high titer (suitable for *in vivo* applications) and infect a broad range of both dividing and nondividing cells. Experiments have demonstrated efficient adenovirus-mediated gene transfer to the vasculature using lacZ, alkaline phosphatase, and luciferase marker genes (Guzman et al., 1993a; Lee et al., 1993; Lemarchand et al., 1993; French et al., 1994; Steg et al., 1994). These studies have shown that gene transfer mediated by recombinant adenovirus is more efficient than recombinant retrovirus or DNA–liposomes, but is transient, peaking at 7–14 days, declining by 21 days, and becoming undetectable by 28 days. Infection and gene transfer appears to be limited to endothelium if the virus is delivered to uninjured arteries, and infection of medial smooth muscle cells requires denudation of the endothelium and either balloon overstretch or hydrostatic pressure distention. There is evidence that the internal elastic lamina presents a physical barrier to the delivery of viral particles to the media (Rome et al., 1994), which may explain the need for mechanical disruption (if only transient) of normal tissue architecture. The lack of persistence of transgene expression is thought to be mediated by the same or similar immune clearance mechanisms as those discussed above, and at least one group has reported finding a perivascular mononuclear infiltrate following successful adenoviral gene transfer to porcine coronary arteries (French et al., 1994). Interestingly, this group found that intracoronary gene transfer of recombinant adenovirus delivered intramurally by the Wolinsky balloon was equally efficient in normal, balloon overstretch–injured, atherosclerotic, and oversized stent–injured arteries. A number of similar studies have confirmed these observations regarding the efficiency and degree of persistence of recombinant adenovirus in the myocardium. More recently, recombinant AAV has shown promise as an efficient, persistent myocardial gene transfer agent (Kaplitt et al., 1996), but the experience with this vector system is currently limited. Similar approaches using other viruses are discussed in the section on gene transfer vectors.

APPROACHES TO GENE THERAPY OF CARDIOVASCULAR DISEASES

Postangioplasty Restenosis

Restenosis remains the primary limitation of the long-term success of angioplasty. Most of the strategies for gene therapy of restenosis target smooth muscle cell proliferation. Many studies have documented the importance of a variety of growth factors, cytokines, cell surface receptors, cytoplasmic second messengers, and cell cycle regulatory factors in controlling smooth muscle cell proliferation (Casscells, 1991). For example, local expression of acidic fibroblast growth factor (FGF) (Nabel et al., 1993c), platelet-derived growth factor-B (Nabel et al., 1993b), transforming growth factor-β1 (Nabel et al., 1993a), and angiotensin-converting enzyme (Morishita et al., 1994a) can all produce neointimal hyperplasia or hypertrophy (either *de novo* or in response to vascular injury) in animal models.

Basic FGF (bFGF), a potent smooth muscle cell mitogen, stimulates proliferation through activation of cell surface FGF receptors. Antibodies targeting bFGF (Lindner and Reidy, 1991) or bFGF-conjugated mitotoxins internalized through FGF receptors (Casscells et al., 1992) can attenuate the response to experimental arterial injury. Expression of an antisense bFGF RNA inhibits bFGF expression and triggers apoptosis or programmed cell death (Fox and Shanley, 1996), as does overexpression of a dominant-negative FGF receptor (T. Miyamoto et al., 1997). The roles that expression of bFGF and autocrine FGF signaling play have been investigated *in vivo* by manipulating either bFGF ligand or FGF receptor activity in an arterial injury model. Gene transfer of these vectors attenuates the response to arterial injury, with evidence of apoptosis in the neointimal layer of the vessel wall (A. Hanna, D. Neschis, T. Miyamoto, M. Golden, and J.C. Fox, unpublished results). These results suggest that genetic manipulation of FGF signaling may reduce or prevent the response to arterial injury (Hanna et al., 1997).

Some investigators have pointed out that many growth factors and cytokines are not smooth muscle specific and that the signal transduction pathways of several mitogen-activated receptors converge on a common set of cell cycle regulatory factors. Another approach has been to target such nuclear regulatory factors that are expressed or are required for progression through the cell cycle in all cells as a means of targeting a central pathway common to the actions of many growth factors and cytokines. A number of previously recognized protooncogenes like c-*myc* and c-*myb* are expressed by many cell types in response to mitogens, including smooth muscle cells, and participate in normal progression through the cell cycle. Antisense oligodeoxynucleotides directed against these genes are effective in inhibiting smooth muscle cell proliferation *in vitro*. Delivered either to the luminal or adventitial surface, antisense oligodeoxynucleotides directed against c-*myb* (Simons et al., 1992a) or c-*myc* (Bennett et al., 1994; Shi et al., 1994) can attenuate the neointimal response to vascular injury in animal models. Other genes known to be expressed by actively dividing cells that can be targeted to attenuate neointimal thickening in animal models include proliferating cell nuclear antigen (Simons et al., 1994) and nonmuscle myosin heavy chain (Simons, 1992b).

Recent progress with genes that more directly control progression through the cell cycle has identified cyclins and their associated cyclin-dependent kinases (CDKs) whose activities in concert are required for proper cell cycling. These genes have been targeted by antisense oligodeoxynucleotides as well, and it was recently demonstrated that inhibition of at least two of these essential (though potentially redundant) components was required to achieve nearly complete inhibition of proliferation whereas each oligodeoxynucleotide alone was partially effective (Morishita et al., 1993a, 1994). All of these oligodeoxynucleotide approaches are subject to the technical caveats and limitations discussed in the earlier section on the use of these agents to modify gene expression. Recent data regarding the nonspecific, nonsequence-dependent or aptameric effects of oligonucleotides have fueled continued controversy regarding the ultimate utility of oligos as therapeutic agents, dampening enthusiasm for their use in this context.

Other recent gene transfer experiments have been reported that demonstrate effective inhibition of neointimal thickening in the rat carotid injury model by targeting cell cycle regulatory proteins. A nonphosphorylatable analog of the cell cycle regulatory protein Rb (the retinoblastoma gene product) can inhibit smooth muscle cell proliferation *in vitro* and neointimal thickening *in vivo* (Chang et al., 1995). Phosphorylation of Rb is required for normal cell cycle progression, and the presence of the nonphosphorylatable analog serves as a dominant-negative regulator preventing cell cycle progression. Similarly, overexpression of the CDK inhibitor p21 induces smooth muscle cell cycle arrest *in vitro* and attenuates neointimal thickening *in vivo* (Yang et al., 1996).

In other recent studies, nitric oxide synthase (NOS) expression targeted to vascular smooth muscle cells following arterial injury has been demonstrated to inhibit neointimal thickening (Von der Leyen et al., 1995). Prior to this study, a number of observations suggested that local production of NO serves as a negative growth regulatory signal for vascular smooth muscle cells. It was postulated that the loss of endothelium following vascular injury, in addition to exposing the vascular smooth muscle cells directly to the circulation, leads to the loss of this negative growth regulatory influence. This gene transfer experiment suggests that the loss of local NO production contributes to the hyperplastic response following injury and that restoration of NO production to physiological levels may partially ameliorate this response.

All of these strategies that target growth factors, protooncogenes, cell cycle regulatory proteins, and signaling molecules that are common to many if not all cells means that specificity is generated strictly by delivery techniques or the use of tissue-restricted expression vectors. Further refinements in the design and delivery of gene therapy vectors suggests that several of these experimental approaches may well turn out to be practical for routine clinical use. These experiments also demonstrate the important principle that the delivery and expression of gene-enhancing or -inhibiting vectors may be efficient enough at the present stage of development to be physiologically meaningful. Further efforts are underway to continue to improve the technology for vascular cell gene transfer and drug delivery, and research on the molecular mechanisms controlling smooth muscle cell behavior continues to provide the rationale for choosing appropriate molecular targets for future intervention.

Myocardial Gene Therapy

The development of gene transfer methods for the heart has followed a parallel track to that of gene transfer to the vasculature. Approaches have included direct application of naked DNA, DNA complexed to liposomes, and recombinant adenovirus. Transgenic mice have already demonstrated the potential of transferred genes to alter myocardial cellularity, mass, proliferative capacity, and contractile function (Field, 1988; Jackson et al., 1990; Gruver et al., 1993; Milano et al., 1994a,b). These elegant studies give a glimpse of the potential for gene therapy of the heart to treat ischemic, congestive, or hypertrophic cardiomyopathies. As the fields of disease gene identification, molecular pathophysiology, and *in vivo* gene transfer continue to converge, the potential to intervene in the diseases of the heart continues to grow.

Direct Myocardial Injection of Plasmid DNA

Direct injection of DNA into the myocardium leads to measurable gene expression (Lin et al., 1990; Acsadi et al., 1991), which is patchy in distribution and limited to cardiomyocytes at the injection site. This suggests that uptake and expression of foreign DNA is a low frequency event despite exposure of a large number of cells to many copies of the DNA. Direct myocardial injection of plasmid DNA produces an acute inflammatory cell infiltrate along the needle track, as well as subsequent foci of fibrosis at the injection site, representing an inflammatory host reaction to plasmid DNA independent of transgene expression. The issues of efficiency and host tissue responses limit the clinical usefulness of this approach, but direct injection of plasmid DNA into the beating heart *in situ* demonstrated the feasibility of gene transfer *in vivo* to nondividing, functioning cardiomyocytes. This approach has been used to study the tissue-specific (Buttrick et al., 1992, 1993; von Harsdorf et al., 1993) or hormone-responsive (Kitsis et al., 1991) expression of transferred genes in the myocardium. These studies have documented important differences in the regulation of tissue-specific gene expression *in vivo* as opposed to *in vitro* and illustrate a major contribution of *in vivo* gene transfer to the study of molecular pathophysiology. Such an approach could conceivably be used to define the molecular mechanisms whereby altered gene expression or mutated gene products contribute to cardiac disease, as in hypertrophic cardiomyopathy or heart failure. Direct myocardial injection of DNA is, therefore, potentially useful for such analytical approaches. Because of low efficiency and lack of persistent expression, however, this technique is probably not of any significant therapeutic value. It has been largely superseded by other approaches, chiefly the use of viral gene transfer vectors such as recombinant adenovirus and AAV.

Myocardial Gene Transfer Using Recombinant Adenovirus

An early report of myocardial gene transfer using recombinant adenovirus used intravenous delivery, resulting in gene transfer to a variety of tissues including the heart (Stratford-Perricaudet et al., 1992). One of the most intriguing findings was the persistence of transgene expression in mice injected as neonates. It has been speculated that the ability of the adenovirus to remain stable in the mouse tissues for many months reflects the lack of complete immune competence in neonates, especially of cell-mediated immunity (see below). Selective adenoviral gene transfer to myocardium has been achieved through the use of intramyocardial injection (Guzman et al., 1993b; French et al., 1994) and via percutaneous, catheter-based approaches targeting either the coronary vasculature (French et al., 1994) or the myocardium (or both)

(Barr et al., 1994). These studies reflect the great therapeutic potential of vascular and myocardial gene therapy, but are all limited by the rapid disappearance of genetically modified cells secondary to cellular immune responses. Despite vastly improved gene transfer efficiency, these studies noted that adenoviral gene transfer was limited in duration, similar to gene transfer using plasmid DNA. Direct injection produces a local inflammatory cell infiltrate, consistent with the studies using plasmid DNA. Although partially due to a nonspecific reaction to the mechanical trauma of direct injection (as measured by the presence of $CD44^+$ leukocytes), a significant component of the inflammatory infiltrate consisted of $CD8^+$ T lymphocytes. Percutaneous, transcoronary delivery also leads to complete loss of expression over several weeks and correlates with physical loss of the vector DNA. However, the lack of transgene expression in this study was not associated with an inflammatory infiltrate or any evidence of scarring or fibrosis of the target tissue (Barr et al., 1994). The immunological basis for these different observations has not been clarified.

Recent studies designed to specifically address the lack of persistence of adenoviral gene transfer in immunocompetent animals have documented a cell-mediated immune response directed against adenovirally infected cells (Engelhardt et al., 1994a; Kass-Eisler et al., 1994; Yang et al., 1994). These studies point out the weaknesses of the currently available adenoviral vectors and have helped spur development of improved vector designs based on a better understanding of the biology of the recombinant adenovirus and the associated host response to infection. These results also emphasize the role that careful monitoring of local tissue reactions and systemic immune responses play in evaluating these emerging gene transfer technologies. The recent progress made in the application of recombinant AAV to myocardial gene transfer may supersede the routine use of adenovirus (Kaplitt et al., 1996), but there are other technical challenges to establishing AAV as a routine gene transfer vector that were discussed above.

The issues of persistent expression and immune clearance aside, recombinant adenoviral vectors have been used to demonstrate some of the therapeutic potential of direct genetic modification of the myocardium. By analogy to some of the transgenic mouse studies cited above (Milano et al., 1994a,b), adenoviral gene transfer of either a β-adrenergic receptor gene or a vector encoding a peptide antagonist for the desensitizing receptor kinase can potentiate β-adrenergic signaling in cultured adult cardiomyocytes (Drazner et al., 1997). Treatment of cardiac allografts with recombinant adenovirus encoding the immunomodulatory cytokine transforming growth factor-β appears to prolong allograft survival by interfering with T lymphocyte–mediated immune functions (Qin et al., 1996). These published experiments offer a glimpse of the types of problems investigators are targeting by *in vivo* gene transfer techniques. Ischemic, hypertrophic, and congestive cardiomyopathies, transplant myopathy and arteriopathy, and metabolic storage diseases are all under intense investigation as targets for gene therapy.

Angiogenic Therapy for Arterial Insufficiency

Both angioplasty and reconstructive surgery are effective treatment options for many patients with either coronary or peripheral arterial insufficiency, restoring perfusion and resulting in the alleviation of symptoms. However, these procedures are associated with considerable risks, notably restenosis after angioplasty, complications of coronary bypass surgery, and cardiac complications of peripheral vascular surgery. In addition, the severity and progressive nature of atherosclerotic vascular disease often limit these treatment options. A major advance in treating arterial insufficiency would be an effective means of promoting collateralization of ischemic tissue through angiogenesis. Approaches to promoting angiogenesis have focused primarily on bFGF and vascular endothelial growth factor (VEGF).

Basic FGF and related proteins are potent angiogenic agents when administered as peptides (Unger et al., 1994; Lazarous et al., 1995). Exogenous FGFs stimulate the expression of endogenous FGFs as well as VEGF. Overexpression of bFGF *in vitro* can promote mitogenesis and lead to a transformed phenotype, or it can promote differentiation. bFGF is synthesized without a secretory signal peptide, so it does not exit cells via the classic secretory pathway but is instead exported by an alternate pathway (Florkiewicz et al., 1995). Physiological studies suggest that an increased local expression and cellular release of bFGF mediates angiogenesis selectively in response to ischemia (Walgenbach et al., 1995). When ectopically expressed in transgenic mice, bFGF has only subtle effects on cell growth (Coffin et al., 1995), with its protective influence perhaps brought out only in the setting of ischemic tissue damage (MacMillan et al., 1993). These observations provide the basis for bFGF as a candidate for gene therapy of ischemic arterial insufficiency.

VEGF promotes endothelial cell proliferation, microvascular hyperpermeability, and angiogenesis (Neufeld et al., 1994; Thomas, 1996). VEGF is essentially an endothelial cell–specific cytokine since expression of VEGF receptors is largely restricted to this cell type. In addition to promoting new vessel formation in areas of normal perfusion, the expression of both VEGF and its receptors is enhanced by tissue ischemia or hypoxia, thus rendering ischemic capillary beds more responsive to VEGF. Available data implicate VEGF as a potential mediator of a spontaneous angiogenic response to ischemia and have led to experiments documenting its efficacy in enhancing this process in experimental models and provided rationale for gene therapy strategies for delivering the VEGF gene to ischemic tissue beds. Gene transfer of a plasmid DNA encoding VEGF can promote the enhanced formation of collaterals in an ischemic hindlimb model (Tsurumi et al., 1996). A recombinant adenovirus encoding VEGF administered as a suspension in basement membrane extract (Matrigel) and injected subcutaneously into mice results in detectable VEGF protein in the surrounding tissues up to 3 weeks after injection, with histological evidence of neovascularization (Muhlhauser et al., 1995). Similarly, cultured fibroblasts transduced *in vitro* with a recombinant herpes viral vector encoding VEGF suspended in Matrigel and injected subcutaneously into syngeneic mice resulted in a detectable angiogenic response (Mesri et al., 1995). The area of therapeutic angiogenesis is generating great interest, and both preclinical and clinical investigations using plasmid DNA encoding VEGF are currently in progress (Isner et al., 1995, 1996).

Gene Therapy of Lipid Disorders

A discussion of gene therapy of cardiovascular diseases would be incomplete without consideration of the approaches being developed to alter noncardiac diseases with serious cardiovascular consequences. The dyslipidemias are a heterogeneous group of syndromes caused by a complex array of genetic and environmental factors. The existence of monogenic diseases (such as familial hypercholesterolemia [FH]) and the generation of transgenic or knockout mice corresponding to individual components of lipid metabolism have helped identify potential targets for gene therapy of lipid disorders.

The liver is one of the major sites of both synthesis and catabolism of lipoproteins with cardiovascular consequences. Most of the current efforts in the area of gene therapy for lipid disorders are therefore focused on liver-directed gene therapy. The liver participates in a wide spectrum of other disorders, so a full discussion of liver-directed gene therapy is beyond the scope of this discussion (for recent reviews, see Wilson and Grossman, 1993; Strauss, 1994). Several important concepts are worth mentioning. First, hepatocytes retain the capacity to proliferate *in vivo* and *in vitro,* and this means that recombinant retroviruses can be suitable vectors for hepatic gene transfer. Second, the capacity to proliferate and therefore regenerate *in vivo* provides a ready source of autologous hepatocytes for schemes using *ex vivo* genetic modification and reimplantation. Third, hepatocytes have the ability, like bone marrow cells, to "home in" on their organ of origin when reinfused into the circulation (Ponder et al., 1991), another advantage for *ex vivo* modification and reimplantation strategies. Fourth, the liver is naturally a site of filtration of the circulation, with a fenestrated endothelium that permits direct contact between the circulation and a large surface area of hepatic parenchyma, rendering the liver perhaps more susceptible than many other organs to gene therapy vectors. In fact, many studies of adenoviral gene transfer targeted to other organs have noted prominent (and unintended) gene transfer to the liver.

Liver-directed gene therapy for lipid disorders has focused mainly on FH, caused by mutations in the LDL-R. Replacement of the defective gene can potentially cure this progressive, fatal disease, as the metabolic defect is completely corrected by orthotopic liver transplantation (Bilheimer et al., 1984). Using the WHHL rabbit model, it has been demonstrated that autologous hepatocytes can be harvested, modified *ex vivo* with a recombinant retrovirus, and reimplanted, resulting in at least temporary improvement in hyperlipidemia (Wilson et al., 1990). This approach was eventually developed for one of the first clinical applications of gene therapy, and the initial experience with an FH patient has been reported (Grossman et al., 1994). Although an important milestone, this approach still requires surgical harvest (partial hepatectomy), considerable *ex vivo* manipulation of the patient's hepatocytes, and the use of recombinant retroviral vectors (with the caveats discussed in the appropriate section above). For these reasons, future approaches to gene therapy for FH will likely rely on strategies that can achieve high efficiency gene transfer *in vivo,* such as recombinant adenovirus.

The principle of adenoviral gene transfer to the liver *in vivo* has been well demonstrated (see reviews cited above). With respect to FH, delivery of a recombinant adenovirus encoding the LDL-R acutely accelerates the clearance of circulating LDL in normal mice (Herz and Gerard, 1993). In a murine charicature of gene therapy for the human disease, adenovirally mediated transfer of the LDL-R gene to the liver of LDL-R knockout mice corrects the hypercholesterolemia associated with a complete lack of functional LDL-R (Ishibashi et al., 1993). This same approach has been successful when applied to the WHHL rabbit (Kozarsky et al., 1994), and has been applied to hepatocytes from FH patients *in vitro* (Kozarsky et al., 1993). Interestingly, gene transfer of the recently cloned very low density lipoprotein-R can also ameliorate the hyperlipidemia characteristic of LDL-R deficiency (Kozarsky et al., 1996). Although the usefulness of the current generation of adenoviral gene transfer vectors in the setting of liver-directed gene therapy is limited by the lack of persistence due to immunological clearance mechanisms, improvements in vector design will eventually make *in vivo* gene transfer the method of choice for genetic correction of FH. In addition, this entire effort provides an instructive paradigm for the treatment of a range of disorders of hepatic origin or amenable to the hepatogenous synthesis and delivery of therapeutic proteins to the circulation.

CONCLUSIONS

Gene therapy for cardiovascular disease, though early in its development, offers great potential for providing novel therapeutic approaches to difficult clinical problems. Since the late 1980s there has been an explosive growth in gene transfer technologies, with many variations in both synthetic DNA and recombinant viral technologies being successfully developed and applied to cellular and animal models of gene transfer *in vitro* and *in vivo.* Improvements in the structure and function of expression vectors, especially in the ability to control expression in a spatial, temporal, and physiological manner, are being aggressively investigated. Basic investigations of how altered gene expression triggers or contributes to disease processes, and the cloning of specific disease-related genes, will help to identify gene targets appropriate for therapeutic manipulation. Expanding knowledge of the molecular mechanisms of disease, further discovery and refinement of gene transfer vectors, and the engineering of improved local drug delivery devices will all contribute to future cardiovascular disease gene therapy strategies. As improvements in technology parallel advances in our knowledge of basic mechanisms of disease, gene therapy strategies for cardiovascular diseases will soon become a clinical reality.

REFERENCES

Acsadi G, et al. (1991): Direct gene transfer and expression into rat heart *in vivo*. *New Biologist* 3:71–81.

Acsadi G, et al. (1994): Cultured human myoblasts and myotubes show markedly different transducibility by replication-detective adenovirus recombinants. *Gene Ther* 1:338–340.

Ali M, et al. (1994): The use of DNA viruses as vectors for gene therapy. *Gene Ther* 1:367–384.

Altman S (1993): RNA enzyme-directed gene therapy. *Proc Natl Acad Sci USA* 90):10898–10900.

Armentano D, et al. (1997): Effect of the E4 region on the persistence of transgene expression from adenovirus vectors. *J Virol* 71:2408–2416.

Arnold TE et al. (1997): *In vivo* gene transfer into rat arterial walls with novel adeno-associated virus vectors. *J Vasc Surg* 25:347–355.

Barr E, et al. (1991): Systemic delivery of recombinant proteins by genetically modified myoblasts. *Science* 254:1507–1509.

Barr E, et al. (1994): Efficient catheter-mediated gene transfer into the heart using replication-defective adenovirus. *Gene Ther* 1:51–58.

Bennett MR, et al. (1994): Inhibition of vascular smooth muscle cell proliferation *in vitro* and *in vivo* by c-myc antisense oligodeoxynucleotides. *J Clin Invest* 93:820–828.

Bergan RC, et al. (1995): Inhibition of protein–tyrosine kinase activity in intact cells by the aptameric action of oligodeoxynucleotides. *Antisense Res Dev* 5:33–38.

Bergelson JM, et al. (1997): Isolation of a common receptor for Coxsackie B viruses and adenoviruses 2 and 5. *Science* 275:1320–1323.

Berns KI, et al. (1996): Biology of adeno-associated virus. *Curr Top Microbiol Immunol* 218:1–23.

Bilheimer DW, et al. (1984): Liver transplantation to provide low-density-lipoprotein receptors and lower plasma cholesterol in a child with homozygous familial hypercholesterolemia. *N Engl J Med* 311:1658–1664.

Blau HM, et al. (1993): Myoblasts in pattern formation and gene therapy. *Trends Genet* 9:269–274.

Bock LC, et al. (1992): Selection of single-stranded DNA molecules that bind and inhibit human thrombin. *Nature* 355:564–566.

Boris LK, et al. (1993): Recent advances in retrovirus vector technology. *Curr Opin Genet Dev* 3:102–109.

Boris LK, et al. (1994): The retroviral vector. Replication cycle and safety considerations for retrovirus-mediated gene therapy. *Ann NY Acad Sci* 716:59–70.

Boyce FM, et al. (1996): Baculovirus-mediated gene transfer into mammalian cells. *Proc Natl Acad Sci USA* 93:2348–2355.

Brown DA, et al. (1994): Effect of phosphorothioate modification of oligodeoxynucleotides on specific protein binding. *J Biol Chem* 269:26801–26805.

Burgess TL, et al. (1995): The antiproliferative activity of c-myb and c-myc antisense oligonucleotides in smooth muscle cells is caused by a nonantisense mechanism. *Proc Natl Acad Sci USA* 92:4051–4055.

Burns JC, et al. (1993): Vesicular stomatitis virus G glycoprotein pseudotyped retroviral vectors: Concentration to very high titer and efficient gene transfer into mammalian and nonmammalian cells. *Proc Natl Acad Sci USA* 90:8033–8037.

Burns JC, et al. (1994): Pantropic retroviral vector-mediated gene transfer, integration, and expression in cultured newt limb cells. *Dev Biol* 165:285–289.

Buttrick PM, et al. (1992): Behavior of genes directly injected into the rat heart *in vivo*. *Circ Res* 70:193–198.

Buttrick PM, et al. (1993): Distinct behavior of cardiac myosin heavy chain gene constructs *in vivo*. Discordance with *in vitro* results. *Circ Res* 72:1211–1217.

Casscells W (1991): Smooth muscle cell growth factors [review]. *Prog Growth Factor Res* 3:177–206.

Casscells W, et al. (1992): Elimination of smooth muscle cells in experimental restenosis: Targeting of fibroblast growth factor receptors. *Proc Natl Acad Sci USA* 89:7159–7163.

Chang MW, et al. (1995): Cytostatic gene therapy for vascular proliferative disorders with a constitutively active form of the retinoblastoma gene product. *Science* 267:518–522.

Chapman GD, et al. (1992): Gene transfer into coronary arteries of intact animals with a percutaneous balloon catheter. *Circ Res* 71:27–33.

Chen HH, et al. (1997): Persistence in muscle of an adenoviral vector that lacks all viral genes. *Proc Natl Acad Sci USA* 94:1645–1650.

Chen L, et al. (1997): Comparison of retroviral and adeno-associated viral vectors designed to express human clotting Factor IX. *Hum Gene Ther* 8:125–135.

Coffin JD, et al. (1995): Abnormal bone growth and selective translational regulation in basic fibroblast growth factor (FGF-2) transgenic mice. *Mol Biol Cell* 6:1861–1873.

Coffin RS, et al. (1996): Gene delivery to the heart *in vivo* and to cardiac myocytes and vascular smooth muscle cells *in vitro* using herpes virus vectors. *Gene Ther* 3:560–566.

Conrad R, et al. (1994): Isozyme-specific inhibition of protein kinase C by RNA aptamers. *J Biol Chem* 269:32051–32054.

Cornetta K, et al. (1991): Safety issues related to retroviral-mediated gene transfer in humans. *Hum Gene Ther* 2:5–14.

Cotten M, et al. (1992): High-efficiency receptor-mediated delivery of small and large (48 kilobase gene constructs using the endosome-disruption activity of defective or chemically inactivated adenovirus particles. *Proc Natl Acad Sci USA* 89:6094–6098.

Cotten M, et al. (1993): Chicken adenovirus (CELO virus) particles augment receptor-mediated DNA delivery to mammalian cells and yield exceptional levels of stable transformants. *J Virol* 67:3777–3785.

Davis HL, et al. (1993): Plasmid DNA is superior to viral vectors for direct gene transfer into adult mouse skeletal muscle. *Hum Gene Ther* 4:733–740.

Delafontaine P, et al. (1996): G-protein coupled and tyrosine kinase receptors: Evidence that activation of the insulin-like growth factor I receptor is required for thrombin-induced mitogenesis of rat aortic smooth muscle cells. *J Clin Invest* 97:139–145.

Dhawan J, et al. (1991): Systemic delivery of human growth hormone by injection of genetically engineered myoblasts. *Science* 254:1509–1512.

Drazner MH, et al. (1997): Potentiation of beta-adrenergic signaling by adenoviral-mediated gene transfer in adult rabbit ventricular myocytes. *J Clin Invest* 99:288–296.

Du J, et al. (1995): Inhibition of vascular smooth muscle cell growth through antisense transcription of a rat insulin-like growth factor I receptor cDNA. *Circ Res* 76:963–972.

Dubensky TW Jr, et al. (1996): Sindbis virus DNA-based expression vectors: Utility for *in vitro* and *in vivo* gene transfer. *J Virol* 70:508–519.

Dzau VJ, et al. (1996): Fusigenic viral liposome for gene therapy in cardiovascular diseases. *Proc Natl Acad Sci USA* 93:11421–11425.

Emerman M (1996): From curse to cure: HIV for gene therapy? *Nat Biotechnol* 14: 943.

Engelhardt JF, et al. (1994a): Prolonged transgene expression in cotton rat lung with recombinant adenoviruses defective in E2a. *Hum Gene Ther* 5:1217–1229.

Engelhardt JF, et al. (1994b): Ablation of E2A in recombinant adenoviruses improves transgene persistence and decreases inflammatory response in mouse liver. *Proc Natl Acad Sci USA* 91:6196–6200.

Faruqi AF, et al. (1996): Recombination induced by triple-helix-targeted DNA damage in mammalian cells. *Mol Cell Biol* 16:6820–6828.

Field LJ (1988): Atrial natriuretic factor-SV40 T antigen transgenes produce tumors and cardiac arrhythmias in mice. *Science* 239:1029–1033.

Fisher KJ, et al. (1994): Biochemical and functional analysis of an adenovirus-based ligand complex for gene transfer. *Biochem J* 299:49–58.

Fisher KJ, et al. (1997): Recombinant adeno-associated virus for muscle directed gene therapy. *Nat Med* 3:306–312.

Florkiewicz RZ, et al. (1995): Quantitative export of FGF-2 occurs through an alternative, energy-dependent, non-ER/Golgi pathway. *J Cell Physiol* 162:388–399.

Flugelman MY, et al. (1992): Low level *in vivo* gene transfer into the arterial wall through a perforated balloon catheter. *Circulation* 85:1110–1117.

Fox JC, et al. (1996): Antisense inhibition of basic fibroblast growth factor induces apoptosis in vascular smooth muscle cells. *J Biol Chem* 271:12578–12584.

French BA, et al. (1994a): Percutaneous transluminal *in vivo* gene transfer by recombinant adenovirus in normal porcine coronary arteries, atherosclerotic arteries, and two models of coronary restenosis. *Circulation* 90:2402–2413.

French BA, et al. (1994b): Direct *in vivo* gene transfer into porcine myocardium using replication-deficient adenoviral vectors. *Circulation* 90:2414–2424.

Frolov I, et al. (1996): Alphavirus-based expression vectors: Strategies and applications. *Proc Natl Acad Sci USA* 93:11371–11377.

Garcia JV, et al. (1994): Retrovirus vector-mediated transfer of functional HIV-1 regulatory genes. *AIDS Res Hum Retroviruses* 10:47–52.

Ghosh CG, et al. (1986): Human adenovirus cloning vectors based on infectious bacterial plasmids. *Gene* 50:161–171.

Goodman JL (1989): Possible transmission of herpes simplex virus by organ transplantation. *Transplantation* 47:609–613.

Graham FL, et al. (1992): Adenovirus-based expression vectors and recombinant vaccines. *Biotechnology* 20:363–390.

Grodums EI, et al. (1981): Experimental studies of acute and recurrent herpes simplex virus infections in the murine heart and dorsal root ganglia. *J Med Virol* 7:163–169.

Grossman M, et al. (1994): Successful *ex vivo* gene therapy directed to liver in a patient with familial hypercholesterolaemia. *Nat Genet* 6:335–341.

Gruver CL, et al. (1993): Targeted developmental overexpression of calmodulin induces proliferative and hypertrophic growth of cardiomyocytes in transgenic mice. *Endocrinology* 133:376–388.

Gu JL, et al. (1995): Ribozyme-mediated inhibition of expression of leukocyte-type 12-lipoxygenase in porcine aortic vascular smooth muscle cells. *Circ Res* 77:14–20.

Gussoni E, et al. (1992): Normal dystrophin transcripts detected in Duchenne muscular dystrophy patients after myoblast transplantation. *Nature* 356:435–438.

Guzman RJ, et al. (1993a): Efficient and selective adenovirus-mediated gene transfer into vascular neointima. *Circulation* 88:2838–2848.

Guzman RJ, et al. (1993b): Efficient gene transfer into myocardium by direct injection of adenovirus vectors. *Circ Res* 73:1202–1207.

Haj AY, et al. (1986): Development of a helper-independent human adenovirus vector and its use in the transfer of the herpes simplex virus thymidine kinase gene. *J Virol* 57:267–274.

Hanna AK, et al. (1997): Antisense basic fibroblast growth factor gene transfer reduces neointimal thickening after arterial injury. *J Vasc Surg* 25:320–325.

Hanvey JC, et al. (1992): Antisense and antigene properties of peptide nucleic acids. *Science* 258:1481–1485.

Herweijer H, et al. (1995): A plasmid-based self-amplifying Sindbis virus vector. *Hum Gene Ther* 6:1161–1167.

Herz J, et al. (1993): Adenovirus-mediated transfer of low density lipoprotein receptor gene acutely accelerates cholesterol clearance in normal mice. *Proc Natl Acad Sci USA* 90:2812–2816.

Hofmann C, et al. (1995): Efficient gene transfer into human hepatocytes by baculovirus vectors. *Proc Natl Acad Sci USA* 92:10099–10103.

Ilan Y, et al. (1997a): Insertion of the adenoviral E3 region into a recombinant viral vector prevents antiviral humoral and cellular immune responses and permits long-term gene expression. *Proc Natl Acad Sci USA* 94:2587–2592.

Ilan Y, et al. (1997b): Oral tolerization to adenoviral antigens permits long-term gene expression using recombinant adenoviral vectors [in process citation]. *J Clin Invest* 99:1098–1106.

Ishibashi S, et al. (1993): Hypercholesterolemia in low density lipoprotein receptor knockout mice and its reversal by adenovirus-mediated gene delivery. *J Clin Invest* 92:883–893.

Isner JM, et al. (1995): Arterial gene therapy for therapeutic angiogenesis in patients with peripheral artery disease. *Circulation* 91:2687–2692.

Isner JM, et al. (1996): Arterial gene transfer for therapeutic angiogenesis in patients with peripheral artery disease. *Hum Gene Ther* 7:959–988.

Jackson T, et al. (1990): The c-myc proto-oncogene regulates cardiac development in transgenic mice. *Mol Cell Biol* 10:3709–3716.

Jarvis TC, et al. (1996): Inhibition of vascular smooth muscle cell proliferation by ribozymes that cleave c-myb mRNA. *RNA* 2:419–428.

Jiao S, et al. (1992): Direct gene transfer into nonhuman primate myofibers *in vivo. Hum Gene Ther* 3:21–33.

Jolly D (1994): Viral vector systems for gene therapy. *Cancer Gene Ther* 1:51–64.

Kaplitt MG, et al. (1996): Long-term gene transfer in porcine myocardium after coronary infusion of an adeno-associated virus vector. *Ann Thorac Surg* 62:1669–1676.

Kass-Eisler A, et al. (1994): The impact of developmental stage, route of administration and the immune system on adenovirus-mediated gene transfer. *Gene Ther* 1:395–402.

Kearns WG, et al. (1996): Recombinant adeno-associated virus (AAV-CFTR) vectors do not integrate in a site-specific fashion in an immortalized epithelial cell line. *Gene Ther* 3:748–755.

Kessler PD, et al. (1996): Gene delivery to skeletal muscle results in sustained expression and systemic delivery of a therapeutic protein. *Proc Natl Acad Sci USA* 93:14082–14087.

Ketner G, et al. (1994): Efficient manipulation of the human adenovirus genome as an infectious yeast artificial chromosome clone. *Proc Natl Acad Sci USA* 91:6186–6190.

Kinniburgh AJ, et al. (1994): DNA triplexes and regulation of the c-myc gene. *Gene* 149:93–100.

Kitsis RN, et al. (1991): Hormonal modulation of a gene injected into rat heart *in vivo. Proc Natl Acad Sci USA* 88:4138–4142.

Koeberl DD, et al. (1997): Persistent expression of human clotting Factor IX from mouse liver after intravenous injection of adeno-associated virus vectors. *Proc Natl Acad Sci USA* 94:1426–1431.

Kozarsky K, et al. (1993): Adenovirus-mediated correction of the genetic defect in hepatocytes from patients with familial hypercholesterolemia. *Somat Cell Mol Gen* 19:449–458.

Kozarsky KF, et al. (1994): *In vivo* correction of low density lipoprotein receptor deficiency in the Watanabe heritable hyperlipidemic rabbit with recombinant adenoviruses. *J Biol Chem* 269:13695–13702.

Kozarsky KF, et al. (1996): Effective treatment of familial hypercholesterolaemia in the mouse model using adenovirus-mediated transfer of the VLDL receptor gene. *Nat Genet* 13:54–62.

Krieg AM, et al. (1993): Modification of antisense phosphodiester oligodeoxynucleotides by a 5′ cholesteryl moiety increases cellular association and improves efficacy. *Proc Natl Acad Sci USA* 90:1048–1052.

Lazarous DF, et al. (1995): Effects of chronic systemic administration of basic fibroblast growth factor on collateral development in the canine heart. *Circulation* 91:145–153.

Leclerc G, et al. (1992): Percutaneous arterial gene transfer in a rabbit model. Efficiency in normal and balloon-dilated atherosclerotic arteries. *J Clin Invest* 90:936–944.

Lee SW, et al. (1993): *In vivo* adenoviral vector-mediated gene transfer into balloon-injured rat carotid arteries. *Circ Res* 73:797–807.

Lemarchand P, et al. (1993): *In vivo* gene transfer and expression in normal uninjured blood vessels using replication-deficient recombinant adenovirus vectors. *Circ Res* 72:1132–1138.

Lim CS, et al. (1991): Direct *in vivo* gene transfer into the coronary and peripheral vasculatures of the intact dog. *Circulation* 83:2007–2011.

Lin H, et al. (1990): Expression of recombinant genes in myocardium *in vivo* after direct injection of DNA. *Circulation* 82:2217–2221.

Lin S, et al. (1994): Integration and germ-line transmission of a pseudotyped retroviral vector in zebrafish. *Science* 265:666–669.

Linden RM, et al. (1996): Site-specific integration by adeno-associated virus. *Proc Natl Acad Sci USA* 93:11288–11294.

Lindner V, et al. (1991): Proliferation of smooth muscle cells after vascular injury is inhibited by an antibody against basic fibroblast growth factor. *Proc Natl Acad Sci USA* 88:3739–3743.

Lu CY, et al. (1996): Adenovirus-mediated increase of exogenous and inhibition of endogenous fosB gene expression in cultured pulmonary arterial smooth muscle cells. *J Mol Cell Cardiol* 28:1703–1713.

MacMillan V, et al. (1993): Mice expressing a bovine basic fibroblast growth factor transgene in the brain show increased resistance to hypoxemic-ischemic cerebral damage. *Stroke* 24:1735–1739.

Malik P, et al. (1997): Recombinant adeno-associated virus mediates a high level of gene transfer but less efficient integration in the K562 human hematopoietic cell line. *J Virol* 71:1776–1783.

McFarlane ES, et al. (1985): Restriction endonuclease digestion analysis of DNA from viruses isolated from different sites of two fatal cases of herpes simplex virus type-1 infection. *J Med Microbiol* 20:27–32.

Mesri EA, et al. (1995): Expression of vascular endothelial growth factor from a defective herpes simplex virus type 1 amplicon vector induces angiogenesis in mice. *Circ Res* 76:161–167.

Michael SI, et al. (1994): Strategies to achieve targeted gene delivery via the receptor-mediated endocytosis pathway. *Gene Ther* 1:223–232.

Milano CA, et al. (1994): Enhanced myocardial function in transgenic mice overexpressing the beta 2–adrenergic receptor. *Science* 264:582–586.

Milano CA, et al. (1994): Myocardial expression of a constitutively active alpha(1b)-adrenergic receptor in transgenic mice induces cardiac hypertrophy. *Proc Natl Acad Sci USA* 91:10109–10113.

Miller AD, et al. (1993): Use of retroviral vectors for gene transfer and expression. *Methods Enzymol* 217:581–599.

Miller DG, et al. (1990): Gene transfer by retrovirus vectors occurs only in cells that are actively replicating at the time of infection. *Mol Cell Biol* 10:4239–4242.

Miyamoto T, Leconte I, Swain JL, and Fox JC (1998): Autocrine FGF signaling is required for vascular smooth muscle cell survival *in vitro*. *J Cell Physiol* (in press).

Morgan JE, et al. (1992): Formation of skeletal muscle *in vivo* from the mouse C2 cell line. *J Cell Sci* 102:779–787.

Morishita R, et al. (1993a): Single intraluminal delivery of antisense cdc2 kinase and proliferating-cell nuclear antigen oligonucleotides results in chronic inhibition of neointimal hyperplasia. *Proc Natl Acad Sci USA* 90:8474–8478.

Morishita R, et al. (1993b): Novel and effective gene transfer technique for study of vascular renin angiotensin system. *J Clin Invest* 91:2580–2585.

Morishita R, et al. (1994a): Evidence for direct local effect of angiotensin in vascular hypertrophy. *In vivo* gene transfer of angiotensin converting enzyme. *J Clin Invest* 94:978–984.

Morishita R, et al. (1994b): Intimal hyperplasia after vascular injury is inhibited by antisense cdk 2 kinase oligonucleotides. *J Clin Invest* 93:1458–1464.

Muhlhauser J, et al. (1995): VEGF165 expressed by a replication-deficient recombinant adenovirus vector induces angiogenesis *in vivo. Circ Res* 77:1077–1086.

Nabel EG, et al. (1990): Site-specific gene expression *in vivo* by direct gene transfer into the arterial wall. *Science* 249:1285–1288.

Nabel EG, et al. (1992): Gene transfer *in vivo* with DNA-liposome complexes: Lack of autoimmunity and gonadal localization. *Hum Gene Ther* 3:649–656.

Nabel EG, et al. (1993a): Direct transfer of transforming growth factor beta 1 gene into arteries stimulates fibrocellular hyperplasia. *Proc Natl Acad Sci USA* 90:10759–10763.

Nabel EG, et al. (1993b): Recombinant platelet-derived growth factor B gene expression in porcine arteries induce intimal hyperplasia in vivo. *J Clin Invest* 91:1822–1829.

Nabel EG, et al. (1993c): Recombinant fibroblast growth factor-1 promotes intimal hyperplasia and angiogenesis in arteries *in vivo. Nature* 362:844–846.

Neufeld G, et al. (1994): Vascular endothelial growth factor and its receptors. *Prog Growth Factor Res* 5:89–97.

Palese P, et al. (1996): Negative-strand RNA viruses: Genetic engineering and applications. *Proc Natl Acad Sci USA* 93:11354–11358.

Phillips MI (1997): Antisense inhibition and adeno-associated viral vector delivery for reducing hypertension. *Hypertension* 29:177–187.

Phillips MI, et al. (1997): Prolonged reduction of high blood pressure with an *in vivo,* nonpathogenic, adeno-associated viral vector delivery of AT(1)-R mRNA antisense. *Hypertension* 29:374–380.

Plank C, et al. (1992): Gene transfer into hepatocytes using asialoglycoprotein receptor mediated endocytosis of DNA complexed with an artificial tetra-antennary galactose ligand. *Bioconjugate Chem* 3:533–539.

Ponder KP, et al. (1991): Mouse hepatocytes migrate to liver parenchyma and function indefinitely after intrasplenic transplantation. *Proc Natl Acad Sci USA* 88:1217–1221.

Ponnazhagan S, et al. (1997): Lack of site-specific integration of the recombinant adeno-associated virus 2 genomes in human cells. *Hum Gene Ther* 8:275–284.

Qin LH, et al. (1996): Gene transfer of transforming growth factor-beta 1 prolongs murine cardiac allograft survival by inhibiting cell-mediated immunity. *Hum Gene Ther* 7:1981–1988.

Quantin B, et al. (1992): Adenovirus as an expression vector in muscle cells *in vivo. Proc Natl Acad Sci USA* 89:2581–2584.

Ragot T, et al. (1993): Efficient adenovirus-mediated transfer of a human minidystrophin gene to skeletal muscle of mdx mice. *Nature* 361:647–650.

Rando TA, et al. (1994): Primary mouse myoblast purification, characterization, and transplantation for cell-mediated gene therapy. *J Cell Biol* 125:1275–1287.

Remy JS, et al. (1995): Targeted gene transfer into hepatoma cells with lipopolyamine-condensed DNA particles presenting galactose ligands: A stage toward artificial viruses. *Proc Natl Acad Sci USA* 92:1744–1748.

Reynolds MA, et al. (1994): Triple-strand-forming methylphosphonate oligodeoxy-nucleotides targeted to messenger-RNA efficiently block protein synthesis. *Proc Natl Acad Sci USA* 91:12433–12437.

Riessen R, et al. (1993): Arterial gene transfer using pure DNA applied directly to a hydrogel-coated angioplasty balloon. *Hum Gene Ther* 4:749–758.

Rome JJ, et al. (1994): Anatomic barriers influence the distribution of *in vivo* gene transfer into the arterial wall. Modeling with microscopic tracer particles and verification with a recombinant adenoviral vector. *Arterioscler Thromb* 14:148–161.

Samulski RJ (1993): Adeno-associated virus: Integration at a specific chromosomal locus. *Curr Opin Genet Dev* 3:74–80.

Schulick AH, et al. (1997): Established immunity precludes adenovirus-mediated gene transfer in rat carotid arteries—Potential for immunosuppression and vector engineering to overcome barriers of immunity. *J Clin Invest* 99:209–219.

Shi Y, et al. (1994): Transcatheter delivery of c-myc antisense oligomers reduces neointimal formation in a porcine model of coronary artery balloon injury. *Circulation* 90:944–951.

Simons M, et al. (1992a): Antisense c-myb oligonucleotides inhibit intimal arterial smooth muscle cell accumulation *in vivo. Nature* 359:67–70.

Simons M, et al. (1992b): Antisense nonmuscle myosin heavy chain and c-myb oligonucleotides suppress smooth muscle cell proliferation in vitro. *Circ Res* 70:835–843.

Simons M, et al. (1994): Antisense proliferating cell nuclear antigen oligonucleotides inhibit intimal hyperplasia in a rat carotid artery injury model. *J Clin Invest* 93:2351–2356.

Smith AE, (1995): Viral vectors in gene therapy. *Annu Rev Microbiol* 49:807–838.

Steg PG, et al. (1994): Arterial gene-transfer to rabbit endothelial and smooth-muscle cells using percutaneous delivery of an adenoviral vector. *Circulation* 90:1648–1656.

Stein CA, et al. (1993): Antisense oligonucleotides as therapeutic agents—is the bullet really magical?" *Science* 261:1004–1012.

Stratford-Perricaudet LD, et al. (1992): Widespread long-term gene transfer to mouse skeletal muscles and heart. *J Clin Invest* 90:626–630.

Strauss M (1994): Liver-directed gene therapy: Prospects and problems. *Gene Ther* 1:156–164.

Strayer DS, et al. (1996): SV40 mediates stable gene transfer *in vivo. Gene Ther* 3:581–587.

Sullenger BA, et al. (1994): Ribozyme-mediated repair of defective mRNA by targeted, trans-splicing. *Nature* 371:619–622.

Sullivan SM (1994): Development of ribozymes for gene therapy. *J Invest Dermatol* 103 (5 suppl):85S-89S.

Thomas KA (1996): Vascular endothelial growth factor, a potent and selective angiogenic agent. *J Biol Chem* 271:603–606.

Tripathy SK, et al. (1994): Stable delivery of physiologic levels of recombinant erythropoietin to the systemic circulation by intramuscular injection of replication-defective adenovirus. *Proc Natl Acad Sci USA* 91:11557–11561.

Tsurumi Y, et al. (1996): Direct intramuscular gene transfer of naked DNA encoding vascular endothelial growth factor augments collateral development and tissue perfusion [see comments]. *Circulation* 94:3281–3290.

Unger EF, et al. (1994): Basic fibroblast growth factor enhances myocardial collateral flow in a canine model. *Am J Physiol* 266:H1588–H1595.

Villa AE, et al. (1995): Effects of antisense c-myb oligonucleotides on vascular smooth muscle cell proliferation and response to vessel wall injury. *Circ Res* 76:505–513.

Von der Leyen HE, et al. (1995): Gene-therapy inhibiting neointimal vascular lesion—*in vivo* transfer of endothelial-cell nitric-oxide synthase gene. *Proc Natl Acad Sci USA* 92:1137–1141.

von Harsdorf R, et al. (1993): Gene injection into canine myocardium as a useful model for studying gene expression in the heart of large mammals. *Circ Res* 72:688–695.

Wagner E, et al. (1992a): Influenza virus hemagglutinin HA-2 N-terminal fusogenic peptides augment gene transfer by transferrin–polylysine–DNA complexes: Toward a synthetic virus-like gene-transfer vehicle. *Proc Natl Acad Sci USA* 89:7934–7938.

Wagner E, et al. (1992): Coupling of adenovirus to transferrin–polylysine/DNA complexes greatly enhances receptor-mediated gene delivery and expression of transfected genes. *Proc Natl Acad Sci USA* 89:6099–6103.

Walgenbach KJ, et al. (1995): Ischaemia-induced expression of bFGF in normal skeletal muscle: A potential paracrine mechanism for mediating angiogenesis in ischaemic skeletal muscle. *Nat Med* 1:453–459.

Wang G, et al. (1995): Targeted mutagenesis in mammalian cells mediated by intracellular triple helix formation. *Mol Cell Biol* 15:1759–1768.

Wang W, et al. (1996): Sequence-independent inhibition of *in vitro* vascular smooth muscle cell proliferation, migration, and *in vivo* neointimal formation by phosphorothioate oligodeoxynucleotides. *J Clin Invest* 98:443–450.

Wilson JM, et al. (1990): Temporary amelioration of hyperlipidemia in low density lipoprotein receptor-deficient rabbits transplanted with genetically modified hepatocytes. *Proc Natl Acad Sci USA* 87:8437–8441.

Wilson JM, et al. (1993): Therapeutic strategies for familial hypercholesterolemia based on somatic gene transfer. *Am J Cardiol* 72:59D-63D.

Wolff JA, et al. (1992): Long-term persistence of plasmid DNA and foreign gene expression in mouse muscle. *Hum Mol Genet* 1:363–369.

Xiao X, et al. (1996): Efficient long-term gene transfer into muscle tissue of immunocompetent mice by adeno-associated virus vector. *J Virol* 70:8098–8108.

Yang Y, et al. (1994): Cellular immunity to viral antigens limits E1–deleted adenoviruses for gene therapy. *Proc Natl Acad Sci USA* 91:4407–4411.

Yang ZY, et al. (1996): Role of the p21 cyclin-dependent kinase inhibitor in limiting intimal cell proliferation in response to arterial injury. *Proc Natl Acad Sci USA* 93:7905–7910.

Yao S-N, et al. (1994): Primary myoblast-mediated gene transfer: Persistent expression of human factor IX in mice. *Gene Ther* 1:99–107.

Yee JK, et al. (1994): A general-method for the generation of high-titer, pantropic retroviral vectors—highly efficient infection of primary hepatocytes. *Proc Natl Acad Sci USA* 91:9564–9568.

Zhang Y, et al. (1996): Intravenous somatic gene transfer with antisense tissue factor restores blood flow by reducing tumor necrosis factor-induced tissue factor expression and fibrin deposition in mouse meth-A sarcoma. *J Clin Invest* 97:2213–2224.

20

APPLICATIONS OF GENE THERAPY TO NEUROLOGICAL DISEASES AND INJURIES

DEREK L. CHOI-LUNDBERG AND MARTHA C. BOHN

Department of Neurobiology and Anatomy, University of Rochester School of Medicine and Dentistry, Rochester, NY 14642 (D.L.C-L), Children's Memorial Institute for Education and Research, Department of Pediatrics, Northwestern University Medical School, Chicago, IL 60614 (M.C.B.)

INTRODUCTION

Gene therapy was first envisaged as a method to correct inherited, single-gene defects. The concept of gene therapy has subsequently been expanded to include molecular genetic techniques to deliver therapeutic molecules for the treatment of any disease, genetic or nongenetic. Potential applications of gene therapy for the nervous system include correction of inherited errors of metabolism that affect the nervous system, delivery of toxic gene products to brain tumors, delivery of neuroprotective molecules to neuronal populations undergoing degeneration, and replacement of neurotransmitter-synthesizing enzymes. A number of inherited diseases resulting from single-gene defects cause pathology in the nervous system and, in many cases, severe mental and behavioral deficits. Examples include disorders of metabolism of amino acids (phenylketonuria); purines (Lesch-Nyhan syndrome); glycolipids, phospholipids, or glycosaminoglycans (Tay-Sachs, Fabry's, Gaucher's, and Niemann-Pick diseases and Sly's syndrome); metals (Wilson's disease); and peroxisomal

Stem Cell Biology and Gene Therapy, Edited by Peter J. Quesenberry, Gary S. Stein, Bernard G. Forget, and Sherman M. Weissman
ISBN 0-471-14656-0

disorders (Zellweger's syndrome). While some of these disorders are manageable by dietary restriction (reduction of phenylalanine intake for phenylketonuria) or by removal of the toxic compound (penicillamine for elimination of copper in Wilson's disease), effective therapies do not exist for most, and curing or managing the disease will likely require replacement of the defective enzyme in peripheral tissues and/or in the nervous system. For a discussion of gene therapy approaches for these disorders, the reader is referred to other reviews (Jinnah and Friedmann, 1995; Jinnah et al., 1993; Suhr and Gage, 1993; and elsewhere in this volume). Brain tumors are often inaccessible to surgical resection and refractory to conventional chemotherapeutics or radiation. Gene therapy vectors that are only capable of infecting and expressing their transgene in mitotically active cells can be used to target expression of a toxic gene to dividing tumor cells while leaving predominantly postmitotic neurons and glia intact. Gene therapy for brain tumors is discussed elsewhere in this volume and in other reviews (Jinnah and Friedmann, 1995; Kramm et al., 1995).

The causative genes of several inherited neurodegenerative disorders have been identified, including Huntington's disease, familial amyotrophic lateral sclerosis (ALS), familial Alzheimer's disease, and spinal muscular atrophy. As the pathogenesis of these disorders is elucidated, their causative genes may become targets for gene therapy. The potential applications of gene therapy in the nervous system are more far-reaching, however, and include the delivery of neuroprotective molecules to specific populations of neurons affected in sporadic neurodegenerative diseases such as the majority of Parkinson's, Alzheimer's, and ALS cases, as well as in anatomically localized stroke and central nervous system (CNS) trauma. Gene therapy approaches are ideally suited to target delivery of therapeutic molecules to subpopulations of cells in the cellularly complex nervous system.

Advances in neuroscience during the last decade have led to the elucidation of many factors and mechanisms involved in neuronal development, plasticity, regeneration, and cell death. These advances are likely to result in novel therapeutic approaches to neurodegenerative diseases. Even in the mature brain, neurons retain incredible plasticity that if provided with a fortuitous microenvironment, are capable of regeneration, sprouting of new fibers, and restoration of function. For example, sprouting of fibers from dopaminergic (DA) neurons damaged by the drug 1-methyl-4-phenyl-1,2,3,6 tetrahydropyridine (MPTP) has been observed in both rodent and nonhuman primate models of Parkinson's disease (Bankiewicz et al., 1993; Bohn et al., 1987; Kordower et al., 1991). Damaged axons in the adult CNS can also grow long distances and establish synapses when provided with a bridging substrate favorable to growth (Benfry and Aguayo, 1982; David and Aguayo, 1981; Xu et al., 1994; Zwimpfer et al., 1992). Furthermore, neurotrophic factors that promote the survival and differentiation of specific types of neurons and molecules that protect neurons from cell death are continually being identified. Genes, cell surface, and extracellular matrix molecules that guide, attract, or promote growing axons are also under intensive study. Consequently, future therapies for neurodegenerative diseases are likely to involve the use of neuroprotective, neurotrophic, and neuronal guidance molecules to reverse or slow the progression of diseases, such as Alzheimer's, Parkinson's, ALS, and retinal degeneration, as well as to ameliorate the outcomes of neuronal injuries following stroke, spinal cord injury, and head trauma. The cellular complexity of the nervous system presents a challenge to gene therapy research in generating approaches to restore or maintain the function of specific neuronal systems, as well as to target transgene expression to specific cell types in a regulatable fashion.

POTENTIAL OF NEUROTROPHIC FACTOR THERAPIES FOR NEURODEGENERATION

Neurotrophic factors promote the survival, neurite outgrowth, and phenotypic differentiation of neurons during development. Furthermore, neurotrophic factors protect neurons from a variety of insults, including axotomy, free radicals, inhibitors of mitochondrial electron transport, glutamatergic excitotoxins, hypoglycemia, and hypoxia/ischemia. The mechanisms by which protection occurs are being elucidated. Several neurotrophic factors increase the activity of superoxide dismutase, glutathione reductase, catalase, and/or glutathione peroxidase, thereby suppressing the accumulation of reactive oxygen species (Cheng and Mattson, 1995; Mattson et al., 1995; Spina et al., 1992). Neurotrophic factors also attenuate the rise in intracellular calcium induced by glutamatergic excitotoxins or hypoglycemia (Cheng et al., 1994a,b; Cheng and Mattson, 1992, 1994). The multiple effects of neurotrophic factors on neurons suggest many strategies in the use of neurotrophic factors for neurodegeneration (see Table 1); however, few of these potentials have been investigated using gene therapy paradigms.

TABLE 1 Potential Applications of Neurotrophic Factor Gene Therapy

Neuroprotection
Protect neurons in aged, diseased, or injured CNS or PNS
Protect neurons in grafts of fetal tissue
Reduce levels of oxidative stress in neurons
Increase number of support cells
Regeneration or sprouting
Stimulate regeneration of damaged, diseased neurons
Stimulate sprouting in healthy, neighboring neurons
Block inhibitory growth molecules
Stimulation of neurotransmitter expression, synthesis, and release
Regulation of neuronal function directly or indirectly

The identification of neurotrophic factors has been achieved initially through *in vitro* studies using specific types of embryonic neurons. In Table 2 are examples of phenotypically well-characterized neurons that are affected by neurodegenerative diseases and those factors that have been reported to promote survival and/or differentiation of these neurons *in vitro* or *in vivo*. The effects of some neurotrophic factors are mediated by glia (Engele and Bohn, 1991). For some factors, *in vivo* studies confirm the *in vitro* results, while for other factors effects reported on neurons in the adult brain have been found to be different from those expected. For example, brain-derived neurotrophic factor (BDNF) promotes survival and neurite outgrowth from embryonic DA neurons *in vitro* (Hyman et al., 1991), but affects the function of DA neurons in the adult rodent brain in the absence of any apparent effect on DA neuronal fiber growth or sprouting (Altar et al., 1992; Yoshimoto et al., 1995).

Delivery of Neurotrophic Factors to the Nervous System

While neurotrophic factors are promising therapeutic agents, there are obvious problems for delivering these factors to specific cell types in the cellularly complex brain at levels that are efficacious. It is important to develop methods for providing a factor or a combination of factors to specific cell types whereby continuous or regulatable delivery can be achieved in the absence of CNS tissue damage, effects in nontargeted cells, and effects in the periphery. The main routes of administration are summarized in Table 3. Peripheral injection is unlikely to be effective for CNS therapies due to the lability of neurotrophic factors and poor passage through the blood–brain barrier. Although transient modification of the blood–brain barrier or molecular modification of the neurotrophic factors may increase their availability to brain (Friden et al., 1993; Kordower et al., 1993, 1994), peripheral injection is not a promising approach, except where peripheral neurons or motoneuron terminals are the desired target. In this regard, subcutaneous injection of Myotropin (insulin-like growth factor 1 [IGF-1]) is the approach being used in ongoing clinical trials for ALS undertaken by Cephalon, Inc. (Walsh, 1995). Injection or infusion of recombinant neurotrophic proteins directly into brain parenchyma or ventricles has been used in many animal studies and is planned in a clinical trial for Parkinson's disease using glial cell line–derived neurotrophic factor (GDNF) (Walsh, 1995). This therapy may be limited, however, by the need to infuse large amounts of neurotrophic factors to elicit effects in animal models of Parkinson's disease and the limited practicality of applying this to long-term progressive human disease. Moreover, infusion of neurotrophic factor proteins is not ideal for providing continuous neurotrophic or neuroprotective support selectively to specific neurons. For example, clinical trials were conducted in which ciliary neurotrophic factor (CNTF) was administered subcutaneously to patients with ALS, a disease characterized by the degenera-

TABLE 2 Neurotrophic Factors for Specific Neuronal Populations and Neurodegenerative Diseases (Selected References)

- Dopaminergic neurons (Parkinson's disease)
  - TGF-β superfamily
    - GDNF (Lin et al., 1993)
    - GDF-5 (Krieglstein et al., 1995b)
    - TGF-β1 (Krieglstein et al., 1995a)
    - TGF-β2, TGF-β3 (Krieglstein et al., 1995a; Poulsen et al., 1994)
    - Activin A (Krieglstein et al., 1995a)
  - Neurotrophins
    - BDNF (Hyman et al., 1991)
    - NT-3 (Hyman et al., 1994)
    - NT-4/5 (Hyman et al., 1994)
  - Cytokines
    - Cardiotrophin-1 (Pennica et al., 1995)
    - CNTF (Hagg and Varon, 1993; Magal et al., 1993)
    - IL-1β (Akaneya et al., 1995)
    - IL-6 (Hama et al., 1991; von Coelln et al., 1995)
    - IL-7 (von Coelln et al., 1995)
  - Mitogenic growth factors
    - aFGF, bFGF (Beck et al., 1993; Engele and Bohn, 1991; Ferrari et al., 1989; Otto and Unsicker, 1993)
    - EGF (Casper et al., 1991)
    - Insulin (Knusel et al., 1990)
    - IGF-1 (Beck et al., 1993)
    - IGF-2 (Liu and Lauder, 1992)
  - Midkine (Kikuchi et al., 1993)
  - PDGF (Othberg et al., 1995)
  - TGF-α (Alexi and Hefti, 1992)
- Spiral ganglion neurons (hearing loss, e.g., aminoglycoside toxicity)
  - NT-3 (Avila et al., 1993; Ernfors et al., 1996)
  - BDNF (Avila et al., 1993; Geschwind et al., 1996)
  - NT-4/5 (Zheng et al., 1995)
  - bFGF (Lefebvre et al., 1991)
- Photoreceptor cells (vision loss)
  - bFGF (LaVail et al., 1992; Mansour-Robaey et al., 1994; Masuda et al., 1995)
  - CNTF (LaVail et al., 1992)
  - BDNF (LaVail et al., 1992)
  - IL-1β (LaVail et al., 1992)
  - aFGF (LaVail et al., 1992)
  - Midkine (Masuda et al., 1995)
- Retinal ganglion cells (vision loss)
  - BDNF (Peinadoramon et al., 1996; Unoki and LaVail, 1994)
  - NT-4/5 (Peinadoramon et al., 1996)
  - CNTF (Unoki and LaVail, 1994)
  - bFGF (Unoki and LaVail, 1994)

TABLE 2 *(Continued).*

Noradrenergic neurons of locus coeruleus (Alzheimer's, Parkinson's)
- GDNF (Arenas et al., 1995)
- NT-3 (Arenas and Persson, 1994; Friedman et al., 1993)
- NT-4/5 (Friedman et al., 1993)
- BDNF (Sklairtavron and Nestler, 1995)

Serotoninergic neurons of raphe nuclei (Alzheimer's disease)
- GDNF (Beck et al., 1996)
- BDNF (Altar et al., 1994b; Mamounas et al., 1995a; Martin-Iverson et al., 1994)
- NT-3 (Martin-Iverson et al., 1994)
- NT-4/5 (Altar et al., 1994b)
- FGF-5 (Lindholm et al., 1994)

Cholinergic neurons of basal forebrain (Alzheimer's disease)
- NGF (Dreyfus, 1989)
- BDNF (Nonomura and Hatanaka, 1992)
- NT-4/5 (Friedman et al., 1993)
- GDNF (Williams et al., 1996)
- bFGF (Enokido et al., 1992)
- FGF-5 (Lindholm et al., 1994)
- G-CSF (Konishi et al., 1993)
- M-CSF (Konishi et al., 1993)
- GM-CSF (Konishi et al., 1993)
- Erythropoietin (Konishi et al., 1993)

Hippocampal neurons (Alzheimer's disease)
- CNTF (Ip et al., 1991)
- IGF-I and -II (Cheng and Mattson, 1992)
- aFGF (Sasaki et al., 1992)
- bFGF (Lowenstein and Arsenault, 1996; Maiese et al., 1993)
- EGF (Maiese et al., 1993)
- NT-3 (Collazo et al., 1992; Ip et al., 1993)
- BDNF (Cheng and Mattson, 1994; Ip et al., 1993; Lowenstein and Arsenault, 1996)
- NT-4/5 (Cheng et al., 1994b; Ip et al., 1993)
- IL-2 (Awatsuji et al., 1993)
- IL-6 (Yamada and Hatanaka, 1994)
- TGF-β1 (Ishihara et al., 1994; Prehn et al., 1994)
- TGF-β2 (Ishihara et al., 1994)
- TNF-α (Cheng et al., 1994a)
- TNF-β (Cheng et al., 1994a)
- PDGF (Cheng and Mattson, 1995)

Cortical neurons (Alzheimer's disease)
- bFGF (Cummings et al., 1992; Peterson et al., 1996)
- IL-2 (Awatsuji et al., 1993)
- TNF-α (Cheng et al., 1994a)
- TNF-β (Cheng et al., 1994a)
- NT-4/5 (Cheng et al., 1994b)

TABLE 2 *(Continued).*

Motoneurons (amyotrophic lateral sclerosis, spinal muscular atrophy, motoneuron diseases)
GDNF (Henderson et al., 1994)
TGF-β1 (Martinou et al., 1992)
BDNF (Henderson et al., 1993; Hughes et al., 1993b; Mitsumoto et al., 1994; Sendtner et al., 1992; Wong et al., 1993; Yan and Miller, 1993)
NT-4/5 (Henderson et al., 1993; Hughes et al., 1993b; Wong et al., 1993)
NT-3 (Henderson et al., 1993; Hughes et al., 1993b; Wong et al., 1993; Yan and Miller, 1993)
CDF/LIF (Hughes et al., 1993b; Martinou et al., 1992)
CNTF (Mitsumoto et al., 1994; Sendtner et al., 1990, 1994)
IGF-1 (Hughes et al., 1993b)
aFGF (Cuevas et al., 1995)
FGF-5 (Hughes et al., 1993a,b)
Cardiotrophin-1 (Henderson, 1995)
Corticospinal neurons (amyotrophic lateral sclerosis)
CNTF (Dale et al., 1995)

TGF, transforming growth factor; GDNF, glial cell line–derived neurotrophic factor; GDF-5, growth differentiation factor-5; BDNF, brain-derived neurotrophic factor; NT, neurotrophin; CNTF, ciliary neurotrophic factor; IL, interleukin; aFGF and bFGF, acidic and basic fibroblast growth factor; EGF, epidermal growth factor; IGF, insulin-like growth factor; PDGF, platelet-derived growth factor; NGF, nerve growth factor; G, M, granulocyte, monocyte colony-stimulating factor (CSF); TNF, tumor necrosis factor; CDF/LIF, cholinergic differentiation factor/leukemia inhibitory factor.

TABLE 3 Delivery of Neurotrophic Factors to the Nervous System

Peripheral injection
Infusion or injection into CNS
Gene therapy
Ex vivo gene therapy
Retroviral vectors and other gene transfer methods
Astrocytes, myoblasts, fibroblasts, cell lines, or progenitor cells as carriers
Encapsulated genetically modified cells
In vivo gene therapy
Herpes simplex recombinant or amplicon viral vectors
Adenoviral vectors
Adenoassociated viral vectors
Lentiviral vectors
DNA–liposome complexes
Novel viral or synthetic vectors
Encapsulated viral vectors

tion of lower and upper motoneurons. Unfortunately, large numbers of patients receiving CNTF experienced side effects, including coughing, fever, pain, and significant weight loss resulting in cessation of these trials (Barinaga, 1994). The implantation of cells encapsulated in polymers, a specialized *ex vivo* gene therapy approach developed by Aebischer and colleagues (1988), is an improvement over infusion or injection of factors since these polymers release factors in a continuous fashion and act as selective molecular size dialysis membranes, thereby minimizing host cellular responses to the implanted cells. In an ongoing clinical trial in Switzerland, baby hamster kidney (BHK) cells genetically engineered to express a secreted form of CNTF and encapsulated in polymers are being implanted intrathecally in ALS patients. Preliminary data from this study suggest that the CNTF-secreting capsules are more likely to succeed than clinical trials using peripheral injections of CNTF (Aebischer et al., 1996; Walsh, 1995).

Genetic approaches, however, are ideal for delivering neurotrophic factors to the nervous system. Viral vectors, genetically modified cells, DNA–liposome complexes, synthetic vectors, or even naked DNA can be injected into specific areas of the nervous system with the potential of producing neurotrophic factors in a continuous, stable fashion (for reviews, see Ridet and Privat, 1995; Tuszynski and Gage, 1995). Furthermore, the inclusion of cell-specific promoters and regulatable elements into vector design will produce biologically synthesized and processed neurotrophic factors in a specific cell type in a manner that can be therapeutically regulated by peripheral administration of drugs or hormones. Moreover, the use of cellular promoters may offer more stable transgene expression than that achieved with viral promoters (Kaplitt et al., 1994a).

GENE THERAPY DELIVERY SYSTEMS: ADVANTAGES AND DISADVANTAGES FOR THE CNS

As in peripheral tissues, there are two main approaches for gene delivery to the CNS: *in vivo* modification of host tissue or *ex vivo* modification of cells with subsequent transplantation (Table 3). *In vivo* approaches include direct injection of naked DNA, DNA–liposome complexes, or one of a variety of viral vectors, including adenovirus (Ad), adeno-associated virus (AAV), herpes simplex virus (HSV), retrovirus (RV), human immunodeficiency virus (HIV), or combinations of approaches (Abdallah et al., 1995; Akli et al., 1993; Andersen and Breakefield, 1995; Geller, 1993; Glorioso et al., 1995; Le Gal La Salle et al., 1993; Mamounas et al., 1995b; McCown et al., 1996; Naldini et al., 1996; Short et al., 1990). Each vector has inherent advantages and disadvantages for use in the CNS, reviewed by several authors (Andersen and Breakefield, 1995; Glorioso et al., 1995; Karpati et al., 1996; Neve, 1993; Smith et al., 1995; see also pertinent chapters in this book). *Ex vivo* approaches involve modification of primary cells or cell lines by conventional transfection methods (calcium phosphate precipitation or cationic lipid complexes, for example) or by infection with any of the viral vectors listed above with subsequent transplantation of genetically modified cells into the host brain (Gage et al., 1987; Ridet and Privat, 1995; Smith et al., 1995).

A wide variety of cell types have been investigated for *ex vivo* gene delivery to the CNS. Tissue may be removed from the host, genetically modified, and subsequently reimplanted (autografts). This approach minimizes immune rejection. The choice of tissue is limited to cells that divide in culture, allowing generation of adequate numbers of genetically modified cells. Cells that have been used for autografting into CNS include fibroblasts, myoblasts, and Schwann cells (isolated from skin, muscle, and nerve biopsies, respectively). Tissues or cell lines derived from other humans (allografts) or species (xenografts) are also potential sources of cells for genetic modification and transplantation. For retina, allografts of genetically modified pigmented epithelial cells have been made into the subretinal space following retroviral transduction with the lacZ gene and shown to have a protective effect on photoreceptor cells in a rat model of retinal degeneration (Dunaief et al., 1995). Dividing neuronal and glial progenitor cells, as well as microglia, are easily isolated from brain, including human brain (Hao et al., 1991; Reynolds and Weiss, 1992; Whittemore and Synder, 1996). Human neuronal progenitor cells have been grafted into brain following adenoviral infection and shown to persist at least 3 weeks (Sabate et al., 1995). In addition, immortalized cell lines, of both neural and non-neural origin, have been generated from tumors or through the introduction of oncogenes. Potential problems with these sources of cells include ethical issues surrounding the use of human abortion material and animal tissues. There is also increased risk of immune rejection, which can be modified by immunosuppression or by encapsulation of cells in semipermeable membranes. The use of cell lines, derived from spontaneous tumors or immortalized with oncogenes, carries the risk of tumor formation. The use of conditionally immortalized cells through regulatable oncogene expression (Hoshimaru et al., 1996; Snyder, 1994) or encapsulation of cells can significantly reduce this risk (Aebischer et al., 1988). However, the use of neuronal progenitor cells obtained without the use of oncogene immortalization are preferable for cell replacement and gene delivery (Reynolds and Weiss, 1992; Whittemore and Synder, 1996). Such cells could be genetically modified using cell type specific promoters that would limit transgene expression to a phenotypically defined cell type that differentiates from the "multipotential" progenitors following their transplantation into the brain. Moreover, as those genes underlying the development of neuronal specificity in mammals are identified, they could be inserted into progenitor cells to drive the differentiation of a specific cell type for neuronal replacement.

The distinguishing features of vectors fall into several areas: the maximal size of DNA that can be inserted, the levels of cytotoxicity elicited following injection into the CNS, cellular tropism, capacity to direct stable expression, and provocation of immune responses. Direct injection of RV is unlikely to be successful, as RV integrates only in dividing cells, and most neurons and glia in the adult brain are postmitotic. However, RV has specialized applications such as in tumor therapies in the CNS (Chen et al., 1994; Gordon and Anderson, 1994; Oldfield et al., 1993; Shinoura et al., 1996; see also pertinent chapters in this book) and may be appropriate for modifying responses to injury by directing gene expression specifically to dividing microglia or astrocytes.

HSV is naturally neurotropic and efficiently infects both neurons and glia. Long-term expression following injection of HSV vectors into CNS has been reported (Andersen et al., 1992; During et al., 1994; Wolfe et al., 1992). Methods for generating helper-free stocks of HSV have been developed; however, at present these do not give high titer (Fraefel et al., 1996). Further development along this line or through genetic modification of recombinant HSV vectors is likely to result in useful vectors (Glorioso, et al., 1995). Since HSV vectors can potentially harbor up to 150 kb of transgene, development of these vectors is crucial for gene therapy in the nervous system, especially with regard to the potential of including large cellular promoters and other regulatory elements into these vectors. HSV vectors are also transported retrogradely in neurons, suggesting another strategy for specific neuronal targeting (Davar et al., 1994; Jin et al., 1996; Keir et al., 1995).

AAV appears to be the safest of the current vector systems, as wild-type AAV is nonpathogenic and requires Ad or HSV to replicate. Moreover, AAV vectors can be generated free of Ad since AAV is heat stable. Transgene expression in neurons and glia for at least 4 months has been reported following AAV injection into striatum (Kaplitt et al., 1994b). However, a recent study suggests that stability of expression from AAV vectors is dependent on brain region (McCown et al., 1996). The main disadvantage of AAV is its small genome, which limits the size of the transgene to about 4.5 kb. Recent advances in generating higher titer AAV have been made. In addition, transduction efficiency of AAV is increased by DNA synthesis and topoisomerase inhibitors (Flotte, 1995; Russell et al., 1994, 1995), although this is not particularly relevant in the CNS, where most cells are postmitotic.

Ad vectors can be grown and concentrated to high titers (up to 10^{12} pfu/ml), enabling large numbers of cells to be infected. Ad vectors can accommodate up to 8–10 kb of transgene and remain episomal, avoiding potential activation or inactivation of cellular genes that can occur with integration. Despite deletions in E1A, E1B, and E3 regions, Ad vectors are still capable of directing the synthesis of viral proteins with the potential of eliciting cytotoxicity. In peripheral tissues a major drawback of current Ad vectors is a strong immune response resulting in destruction of infected cells and loss of transgene expression (Yang et al., 1994). Inflammation also occurs in response to Ad delivery to the brain, although responses are less severe than in peripheral tissues (Byrnes et al., 1995). While this has been presumed to be a response to viral proteins, a recent study also suggests a role for transgene protein (Tripathy et al., 1996). However, several studies have shown that Ad vectors can direct transgene expression in many neuronal and glial cell types in the brain for at least 2 months, although expression declines over time (Akli et al., 1993; Byrnes et al., 1995; Choi-Lundberg et al., 1997; Davidson et al., 1993; Le Gal La Salle et al., 1993). Ad vectors injected into the lateral ventricle directed expression in ependymal cells of all four ventricles and the leptomeningies for up to 1 week, while injection into the cisterna magna led to expression in the leptomeningies and adventitial cells of blood vessels (Bajocchi et al., 1993; Ooboshi et al., 1995). Ad vectors, like HSV vectors, are retrogradely transported by neurons (Ghadge et al., 1995)

Direct *in vivo* modification of host brain tissue may also be accomplished by injection of naked DNA or DNA–liposome complexes. Nonviral delivery of DNA is promising in some peripheral tissues such as muscle (Davis et al., 1993). Although high levels of transgene expression have not been reported in brain using such approaches, neurons are transduced (Abdallah et al., 1995; Boussif et al., 1995; Cao et al., 1995; Mamounas et al., 1995b), and biological effects have been observed following injection of a plasmid DNA–lipofectamine complex carrying the tyrosine hydroxylase (TH) gene into the striatum of a rat model of Parkinson's disease (Cao et al., 1995). The combination of AAV viral vector DNA with liposomes or polycations is also promising and is being used in a clinical trial for Canavan's disease (Levine, 1996).

PARKINSON'S DISEASE

Parkinson's disease, or paralysis agitans, is a progressive neurodegenerative disorder, affecting 1% of people over age 50, with approximately 500,000 patients in the United States. The majority of Parkinson's disease cases are sporadic. Inherited forms usually involve degeneration in other systems, such as in the parkinsonism-dementia-ALS complex of Guam. The main symptoms of Parkinson's disease include resting tremor, rigidity, postural instability, and bradykinesia (slowness of movement). The neuropathological hallmark of Parkinson's disease is loss of DA neurons, with resultant depigmentation, in the substantia nigra pars compacta (SNpc). In addition to loss of DA neurons, noradrenergic neurons of the locus coeruleus and serotonergic neurons of the raphe nuclei are also reduced in number. The DA neuronal loss is most pronounced in the ventrolateral SNpc, which projects mainly to the putamen, the portion of the striatum concerned primarily with motor functions. While a loss of DA neurons occurs in all humans, from about 400,000 at birth to 200,000 by age 80, greater loss occurs in persons with Parkinson's disease, with symptoms becoming evident after approximately 80% of DA neurons have been lost (Adams and Victor, 1993; Mark and Duvoisin, 1995).

Early in the course of Parkinson's disease, medical therapy is based on increasing stimulation of DA receptors in the striatum with DA agonists such as bromocriptine and pergolide; inhibiting breakdown of endogenous DA with selegiline (Deprenyl, a monoamine oxidase-b inhibitor); and the use of anticholinergics. As symptoms worsen, L-dihydroxyphenylalanine (L-dopa) with carbidopa (a peripheral dopa-decarboxylase inhibitor) is added. L-dopa is converted to DA by remaining DA nerve terminals and other cells that express dopa-decarboxylase (also known as aromatic amino acid decarboxylase, AADC) in the striatum. Side effects of the medications include tardive dyskinesia (involuntary movements), psychosis, and autonomic dysfunction. The "on–off" phenomenon develops in many patients, with rapid and unpredictable switching from a state of relative mobility to paralysis. While these medical therapies are effective in most patients for several years, continuing degeneration of DA neurons leads to disability in about 66% and 80% of patients within 5 and 10 years of onset, respectively (Adams and Victor, 1993).

A variety of experimental therapies for Parkinson's disease are in development (for review, see Mizuno et al., 1994). Surgical ablation of the ventroposterior pallidum, which creates a compensatory lesion in basal ganglia circuitry, has resulted in improvement in patients for up to 18 months (Goetz and Diederich, 1996; Obeso et al., 1996). However, this approach creates an additional lesion in the brain to compensate for an existing one and does not slow the continuing degeneration of DA neurons. Other approaches attempt to enhance striatal DA levels by grafting fetal ventral mesencephalon containing the DA neuron cell groups. Eventually, it may be possible to reconstruct the nigrostriatal pathway by grafting ventral mesencephalon tissue into the substantia nigra (SN) and stimulating fiber growth and innervation of the striatum.

Animal Models Of Parkinson's Disease

Several well-characterized animal models of Parkinson's disease are available for testing gene therapy approaches. The weaver mutant in mouse is an autosomal recessive trait that arose spontaneously and exhibits two pathological features: failure of proper cerebellar granule cell migration and partial degeneration of DA neurons. The animals have ataxia and fine tremors. DA innervation of the striatum occurs in a normal developmental pattern, but in the adult, striatal DA levels are reduced 75% and DA neuron numbers in the SN are reduced 70%. As in Parkinson's disease, the mesolimbic DA system is relatively spared (Bankiewicz et al., 1993; Simon and Ghetti, 1994).

The neurotoxin 6-hydroxydopamine (6-OHDA), when injected into the SN, striatum, or the medial forebrain bundle (MFB), which contains DA axons from the ventral tegmental area and SN, causes a specific loss of DA neurons and fibers. 6-OHDA is a dopamine analog that is specifically taken up by high-affinity DA transporters expressed on DA neurons. 6-OHDA undergoes auto-oxidation, generating hydroxyl radical ($^{\cdot}OH$), hydrogen peroxide (H_2O_2), and superoxide anion (O_2^-), which cause damage to various cellular compartments, including membranes, resulting in death of DA neurons (Bankiewicz et al., 1993; Cohen et al., 1974). When 6-OHDA is injected into the SN or MFB, DA neuron death occurs rapidly, within 48 h. However, when 6-OHDA is injected into the striatum, progressive degeneration of the nigrostriatal system occurs over several weeks, with atrophy and decrease of TH preceding cell death (Sauer and Oertel, 1994). Unilateral 6-OHDA lesions lead to an imbalance in DA that can easily be quantified using the Ungerstedt drug-induced rotation test (Ungerstedt, 1971), as well as by assessing deficits in contralateral limb use in several spontaneous behaviors (Olsson et al., 1995; Schallert, 1995). The effects of experimental interventions on these behaviors, as well as on DA phenotypic characteristics, are commonly analyzed.

MPTP (1-methyl-4-phenyl-1,2,3,6-tetrahydropyridine) is toxic to DA neurons of several species, including humans, following systemic administration. MPTP is oxidized by the monoamine oxidase system, generating the toxic molecule MPP^+. In susceptible strains of mice, MPTP causes most TH-positive neurons to disappear, with 80%–95% reduction in striatal DA and DA uptake sites. The mice develop akinesia, but all of the above changes spontaneously recover over time. Thus, MPTP does not kill the DA neurons, but rather severely downregulates their phenotype or damages their terminals. Nonhuman primates are also susceptible to damage by MPTP and following systemic administration develop many of the symptoms of Parkinson's disease, including bradykinesia, resting tremor, and rigidity. An asymmetrical bilateral lesion can be made by combining unilateral internal carotid artery and systemic MPTP (Bankiewicz et al., 1993; Miletich et al., 1994). These well-characterized animal models of Parkinson's disease with their sophisticated tests of behavioral recovery make Parkinson's one of the diseases of choice for developing strategies to prevent, reverse, or compensate for neurodegeneration.

Neurotrophic Factor Protein Infusion in Animal Models of Parkinson's Disease

Many experiments have demonstrated the efficacy of several neurotrophic factors for DA neurons in animal models of Parkinson's disease. These studies have utilized infusion of the proteins into the striatum, near the SN, or into the lateral ventricles. Large quantities of proteins are infused in these studies (typically 10 μg or more), and repeated injections are typically required to maintain effects. For example, epidermal growth factor (EGF), acidic fibroblast growth factor (aFGF), and basic fibroblast growth factor (bFGF) have been reported to promote moderate recovery of DA neurons in 6-OHDA lesioned rat and MPTP lesioned mice (Otto and Unsicker, 1994). bFGF is expressed in DA neurons in the ventral midbrain (Cintra et al., 1991), and ^{125}I-bFGF is retrogradely transported by DA neurons in the SN, suggesting that DA neurons have receptors for bFGF (Ferguson and Johnson, 1991). BDNF and, to a lesser extent, neurotrophin-3 (NT-3), delivered above the SN prevent induction of turning behavior following a unilateral injection of 6-OHDA into the striatum (Altar et al., 1994b). BDNF also has been reported to increase DA turnover in the striatum, the electrical activity of DA neurons, and the function of grafted fetal mesencephalon (Altar et al., 1994a,b; Martin-

Iverson et al., 1994; Sauer et al., 1993; Shen et al., 1994; Shults et al., 1994; Yurek et al., 1996). DA neurons express both BDNF and NT-3 mRNAs, suggesting an autocrine role for these factors in DA neurons (Seroogy et al., 1994). BDNF does not penetrate brain parenchyma following intracerebroventricular administration, likely due to the presence of truncated trkB receptors on ependymal cells that line the ventricles (Klein et al., 1990; Yan et al., 1994b). Thus, BDNF must be delivered within brain parenchyma to reach DA neurons or terminals. In this regard, BDNF and NT-3 are retrogradely transported by DA neurons (DiStefano et al., 1992; Mufson et al., 1994). CNTF has also been reported to rescue axotomized DA neurons from cell death, but not to prevent the loss of TH (Hagg and Varon, 1993).

GDNF, a member of the transforming growth factor-β (TGF-β) family, is the most potent neurotrophic factor for DA neurons described to date, with an EC_{50} for promoting DA uptake in embryonic ventral mesencephalon cultures in the range of 40 pg/ml (1 pM) compared with 1–100 ng/ml for other classes of neurotrophic factors (Lin et al., 1993, 1994), although other members of the TGF-β family have been reported to have comparable potency to GDNF (Poulsen et al., 1994). The highest levels of GDNF in the CNS are observed in the developing striatum (Choi-Lundberg and Bohn, 1995; Strömberg et al., 1993), and GDNF is transported retrogradely from the striatum by DA neurons (Tomac et al., 1995b). While most members of the TGF-β family signal through receptor serine-threonine kinases, the receptor for GDNF is a complex composed of the tyrosine kinase c-ret receptor and GDNF Family Receptor Alpha-1 (GFRα-1), a glycosyl phosphatidylinositol membrane-linked receptor (Durbec et al., 1996; Treanor et al., 1996; Trupp et al., 1996). GDNF has been shown to protect and restore DA neurons in several animal models of Parkinson's disease. In the rat, 60 μg GDNF, delivered near the SN at 4 μg/day beginning the day of MFB transection and continuing for 2 weeks, increased survival of DA neurons to 85% compared with 53% in control treated rats (Beck et al., 1995). In addition, 10 μg GDNF injected into the SN protected approximately 30% of DA neurons from 6-OHDA injected 24 h later into the SN or striatum (Kearns and Gash, 1995). Furthermore, 100 μg (but not 0.1–10 μg) GDNF injected into the SN 4 weeks following unilateral 6-OHDA injection into the MFB partially restored the DA phenotype. Although striatal DA and TH-immunoreactive (IR) levels in the SN remained depleted, GDNF normalized SN DA and induced TH-IR fiber sprouting. TH-IR neuron numbers were restored from 0.6% to 10% of control values by GDNF (Bowenkamp et al., 1995; Hoffer et al., 1994). GDNF also modestly protected and restored DA phenotypic characteristics, including striatal DA levels and TH-IR fiber density, from damage caused by MPTP (Hou et al., 1996; Tomac et al., 1995a). GDNF, but not TGF-β3, injected near the SN at a dose of 10 μg/day every other day for 4 weeks beginning the day of intrastriatal 6-OHDA injection completely protected DA neuron cell number, TH phenotype, and cell size (Sauer et al., 1995). The protective and restorative effects of GDNF in rodent models of Parkinson's disease have recently been extended to nonhuman primates. GDNF delivered into the SN, caudate, or ventricles of MPTP lesioned monkeys resulted in improvements in bradykinesia, rigidity, posture, and balance lasting for 4 weeks. GDNF increased DA neuronal cell size, TH-IR processes, and DA levels in the SN, but did not affect striatal DA or DA metabolite levels (Gash et al., 1996). These results demonstrate the efficacy of neurotrophic factors, especially GDNF, to protect or restore DA neurons and their phenotypes in several animal models of Parkinson's disease. With efficacy of infusion of large quantities of recombinant proteins thus established, it is logical to evaluate the effects of neurotrophic factors delivered by gene therapy approaches.

Neurotrophic Factor Gene Therapy in Animal Models of Parkinson's Disease

To date, published studies of neurotrophic factor gene therapy have been primarily limited to the use of *ex vivo* approaches (Table 4). In one study, BHK cells were transfected with an expression vector containing GDNF and encapsulated in a polymer. Conditioned medium from these cells increased survival and neurite outgrowth of embryonic ventral mesencephalon DA neurons, demonstrating release of bioactive GDNF. When implanted into the striatum of rats previously lesioned with 6-OHDA, the encapsulated cells survived for 90 days and induced ingrowth of TH-IR fibers into the capsules. However, behavioral effects were not observed (Lindner et al., 1995). GDNF-expressing fibroblasts but not control fibroblasts grafted near the SN protected c-ret positive (DA) neurons from 6-OHDA injected into the MFB (Trupp et al., 1996). Several studies have utilized fibroblasts genetically modified with retrovirus to produce BDNF. These fibroblasts secreted bioactive BDNF as demonstrated in bioassays of DA neurons in embryonic ventral mesencephalic cultures. RatI fibroblasts engineered to express BDNF and transplanted near the SN protected 83% ± 22% of DA neurons from a striatal injection of MPP^+ compared with 35% ± 15% for control fibroblasts 7 days postlesion (Frim et al., 1994). Furthermore, DA levels in the SN were enhanced, although DA turnover was unaltered (Galpern et al., 1996). Primary rat fibroblasts modified to express BDNF and grafted into the striatum expressed BDNF mRNA for at least 2 weeks. Increased HVA/DA ratio, indicative of increased DA turnover, and sprouting of TH-IR fibers were noted in normal rats grafted near the SN. However, BDNF fibroblasts grafted into the striatum or SN did not protect DA neurons from 6-OHDA injected into the MFB (Lucidi-Phillipi et al., 1995). Similarly transduced primary fibroblasts transplanted into the striatum prior to striatal 6-OHDA partially protected DA nerve terminals in the striatum and completely protected DA cell bodies in the SN 3 weeks after the lesion (Levivier et al., 1995). Neonatal astrocytes retrovirally transduced to express BDNF and transplanted to the striatum following a partial 6-OHDA lesion of the SN led to reduced apomorphine-induced rotational behavior

TABLE 4 ***Ex Vivo*** **Gene Therapy With Neurotrophic Factors**

Paradigm	Cell type, Factor	Biological Effects	Reference
Dopaminergic neurons (substantia nigra)			
MPP^+	Rat I, BDNF	Protection of soma	Frim et al. (1994)
MPP^+	Fibroblast, BDNF	↑ SN DA levels	Galpern et al. (1996)
6-OHDA, parital	BHK, GDNF	Sprouting of fibers	Lindner et al. (1995)
	Astrocyte, BDNF	Improvement in rotation	Yoshimoto et al. (1995)
6-OHDA, striatal	Fibroblast, BDNF	Protection of soma, fibers	Levivier et al. (1995)
6-OHDA, MFB	Fibroblast, GDNF	Protection of soma	Trupp et al. (1996)
	Fibroblast, BDNF	No effect	Lucidi-Phillipi et al. (1995)
Unlesioned	Fibroblast, BDNF	↑ DA turnover, sprouting	Lucidi-Phillipi et al. (1995)
Noradrenergic neurons (locus coeruleus)			
6-OHDA	Fibroblast, GDNF	Protection, sprouting, ↑ TH and somal size	Arenas et al. (1995)
6-OHDA	Fibroblasts, NT-3	Protection	Arenas and Persson (1994)
Chromaffin cell grafts		Protective/stimulatory	Cunningham et al. (1991, 1994)
Motoneurons			
pmn/pmn	BHK, CNTF	↑ Life span, protection of soma, axons	Sagot et al. (1995b)
pmn/pmn	BHK, GDNF	Protection of soma	Sagot et al. (1996)
ALS patients	BHK, CNTF	No deleterious effects	Aebischer et al. (1996)
Basal forebrain cholinergic neurons			
Fimbria-fornix	208F, NGF	Phenotypic rescue and sprouting	Rosenberg et al. (1988)
Fimbria-fornix	208F, NGF	Phenotypic rescue	Hoffman et al. (1993)
Fimbria-fornix	BHK, NGF	Phenotypic rescue and sprouting	Emerich et al. (1994b)
Fimbria-fornix	BHK, NGF	Phenotypic rescue	Winn et al. (1994)
Fimbria-fornix	Progenitor, NGF	Phenotypic rescue and sprouting	Martinez-Serrano et al. (1995b)

Fimbria-fornix	Fibroblast, NGF	Phenotypic rescue and sprouting	Tuszynski et al. (1996b)
Fimbria-fornix	Fibroblast, NGF	Phenotypic rescue and sprouting	Kawaja et al. (1992)
Fimbria-fornix	Fibroblast, NGF	Phenotypic rescue and sprouting	Lucidi-Phillipi et al. (1995)
Fimbria-fornix	Fibroblast, BDNF	No effect	Lucidi-Phillipi et al. (1995)
Aged rats	Progenitor, NGF	Improved spatial learning, reversal of soma atrophy	Martinez-Serrano et al. (1995a)
Ibotenic acid	Fibroblast, NGF	Improved spatial learning, reversal of soma atrophy	Dekker et al. (1994)
Striatal cholinergic neurons			
QA	RatI, NGF	↓ Lesion size, protect cholinergic neurons	Schumacher et al. (1991)
QA	RatI, NGF	↓ Lesion, size, protect cholinergic neurons	Frim et al. (1993a)
3-NPA	RatI, NGF	↓ Lesion size	Frim et al. (1993b)
QA	BHK, NGF	↓ Lesion size, protect cholinergic neurons	Emerich et al. (1994a)
Unlesioned	BHK, NGF	Hypertrophy of cholinergic, neuropeptide Y neurons	Kordower et al. (1996)
Hippocampal neurons			
Stroke	RatI, NGF	Protection of CA1 neurons	Pechan et al. (1995)
Spinal cord			
Normal	Fibroblast, NGF	Sprouting of sensory neurons	Tuszynski et al. (1994)
Hemisection	Fibroblast, NGF	Sprouting of sensory neurons Sprouting of noradrenergic neurons	Tuszynski et al. (1996a)
Normal	Fibroblast, NT-3	Sprouting of sensory neurons	Nakahara et al. (1996)
	Fibroblast, bFGF	Sprouting of sensory neurons	Nakahara et al. (1996)
Normal	Fibroblast, PDGF	↑ Glial precursors	Ijichi et al. (1996)
Glutamatergic/calbindin-negative neurons of entorhinal cortex			
Lesion of perforant path	Fibroblasts, FGF-2	Protection	Peterson et al. (1996)

without significantly affecting the density of TH-IR fibers in the striatum (Yoshimoto et al., 1995).

Three studies utilizing *in vivo* gene therapy approaches have demonstrated protective or restorative effects of GDNF on DA neurons in a rat model of PD developed by Sauer and Oertel (1994) (see Table 5). In one study, a subpopulation of DA neurons were prelabeled by bilateral intrastriatal injection of the retrograde tracer fluorogold (FG). Rats received injections of Ad vectors encoding GDNF, a mutant form of GDNF (mGDNF) lacking bioactivity, lacZ, or no injection dorsal to the SN 1 week prior to unilateral striatal 6-OHDA. Six weeks following 6-OHDA lesion, 69% ± 3% of the FG-labeled DA neurons had degenerated on the lesioned side in the Ad-mGDNF, Ad-lacZ, and no injection groups compared with only 21% ± 5% in rats receiving Ad-GDNF (Fig. 1). The protective effect of *in vivo* GDNF gene therapy observed in this study in which GDNF protein was biologically synthesized and secreted in the range of nanograms per day was comparable with that reported after injection of microgram quantities of recombinant GDNF

TABLE 5 ***In Vivo* Gene Therapy With Neuroprotective Factors**

Paradigm	Vector, Factor	Biological Effects	Reference
Sympathetic neurons			
Axotomy	HSV, NGF	Protection of TH phenotype	Federoff et al. (1992)
Dopaminergic neurons			
6-OHDA	Ad, GDNF	Protection of soma	Choi-Lundberg et al. (1997)
6-OHDA	Ad, GDNF	Protection of soma, behavior	Choi-Lundberg et al. (1998)
6-OHDA	Ad, GDNF	Protection of soma, fiber density, behavior	Bilang-Bleuel et al. (1997)
Cholinergic neurons			
Aged rats	Ad, NGF	↑ Cell size	Castel-Barthe, et al. (1996)
Spiral ganglion (auditory) neuron			
Explant	HSV, BDNF	↑ Neurite outgrowth	Geschwind et al. (1996)
Telencephalic neurons			
Focal ischemia	HSV, bcl-2	Area of viable tissue around injection site	Linnik et al. (1995)
Focal ischemia	HSV, bcl-2	Protection of transgene expressing cells	Lawrence et al. (1996)
Focal ischemia	Ad, IL-1ra	Reduced infarct volume	Betz et al. (1995)

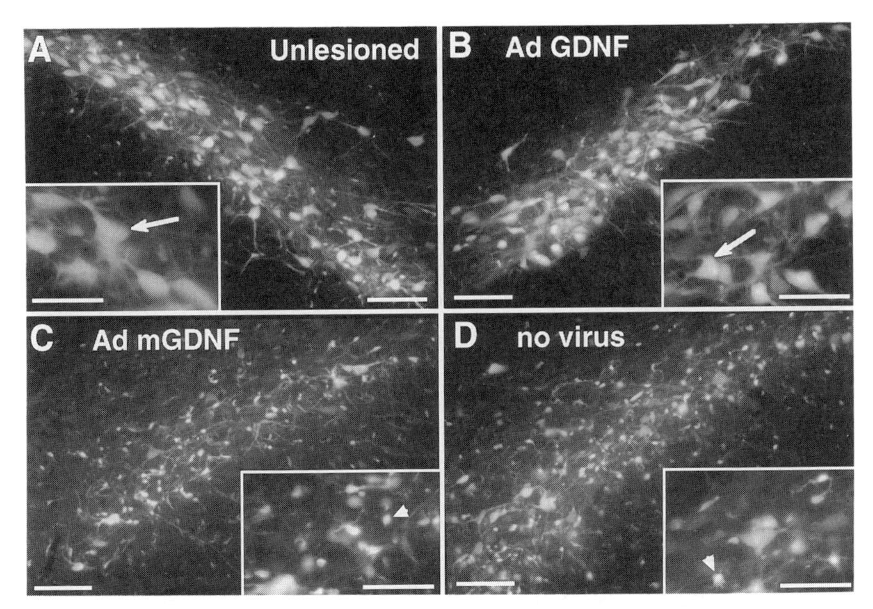

FIGURE 1. *In vivo* gene therapy with Ad GDNF prevents progressive degeneration of fluorogold (FG)-prelabeled DA neurons. Six weeks following intrastriatal 6-OHDA, many large, FG-positive cells (DA neurons, arrows) are observed in the SN on the unlesioned side (**A**) and on the lesioned side of a rat treated with Ad GDNF (**B**). In contrast, fewer large FG-positive cells, but many small, secondarily labeled FG-positive cells (microglia and other non-neuronal cells, arrowheads) are noted in rats treated with Ad mGDNF (**C**) or untreated (**D**). Bars in A–D are 100 μm and 50 μm for the insets. Reprinted with permission from Choi-Lundberg et al., 1997. Dopaminergic neurons protected from degeneration by GDNF gene therapy. *Science 275,* 838–841. Copyright 1997 American Association for the Advancement of Science.

protein (Choi-Lundberg et al., 1997; Sauer et al., 1995). Two additional studies demonstrated the protective effects of Ad-GDNF injected into the rat striatum, the site of DA nerve terminals and a more surgically accessible site than the SN in humans. Ad-GDNF decreased the loss of DA neuron soma, accessed by counts of FG-positive or TH-IR cells in the SN. In one of the studies, TH-IR fiber density in the striatum was partially protected (Bilang-Bleuel et al., 1997). Furthermore, Ad-GDNF prevented the development of limb use asymmetry during spontaneous movements along vertical surfaces and prevented amphetamine-induced rotation behavior, which occurred in control-treated, lesioned rats (Bilang-Bleuel et al., 1997, Choi-Lundberg et al., 1998).

Other Gene Therapy Approaches for Parkinson's Disease

Enhancing striatal DA levels by introducing DA synthesizing enzymes by modifying host tissue or the transplantation of L-dopa or DA producing cells is an alternative gene therapy strategy. It is hypothesized that the local production of dopamine in a continuous fashion within the striatum may avoid the problems associated with oral L-dopa administration, especially the on–off phenomenon. DA is synthesized by a two-step process. First, tyrosine is converted to L-dopa by the rate-limiting enzyme TH, which requires the cofactor tetrahydrobiopterin (BH4). Next, L-dopa is decarboxylated to DA by the enzyme dopa-decarboxylase (or AADC). Cells that normally do not make L-dopa can be made to do so by the introduction of genes encoding TH. Providing exogenous BH4 or co-expressing GTP-cyclohydrolase 1 (GTP-CH1, an enzyme required for BH4 synthesis) may enhance L-dopa production, and co-expressing AADC may enhance DA production.

Ex vivo modification of fibroblast cell lines with a TH gene resulted in production and release of L-dopa *in vitro* and *in vivo,* which in some instances depended on exogenous biopterin or BH4 (Uchida et al., 1992; Wolff et al., 1989). TH-expressing fibroblasts did not produce DA *in vitro,* but *in vivo* they released L-dopa that was subsequently converted to DA by host striatal AADC (Horellou et al., 1990). Co-culture of primary fibroblasts modified to express TH with fibroblasts modified to express AADC produced DA *in vitro* (Kang et al., 1993), and it would be expected that a fibroblast modified to express both TH and AADC would release DA as well. Transplantation of TH-fibroblast cell lines into the striatum of rats with unilateral 6-OHDA lesions resulted in a 34%–50% reduction in apomorphine-induced turning behaviors (Horellou et al., 1990; Uchida et al., 1992; Wolff et al., 1989). A study that demonstrated similar results with primary myoblasts transfected with a TH gene was retracted (Jiao et al., 1993; Jiao et al., 1996). In another approach, a temperature-sensitive immortalized neural cell line derived from embryonic day 14 mesencephalon was modified to produce additional TH by infection with a retrovirus or transfection with a plasmid DNA encoding a TH gene and shown to produce increased amounts of L-dopa *in vitro.* Following grafting into 6-OHDA lesioned rats and two immunosuppressed MPTP lesioned monkeys, the TH-modified but not the parental cell line improved apomorphine-induced rotational behavior 43% in rats and 54% in monkeys (Anton et al., 1994). Astrocytes genetically engineered to express TH secreted L-dopa *in vitro* and 2 weeks after transplantation into the unilaterally 6-OHDA lesioned rat improved apomorphine-induced rotations by 50% (Lundberg et al., 1996).

Long-term efficacy of direct *in vivo* modification of striatum by AAV and HSV vectors encoding the TH gene, but not lacZ, have been reported in rat models of Parkinson's disease (During et al., 1994; Kaplitt et al., 1994b). Injection of AAV-lacZ resulted in expression in the striatum for at least 3 months and AAV-TH for at least 4 months. AAV-TH improved apomorphine-induced rotational behavior by about 35% in unilateral 6-OHDA lesioned rats for 9 weeks. HSV-TH improved rotation rate 65% compared with HSV-lacZ and saline treatment from 2 weeks through 1 year following HSV injection. *In vivo* microdialysis in these HSV-TH injected rats at 4 or 6 months after gene transfer showed 300% increase in DA levels compared with HSV-lacZ and saline rats (although unlesioned controls had 1,150% higher levels of DA compared with lesioned controls). Thus, *in vivo* gene transfer of TH to the striatum led to increased DA production in 6-OHDA lesioned rats, which was manifested by partial behavioral recovery. Similarly, Ad-TH has also been observed to decrease sensorimotor asymmetry in rats following intracerebral injection (Horellou et al., 1994).

Another *in vivo* gene therapy approach applicable to Parkinson's disease is to modify DA receptor expression. An initial study toward this goal used Ad to deliver the gene encoding the D2 receptor to the striatum (Ikari et al., 1995). Increased D2 receptor expression in the striatum was confirmed by autoradiographic analysis of ^{3}H-spiperone binding.

MOTONEURON DEGENERATIVE DISORDERS

ALS

ALS (also known as Lou Gehrig's disease or motoneuron disease) is characterized by progressive degeneration of upper motoneurons (corticospinal neurons) and lower motoneurons in the spinal cord and brain stem, resulting in progressive weakness, with paralysis and death within 3–4 years of diagnosis on average. The annual incidence of ALS is 1 per 100,000. The only currently available therapy for ALS approved by the FDA is riluzole, a glutamate release inhibitor, which was shown to increase survival 17% and 14% at 12 and 18 months, respectively, compared with placebo (discussed by Festoff, 1996). Approximately 5%–10% of cases are transmitted as autosomal dominant mutations, and 25% of these contain mutations in Cu/Zn superoxide dismutase (SOD1), which catalyzes the conversion of two superoxide anions to hydrogen peroxide and molecular oxygen (Deng et al., 1993; Rosen et al., 1993). Transgenic mice overexpressing mutant forms of SOD1, but not wild-type SOD1, demonstrate motoneuron degeneration (Gurney et al., 1994), suggesting a toxic gain of function rather than partial reduction in SOD1 activity in the pathogenesis of ALS. Mutant forms of SOD1 promote apoptosis when expressed in a neural cell line (Rabizadeh et al., 1995). The mutant forms of SOD1 have increased peroxidase activity, which may lead to enhanced oxidation of cellular components with resultant death (Wiedau-Pazos et al., 1996). Neurotrophic factors are able to modulate the oxidation-reduction potential of cells (Cheng and Mattson, 1995; Mattson et al., 1995; Spina et al., 1992) and may thus protect neurons from this mode of damage.

In addition to mutant SOD1 transgenic animals, several other animal models of motoneuron degeneration exist, demonstrating spontaneous degeneration of motoneurons and axons, including progressive motor neuronopathy (pmn/pmn), mnd, and wobbler mice. Other models for motoneuron degeneration include peripheral axotomy, which in the neonate results in death of a large percentage of axotomized motoneurons and in the adult results in a decrease of choline acetyltransferase (ChAT) expression. Several neurotrophic factors for motoneurons have been tested in several of these models, as well as during developmentally programmed cell death. Increased motoneuron survival has been observed with BDNF (Chiu et al., 1994; Koliatsos et al., 1993; Sendtner et al., 1992; Yan et al., 1994a), NT-3 (Yan et al., 1993), NT-4/5 (Hughes et al., 1993b; Koliatsos et al., 1994), IGF-1 (Festoff et al., 1995; Hughes et al., 1993b), GDNF (Henderson et al., 1994; Li, L.X. et al., 1995; Oppenheim et al., 1995; Sagot et al., 1996; Yan et al., 1995), leukemia inhibitory factor (LIF) (Hughes et al., 1993b), and CNTF (Chiu et al., 1994; Forger et al., 1993; Sendtner et al., 1990), as well as with combined administration of CNTF or GDNF with BDNF (Mitsumoto et al., 1994; Vejsada et al., 1995). The protective effects of IGF-1 and CNTF on motoneurons have led to clinical trials in patients with ALS. A phase III clinical trial of Myotrophin (IGF-1) by Cephalon involving 266 patients has demonstrated slower progression of disease and improved function with 50 or 100 μg IGF-1/kg/day compared with placebo (Festoff, 1996; Walsh, 1995). In contrast, two large (570 and 730 patients) phase III double-blind, placebo-controlled clinical trials with subcutaneous injections of 0.5, 2, or 5 μg CNTF/kg/day, or 15 or 30 μg CNTF/kg three times per week, failed to show beneficial effects. Side effects included anorexia, weight loss, and cough, and in many patients they were severe enough to limit dosing (ALS CNTF Treatment Study Group, 1996; Miller et al., 1996).

The severe side effects observed in the CNTF trials underscore the need to target delivery of neurotrophic factors to the cells of interest, while minimizing exposure to other systems. In addition, continuous delivery of trophic factor may be required. CNTF has a half life of 2–6 h in the human bloodstream (ALS CNTF Treatment Study [ACTS] Phase I–II Study Group, 1995) and daily or three times per week subcutaneous injections may not have been able to deliver adequate levels of CNTF to motoneurons. Gene therapy approaches have the potential to provide continuous release of neurotrophic factor targeted to cells of interest. In pmn/pmn mice, encapsulated cells engineered to express CNTF increased life span 40% and reduced motoneuron loss (Sagot et al., 1995b). Encapsulated BHK cells modified to express GDNF reduced motoneuron loss 50% in the facial nucleus; however, there were no beneficial effects on axon degeneration and life span of pmn/pmn mice (Sagot et al., 1996). In six ALS patients, intrathecal implantation of encapsulated BHK cells engineered to produce about 0.5 μg CNTF/day resulted in 0.2–6 ng/ml CNTF in the CSF over 17 weeks and did not cause side effects observed in the systemic delivery trials (Aebischer et al., 1996).

ALZHEIMER'S DISEASE

Alzheimer's disease (AD) is the most prevalent neurodegenerative disorder, affecting 5%–10% of persons over age 65 and >20% over age 75. Persons with AD suffer progressively worsening dementia, with deficits in memory, spatial orientation, problem-solving, judgment, and language. Pathological features of AD brains include atrophy due to widespread neuronal loss, especially in the hippocampus, association cortex, entorhinal cortex, basal forebrain, and locus coeruleus. Cholinergic neurons of the basal forebrain are affected early in the disease. The neuropathological hallmarks of AD are neurofibrillary tangles composed of paired helical filaments of hyperphosphorylated tau protein and neuritic plaques containing depositions of amyloid. At present, the only FDA-approved drug for AD is tacrine, a cholinesterase inhibitor, which modestly improves cognitive function in a subpopulation of AD patients. Observational epidemiological studies have suggested a role for anti-inflammatory drugs in delaying the age of onset of AD.

Approximately 5%–10% of AD cases are inherited in an autosomal dominant fashion. Mutations in three genes have been identified. The first gene identified encodes amyloid precursor protein (APP) (Yankner, 1996). APP is a precursor of Aβ, the main constituent of neuritic plaques. The identification of mutations in APP has led to the development of animal models that exhibit some of the neuropathological features of AD. For example, mice overexpressing a mutation of APP (V717F) develop amyloid plaques and diffuse synaptic and dendritic loss, although neurofibrillary tangles are not observed (Games et al., 1995). These and other related models may be useful for testing the efficacy of therapies and unraveling the pathogenesis of AD. The other two AD genes identified are presenilin 1 (S182) and presenilin 2 (STM2), which are related seven transmembrane spanning domain proteins (Yankner, 1996). A partial clone of the mouse homolog of presenilin 2, designated ALG-3 (apoptosis-linked gene), was recently identified by a "death trap" assay, based on its ability to rescue a T-cell hybridoma from T-cell receptor crosslinking induced apoptosis (Vito et al., 1996). This finding suggests that the presenilins may be involved in signaling pathways in programmed cell death. Alternatively, they may be important in protein trafficking and might affect APP processing based on homology with the *Caenorhabditis elegans* protein SPE-4, which functions in the Golgi apparatus to segregate membrane proteins (Yankner, 1996). As the roles these proteins play in the normal brain and in the pathogenesis of AD are elucidated, they may become targets for gene therapy approaches.

Cholinergic neurons of the basal forebrain (medial septum, vertical limb of the diagonal band of Broca, and nucleus basalis of Meynert) that project to the hippocampus and cortex undergo extensive degeneration in AD, and the degree of cholinergic cell loss correlates with cognitive decline. In addition, learning deficits in aged animals are correlated with decreases in acetylcholine and can be reversed by treatment with cholinergic agonists or cholinesterase inhibitors. Loss of cholinergic neurons can be induced by transection of the fimbria-fornix, which contains projection axons from cholinergic cell groups in the basal forebrain to the hippocampus. Numerous studies have demonstrated the efficacy of nerve growth factor (NGF) in rescuing cholinergic neurons from degeneration, atrophy, and downregulation of neurotransmitter phenotype following axotomy in rodents and primates (for example, see Hefti, 1986; Tuszynski et al., 1990; Williams et al., 1986), and improvements in memory tasks have been noted as well in aged rodents (Fischer et al., 1987). An AD patient has been infused intraventricularly with 6.6 mg NGF over 3 months (75 μg/day). While some potentially beneficial effects were observed, including increased nicotine-binding sites in the cortex, increased cerebral blood flow and improvement on a verbal memory test (although other tests were unchanged), side effects including weight loss, pain, and sleep disturbances were noted (Olson et al., 1992, 1994).

Ex vivo gene therapy approaches, using a variety of cell types genetically modified by classic transfection or retroviral infection methods to express NGF are also effective in protecting rodent and primate cholinergic neurons from damage induced by fimbria-fornix transection or ibotenic acid lesion. They also enhance function and increase cell size in aged rats. Cells transplanted in these studies include fibroblast tumor cell lines 208F (Hoffman et al., 1993; Rosenberg et al., 1988) and BHK (Emerich et al., 1994b; Winn et al., 1994), primary fibroblasts (Dekker et al., 1994; Kawaja et al., 1992; Tuszynski et al., 1996b), and conditionally immortalized neural progenitors (Martinez-Serrano et al., 1995a,b). In the majority of the studies, the NGF gene, under the Moloney murine leukemia virus long terminal repeat (MMLV-LTR) promoter, was delivered via retroviral infection, although $CaPO_4$ transfection of NGF under the metallothionein promoter was also used (Emerich et al., 1994b; Winn et al., 1994). Cells were grafted into the basal forebrain, the fimbria-fornix lesion cavity, or lateral ventricle. Prior to transplantation, the cells secreted 1–135 ng NGF/10^5 cells/day as determined by ELISA. In two studies in which encapsulated cells were used, capsules were retrieved at the time of sacrifice, and NGF secretion fell from 27 to 5 ng NGF/capsule/day 3 weeks post-transplantation in the rat (Winn et al., 1994) and from 21 to 9 ng NGF/capsule/day 5 weeks post-transplantation in the monkey (Emerich et al., 1994b). Three or 6 months following transplantation into the unlesioned striatum, capsules released 3–22 ng NGF/capsule/day, suggesting long-term expression is possible with the metallothionein promoter (Winn et al., 1994). Animals were sacrificed at 2–4 weeks, or in some cases at 6–10 weeks, following lesioning and grafting. In fimbria-fornix models, the biological effect studied was the percentage of septal cholinergic neurons on the lesioned side compared with the unlesioned side, with septal cholinergic neurons identified by p75 (low-affinity NGF receptor) or (ChAT) immunoreactivity. NGF-secreting cells maintained 68%–93% of cholinergic neurons, compared with 14%–49% in control cell grafts. In aged and ibotenic acid lesioned rats, NGF-secreting cells improved deficits in spatial learning (water maze task) and reversed cholinergic neuron atrophy (Dekker et al., 1994; Martinez-Serrano et al., 1995a).

Modest effects on atrophied basal forebrain cholinergic neurons also have been noted with an *in vivo* gene therapy approach in aged rats. An Ad vector encoding NGF under control of the Rous sarcoma virus LTR promoter injected into the nucleus basalis magnocellularis increased the cell soma area of cholinergic neurons 3 weeks after injection. Aged rats receiving 2×10^6 or 5×10^6 pfu of Ad-NGF had acetylcholinesterase-positive neurons that were 6.7%–11.1% or 6%–23% larger than neurons on the contralateral side, respectively, compared with −7.4%–1.6% for animals injected with 2 x 10^6 pfu of Ad-lacZ (Castel-Barthe et al., 1996).

These studies demonstrate the efficacy of NGF delivered via gene therapy approaches to protect cholinergic neurons of the basal forebrain from axotomy-induced degeneration, as well as to improve spatial learning in aged rats. Although cholinergic neurons are severely affected in AD, many other neuronal populations are affected as well, and it is unclear if protecting cholinergic neurons alone will be adequate to prevent or reverse the progressive dementia of AD. Additional trophic factors aimed at other neuronal populations, including cortical, hippocampal, and noradrenergic locus coeruleus neurons, might also be required.

CAG TRINUCLEOTIDE REPEAT EXPANSION DISORDERS

Several progressive neurodegenerative diseases, including Huntington's disease (HD), spinal and bulbar muscular atrophy (SBMA, or Kennedy's disease), spinocerebellar ataxia type 1 (SCA1), Machado-Joseph disease (MJD), dentatorubral-pallidoluysian atrophy (DRPLA), and Haw River syndrome (HRS) are caused by CAG repeat expansions that encode polyglutamine within proteins (Paulson and Fischbeck, 1996; Sharp and Ross, 1996). Except for DRPLA and HRS, each disease is caused by a mutation in a distinct gene. The proteins encoded by the mutated genes are as follows: HD, huntingtin; SBMA, androgen receptor; SCA1, ataxin-1; MJD, MJD1 gene product; and DRPLA and HRS, atrophin. The normal numbers of CAG repeats in these genes range from 6 to 40, while 36 to 121 repeats are associated with disease. Longer CAG expansions are associated with earlier onset and more rapid progression.

Although each disease shares the common genetic mechanism of CAG expansion, the pattern of neuronal degeneration, symptoms, and signs differs. In SBMA, degeneration of lower motoneurons (LMNs) in the spinal cord and brain stem leads to LMN signs, including muscle atrophy, weakness, fasciculations, and hyporeflexia. Patients also have androgen insensitivity. SCA1 is characterized by neuronal loss in the cerebellum, especially Purkinje neurons, as well as in various nuclei in the brain stem, resulting in ataxia, dysarthria, paralysis of eye muscles, and weakness. In DRPLA and HRS, degeneration occurs in the globus pallidus, dentatorubral system, and subthalamic nucleus, leading to ataxia, choreathetosis, and myoclonus, as well as dementia and epilepsy. In MJD, spinocerebellar systems degenerate, resulting in ataxia. HD is the most common of the polyglutamine expansion disorders. Degeneration in the striatum, especially GABAergic medium spiny neurons, typically occurs first, and later involves widespread areas of cortex and cerebellum. Clinical signs include involuntary movements (chorea and dystonia), as well as cognitive, emotional, and behavioral disturbances.

With the exception of SBMA, which is X-linked, inheritance is autosomal dominant, suggesting a toxic gain of function. As the causative genes have no significant homology to each other, and all result in progressive neurodegeneration, albeit in different neuronal populations, it has been proposed that the expanded polyglutamine repeats may interact with a related set of proteins, possibly interfering with their function. Two proteins that preferentially interact with expanded polyglutamine repeats have been described. In a yeast two-hybrid screening system, the huntingtin protein interacts with a protein designated HAP-1 (huntingtin-associated protein), and HAP-1 binds to polyglutamine-expanded huntingtin with greater affinity than normal huntingtin (Li et al., 1995). Glyceraldehyde-3-phosphate dehydrogenase (GAPDH), an enzyme in the glycolysis pathway, binds a 60-glutamine peptide (representing a mutant, expanded CAG repeat), but not a 20-glutamine peptide (representing normal). Furthermore, both atrophin and huntingtin bind GAPDH (Burke et al., 1996). In addition, expanded glutamine repeats reduce GAPDH activity (Roses, 1996). This finding is particularly interesting, as persons with an HD mutant allele, even those not yet showing clinical signs, have decreased glucose metabolism in the striatum, suggesting an impairment in energy generation. Furthermore, striatal degeneration similar to that seen in HD patients can be created in animals by injection of mitochondrial toxins, such as malonate and 3-nitropropionic acid (3-NPA) (Sharp and Ross, 1996).

Transgenic animals overexpressing or heterozygous for the mutant genes with expanded CAG repeats should be good animal models for these diseases and should assist in elucidating pathogenesis and testing therapies. A possible gene therapy approach would be to deliver a gene encoding an antisense sequence to the expanded polyglutamine stretch to prevent its translation while not interfering with translation of the normal allele. This may prove technically challenging as the normal allele has from 6 to 40 CAG repeats and the disease allele 36 to 121, and such an antisense RNA might bind both transcripts. Additionally, such a gene would have to be expressed in a large percentage of at-risk neurons. Another possibility would be to target those proteins interacting with polyglutamine.

Other animal models of HD induce striatal degeneration by the injection of excitotoxins, such as the glutamate agonist quinolinic acid (QA), or mitochondrial toxins, including 3-NPA. NGF protects cholinergic striatal neurons from degeneration induced by QA. Infusion of NGF (1 μg/day) for 1 week into the striatum beginning the day of QA lesion decreased the loss of ChAT-IR neurons from 40% to 6%. ChAT mRNA per neuron was increased 180% in NGF-treated animals compared with a 40% decline in controls. However, GABAergic neurons were not protected (Venero et al., 1994). In a similar QA lesion paradigm, intravenous injection of 20 μg/day NGF conjugated to OX-26 (an antibody to the transferrin receptor, allowing passage across the blood–brain barrier) beginning 2 days before QA lesion and continuing for 2 weeks modestly protected cholinergic neurons (24% cell loss in NGF-OX-26–treated animals vs. 40% in controls), while somatostatin and NADPH-diaphorase neurons were unaffected (Kordower et al., 1994).

Ex vivo gene therapy with NGF-secreting cells protects cholinergic and, in some cases, NADPH-diaphorase striatal neurons from QA– and/or 3-NPA–induced degeneration. RatI fibroblasts, modified to express 4 ng NGF/10^5 cells/day by retroviral infection and implanted 1 week before QA, quisqualate, or 3-NPA lesion reduced the lesion size 60%–80% compared with control RatI fibroblasts. In QA and quisqualate, but not 3-NPA lesioned rats, cholinergic neurons within the lesioned area were protected (Frim et al., 1993a,b; Schumacher et al., 1991). Polymer-encapsulated BHK cells, modified to express 34 ng NGF/capsule/day by transfection of NGF under the metallothionein promoter, protected cholinergic and NADPH diaphorase neurons and reduced lesion size, although no quantitative analysis was performed. NGF-treated rats also rotated less following apomorphine injection, suggesting a preservation of striatal neurons with dopamine receptors (Emerich et al., 1994a). In the intact striatum, polymer-encapsulated BHK cells expressing NGF under the metallothionein promoter induced hypertrophy of cholinergic and neuropeptide Y striatal neurons, but not GABAergic neurons, 1, 2, and 4 weeks post grafting, with the hypertrophy reversible following removal of implants (Kordower et al., 1996). While these studies clearly demonstrate the ability of NGF to protect cholinergic neurons of the striatum, GABAergic medium spiny neurons, which are the most severely affected in HD, were not protected. Future studies to identify trophic factors for these neurons, as well as cortical and cerebellar neurons, which also degenerate in HD, and to demonstrate protective effects *in vivo* are needed before neurotrophic factor therapy can be applied to HD.

STROKE AND APOPTOSIS

The lack of blood flow in ischemic stroke results in hypoxic and hypoglycemic conditions in the neuronal environment, resulting in reduced energy production. With reduced energy available, ionic gradients are not maintained. Elevated levels of intracellular calcium play a key role in neuronal damage and death. A variety of neurotrophic factors protect neurons from hypoglycemic and hypoxic conditions *in vitro* (Mattson and Cheng, 1993). Infusion of several neurotrophic factors, including BDNF (Beck et al., 1994), bFGF (Koketsu et al., 1994), insulin, and IGF-I (Zhu and Auer, 1994), into the brain has been shown to reduce lesion size following experimental ischemia. *Ex vivo* gene therapy with RatI fibroblasts, modified by retroviral vector to produce NGF and implanted into the hippocampus 7 days prior to four-vessel occlusion, protected 61% of CA1 neurons from damage within 400 μm of the graft. In contrast, in rats that received control RatI fibroblasts or no grafts, 97% of CA1 neurons were damaged (Pechan et al., 1995).

There is a growing body of evidence that apoptosis plays a role in neuronal death in the penumbra region of strokes (Bredesen, 1995). *In vivo* gene therapy designed to interfere with the expression of pro-apoptotic genes by directing expression of an antisense message or overexpressing anti-apoptotic genes may thus protect neurons from degeneration. Transgenic animals over expressing bcl-2, an antiapoptotic protein, were found to have a 50% reduction in infarct volume after experimentally induced ischemia (Martinou et al., 1994). Using an *in vivo* gene therapy approach, Linnik and colleagues (1995) injected an HSV amplicon vector directing the expression of bcl-2 24 h prior to focal ischemia in rats. There was an area of viable tissue around the injection site with HSV-bcl-2, whereas HSV-lac had no effect. In another experiment, HSV vectors encoding a bicistronic lacZ and bcl-2 or a bicistronic lacZ and a truncated, nonfunctional bcl-2 (designated bst) were injected into the striatum 6 h before middle cerebral artery occlusion. At 48 h after ischemia, the percentage of lacZ expressing cells on the lesioned side compared with the unlesioned side was 107% ± 17% for lacZ-bcl-2 and 64% ± 11% for lacZ-bst, suggesting some protection of neurons expressing bcl-2; however, there was no effect on total lesion volume (Lawrence et al., 1996). Gene therapies designed to interfere with the interleukin-1b–converting enzyme (ICE) proteases, which are pro-apoptotic, may also prevent degeneration (Milligan et al., 1995).

Delivery of interleukin-1 receptor antagonist (IL-1ra) protein into the CNS reduces ischemic brain damage (see Betz et al., 1995). *In vivo* gene therapy with an Ad vector encoding IL-1ra injected into the rat lateral ventricle 5 days before 24 h of middle cerebral artery occlusion reduced infarct volume 64% compared with Ad-lacZ (Betz et al., 1995). In this study, rats injected with Ad IL-1ra had cerebrospinal fluid and brain levels of IL-1ra that were 50- and 5-fold higher than Ad-lacZ rats, respectively.

NEURODEGENERATIVE DISEASES AND APOPTOSIS

In addition to its role in stroke, apoptosis may be involved in neuronal death in neurodegenerative diseases (Bredesen, 1995). Overexpression of the anti-apoptotic gene bcl-2 under control of the neuron-specific enolase promoter in mice protected neurons in several models of neurodegeneration. For example, these transgenic mice exhibited no motor neuron cell death following neonatal peripheral nerve axotomy, whereas in wild-type mice, 77% of neurons degenerated (De Bilbao and Dubois-Dauphin, 1996; Dubois-Dauphin et al., 1994). Similarly, optic nerve transection caused a 50% reduction in the total number of retinal ganglion cells (RGC) 24 h after transection in wild-type mice, whereas no significant RGC loss occurred in bcl-2 transgenic mice. The number of pyknotic RGC cells was reduced 94% in bcl-2 transgenic compared with wild-type mice (Bonfanti et al., 1996). Additionally, overexpression of bcl-2 in pmn/pmn mice, obtained by crossing pmn/pmn mice with bcl-2 transgenic mice, prevented loss of motoneuron soma, but not axonal degeneration (Sagot et al., 1995a). These studies were performed in mice overexpressing bcl-2 in neurons and represent a germ line gene therapy. Future studies will need to address somatic gene therapy with bcl-2 or other apoptosis-related proteins using *in vivo* modification of host tissue.

Some persons with autosomal recessive spinal muscular atrophy (SMA, not to be confused with the autosomal dominant trinucleotide repeat expansion disorder spinal and bulbar muscular atrophy, SBMA) have mutations in the naip (neuronal apoptosis inhibitor protein) gene, suggesting that loss of function of an anti-apoptotic gene can lead to motoneuron degeneration (Roy et al., 1995). Overexpressing naip or other anti-apoptotic genes in motoneurons by gene therapy approaches might prevent motoneuron loss in these patients. The development of naip knockout mice should be helpful in understanding the pathogenesis and testing therapies of this disease.

DISORDERS OF THE PERIPHERAL NERVOUS SYSTEM

There are numerous disorders of the peripheral nervous system (PNS) with diverse etiologies, including inflammatory demyelination and neuropathies caused by injury, compression, diabetes, or other endocrine abnormalities, toxic agents (including some pharmaceutical agents), leprosy, and inherited neuropathies. Neuropathy is a frequent complication of diabetes, and its chronic, progressive nature makes it a potential target for gene therapy approaches. Metabolic abnormalities and damage to the vasa nervorum (blood vessels supplying nerves) likely contribute to peripheral nerve damage in diabetes. In addition, decreased expression of trkA, the high-affinity NGF receptor, by nociceptive sensory neurons and decreased expression of NGF in target organs, such as skin, contribute to degeneration and dysfunction of sensory neurons (Anand, 1996; Tomlinson et al., 1996). Administration of NGF has been shown to prevent behavioral and biochemical changes associated with streptozocin-induced diabetes in rats (Apfel et al., 1994). However, local or systemic injection of NGF is associated with hyperalgesia in rodents and humans (McMahon, 1996). Modifying tissues to produce modest amounts of NGF by gene therapy approaches could potentially prevent these toxic side effects. Neurotrophic factors can prevent PNS damage induced by some therapeutic agents. For example, NT-3 protected large sensory neurons from cisplatin-induced neuropathy (Gao et al., 1995) and spiral ganglion auditory neurons from aminoglycoside-induced degeneration (Ernfors et al., 1996). In addition, explant cultures of spiral ganglion infected with HSV BDNF exhibited robust neurite outgrowth (Geschwind et al., 1996). Injection of HSV NGF, but not HSV lacZ, into the superior cervical ganglion prevented the decline in TH activity that normally occurs following axotomy (Federoff et al., 1992).

SPINAL CORD TRAUMA

Spinal cord trauma can damage ascending sensory pathways, descending motor pathways, local connections, and neurons. Axonal regrowth in the spinal cord is inhibited by glial scar formation, myelin-associated inhibitors of neurite outgrowth, lack of growth-promoting extracellular matrix and cell surface molecules, and perhaps a limited supply of trophic support. Grafts of fibroblasts genetically modified to secrete neurotrophic factors promote neurite growth into grafts. For example, primary fibroblasts modified to express NGF by a retroviral vector and grafted into the unlesioned or dorsal hemisected spinal cord promoted robust ingrowth of sensory and noradrenergic fibers, limited ingrowth of motor fibers in lesioned cord only, and no ingrowth of serotonergic fibers (Tuszynski et al., 1994, 1996a). Transgene expression, analyzed by RT-PCR, was still detectable at 14 months after grafting (Tuszynski et al., 1996a). Sensory neurites were also observed to penetrate grafts of fibroblasts expressing NT-3 and bFGF, but not BDNF (Nakahara et al., 1996). In another *ex vivo* gene therapy study, fibroblasts genetically modified to express PDGF and injected into the cisterna magna increased the number of O-2A progenitors in the adult rat cervical spinal cord (Ijichi et al., 1996). For regeneration to occur in the spinal cord or brain, several strategies will likely be required in addition to providing neurotrophic factors, such as expression of substrates that promote axon outgrowth or block the action of inhibitory molecules.

CONCLUSIONS AND FUTURE GOALS

The potential clinical application of gene therapy to neurological disorders is promising, but presently limited by lack of technological advances at three levels. The first level is common to the goals of gene therapy in general, regardless of tissue type (see Table 6). The most important of these goals is to develop vectors or DNA transfer techniques that result in long-term sustainable expression. While advances in vector design may enhance the durability of transgene expression, this might also be enhanced through the use of cell-specific promoters. The cellular complexity of the nervous system presents unique challenges to gene therapy technology and is an ideal tissue for studying the incorporation of cell-specific promoters on the duration of transgene expression. The cellular context of transgene also may be a contributing factor to cellular and immune reactions to vector. Although cell-specific promoters have been tried in several studies in neural tissue *in vitro* and *in vivo* (Andersen et al., 1992; Jin et al., 1996; Kaplitt et al., 1994a; Lu and Federoff, 1995), this remains an area ripe for investigation, especially for making rigorous, side-by-side comparisons on level and duration of transgene expression with specific cellular promoters and different viral vectors.

TABLE 6 Clinical Application of Gene Therapy to the Nervous System: Goals to be Met

- Level 1: Vector development in general
  - High titer, low particle ratio, or absence of helper virus
  - Low cytotoxicity
  - High, sustained levels of transgene expression
  - Regulatable expression
  - Modulation of immune responses
  - Vector safety
- Level 2: Vector development specific to the nervous system
  - Specific cell targeting
    - Viral capsid or fiber modification
    - Cell specific promoters
    - Retrograde transport
    - Route of administration
  - Focal vs. widespread distribution of vectors in brain
  - Comparative testing *in vivo*
- Level 3: Biological issues of neurotrophic/neuroprotective factor gene therapy
  - Differential effects of factors on embryonic, adult, aged, and diseased neurons
  - Identify novel molecules that promote or repress survival and growth of specific classes of neurons
  - Determine the optimal cell type(s) for gene targeting
  - Identify optimal vector(s)
  - Improve animal models of progressive degeneration
  - Design more sophisticated behavioral tests of neuronal recovery
  - Interfere directly with apoptosis program in neurons

Ideally, the therapeutic gene should be delivered to the site where its product can mimic, substitute, or block endogenous mechanisms in a specific class of neurons and not produce deleterious side effects in nontargeted neurons. There are several strategies for targeted delivery. The cell type to be targeted could be the diseased or injured neuron itself where the factor would act in an autocrine or paracrine manner. Alternatively, the vector could target the cells upon which the diseased or injured neuron normally synapses where the factor would be released, taken up by neuronal terminals, and be retrogradely transported or act through terminal transduction mechanisms. Finally, the vector could be targeted to glial cells, such as astrocytes, oligodendrocytes, or microglia. These cells would then release the factor and increase its level in the micromileu of the diseased neuron. The particular strategy employed would be based on the biological nature of the neuronal system to be treated and on the molecular mechanisms involved.

Another issue is the route of delivery to the CNS. A single injection of virus in microliter quantities into rat brain produces a focal region of infected cells. Slow perfusion of virus into brain parenchyma has been reported to give fairly widespread distribution with up to 9 mm reported for Ad and 14 mm for HSV (Davidson et al., 1993; Muldoon et al., 1995). Another method for focal delivery would be to utilize retrograde transport. Combining retrograde transport with a cell-specific promoter driving transgene expression would give this approach an even higher degree of specificity. One study applying this approach has demonstrated specific expression of transgene in DA neurons in the SN following injection of an HSV amplicon vector harboring lacZ driven by the TH promoter into the striatum (Jin et al., 1996). Such focal delivery might be optimal for diseases affecting a single group of cells such as in Parkinson's disease, whereas other diseases or injuries to the CNS might best be served by a more global delivery of vector. For example, vector could be delivered fairly noninvasively through injection into cerebral vessels following disruption of the blood–brain barrier or injection into the cerebrospinal fluid. Transient osmotic disruption of the blood–brain barrier results in a larger spread of transduced cells following administration of Ad or HSV, with labeled cells found throughout the disrupted cerebral cortex (Doran et al., 1995; Muldoon et al., 1995; Neuwelt et al., 1991). Viral vectors have also been injected into the cisterna magna and lateral ventricle with transgene expression observed in ependymal cells, leptomeninges, and cells lining cerebral vessels (Bajocchi et al., 1993; Ooboshi et al., 1995). Again, the optimal route of delivery would depend on the disease or injury to be targeted.

Another goal should be the use of regulatable promoters so that transgene expression can be induced or eliminated. In some instances, such as in stroke, transient production of transgene might be sufficient to offer protection to neurons, whereas long-term expression might not be needed or might be deleterious. In other instances, long-term expression might be theoretically suitable. However, side effects, such as affective, cognitive, and behavioral disorders produced by long-term transgene expression in the human, might not be predicted from animal studies. The incorporation of a regulatable promoter would be advantageous if such side effects were to occur. To date, only preliminary studies on the use of regulatable promoters in neural cells have been done (Emerich et al., 1994a,b; Paulus et al., 1996; Winn et al., 1994).

The third level of consideration for applying gene therapy to the CNS involves issues basic to neurobiology and behavior. The more fundamental knowledge we have on the molecular mechanisms involved in development, aging, and disease of specific neuronal systems, the more genes will be identified as candidates for gene therapy. Likewise, means for acting on downstream pathways involved in apoptosis and neurotrophic factor action through gene therapy are likely to be fruitful areas of investigation. The development of animal models that more closely mimic the morphological, neurochemical, and behavioral consequences of neurological diseases in humans are also critically important to evaluating preclinical gene therapy studies. Finally, it is important to include nonhuman primates in the preclinical studies before gene therapy is applied to the human CNS.

ACKNOWLEDGMENTS

The authors are grateful for the secretarial assistance of Ms. Kim Coene.

REFERENCES

Abdallah B, Sach L, Demeneix BA (1995): Non-viral gene transfer—Applications in developmental biology and gene therapy. *Biol Cell* 85:1–7.

Adams RD, Victor M (1993): Degenerative diseases of the nervous system. In *Principles of Neurology.* New York: McGraw-Hill, Inc., pp 975–982.

Aebischer P, Schluep M, Deglon N, Joseph M, Hirt L, Heyd B, Hammang JP, Zurn AD, Kato AC, Regli F, Baetge EE (1996): Intrathecal delivery of CNTF using encapsulated genetically modified xenogeneic cells in amyotrophic lateral sclerosis patients. *Nat Med* 2:696–699.

Aebischer P, Winn SR, Galletti PM (1988): Transplantation of neural tissue in polymer capsules. *Brain Res* 448:36–48.

Akaneya Y, Takahashi M, Hatanaka H (1995): Interleukin-1 beta enhances survival and interleukin-6 protects against MPP^+ neurotoxicity in cultures of fetal rat dopaminergic neurons. *Exp Neurol* 136:44–52.

Akli S, Caillaud C, Vigne C, Stratford-Perricaudet LD, Poenaru L, Perricaudet M, Kahn A, Peschanski MR (1993): Transfer of a foreign gene into the brain using adenovirus vectors. *Nat Genet* 3:224–228.

Alexi T, Hefti F (1992): TGFα selectively increases dopaminergic cell survival in ventral mesencephalic cultures. *Soc Neurosci Abst* 18:1292.

ALS CNTF Treatment Study (ACTS) Phase I–II Study Group (1995): The pharmacokinetics of subcutaneously administered recombinant human ciliary neurotrophic factor (rHCNTF) in patients with amyotrophic lateral sclerosis: Relation to parameters of the acute-phase response. *Clin Neuropharmacol* 18:500–514.

ALS CNTF Treatment Study Group (1996): A double-blind placebo-controlled clinical trial of subcutaneous recombinant human ciliary neurotrophic factor (rHCNTF) in amyotrophic lateral sclerosis. *Neurology* 46:1244–1249.

Altar CA, Boylan CB, Fritsche M, Jackson C, Hyman C, Lindsay RM (1994a): The neurotrophins NT-4/5 and BDNF augment serotonin, dopamine, and GABAergic systems during behaviorally effective infusions to the substantia nigra. *Exp Neurol* 130:31–41.

Altar CA, Boylan CB, Fritsche M, Jones BE, Jackson C, Wiegand SJ, Lindsay RM, Hyman C (1994b): Efficacy of brain-derived neurotrophic factor and neurotrophin-3 on neurochemical and behavioral deficits associated with partial nigrostriatal dopamine lesions. *J Neurochem* 63:1021–1032.

Altar CA, Boylan CB, Jackson C, Hershenson S, Miller J, Wiegand SJ, Lindsay RM, Hyman C (1992): Brain-derived neurotrophic factor augments rotational behavior and nigrostriatal dopamine turnover *in vivo. Proc Natl Acad Sci USA* 89:11347–11351.

Anand P (1996): Neurotrophins and peripheral neuropathy. *Philos Trans R Soc Lond B Biol Sci* 351:449–454.

Andersen J, Garber D, Meaney C, Breakefield X (1992): Gene transfer into mammalian central nervous system using herpes virus vectors: Extended expression of bacterial lacZ in neurons using the neuron-specific enolase promoter. *Hum Gene Ther* 3:487–499.

Andersen JK, Breakefield XO (1995): Gene delivery to neurons of the adult mammalian nervous system using herpes and adenovirus vectors. In Chang PL (ed): *Somatic Gene Therapy.* Boca Raton: CRC Press, Inc., pp 135–160.

Anton R, Kordower JH, Maidment NT (1994): Neural-targeted gene therapy for rodent and primate hemiparkinsonism. *Exp Neurol* 127:207–218.

Apfel SC, Arezzo JC, Brownlee M, Federoff H, Kessler JA (1994): Nerve growth factor administration protects against experimental diabetic sensory neuropathy. *Brain Res* 634:7–12.

Arenas E, Persson H (1994): Neurotrophin-3 prevents the death of adult central noradrenergic neurons *in vivo. Nature* 367:368–371.

Arenas E, Trupp M, Akerud P, Ibanez CF (1995): GDNF prevents degeneration and promotes the phenotype of brain noradrenergic neurons *in vivo. Neuron* 15:1465–1473.

Avila MA, Varela-Nieto I, Romero G, Mato JM, Giraldez F, Van De Water TR, Represa J (1993): Brain-derived neurotrophic factor and neurotrophin-3 support survival and neuritogenesis response of developing cochleovestibular ganglion neurons. *Dev Biol* 159:266–275.

Awatsuji H, Furukawa Y, Nakajima M, Furukawa S, Hayashi K (1993): Interleukin-2 as a neurotrophic factor for supporting the survival of neurons cultured from various regions of fetal rat brain. *J Neurosci Res* 35:305–311.

Bajocchi G, Feldman SH, Crystal RG, Mastrangeli A (1993): Direct *in vivo* gene transfer to ependymal cells in the central nervous system using recombinant adeovirus vectors. *Nat Genet* 3:224–228.

Bankiewicz K, Mandel RJ, Sofroniew MV (1993): Trophism, transplantation, and animal models of Parkinson's disease. *Exp Neurol* 124:140–149.

Barinaga M (1994): Neurotrophic factors enter the clinic. *Science* 264:772–774.

Beck KD, Irwin I, Valverde J, Brennan TJ, Langston JW, Hefti F (1996): GDNF induces a dystonia-like state in neonatal rats and stimulates dopamine and serotonin synthesis. *Neuron* 16:665–673.

Beck KD, Knüsel B, Hefti F (1993): The nature of the trophic action of brain-derived neurotrophic factor, des(1-3)-insulin-like growth factor-1, and basic fibroblast growth factor on mesencephalic dopaminergic neurons developing in culture. *Neuroscience* 52:855–866.

Beck KD, Valverde J, Alexi T, Poulsen K, Moffat B, Vendlen RA, Rosenthal A, Hefti F (1995): Mesencephalic dopaminergic neurons protected by GDNF from axotomy-induced degeneration in the adult brain. *Nature* 373:339–341.

Beck T, Lindholm D, Castren E, Wree A (1994): Brain-derived neurotrophic factor protects against ischemic cell damage in rat hippocampus. *J Cereb Blood Flow Metab* 14:689–692.

Benfry M, Aguayo AJ (1982): Extensive elongation of axons from rat brain into peripheral nerve grafts. *Nature* 296:150–152.

Betz AL, Yang GY, Davidson BL (1995): Attenuation of stroke size in rats using an adenoviral vector to induce overexpression of interleukin-receptor antagonist in brain. *J Cereb Blood Flow Metab* 15:547–551.

Bilang-Bleuel A, Reuah F, Colin P, Locquet I, Robert J-J, Mallet J, Horellou P (1997): Intrastriatal injection of an adenoviral vector expressing glial-cell-line-derived neurotrophic factor prevents dopaminergic neuron degeneration and behavioral impairment in a rat model of parkinson disease. *Proc Natl Acad Sci USA* 94:8818–8823.

Bohn MC, Marciano F, Cupit L, Gash DM (1987): Adrenal medulla grafts enhance recovery of striatal dopaminergic fibers. *Science* 237:913–916.

Bonfanti L, Strettoi E, Chierzi S, Cenni MC, Liu X-H, Martinou J-C, Maffei L, Rabacchi SA (1996): Protection of retinal ganglion cells from natural and axotomy-induced cell death in neonatal transgenic mice overexpressing bcl-2. *J Neurosci* 16:4186–4194.

Boussif O, Lezoualc'h F, Zanta MA, Mergny MD, Scherman D, Demeneix B, Behr J-P (1995): A versatile vector for gene and oligonucleotide transfer into cells in culture and *in vivo:* Polyethylenimine. *Proc Natl Acad Sci USA* 92:7297–7301.

Bowenkamp KE, Hoffman AF, Gerhardt GA, Henry MA, Biddle PT, Hoffer BJ, Granholm A-CE (1995): Glial cell line-derived neurotrophic factor supports survival of injured midbrain dopaminergic neurons. *J Comp Neurol* 355:479–489.

Bredesen DE (1995): Neural apoptosis [review]. *Ann Neurol* 38:839–851.

Burke JR, Enghild JJ, Martin ME, Jou Y-S, Myers RM, Roses AD, Vance JM, Strittmatter WJ (1996): Huntingtin and DRPLA proteins selectively interact with the enzyme GAPDH. *Nat Med* 2:347–350.

Byrnes AP, Rusby JE, Wood MJA, Charlton HM (1995): Adenovirus gene transfer causes inflammation in the brain. *Neuroscience* 66:1015–1024.

Cao L, Zheng ZC, Zhao YC, Jiang ZH, Liu ZG, Chen SD, Zhou CF, Liu XY (1995): Gene therapy of Parkinson disease model rat by direct injection of plasmid DNA-lipofectin complex. *Hum Gene Ther* 6:1497–1501.

Casper D, Mytilineus C, Blum M (1991): EGF enhances the survival of dopaminergic neurons in rat embryonic mesencephalon primary cell culture. *J Neurosci Res* 30:372–381.

Castel-Barthe MN, Jazat-Poindessous F, Barneoud P, Vigne E, Revah F, Mallet J, Lamour Y (1996): Direct intracerebral nerve growth factor gene transfer using a recombinant adenovirus: Effect on basal forebrain cholinergic neurons during aging. *Neurobiol Dis* 3:76–86.

Chen S-H, Shine HD, Goodman JC, Grossman RG, Woo SLC (1994): Gene therapy for brain tumors: Regression of experimental gliomas by adenovirus-mediated gene transfer *in vivo. Proc Natl Acad Sci USA* 91:3054–3057.

Cheng B, Christakos S, Mattson MP (1994a): Tumor necrosis factors protect neurons against metabolic-excitotoxic insults and promote maintenance of calcium homeostasis. *Neuron* 12:139–153.

Cheng B, Goodman Y, Begley JG, Mattson MP (1994b): Neurotrophin-4/5 protects hippocampal and cortical neurons against energy deprivation- and excitatory amino acid-induced injury. *Brain Res* 650:331–335.

Cheng B, Mattson MP (1992): IGF-I and IGF-II protect cultured hippocampal and septal neurons against calcium-mediated hypoglycemic damage. *J Neurosci* 12:1558–1566.

Cheng B, Mattson MP (1994): NT-3 and BDNF protect CNS neurons against metabolic/excitotoxic insults. *Brain Res* 640:56–67.

Cheng B, Mattson MP (1995): PDGFs protect hippocampal neurons against energy deprivation and oxidative injury: Evidence for induction of antioxidant pathways. *J Neurosci* 15:7095–7104.

Chiu AY, Chen EW, Loera S (1994): Distinct neurotrophic responses of axotomized motor neurons to BDNF and CNTF in adult rats. *Neuroreport* 5:693–696.

Choi-Lundberg DL, Bohn MC (1995): Ontogeny and distribution of glial cell line-derived neurotrophic factor (GDNF) mRNA in rat. *Brain Res Dev Brain Res* 85:80–88.

Choi-Lundberg DL, Lin Q, Chang Y-N, Chiang YL, Hay CM, Mohajeri H, Davidson BL, Bohn MC (1997): Dopaminergic neurons protected from degeneration by GDNF gene therapy. *Science* 275:838–841.

Choi-Lundberg DL, Lin Q, Schallert T, Crippens D, Davidson BL, Chang Y-N, Chiang YL, Qian J, Bardwaj L, Bohn MC (1998): Behavioral and cellular protection of rat dopaminergic neurons by an adenoviral vector encoding glial cell line-derived neurotrophic factor (GDNF). Submitted.

Cintra A, Cao Y, Oelling C, Tinner B, Bortolotti F, Goldstein M, Pettersson RF, Fuxe K (1991): Basic FGF is present in dopaminergic neurons of the ventral midbrain of the rat. *Neuroreport* 2:597–600.

Cohen G, Heikkila RE, MacNamee D (1974): The generation of hydrogen peroxide superoxide radical, and hydroxyl radical by 6-hydroxydopamine, dialuric acid, and related cytotoxic agents. *J Biol Chem* 249:2447–2452.

Collazo D, Takahashi H, McKay RDG (1992): Cellular targets and trophic functions of neurotrophin-3 in the developing rat hippocampus. *Neuron* 9:643–656.

Cuevas P, Carceller F, Gimenez-Gallego G (1995): Acidic fibroblast growth factor prevents post-axotomy neuronal death of the newborn rat facial nerve. *Neurosci Lett* 197:183–186.

Cummings BJ, Yee GJ, Cotman CW (1992): bFGF promotes the survival of entorhinal layer II neurons after perforant path axotomy. *Brain Res* 591:271–276.

Cunningham LA, Hansen JT, Short MP, Bohn MC (1991): The use of genetically altered astrocytes to provide nerve growth factor (NGF) to adrenal chromaffin cells grafted into the striatum. *Brain Res* 561:192–202.

Cunningham LA, Short MP, Breakefield XO, Bohn MC (1994): Nerve growth factor released by transgenic astrocytes enhances the function of adrenal chromaffin cell grafts in a rat model of Parkinson's disease. *Brain Res* 658:219–231.

Dale SM, Kuang RZ, Wei X, Varon S (1995): Corticospinal motor neurons in the adult rat: Degeneration after intracortical axotomy and protection by ciliary neurotrophic factor (CNTF). *Exp Neurol* 135:67–73.

Davar G, Kramer MF, Garber D, Roca AL, Andersen JK, Bebtin W, Coen DM, Kosz-Vnenchak M, Knipe DM, Breakfield XO, Isacson O (1994): Comparative efficacy of expression of genes delivered to mouse sensory neurons with herpes virus vectors. *J Comp Neurol* 339:3–11.

David S, Aguayo AJ (1981): Axonal elongation into peripheral nervous system "bridge" after central nervous system injury in adult rats. *Science* 214:931–933.

Davidson BL, Allen ED, Kozarsky KF, Wilson JM, Roessler BJ (1993): A model system for *in vivo* gene transfer into the central nervous system using an adenoviral vector. *Nat Genet* 3:219–223.

Davis HL, Demeneix BA, Quantin B, Coulombe J, Whalen RG (1993): Plasmid DNA is superior to viral vectors for direct gene transfer into adult mouse skeletal muscle. *Hum Gene Ther* 4:733–740.

De Bilbao F, Dubois-Dauphin M (1996): Time course of axotomy-induced apoptotic cell death in facial motoneurons of neonatal wild type and bcl-2 transgenic mice. *Neuroscience* 71:1111–1119.

Dekker AJ, Winkler J, Ray J, Thal LJ, Gage FH (1994): Grafting of nerve growth factor-producing fibroblasts reduces behavioral deficits in rats with lesions of the nucleus basalis magnocellularis. *Neuroscience* 60:299–309.

Deng HX, Hentati A, Rainer JA, Iqbal Z, Cayabyab Z, Hung W-Y, Getzoff ED, Hu P, Herzfeldt B, Roos RP, Warner C, Deng G, Soriano E, Smyth C, Parge HE, Ahmed A, Roses AD, Hallewell RA, Pericak-Vance MA, Siddique T (1993): Amyotrophic lateral sclerosis and structural defects in Cu,Zn superoxide dismutase. *Science* 261:1047–1051.

DiStefano PS, Friedman B, Radziejewski C, Alexander C, Boland P, Schick CM, Lindsay RM, Wiegand SJ (1992): The neurotrophins BDNF, NT-3, and NGF display distinct patterns of retrograde axonal transport in peripheral and central neurons. *Neuron* 8:983–993.

Doran SE, Ren XD, Betz AL, Pagel MA, Neuwelt EA, Roessler BJ, Davidson BL (1995): Gene expression from recombinant viral vectors in the central vervous system after blood–brain barrier disruption. *Neurosurgery* 36:965–970.

Dreyfus CF (1989): Effects of nerve growth factor on cholinergic brain neurons. *Trends Pharmacol Sci* 10:145–149.

Dubois-Dauphin M, Frankowski H, Tsujimoto Y, Huarte J, Martinou J-C (1994): Neonatal motoneurons overexpressing the bcl-2 protooncogene in transgenic mice are protected from axotomy-induced cell death. *Proc Natl Acad Sci USA* 91:3309–3313.

Dunaief JL, Kwun RC, Bhardwaj N, Lopez R, Gouras P, Goff SP (1995): Retroviral gene transfer into retinal pigment epithelial cells followed by transplantation into rat retina. *Hum Gene Ther* 6:1225–1229.

Durbec P, Marcos-Gutierrez CV, Kilkenny C, Grigoriou M, Wartiowaara K, Suvanto P, Smith D, Ponder B, Costantini F, Saarma M, Sariola H, Pachnis V (1996): GDNF signalling through the ret receptor tyrosine kinase. *Nature* 381:789–793.

During MJ, Naegele JR, O'Malley KL, Geller AI (1994): Long-term behavioral recovery in parkinsonian rats by an HSV vector expressing tyrosine hydroxylase. *Science* 266:1399–1403.

Emerich DF, Hammang JP, Baetge EE, Winn SR (1994a): Implantation of polymer-encapsulated human nerve growth factor-secreting fibroblasts attenuates the behavioral and neuropathological consequences of quinolinic acid injections into rodent striatum. *Exp Neurol* 130:141–150.

Emerich DF, Winn SR, Harper J, Hammang JP, Baetge EE, Kordower JH (1994b): Implants of polymer-encapsulated human NGF-secreting cells in the nonhuman primate: Rescue and sprouting of degenerating cholinergic basal forebrain neurons. *J Comp Neurol* 349:148–164.

Engele J, Bohn MC (1991): The neurotrophic effects of fibroblast growth factors on dopaminergic neurons *in vitro* are mediated by mesencephalic glia. *J Neurosci* 11:3070–3078.

Enokido Y, Akaneya Y, Niinobe M, Mikoshiba K, Hatanaka H (1992): Basic fibroblast growth factor rescues CNS neurons from cell death caused by high oxygen atmosphere in culture. *Brain Res* 599:261–271.

Ernfors P, Duan ML, ElShamy WM, Canlon B (1996): Protection of auditory neurons from aminoglycoside toxicity by neurotrophin-3. *Nat Med* 2:463–467.

Fraefel C, Song S, Lim F, Lang P, Yu L, Wang Y, Wild P, Geller AI (1996): Helper virus-free transfer of herpes simplex virus type 1 plasmid vectors into neural cells. *J Virol* 70:7190–7197.

Federoff HJ, Geschwind MD, Geller AI, Kessler JA (1992): Expression of nerve growth factor *in vivo* from a defective herpes simplex virus 1 vector prevents effects of axotomy on sympathetic ganglia. *Proc Natl Acad Sci USA* 89:1636–1640.

Ferguson IA, Johnson EM (1991): Fibroblast growth factor bearing neurons in the CNS: Identification by receptor mediated retrograde transport. *J Comp Neurol* 313:693–706.

Ferrari G, Minozzi M-C, Toffano G, Leon A, Skaper SB (1989): Basic fibroblast growth factor promotes the survival and development of mesencephalic neurons in culture. *Dev Biol* 133:140–147.

Festoff BW (1996): Amyotrophic lateral sclerosis. *Drugs* 51:28–44.

Festoff BW, Yang SX, Vaught J, Bryan C, Ma JY (1995): The insulin-like growth factor signaling system and ALS neurotrophic factor treatment strategies [review]. *J Neurol Sci* 129(suppl):114–121.

Fischer W, Wictorin K, Bjorklund A, Williams LR, Varon S, Gage FH (1987): Amelioration of cholinergic neuron atrophy and spatial memory impariment in aged rats by nerve growth factor. *Nature* 329:65–68.

Flotte TR (1995): Adeno-associated virus vectors for gene therapy. *Gene Ther* 2:357–362.

Forger NG, Roberts SL, Wong V, Breedlove SM (1993): Ciliary neurotrophic factor maintains motoneurons and their target muscles in developing rats. *J Neurosci* 13:4720–4726.

Friden PM, Walus LR, Watson P, Doctrow SR, Kozarich JW, Backman C, Bergman H, Hoffer B, Bloom F, Granholm A-C (1993): Blood-brain barrier penetration and *in vivo* activity of an NGF conjugate. *Science* 259:373–377.

Friedman WJ, Ibanez CF, Hallböök F, Persson H, Cain LD, Dreyfus CF, Black IB (1993): Differential actions of neurotrophins in the locus coeruleus and basal forebrain. *Exp Neurol* 119:72–78.

Frim DM, Short MP, Rosenberg WS, Simpson J, Breakefield XO, Isacson O (1993a): Local protective effects of nerve growth factor-secreting fibroblasts against excitotoxic lesions in the rat striatum. *J Neurosurg* 78:267–273.

Frim DM, Simpson J, Uhler TA, Short MP, Bossi SR, Breakefield XO, Isacson O (1993b): Striatal degeneration induced by mitochondrial blockade is prevented by biologically delivered NGF. *J Neurosci Res* 35:452–458.

Frim DM, Uhler TA, Galpern WR, Beal MF, Breakefield XO, Isacson O (1994): Implanted fibroblasts genetically engineered to produce brain-derived neurotrophic factor prevent 1-methyl-4-phenylpyridinium toxicity to dopaminergic neurons in the rat. *Proc Natl Acad Sci USA* 91:5104–5108.

Gage FH, Wolff JA, Rosenberg MB, Xu L, Yee J-K, Shults C, Friedmann T (1987): Grafting genetically modified cells into the brain: Possibilities for the future. *Neuroscience* 23:795–807.

Galpern WR, Frim DM, Tatter SB, Altar CA, Beal MF, Isacson O (1996): Cell-mediated delivery of brain-derived neurotrophic factor enhances dopamine levels in an MPP^+ rat model of substantia nigra degeneration. *Cell Transplant* 5:225–232.

Games D, Adams D, Alseeandrini R, Barbour R, Berthelette P, Blackwell C, Carr T, Clemens J, Donaldson T, Gillespie F, Guido T, Hagopian S, Johnson-Wood K, Khan K, Lee M, Leibowitz P, Lieberburg I, Little S, Masliah E, McConlogue L, Montoya-Zavala M, Mucke L, Paganini L, Penniman E, Power M, Schenk D, Seubert P, Snyder B, Soriano F, Tan H, Vitale J, Wadsworth S, Wolozin B, Zhao J (1995): Alzheimer-type neuropathology in transgenic mice overexpressing V717F β-amyloid precursor protein. *Nature* 373:523–527.

Gao WQ, Dybdal N, Shinsky N, Murnane A, Schmelzer C, Siegel M, Keller G, Hefti F, Phillips HS, Winslow JW (1995): Neurotrophin-3 reverses experimental cisplatin-induced peripheral sensory neuropathy. *Ann Neurol* 38:30–37.

Gash DM, Zhang Z, Ovadia A, Cass WA, Yi A, Simmerman L, Russell D, Martin D, Lapchak PA, Collins F, Hoffer BJ, Gerhardt GA (1996): Functional recovery in parkinsonian monkeys treated with GDNF. *Nature* 380:252–255.

Geller AI (1993): Herpesviruses: Expression of genes in postmitotic brain cells [review]. *Curr Opin Genet Dev* 3:81–85.

Geschwind MD, Hartnick CJ, Liu W, Amat J, Vandewater TR, Federoff HJ (1996): Defective HSV-1 vector expressing BDNF in auditory ganglia elicits neurite outgrowth—Model for treatment of neuron loss following cochlear degeneration. *Hum Gene Ther* 7:173–182.

Ghadge GD, Roos RP, Kang UJ, Wollmann R, Fishman PS, Kalynych AM, Barr E, Leiden JM (1995): CNS gene delivery by retrograde transport of recombinant replication-defective adenoviruses. *Gene Ther* 2:132–137.

Glorioso JC, DeLuca NA, Fink DJ (1995): Development and application of herpes simplex virus vectors for human gene therapy. *Annu Rev Microbiol* 49:675–710.

Goetz CG, Diederich NJ (1996): There is a renaissance of interest in pallidotomy for Parkinson's disease. *Nat Med* 2:510–514.

Gordon EM, Anderson WF (1994): Gene therapy using retroviral vectors. *Curr Opin Biotechnol* 5:611–616.

Gurney ME, Pu H, Chiu AY, Dal Canto MC, Polchow CY, Alexander DD, Caliendo J, Hentati A, Kwon YW, Deng H-X, Chen W, Zhai P, Sufit RL, Siddique T (1994): Motor neuron degeneration in mice that express a human Cu,Zn superoxide dismutase mutation. *Science* 264:1772–1775.

Hagg T, Varon S (1993): Ciliary neurotrophic factor prevents degeneration of adult rat substantia nigra dopaminergic neurons *in vivo*. *Proc Natl Acad Sci USA* 90:6315–6319.

Hama T, Kushima Y, Miyamoto M, Kubota M, Takei M, Hatanaka H (1991): Interleukin-6 improves the survival of mesencephalic catecholaminergic and septal cholinergic neurons from post-natal two-week old rats in cultures. *Neuroscience* 40:445–452.

Hao C, Richardson A, Fedoroff S (1991): Macrophage-like cells originate from neuro-epithelium in culture: Characterization and proterties of the macrophage-like cells. *Int J Dev Neurosci* 9:1–14.

Hefti F (1986): Nerve growth factor promotes survival of septal cholinergic neurons after fimbrial transections. *J Neurosci* 6:2155–2162.

Henderson CE (1995): Symposium. GDNF: A new neurotrophic factor with multiple roles. *Soc Neurosci Abst* 21:1259.

Henderson CE, Camu W, Mettling C, Gouin A, Poulsen K, Karihaloo M, Rullamas J, Evans T, McMahon SB, Armanini MP, Berkemeier L, Phillips HS, Rosenthal A (1993): Neurotrophins promote motor neuron survival and are present in embryonic limb bud. *Nature* 363:266–270.

Henderson CE, Phillips HS, Pollock RA, Davies AM, Lemeulle C, Armanini M, Simpson LC, Moffet B, Vandlen RA, Koliatsos E, Rosenthal A (1994): GDNF: A potent survival factor for motoneurons present in peripheral nerve and muscle. *Science* 266:1062–1064.

Hoffer BJ, Hoffman A, Bowenkamp K, Huettl P, Hudson J, Martin D, Lin LF, Gerhardt GA (1994): Glial cell line-derived neurotrophic factor reverses toxin-induced injury to midbrain dopaminergic neurons in vivo. *Neurosci Lett* 182:107–111.

Hoffman D, Breakefield XO, Short MP, Aebischer P (1993): Transplantation of a polymer-encapsulated cell line genetically engineered to release NGF. *Exp Neurol* 122:100–103.

Horellou P, Brundin P, Kalén P, Mallet J, Björklund A (1990): *In vivo* release of DOPA and dopamine from genetically engineered cells grafted to the denervated rat striatum. *Neuron* 5:393–402.

Horellou P, Vigne E, Castel MN, Barneoud P, Colin P, Perricaudet M, Delaere P, Mallet J (1994): Direct intrcerebral gene transfer of an adenoviral vector expressing tyrosine hydroxylase in a rat model of Parkinson's disease. *Neuroreport* 6:49–53.

Hoshimaru M, Ray J, Sah DWY, Gage FH (1996): Differentiation of the immortalized adult neuronal progenitor cell line HC2S2 into neurons by regulatable suppression of the v-myc oncogene. *Proc Natl Acad Sci USA* 93:1518–1523.

Hou J-G, G., Lin L-F, H., Mytilineou C (1996): Glial cell line-derived neurotrophic factor exerts neurotrophic effects on dopaminergic neurons in vitro and promotes their survival and regrowth after damage by 1-methyl-4-phenylpyridinium. *J Neurochem* 66:74–82.

Hughes RA, Sendtner M, Goldfarb M, Lindholm D, Thoenen H (1993a): Evidence that fibroblast growth factor 5 is a major muscle-derived survival factor for cultured spinal motoneurons. *Neuron* 10:369–377.

Hughes RA, Sendtner M, Thoenen H (1993b): Members of several gene families influence survival of rat motoneurons *in vitro* and *in vivo*. *J Neurosci Res* 36:663–671.

Hyman C, Hofer M, Barde Y-A, Juhasz M, Yancopoulos GD, Squinto SP, Lindsay RM (1991): BDNF is a neurotrophic factor for dopaminergic neurons of the substantia nigra. *Nature* 350:230–232.

Hyman C, Juhasz M, Jackson C, Wright P, Ip NY, Lindsay RM (1994): Overlapping and distinct actions of the neurotrophins BDNF, NT-3, and NT-4/5 on cultured dopaminergic and GABAergic neurons of the ventral mesencephalon. *J Neurosci* 14:335–347.

Ijichi A, Noel F, Sakuma S, Weil MM, Tofilon PJ (1996): *Ex vivo* gene delivery of platelet-derived growth factor increases O-2A progenitors in adult rat spinal cord. *Gene Ther* 3:389–395.

Ikari H, Zhang L, Chernak JM, Mastrangeli A, Kato S, Kuo H, Crystal RG, Ingram DK, Roth GS (1995): Adenovirus-mediated gene transfer of dopamine D2 receptor cDNA into rat striatum. *Brain Res Mol Brain Res* 34:315–320.

Ip NY, Li T, Yancopoulos GD, Lindsay RM (1993): Cultured hippocampal neurons show responses to BDNF, NT-3, and NT-4, but not NGF. *J Neurosci* 13:3394–3405.

Ip NY, Li YP, van de Stadt I, Panayotatos N, Alderson RF, Lindsay RM (1991): Ciliary neurotrophic factor enhances neuronal survival in embryonic rat hippocampal cultures. *J Neurosci* 11:3124–3134.

Ishihara A, Saito H, Abe K (1994): Transforming growth factor-β1 and -β2 promote neurite sprouting and elongation of cultured rat hippocampal neurons. *Brain Res* 639:21–25.

Jiao S, Gurevich V, Wolff JA (1993): Long-term correction of rat model of Parkinson's disease by gene therapy. *Nature* 362:450–455.

Jiao S, Gurevich V, Wolff JA (1996): Long-term correction of rat model of Parkinson's disease by gene therapy. *Nature* 380:734.

Jin BK, Belloni M, Conti B, Federoff HJ, Starr R, Son H, Baker H, Joh TH (1996): Long-term *in vivo* gene expression driven by a tyrosine hydroxylase promoter in a defective herpes simplex virus amplicon vector. *Human Gene Ther* 7:2015–2027.

Jinnah HA, Friedmann T (1995): Gene therapy and the brain. *Br Med Bull* 51:138–148.

Jinnah HA, Gage FH, Friedmann T (1993): Gene therapy and neurologic disease. In Rosenberg RN, Prusiner SB, DiMauro S, Barchi RL, Kunkel LM (eds): *The Molecular and Genetic Basis of Neurological Disease.* Boston: Butterworth-Heinemann, pp 969–976.

Kang UJ, Fisher LJ, Joh TH, O'Malley KL, Gage FH (1993): Regulation of dopamine production by genetically modified primary fibroblasts. *J Neurosci* 13:5203–5211.

Kaplitt M, Kwong A, Leopoulos S, Mobbs C, Rabkin S, Pfaff D (1994a): Preproenkephalin promoter yields region-specific and long-term expression in adult brain after direct *in vivo* gene transfer via a defective herpes simplex viral vector. *Proc Natl Acad Sci USA* 91:8979–8983.

Kaplitt MG, Leone P, Samulski RJ, Xiao X, Pfaff DW, O'Malley KL, During MJ (1994b): Long-term gene expression and phenotypic correction using adeno-associated virus vectors in the mammalian brain. *Nat Genet* 8:148–154.

Karpati G, Lochmuller H, Nalbantoglu J, Durham H (1996): The principles of gene therapy for the nervous system. *Trends Neurosci* 19:49–54.

Kawaja MD, Rosenberg MB, Yoshida K, Gage FG (1992): Somatic gene transfer of nerve growth factor promotes the survival of axtomized septal neurons and the regeneration of their axons in adult rats. *J Neurosci* 12:2849–2864.

Kearns CM, Gash DM (1995): GDNF protects nigral dopamine neurons against 6-hydroxydopamine *in vivo. Brain Res* 672:104–111.

Keir SD, Mitchell WJ, Feldman LT, Martin JR (1995): Targeting and gene expression in spinal cord motor neurons following intramuscular inoculation of an HSV-1 vector. *J Neurovirol* 1:259–267.

Kikuchi S, Muramatsu H, Muramatsu T, Kim SU (1993): Midkine, a novel neurotrophic factor, promotes survival of mesencephalic neurons in culture. *Neurosci Lett* 160:9–12.

Klein R, Conway D, Parade LF, Barbacid M (1990): The trkB tyrosine protein kinase gene codes for a second neurogenic receptor that lacks the catalytic kinase domain. *Cell* 61:647–656.

Knusel B, Michel PP, Schwaber JS, Hefti F (1990): Selective and nonselective stimulation of central cholinergic and dopaminergic development *in vitro* by nerve growth factor, basic fibroblast growth factor, epidermal growth factor, insulin and the insulin-like growth factors I and II. *J Neurosci* 10:558–570.

Koketsu N, Berlove DJ, Moskowitz MA, Kowall NW, Caday CG, Finklestein SP (1994): Pretreatment with intraventricular basic fibroblast growth factor decreases infarct size following focal cerebral ischemia in rats. *Ann Neurol* 35:451–457.

Koliatsos VE, Cayouette MH, Berkemeier LR, Clatterbuck RE, Price DL, Rosenthal A (1994): Neurotrophin 4/5 is a trophic factor for mammalian facial motor neurons. *Proc Natl Acad Sci USA* 91:3304–3308.

Koliatsos VE, Clatterbuck RE, Winslow JW, Cayouette MH, Price DL (1993): Evidence that brain-derived neurotrophic factor is a trophic factor for motor neurons *in vivo. Neuron* 10:359–367.

Konishi Y, Chui DH, Hirose H, Kunishita T, Tabira T (1993): Trophic effects of erythropoietin and other hematopoietic factors on central cholinergic neurons *in vitro* and *in vivo. Brain Res* 609:29–35.

Kordower J, Mufson E, Granholm A, Hoffer B, Friden P (1993): Delivery of trophic factors to the primate brain [review]. *Exp Neurol* 124:21–30.

Kordower JH, Charles V, Bayer R, Bartus RT, Putney S, Walus LR, Friden PM (1994): Intravenous administration of a transferrin receptor antibody-nerve growth factor conjugate prevents the degeneration of cholinergic striatal neurons in a model of Huntington disease. *Proc Natl Acad Sci USA* 91:9077–9080.

Kordower JH, Chen E-Y, Mufson EJ, Winn SR, Emerich DF (1996): Intrastriatal implants of polymer encapsulated cells genetically modified to secrete human nerve growth factor: Trophic effects upon cholinergic and noncholinergic striatal neurons. *Neuroscience* 72:63–77.

Kordower JH, Cochran E, Penn RD, Goetz CG (1991): Putative chromaffin cell survival and enhanced host-derived TH-fiber innervation following a functional adrenal medulla autograft for Parkinson's disease. *Ann Neurol* 29:405–412.

Kramm CM, Senaestevers M, Barnett FH, Rainov NG, Schuback DE, Yu JS, Pechan PA, Paulus W, Chiocca EA, Breakefield XO (1995): Gene therapy for brain tumors [review]. *Brain Pathol* 5:345–381.

Krieglstein K, Suter-Crazzolara C, Fischer WH, Unsicker K (1995a): TGF-β superfamily members promote survival of midbrain dopaminergic neurons and protect them against MPP^+ toxicity. *EMBO J* 14:736–742.

Krieglstein K, Suter-Crazzolara C, Hotten G, Pohl J, Unsicker K (1995b): Trophic and protective effects of growth/differentiation factor 5, a member of the transforming growth factor-B superfamily, on midbrain dopaminergic neurons. *J Neurosci Res* 42:724–732.

LaVail MM, Unoki K, Yasumura D, Matthes MT, Yancopoulos GD, Steinberg RH (1992): Multiple growth factors, cytokines, and neurotrophins rescue photoreceptors from the damaging effects of constant light. *Proc Natl Acad Sci USA* 89:11249–11253.

Lawrence MS, Ho DY, Sun GH, Steinberg GK, Sapolsky RM (1996): Overexpression of bcl-2 with herpes simplex virus vectors protects CNS neurons against neurological insults *in vitro* and *in vivo. J Neurosci* 16:486–496.

Lefebvre PP, Van De Water TR, Weber T, Rogister B, Monnen G (1991): Growth factor interactions in cultures of dissociated adult acoustic ganglia: Neuronotrophic effects. *Brain Res* 567:306–312.

Le Gal La Salle G, Robert JJ, Berrard S, Ridoux V, Stratford-Perridaudet LD, Perricaudet M, Mallet J (1993): An adenovirus vector for gene transfer into neurons and glia in the brain. *Science* 259:988–990.

Levine RJ (1996): Canavan gene therapy protocol. *Science* 272:1085.

Levivier M, Przedborski S, Bencsics C, Kang UJ (1995): Intrastriatal implantation of fibroblasts genetically engineered to produce brain-derived neurotrophic factor prevents degeneration of dopaminergic neurons in a rat model of Parkinson's disease. *J Neurosci* 15:7810–7820.

Li LX, Wu WT, Lin LFH, Lei M, Oppenheim RW, Houenou LJ (1995): Rescue of adult mouse motoneurons from injury-induced cell death by glial cell line-derived neurotrophic factor. *Proc Natl Acad Sci USA* 92:9771–9775.

Li X-J, Li S-H, Sharp AH, Nucifora FC, Schilling G, Lanahan A, Worley P, Snyder SH, Ross CA (1995): A huntingtin-associated protein enriched in brain with implications for pathology. *Nature* 378:398–402.

Lin L-F, Zhang TJ, Collins F, Armes LG (1994): Purification and initial characterization of rat B49 glial cell line-derived neurotrophic factor. *J Neurochem* 63:758–768.

Lin L-FH, Doherty DH, Lile JD, Bektesh S, Collins F (1993): GDNF: A glial cell line-derived neurotrophic factor for midbrain dopaminergic neurons. *Science* 260:1130–1132.

Lindholm D, Hartikka J, da Penha Berzaghi M, Castrén E, Tzimagiorgis G, Hughes RA, Thoenen H (1994): Fibroblast growth factor-5 promotes differentiation of cultured rat septal cholinergic and raphe serotonergic neurons: Comparison with the effects of neurotrophins. *Eur J Neurosci* 6:2–13.

Lindner MD, Winn SR, Baetge EE, Hammang JP, Gentile FT, Doherty E, McDermott PE, Frydel B, Ullman MD, Schallert T, Emerich DF (1995): Implantation of encapsulated catecholamine and GDNF producing cells in rats with unilateral dopamine depletions and Parkinsonian symptoms. *Exp Neurol* 132:62–76.

Linnik MD, Zahos P, Geschwind MD, Federoff HJ (1995): Expression of BCL-2 from a defective herpes simplex virus-1 vector limits neuronal death in focal cerebral ischemia. *Stroke* 26:1670–1675.

Liu JP, Lauder JM (1992): S-100β and insulin-like growth factor-II differentially regulate growth of developing serotonin and dopamine neurons *in vitro*. *J Neurosci Res* 33:248–256.

Lowenstein DH, Arsenault L (1996): The effects of growth factors on the survival and differentiation of cultured dentate gyrus neurons. *J Neurosci* 16:1759–1769.

Lu B, Federoff HJ (1995): Herpes simplex virus type 1 amplicon vectors with glucocorticoid-inducible gene expression. *Hum Gene Ther* 6:421–430.

Lucidi-Phillipi CA, Gage FH, Shults CW, Jones KR, Reichardt LF, Kang UJ (1995): Brain-derived neurotrophic factor-transduced fibroblasts: Production of BDNF and effects of grafting to the adult rat brain. *J Comp Neurol* 354:361–376.

Lundberg C, Horellou P, Mallet J, Bjorklund A (1996): Generation of DOPA-producing astrocytes by retroviral transduction of the human tyrosine hydroxylase gene: *In vitro* characterization and *in vivo* effects in the rat Parkinson model. *Exp Neurol* 139:39–53.

Magal E, Burnham P, Varon S, Louis J-C (1993): Convergent regulation by ciliary neurotrophic factor and dopamine of tyrosine hydroxylase expression in cultures of rat substantia nigra. *Neuroscience* 52:867–881.

Maiese K, Boniece I, DeMeo D, Wagner JA (1993): Peptide growth factors protect against ischemia in culture by preventing nitric oxide toxicity. *J Neurosci* 13:3034–3040.

Mamounas LA, Blue ME, Sluclak JA, Altar CA (1995a): Brain-derived neurotrophic factor promotes the survival and sprouting of serotonergic axons in rat brain. *J Neurosci* 15:7929–7939.

Mamounas M, Leavitt M, Yu M, Wongstaal F (1995b): Increased titer of recombinant AAV vectors by gene transfer with adenovirus coupled to DNA–polylysine complexes. *Gene Ther* 2:429–432.

Mansour-Robaey S, Clarke DB, Wang Y-C, Bray GM, Aguayo AJ (1994): Effects of ocular injury and administration of brain-derived neurotrophic factor on survival and regrowth of axotomized retinal ganglion cells. *Proc Natl Acad Sci USA* 91:1632–1636.

Mark MH, Duvoisin RC (1995): The history of the medical therapy of Parkinson's disease. In Koller WC, Paulson G (eds): *Therapy of Parkinson's Disease.* New York: Marcel Dekker,

Martin-Iverson MT, Todd KG, Altar CA (1994): Brain-derived neurotrophic factor and neurotrophin-3 activate striatal dopamine and serotonin metabolism and related behaviors: Interactions with amphetamine. *J Neurosci* 14:1261–1270.

Martinez-Serrano A, Fischer W, Bjorklund A (1995a): Reversal of age-dependent cognitive impairments and cholinergic neuron atrophy by NGF-secreting neural progenitors grafted to the basal forebrain. *Neuron* 15:473–484.

Martinez-Serrano A, Lundberg C, Horellou P, Fischer W, Bentlage C, Campbell K, McKay RDG, Mallet J, Bjorklund A (1995b): CNS-derived neural progenitor cells for gene transfer of nerve growth factor to the adult rat brain: Complete rescue of axotomized cholinergic neurons after transplantation into the septum. *J Neurosci* 15:5668–5680.

Martinou J-C, Dubois-Douphin M, Staple JK, Rodriguez I, Frankowski H, Missotten M, Albertini P, Talabot D, Catsicas S, Pietra C, Huarte J (1994): Overexpression of BCL-2 in transgenic mice protects neurons from naturally occurring cell death and experimental ischemia. *Neuron* 13:1017–1030.

Martinou JC, Martinou I, Kato AC (1992): Cholinergic differentiaion factor (CDF/LIF) promotes survival of isolated rat embryonic motoneurons *in vitro. Neuron* 8:737–744.

Masuda K, Watanabe I, Unoki K, Ohba N, Muramatsu T (1995): Functional rescue of photoreceptors from the damaging effects of constant light by survival-promoting factors in the rat. *Invest Ophthalmol Vis Sci* 36:2142–2146.

Mattson MP, Cheng B (1993): Growth factors protect neurons against excitotoxic/ischemic damage by stabilizing calcium homeostasis. *Stroke* 24:I136–140.

Mattson MP, Lovell MA, Furukawa K, Markesbery WR (1995): Neurotrophic factors attenuate glutamate-induced accumulation of peroxides, elevation of intracellular $Ca2^+$ concentration, and neurotoxicity and increase antioxidant enzyme activities in hippocampal neurons. *J Neurochem* 65:1740–1751.

McCown TJ, Xiao X, Li J, Breese GR, Samulski RJ (1996): Differential and persistent expression patterns of CNS gene transfer by an adeno-associated virus (AAV) vector. *Brain Res* 713:99–107.

McMahon SB (1996): NGF as a mediator of inflammatory pain. *Philos Trans R Soc Lond B Biol Sci* 351:431–440.

Miletich RS, Bankiewicz KS, Quarantelli M, Plunkett RJ, Frank J, Kopin IJ, Di Chiro G (1994): MRI detects acute degeneration of the nigrostriatal dopamine system after MPTP exposure in hemiparkinsonian monkeys. *Ann Neurol* 35:689–697.

Miller RG, Petajan JH, Bryan WW, Armon C, Barohn RJ, Goodpasture JC, Hoagland RJ, Parry GJ, Ross MA, Stromatt SC, rhCNTF ALS Study Group (1996): A placebo-controlled trial of recombinant human ciliary neurotrophic (rhCNTF) factor in amyotrophic lateral sclerosis. *Ann Neurol* 39:256–260.

Milligan CE, Prevette D, Yaginuma H, Homma S, Cardwell C, Fritz LC, Tomaselli KJ, Oppenheim RW, Schwartz LM (1995): Peptide inhibitors of the ICE protease family arrest programmed cell death of motoneurons *in vivo* and *in vitro*. *Neuron* 15:385–393.

Mitsumoto H, Kieda K, Klinkosz B, Cedarbaum JM, Wong V, Lindsay RM (1994): Arrest of motor neuron disease in wobbler mice cotreated with CNTF and BDNF. *Science* 265:1107–1109.

Mizuno Y, Mori H, Kondo T (1994): Potential of neuroprotective therapy in Parkinson's disease. *CNS Drugs* 1:45–56.

Mufson EJ, Kroin JS, Sobreviela T, Burke MA, Kordower JH, Penn RD, Miller JA (1994): Intrastriatal infusions of brain-derived neurotrophic factor: retrograde transport and colocalization with dopamine-containing substantia nigra neurons in rat. *Exp Neurol* 129:15–26.

Muldoon LL, Nilaver G, Kroll RA, Pagel MA, Breakefield XO, Chiocca EA, Davidson BL, Weissleder R, Neuwelt EA (1995): Comparison of intracerebral inoculation and osmotic blood-brain barrier disruption for delivery of adenovirus, herpesvirus and iron oxide particles to normal rat brain. *Am J Pathol* 147:1840–1851.

Nakahara Y, Gage FH, Tuszynski MH (1996): Grafts of fibroblasts genetically modified to secrete NGF, BDNF, NT-3 or basic FGF elicit differential responses in the adult spinal cord. *Cell Transplant* 5:191–204.

Naldini L, Blomer U, Gallay P, Ory D, Mulligan R, Gage FH, Verma IM, Trono D (1996): *In vivo* gene delivery and stable transduction of nondividing cells by a lentiviral vector. *Science* 272:263–268.

Neuwelt EA, Pagel MA, Bix RD (1991): Delivery of ultraviolet-inactivated 35S-herpesvirus across an osmotically modified blood–brain barrier. *J Neurosurg* 74:475–479.

Neve RL (1993): Adenovirus vectors enter the brain. *Trends Neurosci* 16:251–253.

Nonomura T, Hatanaka H (1992): Neurotrophic effect of brain-derived neurotrophic factor on basal forebrain cholinergic neurons in culture from postnatal rats. *Neurosci Res* 14:226–233.

Obeso JA, Linazasoro G, Rothwell JC, Jahanshahi M, Brown R (1996): Assessing the effects of pallidotomy in Parkinson's disease. *Lancet* 347:1490.

Oldfield EH, Ram Z, Culver KW, Blaese RM, DeVroom HL, Anderson WF (1993): Gene therapy for the treatment of brain tumors using intra-tumoral transduction with the thymidine kinase gene and intravenous ganciclovir. *Hum Gene Ther* 4:39–69.

Olson L, Backman L, Ebendal T, Eriksdotter-Jonhagen M, Hoffer B, Humpel C, Freedman R, Giacobini M, Meyerson B, Nordberg A, Seiger A, Stromberg I, Sydow

O, Tomac A, Trok K, Winblad B (1994): Role of growth factors in degeneration and regeneration in the central nervous system; clinical experiences with NGF in Parkinson's and Alzheimer's diseases. *J Neurol* 241:S12–S15.

Olson L, Nordberg A, vonHolst H, Bäckman L, Ebendal T, Alafuzoff I, Amberla K, Hartvig P, Herlitz A, Lilja A, Viitanen M, Winblad B, Seiger Å (1992): Nerve growth factor affects ^{11}C-nicotine binding, blood flow, EEG and verbal episodic memory in an Alzheimer patient (case report). *J Neural Transm Park Dis Dement Sect* 4:79–95.

Olsson M, Nikkhah G, Bentlage C, Bjorklund A (1995): Forelimb akinesia in the rat parkinson model—Differential effects of dopamine agonists and nigral transplants as assessed by a new stepping test. *J Neurosci* 15:3863–3875.

Ooboshi H, Welsh MJ, Rios CD, Davidson BL, Heistad DD (1995): Adenovirus-mediated gene transfer *in vivo* to cerebral blood vessels and perivascular tissue. *Circ Res* 77:7–13.

Oppenheim RW, Housenou LJ, Johnson JE, Lin L-FH, Li L, Lo AC, Newsome AL, Prevette DM, Wang S (1995): Developing motor neurons rescued from programmed and axotomy-induced cell death by GDNF. *Nature* 373:344–346.

Othberg A, Odin P, Ballagi A, Ahgren A, Funa K, Lindvall O (1995): Specific effects of platelet derived growth factor (PDGF) on fetal rat and human dopaminergic neurons *in vitro. Exp Brain Res* 105:111–122.

Otto D, Unsicker K (1993): FGF-2-mediated protection of cultured mesencephalic dopaminergic neurons against MPTP and MPP^+: Specificity and impact of culture conditions, non-dopaminergic neurons, and astroglial cells. *Neurosci Res* 34:382–393.

Otto D, Unsicker K (1994): FGF-2 in the MPTP model of Parkinsons's disease: Effects on astroglial cells. *Glia* 11:47–56.

Paulson HL, Fischbeck KH (1996): Trinucleotide repeats in neurogenetic disorders. *Annu Rev Neurosci* 19:79–107.

Paulus W, Baur I, Boyce FM, Breakefield XO, Reeves SA (1996): Self-contained, tetracycline-regulated retroviral vector system for gene delivery to mammalian cells. *J Virol* 70:62–67.

Pechan PA, Yoshida T, Panahian N, Moskowitz MA, Breakefield XO (1995): Genetically modified fibroblasts producing NGF protect hippocampal neurons after ischemia in the rat. *Neuroreport* 6:669–672.

Peinadoramon P, Salvador M, Villegasperez MP, Vidalsanz M (1996): Effects of axotomy and intraocular administration of NT-4, NT-3, and brain-derived neurotrophic factor on the survival of adult rat retinal ganglion cells—A quantitative *in vivo* study. *Invest Ophthalmol Vis Sci* 37:489–500.

Pennica D, Shaw KJ, Swanson TA, Moore MW, Shelton DL, Zioncheck KA, Rosenthal A, Taga T, Paoni NF, Wood WI (1995): Cardiotrophin-1: Biological activities and binding to the leukemia inhibitory factor receptor/gp130 signaling complex. *J Biol Chem* 270:10915–10922.

Peterson DA, Lucici-Phillipi CA, Murphy DP, Ray J, Gage FH (1996): Fibroblast growth factor-2 protects entorhinal layer II glutamatergic neurons from axotomy-induced death. *J Neurosci* 16:886–898.

Poulsen KT, Armanini MP, Klein RD, Hynes MA, Phillips HS, Rosenthal A (1994): TGF beta 2 and TGF beta 3 are potent survival factors for midbrain dopaminergic neurons. *Neuron* 13:1245–1252.

Prehn JH, Bindokas VP, Marcuccilli CJ, Krajewski S, Reed JC, Miller RJ (1994): Regulation of neuronal Bcl2 protein expression and calcium homeostasis by transforming growth factor type beta confers wide-ranging protection on rat hippocampal neurons. *Proc Natl Acad Sci USA* 91:12599–12603.

Rabizadeh S, Graller EB, Borchelt DR, Gwinn R, Valentine JS, Sisodia S, Wong P, Lee M, Hahn H, Bredesen DE (1995): Mutations associated with amyotrophic lateral sclerosis convert superoxide dismutase from an antiapoptotic gene to a proapoptotic gene: Studies in yeast and neural cells. *Proc Natl Acad Sci USA* 92:3024–3028.

Reynolds BA, Weiss S (1992): Generation of neurons and astrocytes from isolated cells of the adult mammalian central nervous system. *Science* 255:1707–1710.

Ridet J-L, Privat A (1995): Gene therapy in the central nervous system: Direct versus indirect gene delivery. *J Neurosci Res* 42:287–293.

Rosen DR, Siddique T, Patterson D, Figlewicz DA, Sapp P, Hentati A, Donaldson D, Goto J, O'Regan GP, Deng HX, Rahmani Z, Krizus A, McKenna-Yasek D, Cayabyab A, Gaston SM, Berger R, Tanzi RE, Halperin JJ, Herzfeldt B, Van den Bergh R, Hung W-Y, Bird T, Deng G, Mulder DW, Smyth C, Laing NG, Soriano E, Pericak-Vance MA, Haines J, Rouleau GA, Gusella JS, Horvitz HR, Brown RH Jr (1993): Mutations in Cu/Zn superoxide dismutase gene are associated with familial amyotrophic lateral sclerosis. *Nature* 362:59–62.

Rosenberg MB, Friedmann T, Robertson RC, Tuszynski M, Wolff JA, Breakefield XO, Gage FH (1988): Grafting genetically modified cells to the damaged brain: Restorative effects of NGF expression. *Science* 242:1575–1578.

Roses AD (1996): From genes to mechanisms to therapies: Lessons to be learned from neurological disorders. *Nat Med* 2:267–269.

Roy N, Mahadevan MS, McLean M, Shutler G, Yaraghi Z, Farahani R, Baird S, Besner-Johnson A, Lefebvre C, Kang X, Xalih M, Aubry H, Tamai K, Guan X, Ioannou P, Crawford TO, de Jong PJ, Surh L, Ikeda JE, Korneluk RG, MacKenzie A (1995): The gene for neuronal apoptosis inhibitory protein is partially deleted in individuals with spinal muscular atrophy. *Cell* 80:167–178.

Russell D, Miller A, Alexander I (1994): Adeno-associated virus vectors preferentially transduce cells in S phase. *Proc Natl Acad Sci USA* 91:8915–8919.

Russell DW, Alexander IE, Miller AD (1995): DNA synthesis and topoisomerase inhibitors increase transduction by adeno-associated virus vectors. *Proc Natl Acad Sci USA* 92:5719–5723.

Sabate O, Horellou P, Vigne E, Colin P, Perricaudet M, Buc-Caron MH, Mallet J (1995): Transplantation to the rat brain of human neural progenitors that were genetically modified using adenoviruses. *Nat Genet* 9:256–260.

Sagot Y, Dubois-Dauphin M, Tan SA, de Bilbao F, Aebischer P, Martinou J-C, Kato AC (1995a): Bcl-2 overexpression prevents motoneuron cell body loss but not axonal degeneration in a mouse model of a neurodegenerative disease. *J Neurosci* 15:7727–7733.

Sagot Y, Tan SA, Baetge E, Schmalbruch H, Kato AC, Aebischer P (1995b): Polymer encapsulated cell lines genetically engineered to release ciliary neurotrophic factor can slow down progressive motor neuronopathy in the mouse. *Eur J Neurosci* 7:1313–1322.

Sagot Y, Tan SA, Hammang JP, Aebischer P, Kato AC (1996): GDNF slows loss of motoneurons but not axonal degeneration or premature death of pmn/pmn mice. *J Neurosci* 16:2335–2341.

Sasaki K, Oomura Y, Suzuki K, Hanai K, Yagi H (1992): Acidic fibroblast growth factor prevents death of hippocampal CA1 pyramidal cells following ischemia. *Neurochem Int* 21:397–402.

Sauer H, Fischer W, Nikkhah G, Wiegand SJ, Brundin P, Lindsay RM, Bjorklund A (1993): Brain-derived neurotrophic factor enhances function rather than survival of intrastriatal dopamine cell-rich grafts. *Brain Res* 616:37–44.

Sauer H, Oertel WH (1994): Progressive degeneration of nigrostriatal dopamine neurons following intrastriatal terminal lesions with 6-hydroxydopamine: A combined retrograde tracing and immunocytochemical study in the rat. *Neuroscience* 59:401–415.

Sauer H, Rosenblad C, Björklund A (1995): Glial cell line-derived neurotrophic factor but not transforming growth factor $\beta3$ prevents delayed degeneration of nigral dopaminergic neurons following striatal 6-hydroxydopamine lesion. *Proc Natl Acad Sci USA* 92:8935–8939.

Schallert T (1995): Models of neurological defects and defects in neurological models. *Behav Brain Sci* 18:68–69.

Schumacher JM, Short MP, Hyman BT, Breakefield XO, Isacson O (1991): Intracerebral implantation of nerve growth factor-producing fibroblasts protects striatum against neurotoxic levels of excitatory amino acids. *Neuroscience* 45:561–570.

Sendtner M, Carroll P, Holtmann B, Hughes RA, Thoenen H (1994): Ciliary neurotrophic factor. *J Neurobiol* 25:1436–1453.

Sendtner M, Hotmann B, Kolbeck R, Thoenen H, Barde Y-A (1992): Brain-derived neurotrophic factor prevents the death of motoneurons in newborn rats after nerve section. *Nature* 360:757–759.

Sendtner M, Kretzber GW, Thoenen H (1990): Ciliary neurotrophic factor prevents the degeneration of motor neurons after axotomy. *Nature* 345:440–441.

Seroogy KB, Lundgren KH, Tran TMD, Guthrie KM, Isackson PJ, Gall CM (1994): Dopaminergic neurons in rat ventral midbrain express brain-derived neurotrophic factor and neurotrophin-3 mRNAs. *J Comp Neurol* 342:321–334.

Sharp AH, Ross CA (1996): Neurobiolgy of Huntington's disease. *Neurobiol Disease* 3:3–15.

Shen RY, Altar CA, Chiodo LA (1994): Brain-derived neurotrophic factor increases the electrical activity of pars compacta dopamine neurons *in vivo*. *Proc Natl Acad Sci USA* 91:8920–8924.

Shinoura N, Chen L, Wani MA, Kim YG, Larson JJ, Warnick RE, Simon M, Menon AG, Bi WL, Stambrook PJ (1996): Protein and messenger RNA expression of connexin43 in astrocytomas: Implications in brain tumor gene therapy. *J Neurosurg* 84:839–846.

Short MP, Choi BC, Lee JK, Malick A, Breakefield XO, Martuza RL (1990): Gene delivery to glioma cells in rat brain grafting of a retrovirus packaging cell line. *J Neurosci Res* 27:427–439.

Shults CW, Matthews RT, Altar CA, Hill LR, Langlais PJ (1994): A single intramesencephalic injection of brain-derived neurotrophic factor induces persistent rotational asymmetry in rats. *Exp Neurol* 125:183–194.

Simon JR, Ghetti B (1994): The weaver mutant mouse as a model of nigrostriatal dysfunction. *Mol Neurobiol* 9:183–189.

Sklairtavron L, Nestler EJ (1995): Opposing effects of morphine and the neurotrophins, NT-3, NT-4, and BDNF, on locus coeruleus neurons *in vitro. Brain Res* 702:117–125.

Smith F, Jacoby D, Breakfield XO (1995): Virus vectors for gene delivery to the nervous system. *Restorative Neurol Neurosci* 8:21–34.

Snyder EY (1994): Grafting immortalized neurons to the CNS. *Curr Opin Neurobiol* 4:742–751.

Spina MB, Squinto SP, Miller J, Lindsay RM, Hyman C (1992): Brain-derived neurotrophic factor protects dopamine neurons against 6-hydroxydopamine and *N*-methyl-4-phenylpyridinium ion toxicity: Involvement of the glutathione system. *J Neurochem* 59:99–106.

Strömberg I, Björklund L, Johansson M, Tomac A, Collins F, Olson L, Hoffer B, Humpel C (1993): Glial cell line-derived neurotrophic factor is expressed in the developing but not adult striatum and stimulates developing dopamine neurons *in vivo. Exp Neurol* 124:401–412.

Suhr ST, Gage FH (1993): Gene therapy for neurologic diseases. *Arch Neurol* 50:1252–1268.

Tomac A, Lindqvist E, Lin L-FH, Ögren SO, Young D, Hoffer B, Olson L (1995a): Protection and repair of the nigrostriatal dopaminergic system by GDNF *in vivo. Nature* 373:335–339.

Tomac A, Widenflak J, Lin L-F, Kohno T, Ebendal R, Hoffer BJ, Olson L (1995b): Retrograde axonal transport of glial cell line-derived neurotrophic factor in the adult nigrostriatal system suggests a trophic role in the adult. *Proc Natl Acad Sci USA* 92:8274–8278.

Tomlinson DR, Fernyhough P, Diemel LT (1996): Neurotrophins and peripheral neuropathy. *Philos Trans R Soc Lond B Biol Sci* 351:455–462.

Treanor JJS, Goodman L, de Sauvage F, Stone DM, Poulsen KT, Beck CD, Gray C, Armanini MP, Pollock RA, Hefti F, Phillips HS, Goddard A, Moore MW, Buj-Bello A, Davies AM, Asai N, Takahaski M, Vandlen R, Henderson CE, Rosenthal A (1996): Characterization of a multicomponent receptor for GDNF. *Nature* 382:80–83.

Tripathy SK, Black HB, Goldwasser E, Leiden JM (1996): Immune responses to transgene-encoded proteins limit the stability of gene expression after injection of replication-defective adenovirus vectors. *Nat Med* 2:545–550.

Trupp M, Arenas E, Fainzilber M, Nilsson A-S, Sieber B-A, Grigoriou M, Kilkenny C, Salazar-Grueso E, Pachnis V, Arumae U, Sariola H, Saarma M, Ibanez CF (1996): Functional receptor for GDNF encoded by the c-ret proto-oncogene. *Nature* 381:785–789.

Tuszynski MH, Gabriel K, Gage FH, Suhr S, Meyer S, Rosetti A (1996a): Nerve growth factor delivery by gene transfer induced differential outgrowth of sensory, motor and noradrenergic neurites after adult spinal cord injury. *Exp Neurol* 137:157–173.

Tuszynski MH, Gage FH (1995): Maintaining the neuronal phenotype after injury in the adult CNS—Neurotrophic factors, axonal growth substrates, and gene therapy. *Mol Neurobiol* 10:151–167.

Tuszynski MH, Peterson DA, Ray J, Baird A, Nakahara Y, Gage FH (1994): Fibroblasts genetically modified to produce nerve growth factor induce robust neuritic ingrowth after grafting to the spinal cord. *Exp Neurol* 126:1–14.

Tuszynski MH, Roberts J, Senut MC, U HS, Gage FH (1996b): Gene therapy in the adult primate brain—Intraparenchymal grafts of cells genetically modified to produce nerve growth factor prevent cholinergic neuronal degeneration. *Gene Ther* 3:305–314.

Tuszynski MH, U HS, Amaral DG, Gage FH (1990): Nerve growth factor infusion in the primate brain reduces lesion-induced cholinergic neuronal degeneration. *J Neurosci* 10:3604–3614.

Uchida K, Tsuzaki N, Nagatsu T, Kohsaka S (1992): Tetrahydrobiopterin-dependent functional recovery in 6-hydroxydopamine-treated rats by intracerebral grafting of fibroblasts transfected with tyrosine hydroxylase cDNA. *Dev Neurosci* 14:173–180.

Ungerstedt U (1971): Post-synaptic supersensitivity after 6-OHDA induced degeneration of the nigrostriatal dopamine system. *Acta Physiol Scand* 82:69–93.

Unoki K, LaVail MM (1994): Protection of the rat retina from ischemic injury by brain-derived neurotrophic factor, ciliary neurotrophic factor, and basic fibroblast growth factor. *Invest Ophthalmol Vis Sci* 35:907–915.

Vejsada R, Lindsay RM, Kato AC (1995): Additive rescue effects of GDNF and BDNF on axotomized sciatic motoneurons in newborn rats. *Soc Neurosci Abs* 21:1536.

Venero JL, Beck KD, Hefti F (1994): Intrastriatal infusion of nerve growth factor after quinolinic acid prevents reduction of cellular expression of choline acetyltransferase messenger RNA and trkA messenger RNA, but not glutamate decarboxylase messenger RNA. *Neuroscience* 61:257–268.

Vito P, Lacana E, D'Adamio L (1996): Interfering with apoptosis: $Ca2^+$-binding protein ALG-2 and Alzheimer's disease gene ALG-3. *Science* 271:521–525.

von Coelln R, Unsicker K, Krieglstein K (1995): Screening of interleukins for survival-promoting effects on cultured mesencephalic dopaminergic neurons from embryonic rat brain. *Brain Res Dev Brain Res* 89:150–4.

Walsh (1995): Nervous excitement over neurotrophic factors. *Biotechnology* 13:1167–1171.

Whittemore SR, Synder EY (1996): Physiological relevance and functional potential of central nervous system-derived cell lines. *Mol Neurobiol* 12:13–38.

Wiedau-Pazos M, Goto JJ, Rabizadeh S, Gralla EB, Roe JA, Lee MK, Valentine JS, Bredesen DE (1996): Altered reactivity of superoxide dismutase in familial amyotrophic lateral sclerosis. *Science* 271:515–158.

Williams L, Varon S, Peterson GM, Wictorin K, Fischer W, Bjorklund A, Gage FH (1986): Continous infusion of nerve growth factor prevents basal forebrain neuronal death after fimbria fornix transection. *Proc Natl Acad Sci USA* 91:2324–2328.

Williams LR, Inouye G, Cummins V, Pelleymounter MA (1996): Glial cell line-derived neurotrophic factor sustains axotomized basal forebrain cholinergic neurons *in vivo:* Dose–response comparison to nerve growth factor and brain-derived neurotrophic factor. *Am Soc Pharmacol Exp Ther* 277:1140–1151.

Winn SR, Hammang JP, Emerich DF, Lee A, Palmiter RD, Baetge EE (1994): Polymer-encapsulated cells genetically modified to secrete human nerve growth factor promote the survival of axotomized septal cholinergic neurons. *Proc Natl Acad Sci USA* 91:2324–2328.

Wolfe J, Deshmane S, Fraser N (1992): Herpes virus vector gene transfer and expression of beta-glucuronidase in the central nervous system of MPS VII mice. *Nat Genet* 1:379–384.

Wolff JA, Fisher LJ, Xu L, Jinnah HA, Langlais PJ, Iuvone PM, O'Malley KL, Rosenberg MB, Shimohama S, Friedmann T, Gage FH (1989): Grafting fibroblasts genetically modified to produce L-dopa in a rat model of Parkinson's disease. *Proc Natl Acad Sci USA* 86:9011–9014.

Wong V, Arriaga R, Ip NY, Lindsay RM (1993): The neurotrophins BDNF, NT-3 and NT-4/5, but not NGF, up-regulate the cholinergic phenotype of developing motor neurons. *Eur J Neurosci* 5:466–474.

Xu XM, Guenard V, Kleitman N, Bunge MB (1994): Axonal regeneration into Schwann cell-seeded guidance channels grafted into transected adult rat spinal cord. *J Comp Neurol* 351:145–160.

Yamada M, Hatanaka H (1994): Interleukin-6 protects cultured rat hippocampal neurons against glutamate-induced cell death. *Brain Res* 643:173–180.

Yan Q, Elliott JL, Matheson C, Sun J, Zhang L, Mu X, Rex KL, Snider WD (1993): Influences of neurotrophins on mammalian motoneurons *in vivo*. *J Neurobiol* 24:1555–1577.

Yan Q, Matheson C, Lopez OT, Miller JA (1994a): The biological responses of axotomized adult motoneurons to brain-derived neurotrophic factor. *J Neurosc* 14:5281–5291.

Yan Q, Matheson C, Sun J, Radeke MJ, Feinstein SC, Miller JA (1994b): Distribution of intracerebral ventricularly administered neurotrophins in rat brain and its correlation with trk receptor expression. *Exp Neurol* 127:23–36.

Yan Q, Matheson C, Lopez OT (1995): *In vivo* neurotrophic effects of GDNF on neonatal and adult facial motor neurons. *Nature* 373:341–344.

Yan Q, Miller JA (1993): The use of trophic factors in degenerative motoneuron disease. *Exp Neurol* 124:60–63.

Yang Y, Nunes FA, Berensci K, Furth EA, Gönczöl E, Wilson JM (1994): Cellular immunity to viral antigens limits E1-deleted adenoviruses for gene therapy. *Proc Natl Acad Sci USA* 91:4407–4411.

Yankner BA (1996): Mechanisms of neuronal degeneration in Alzheimer's disease. *Neuron* 16:921–932.

Yoshimoto Y, Lin Q, Collier T, Frim D, Breakefield XO, Bohn MC (1995): Astrocytes retrovirally transduced with BDNF elicit behavioral improvement in a rat model of Parkinson's disease. *Brain Res* 691:25–36.

Yurek DM, Lu W, Hipkens S, Wiegand SJ (1996): BDNF enhances the functional reinnervation of the striatum by grafted fetal dopamine neurons. *Exp Neurol* 127:105–118.

Zheng JL, Stewart RR, Gao WQ (1995): Neurotrophin-4/5 enhances survival of cultured spiral ganglion neurons and protects them from cisplatin neurotoxicity. *J Neurosci* 15:5079–5087.

Zhu CZ, Auer RN (1994): Intraventricular administration of insulin and IGF-1 in transient forebrain ischemia. *J Cereb Blood Flow Metab* 14:237–242.

Zwimpfer TJ, Aguayo AJ, Bray GM (1992): Synapse formation and preferential distribution in the granule cell layer by regenerating retinal ganglion cell axons guided to the cerebellum of adult hamster. *J Neurosci* 12:1144–1159.

INDEX